Engineering Graphics with AutoCAD® 2008

Engineering Graphics with AutoCAD® 2008

James D. Bethune
Boston University

PEARSON
Prentice
Hall

Upper Saddle River, New Jersey
Columbus, Ohio

Library of Congress Control Number: 2007923738

Editor in Chief: Vernon Anthony
Acquisitions Editor: Jill Jones-Renger
Editorial Assistant: Doug Greive
Development Editor: Lisa S. Garboski, bookworks publishing services
Production Editor: Louise N. Sette
Production Supervision: Lisa S. Garboski, bookworks publishing services
Design Coordinator: Diane Ernsberger
Cover Designer: Thomas Mack
Art Coordinator: Janet Portisch
Production Manager: Deidra M. Schwartz
Director of Marketing: David Gesell
Marketing Manager: Jimmy Stephens
Marketing Coordinator: Alicia Dysert

This book was set by Aptara, Inc. It was printed and bound by Courier Kendallville, Inc. The cover was printed by Coral Graphic Services, Inc.

Certain images and materials contained in this publication were reproduced with the permission of Autodesk, Inc. © 2006. All rights reserved. Autodesk and AutoCAD are registered trademarks of Autodesk, Inc., in the U.S.A. and certain other countries.

Disclaimer:
The publication is designed to provide tutorial information about AutoCAD® and/or other Autodesk computer programs. Every effort has been made to make this publication complete and as accurate as possible. The reader is expressly cautioned to use any and all precautions necessary, and to take appropriate steps to avoid hazards, when engaging in the activities described herein.

Neither the author nor the publisher makes any representations or warranties of any kind, with respect to the materials set forth in this publication, express or implied, including without limitation any warranties of fitness for a particular purpose or merchantability. Nor shall the author or the publisher be liable for any special, consequential or exemplary damages resulting, in whole or in part, directly or indirectly, from the reader's use of, or reliance upon, this material or subsequent revisions of this material.

Pearson Education Ltd.   Pearson Education Australia Pty. Limited
Pearson Education Singapore Pte. Ltd.   Pearson Education North Asia Ltd.
Pearson Education Canada, Ltd.   Pearson Educación de Mexico, S.A. de C.V.
Pearson Education—Japan   Pearson Education Malaysia Pte. Ltd.

10 9 8 7 6 5 4 3
ISBN-13: 978-0-13-159233-9
ISBN-10:   0-13-159233-5

Preface

This book teaches technical drawing and uses AutoCAD® as its drawing instrument. This book updates *Engineering Graphics with AutoCAD® 2007* for AutoCAD® 2008. It follows the general format of many technical drawing texts and presents much of the same material about drawing conventions and practices, with emphasis on creating accurate, clear drawings. For example, the book shows how to locate dimensions on a drawing so that they completely define the object in accordance with current national standards, but the presentation centers on the **Dimension** toolbar and its associated tools and options. The standards and conventions are presented and their applications are shown using AutoCAD® 2008. This integrated teaching concept is followed throughout the book.

Most chapters include design problems. The problems are varied in scope and are open-ended, which means that there are several correct solutions. This is intended to encourage student creativity and increase their problem-solving abilities.

Chapters 1 through 3 cover AutoCAD **Draw** and **Modify** toolbars and other commands needed to set up and start drawings. The text starts with simple **Line** commands and proceeds through geometric constructions. The final sections of Chapter 3 describe how to bisect a line and how to draw a hyperbola, a parabola, a helix, and an ogee curve. Redrawing many of the classic geometric shapes will help students learn how to use the **Draw** and **Modify** toolbars, along with other associated commands, with accuracy and creativity.

Chapters 2 and 3 also show how to use the dynamic input options.

Chapter 4 presents freehand sketching. Simply stated, there is still an important place for sketching in technical drawing. Many design ideas start as freehand sketches and are then developed on the computer. This chapter now includes extensive exercise problems associated with object orientation.

Chapter 5 presents orthographic views. Students are shown how to draw three views of an object using AutoCAD. The discussion includes projection theory, hidden lines, compound lines, oblique surfaces, rounded surfaces, holes, irregular surfaces, castings, and thin-walled objects. The chapter ends with several intersection problems. These problems serve as a good way to pull together orthographic views and projection theory. Several new, more difficult, exercise problems have been added to this edition.

Chapter 6 presents sectional views and introduces the **Hatch** and **Gradient** commands. The chapter includes multiple, broken-out, and partial sectional views and shows how to draw an S-break for a hollow cylinder.

Chapter 7 covers auxiliary views and shows how to use the **Snap, Rotate** command to create axes aligned with slanted surfaces. Secondary auxiliary views are also discussed. Solid modeling greatly simplifies the determination of the true shape of a line or plane, but a few examples of secondary auxiliary views help students refine their understanding of orthographic views and eventually the application of UCSs.

Chapter 8 shows how to dimension both two-dimensional shapes and orthographic views. The **Dimension** command and its associated commands are demonstrated, including how to use the **Dimension Styles** tool. The commands are presented as needed to create required dimensions. The conventions demonstrated are in compliance with ANSI Y32.

Chapter 9 introduces tolerances. First, the chapter shows how to draw dimensions and tolerances using the **Dimension** and **Tolerance** commands, among others. The chapter ends with an explanation of fits and shows how to use the tables included in the appendix to determine the maximum and minimum tolerances for matching holes and shafts.

Chapter 10 continues the discussion of tolerancing using geometric tolerances and explains how AutoCAD® 2008 can be used to create geometric tolerance symbols directly from dialog boxes. Both profile and positional tolerances are explained. The overall intent of the chapter is to teach students how to make parts fit together. Fixed and floating fastener applications are discussed, and design examples are given for both conditions.

Chapter 11 covers how to draw and design using standard fasteners, including bolts, nuts, machine screws, washers, hexagon heads, square heads, setscrews, rivets, and springs. Students are shown how to create wblocks of the individual thread representations and how to use them for different size requirements.

Chapter 12 discusses assembly drawings, detail drawings, and parts lists. Instructions for drawing title blocks, tolerance blocks, release blocks, and revision blocks, and for inserting drawing notes, are also included to give students better preparation for industrial practices. Several new exercise problems have been added to the chapter.

Chapter 13 presents gears, cams, and bearings. The intent of the chapter is to teach how to design using gears selected from a manufacturer's catalog. The chapter shows how to select bearings to support gear shafts and how to tolerance holes in support plates to maintain the desired center distances of meshing gears. The chapter also shows how to create a displacement diagram and then draw the appropriate cam profile. Two new assembly problems involving gears have been added.

Chapter 14 introduces AutoCAD 3D capabilities. Both parallel (isometric) and the new perspective grids are demonstrated as well as both WCS and UCS coordinate systems. The intent is to learn the fundamentals of 3D drawings before drawing objects.

Chapter 15 extends the discussion of Chapter 14 to cover surfaces. The surface primitive commands are covered as well as **3D face** and **3D mesh.** New exercise problems have been added.

Chapter 16 shows how to draw three-dimensional solid models. It includes examples of both parallel and perspective grids using all the different **Visual Style** options. The chapter shows how to union primitive shapes to create more complex models. It also shows how to create orthographic views from those models. Several new, more complex exercise problems have been added.

Chapter 17 presents a solid modeling approach to descriptive geometry. For example, a plane is drawn as a solid that is 0.00001 inch thick. AutoCAD®'s solid modeling and other commands are then used to manipulate the plane. The true lengths of lines and shapes of planes, point and plane locations, and properties between lines and planes are discussed. Piercing points and line visibility are also covered.

Online Instructor Materials

To access supplementary materials online, instructors need to request an instructor access code. Go to **www.prenhall.com,** click the **Instructor Resource Center** link, and then click **Register Today** for an instructor access code. Within 48 hours after registering you will receive a confirming e-mail including an instructor access code. Once you have received your code, go to the site and log on for full instructions on downloading the materials you wish to use.

Autodesk Learning License

Through a recent agreement with AutoCAD publisher Autodesk®, Prentice Hall now offers the option of purchasing *Engineering Graphics with AutoCAD 2008* with either a 180-day or a 1-year student software license agreement. This provides adequate time for a student to complete all the activities in this book. The software is functionally identical to the professional license, but is intended **for student personal use only.** It is not for professional use. For more information about this book and the Autodesk Learning License, contact your local Pearson Prentice Hall sales representative, or contact our National Marketing Manager, Jimmy Stephens at 1 (800) 228-7854 x3725 or at Jimmy_Stephens@prenhall.com. For the name and number of your sales rep, please contact Prentice Hall Faculty Services at 1 (800) 526-0485.

ACKNOWLEDGMENTS

I would like to thank the following reviewers. Their comments and suggestions were most helpful in creating this and previous editions: Anthony Duva, Wentworth Institute of Technology; Dale M. Gerstenecker, St. Louis Community College at Florissant; John Loebach, Spoon River College; N. S. Malladi, Ph.D., University of South Alabama; Jack Zhou, Drexel University; Kent Eddy, Kettering University;

Dennis Maas, James A. Rhodes State College; Jill Palmer-Wood, Hudson Valley Community College; Charles Bales, Moraine Valley Community College; and Andy S. Zhang, New York City College of Technology.

Thanks to Jill Jones-Renger, the editor who pulled it all together; to Lisa Garboski; and Barbara Liguori; also thanks to David, Maria, Randy, Lisa, Hannah, Wil, Madison, Jack, Luke, Sam, and Ben for their continued support. A special thanks to Cheryl.

James D. Bethune
Boston University

Contents

Chapter 3—More Advanced Commands

Chapter 4—Sketching

Chapter 5—Orthographic Views

Chapter 6—Sectional Views

Chapter 7—Auxiliary Views

Chapter 8—Dimensioning

Chapter 9—Tolerancing

Chapter 10—Geometric Tolerances

Chapter 11—Threads and Fasteners

Chapter 12—Working Drawings

Chapter 13—Gears, Bearings, and Cams

Chapter 14—Fundamentals of 3D Drawing

Chapter 15—3D Surfaces

Chapter 16—Modeling

Chapter 17—Descriptive Geometry

Appendix

Engineering Graphics with AutoCAD® 2008

C H A P T E R 1

Getting Started

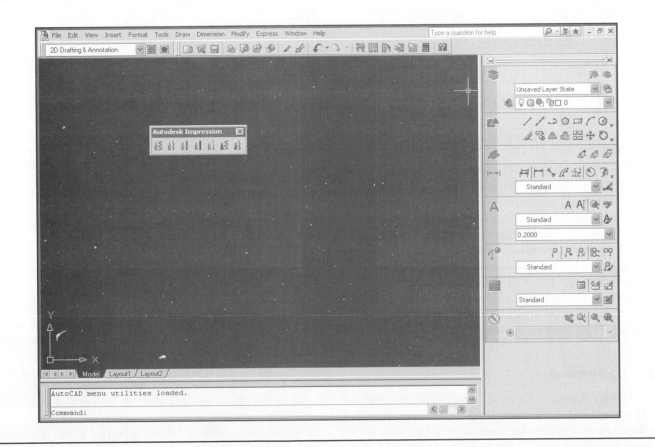

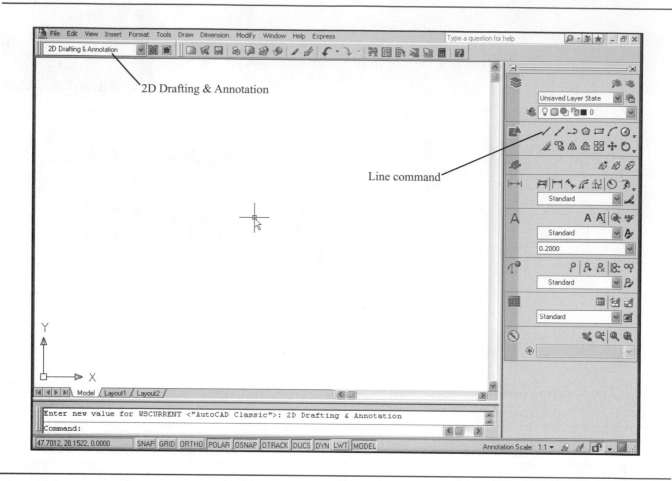

Figure 1-1

1-1 INTRODUCTION

The figure on the previous page and on the book's cover shows the initial AutoCAD drawing screen. It will appear when the program is first accessed. Figure 1-1 shows an AutoCAD screen that uses the **2D Drafting & Annotation** format. AutoCAD has three screen formats. Figure 1-2 shows the **AutoCAD Classic** format. The third format, **3D modeling,** will be introduced in the chapter on 3D modeling.

The different formats are intended to make creating a drawing easier. Each format includes commands most often used with the type of drawing being created. The commands function the same way regardless of how they are accessed. For example, the **Line** command is available in both the **2D Drafting & Annotation** and **AutoCAD Classic** formats.

Color background

If your screen background is black and you prefer a white background, click **Tools, Drafting Settings, Drafting Tooltip Appearance,** and **Colors.** Select **Uniform Background** and set the color for **white.** Click **Apply** and **Close.**

To start a new drawing

1. Select the **File** menu at the top of the screen.

   A listing of commands will cascade down. See Figure 1-3.

2. Select the **New** command.

   The **Select Template** dialog box will appear. See Figure 1-4.

3. Select the **acad** template by double-clicking the **acad** option.

   The **2D AutoCAD** screen will appear. See Figure 1-5.

   The top line of Figure 1-5 displays the pull-down menus for exiting a program and changing a program. It is assumed that the reader is familiar with basic Windows operations.

   The second line is the **Standard** and **Styles** toolbars and contains a group of the most commonly used commands. The third line contains the **Layers** and **Properties** toolbars.

   The line above the drawing portion of the screen displays the name of the current drawing. Since no drawing has

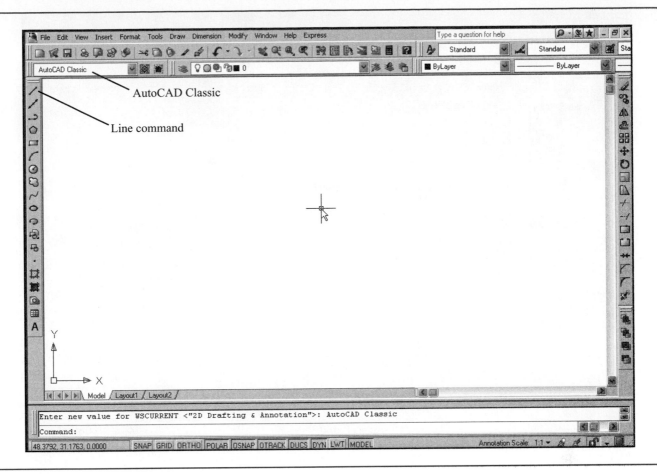

Figure 1-2

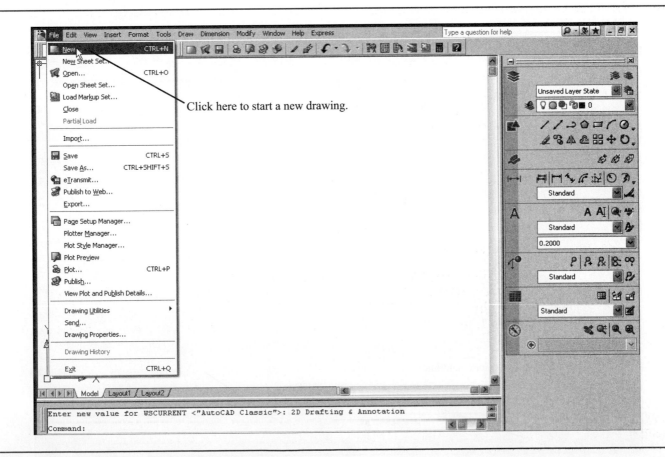

Figure 1-3

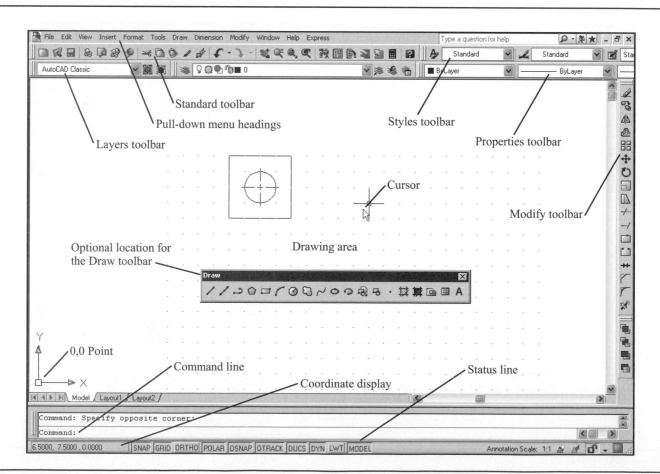

Figure 1-4

Figure 1-5

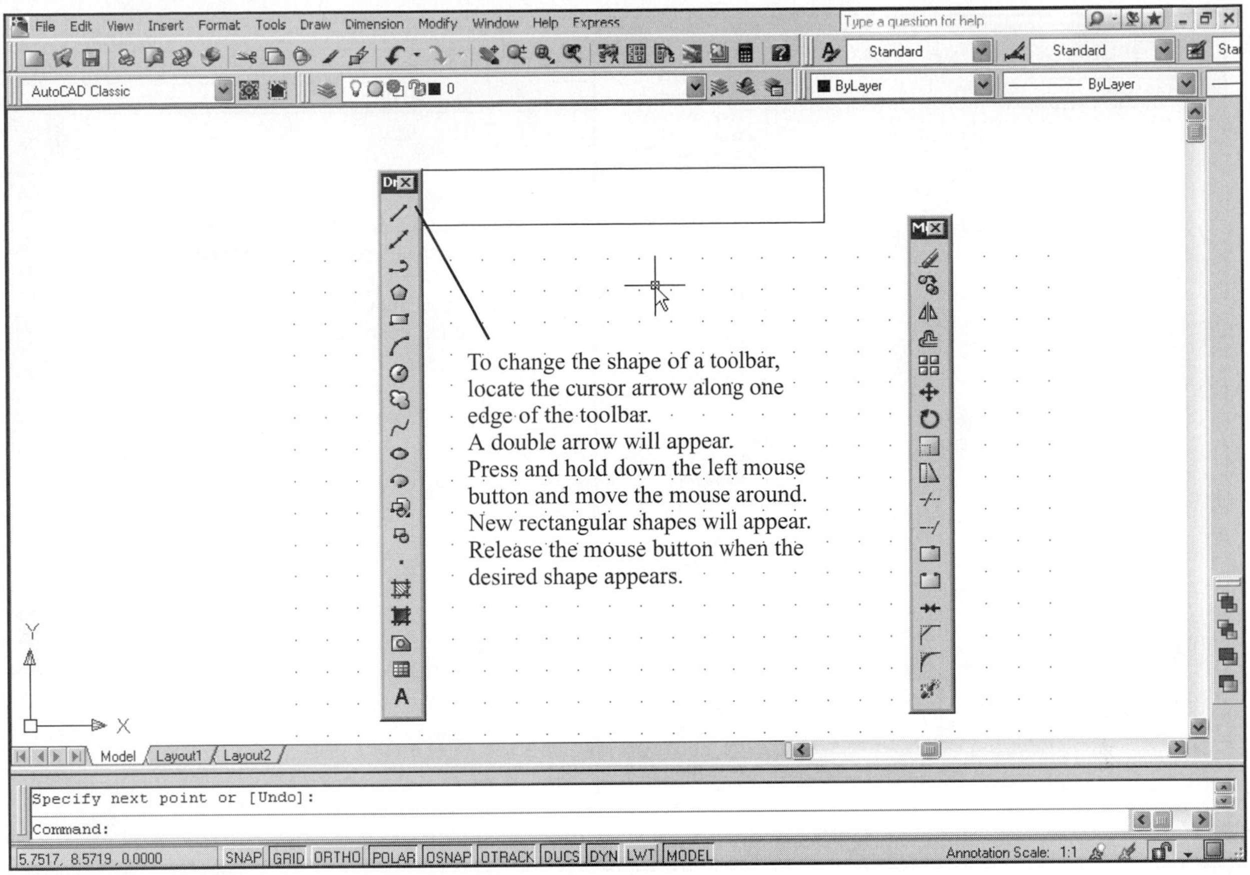

To change the shape of a toolbar, locate the cursor arrow along one edge of the toolbar.
A double arrow will appear.
Press and hold down the left mouse button and move the mouse around.
New rectangular shapes will appear.
Release the mouse button when the desired shape appears.

Figure 1-6

been named, the line reads **Drawing1.dwg.** Once a drawing name has been defined, it will appear at the top of the screen.

The bottom left corner of the drawing screen shows the coordinate display position of the horizontal, vertical crosshairs in terms of an X,Y coordinate value, whose origin is the lower left corner of the drawing screen.

The commands listed at the bottom of the screen in the status line **(SNAP, GRID, . . .)** are activated by clicking them. When they are active, they appear recessed.

The large open area in the center of the screen is called the *drawing screen,* or *drawing editor*. The **Modify** toolbar is located along the left edge of the showing screen. The **Draw** toolbar has been moved into the drawing area.

1-2 TOOLBARS

An AutoCAD toolbar contains a group of command icons located under a common heading. The initial AutoCAD as screen, shown in Figure 1-3, contains six toolbars: **Layers, Standard, Properties, Styles, Draw,** and **Modify.** There are 33 additional predefined toolbars, and you can create your own user-specific toolbars as needed.

To move a toolbar

1. Locate the cursor arrow on the top of the **Modify** toolbar.
2. Press and hold down the left mouse button.

A light gray broken-line box appears around the edge of the toolbar.

3. Still holding the left mouse button down, move the gray outline box to a new location on the screen.
4. Release the left button.

The toolbar will appear in the new location.

To change the shape of a toolbar

See Figure 1-6.

1. Locate the cursor arrow along the right edge of the **Modify** toolbar.

A double arrow will appear.

2. Press and hold the left mouse button.

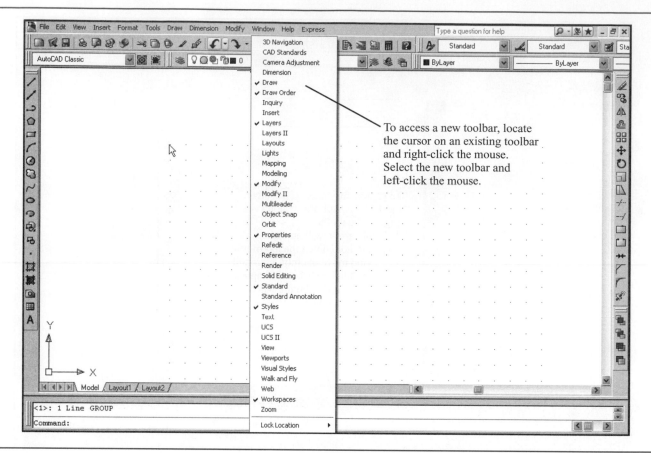

Figure 1-7

A light gray broken-line box will appear around the outside of the toolbar.

3. Still holding the left mouse button down, move the mouse around and watch how the gray box changes shape.
4. When the gray toolbar shape is a long, vertical rectangle, release the left mouse button.

A reshaped toolbar will appear.

To return the toolbar to its original location and shape

1. Locate the cursor arrow along the bottom or edge lines of the toolbar and return the toolbar to its original shape using the procedure outlined in Figure 1-6.
2. Dock the **Draw** toolbar on the left side of the drawing screen and the modify toolbar on the right side.

To add a new toolbar to the screen

See Figures 1-7 and 1-8.

1. Locate the cursor on any one of the existing toolbars and press the right mouse button. In this example the **Standard** toolbar was selected.

A list of the available toolbars will cascade down the screen.

2. Select a new toolbar and click the left mouse button. In this example, the **Dimension** toolbar was selected.

The selected toolbar will appear on the drawing screen. See Figure 1-8. You can move any toolbar or change its shape as described in Figure 1-6.

To remove a toolbar from the screen

1. Locate the cursor arrow on the check mark located in the upper right corner of the toolbar and press the left mouse button.

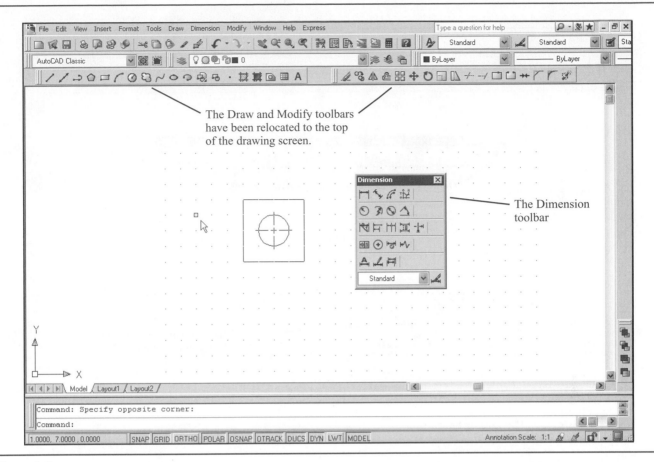

The Draw and Modify toolbars have been relocated to the top of the drawing screen.

The Dimension toolbar

Figure 1-8

Figure 1-8 shows the **Draw** and **Modify** toolbars docked horizontally at the top of the drawing screen. The **Dimension** toolbar will initally appear within the drawing area of the screen, as will any other toolbar activated. Toolbars can then be moved to different locations, as explained. If the toolbar is located close to either the top or sides of the drawing screen, it will blend into the area surrounding the drawing area and will no longer be within the drawing area.

To relocate the toolbars from positions outside the drawing area

1. Locate the cursor arrow above the icons, but still below the horizontal line that defines the toolbar area, and press and hold down the left mouse button. A rectangular box will appear around the toolbars that will move with the cursor arrow. Release the left button when the new location is reached.

1-3 THE COMMAND LINE BOX

The size of the **Command** window, located at the bottom of the screen, may be changed to display more or fewer command lines. It is recommended that at least two command lines be visible at all times.

To resize the command line box

See Figure 1-9.

1. Locate the cursor arrow along the top edge of the command line box. A double arrow will appear.
2. Still holding the left mouse button down, move the cursor arrow to a new location on the drawing screen.

The command line box will expand as the double arrow is moved upward.

3. Release the left mouse button to relocate the command line box.

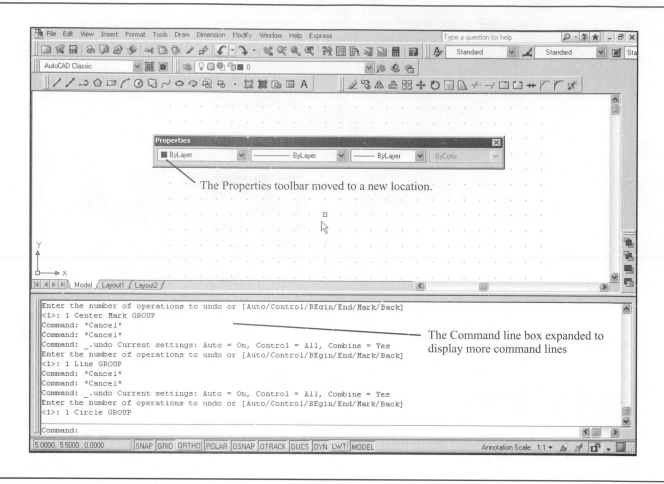

Figure 1-9

The command line box may now be moved and reshaped just like a toolbar. Figure 1-9 also shows the **Properties** toolbar moved from its original position above the drawing area to a location within the drawing area. AutoCAD allows you to customize the screen to whatever configuration suits your work best.

1-4   COMMAND TOOLS

A *tool* is a picture (icon) that represents an AutoCAD command. Most commands have equivalent tools.

To determine the command a tool represents

See Figure 1-10.

1. Locate the cursor arrow on the selected icon (tool).

   In the example shown, the **Line** command tool within the **Draw** toolbar was selected.

2. Hold the arrow still without pressing any mouse buttons.

The command name will appear below the tool. This name is referred to as a *tooltip*.

1-5   STARTING A NEW DRAWING

When a new drawing is started a drawing name is assigned, the drawing units are specified, the drawing limits are modified, if needed, and **Grid** and **Snap** values are defined. The following four sections will show how to start a new drawing.

1-6   NAMING A DRAWING

Any combination of letters and numbers may be used as a file name. Either upper- or lowercase letters may be used, since AutoCAD file names are not case-sensitive. The symbols $, -, and _ (underscore) may also be used. Other symbols, such as % and *, may not be used. See Figure 1-11.

All AutoCAD drawing files will automatically have the extension *.dwg* added to the given file name. If you name

A toolbar contains tools (icons that allow access to commands).

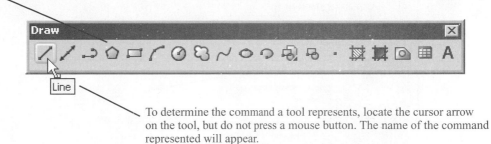

To determine the command a tool represents, locate the cursor arrow on the tool, but do not press a mouse button. The name of the command represented will appear.

Figure 1-10

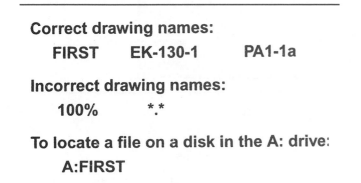

Correct drawing names:

FIRST   EK-130-1   PA1-1a

Incorrect drawing names:

100%   *.*

To locate a file on a disk in the A: drive:

A:FIRST

Figure 1-11

a drawing **FIRST,** it will appear in the files as **FIRST.dwg.** Other extensions can be used, but the .dwg is the default setting. (A default setting is one that AutoCAD will use unless specifically told to use some other value.)

Drawings may be saved to any drive and directory. The defaults for Windows XP Professional are the C: drive and the My Documents directory.

If you want to locate a file on another drive, specify the drive letter followed by a colon in front of the drawing name. For example, in Figure 1-11 the file specified **A:FIRST** will locate the drawing file FIRST on the A: drive.

For large drawings, that is, drawings that have very large databases likely to exceed the capacity of the specified drive, it is better to work on a drive that you are sure has sufficient space to accept the drawing file and then transfer the drawing to another drive as part of the saving process.

To start a new drawing

There are four ways to access the **Create New Drawing** dialog box that is used to name a new drawing:

1. Select the **New Drawing** tool in the **Standard** toolbar (see Figure 1-12).
2. Select the **File** pull-down menu, then select **New** (see Figure 1-13).

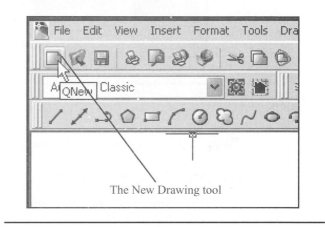

The New Drawing tool

Figure 1-12

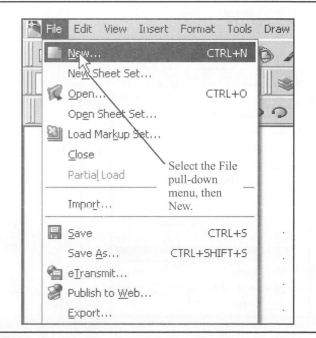

Select the File pull-down menu, then New.

Figure 1-13

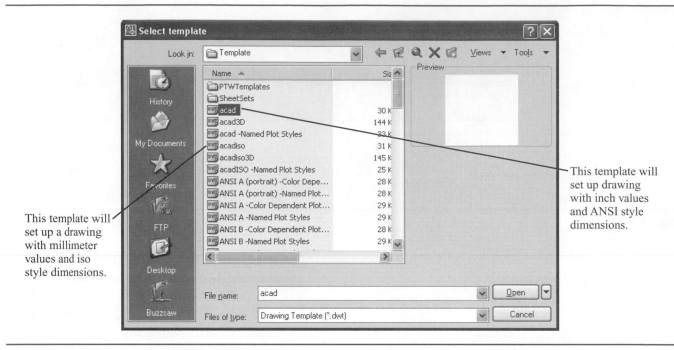

Figure 1-14

3. Type the word **new** in response to a Command prompt.
4. Hold down the **<Ctrl>** key and press **N**.

Any of these methods will cause the **Select template** dialog box to appear on the screen. See Figure 1-14. The **acad** template will set up a drawing using inch values and ANSI style dimensions. The **acadiso** template will set up a drawing using millimeter values and ISO style dimensions.

To save a new drawing file

A drawing name is entered as a file name by using the **Save As** option under the **File** pull-down menu. See Figure 1-15. It is recommended that a drawing name be assigned before a new drawing is started.

1. Select the **File** pull-down menu.
2. Select **Save As.**

The **Save Drawing As** dialog box will appear. See Figure 1-16. Note that the operating system for the computer

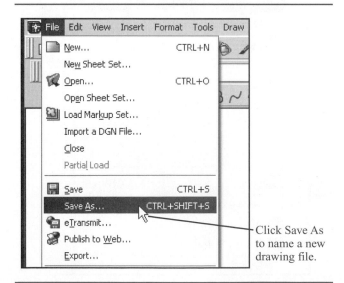

Figure 1-15

Figure 1-16

Drawing name will appear at the top of the screen.

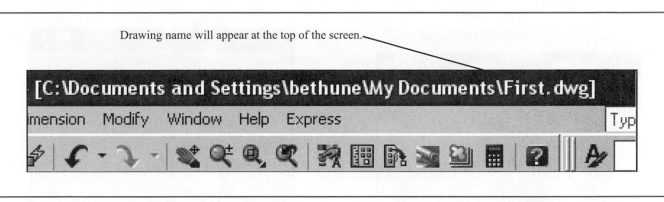

Figure 1-17

used to create Figure 1-16 was Microsoft Windows XP Pro. This dialog box may look different if a different operating system is used.

The **Save Drawing As** dialog box will list all existing drawings. Click on the thumbnail option under the **Views** heading to change the list to thumbnail drawings.

3. Enter the drawing name.

   In this example, the drawing name **First** was used.

4. Select **OK.**

   The name of the drawing will appear at the top of the screen. See Figure 1-17.

1-7 DRAWING UNITS

AutoCAD 2008's **Select template** dialog box allows for either English or metric units to be used as default values; however, AutoCAD can work in any of five different unit systems: Scientific, Decimal, Engineering, Architectural, or Fractional. The default system is the Decimal system and can be applied to either English (inches) or metric values (millimeters).

To specify or to change the drawing units

1. Select the **Format** pull-down menu.
2. Select **Units** (see Figure 1-18).

   The **Drawing Units** dialog box will appear. See Figure 1-19.

3. Select architectural units by clicking the arrow to the right of the **Type** box.

   A listing of the five unit options will cascade down.

4. Select **Architectural.**
5. Select **OK.**

The original drawing screen will reappear. Note that the coordinate display box now displays the cursor location in terms of feet and fractional inches.

6. Repeat the procedure and set the units back to decimal.

To specify or change the precision of the units system

Unit values can be expressed with zero to eight decimal places or from 0 to 1/256 of an inch.

1. Access the **Drawing Units** dialog box as explained previously.

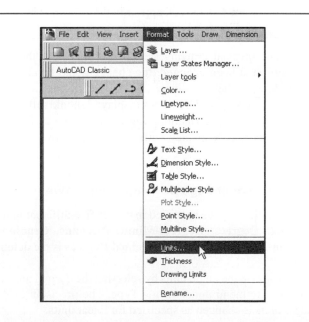

Figure 1-18

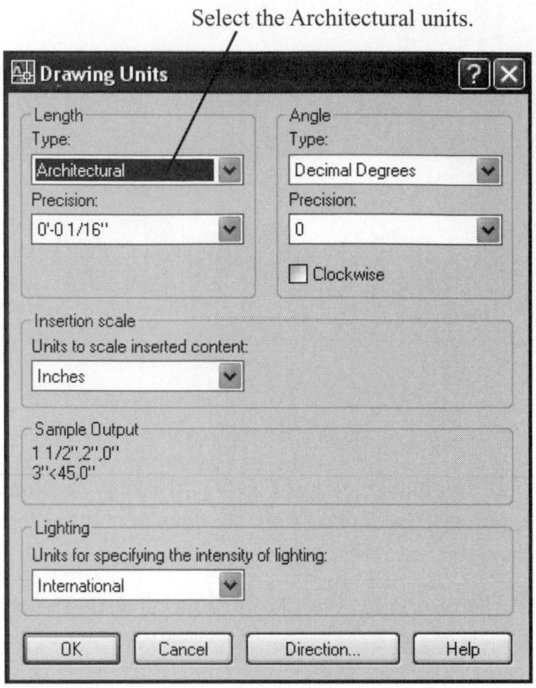

Select the Architectural units.

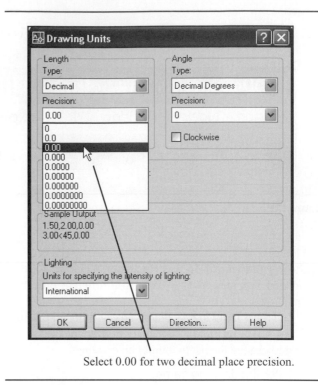

Figure 1-19

Select 0.00 for two decimal place precision.

Figure 1-20

2. Select the arrow to the right of the current precision value display box below the word **Precision.**

A listing of the possible decimal precision values will cascade from the box. See Figure 1-20.

3. Select **0.0000.**

The value 0.0000 will appear in the **Precision** box.

4. Select **OK.**

The original drawing screen will appear. If the unit precision had been changed, the change could be checked by looking at the coordinate display box in the lower left corner of the screen. The values displayed should reflect the new precision specification. Figure 1-20 shows how to select a coordinate display with a precision of two decimal places.

To specify or change the Angle units value

Angles may be specified in one of five different units: **Decimal Degrees, Degrees/Minutes/Seconds, Gradians, Radians,** or **Surveyor** units. Decimal Degrees is the default value.

Change the angle units by selecting the desired units in the cascade menu under **Angle Type.** The precision of the angle units is changed as specified for linear units.

1-8   DRAWING LIMITS

Drawing limits are used to set the boundaries of the drawing. The drawing boundaries are usually set to match the size of a sheet of drawing paper. This means that when the drawing is plotted and a hard copy is made, it will fit on the drawing paper.

Figure 1-21 shows a listing of standard flat-size drawing papers for engineering applications, Figure 1-22 shows standard metric sizes, and Figure 1-23 shows standard architectural sizes.

Standard Drawing Sheet Sizes—Inches

A = 8.5 × 11
B = 11 × 17
C = 17 × 22
D = 22 × 34
E = 34 × 44

Figure 1-21

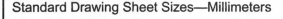

Standard Drawing Sheet Sizes—Millimeters

A4 = 210 × 297
A3 = 297 × 420
A2 = 420 × 594
A1 = 594 × 841
A0 = 841 × 1189

Figure 1-22

A standard 8.5″ × 11″ letter-size sheet of paper as used by most ink-jet and laser printers is referred to as an *A-size* sheet of drawing paper.

To align the drawing limits with a standard A4 (metric) paper size

1. Create an **acadiso** template, then select the **Format** pull-down menu.
2. Select **Drawing Limits.**

See Figure 1-24. The following prompts will appear in the **Command Line** box.

Reset Model space limits:
Specify lower left corner or [ON/OFF] <0.0000,0.0000>:

3. Press the **Enter** key or the right button on the mouse.

This means that you have accepted the default value of 0.0000,0.0000, or that the lower left of the drawing screen is the origin of an X,Y axis.

Standard Drawing Sheet Sizes—Architectural USA

A = 9 × 12
B = 12 × 18
C = 18 × 24
D = 24 × 36
E = 36 × 48

Figure 1-23

Create an acadiso template.

Figure 1-24

Specify upper right corner <420.0000,297.0000>:

4. Type **297,210;** press **Enter.**

This changes the upper right corner of the drawing screen to an X,Y value of 297,210. There will be no visible change to the screen. The new limit extends beyond the current screen calibration. The screen must be recalibrated to align with the new drawing limits.

5. Type **Zoom.**

Specify corner of window, enter a scale factor (nX or nXP) or [All/Center/Dynamic/Extents/Previous/Scale/Window] <real time>:

6. Type **A;** press **Enter.**

The **Zoom All** command will align whatever drawing limits have been defined to the drawing screen. You can verify that the new drawing limits are in place by moving the cursor around the screen and noting the coordinate display box in the lower left corner of the screen. The values should be much larger than they were with the original inch default screen settings.

1-9 GRID AND SNAP

The **Grid** command is used to place a dotted grid background on the drawing screen. This background grid is helpful for establishing visual reference points for sizing and locating points and lines.

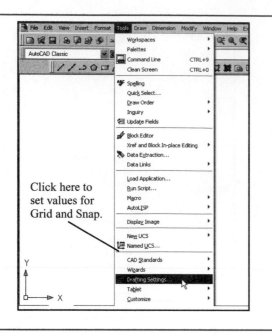

Click here to
set values for
Grid and Snap.

Figure 1-25

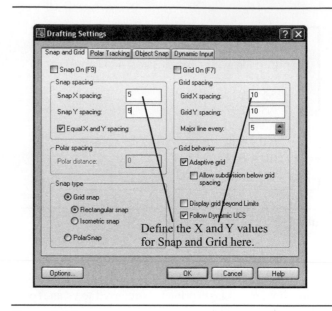

Define the X and Y values
for Snap and Grid here.

Figure 1-26

The **Snap** command limits the movement of the cursor to predefined points on the screen. For example, if the **Snap** command values are set to match the **Grid** values, the cursor will snap from grid point to grid point. It cannot be located at a point between the grid points.

The default **Grid** and **Snap** setting for the acad template is **0.50** inch, and the default setting for **Grid** and **Snap** for the acadiso template is **10** millimeters.

To set the Grid and Snap values

1. Create an **acadiso** template with **AutoCAD Classic** format, then select the **Tools** pull-down menu.
2. Select **Drafting Settings.**

See Figure 1-25. The **Drafting Settings** dialog box will appear. See Figure 1-26. If necessary, click the **Snap and Grid tab.**

3. Click on the **Grid** and **Snap** boxes.
4. Place the cursor on the **Snap X** spacing box to the right of the given value under the **Snap On** heading.

A vertical flashing cursor will appear.

5. Backspace out the existing value and type in **5.**
6. Click the **Snap Y** spacing box.

The Y spacing will automatically be made equal to the X spacing value. Nonrectangular snap spacing can be created by specifying different X and Y spacing values.

7. Select the **Grid X** spacing box under the **Grid On** heading.
8. Backspace out the existing value and type in **10** (just 10).
9. Click the **Grid Y** spacing box to make the X and Y values equal.
10. Select **OK.**

The original drawing screen will appear with a dotted grid. See Figure 1-27. Since the **Snap** values have been set to exactly half the **Grid** values, the cursor can be located either directly on grid points or halfway between them.

The grid will first appear in the lower left corner of the drawing screen. Use the **All** option of the **Zoom** command to increase the size of the grid to match the drawing screen. See Figure 1-27.

The grid can be turned on and off by either double-clicking the word **GRID** at the bottom of the screen or by pressing the **<F7>** key on the keyboard. Snap can be turned on and off by double-clicking the word **SNAP** at the bottom of the screen or by pressing the **<F9>** key on the keyboard.

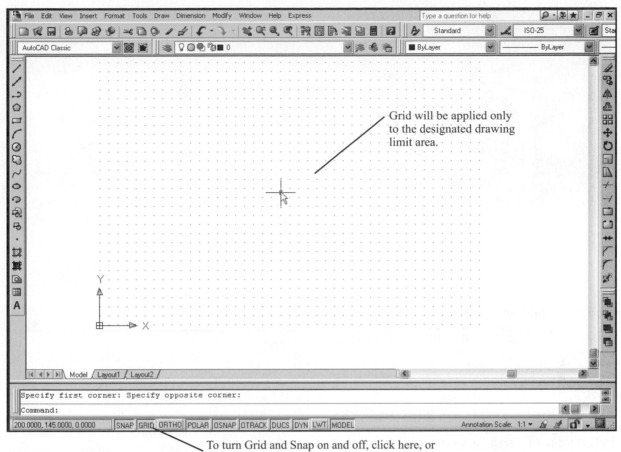

Grid will be applied only
to the designated drawing
limit area.

Specify first corner: Specify opposite corner:
Command:

200.0000, 145.0000, 0.0000   SNAP GRID ORTHO POLAR OSNAP OTRACK DUCS DYN LWT MODEL   Annotation Scale: 1:1

To turn Grid and Snap on and off, click here, or
press F7 for Grid and F9 for Snap.

Figure 1-27

1-10 SAMPLE PROBLEM SP1-1

Set up a drawing that will use millimeter dimensions
and the following parameters:

Sheet size = **297,210**
Grid = **10** spacing
Snap = **5** spacing
Whole-number precision

To specify the drawing units

1. Select the **New** tool from the **Standard** toolbar.

The **Select template** dialog box will appear. See
Figure 1-28.

2. Select the **acadiso** template, then **OK.**

To define the drawing precision

1. Select the **Format** pull-down menu, then **Units.**

The **Drawing Units** dialog box will appear. See Figure 1-29. In this example only whole numbers will be used,
so the "0" option is selected.

2. Select the desired precision, then **OK.**

To calibrate the sheet size

1. Select the **Format** pull-down menu, then **Drawing Limits.**
2. Set the lower limit for **0.0000,0.0000** (accept the default values), and the upper right limit for **297,210.**

The limits are given in terms of their X,Y coordinate
values.

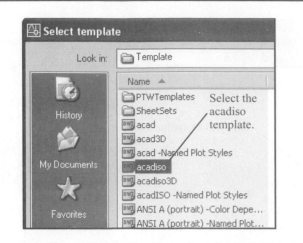

Figure 1-28

To set the Grid and Snap values

1. Select the **Tools** pull-down menu, then **Drafting Settings.**

   The **Drafting Settings** dialog box will appear. See Figure 1-30.

2. Enter the designated **Grid** and **Snap** values.

   The screen is now ready for starting a drawing using millimeter values. Note that the grid fills only the 297 × 210 area specified as the drawing limits.

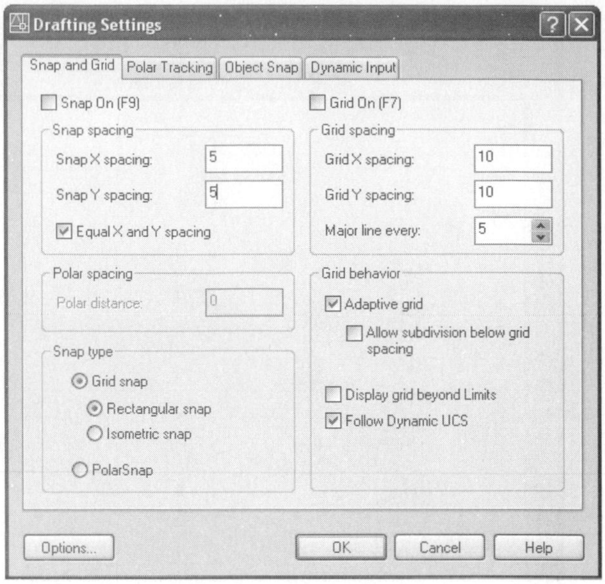

Define the X and Y values for the Grid and Snap.

Figure 1-30

1-11 SAVE AND SAVE AS

The **Save** and **Save As** commands are used to save a drawing. The Save command is also called the *quick save* command. It is used primarily while you are working on a drawing to save your work as you go. **Save** can also be used to save a finished drawing, but only under the original file name of the drawing. The **Save As** command allows you to save a drawing under a different file name.

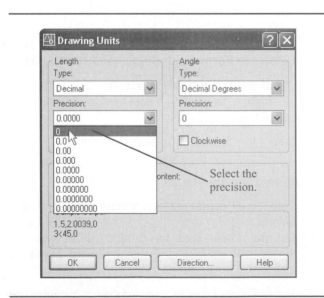

Figure 1-29

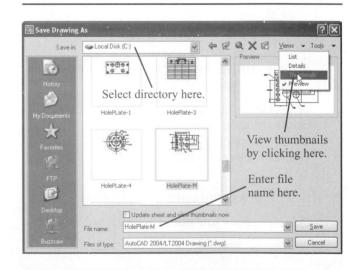

Figure 1-31

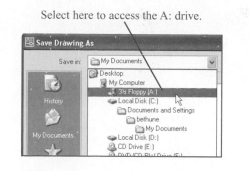

Select here to access the A: drive.

Figure 1-32

To use the Save command

The **Save** command can be accessed in one of three ways:

1. Click the **Save** icon on the **Standard** toolbar.
2. Select the **File** pull-down menu and select **Save**.
3. Press **<Ctrl> + <S>**.

The **Save Drawing As** dialog box will appear with the drawing name listed as the **File name.** See Figure 1-31.

4. Click the **Save** box.

To use the Save As command

The **Save As** command is used to save a drawing just as the **Save** command does, but under a name that is different from the original drawing name. For example, you named a drawing **FIRST** and then decided to save it on a disk on the A: drive under the name **BOX.**

1. Select the **File** pull-down menu, then **Save As.**

The **Save Drawing As** dialog box will appear. See Figure 1-32.

2. Scroll down the **Save** box and select a drive.
3. Type the new file name in the **File name** box.
4. Click the **Save** box.

1-12 OPEN

The **Open** command is used to call up an existing drawing so that you may continue working on it, or revise it.

To use the Open command

The **Open** command may be accessed in one of three ways:

1. Use the **Open** tool on the **Standard** toolbar.
2. Select the **File** pull-down menu, then select **Open**.
3. Press **<Ctrl> + <O>**.

The **Select File** dialog box will appear. Select the **Preview** option under the **Views** heading. See Figure 1-33. Drawings will be listed and a preview of each presented. Click the desired drawing's preview, and the file name will appear in the **File name** box, then click **Open.**

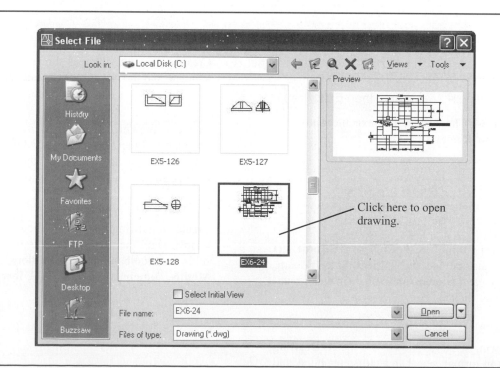

Click here to open drawing.

Figure 1-33

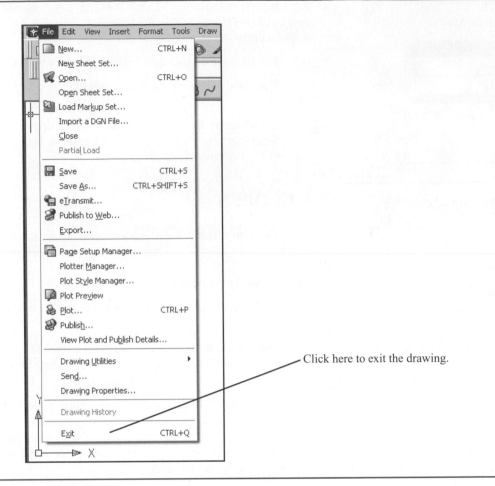

Click here to exit the drawing.

Figure 1-34

1-13 EXIT

The **Exit** command allows you to exit AutoCAD.

1. Select the **File** pull-down menu, then **Exit.**

See Figure 1-34. The system will exit the AutoCAD program and return to Windows.

1-14 EXERCISE PROBLEMS

EX1-1

Create a drawing screen as shown. Use the **AutoCAD Classic** format. Include the **Object Snap, Zoom, Draw, Modify,** and **Dimension** toolbars in the orientation and format shown.

EX1-2

Create a drawing screen as shown. Use the **2D Drafting & Annotation** format, select an **acadiso** template, and set the **Grid** spacing for **10** and **Snap** for **5.**

EX1-3

Create a drawing screen as shown. Select an **acad 3D** template and the **AutoCAD Classic** format.

EX1-4

Create a drawing screen as shown. Use the **acad** template and the **2D Drafting & Annotation** format. Add the following toolbars in the locations shown: **Object Snap, Modify II, Zoom,** and **Text.**

EX1-5

Create a drawing screen as shown. Use the **acad** template and the **AutoCAD Classic** format. Add the following toolbars in the locations shown: **Draw, Modify, Dimension, Zoom, USC, Text,** and **Orbit.**

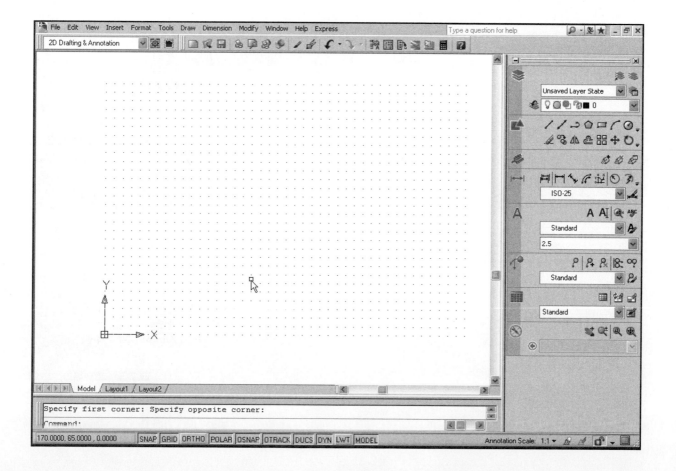

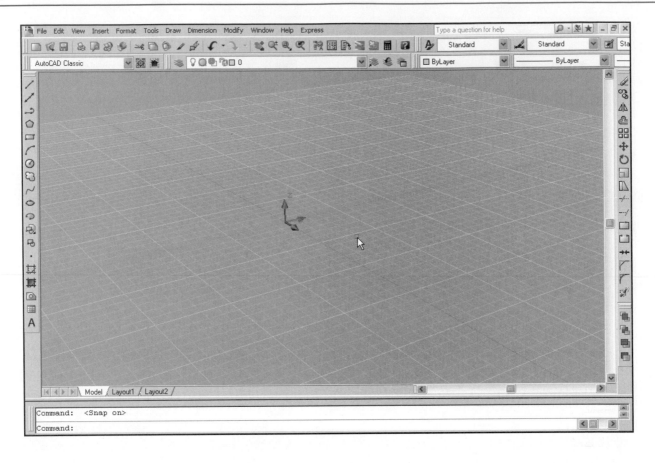

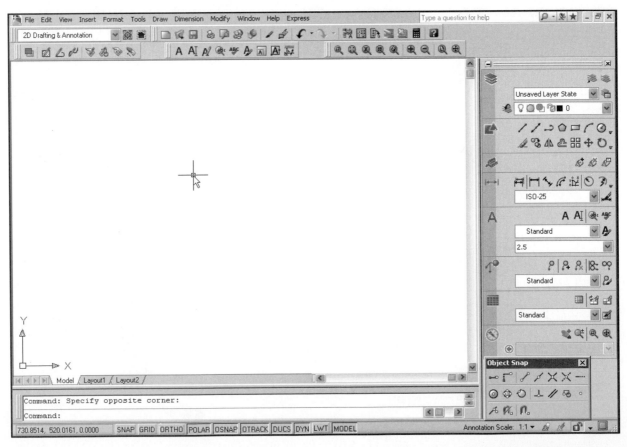

EX1-6

Create the following screen setup.
Decimal Units
Drawing Limits = **594,420**
Grid = **25**
Snap = **12.5**

EX1-7

Create the following screen setup.
Architectural Units
Drawing Limits = **48,36**
Grid = **6**
Snap = **3**

EX1-8

Create the following screen setup.
Fractional Units
Drawing Limits = **36,24**
Grid = **1**
Snap = **1/2**

CHAPTER 2

Fundamentals of 2D Construction

2-1 INTRODUCTION

This chapter demonstrates how to work with AutoCAD commands. AutoCAD commands are executed by using command tools, toolbars, and a series of prompts. The prompts appear in the command line box and ask for a selection or numeric input so that a command sequence can be completed. The **Line** command was chosen to demonstrate the various input and prompt sequences that are typical with AutoCAD commands.

Most of the commands contained in the **Draw** and **Modify** toolbars will be demonstrated in this chapter. The purpose is to present enough commands for the reader to be able to create simple 2D shapes. The next chapter will present more advanced applications for these commands and will present some new commands as well.

2-2 LINE—RANDOM POINTS

The **Line** command is used to draw straight lines between two defined points. Figure 2-1 shows the **Line** tool on the **Draw** toolbar.

There are four ways to define the length and location of a line: randomly select points; set a specific value for the **Snap** function and select points using the **Snap** spacing

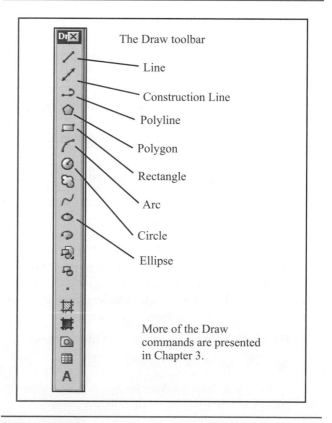

The Draw toolbar

Line

Construction Line

Polyline

Polygon

Rectangle

Arc

Circle

Ellipse

More of the Draw commands are presented in Chapter 3.

Figure 2-1

23

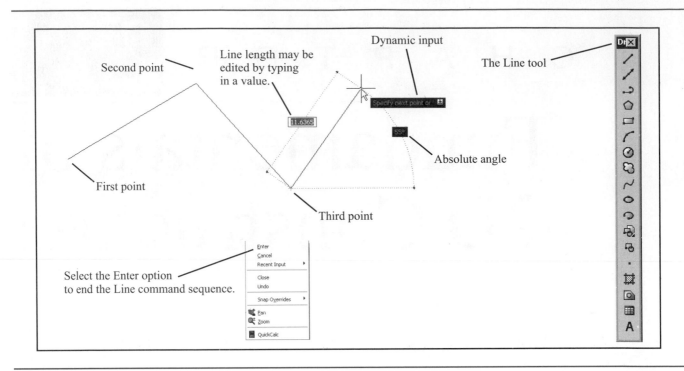

Figure 2-2

values; enter the coordinate values for the start and end points; or use relative inputs and specify the starting point, the length, and the direction of the line.

To randomly select points (See Figure 2-2.)

1. Select the **Line** tool in the **Draw** toolbar.

You can also access the **Line** command by typing the word **line** in response to a command prompt.

The following command sequence will appear in the command line box.

Command: _line Specify first point:

2. Place the cursor randomly on the drawing screen and press the left mouse button.

Specify next point or [Undo]:

As you move the cursor from point to point dynamic inputs will appear on the screen. See Figure 2-2. Dynamic inputs include an absolute angle and length change box. Dynamic inputs will be discussed in Section 2-6.

3. Randomly pick another point on the screen.

Specify next point or [Undo]:

AutoCAD will keep asking for another point until you press either the **Enter** key or the right mouse button.

4. When the right mouse button is pressed a small dialog box will appear. See Figure 2-2.

5. Click the **Enter** option to end the **Line** command sequence.

The **Line** command may be reactivated by pressing the right mouse button immediately after selecting the **Enter** option. A dialog box will appear. See Figure 2-3. Select the **Repeat Line** option to restart the **Line** command.

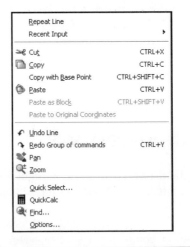

Figure 2-3

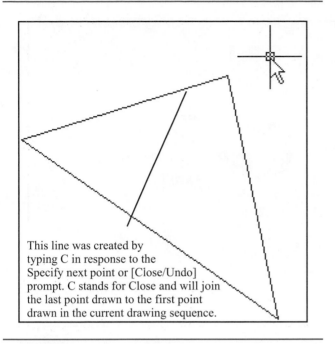

This line was created by
typing C in response to the
Specify next point or [Close/Undo]
prompt. C stands for Close and will join
the last point drawn to the first point
drawn in the current drawing sequence.

Figure 2-4

To exit a command sequence

If, as you work, you need to exit a command sequence, press the **<Esc>** key, which will return you to a command prompt. The **<Esc>** key allows you to exit almost all AutoCAD commands.

To create a closed area (See Figure 2-4.)

1. Select the **Line** tool from the **Draw** toolbar.

   Command: _line Specify first point:

2. Select a random point.

   Specify next point or [Undo]:

3. Select a second random point.

   Specify next point or [Undo]:

4. Select a third random point.

   Specify next point or [Close/Undo]:

5. Type **c,** then press **Enter** to activate the **Close** option.

   A line will be drawn from the third point to the starting point, creating a closed area.

2-3 ERASE

You can erase any line by using the **Erase** command. There are two ways to erase lines: select individual lines or window a group of lines. The **Erase** tool is located on the **Modify** toolbar. See Figure 2-5.

To erase individual lines

1. Select the **Erase** tool on the **Modify** toolbar.

   The following prompt will appear in the command line box.

   Command: _erase
   Select objects:

   The cursor will be replaced by a rectangular cursor. This is the select cursor (*pickbox*). Any time you see this

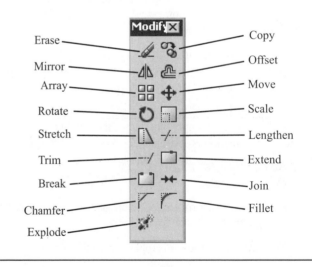

Figure 2-5

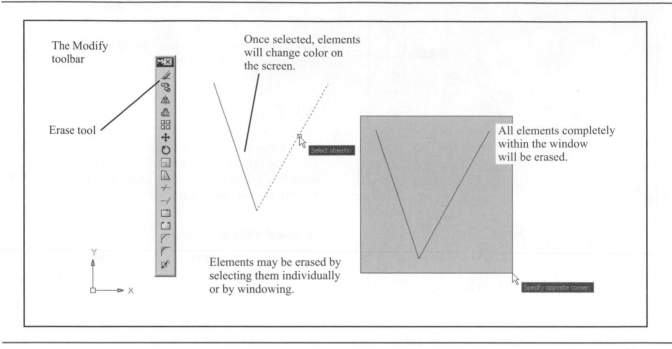

The Modify toolbar

Erase tool

Once selected, elements will change color on the screen.

Select objects:

All elements completely within the window will be erased.

Elements may be erased by selecting them individually or by windowing.

Specify opposite corner:

Figure 2-6

cursor, AutoCAD expects you to select an entity—in this example, a line. See Figure 2-6.

If you change your mind and do not want to erase anything, select the next command you want to use, and the Erase command will be terminated.

2. Select the two open lines by placing the rectangular cursor on each line, one at a time, and pressing the left mouse button.

The lines will change color from black to a broken gray pattern. This color change indicates that the line has been selected. The selection is confirmed by a change in the prompt.

Select objects: 1 found
Select objects:

3. Press the right mouse button or the **Enter** key to complete the **Erase** sequence.

The two lines should disappear from the screen.

To erase a group of lines simultaneously (See Figure 2-6.)

1. Select the **Erase** tool.

Command: _erase
Select objects:

2. Place the rectangular select cursor above and to the left of the lines to be erased and press the left mouse button.

3. Move the mouse, and a shaded window will drag from the selected first point.

4. When all the lines to be erased are completely within the window, press the left mouse button.

All the lines completely within the window will be selected and will change to broken gray lines. The number of lines selected will be referenced in the command line box.

5. Press the right mouse button or the **Enter** key to complete the command sequence.

The lines will disappear from the screen. The **Undo** tool will return all the lines, if needed. If the **Erase** window was created from right to left, any line even partially within the **Erase** window would be erased.

2-4 LINE—SNAP POINTS

Lines can be drawn using calibrated **Snap** spacing. This technique is similar to drawing points randomly, but because the cursor is limited to specific points, the length of lines can be determined accurately. See Figure 2-7.

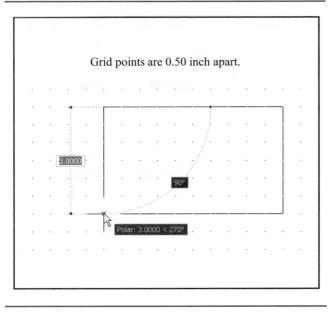

Grid points are 0.50 inch apart.

Figure 2-7

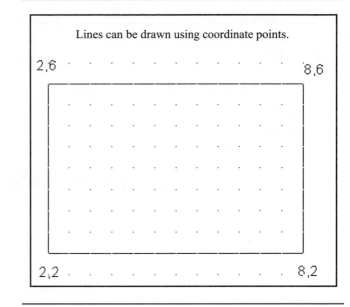

Lines can be drawn using coordinate points.

2,6   8,6

2,2   8,2

Figure 2-8

Problem: Draw a 3″ × 5″ rectangle.

1. Set the **Grid** spacing for **.5** and the **Snap** spacing for **.50,** and turn both commands on. Zoom the grid to fit the screen.
2. Select the **Line** tool from the **Draw** toolbar.

   Command: _line Specify first point:

3. Select a grid point.

   Specify next point or [Undo]:

4. Move the cursor horizontally to the right **10** grid spaces and press the left mouse button.

   The **Grid** spacing has been set at .5, so 10 spaces equals 5 units. Watch the dynamic input readings as you move the cursor. The X value should increase by 5, and the Y value should stay the same.

   Specify next point or [Undo]:

5. Move the cursor vertically **6** spaces and press the left mouse button.

   Specify next point or [Close/Undo]:

6. Move the cursor horizontally to the left **10** spaces and press the left mouse button.

   Specify next point or [Close/Undo]:

7. Type **c**; press **Enter.**

   Save or erase the drawing as desired.

2-5 LINE—COORDINATE VALUES

AutoCAD will also accept locational inputs, the start and end points of a line, in terms of X,Y coordinate values. The 0,0 origin is located in the lower left corner of the drawing screen.

Problem: Draw a 4″ × 6″ rectangle starting at the 2,2 coordinate point. See Figure 2-8.

1. Select the **Line** tool from the **Draw** toolbar.

   Command: _line Specify first point:

2. Move the cursor to the right of the word **point:** in the command line and left-click the mouse. Type **2,2**; press **Enter.**

   Specify next point or [Undo]:

3. Move the cursor to the command line and type **8,2**; press **Enter.**

   The X value has been increased 6 inches.

   Specify next point or [Close/Undo]:

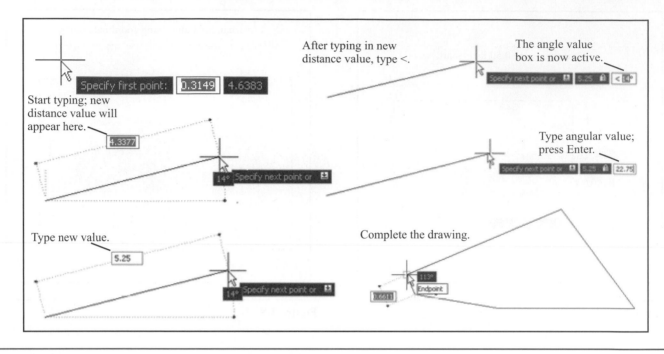

After typing in new
distance value, type <.

The angle value
box is now active.

Specify first point: 0.3149 4.6383

Start typing; new
distance value will
appear here.

4.3377

Type angular value;
press Enter.

14° Specify next point or

Type new value.

5.25

Complete the drawing.

14° Specify next point or

Figure 2-9

4. Move the cursor to the command line and type **8,6**; press **Enter**.

The Y value has been increased 4 inches.

Specify next point or [Close/Undo]:

5. Move the cursor to the command line and type **2,6**; press **Enter**.

Specify next point or [Close/Undo]:

6. Type **c**; press **Enter**.

Save or erase the drawing as desired.

2-6 LINE—DYNAMIC INPUTS

Dynamic Input allows both angular and distance values to be added to a drawing using screen prompts.

To create lines using Dynamic Input

This example was created using inch values.

1. Click the **Line** command tool and select a starting point.

Figure 2-9 shows the dynamic input for the selected first point.

2. Move the cursor to the approximate location of the second line point.

The distance and angular values for this new point will be displayed in boxes.

Angular values assume that 0° is a horizontal line to the right of the first line point.

3. Type in the desired distance value.

The new value will appear in the distance value box.

4. Continue typing (do not press **Enter**) and type the < symbol.

The line distance will change to the typed value, and the angle box will become active.

5. Type the desired angular value; press **Enter**.

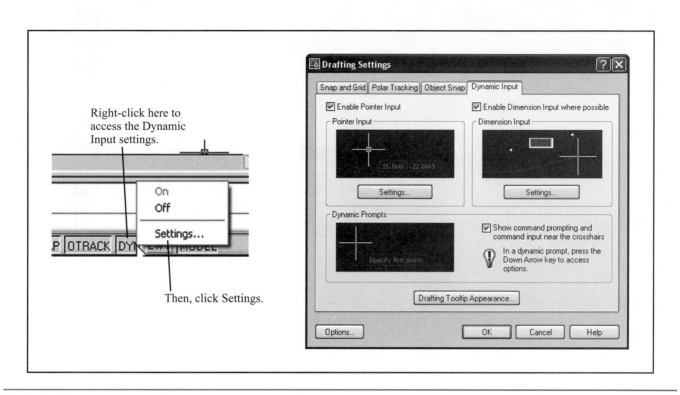

Figure 2-10

The angular value will be applied to the line, and the dynamic input command will prepare for the next line point input.

6. Complete the desired shape; press **Enter.**

To access the Dynamic Input settings

See Figure 2-10.

1. Right-click the **DYN** box located at the bottom of the screen.

   A small dialog box will appear. See Figure 2-10.

2. Click the **Settings** option.

   The **Dynamic Input** settings box will appear.

2-7 CONSTRUCTION LINE

The **Construction Line** command is used to draw lines of infinite length. Construction lines are very helpful during the initial layout of a drawing. They can be trimmed during the creation of the drawing as needed.

1. Select the **Construction Line** tool from the **Draw** toolbar.

   Command:_ xline Specify a point or [Hor/Ver/ Ang/Bisect/Offset]:

   The **Specify a point** command is the default setting.

2. Select or define a starting point.

   You can position the direction of the line by moving the cursor or by using dynamic input, as defined in Section 2-6, to define the line's direction.

3. Select or use dynamic input to define a through point.

   Specify through point:

   A line will pivot about the designated starting point and extend an infinite length in a direction through the cursor. You can position the direction of the line by moving the cursor. See Figure 2-11.

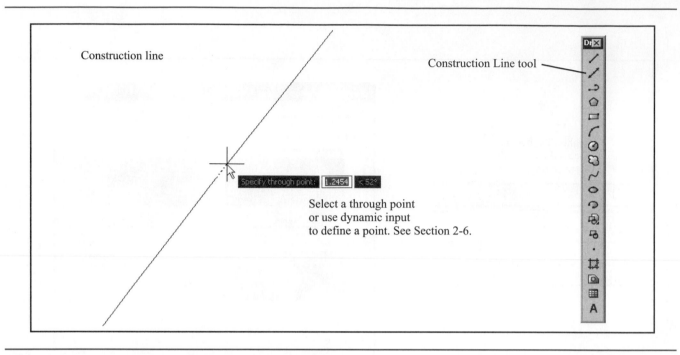

Construction line

Construction Line tool

Specify through point: 1.2454 < 52°

Select a through point
or use dynamic input
to define a point. See Section 2-6.

Figure 2-11

4. Select or define a through point.

An infinitely long line will be drawn through the two designated points.

Specify through point:

Another infinite line will appear through the starting point through the cursor.

5. Press **Enter.**

This ends the **Construction Line** command sequence. The sequence may be reactivated and another construction line drawn by pressing the **Enter** key a second time.

Other Construction Line commands: Hor/Ver/Ang

The lines shown in Figure 2-12 were created using the **Hor** (horizontal), **Ver** (vertical), and **Ang** (angular) options. A grid was created and the **Snap** option was turned on so the lines could be drawn through known points. Figure 2-12 was created as follows:

1. Set up the drawing screen with **Grid** and **Snap** spacings of **.5.**

2. Select the **Construction Line** tool in the **Draw** toolbar.

   Command:_ xline Specify a point or [Hor/Ver/Ang/ Bisect/Offset]:

3. Type **H**; press **Enter.**

   A horizontal line will appear through the cursor.

   Specify through point:

4. Select or define a point on the drawing screen.

   A horizontal line will appear through the point. As you move the mouse another horizontal line will appear through the cursor.

   Specify through point:

5. Select or define a second point.

   Specify through point:

6. Select or define a third point.

   Specify through point:

7. Double-click the right mouse button and select the **Repeat Construction Line** option from the dialog box that appears on the screen.

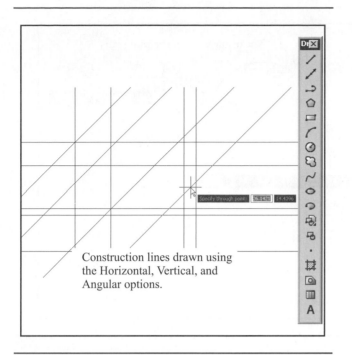

Figure 2-12

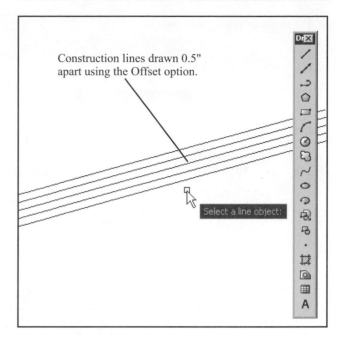

Figure 2-13

Command:_ xline Specify a point or [Hor/Ver/ Ang/Bisect/Offset]:

8. Type **V**; press **Enter.**

Specify through point:

9. Draw vertical lines, then double-click the **Enter** key and select the **Repeat Construction Line** option.

Command:_ xline Specify a point or [Hor/Ver/ Ang/Bisect/Offset]:

10. Type **A**; press **Enter.**

Enter angle of xline (0) or [Reference]:

11. Type **45**; press **Enter.**

Specify through point:

An infinite line at 45° will appear through the crosshairs.

12. Draw **45°** lines.

Specify through point:

13. Press **Enter.**

Your drawing should look approximately like Figure 2-12.

Other Construction Line command: Offset

The **Offset** option allows you to draw a line parallel to an existing line, regardless of the line's orientation, at a predefined distance. See Figure 2-13.

1. Select the **Construction Line** tool in the **Draw** toolbar.

Command:_ xline Specify a point or [Hor/Ver/ Ang/Bisect/Offset]:

2. Draw a line approximately **15°** to the horizontal.

Specify through point:

3. Double-click the **Enter** key to restart the **Construction Line** command sequence.

Command:_ xline Specify a point or [Hor/Ver/ Ang/Bisect/Offset]:

4. Type **O**; press **Enter.**

The O will appear in the dynamic input box.

Offset distance or through <Through>:

5. Type **.5**; press **Enter.**

Select a line object:

6. Select the line.

Select side to offset:

7. Select a point above the line.

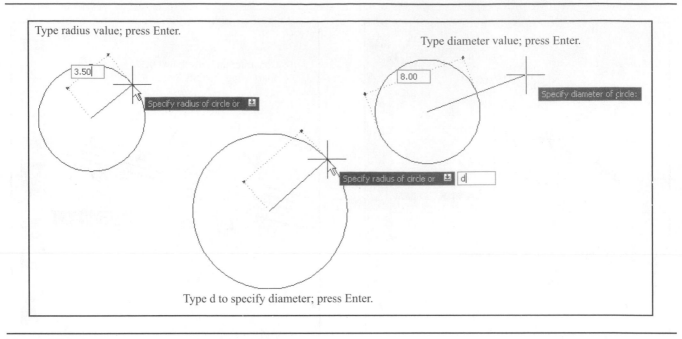

Figure 2-14

Select a line object:

8. Select the line just created by the **Offset** option.

Select side to offset:

9. Again select a point above the selected line.

Select a line object:

10. Draw several more lines below the original line.

Select a line object:

11. Press the right mouse button.

This will end the **Offset** command sequence and return a command prompt to the command prompt line.

2-8 CIRCLE

The **Circle** tool is on the **Draw** toolbar. Circles may be defined by a center point and either a radius or diameter; by two or three points on the diameter; or by the tangents to two existing lines or arcs and a radius value.

To draw a circle—radius (See Figure 2-14.)

1. Select the **Circle** tool from the **Draw** toolbar.

Command: _circle Specify a center point for circle or [3P/2P/Ttr (tan tan radius)]:

2. Select or use dynamic input to define a center point.

Specify radius of circle or [Diameter]:

3. Type **3.50**; press **Enter.**

The radius value will appear in the **Dynamic Input** box.

To draw a circle—diameter (See Figure 2-14.)

1. Select the **Circle** tool from the **Draw** toolbar.

Command: _circle Specify a center point for circle or [3P/2P/Ttr (tan tan radius)]:

2. Select or define a center point.

Specify radius of circle or [Diameter]:

3. Type **d**; press **Enter.**

Specify diameter of circle <value>:

4. Type **8.00**; press **Enter.**

To draw a circle—2 points (See Figure 2-15.)

1. Select the **Circle** tool from the **Draw** toolbar.

Command: _circle Specify a center point for circle or [3P/2P/Ttr (tan tan radius)]:

2. Type **2p**; press **Enter.**

Specify first end point of circle's diameter:

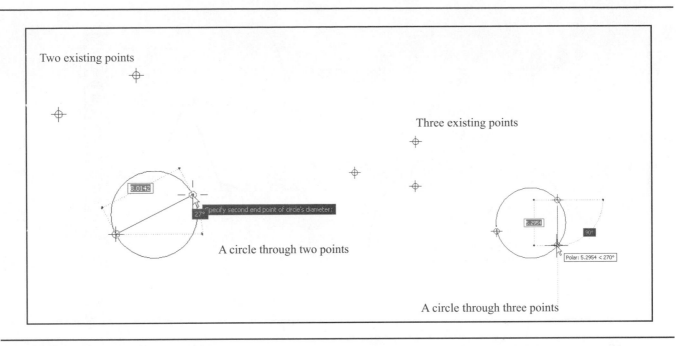

Two existing points

A circle through two points

Three existing points

A circle through three points

Figure 2-15

3. Select or define a first point.

   Select second end point of circle's diameter:

4. Select or define a second point.

   The circle will automatically be drawn through the first and second points.

To draw a circle—3 points (See Figure 2-15.)

1. Select the **Circle** tool from the **Draw** toolbar.

   Command: _circle Specify a center point for circle or [3P/2P/Ttr (tan tan radius)]:

2. Type **3p;** press **Enter.**

   Specify first point on circle:

3. Select or define a first point.

   Specify second point on circle:

4. Select or define a second point.

   Spccify third point on circle:

5. Select or define a third point.

To draw a circle—tangent tangent radius

   The **Tangent Tangent Radius** option allows you to draw a circle tangent to two other entities. Figure 2-16 shows a circle drawn tangent to two lines.

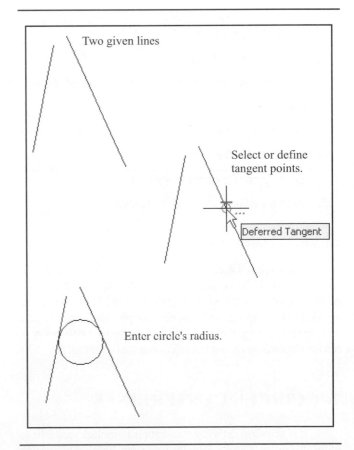

Two given lines

Select or define tangent points.

Deferred Tangent

Enter circle's radius.

Figure 2-16

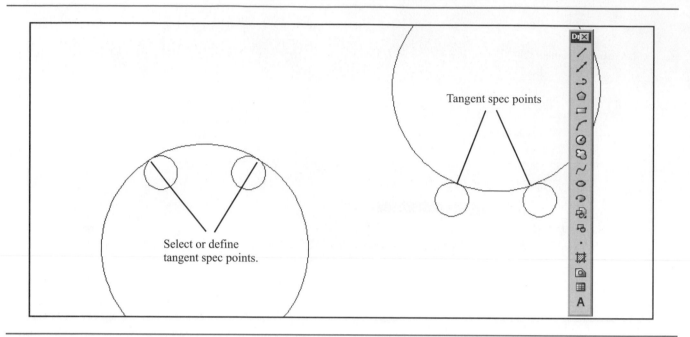

Figure 2-17

1. Select the **Circle** tool from the **Draw** toolbar.

   Command: _circle Specify a center point for circle or [3P/2P/Ttr (tan tan radius)]:

2. Type **ttr;** press **Enter.**

   Specify point on object for first tangent of circle:

3. Select an entity.

   Specify point on object for second tangent of circle:

4. Select the other entity.

   Specify radius of circle <value>:

5. Type a radius value; press **Enter.**

Quadrant-sensitive applications

The circle TTR option is quadrant-sensitive, that is, the final location of the tangent circle will depend on the location of the tangent spec points. In Figure 2-17 the TTR option was applied to two circles, and in both examples the tangent circle was created using the same radius. Note the difference in results depending on the circle's quadrants selected.

2-9 CIRCLE CENTERLINES

It is standard drawing convention to include centerlines with all circles. Circles usually represent holes. The circle's center point is used to locate the circle and, during manufac-

ture, serves as the location point for a drill. There are two ways to set up AutoCAD to add centerlines easily to a given circle: redefine the **Dimcen** system variable, or use the **Center Mark** option found on the **Dimension** toolbar.

To change the Dimcen system variable

There are no icons for **Dimcen;** the change is made using keyboard input entirely.

1. Type **setvar** next to the command prompt; press **Enter.**

   Enter variable name or? [?] <WSCURRENT>;

2. Type **dimcen;** press **Enter.**

   New value for DIMCEN <0.09>:

3. Type **−.1** for inch units or **−2** for metric units; press **Enter.**

   Command:

   The system variable is now changed but must be applied to the circle.

4. Type **dim;** press **Enter.**

   Dim:

5. Type **center;** press **Enter.**

   Select arc or circle:

6. Select the given circle.

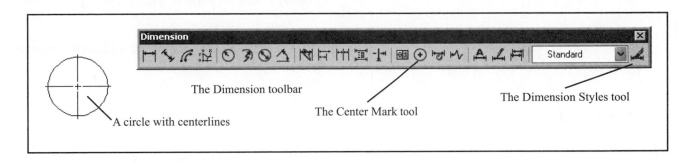

Figure 2-18

A centerline and center point will be added as shown in Figure 2-18.

To use the Center Mark tool

1. Access the **Dimension** toolbar by locating the cursor on an existing toolbar and right-clicking the mouse.
2. Select the **Dimension Styles** tool from the **Dimension** toolbar.

   The **Dimension Style Manager** dialog box will appear. See Figure 2-19.

3. Select the **Modify** box.

The **Override Current Style: Standard** dialog box will appear.

4. Select the **Symbols and Arrows** tab, then the **Center Marks** option, then **Line.**

   The preview screen will display centerlines.

5. Select **OK** and return to the drawing screen.

   Change the size value to create a clear center point and lines.

6. Click the **Center Mark** tool on the **Dimension** toolbar, then the given circle.

   Centerlines will appear on the circle.

The Dimension Style Manager dialog box

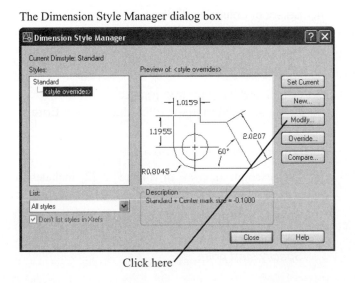

Click here

The Override Current Style: Standard dialog box

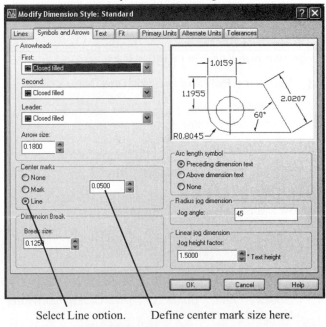

Select Line option.   Define center mark size here.

Figure 2-19

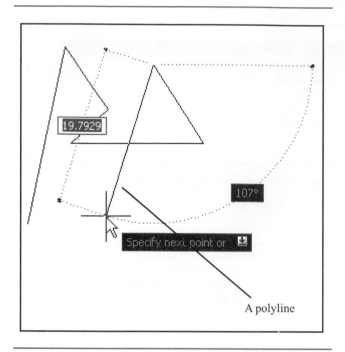

Figure 2-20

2-10 POLYLINE

A *polyline* is a line made from a series of individual, connected line segments that act as a single entity. Polylines are used to generate curves and splines and can be used in three-dimensional applications to produce solid objects.

Polylines can be entered using dynamic input. See Section 2-6.

To draw a polyline (See Figure 2-20.)

1. Select the **Polyline** tool from the **Draw** toolbar.

   Command: _pline
   Specify start point:

2. Select or define a start point.

   Specify next point or [Arc/Close/Halfwidth/ Length/Undo/Width]:

3. Select or define a second point.

   Specify next point or [Arc/Close/Halfwidth/ Length/Undo/Width]:

4. Select or define several more points.

   Specify next point or [Arc/Close/Halfwidth/ Length/Undo/Width]:

5. Press the right mouse button, then select **Enter.**

To verify that a polyline is a single entity

Figure 2-20 shows a polyline.

1. Select the **Erase** tool from the **Modify** toolbar or type the word **erase.**

   Select objects:

2. Select any one of the line segments in the polyline.

   Select objects:

The entire polyline, not just the individual line segment, will be selected, because in AutoCAD the polyline is a single entity.

3. Press **Enter.**

   The entire polyline will disappear.

4. Select the **Undo** tool from the **Standard** toolbar.

   The object will reappear.

To draw a polyline: Arc (See Figure 2-21.)

1. Select the **Polyline** tool from the **Draw** toolbar.

   From point:

2. Select or define a start point.

   Specify next point or [Arc/Close/Halfwidth/ Length/Undo/Width]:

3. Type **A;** press **Enter.**

   Specify endpoint of arc or [Angle/CEnter/ CLose/Direction/Halfwidth/Line/Radius/Second pt/Undo/Width]:

4. Select or define another point.

   Specify endpoint of arc or [Angle/CEnter/ CLose/Direction/Halfwidth/Line/Radius/Second pt/Undo/Width]:

5. Select or define another point.
6. Press the right mouse button, then select **Enter.**

Other options with a polyline arc

The example shown in Figure 2-22 includes a background grid with .5 spacing.

1. Select the **Polyline** tool from the **Draw** toolbar.

   Command: _pline
   Specify start point:

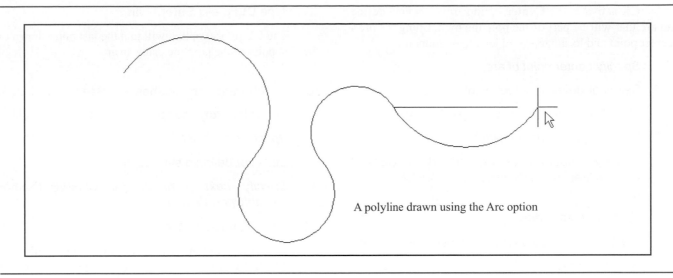

A polyline drawn using the Arc option

Figure 2-21

2. Select or define a start point.

   Specify next point or [Arc/Close/Halfwidth/Length/Undo/Width]:

3. Draw a short horizontal line segment.

   Specify next point or [Arc/Close/Halfwidth/Length/Undo/Width]:

4. Type **A**; press **Enter.**

   Specify endpoint of arc or [Angle/CEnter/CLose/Direction/Halfwidth/Line/Radius/Second pt/Undo/Width]:

5. Type **CE**; press **Enter.**

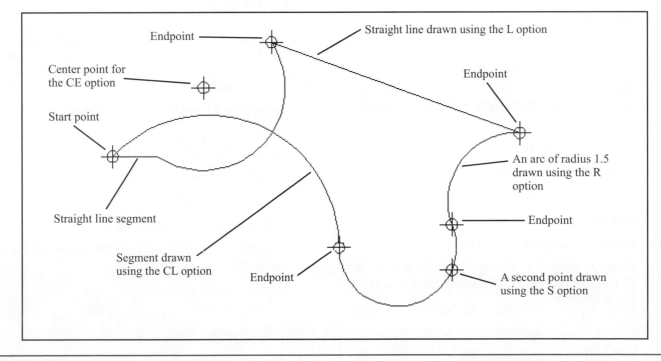

Figure 2-22

CE activates the **Center** option. You can now define an arc that will be part of the polyline by defining the arc's center point and its angle, chord length, or endpoint.

Specify center point of arc:

6. Select or define a center point.

Specify endpoint of arc or [Angle/Width]:

7. Select or define an endpoint.

Specify endpoint of arc or [Angle/CEnter/ CLose/Direction/Halfwidth/Line/Radius/Second pt/ Undo/Width]:

8. Type **L**; press **Enter**.

The **Line** option is used to draw straight-line segments.

Specify next point or [Arc/Close/Halfwidth/ Length/Undo/Width]:

9. Select or define an endpoint.

Specify next point or [Arc/Close/Halfwidth/ Length/Undo/Width]:

10. Type **A**; press **Enter**.

Specify endpoint of arc or [Angle/CEnter/ CLose/Direction/Halfwidth/Line/Radius/Second pt/ Undo/Width]:

11. Type **R**; press **Enter**.

Specify radius of arc:

12. Type **1.5** or other value; press **Enter**.

Specify endpoint of arc or [Angle]:

13. Select or define an endpoint.

Specify endpoint of arc or [Angle/CEnter/ CLose/Direction/Halfwidth/Line/Radius/Second pt/ Undo/Width]:

14. Type **S**; press **Enter**.

Specify second point on arc:

15. Select or define a point.

Specify endpoint of arc:

16. Select or define an endpoint.

Specify endpoint of arc or [Angle/CEnter/ CLose/Direction/Halfwidth/Line/Radius/Second pt/ Undo/Width]:

17. Type **CL**; press **Enter**.

The **CL** (close) option will join the last point drawn to the first point of the polyline using an arc.

To draw different line thicknesses (See Figure 2-23.)

1. Select the **Polyline** tool from the **Draw** toolbar.

Specify start point:

2. Select or define a start point.

Specify next point or [Arc/Close/Halfwidth/ Length/Undo/Width]:

3. Type **W**; press **Enter**.

The **Width** option is used to define the width of a line. The **Halfwidth** option is used to define half the width of a line.

Specify starting width <0.0000>:

4. Type **1.00**; press **Enter**.

Specify ending width <1.0000>:

5. Press **Enter**.

Specify next point or [Arc/Close/Halfwidth/ Length/Undo/Width]:

6. Draw several line segments.

2-11 SPLINE

A *spline* is a curved line created through a series of predefined points. If the curve forms an enclosed area, it is called a *closed spline.* Curved lines that do not enclose an area are called *open splines.* See Figure 2-24.

1. Select the **Spline** tool from the **Draw** toolbar.

Command: _spline
Specify first point or [Object]:

2. Select or define a start point.

Specify next point:

3. Select or define a second point.

Specify next point or [Close/Fit tolerance] <start tangent>:

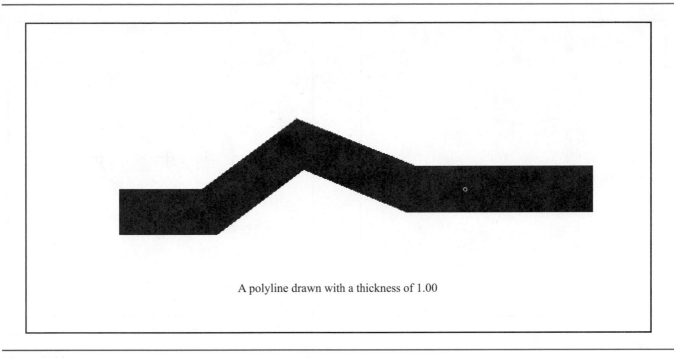

A polyline drawn with a thickness of 1.00

Figure 2-23

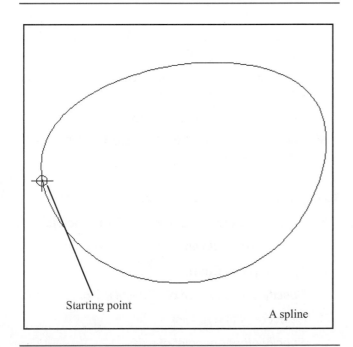

Starting point

A spline

Figure 2-24

4. Select or define two more points.

   Specify next point or [Arc/Close/Halfwidth/ Length/Undo/Width]:

5. Type **C**; press **Enter.**

   Specify tangent.

6. Select or define a point.

   Note how the shape of the spline changes as you move the cursor to locate the tangent point. These changes are based on your selection of a tangent point, which, in turn, affects the mathematical calculations used to create the curve.

2-12 ELLIPSE

   There are three options associated with the **Ellipse** tool. These three options allow you to define an ellipse using the lengths of its major and minor axes, an included angle, or an angle of rotation about the major axis.

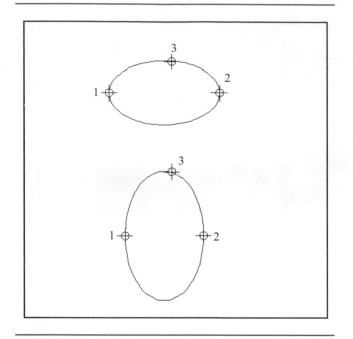

Figure 2-25

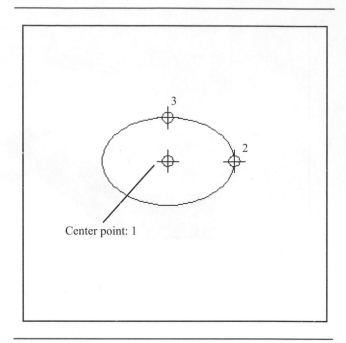

Figure 2-26

To draw an ellipse—axis endpoint (See Figure 2-25.)

1. Select the **Ellipse** tool from the **Draw** toolbar.

   Specify axis endpoint of ellipse or [Arc/Center]:

2. Select or define a start point for one of the axes.

   Specify other endpoint of axis:

3. Select or define an endpoint that defines the length of the axis.

   Specify distance to other axis or [Rotation]:

4. Select or define a point that defines half the length of the other axis.

   This distance is the radius of the axis. In the example shown, points 1 and 2 were used to define the major axes, and point 3 defines the minor axis. The bottom figure in Figure 2-25 shows an object in which points 1 and 2 were used to define the minor axes, and point 3 the major axis.

To draw an ellipse—center (See Figure 2-26.)

1. Select the **Ellipse** tool from the **Draw** toolbar.

   Specify axis endpoint of ellipse or [Arc/Center]:

2. Type **C:** press **Enter.**

   Specify center of ellipse:

3. Select or define the center point of the ellipse.

   Specify axis endpoint:

4. Select or define one of the endpoints of one of the axes.

   The distance between the center point and the endpoint is equal to the radius of the axis.

   Specify distance to other axis or [Rotation]:

5. Select or define a point that defines half the length of the other axis.

To draw an ellipse—arc (See Figure 2-27.)

1. Select the **Ellipse** tool from the **Draw** toolbar.

   Specify axis endpoint of ellipse or [Arc/Center]:

2. Type **A;** press **Enter.**

   Specify axis endpoint of elliptical arc or [Center]:

3. Select or define a point.

   Specify other endpoint of axis:

4. Select or define a point.

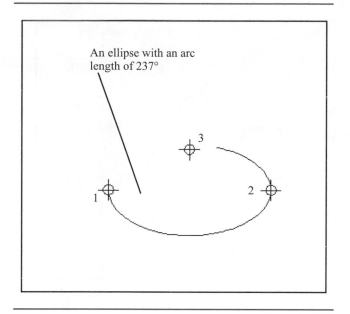

An ellipse with an arc length of 237°

Figure 2-27

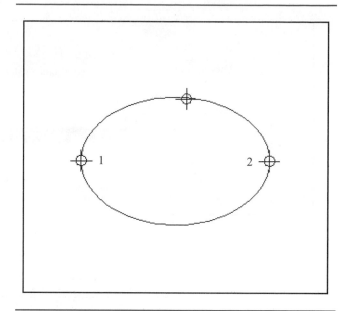

Figure 2-29

The distance between points 1 and 2 defines the length of the major axis.

Specify distance to other axis or [Rotation]:

5. Select or define a point that defines the minor axis.

Specify start angle or [Parameter]:

6. Type **0**; press **Enter.**

Specify end angle or [Parameter/Included angle]:

7. Type **237**; press **Enter.**

An ellipse may also be defined in terms of its angle of rotation about the major axis. An ellipse with 0° rotation is a circle, an ellipse of constant radius. An ellipse with 90° rotation is a straight line, an end view of an ellipse. See Figure 2-28.

To draw an ellipse by defining its angle of rotation about the major axis (See Figure 2-29.)

1. Select the **Ellipse** tool from the **Draw** toolbar.

Specify axis endpoint of ellipse or [Arc/Center]:

2. Select or define an endpoint of the major axis.

Specify other endpoint of axis:

3. Select or define the other endpoint of the major axis.

Specify distance to other axis or [Rotation]:

4. Type **R**; press **Enter.**

Specify rotation around major axis:

5. Type **47**; press **Enter.**

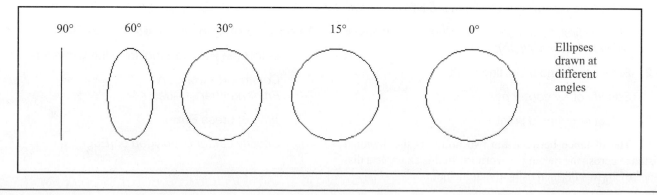

Ellipses drawn at different angles

Figure 2-28

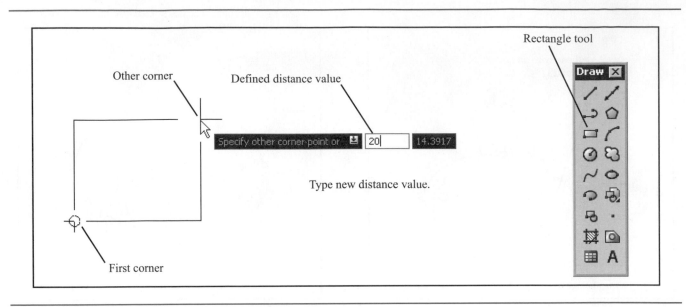

Figure 2-30

The same sequence of commands could have been applied using the **Ellipse Center** tool option. The distance between points 1 and 2 with the **Center** option is half the distance of the major axis.

2-13   RECTANGLE

The **Rectangle** command is used to draw rectangles. The rectangles generated are *blocks,* that is, they are considered to be one entity, not four straight lines. If you try to erase one of the lines that the rectangle comprises, all lines in the rectangle will be erased. The **Explode** command, whose tool is located on the **Modify** toolbar, may be applied to any block to reduce it to its individual elements. If **Explode** is applied to a rectangle, it will be changed from a single entity to four individual straight lines.

To draw a rectangle (See Figure 2-30.)

1. Select the **Rectangle** tool from the **Draw** toolbar.

    Specify first corner point or [Chamfer/Elevation/ Fillet/Thickness/Width]:

2. Select or define a starting point.

    Specify other corner point:

3. Select or define a point.

The distance between the two points is the diagonal distance across the rectangle's corners. In this example a distance of 20 was defined using dynamic input.

To explode a rectangle (See Figure 2-30.)

1. Select the **Explode** tool from the **Modify** toolbar.

    Select object:

2. Select the rectangle.

    Select object:

3. Press **Enter.**

There is no visible change in the rectangle, but it is now composed of four individual straight lines.

2-14   POLYGON

A *polygon* is a closed figure bounded by straight lines. The **Polygon** command will draw only regular polygons, in which all sides are equal. A regular polygon with four equal sides is a square. A six-sided polygon, or hexagon, will be drawn in this example.

To draw a polygon center point (See Figure 2-31.)

1. Select the **Polygon** tool from the **Draw** toolbar.

    Command: _polygon
    Enter number of sides <4>:

2. Type **6**; press **Enter.**

    Specify center of polygon or [Edge]:

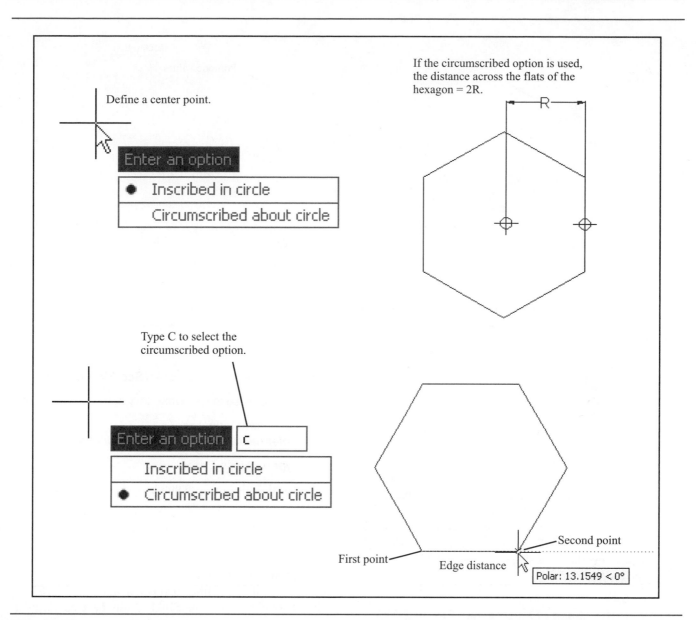

Figure 2-31

3. Select or define a center point.

   Enter an option [Inscribed in circle/Circumscribed about a circle] <I>:

4. Type **C**; press **Enter**.

   Specify radius of circle:

5. Type a value; press **Enter**.

The **Circumscribe** option was selected because the diameter of the designated circle will equal the distance across the flats of the hexagon. The distance across the flats is often used to specify hexagon head bolt sizes and the wrench sizes used to fit them. The **Inscribe** option is used in a similar manner.

To draw a polygon edge distance (See Figure 2-31.)

1. Select the **Polygon** tool from the **Draw** toolbar.

   Command: _polygn
   Enter number of sides <4>:

2. Type **6**; press **Enter**.

   Specify center of polygon or [Edge]:

3. Type **E**; press **Enter**.

   Specify first endpoint of edge:

4. Select or define a point.

   Specify second endpoint of edge:

5. Select or define a point.

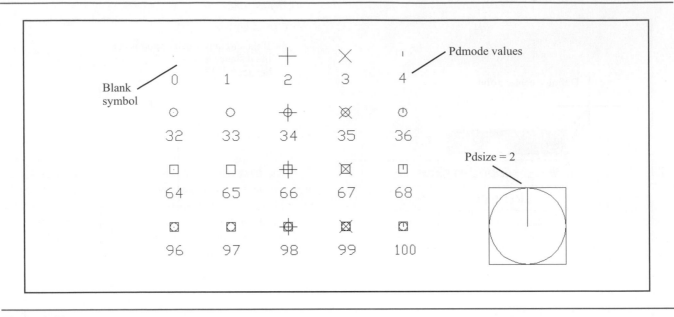

Figure 2-32

2-15 POINT

The **Point** command is used to draw points on a drawing. The default setting for a point's shape is a small dot, the kind used to display background grids. Other point shapes are available. See Figure 2-32. The default small dot point shape is listed as 0. Note that shape 1 is a void. The five available shapes can be enhanced by adding the numbers 32, 64, and 96 to the point's assigned value. For example, shape 2 is crossed horizontal and vertical lines. Shape 34 is shape 2 with a circle around it; shape 66 is shape 2 with a square overlay; and shape 98 is shape 2 with both a circle and a square around it.

To change the shape of a point (See Figure 2-32.)

Start with a command prompt.

1. Type **pdmode** and press **Enter** in response to a command prompt.

    Command: _pdmode
    Enter new value for PDMODE<0>:

2. Type **2**; press **Enter.**

    Command: _point POINT

3. Select several points to verify that the shape has been changed.

An **Enter** command should end the command sequence. If it does not, draw a line and use **Enter** to return to a command prompt. Erase the line and continue working.

To change the size of a point (See Figure 2-32.)

1. Type **pdsize** in response to a command prompt.

    Command: _pdsize
    Enter new value for PDSIZE<0.0000>:

2. Type **2**; press **Enter.**

    The default value is **0.0000.**

2-16 TEXT

There are two ways to apply text to an AutoCAD drawing: use the **Multiline Text** command or use the **Single Line Text** command. The **Single Line Text** command is accessed through the **Draw** pull-down menu, then selecting the **Text** option. The **Single Line Text** command is most useful when entering small amounts of text.

To use the Single Line Text command (See Figure 2-33.)

1. Select the **Single Line Text** command from the **Text** option on the **Draw** pull-down menu.

    Current text style: Standard
    Text height: 0.200
    Specify start point of text or [Justify/Style]:

2. Select or define a start point.

    The text will start at the selected point and run left to right.

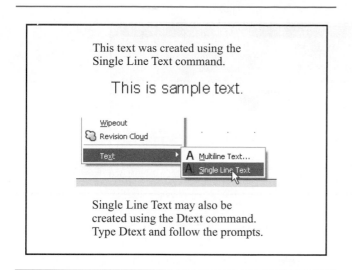

Figure 2-33

Specify height <0.2000>:

3. Type **.25**; press **Enter**.

This changes the text height from .2000 to .2500.

Specify rotation angle of text <0>:

4. Press **Enter**.

A bracket will appear with its lower left corner at the start point.

Enter text:

5. Type: **This is sample text**; press **Enter**.

Enter text:

6. Press **Enter**.

The **Single Line Text** command always assumes you will be typing another line of text regardless of how many lines you type. An **Enter** response signals that you are done entering text.

To use the Multiline Text tool (See Figures 2-34 through 2-39.)

The **Multiline Text** command is used to enter a large amount of text. First, define the area in which the text is to be entered; the **Text Formatting** dialog box will appear on the screen. Text is typed into the dialog box, just as with word processing programs, then transferred back to the drawing screen. Text can be moved to different locations on the drawing screen using the **Move** command.

1. Select the **Text** tool from the **Draw** toolbar.

Command:_mtext
Current text style: Standard
Text height: 0.2000
Specify first corner:

2. Select or define a point.

The area you are about to define is for the entire text entry. See Figure 2-34.

Specify opposite corner or [Height/Justify/Line spacing/Rotation/Style/Width]:

3. Define the other corner of the area.

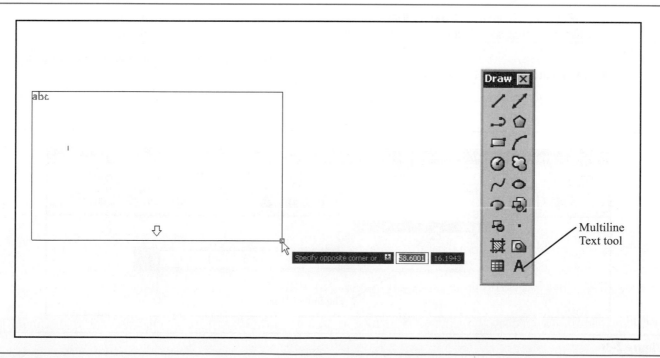

Figure 2-34

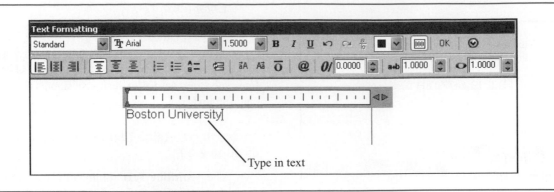

Figure 2-35

The **Text Formatting** dialog box will appear. See Figure 2-35.

4. Type in your text, then select the **OK** box.

The text will be located on the drawing screen within the specified area.

The Text Formatting dialog box options

The **Text Formatting** dialog box can be used to change text height, color, font style, and text style. The default justification is top left. The text will start in the upper left corner of the text area and proceed to the right and down, line by line, from the top to the bottom. Other justifications are available.

To change the text font

1. Select the **Text** tool from the **Draw** toolbar.
2. Designate a text area as specified in the previous section.

The **Text Formatting** dialog box will appear.

3. Select the scroll arrow to the right of the **Font** box.
4. Scroll down the available text listing.
5. Select the **Arial Black** option. (See Figure 2-36.)

Figure 2-37 shows the new text font. Compare the **Arial Black** text font shown in Figure 2-37 with the **Arial** text font shown in Figure 2-35.

To justify text

1. Select the **Text** tool from the **Draw** toolbar.
2. Select the first corner; do not select the other corner.

    Specify opposite corner or [Height/Justify/Line spacing/Rotation/Style/Width]:

3. Type **J**; press **Enter**.

A listing of justification options will appear. Figure 2-38 shows the meaning of the available options.

4. Select an option, then locate the other corner.

Figure 2-36

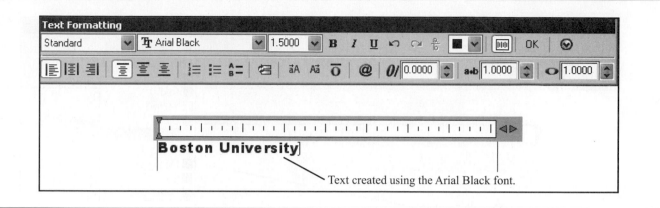

Figure 2-37

The justification options created using AutoCAD's standard text font.

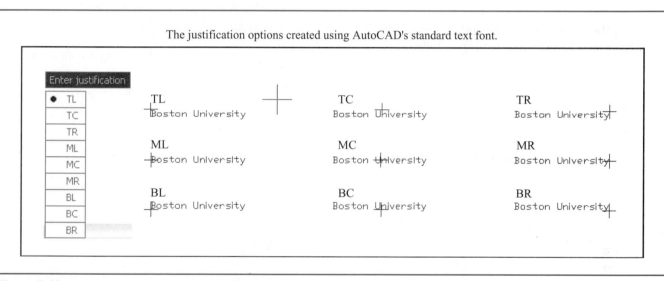

Figure 2-38

The **Text Formatting** dialog box will appear. Text may be entered as described before.

The symbol options

AutoCAD has three symbol commands:

%%C = ∅
%%P = ±
%%D = °

A symbol will initally appear in the %% form in the **Text Formatting** dialog box but will change to the equivalent symbol when the text is transferred to the drawing.

Symbols may also be created using the **Windows Character** chart as follows:

ALT + 0216 = ∅
ALT + 0177 = ±
ALT + 0176 = °

Text color

The color of a text input may be changed using the **Color** option located on the **Text Formatting** dialog box. See Figure 2-39. Select the scroll arrow to the right of the **Color** box, and a listing of available colors will scroll down.

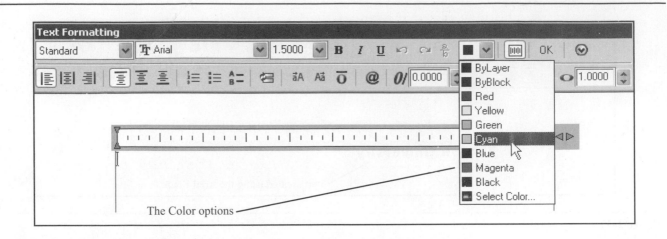

The Color options

Figure 2-39

2-17 MOVE

The **Move** command is used to move a line or object to a new location on the drawing. See Figure 2-40.

To move an object

1. Select the **Move** tool from the **Modify** toolbar.

    Select objects:

2. Window the entire object.

    Select objects:

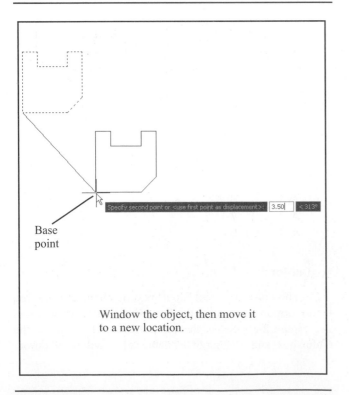

Base point

Window the object, then move it to a new location.

Figure 2-40

3. Press **Enter** or right-click the mouse.

    Specify base point or displacement:

4. Select or define a base point.

    Any point may be selected. Snap points are usually used as base points because they can define precise displacement distances and accurate new locations.

    Specify second point of displacement or <use first point of displacement>:

5. Select or define a second displacement point.

    The object will now be located relative to the second displacement point.

2-18 COPY

The **Copy** command is used to make an exact copy of an existing line or object. The **Copy** command can also be used to create more than one copy without reactivating the command. See Figure 2-41.

To copy an object

1. Select the **Copy** tool from the **Modify** toolbar.

    Select objects:

2. Window the entire object.

    Select objects:

3. Press **Enter** or right-click the mouse.

    Specify base point:

4. Select or define a base point.

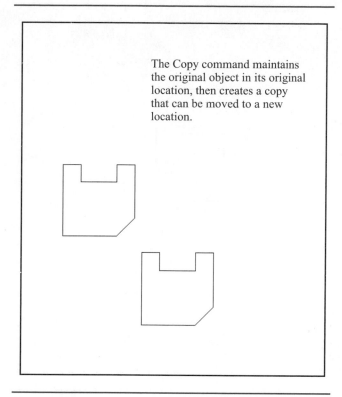

The Copy command maintains the original object in its original location, then creates a copy that can be moved to a new location.

Figure 2-41

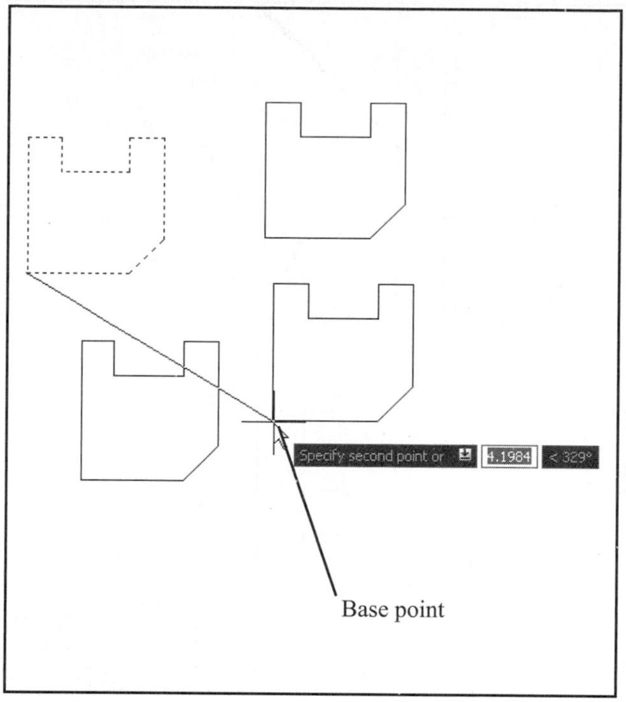

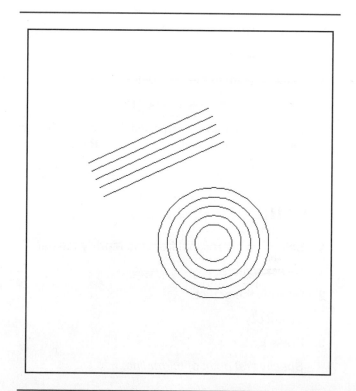

Base point

Figure 2-42

In the example shown in Figure 2-40 the lower left corner of the object was selected as the base point.

Specify second point of displacement or <use first point as displacement>:

5.   Select or define a second displacement point.

The original object will remain in its original location, and a new object will appear at the second displacement point.

6.   Right-click the mouse to enter the copied object.

To draw multiple copies (See Figure 2-42.)

AutoCAD will continue to make copies of a selected object until all the copied objects are entered. After each copy appears the prompt

Select a second displacement point:

Move the cursor to a new base point, or right-click the mouse to enter the objects.

2-19 OFFSET (See Figure 2-43.)

1.   Select the **Offset** tool from the **Modify** toolbar.

Specify offset distance or [Through/Erase/Layer] <Through>:

Figure 2-43

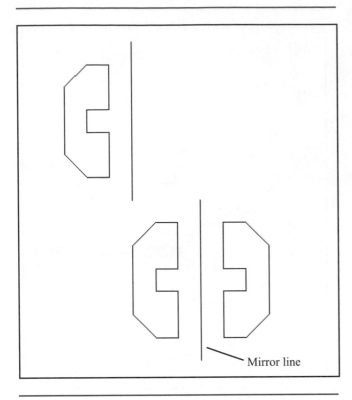

Figure 2-44

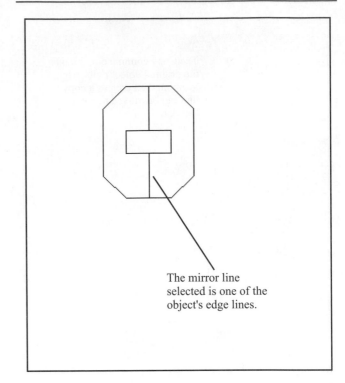

The mirror line selected is one of the object's edge lines.

Figure 2-45

2. Specify the offset distance by typing a value.

In this example, a distance of **.25** was selected.

Select object to offset or <exit>:

3. Select a line.

Specify point on side to offset:

4. Select a point to the right of the line.

Select object to offset or <exit>:

The process may be repeated by double-clicking the right mouse button and selecting the **Repeat Offset** option.

2-20 MIRROR (See Figure 2-44.)

1. Select the **Mirror** tool from the **Modify** toolbar.

Select objects:

2. Window the object.

Select objects:

3. Press **Enter.**

Specify first point of mirror line:

4. Select a point on the mirror line.

Specify second point of mirror line:

5. Select a second point on the mirror line.

Delete source object? [Yes/No]<N>:

6. Press **Enter.**

Any line may be used as a mirror line, including lines within the object. Figure 2-45 shows an object mirrored about one of its edge lines. The **Mirror** command is very useful when drawing symmetrical objects. Only half of the object need be drawn. The second half can be created using the **Mirror** command.

2-21 ARRAY

Objects may be arrayed using either the **Rectangular** or the **Polar** option. The **Rectangular** option is demonstrated in Figure 2-46.

1. Select the **Array** tool from the **Modify** toolbar.

The Array dialog box will appear.

2. Set **Rows** for **2**, Columns for **3, Row offset** for **3.50,** and **Column offset** for **4.00.**

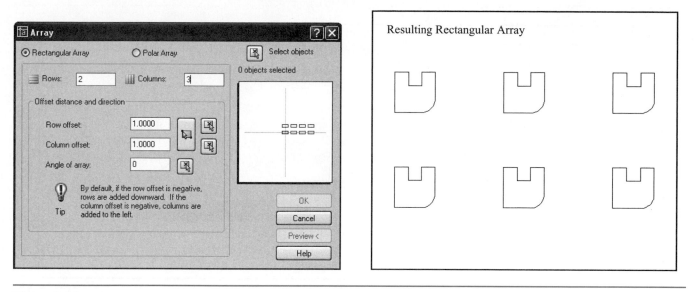

Figure 2-46

3. Click the **Select objects** box and select the object to be arrayed.

   The **Array** dialog box will appear.

4. Select **OK.**

To use the Polar Array option

Objects may also be arrayed using the **Polar Array** option. The **Polar Array** option will array objects about a defined center point at a given radius. See Figure 2-47.

1. Select the **Array** icon from the **Modify** toolbar.

   The **Array** dialog box will appear.

2. Select the **Polar Array** option.

3. Set the **Total number of items** to **8** and the **Angle to fill** to **360,** and turn off the **Rotate items as copied** box.

4. Select the **Center point,** and select the array's center point.

   The **Array** dialog box will appear.

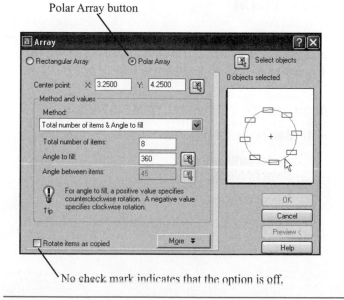

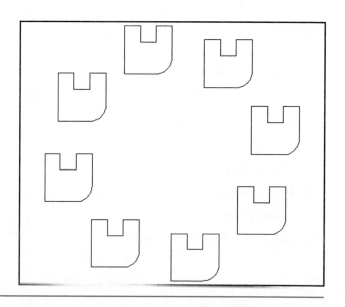

Figure 2-47

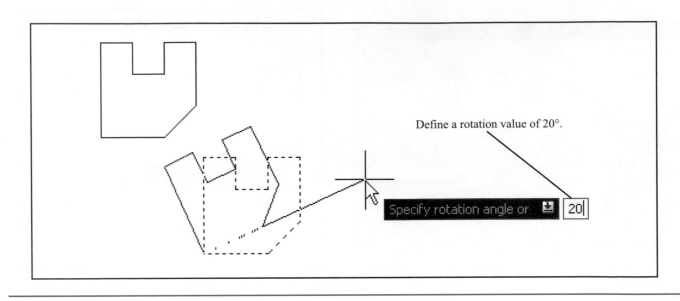

Define a rotation value of 20°.

Specify rotation angle or   20

Figure 2-48

5. Select the **Select objects** box and select the figure.
6. Select **OK.**

2-22 ROTATE

The **Rotate** tool is used to rotate an object about a spec-ified base point. The **3D Rotate** command will be covered later in the book. See Figure 2-48.

AutoCAD defines a horizontal line to the right as **0°.** Angles in the counterclockwise direction are positive angles.

To rotate an object (See Figure 2-48.)

1. Select the **Rotate** tool from the **Modify** toolbar.

   Select objects:

2. Window the object.

   Select objects:

3. Press **Enter.**

   Specify base point:

4. Select or define a base point.

The base point may be anywhere on the screen. In the example shown, the lower left corner of the object was selected as the base point.

   Specify rotation angle or [Reference]:

5. Type **20**; press **Enter.**

The object will rotate about the base point 20° in the counterclockwise direction.

2-23 TRIM

The **Trim** tool is used to cut away excessively long lines. It is a very important AutoCAD command and is used frequently.

To use the Trim command (See Figure 2-49.)

1. Select the **Trim** tool from the **Modify** toolbar.

   Select cutting edges:
   Select objects:

2. Select the circle.

   Select objects:

3. Press **Enter.**

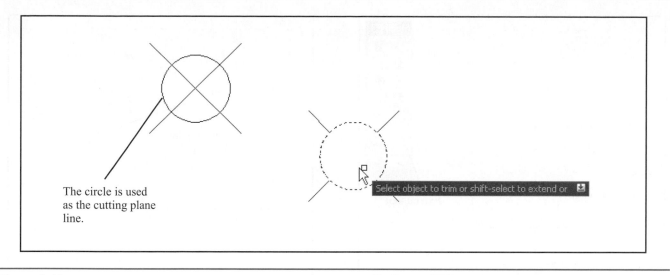

The circle is used as the cutting plane line.

Select object to trim or shift-select to extend or ⬇

Figure 2-49

Select object to trim or [Project/Edge/Undo]:

4. Select the center portions of the lines within the circle.
5. Press **Enter.**

The lines within the circle disappear. Figure 2-50 shows an example similar to that presented in Figure 2-49, but in this example the lines were selected as the cutting edges, and portions of the circle were trimmed.

2-24 EXTEND

The **Extend** command allows given lines to be extended to new lengths. See Figure 2-51.

1. Select the **Extend** tool from the **Modify** toolbar.

   Select boundary edges:
   Select objects:

2. Select a line that can be used as a boundary edge.

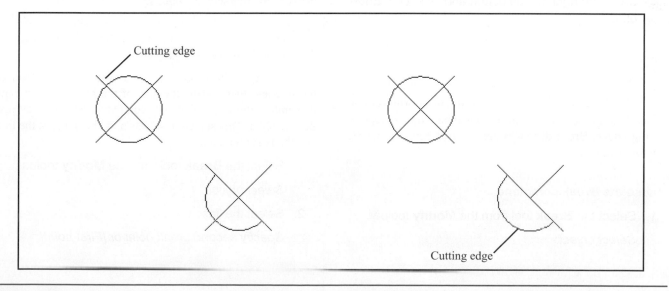

Cutting edge

Cutting edge

Figure 2-50

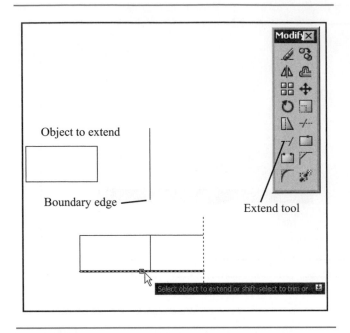

Figure 2-51

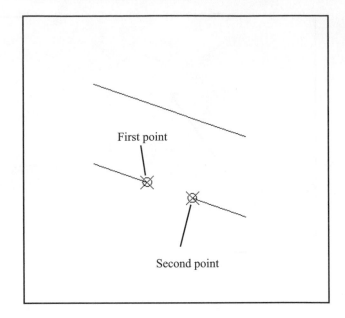

Figure 2-52

A boundary line must be positioned so that all the lines to be extended will intersect it. If you have an object that you want to make longer, draw a boundary line, or use the **Move** command and move one of the object's edge lines to the new position and use it as a boundary line.

Select object to extend or [Project/Edge/Undo]:

3. Select the lines to be extended.

Select object to extend or [Project/Edge/Undo]:

4. Press the right mouse button and select the **Enter** option.

2-25 BREAK

The **Break** command is used to remove portions of an object. It is similar to the **Trim** command but does not require cutting edges. Break distances are defined between points. See Figure 2-52.

To use the Break command

1. Select the **Break** tool from the **Modify** toolbar.

Select object:

2. Select a point on the line.

If needed, press the <**Shift**> key and right mouse button simultaneously to access the **Object Snap** screen menu. Select the **None** option to eliminate the default snap function.

Specify second break point or [First point]:

3. Select a second point.

The length of the break will be equal to the distance between the two selected points.

To use the First point option

The **F** or **First point** option allows you to first select the line and then define the size of the break. This means that unlike the original **Break** command sequence, as just described, the first selection selects only the line, not the line and the first break point.

1. Select the **Break** tool from the **Modify** toolbar.

Select object:

2. Select the line.

Specify second break point or [First point]:

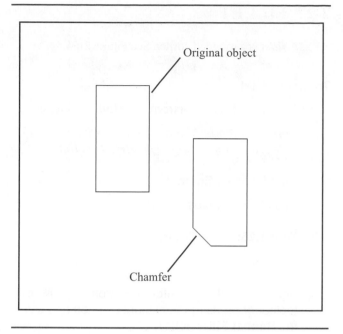

Figure 2-53

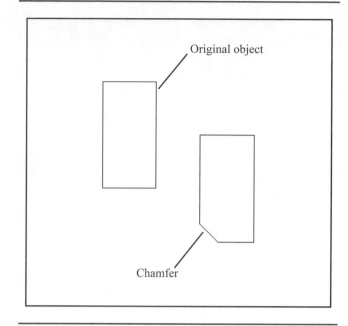

Figure 2-54

3. Type **F**; press **Enter**.

   Specify first break point:

4. Select or define the first point of the break.

   Specify second break point:

5. Select or define the second point of the break.

2-26 CHAMFER

A *chamfer* is a straight-line corner cut, usually cut at 45°. Other angles may be used. See Figure 2-53.

To create a chamfer

1. Select the **Chamfer** tool from the **Modify** toolbar.

   (TRIM mode) Current chamfer Dist1 = 0.0000, Dist2 = 0.5000
   Select first line or [Polyline/Distance/Angle/Trim/ Method]<Select first line>:

2. Type **D**; press **Enter**.

   Specify first chamfer distance <0.0000>:

3. Type **.75**; press **Enter**.

   Specify second chamfer distance <0.7500>:

   AutoCAD assumes that the chamfer will be at 45° and automatically sets the second distance equal to the first.

4. Press **Enter**.

   Command:

   The chamfer has now been set for a 45° chamfer of length .75. The chamfer is now applied to the rectangular figure in Figure 2-53.

5. Again, select the **Chamfer** tool on the **Modify** toolbar or press the **Enter** key and select the **Repeat Chamfer** option.

   (TRIM mode) Current chamfer Dist1 = 0.7500, Dist2 = 0.7500
   Select first line or [Polyline/Distance/Angle/Trim/ Method]<Select first line>:

6. Select a line.

   Select a second line:

7. Select a second line.

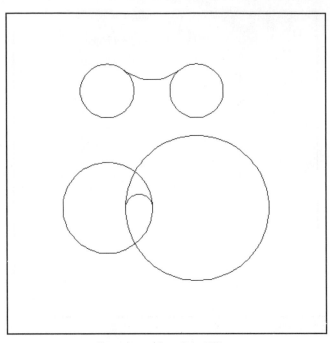

Some examples of the Fillet
command applied to circles

Figure 2-55

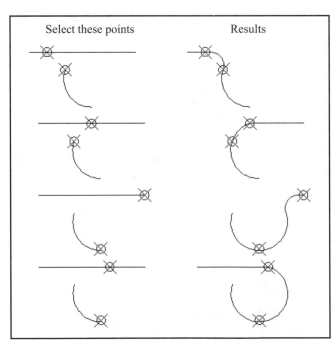

The Fillet command is location sensitive. The final shape
of the fillet will depend on the location of your selection
points. Here are four examples of fillets drawn from the
same initial arc line setup.

Figure 2-56

2-27 FILLET

A *fillet* is a rounded corner. See Figure 2-54.

To draw a fillet

1. Select the **Fillet** tool from the **Modify** toolbar.

   Current settings: Mode = Trim, Radius = 0.5000
   Select first object or [Polyline/Radius/Trim]:

2. Type **R**; press **Enter.**

   Specify fillet radius <0.0000>:

3. Type **1.25**; press **Enter.**

   Command:

4. Again, select the **Fillet** tool from the **Modify**
   toolbar or press the right mouse button and select
   the **Repeat Fillet** option.

   (TRIM mode) Current fillet radius = 1.2500
   Select first object or [Polyline/Radius/Trim]:

5. Select a line.

   Select second object:

6. Select a second line.

   Fillets can also be drawn between circles. See Figure 2-55. The **Fillet** command is location sensitive, meaning
that the location of the selection points will affect the shape
of the resulting fillet. Figure 2-56 shows how selection
points can be used to create different fillet shapes.

2-28 TABLE

The **Table** command is used to create tables such as
the one shown in Figure 2-57. Tables are used on drawings to
define multiple hole dimensions, to define coordinate data,
and to create parts lists, among other uses.

To create a table

1. Click the **Table** command tool on the **Draw** toolbar.

   The **Insert Table** dialog box will appear. See Figure 2-58.

2. Click the **Table style** button.

   The **Table Style** dialog box will appear. See Figure 2-59.

3. Click the **Modify** box.

   The **Modify Table Style: Standard** dialog box will
appear. See Figure 2-60.

LOCATION	X-DIM	Y-DIM	Ø
A1	15.00	85.00	20.00
A2	15.00	15.00	20.00
A3	145	85.00	20.00
A4	145	15.00	20.00
B1	55	70.00	30.00
B2	110	70.00	30.00
C1	80	30.00	40.00

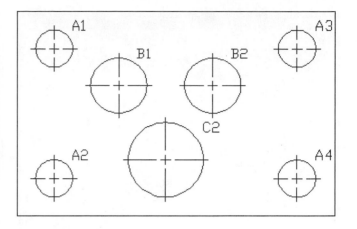

Figure 2-57

4. Click the **Text** tab under the **Cell styles** heading.

   The word **Data** indicates that the modifications will be made on the text used in the data cells.

5. Click the box to the right of **Text style: Standard** as shown in Figure 2-61.

   The **Text Style** dialog box will appear. See Figure 2-62.

6. Select the **Arial** font and set the height for **0.25,** then click **Apply** and **Close.**

   The **Modify Table Style: Standard** dialog box will appear. See Figure 2-63.

7. Click the arrow to the right of **Data** under **Cell styles** and select the **Title** option.

   The table's title can now be modified.

8. Return to the **Text Style** dialog box and set the text font for **Arial** and the text height for **0.375.** Click **Apply** and **Close.**

9. Return to the **Insert Table** dialog box and set the number of data rows and columns and size of boxes, then click **OK.** See Figure 2-58.

   Note that the number of data rows does not include the column headings.
   A blank table will appear on the screen. See Figure 2-64.

10. Click on the appropriate box and type in the required data.

    The arrow keys may be used to move from one box to another. Figure 2-57 shows the resulting table.

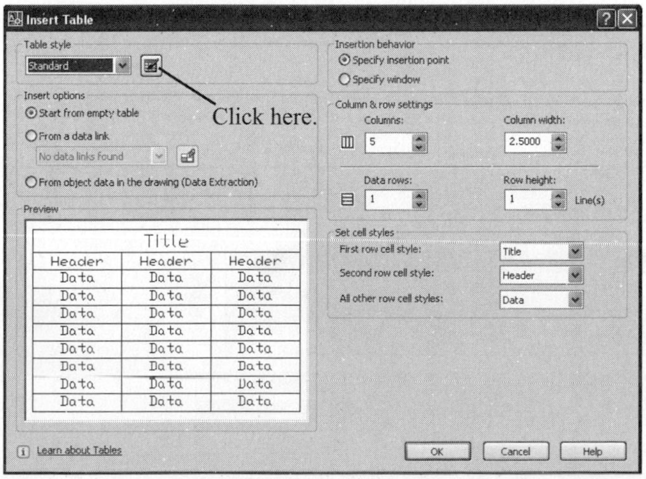

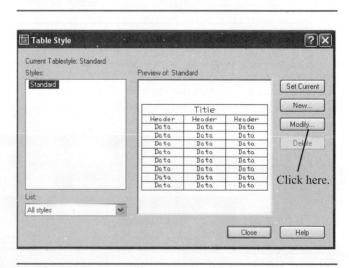

Figure 2-58

Figure 2-59

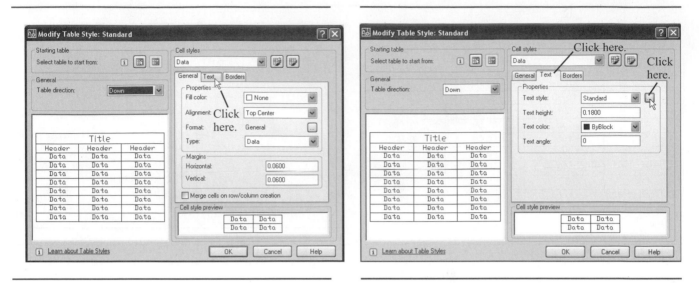

Figure 2-60

Figure 2-61

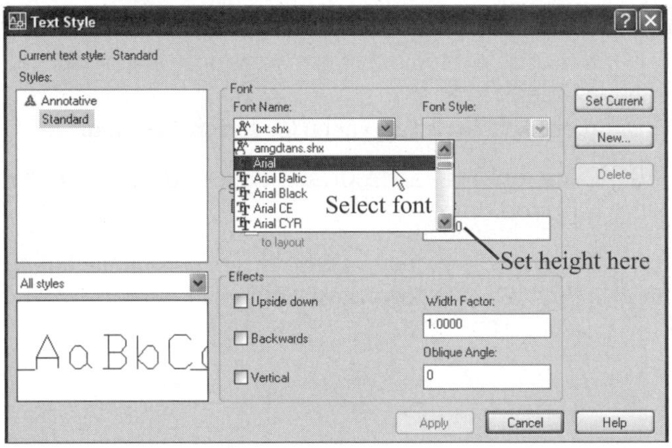

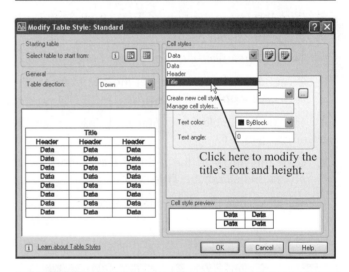

Figure 2-62

Figure 2-63

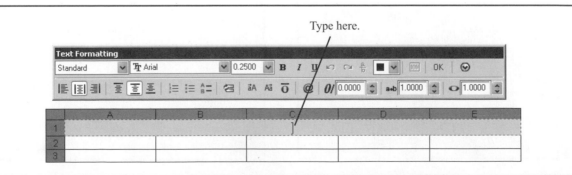

Figure 2-64

2-29 EXERCISE PROBLEMS

Redraw the following figures. Do not include dimensions. Dimensional values are as indicated.

EX2-1 INCHES

GUIDE PLATE

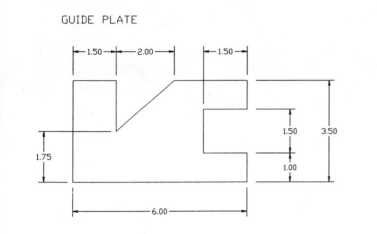

EX2-3 MILLIMETERS

BASE PLATE

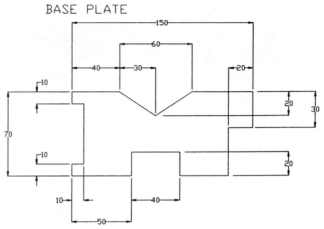

EX2-2 INCHES

TOP GASKET

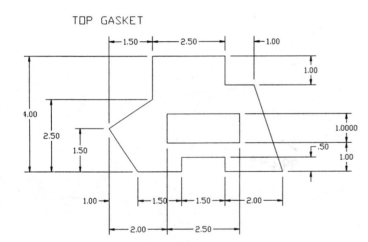

EX2-4 MILLIMETERS

GASKET

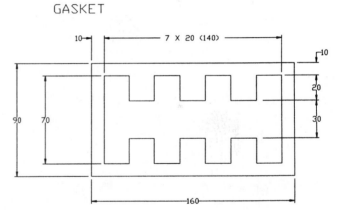

EX2-5 INCHES

AB = 5.0000
BC = 3.8376
CD = 1.5403
DE = ?
EF = 3.4361
FG = 2.3679
GH = 1.7713
HA = 3.0881
Angle CDE = ?

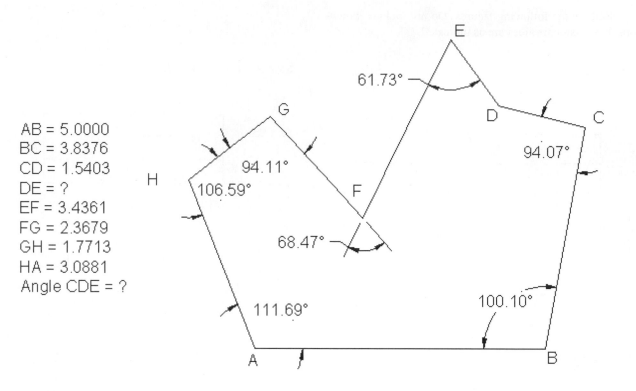

EX2-6 MILLIMETERS

AB = 130.0
BC = 28.1
CD = 40.6
DE = 111.0
EF = 133.9
FG = ?
GH = 114.0
HJ = 70.6
JA = 56.7
Angle EFG = ?

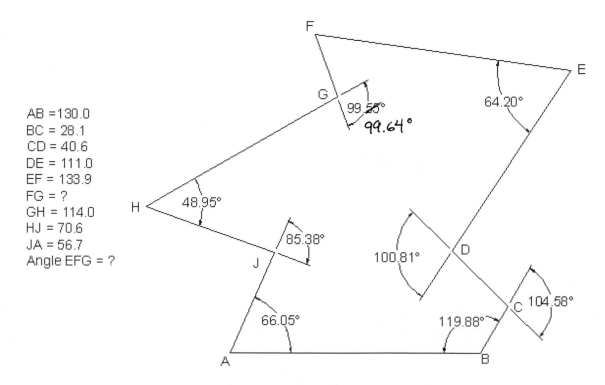

EX2-7 INCHES

SIDE BRACKET

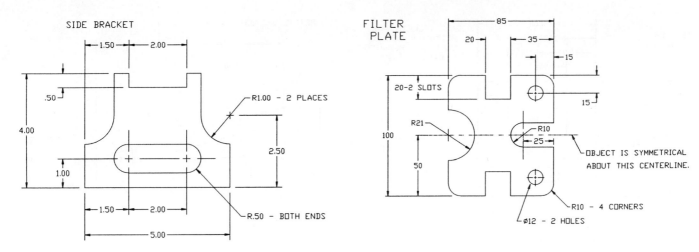

EX2-9 MILLIMETERS

FILTER
PLATE

EX2-8 INCHES

FILTER PLATE

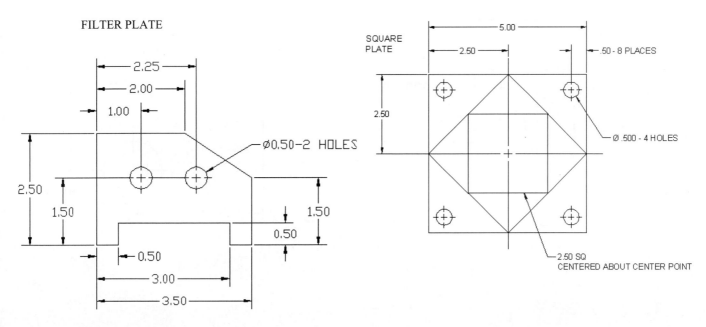

EX2-10 INCHES

EX2-11 INCHES

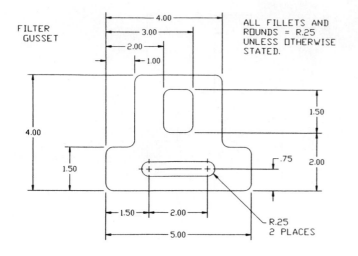

FILTER GUSSET

ALL FILLETS AND ROUNDS = R.25 UNLESS OTHERWISE STATED.

EX2-13 MILLIMETERS

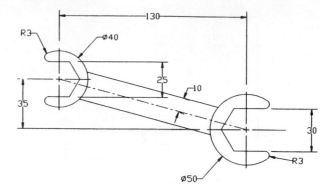

METRIC WRENCH

EX2-12 MILLIMETERS

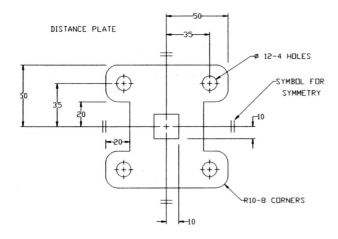

DISTANCE PLATE

EX2-14 INCHES

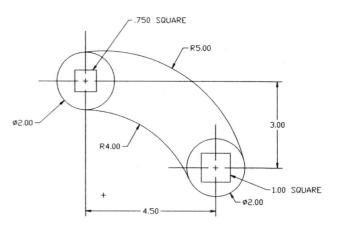

EX2-15 MILLIMETERS

TOP FILTER

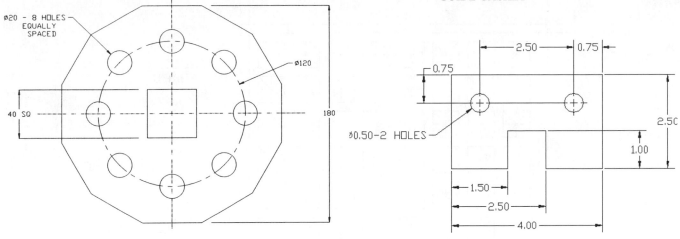

Ø20 - 8 HOLES
EQUALLY
SPACED

Ø120

40 SQ

180

EX2-17 INCHES

GUIDE GASKET

2.50   0.75

0.75

Ø0.50-2 HOLES

2.50

1.00

1.50

2.50

4.00

EX2-16 MILLIMETERS

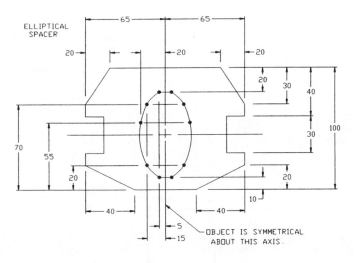

ELLIPTICAL
SPACER

65   65

20   20   20

20   30   40

30   100

70

55

20   20

40   40   10

5   OBJECT IS SYMMETRICAL
15   ABOUT THIS AXIS.

EX2-18 MILLIMETERS

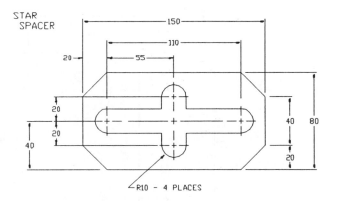

STAR
SPACER

150

110

55

20

20   40   80

20

40   20

R10 - 4 PLACES

EX2-19 INCHES

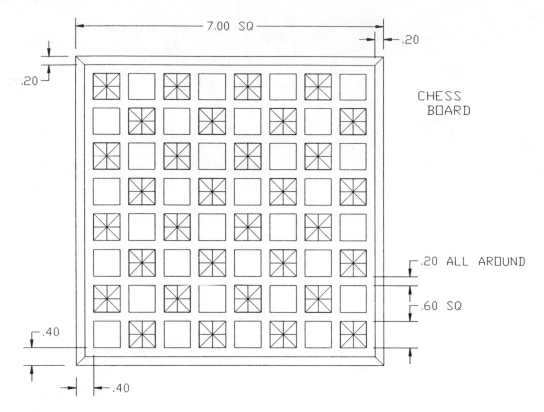

EX2-20 MILLIMETERS

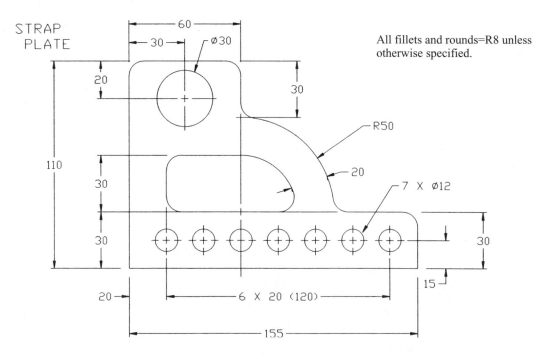

EX2-21 INCHES

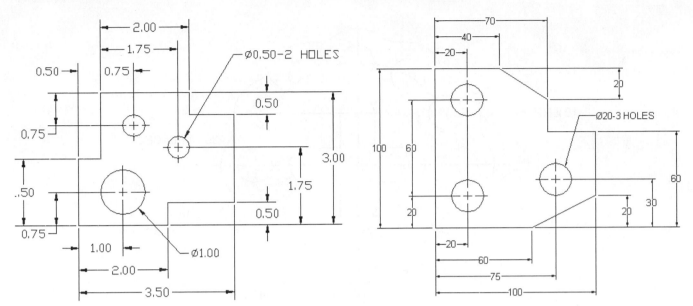

EX2-23 MILLIMETERS

EX2-22 MILLIMETERS

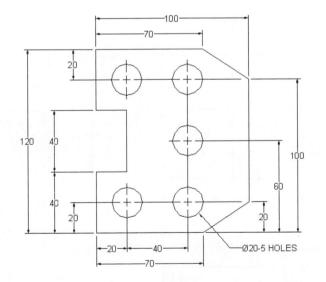

EX2-24 INCHES

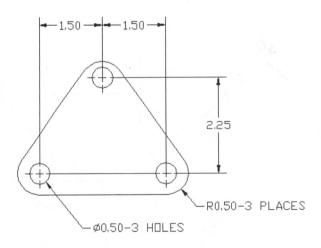

EX2-25 MILLIMETERS

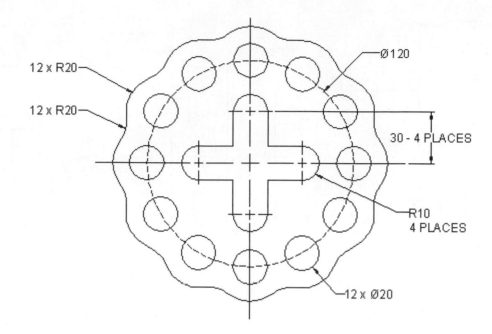

EX2-26 MILLIMETERS

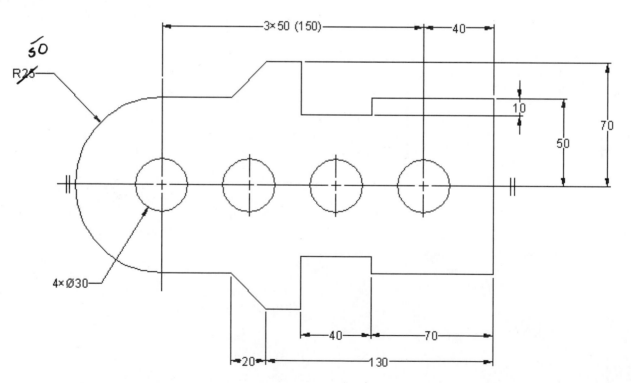

EX2-27 MILLIMETERS

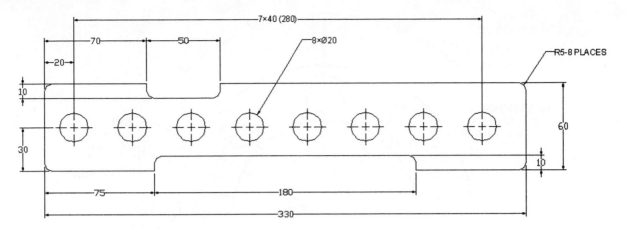

EX2-28 INCHES

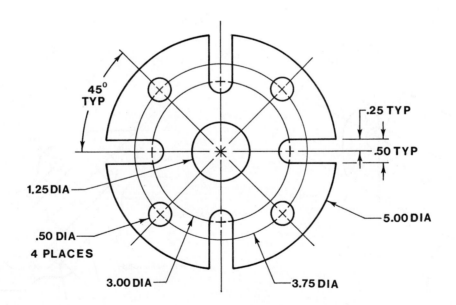

MATL .25 STEEL

EX2-29 MILLIMETERS

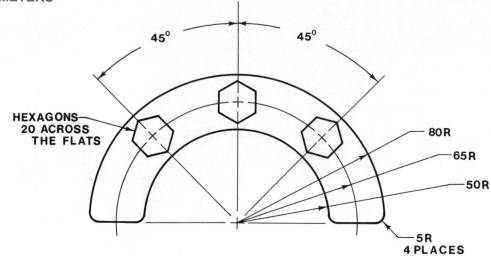

EX2-30 MILLIMETERS

EX2-31 MILLIMETERS

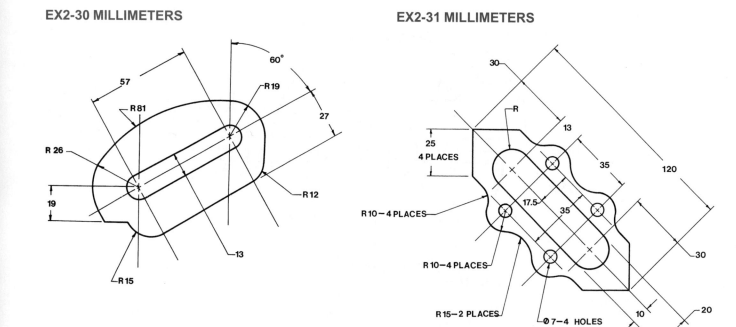

EX2-32 INCHES

Redraw the Hole Plate shown below. Do not include the dimensions. Draw and complete the hole table outlined below. Set the text style for Arial. The data text is .18 high, the heading text is .25 high, and the title text is .375 high.

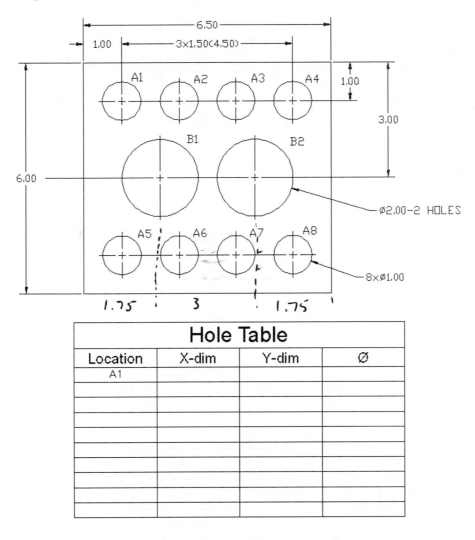

Hole Table			
Location	X-dim	Y-dim	Ø
A1			

EX2-33 MILLIMETERS

Redraw the Hole Plate shown below. Do not include the dimensions. Draw and complete the hole table outlined below. Set the text style for Arial. The data text is 5 high, the heading text is 10 high, and the title text is 15 high.

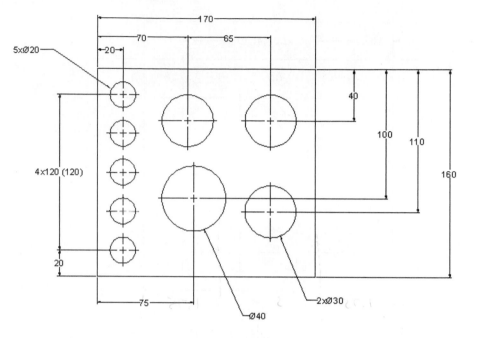

Hole Table			
Location	X-dim	Y-dim	Ø
A1			

C H A P T E R **3**

More Advanced Commands

3-1 INTRODUCTION

This chapter explains how to use some of the more advanced AutoCAD commands. Included are the **Osnap** toolbar, **Grips, Layer, Block,** and **Attribute** commands, and some of the commands on the **Modify II** toolbar.

3-2 OSNAP

The **Object Snap** command allows you to snap to entities on the drawing screen rather than just to points, as does the **Snap** command. **Osnap** can snap to the endpoint of a line, to the intersection of two lines, or to the center point of a circle, as well as to other items. The **Osnap** toolbar allows you to access all **Osnap** commands. See Figure 3-1.

To access the Osnap toolbar

1. Locate the cursor on an existing toolbar and right-click the mouse.
2. Select the **Osnap** toolbar and close the dialog box.
3. Dock the **Osnap** toolbar at a convenient location on the drawing screen.

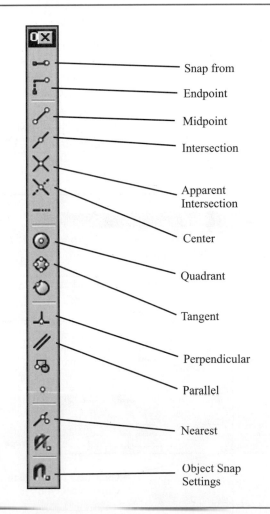

Snap from
Endpoint
Midpoint
Intersection
Apparent Intersection
Center
Quadrant
Tangent
Perpendicular
Parallel
Nearest
Object Snap Settings

Figure 3-1

71

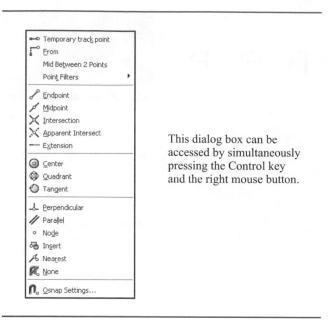

This dialog box can be accessed by simultaneously pressing the Control key and the right mouse button.

Figure 3-2

Osnap can also be accessed by holding down the <Ctrl> key and pressing the right mouse button. This feature can be activated during any command sequence. A dialog box appears on the screen listing the **Osnap** options. See Figure 3-2.

The **Osnap Settings** tool allows you to access the **Osnap Settings** dialog box. See Figure 3-2. The **Osnap**

Settings dialog box may also be accessed using the **Tools** pull-down menu. The **Osnap Settings** dialog box can be used to turn an **Osnap** option on permanently; that is, the option will remain on until you turn it off. If the **Endpoint** option is on, the cursor will snap to the nearest endpoint every time a point selection is made. Turning an **Osnap** option on is very helpful when you know you are going to select several points of the same type; however, in most cases, it is more practical to activate an **Osnap** option as needed using the **Osnap** toolbar or the **Osnap** menu.

To turn Osnap on

1. Select the **Tools** pull-down menu, then **Drafting Settings,** then select the **Object Snap** tab.

   The **Drafting Settings** dialog box will appear. See Figure 3-3.

2. Click the box to the left of the command you wish to activate.

   A check mark will appear, indicating that the command is active.

3. Close the **Drafting Settings** dialog box.

   The **Options** box on the **Drafting Settings** dialog box allows you to access the **Options** dialog box, which can be used to change the color of the cursor box used to identify **Osnap** points.

Figure 3-3

To change the size of the Osnap cursor box

The **Options** dialog box also contains an **Aperture Size** option. This option allows you to change the size of the rectangular box on the cursor. A larger box makes it easier to grab objects, but too large a box may grab more than one object, or the wrong object.

3-3 OSNAP—Endpoint

The **Endpoint** option is used to snap to the endpoint of an existing entity. Figure 3-4 shows an existing line.

To snap to the endpoint of a line

1. Select the **Line** tool from the **Draw** toolbar.

   Command: _line Specify first point:

2. Select the **Endpoint** tool from the **Object Snap** toolbar.

   Specify next point or [Undo]: _endp of

Most of the **Osnap** options do not work independently but in conjunction with other commands. In this example, the **Line** command must be activated first, then the **Osnap** command. Note that a box will appear around the endpoint of the line, indicating that the **Osnap Endpoint** command has selected the line's endpoint. The **Endpoint** command also works with arcs, mlines, and 3D applications.

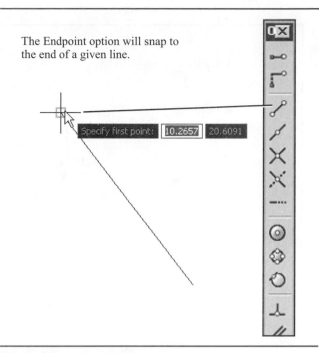

The Endpoint option will snap to the end of a given line.

Figure 3-4

3-4 OSNAP—Snap From

The **Snap From** option allows you to grab the line directly, without having to enter another command. See Section 3-15, Grips.

If you first click the **Snap From** option, then select a line, the endpoints and midpoint of the line are defined by boxes. See Figure 3-5. The points selected in Figure 3-5 are selected at random locations.

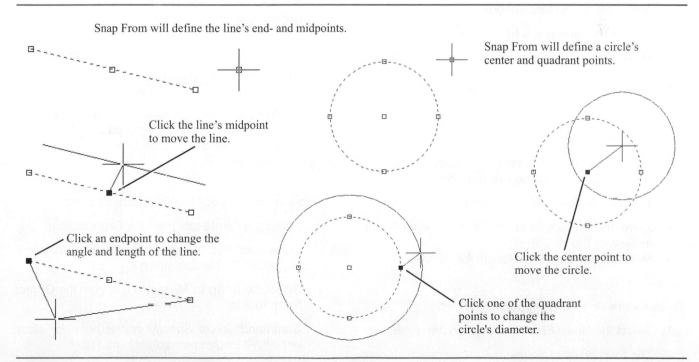

Snap From will define the line's end- and midpoints.

Click the line's midpoint to move the line.

Click an endpoint to change the angle and length of the line.

Snap From will define a circle's center and quadrant points.

Click the center point to move the circle.

Click one of the quadrant points to change the circle's diameter.

Figure 3-5

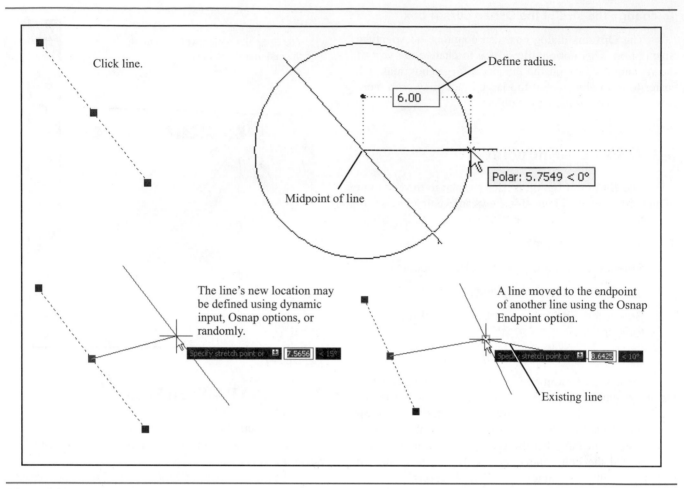

Figure 3-6

To move a line

1. Select the line's midpoint box.

   The box will change to a filled red box.

2. Move the cursor, and the line moves with the cursor.
3. Select a new location for the line.

To change the angle and length of a line

1. Select one of the designated endpoint boxes created using the **Snap From** option.

   The box will change to a filled red box.

2. Move the cursor, and the line's endpoint will move with the cursor.
3. Select a new angle and length for the line.

To apply the Snap From option to a circle

1. Select the **Snap From** option, then the circle.

   The four quadrant points and the circle's midpoint are defined by boxes. See Figure 3-5.

2. Select the circle's center point to move the circle, or select one of the quadrant points to change the circle's diameter.

3-5 OSNAP—Midpoint

The **Midpoint** option is used to snap to the midpoint of an existing entity. In the example presented in Figure 3-6, a circle is to be drawn with its center point at the midpoint of the line.

To draw a circle about the midpoint of a line

1. Select the **Circle** tool from the **Draw** toolbar.

   Command:_circle Specify center point for circle or [3P/2P/Ttr (tan tan radius)]:

2. Select the **Snap to Midpoint** tool from the **Object Snap** toolbar.

   Command:_circle Specify center point for circle or [3P/2P/Ttr (tan tan radius)]:_mid of

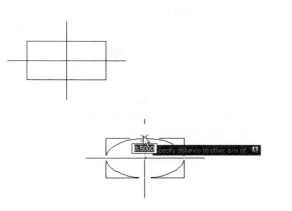

The Intersection option snaps to
the intersection of any two or
more entities.

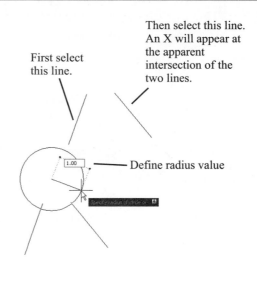

Define radius value

The circle's center point will be located on
the apparent intersection of the two lines.

Figure 3-7

3. Select the line.

   Specify radius of circle or [Diameter]:

4. Select a radius value.

3-6 OSNAP—Intersection

The **Intersection** option is used to snap to the intersection of two or more entities. Figure 3-7 shows a set of projection lines that are to be used to define an ellipse.

To use the Osnap Intersection command to define an ellipse

1. Select the **Ellipse** tool from the **Draw** toolbar.

   Specify axis endpoint of ellipse or [Arc/Center]:

2. Select the **Intersection** tool from the **Object Snap** toolbar.

   Specify axis endpoint of ellipse or [Arc/Center]: _int of

3. Select the first point for the ellipse.

   Specify other endpoint of axis:_int of

4. Select the **Intersection** option and then select the second point.

   Select the intersection that defines the length of one of the axes from the ellipse centerpoint. In this example the

Figure 3-8

distance between the two points selected equals the length of the major axis of the ellipse.

   Specify distance to other axis or [Rotation]:_int of

5. Select the **Intersection** tool from the **Object Snap** toolbar and select the intersection that defines the other axis length.

3-7 OSNAP—Apparent Intersection

The **Apparent Intersection** option is used to snap to an intersection that would be created if the two entities were extended to create an intersection. Figure 3-8 shows two lines that do not intersect but, if continued, would intersect.

To draw a circle centered about an apparent intersection

1. Select the **Circle** tool from the **Draw** toolbar.

   Command: _circle Specify center point for circle or [3P/2P/Ttr (tan tan radius)]:

2. Select the **Apparent Intersection** tool from the **Object Snap** toolbar.

   Command: _circle Specify center point for circle or [3P/2P/Ttr (tan tan radius)]: _appint of

3. Select one of the lines.

   Command: _circle Specify center point for circle or [3P/2P/Ttr (tan tan radius)]: _appint of

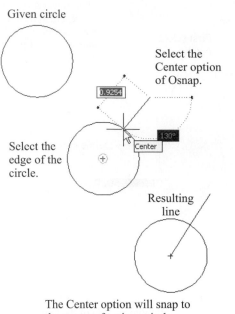

Given circle

Select the Center option of Osnap.

Select the edge of the circle.

Resulting line

The Center option will snap to the center of a given circle.

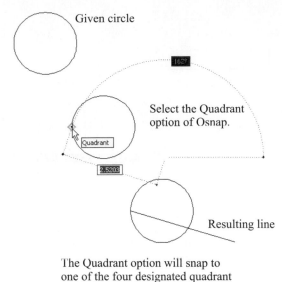

Given circle

Select the Quadrant option of Osnap.

Resulting line

The Quadrant option will snap to one of the four designated quadrant points.

Figure 3-9

4. Select the other line.

   Specify radius of circle or [Diameter]:

5. Define a radius value for the circle.

3-8 OSNAP—Center

The **Center** option is used to draw a line from a given point directly to the center point of a circle. See Figure 3-9.

To draw a line to the center point of a circle

1. Select the **Line** tool from the **Draw** toolbar.

   Command: _line Specify first point:

2. Select the start point for the line.

   Specify next point or [Undo]:

3. Select the **Center** tool from the **Object Snap** toolbar.

   Specify next point or [Undo]: _cen of

4. Select any point on the circle.

   Do *not* try to select the center point directly. Select any point on the edge of the circle or arc; the center point will be calculated automatically.

Figure 3-10

3-9 OSNAP—Quadrant

The **Quadrant** option is used to snap directly to one of the quadrant points of an arc or circle. Figure 3-10 shows the quadrant points for an arc and a circle.

To draw a line to one of a circle's quadrant points

1. Select the **Line** tool from the **Draw** toolbar.

   Command: _line Specify first point:

2. Select a start point for the line.

   Specify next point or [Undo]:

3. Select the **Quadrant** tool from the **Osnap** toolbar.

   Specify next point or [Undo]: _qua of

4. Select a point on the circle near the desired quadrant point.

3-10 OSNAP—Perpendicular

The **Perpendicular** option is used to draw a line perpendicular to an existing entity. See Figure 3-11.

To draw a line perpendicular to a line

1. Select the **Line** tool from the **Draw** toolbar.

   Command: _line Specify first point:

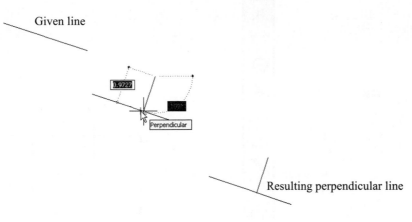

The Perpendicular option can be used to
draw a line perpendicular to another line.

Figure 3-11

2. Select a start point for the line.

   Specify next point or [Undo]:

3. Select the **Perpendicular** tool on the **Osnap** toolbar.

   Specify next point or [Undo]: _per to

4. Select the line or entity that will be perpendicular to the drawn line.

3-11 OSNAP—Tangent

The **Tangent** option is used to draw lines tangent to existing circles and arcs. See Figure 3-12.

To draw a line tangent to a circle

1. Select the **Line** tool from the **Draw** toolbar.

   Command: _line Specify first point:

2. Select the start point for the line.

   Specify next point or [Undo]:

3. Select the **Tangent** tool from the **Osnap** toolbar.

   Specify next point or [Undo]: _tan to

4. Select the circle.

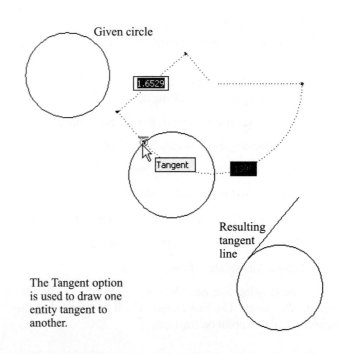

The Tangent option
is used to draw one
entity tangent to
another.

Figure 3-12

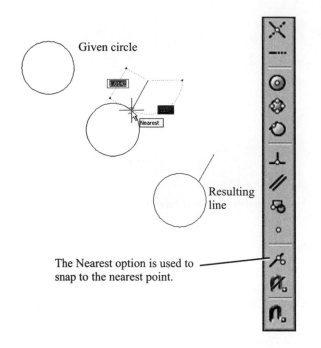

Given circle

Resulting line

The Nearest option is used to snap to the nearest point.

Figure 3-13

3-12 OSNAP—Nearest

The **Nearest** option is used to snap to the nearest available point on an existing entity. See Figure 3-13.

To draw a line from a point to the nearest selected point on an existing line

1. Select the **Line** tool from the **Draw** toolbar.

   Command: _line Specify first point:

2. Select a start point for the line.

   Specify next point or [Undo]:

3. Select the **Nearest** tool from the **Osnap** toolbar.

   Specify next point or [Undo]: _nea to

4. Select the existing line.

The existing line need be only within the rectangular box on the cursor. The line endpoint will be snapped to the nearest available point on the line.

3-13 SAMPLE PROBLEM SP3-1

Redraw the object shown in Figure 3-14. Do *not* include dimensions. Use **Osnap** commands whenever possible.

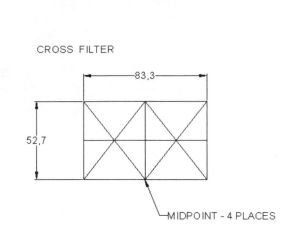

CROSS FILTER

83,3

52,7

MIDPOINT - 4 PLACES

Figure 3-14

1. Create a new drawing called **SP3-1**.
2. Drawing setup:

   Limits: lower left = **<0.0000,0.0000>**
   Limits: upper right = **297,210**
   Zoom All
   Grid = **10**
   Snap = **5**

3. Draw an **83.3 × 52.7** rectangle starting at point **40,40.** Use dynamic input values to draw the rectangle's edge lines. See Figure 3-15.

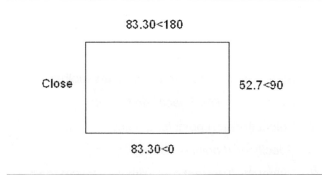

83.30<180

Close

52.7<90

83.30<0

Figure 3-15

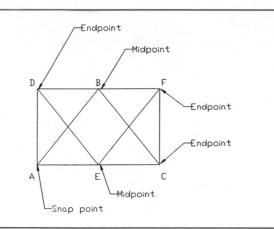

Figure 3-16

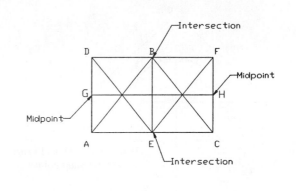

Figure 3-17

4. Draw line **A-B-C**. Point **A** is a grid snap point, so it may be selected directly. Point **B** is selected using **Osnap Midpoint,** and point **C** is selected using **Osnap Endpoint** (or Intersection). See Figure 3-16.
5. Draw line **D-E-F** using **Endpoint, Midpoint,** and **Endpoint,** respectively.
6. Draw line **B-E** using **Intersection.** See Figure 3-17.
7. Draw line **G-H** using **Midpoint.**
8. Save the drawing if desired.

3-14 SAMPLE PROBLEM—SP3-2

Redraw the object shown in Figure 3-18. Do not include dimensions.

1. Create a new drawing called **SP3-2.**
2. Drawing setup:

   Limits: accept the default values
   Grid = **.50**
   Snap = **.25**

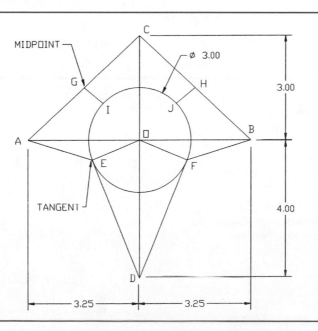

Figure 3-18

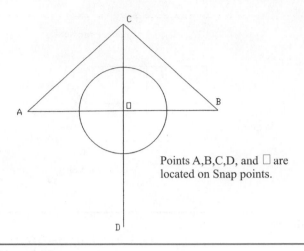

Points A,B,C,D, and ☐ are
located on Snap points.

Figure 3-19

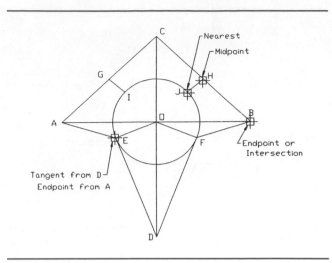

Figure 3-20

3. Draw lines **A-B, C-D, A-C,** and **C-B.** The endpoint of each of these lines is located on a grid snap point, so the points may be selected directly.
4. Draw circle **O.** See Figure 3-19. Respond to the center point prompt by selecting **Intersection** from the **Osnap** pop-up menu.
5. Draw lines **G-I** and **H-J** using **Midpoint** and **Nearest.** See Figure 3-20.
6. Draw lines **D-E** and **D-F** using **Tangent.** Draw lines **A-E** and **A-F** using **Endpoint.** Points **A, B,** and **D** are grid snap points and also are endpoints.
7. Save the drawing if desired.

3-15　GRIPS

The **Grips** function is used to quickly identify and lock onto convenient points on entities such as the endpoints of lines or the center point of a circle. Figure 3-21 shows some examples of grip points.

Grips are helpful when using some **Modify** commands. For example, if a line is to be rotated about its center point, a grip can be used to first identify the line's center point (pickbox) and then lock onto it as the center point (basepoint) for the rotation. The center point of a circle can be used to move the circle to a new location.

To turn the Grips function off

By default, the **Grips** command is automatically on. The **Grips** command may be turned off as follows.

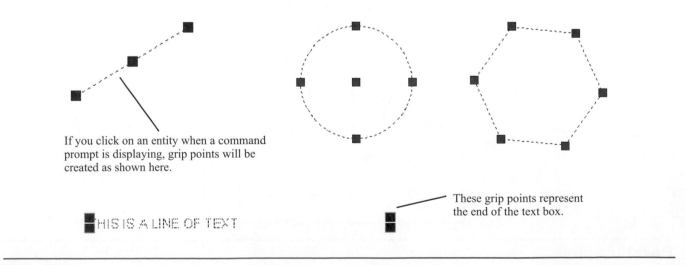

If you click on an entity when a command prompt is displaying, grip points will be created as shown here.

These grip points represent the end of the text box.

Figure 3-21

THE DEFAULT GRIPS VALUE IS 1.

1 = GRIPS ON

0 = GRIPS OFF

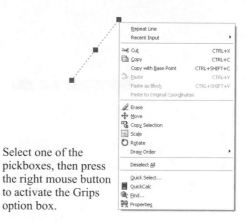

Select one of the pickboxes, then press the right mouse button to activate the Grips option box.

Figure 3-22

Figure 3-23

1. Type **Grips** in response to a command prompt.

   Enter new value for GRIPS <1>:

   There are two possible values: 1 and 0. The **1** value turns the **Grips** command on. The **0** value turns the **Grips** command off. See Figure 3-22.

2. Type **0**; press **Enter**.

   The **Grips** command is now off.

To access the Grips dialog box

1. Grip a line; that is, place the cursor on the line and press the left mouse button.

   A line is gripped when a command prompt is displayed and no other command sequence is active.

2. Click on one of the grip points.

   In this example, the lower endpoint was selected.

   <Stretch to point>/Base point/Copy/Undo/eXit:

3. Press the right mouse button.

   The **Grips** dialog box appears. See Figure 3-23.

3-16 GRIPS—Extend

To extend the length of a line (See Figure 3-24.)

   Given a line, extend it 1.25 inches.

1. Draw an angled line as shown, then draw a circle of radius **1.25** centered on one of the line's endpoints.
2. Press the **<Esc>** key to ensure a command prompt.
3. Select the line.

   Blue pickboxes should appear at the two endpoints and at the midpoint.

4. Select the endpoint that also serves as the circle's center point.

   The blue pickbox should change to a solid red square box.

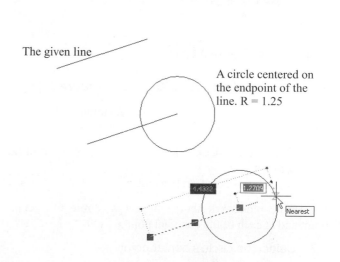

The given line

A circle centered on the endpoint of the line. R = 1.25

Figure 3-24

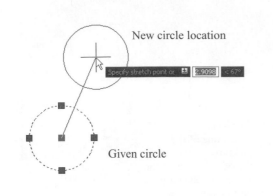

Figure 3-25

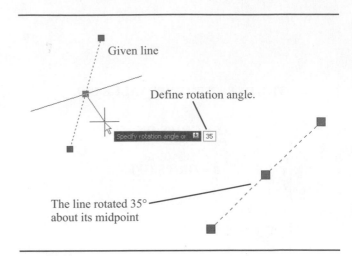

Figure 3-26

5. Press the right mouse button and select the **Stretch** option from the **Grips** menu.

STRETCH
Specify stretch point or [Base point/Copy/Undo/eXit]:

Stretch the endpoint of the line to the edge of the 1.25 circle by accessing the **Osnap** menu (press the **<Ctrl>** key and right mouse button) and selecting the **Nearest** option.

6. Select a point on the circle that aligns the line through the three grip points, then press the left button.
7. Erase the circle.

3-17 GRIPS—Move

To move an object using Grips (See Figure 3-25.)

Move the given circle to a new location on the drawing.

1. Press the **<Esc>** key to ensure a command prompt.
2. Select any point on the circle.

Blue pickboxes should appear at circle O's center point and at each of the quadrant points.

3. Select the circle's center point.

The center point pickbox should change to a red solid square box.

STRETCH
Specify stretch point or [Base_point/Copy/Undo/eXit]:

4. Move the circle to a new location and press the left mouse button.

The circle's new location can be located using dynamic input.

3-18 GRIPS—Rotate

To rotate an object using Grips (See Figure 3-26.)

Given a line, rotate it 35° about its midpoint.

1. Press the **<Esc>** key to ensure a command prompt.
2. Select the line.

Blue pickboxes should appear at the line's two endpoints and at its midpoint.

3. Select the line's midpoint.

The blue pickbox at the midpoint should change to a solid red square box.

4. Press the right mouse button, then select **Rotate** from the **Grips** dialog box.

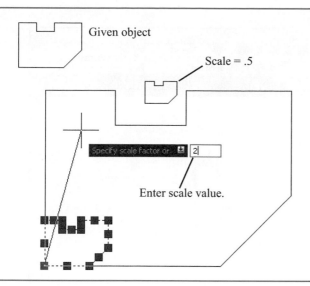

Figure 3-27

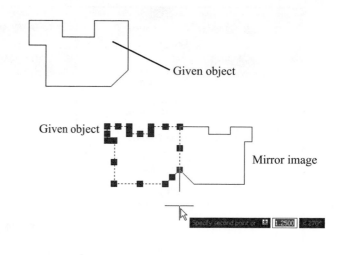

Figure 3-28

ROTATE
Specify rotation angle or [Base point/Copy/ Undo/eXit]:

5. Type **35**; press **Enter.**

The value 35 will appear in the dynamic input box.

3-19 GRIPS—Scale

To change the scale of an object (See Figure 3-27.)

Given an object, reduce it to half its original size.

1. Press the **<Esc>** key to ensure a command prompt.
2. Select the object by first windowing it, then pressing the left mouse button.

Blue pickboxes should appear all around the object.

3. Select any one of the pickboxes.

The selected pickbox should change to a red solid square box.

4. Press the right mouse button to activate the **Grips** options box.
5. Select **Scale** from the menu.

SCALE
Specify scale factor or [Base point/Copy/Undo/ Reference/eXit]:

6. Type **.5**; press **Enter.**

Figure 3-27 also shows the same object scaled to twice its original size, that is, a scale factor of 2.

3-20 GRIPS—Mirror

To mirror an object (See Figure 3-28.)

Given an object, draw a mirror image of the object.

1. Press the **<Esc>** key to ensure a command prompt.
2. Window the object, then press the left mouse button.

Blue pickboxes should appear at each corner intersection of the hexagon.

3. Select the upper right pickbox.

The selected pickbox should change to a red solid square box.

4. Press the right mouse button to activate the **Grips** options box.
5. Select the **Mirror** option.

MIRROR
Specify second point or [Base point/Copy/ Undo/eXit]:

6. Select the pickbox vertically below the selected right upper pickbox.
7. Press the **<Esc>** key twice to fix the mirrored object.

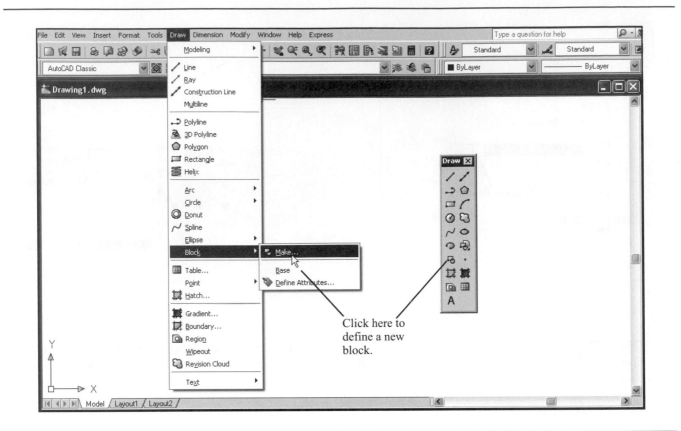

Click here to define a new block.

Figure 3-29

3-21 BLOCKS

Blocks are groups of entities saved as a single unit. Blocks are used to save shapes and groups of shapes that are used frequently when creating drawings. Once created, blocks can be inserted into the drawings, thereby saving drawing time.

The **Block** tool is located on the **Draw** pull-down menu and on the **Draw** toolbar. See Figure 3-29. The **Insert Block** tool is associated with the **Insert** pull-down menu and the **Draw** toolbar.

To create a block

Figure 3-30 shows an object. A block can be made from this existing drawing as follows.

1. Select the **Make Block** tool from the **Draw** toolbar.

   The **Block Definition** dialog box appears. See Figure 3-30.

2. Click the **Select Objects** box.

   Select objects:

3. Window the object; press **Enter.**

   The **Block Definition** dialog box will reappear.

4. Type **SHAPE** in the **Name** box.

   Block names may contain up to 31 characters.

5. Click the **Pick point** box.
6. Define the insertion base point for the block.

   The **Block Definition** dialog box will reappear. **Osnap** may be used to locate an insertion point. The values listed in the **Block Definition** dialog box are XYZ values relative to the current origin.

   The insertion base point will be a point on the object that will align with the screen cursor during the block insertion process.

7. Select the **OK** box.

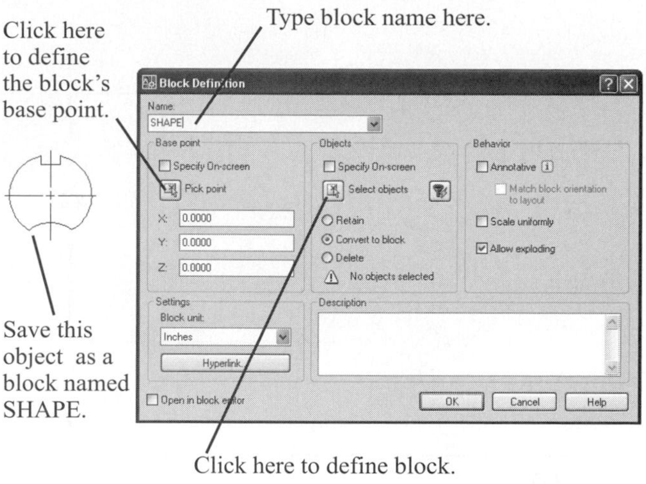

Click here
to define
the block's
base point.

Type block name here.

Save this
object as a
block named
SHAPE.

Click here to define block.

Figure 3-30

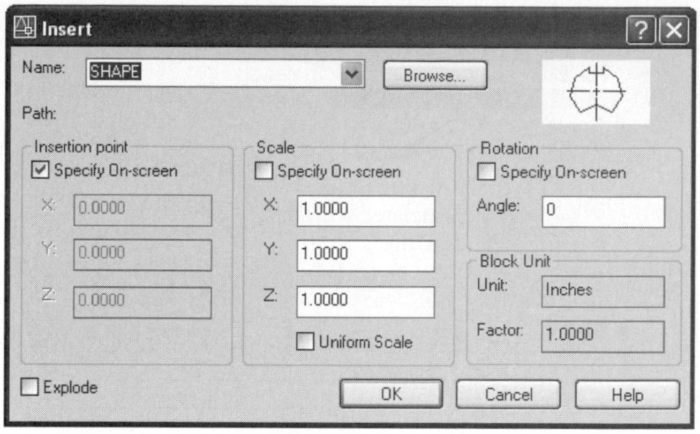

Figure 3-31

To insert a block

1. Select **Insert Block** from the **Draw** toolbar.

    The **Insert** dialog box appears. See Figure 3-31.

2. Select the **Name** box.

    A cascade menu will appear. See Figure 3-32.

3. Select **SHAPE.**

    The word **SHAPE** appears in the **Name** box.

4. Select **OK.**

    The **Insert** dialog box reappears with the word **SHAPE** in the **Name** box.

5. Select **OK.**

    The **SHAPE** block will appear on the screen with its insertion point aligned with the cursor. The shape will move as you move the crosshairs.

    Specify insertion point or [Scale/X/Y/Z/Rotate/ PScale/PX/PY/PZ/PRotate]:

6. Select an insertion point for the shape.

    The default scale factor is 1. This means that if you accept the default value by pressing **Enter,** the shape will be redrawn at its original size. The shape of the object may be changed by using the options available at the **Insert** prompt.

To change the scale of a block

1. Select **Insert Block** from the **Draw** toolbar.
2. Select the **SHAPE** block.

    Specify insertion point or [Scale/X/Y/Z/Rotate/ PScale/PX/PY/PZ/PRotate]:

3. Type **S;** press **Enter.**

    Specify scale factor for XYZ axes:

4. Type **2;** press **Enter.**

    Specify insertion point:

5. Select a point.

    The **Rotation** option is used to rotate a block about its insertion point. AutoCAD assumes that a horizontal line to the right of the insertion point is 0° and that counterclockwise is the positive angular direction.

    The block is now part of the drawing; however, the block in its present form may not be edited. Blocks are treated as a single entity and not as individual lines. You can verify this by trying to erase any one of the lines in the block. The entire object will be erased. The object can be returned to the screen by clicking the **Undo** tool. A block must first be exploded before it can be edited.

To explode a block

1. Select the **Explode** tool from the **Modify** toolbar.

    *Command: _explode
    Select objects:*

2. Window the entire object.

    Select objects:

3. Press **Enter.**

    The object is now exploded and can be edited. There is no visible change in the screen, but there will be a short blink after the **Explode** command is executed.

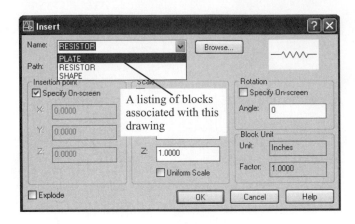

Figure 3-32

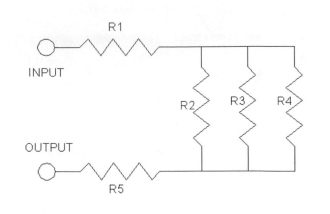

Figure 3-33

3-22 WORKING WITH BLOCKS

Figure 3-33 shows a resistor circuit. It was created from an existing block called RESISTOR. The drawing used **Decimal** units with a **Grid** set to .5 and a **Snap** set to .25. The default drawing limits were accepted. The procedure is as follows.

To insert blocks at different angles

1. Select **Insert Block** from the **Draw** toolbar.

   The **Insert** dialog box will appear.

2. Select the cascade menu on the right side of the **Name** box.

3. Figure 3-34 shows **RESISTOR** in the **Name** box. Select the **RESISTOR** block.

4. Select **OK** to return to the drawing screen.

   The RESISTOR block will appear attached to the crosshairs at the predefined insertion point. See Figure 3-35.

   The resistor is too large for the drawing, so a reduced scale will be used to generate the appropriate size.

   Specify insertion point or [Scale/X/Y/Z/Rotate/ PScale/PX/PY/PZ/PRotate]:

5. Type **S**; press **Enter.**

   Specify scale factor for XYZ axes:

6. Type **.50**; press **Enter.**

   Specify insertion point:

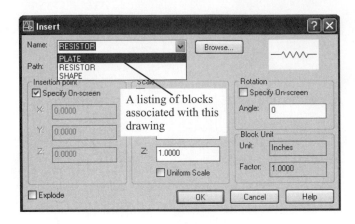

Figure 3-34

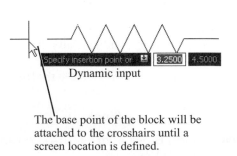

Dynamic input

The base point of the block will be attached to the crosshairs until a screen location is defined.

Figure 3-35

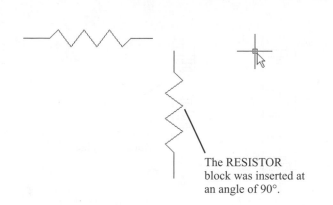

Figure 3-36

The RESISTOR block was inserted at an angle of 90°.

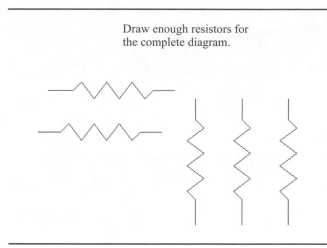

Draw enough resistors for the complete diagram.

Figure 3-37

Locate the resistor on the screen.

7. Press the right mouse button and select the **Repeat Insert Block** option.
8. Select the **RESISTOR** block again and set the X Y Z scale factors for **.50.**

Specify insertion point or [Scale/X/Y/Z/Rotate/PScale/PX/PY/PZ/PRotate]:

9. Type **R**; press **Enter.**

Specify rotation angle:

10. Type **90**; press **Enter.**

Specify insertion point:

11. Select an insertion point as shown in Figure 3-36.

Your screen should look like Figure 3-36.

12. Use the **Copy** command (**Modify** toolbar) to add the additional required resistors.

See Figure 3-37.

13. Use the **Line** command (**Draw** toolbar) to draw the required lines.
14. Select the **Circle** command (**Draw** toolbar) and draw a circle of diameter **.250.**
15. Use the **Copy** command (**Modify** toolbar) to create a second circle.
16. Use the **Move** command to position the circles so that they touch the edges of the two open circles.
17. Use the **Dtext** command to add the appropriate text.

The drawing should now look like the diagram presented in Figure 3-33.

To insert blocks with different scale factors

Figure 3-38 shows four different-sized threads, all created from the same block. The block labeled A used the default scale factor of 1, so it is exactly the same size as the original drawing used to create the block. The block labeled B was created with an X scale factor equal to 1, and a Y scale factor equal to 2. The procedure is as follows.

1. Select the **Insert Block** tool from the **Draw** toolbar.
2. Select the **THREAD** block.

The THREAD block is not an AutoCAD creation. THREAD was created specifically for this example.

Specify insertion point or [Scale/X/Y/Z/Rotate/PScale/PX/PY/PZ/PRotate]:

3. Type **X**; press **Enter.**

Specify X scale factor:

4. Type **1**; press **Enter.**

Specify insertion point:

5. Type **Y**; press **Enter.**

Specify Y scale factor:

6. Type **2**; press **Enter.**

Specify insertion point:

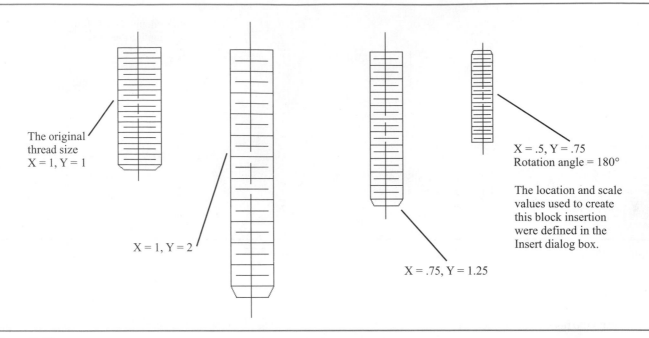

The original
thread size
X = 1, Y = 1

X = 1, Y = 2

X = .5, Y = .75
Rotation angle = 180°

The location and scale
values used to create
this block insertion
were defined in the
Insert dialog box.

X = .75, Y = 1.25

Figure 3-38

7. Specify a point.

The thread labeled C has an X scale factor of .75 and a Y scale factor of 1.25. The thread labeled D has an X scale factor of .5, a Y scale factor of .75, and a rotation angle of 180°.

To use the Insert dialog box to change the shape of a block

The D thread scale factors and rotation angle were defined using the **Insert** dialog box. The procedure is as follows.

1. Select the **Insert** block.

The **Insert** dialog box will appear. See Figure 3-39.

2. Click the box to the left of the words **Specify On-screen.**

The click should remove the check mark in the box, indicating that the option has been turned off. When the check mark is removed, the options labeled **Insertion Point, Scale,** and **Rotation** should change from gray to black in the dialog box.

3. Place the cursor arrow inside the **X Scale** box, backspace to remove the existing value, and type **.50.**
4. Change the **Y Scale** factor to **.75.**
5. Change the **Angle** to **180.**

See Figure 3-39. The insertion point was also defined using the **Insert** dialog box. In this example, the thread

will appear on the screen with its insertion point at the 11.2500,7.5000 point. These values are unique to the example presented.

To combine blocks

Figure 3-40 shows a hex head screw that was created from two blocks. The threaded portion of the screw was created first, then the head portion was added. Note in Figure 3-40 that when the hex head block was created, the insertion point was deliberately selected so that it could

Turn this option off before
entering insertion point, scale,
and rotation values.

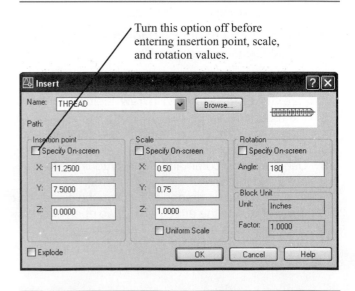

Figure 3-39

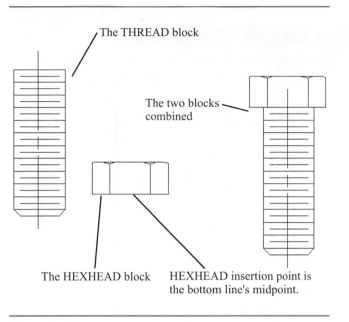

The THREAD block

The two blocks combined

The HEXHEAD block

HEXHEAD insertion point is the bottom line's midpoint.

Figure 3-40

Figure 3-41

easily be aligned with the centerline and the top surface of the thread block.

3-23 WBlock

Wblocks are blocks that can be entered into any drawing. When a block is created, it is unique to the drawing on which it was defined. This means that if you create a block on a drawing, then save and exit the drawing, the block is saved with the drawing but cannot be used on another drawing. If you start a new drawing, the saved block will not be available.

Any block can be defined and saved, however, as a wblock. Wblocks are saved as individual drawing files and can be inserted into any drawing.

To create a wblock

This section shows how to create a wblock for the block called SHAPE shown in Figure 3-30. A wblock may be created directly from an object shown on the screen.

1. At a command prompt, type **wblock** and press **Enter.**

   The **Write Block** dialog box appears. See Figure 3-41. There is no wblock tool.

2. Click the **Block** radio button under **Source.**
3. Select the **SHAPE** file from the listed blocks.

A listing of available blocks is accessed by clicking the arrow on the upper right side of the dialog box.

4. Select **OK.**

To verify that a wblock has been created

1. Select the **Insert block** tool from the **Draw** toolbar.

   The **Insert** dialog box will appear. See Figure 3-42.

Click here to access the drawing files.

Figure 3-42

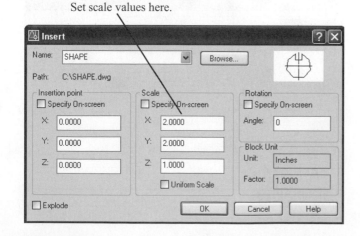

Figure 3-43

2. Click the **Browse** box.

   The **Select Drawing File** dialog box will appear.

3. Scroll down the file listing to locate the **SHAPE** file, then click it.

   See Figure 3-43.

4. Click the **Open** box, then the **OK** box.

   The wblock will appear on the screen.

To change the size of a wblock.

1. Select the **Insert block** tool from the **Draw** toolbar.

   The **Insert** dialog box will appear. See Figure 3-44.

2. Select the **Browse** option.
3. Select the **SHAPE** wblock file, then click the **Open** box.

   The **Insert** dialog box will appear.

4. Set the **Scale** values.
5. Select **OK.**

   See Figure 3-45.

Set scale values here.

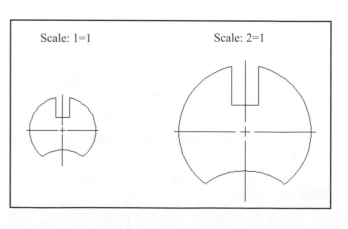

Figure 3-44

Figure 3-45

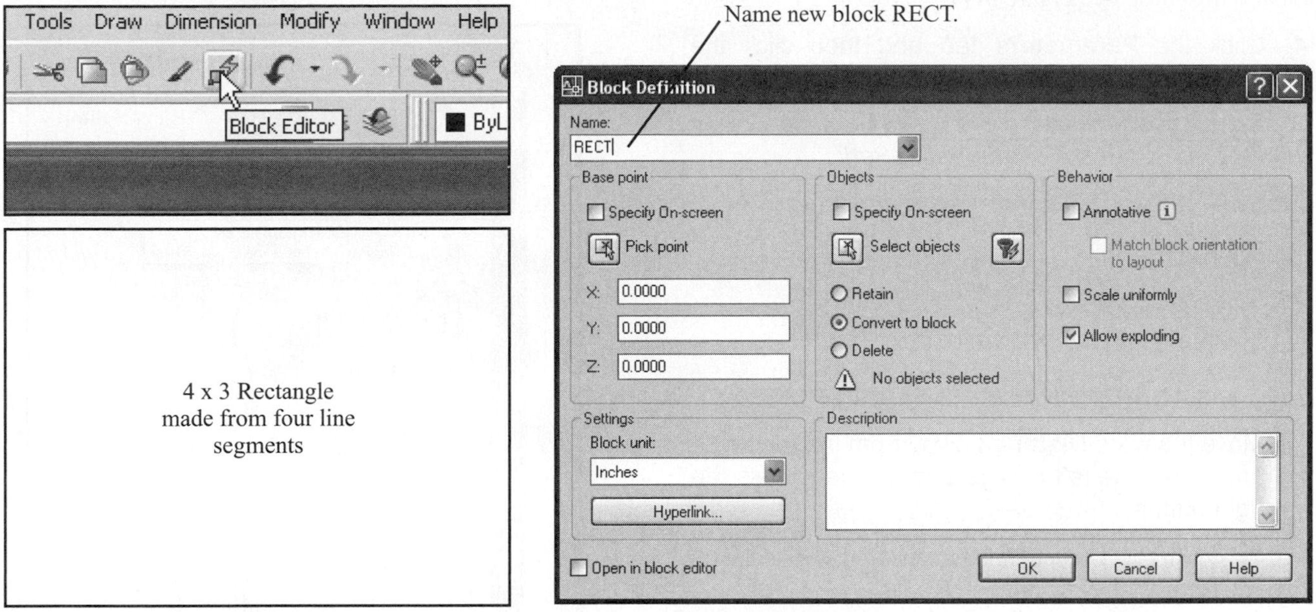

Name new block RECT.

Figure 3-46

3-24 DYNAMIC BLOCKS

Dynamic blocks are blocks that can be changed in real time. For example, a dynamic block can be stretched by locating a point and moving the cursor to a new position. In the following example a 4×3 rectangular block is changed into a 7×3 block.

To create a dynamic block—stretch option

Figure 3-46 shows a 4×3 rectangle. It was created using four line segments.

1. Access the **Make Block** command and define the rectangle as a block. Name the block **RECT**.
2. Click the **Block Editor** tool on the **Standard** toolbar.

The **Edit Definition** dialog box will appear. See Figure 3-47.

3. Select the **RECT** block and click **OK**.

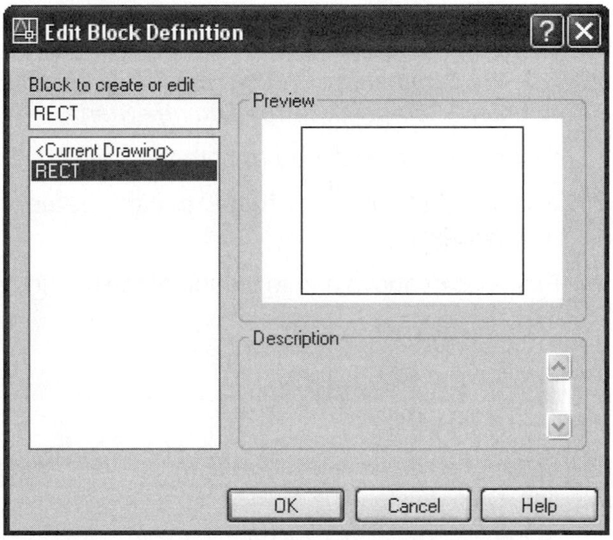

Figure 3-47

The **Block Authoring Palettes** and the **Block Editor** toolbar will appear. See Figure 3-48.

4. Click the **Parameters** tab and then click the **Linear Parameter** option.

   Specify start point or [Name/Label/Chain/ Description/Base/Palette/Value Set]:

5. Select the top left corner of the rectangle.

   See Figure 3-49.

   Specify endpoint:

6. Select the top right corner.

   Specify label location:

7. Move the word **Distance** away from the rectangle and press the left mouse button, then press the right button.

   This defines the block's parameter. Now define the related actions.

8. Click the **Actions** tab on the **Block Authoring Palettes - All Palettes** dialog box and select the **Stretch Action** option.

   See Figure 3-50.

   Select parameter:

9. Select the word **Distance.**

   Specify parameter point to associate with action or enter [sTart point/Second point] <Start>:

10. Select the upper left corner of the rectangle.

    Specify first corner of the stretch frame or [CPolygon]:

11. Pick a point above and to the left of the rectangle.

    Specify opposite corner:

12. Window the **RECT** block.

    Select objects:

13. Select the two horizontal lines and the left vertical line. Do not select the right vertical line.

    See Figure 3-51.

14. Press the right mouse button.

    Specify action location [Multiplier/Offset]:

    The word **Stretch** will appear on the screen.

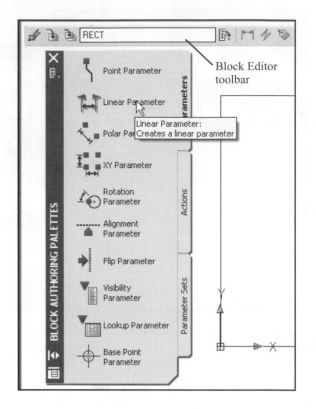

Block Editor toolbar

Figure 3-48

15. Locate the word **Stretch** on the upper left corner of the rectangle.

    See Figure 3-52.

16. Select the **Save Block Definition** and save the block edits, then select the **Close Block Editor** option on the **Block Editor** toolbar.

    The **RECT** block will appear on the screen.

17. Left-click the **RECT** block.

    See Figure 3-53. Two arrowheads and a filled square will appear. These are the dynamic block indicators.

18. Locate the cursor on the upper left corner and, holding down the left mouse button, move the arrow to the left creating a **7 × 3** rectangle. Use the dynamic input boxes to define the new size.

    See Figures 3-54 and 3-55.

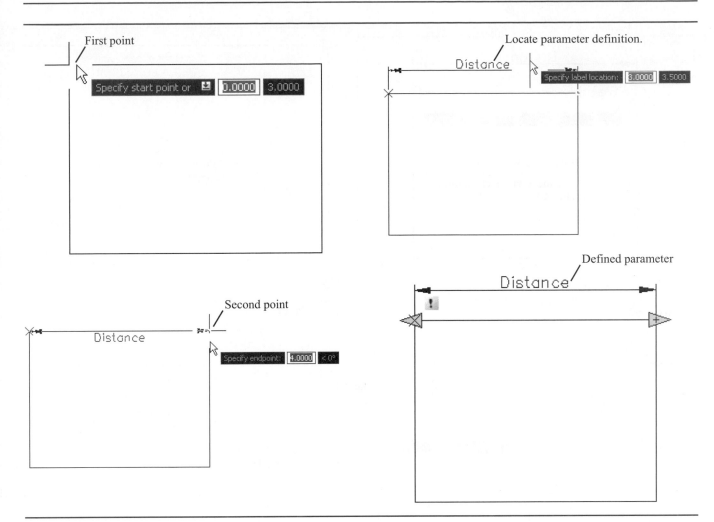

Figure 3-49

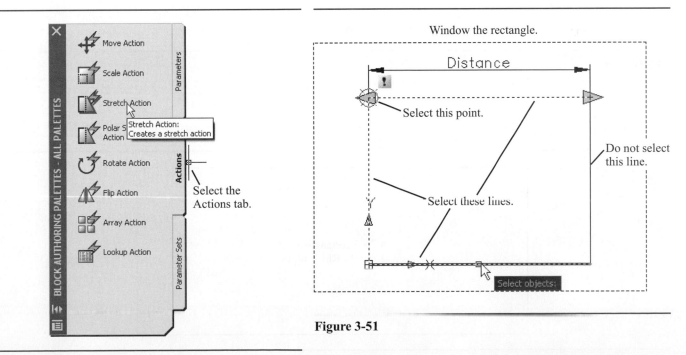

Figure 3-50

Figure 3-51

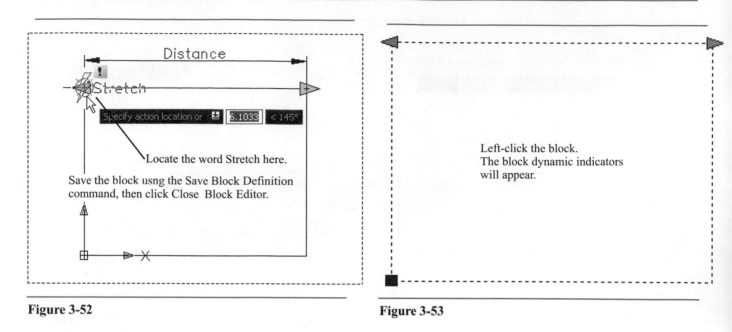

Locate the word Stretch here.

Save the block usng the Save Block Definition command, then click Close Block Editor.

Figure 3-52

Left-click the block.
The block dynamic indicators
will appear.

Figure 3-53

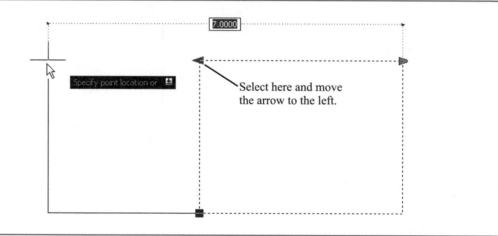

Select here and move
the arrow to the left.

Figure 3-54

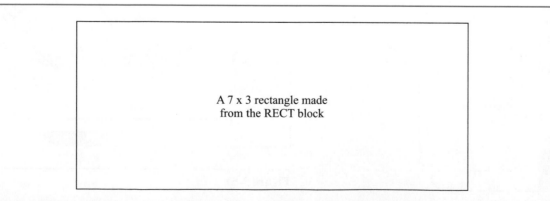

A 7 x 3 rectangle made
from the RECT block

Figure 3-55

To add a second parameter and action

Figure 3-56 shows the RECT block with a stretch action applied to the horizontal direction. This allows dynamic applications in the horizontal direction. A second action will be added to allow dynamic applications in the vertical direction.

1. Add a linear parameter to the right side of the block.
2. Add a stretch action to the vertical linear parameter.

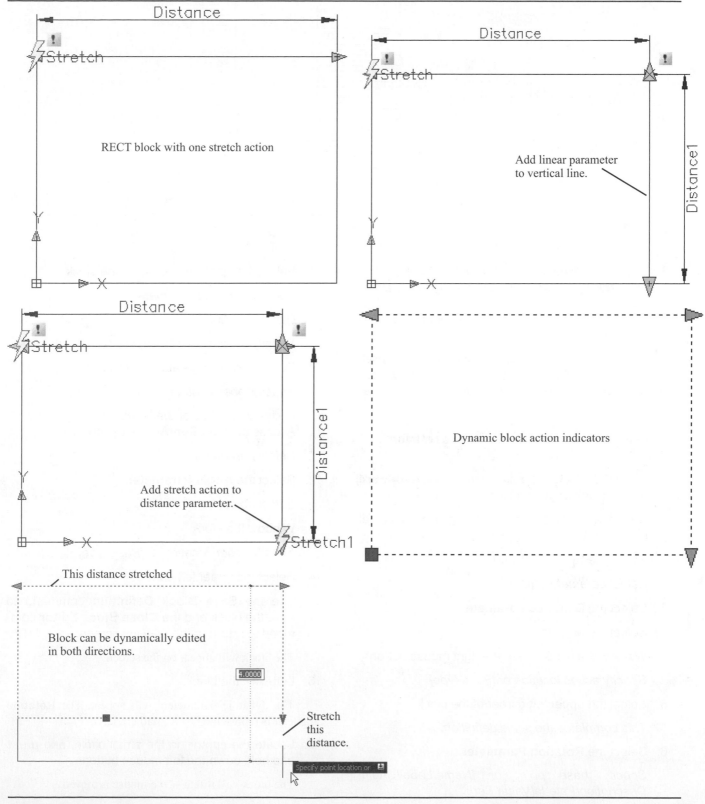

Figure 3-56

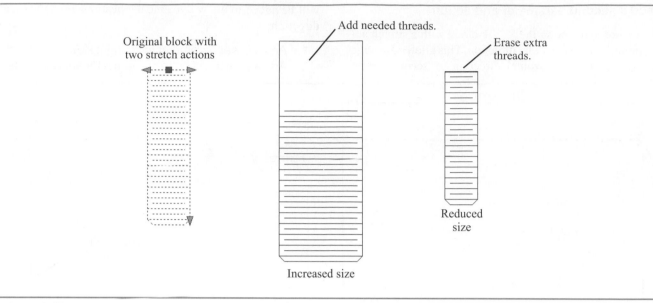

Original block with two stretch actions

Add needed threads.

Erase extra threads.

Increased size

Reduced size

Figure 3-57

3. Save the revised block and close the **Block Editor.**
4. Left-click the block to reveal the dynamic arrow indicators.
5. Stretch the block in both the horizontal and vertical directions.

Figure 3-57 shows horizontal and vertical stretch actions applied to a thread representation. These actions allow the thread to be sized as needed for different applications.

To create a dynamic block—scale and rotation options

Figure 3-58 shows a block named Thread. A scale and rotation action option will be added to the block.

1. Define a linear parameter along the horizontal section of the block.
2. Select the **Scale Action.**

   Select parameter:

3. Select the **Distance Parameter.**

   Select the objects:

4. Window the block; press the right mouse button.

   Specify action location or [Base type]:

5. Select the upper left corner of the block.

   This completes the scale definition.

6. Select the **Rotation Parameter.**

   Specify base point or [Name/Label/Chain/ Description/Palette/Value Set]:

7. Select the upper left corner of the block.

   Specify radius of parameter:

8. Define a radius.

   Specify default rotation angle or [Base angle]<0>:

9. Define an angle greater than 180°.

   Specify label location:

10. Define a location for the label.
11. Access the **Rotation Action** option.

    Select parameter:

12. Select the **Angle Parameter.**

    Select object:

13. Window the block.

    Specify action location or [Base type]:

14. Select the upper left corner.
15. Use the **Save Block Definition** command to save the block and the **Close Block Editor** command to return to the drawing screen.

    The block will appear on the screen.

16. Left-click the block.

    The dynamic parameters will appear. The **Rotation** parameter is represented by a small circle.

17. Locate the cursor on the small circle, and press and hold down the left mouse button.

    The block will rotate as the mouse is moved.

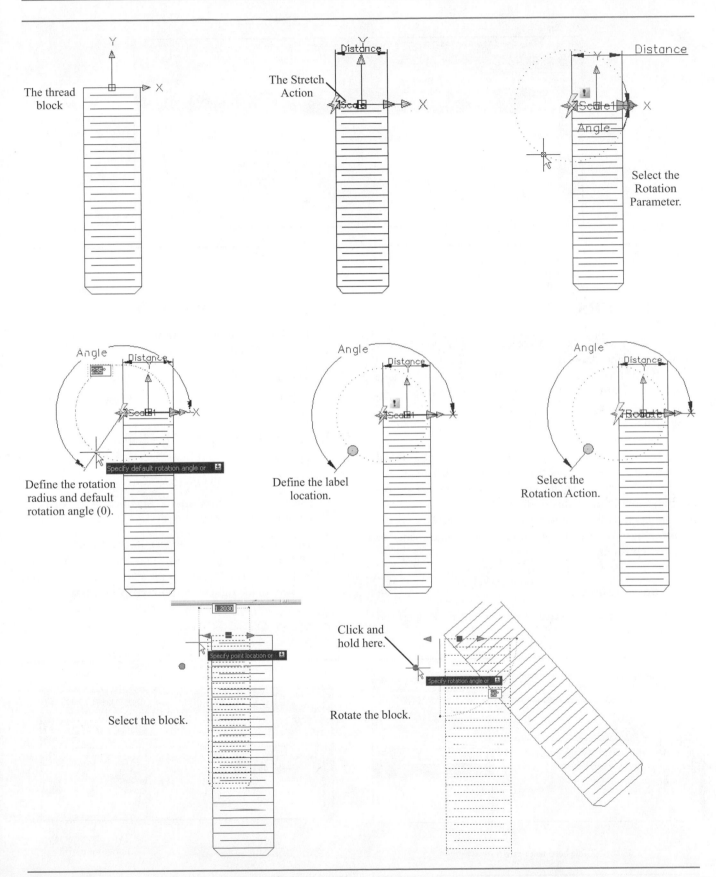

The thread block

The Stretch Action

Select the Rotation Parameter.

Define the rotation radius and default rotation angle (0).

Define the label location.

Select the Rotation Action.

Select the block.

Click and hold here.

Rotate the block.

Figure 3-58

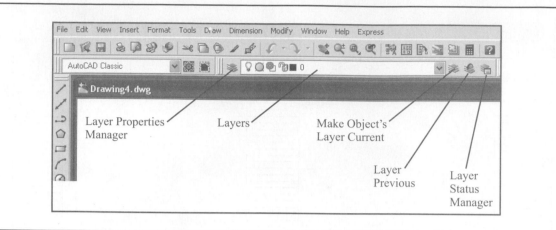

Figure 3-59

3-25 LAYERS

A *layer* is like a clear piece of paper you can lay directly over the drawing. You can draw on the layer and see through it to the original drawing. Layers can be made invisible, and information can be transferred between layers.

In the example shown, a series of layers will be created, then a group of lines will be moved from the initial 0 layer to the other layers.

The **Layers** toolbar is located just below the **Standard** toolbar and includes six associated tools and headings. See Figure 3-59. The associated tools are not command tools but indicators of the status of the **Layer** options. For example, the open padlock shown indicates that the layer is not locked. If it were locked, the **Padlock** tool would change to the closed position.

To create new layers

This exercise will create two new layers: Hidden and Center.

1. Select the **Layer Properties Manager** tool.

The **Layer Properties Manager** dialog box will appear. See Figure 3-60.

2. Click the **New** box.

A new layer listing will appear with the name **Layer1.** See Figure 3-61.

3. Type **Hidden** in place of the **Layer1** name.

The name Hidden should appear under the 0 layer heading in the **Name** column.

4. Click the **New** box again, and add another new layer named **Center.**

The name Center will appear in the **Name** column. There are now three layers associated with the drawing. See Figure 3-62. The 0 layer is the current layer, as indicated by the 0 to the right of the **Current Layer** box.

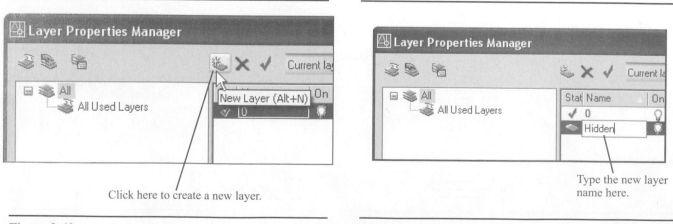

Figure 3-60

Figure 3-61

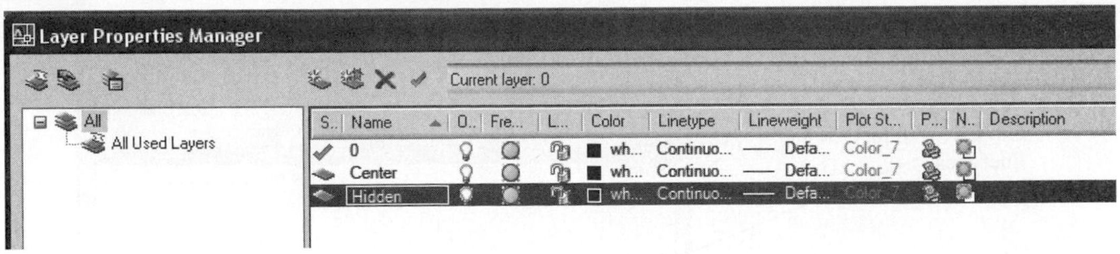

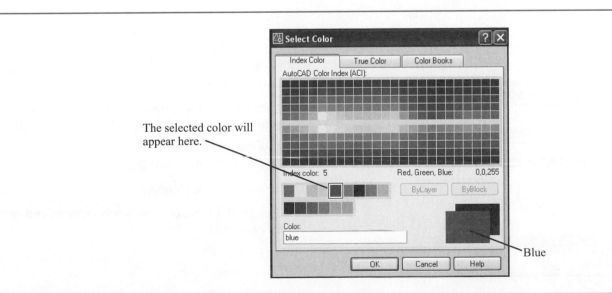

Figure 3-62

To change the color and linetype of a layer

1. Click the **Hidden** layer line on the **Layer Properties Manager** dialog box.
2. Click the word **white** on the **Hidden** layer line.

   See Figure 3-63. The **Select Color** dialog box will appear. See Figure 3-64. The **Select Color** dialog box can also be accessed by clicking the **Show details** box, then scrolling down the options given in the **Color** box.

3. Select the color **blue,** then click **OK.**

   All lines drawn on the Hidden layer will be blue.

4. Select the word **Continuous** on the **Hidden** layer line under the **Linetype** heading.

   The **Select Linetype** dialog box will change. See Figure 3-65.

5. Click the **Load** button.

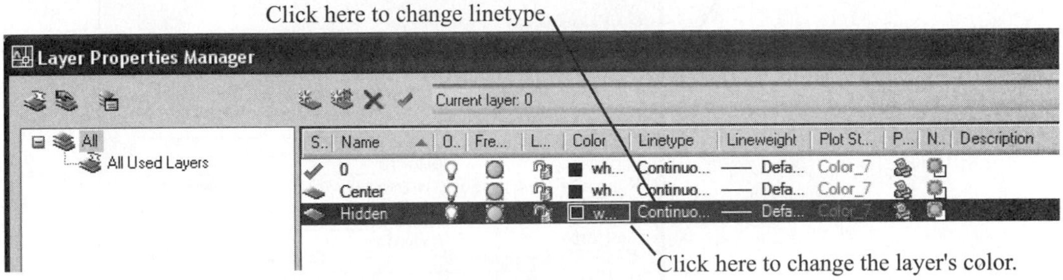

Click here to change linetype

Click here to change the layer's color.

Figure 3-63

The selected color will appear here.

Blue

Figure 3-64

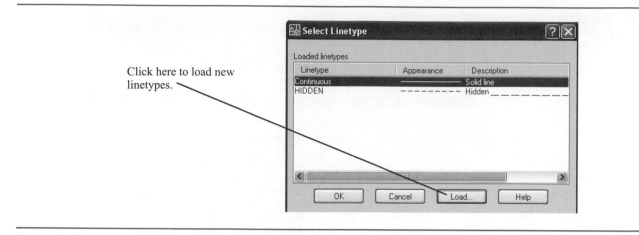

Click here to load new
linetypes.

Figure 3-65

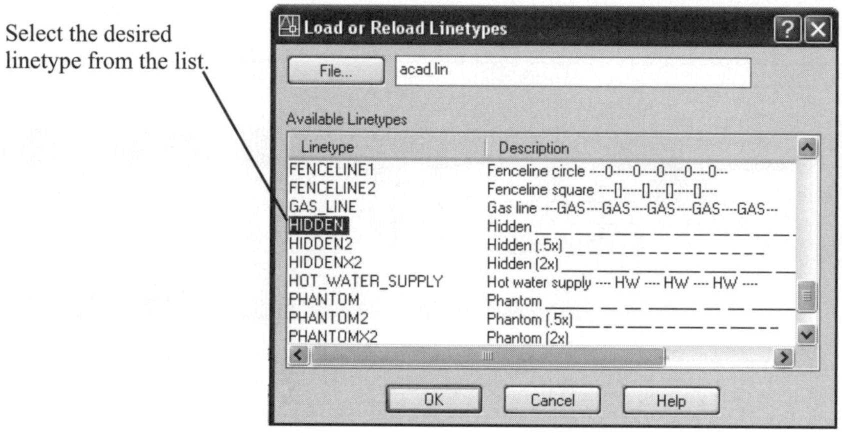

Select the desired
linetype from the list.

Figure 3-66

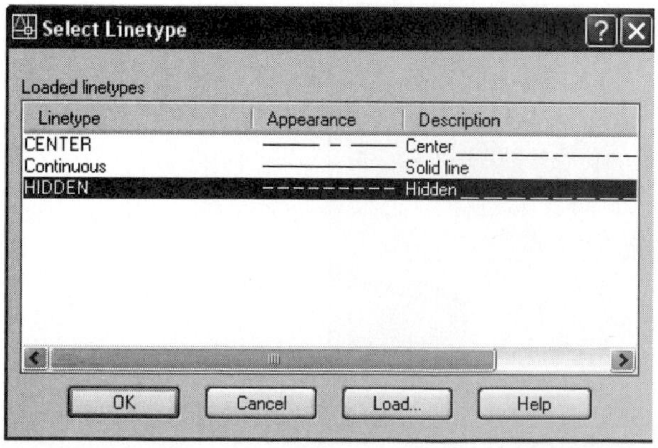

Figure 3-67

The **Load or Reload Linetypes** dialog box will appear. See Figure 3-66.

6. Scroll down the **Linetype** list and select the **Hidden** pattern by clicking on the pattern preview.

7. Click **OK.**

8. Select the **Center** layer line.

   The line will be highlighted.

9. Change the **Center** layer's line color to **red** and its linetype to **Center.**

   See Figure 3-67.

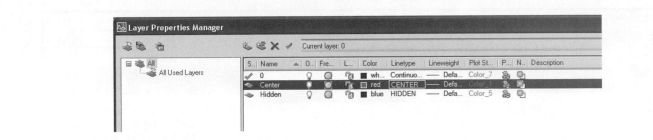

Figure 3-68

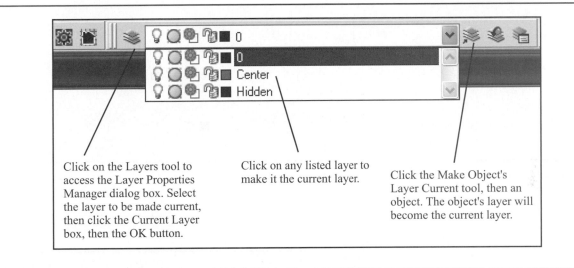

Click on the Layers tool to access the Layer Properties Manager dialog box. Select the layer to be made current, then click the Current Layer box, then the OK button.

Click on any listed layer to make it the current layer.

Click the Make Object's Layer Current tool, then an object. The object's layer will become the current layer.

Figure 3-69

The **Layer Properties Manager** dialog box should look like Figure 3-68.

To draw on different layers

Only the layer designated as the current layer may be worked on, even though other layers may be visible. There are three ways to make a layer the current layer. See Figure 3-69.

1. Use the **Layer control** tool and click the layer name. The layer will become the current layer. The current layer is the name displayed in the **Layer control** box after the cascade disappears.
2. Use the **Layers** tool to access the **Layer Properties Manager** dialog box. Click on a layer name, then click the current box, then **OK.** The layer name will appear in the **Layer control** box, indicating that it is the current layer.
3. Use the **Make Object's Layer Current** tool, by clicking an entity on the screen. The entity's layer will become the current layer.

Once a current layer has been established, objects may be drawn and modified.

To change layers

An object may be drawn on one layer, then moved to another layer. This feature is helpful when designing; for example, an original layout may be created in a single layer, then lines may be transferred to different layers as needed. As a line changes layers, it assumes the color and linetype of the new layer. Figure 3-70 shows a circle and a line drawn on the 0 layer. Change the line to a centerline and the circle to a hidden line by moving the objects to the appropriate layer.

1. Select the circle.

   The grip points will appear.

2. Use the **Layer control** tool and select the **Hidden** layer, making it the current layer.
3. Double-click the **<Esc>** key.

   The circle will be moved to the **Hidden** layer and will appear as a blue hidden line.

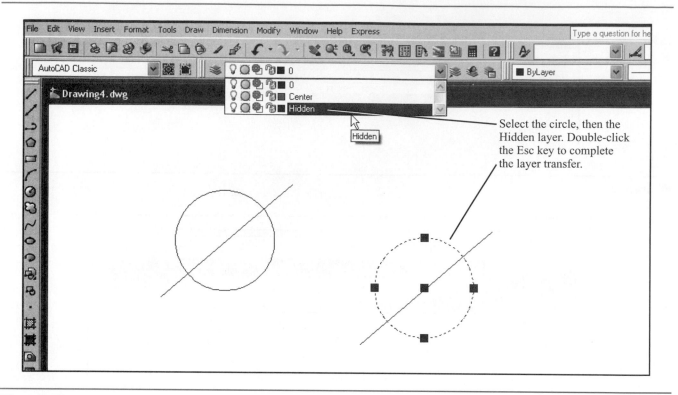

Select the circle, then the Hidden layer. Double-click the Esc key to complete the layer transfer.

Figure 3-70

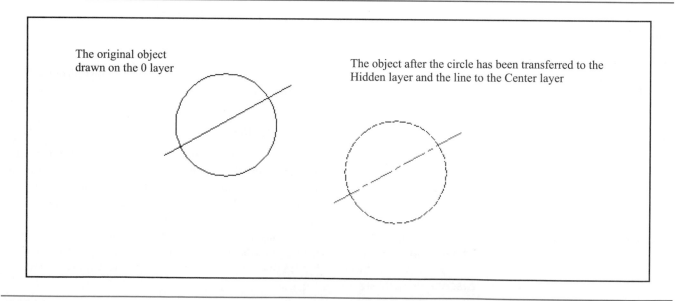

The original object drawn on the 0 layer

The object after the circle has been transferred to the Hidden layer and the line to the Center layer

Figure 3-71

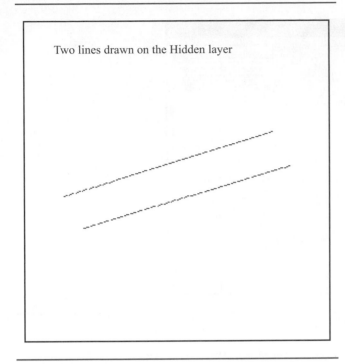

Two lines drawn on the Hidden layer

Figure 3-72

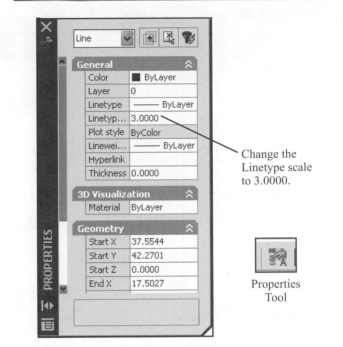

Change the
Linetype scale
to 3.0000.

Properties
Tool

Figure 3-73

4. Move the line to the **Center** layer using the same procedure.

Figure 3-71 shows the resulting changes.

To change the scale of a linetype

Figure 3-72 shows two parallel lines drawn on the Hidden layer.

1. Select the **Modify** pull-down menu, then **Properties,** or click the **Properties** tool on the **Standard** toolbar.

The **Properties** palette will appear. See Figure 3-73.

2. Select the **Linetype scale** and change the value from **1.0000** to **3.0000.**
3. Select the **Exit** button on the dialog box and return to the screen.
4. Draw two more lines.

Figure 3-74 shows the resulting change. The new lines have a different linetype scale and therefore different spacing.

To use the Global Linetype scale factor

All line patterns may be modified by the same scale factor using the **Global Linetype** scale factor.

1. Select the **Format** pull-down menu, then **Linetype,** then click the **Show Details** box.

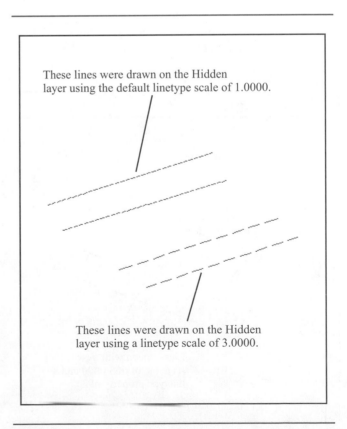

These lines were drawn on the Hidden layer using the default linetype scale of 1.0000.

These lines were drawn on the Hidden layer using a linetype scale of 3.0000.

Figure 3-74

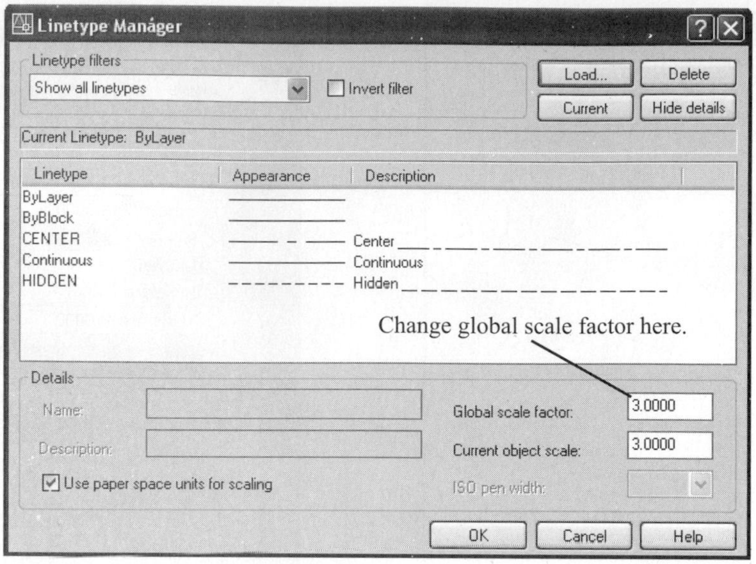

Figure 3-75

The **Linetype Manager** dialog box will appear. See Figure 3-75. The current global scale is **1.0000,** and the current object scale, the one set for the **Hidden** layer, equals **3.0000.**

2. Change the **Global scale factor** to **3.000**; select **OK.**

Figure 3-76 shows the resulting changes. All lines were changed by a factor of 3. The difference between the line scales is still 3.

To use the Match Properties tool

1. Select the **Match Properties** tool from the **Standard** toolbar.

   Select source object:

2. Select one of the lines with the smaller spacing.

   Select destination object(s) or[Settings]:

3. Select one of the lines with the larger spacing.

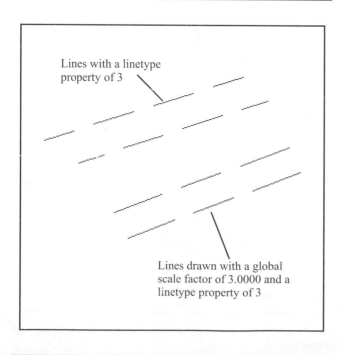

Figure 3-76

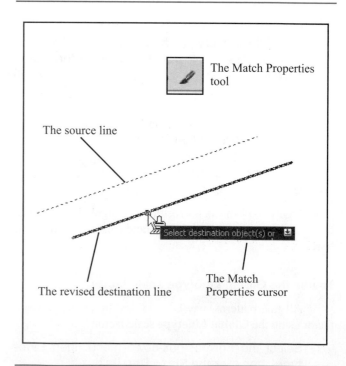

Figure 3-77

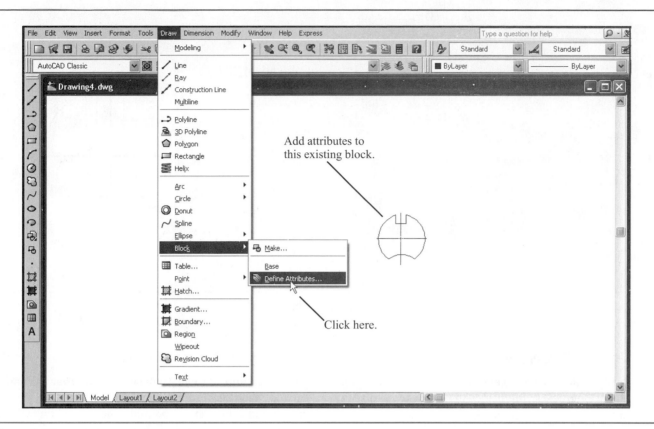

Figure 3-78

Figure 3-77 shows the resulting changes as applied to the line presented in Figure 3-74. The destination line now has the same properties as the source line.

To turn layers off

1. Select the **Layer control** tool.

   A list of current layers will cascade from the box.

2. Click the lightbulb icon on the **Center** layer.

   The lightbulb icon will change from yellow to blue, indicating that the layer is off.

3-26 ATTRIBUTES

Attributes are sections of text added to a block that prompt the drawer to add information to the drawing. For example, a title block may have an attribute that prompts the illustrator to add the date to the title block as it is inserted into a drawing. Attributes have their own toolbar.

To add an attribute to a block

Figure 3-78 shows a block. This example will add attributes that request information about the product's part number, material, quantity, and finish.

1. Select the **Draw** pull-down menu, then **Block,** then **Define Attributes.**

   The **Attribute Definition** dialog box appears. See Figure 3-79.

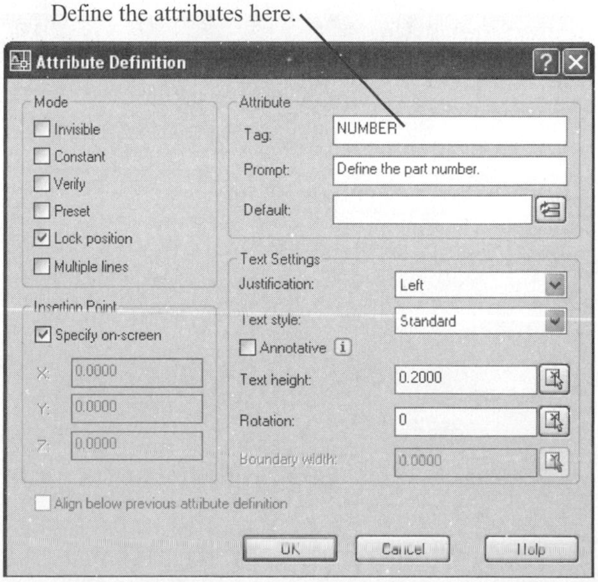

Figure 3-79

Figure 3-80

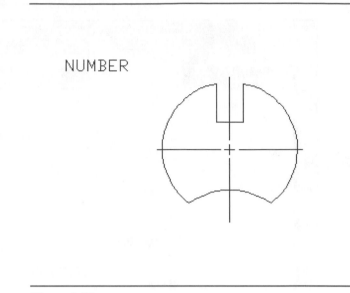

Figure 3-81

2. Select the **Tag** box and type **NUMBER.**

A *tag* is the name of an attribute and is used for filing and reference purposes. Tag names must be one word with no spaces.

3. Select the **Prompt** box and type: **Define the part number.**

The prompt line will eventually appear at the bottom of the screen when a block containing an attribute is inserted into a drawing. If you do not define a prompt, the tag name will be used as a prompt.

4. Leave the **Default** box empty.

The **Value** box entry will be the default value if none is entered when the prompt appears. Based on the defined prompt and value inputs, the following prompt line will appear when the attribute is combined with a block:

Define the part number <>:

5. Select **Specify on-screen.**

The drawing will appear with crosshairs. Select a location for the attribute tag by moving the crosshairs, then press the left mouse button.

Click here to align attributes.

Figure 3-82

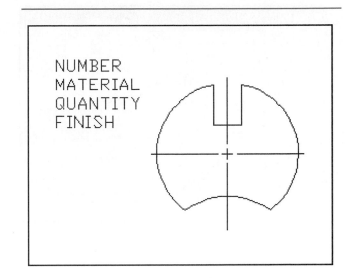

Figure 3-83

Figure 3-84

The **Attribute Definition** dialog box will reappear with the X,Y coordinate values of the selected point listed in the **Insertion Point** box. See Figure 3-80.

6. Select **OK.**

Figure 3-81 shows the attribute tag applied to a drawing. Figures 3-82 and 3-83 show three more **Attribute Definition** dialog boxes. Note that the **Align below previous attribute definition** box has a check mark, indicating it has been turned on. The option is turned on by clicking the box. Figure 3-84 shows the resulting drawing.

To create a new block that includes attributes

1. Select the **Make Block** tool from the **Draw** toolbar.

    The **Block Definition** toolbar will appear.

2. Name the new block **SHAPE-1.**
3. Select a base point.

    In this example the center point was selected.

4. Select the object by windowing the block and all the attribute tags.
5. Select **OK.**
6. The **Edit Attributes** dialog box will appear. If all entries are correct, select **OK.**

    See Figure 3-85.
    Figure 3-86 shows the resulting block.

Figure 3-85

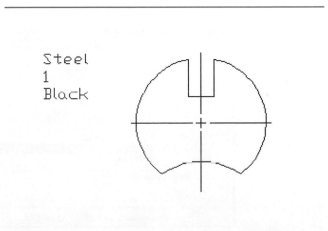

Figure 3-86

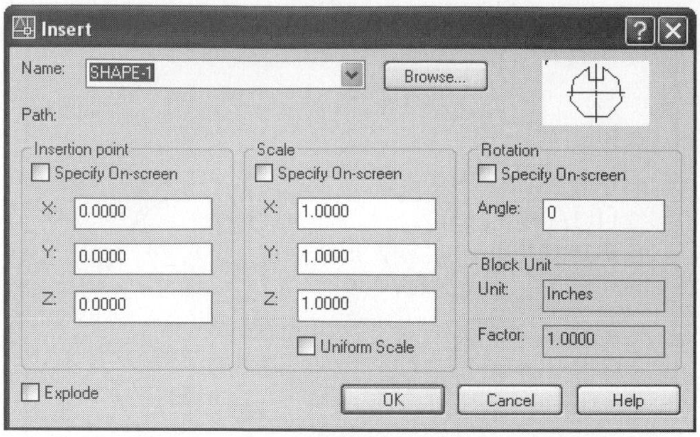

Figure 3-87

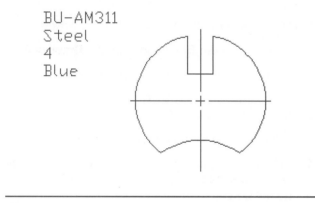

Figure 3-88

To insert an exisiting block with attributes

1. Select the **Insert Block** tool on the **Draw** toolbar.

   The **Insert** dialog box will appear. See Figure 3-87.

2. Select **SHAPE-1,** then click **OK.**

   Specify the insertion point or [Scale/X/Y/Z/Rotate/PScale/PX/PY/PZ/PRotate]:

3. Select an insertion point.

   Enter attribute values
   What material is needed? <Steel>:

4. Press **Enter.**

   What color do you want? <Black>:

5. Select a different color by typing **Blue;** press **Enter.**

   How many are needed?

6. Type **4;** press **Enter.**

   Define the part number <>:

7. Type **BU-AM311;** press **Enter.**

   Figure 3-88 shows the resulting drawing.

To edit an existing attribute

Once a block has been created that includes attributes, it may be edited using the **Edit Attributes** command. The procedure is as follows. The block to be edited must be on the drawing screen.

1. Select the **Edit Attribute** tool from the **Modify II** toolbar, or select the **Modify** pull-down menu, then **Object,** then **Attribute,** then **Single.** See Figure 3-89.

   Command: _attedit
   Select block reference:

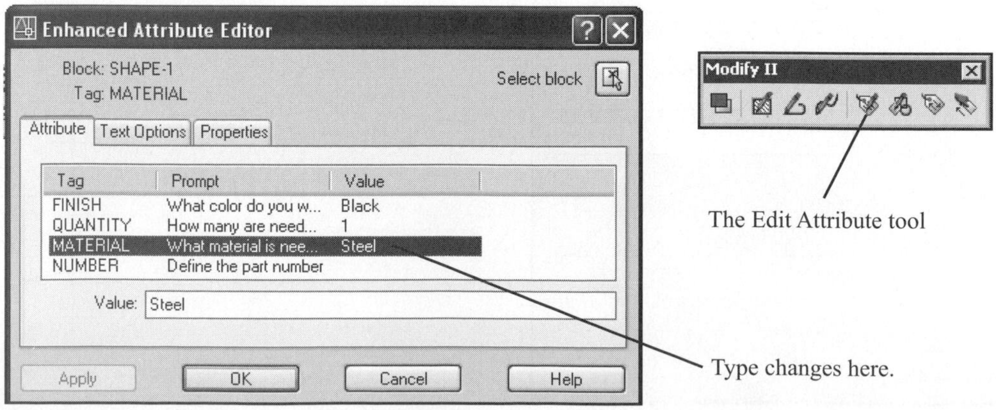

The Edit Attribute tool

Type changes here.

Figure 3-89

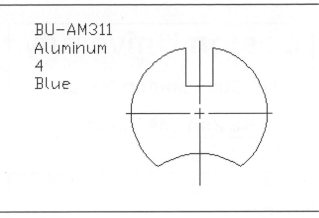

Figure 3-90

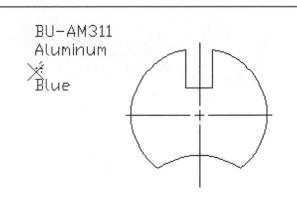

Figure 3-91

2. Select the word **Steel.**

The **Enhanced Attribute Editor** dialog box will appear. See Figure 3-89. In this example, the material will be changed from steel to aluminum. Note that the attribute prompt, tag, and value lines originally entered in the **Attribute Definition** dialog box are listed on the **Enhanced Attribute Editor** dialog box.

3. Locate the cursor to the right of the word **Steel,** backspace to remove **Steel,** then type **Aluminum.**

Figure 3-90 shows the resulting changes.

To use the Edit Global command

The **Edit Global** command is used to change information in a block independent of the block's definitions. The block to be edited must be on the screen.

1. Select the **Modify** pull-down menu, then **Object,** then **Attribute,** then **Global.**

Command: _-attedit
Edit attributes one at a time? [Yes/No] <Y>:

2. Press **Enter.**

Block name specification<>:*

3. Press **Enter.**

Enter attribute tag specification <>:*

4. Press **Enter.**

Enter attribute value specification <>:*

5. Press **Enter.**

Select attributes:

6. Select the number **4** associated with the **SHAPE-1** block on the screen; press **Enter.**

The value 4 will change to a hidden pattern, and a large X will appear on the screen. See Figure 3-91.

Enter an option [Value/Position/Height/Angle/Style/Layer/Color/Next] <N>:

7. Type **V;** press **Enter.**

Enter type of value modification [Change/Replace?] <R>:

The **Change** option is used to modify a few characters of the existing value. The **Replace** option is used to create an entirely new value. The **Replace** option is the default option.

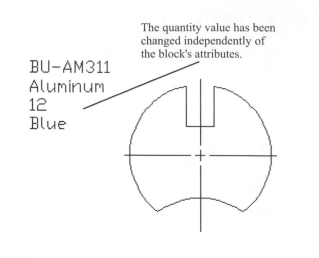

The quantity value has been changed independently of the block's attributes.

BU–AM311
Aluminum
12
Blue

Figure 3-92

8. Press **Enter.**

   Enter new attribute value:

9. Type **12;** press **Enter.**

   Enter an option [Value/Position/Height/Angle/ Style/Layer/Color/Next]<N>:

10. Press **Enter.**

   The new value will replace the former value. See Figure 3-92.

Boston University

110 Cummington Street

Boston, MA 02215

TITLE:		
PART NO.:		REV:
SCALE:	DATE:	

Figure 3-93

3-27 TITLE BLOCKS WITH ATTRIBUTES

Figure 3-93 shows a title block that has been saved as a block. It would be helpful to add attributes to the block so that when it is called up to a drawing, the drawer will be prompted to enter the required information.

Note that text may be added to an AutoCAD template using the **Mtext** command. Create the desired text and use the **Move** command to position it within the title block. The title block shown in Figure 3-93 was custom drawn for a special application.

Figures 3-94, 3-95, 3-96, 3-97, and Figure 3-98 show the five **Attribute Definition** dialog boxes used to define the

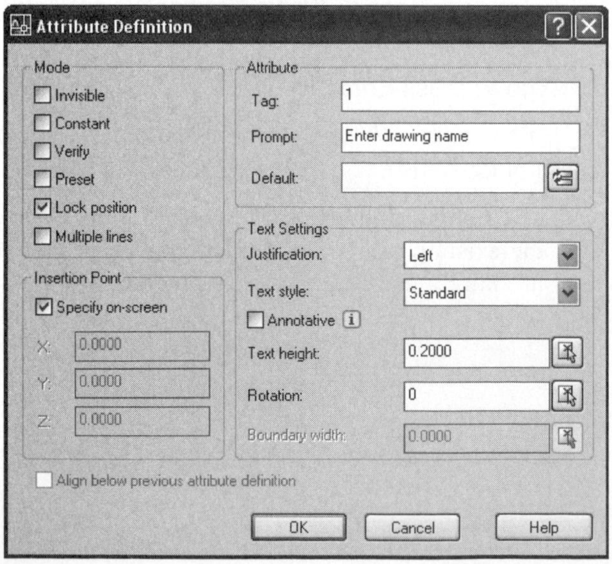

Figure 3-94

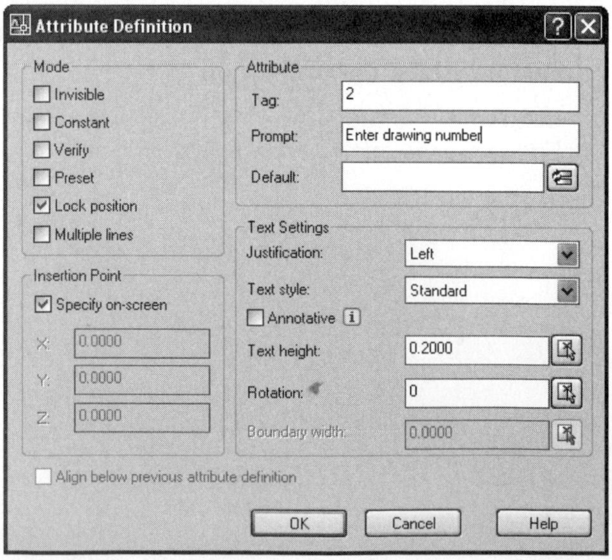

Figure 3-95

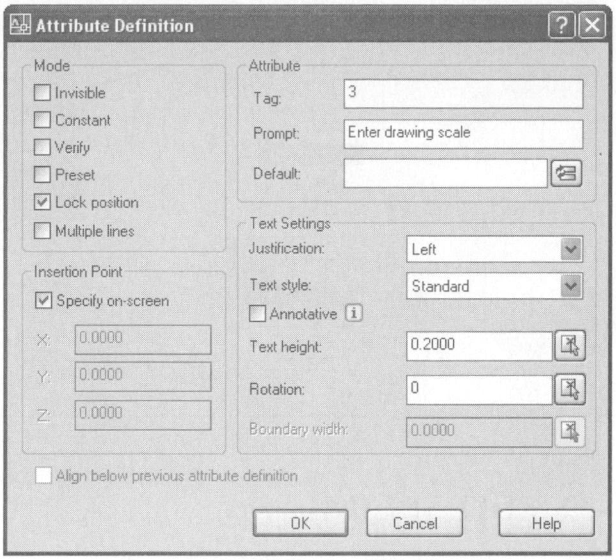

Figure 3-96

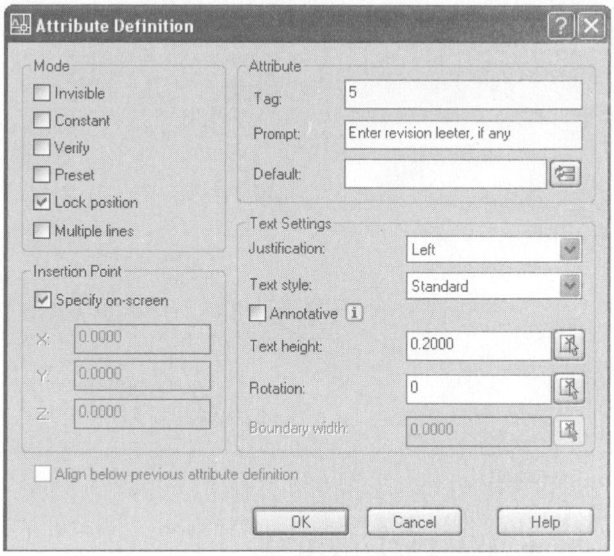

Figure 3-97

Figure 3-98

title block attributes. Note that the text height in Figures 3-94 through 3-98 may be varied if desired. Also note that no attribute defaults were assigned. This means that the space on the drawing will be left blank if no prompt value is entered. Figure 3-99 shows the resulting title block with the attribute tags in place.

The new block with attributes was saved as **TITLE-A**, using the **Block** command. It can now be saved as a wblock and used on future drawings. Figure 3-100 shows a possible title block created from the TITLE-A block.

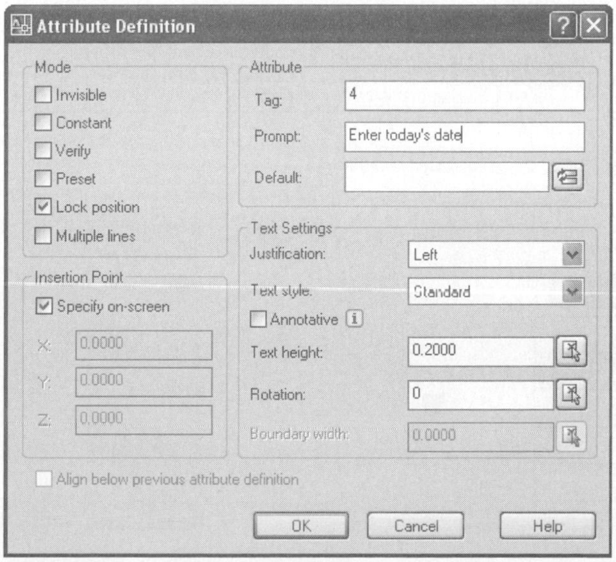

Boston University

110 Cummington Street

Boston, MA 02215

TITLE: 1		
PART NO.: 2		REV: 5
SCALE: 3	DATE: 4	

Figure 3-99

A sample use of the block attributes

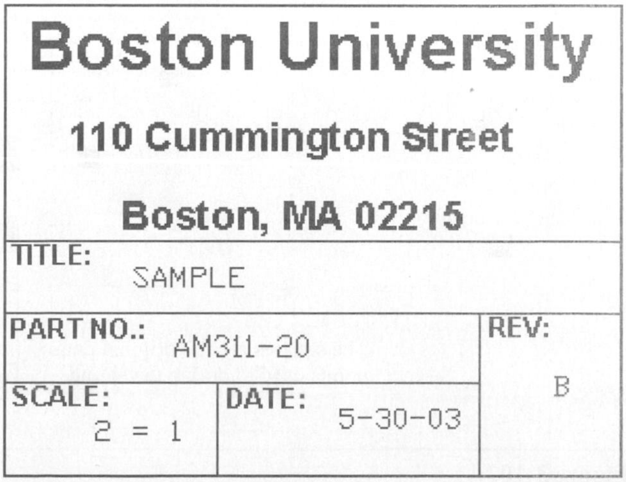

Boston University

110 Cummington Street

Boston, MA 02215

TITLE: SAMPLE		
PART NO.: AM311-20		REV: B
SCALE: 2 = 1	DATE: 5-30-03	

Figure 3-100

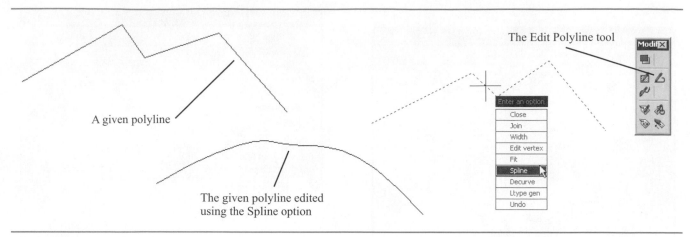

A given polyline

The given polyline edited
using the Spline option

The Edit Polyline tool

Figure 3-101

3-28 EDIT POLYLINE

The **Edit Polyline** tool, found on the **Modify II** tool-bar, is used to change the shape of a given polyline. In the example shown in Figure 3-101 the **Edit Polyline** command was used to create a spline from a given polyline. The **Edit Polyline** tool is discussed further in Section 16-17.

To create a spline from a given polyline

1. Select the **Edit Polyline** tool from the **Modify II** toolbar.

   Command: _pedit Select polyline:

2. Select the polyline.

   Select an option [Close/Join/Width/Edit vertex/Fit/ Spline/Decurve/Ltype gen/Undo]:

3. Select the **Spline** option; press **Enter.**

   Figure 3-101 shows the edited polyline.

3-29 EDIT SPLINE

The **Edit Spline** tool, found on the **Modify II** toolbar, is used to change the shape of a given spline.

To edit a spline

1. Select the **Edit Spline** tool from the **Modify II** toolbar.

   Command: _splinedit
   Select spline:

2. Select the given spline.

   A series of squares will appear on the screen. See Figure 3-102. These squares are the original data points used to define the spline. In this example one of the points will be moved.

3. Select the **Move vertex** option; press **Enter.**

   Specify new location or [Next/Previous/Select point/eXit] <N>:

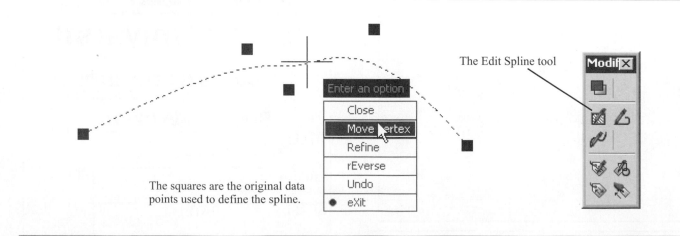

The squares are the original data points used to define the spline.

The Edit Spline tool

Figure 3-102

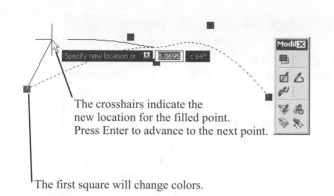

The crosshairs indicate the
new location for the filled point.
Press Enter to advance to the next point.

The first square will change colors.

Figure 3-103

The first square will change colors, and a rubber-band–type line with a large + will appear. See Figure 3-103.

4. Move the first square to a new location, then type **X**; press **Enter.**

Figure 3-104 shows the edited spline. The chosen location will become the new location for the first point. If a point other than the first point needs to be moved, press

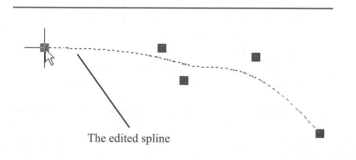

The edited spline

Figure 3-104

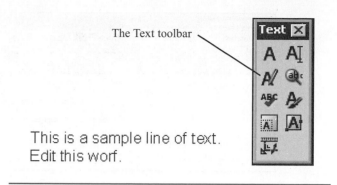

The Text toolbar

This is a sample line of text.
Edit this worf.

Figure 3-105

Enter when the X appears, and the **Edit** command will advance to the next point. Every time the **Enter** key is pressed, the **Edit** command will advance to the next point. Only the **X** option will allow you to exit the command sequence.

3-30 EDIT TEXT

The **Edit** tool on the **Text** toolbar is used to change existing text. Figure 3-105 shows a sample of existing text.

To change existing text

1. Select the **Edit** tool from the **Text** toolbar.

    Command: _ddedit
    Select an annotated object or [Undo]:

2. Select the existing text.

The **Text Formatting** dialog box will appear (see Figure 3-106) with the selected text.

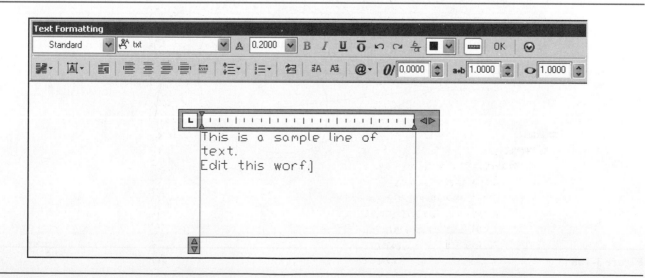

Figure 3-106

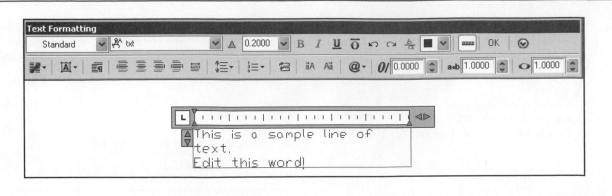

Figure 3-107

3. Edit the text shown in the dialog box.

   See Figure 3-107.

4. Select **OK.**

   The edited text will replace the previous text.

3-31 CONSTRUCTING THE BISECTOR OF AN ANGLE— METHOD I

Given angle A-O-B, bisect it. See Figure 3-108.

1. Select **Draw, Circle.**
2. Draw a circle with a center point located at point **O.** The circle may be of any radius. This is an excellent place to take advantage of AutoCAD's **Drag** mode. Move the cursor until the circle appears to be approximately the same size as that shown.
3. Select **Copy, Multiple** to draw two more circles equal in radius to the one drawn in step 1. Locate the circles' center points on intersections of the first circle and angle **A-O-B.** Use **Osnap, Intersection** to ensure accuracy.
4. Draw a line from point **O** to intersection **C** as shown. Use **Osnap, Intersection** to ensure accuracy.

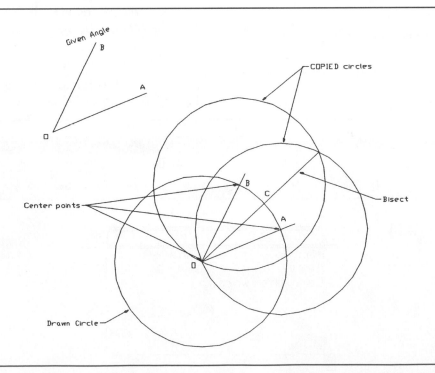

Figure 3-108

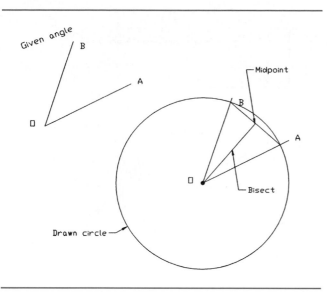

Figure 3-109

3-32 CONSTRUCTING THE BISECTOR OF AN ANGLE— METHOD II

Given angle A-O-B, bisect it. See Figure 3-109.

1. Draw a circle with a center point located at point **O**. The circle may be of any radius. This is an excellent place to take advantage of AutoCAD's **Drag** mode. Move the cursor until the circle

appears to be approximately the same size as that shown.
2. Draw a line between the intersection point created by the circle and angle **A-O-B**. Use **Osnap, Intersection** to ensure accuracy.
3. Draw a line between point **O** and the midpoint of the line. Use **Osnap, Midpoint** to ensure accuracy.

3-33 CONSTRUCTING AN OGEE CURVE (S-CURVE) WITH EQUAL ARCS

Given points B and C, construct an ogee curve between them. See Figure 3-110.

1. Draw a straight line between points **B** and **C**. Use **Osnap, Endpoint** to ensure accuracy.
2. Draw lines perpendicular to lines **A-B** and **C-D** as shown.

Use **Osnap, Endpoint** to accurately start the lines on points B and C, respectively. Use relative coordinates to draw the lines. The lines may be of any length greater than the perpendicular distance between lines A-B and C-D.

3. Construct a perpendicular bisector of line **B-C**.
4. Define the intersections of the perpendicular lines drawn in step 2 with the perpendicular bisector of **B-C** at points **F** and **G**.
5. Draw a circle centered at point **F** that passes points **B** and **E**. Trim the circle using line **B-C**.
6. Draw a circle centered at point **G** that passes points **C** and **E**. Trim the circle using line **B-C**.

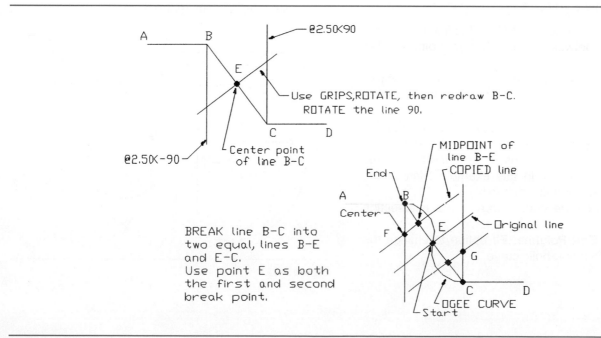

Figure 3-110

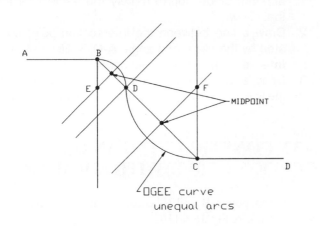

Figure 3-111

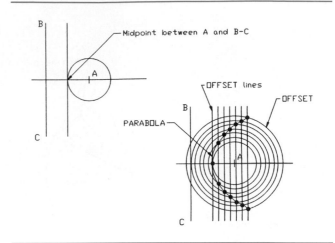

Figure 3-112

The two arcs of steps 5 and 6 define an ogee curve using equal arcs. Figure 3-111 shows the construction of an ogee curve with unequal arcs. The construction techniques are similar to those presented above.

3-34 CONSTRUCTING A PARABOLA

Definition: A *parabola* is the loci of points such that the distances between a fixed point, the *focus,* and a fixed line, the *directrix,* are always equal.

Given a focus point A and a directrix B-C, construct a parabola. See Figure 3-112.

1. Draw a line parallel to **B-C** so that it intersects the midpoint between line **B-C** and point **A.** Use **Offset.**
2. Draw a circle centered about point **A** whose radius is equal to half the distance between line **B-C** and point **A.**
3. Offset lines and a circle **0.2 inch** from the parallel line and the circle created in steps 1 and 2.
4. Identify the intersections of the lines and circles created in step 3. In this example filled circles were used to identify the points.
5. Draw a polyline connecting all the intersection points.
6. Use the **Edit Polyline,** Fit option to change the polyline to a parabolic curve.

3-35 CONSTRUCTING A HYPERBOLA

Definition: A *hyperbola* is the loci of points equidistant between two fixed foci.

Given foci points F1 and F2 equidistant from a vertical line and two other vertical lines drawn through points A and B, draw a hyperbola. See Figures 3-113, 3-114, and 3-115.

1. Draw a circle of arbitrary radius using point **F2** as the center. If possible, set the drawing up so that both points **F1** and **F2** are on snap points. If this is not possible, draw a vertical line through point **F2** so that the **Osnap, Intersection** option may be used to locate the center accurately.

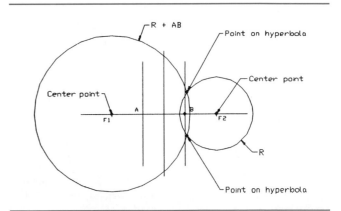

Figure 3-113

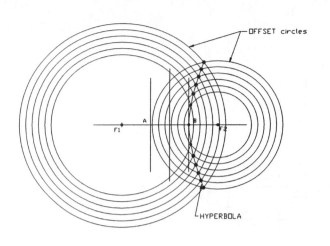

Figure 3-114

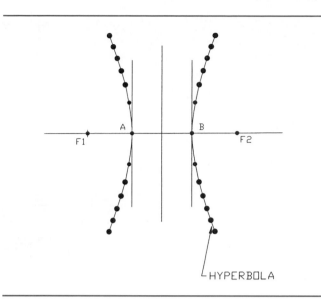

Figure 3-115

2. Draw a circle centered about **F2** of radius equal to the radius of the circle drawn in step 1, plus the distance between points **A** and **B.** See Figure 3-114.

3. Draw concentric circles about the circles centered about **F1** and **F2**. The distance between all the circles should be equal. Use **Offset.**

4. Mark the intersections between the smaller and larger circles as shown in Figure 3-114. Use filled circles to define the points.

5. Draw a polyline connecting points 1 through 12.

6. Use the **Edit Polyline, Fit** option to draw the hyperbola.

7. Use **Mirror** to create the opposing hyperbola. See Figure 3-115.

3-36 CONSTRUCTING A SPIRAL

Construct a spiral of Archimedes. See Figure 3-116.

1. Draw a set of perpendicular centerlines as shown.

2. Draw **12** concentric circles about center point **O.** Use **Offset** to draw circles.

3. Draw **12** equally spaced ray lines as shown. Array either the horizontal or vertical line using the **Polar Array** command. Use point **O** as the center of rotation.

4. Draw a polyline starting at point **O.** The second point is the intersection of the **30°** ray with the first circle. Point 2 is the intersection of the **60°** ray and the second circle. Continue through all 12 rays and circles.

5. Use the **Edit Polyline, Fit** option to change the polyline to a spiral.

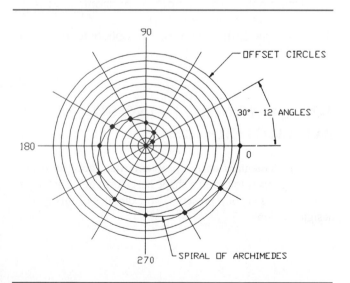

Figure 3-116

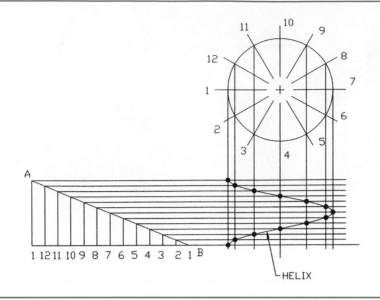

Figure 3-117

3-37 CONSTRUCTING A HELIX

Construct a helix that advances 1.875 inches every 360°. See Figure 3-117.

1. Draw a vertical line **1.875** inches long and a horizontal line as shown.
2. Draw a horizontal line that can easily be divided into **12** equal spaces. The length of the horizontal line is arbitrary. Set up a grid so that the 12 equal spaces can easily be identified.
3. Draw a circle as shown. Use **Array** and divide the circle using **12** equally spaced rays. Use the **Polar Array** command with the circle's center point as the center of rotation.
4. Draw line **A-B** as shown. Draw **12** vertical, equally spaced lines between the horizontal line and line A-B.
5. Label both the circle and horizontal line as shown. The number of divisions must be the same for both the circle and the horizontal line.
6. Draw vertical lines from the intersections of the 12 equally spaced rays and the circle's circumfer-

ence as shown in Figure 3-116. Use the **Osnap, Intersection** option to ensure accuracy.
7. Draw horizontal lines from the intersections on line A-B so that they intersect the vertical lines from step 5. Use the **Osnap, Intersection** option to ensure accuracy.
8. Mark the intersections of the lines from steps 5 and 6 as shown.
9. Use a polyline to connect the horizontal and vertical intersections as shown.
10. Use the **Edit Polyline, Fit** option to draw the helix.

3-38 DESIGNING USING SHAPE PARAMETERS

An elementary design problem often faced by beginning designers is to create a shape based on a given set of parameters. This section presents two examples of this type of design problem.

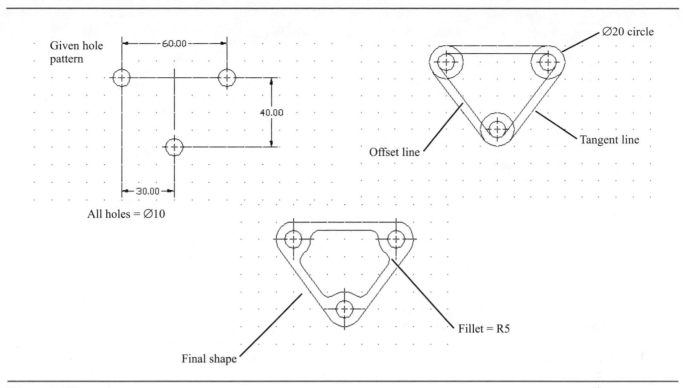

Figure 3-118

Design Problem DP3-1

Design a shape that will support the hole pattern shown in Figure 3-118 subject to the following parameters. Design for the minimum amount of material. All dimensions are in millimeters.

1. The edge of the material may be no closer to the center of a hole than a distance equivalent to the radius of the hole.
2. The minimum distance from any cutout to the edge of the part may not be less than the smallest distance calculated in step 1.
3. The minimum inside radius for a cutout is R = **5** millimeters.

The solution is as follows.

1. Calculate the minimum edge distances for the holes.

In this problem the three holes are all the same diameter: 10 millimeters. Given the parameter that an edge may be no closer to the center of the hole than a distance equivalent to the size of the hole's diameter, the distance from the outside edge of the hole to the edge of the part must be no less than 5 millimeters, or a circular shape of diameter 20 millimeters.

2. Draw **Ø20** circles using the existing circle's center point as shown in Figure 3-118.
3. Draw outside tangent lines between the Ø20 circles.

The second parameter limits the minimum edge distance to the minimum edge distance calculated in step 1, or 5 millimeters.

4. Use the **Offset** command to draw lines **5** millimeters from the outside tangent lines drawn in step 3 to define the internal cutout.
5. Use the inside radius parameter of **5** millimeters and add fillets to the internal cutout.
6. Erase any excess lines.

Figure 3-118 shows the final shape.

Design Problem DP3-2

Design a shape that will support the hole pattern shown in Figure 3-119. Support the center hole with four perpendicular webs. All dimensions are in inches.

1. The edge of the material may be no closer to the center of a hole than a distance equivalent to the radius of the hole.

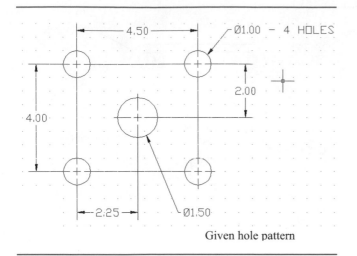

Given hole pattern

Figure 3-119

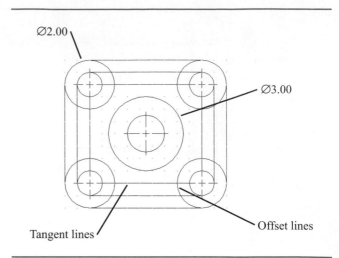

Tangent lines   Offset lines

Figure 3-120

2. The minimum distance from any cutout to the edge of the part may not be less than the smallest distance calculated in step 1.
3. The minimum inside radius for a cutout is R = **.125** inch.

The solution is as follows.

1. Calculate the minimum edge distances for the holes.

In this example, the hole diameters are 1.00 and 1.50 inches.

2. Add circles of radius **2.00** and **3.00** as shown. See Figure 3-120.

The minimum edge distance based on the Ø1.00 of the smaller hole equals .50 inch.

3. Use the minimum edge distance and define the inside edge of the internal cutouts.
4. Use the minimum edge distance to define the four supporting webs for the center hole.
5. Add the inside bend radii.

The inside bend radii were added using the **Fillet** command. This will cause a portion of the internal edge to disappear. Use the **Offset** command again to replace the line defining the edge of the internal cutout. See Figure 3-121.

6. Remove all excess lines.

See Figure 3-122.

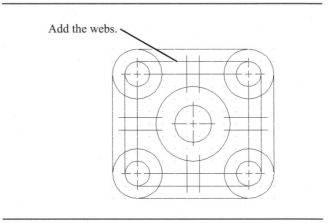

Add the webs.

Figure 3-121

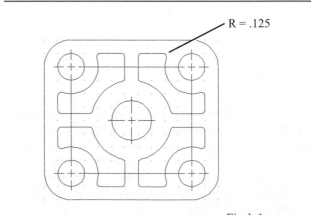

R = .125

Final shape

Figure 3-122

3-39 EXERCISE PROBLEMS

Redraw the following figures to scale using the given dimensions. Do not include dimensions on the drawing.

EX3-1 INCHES (Hint: Use Osnap)

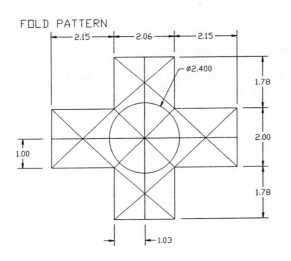

EX3-3 MILLIMETERS (Hint: Use Osnap)

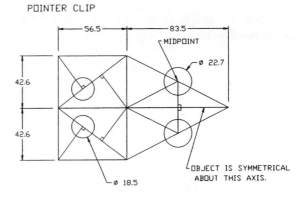

EX3-2 MILLIMETERS (Hint: Use Osnap)

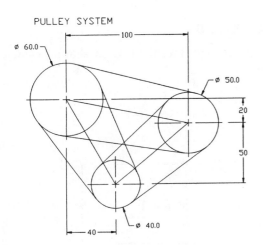

EX3-4 MILLIMETERS (Hint: Use Osnap)

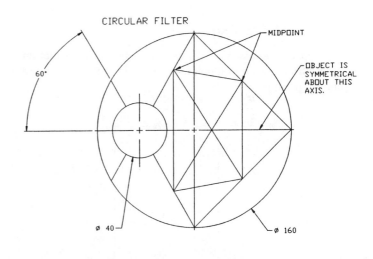

EX3-5

Draw a 54° angle; construct the bisector.

EX3-6

Draw an 89.33° angle; construct the bisector.

EX3-7 INCHES

Given points A and B as shown below, construct an ogee curve of equal arcs.

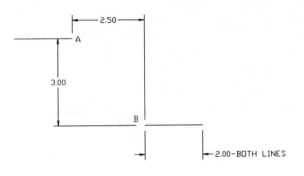

EX3-8 MILLIMETERS

Given points A, B, C, as shown below, construct an ogee curve that starts at point A, passes through point C, and ends at point B.

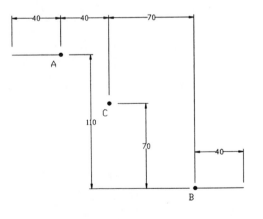

EX3-9

Construct a hexagon inscribed within a 3.63-inch diameter circle.

EX3-10

Construct a hexagon inscribed within a 120-millimeter diameter circle.

EX3-11 INCHES

Given point A and directrix B-C as shown below, construct a parabola.

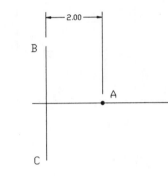

EX3-12 MILLIMETERS

Given point A and directrix B-C as shown below, construct a parabola.

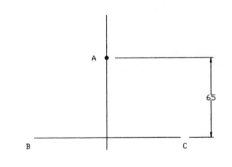

EX3-13 INCHES

Given foci F1 and F2 as shown below, draw a hyperbola.

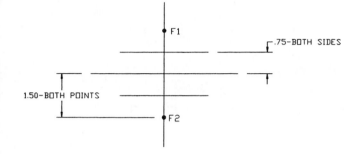

EX3-15 INCHES

Given the concentric circles and rays shown below, construct a spiral.

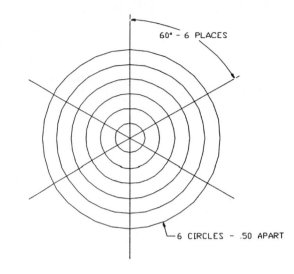

EX3-14 MILLIMETERS

Given foci F1 and F2 as shown below, draw a hyperbola.

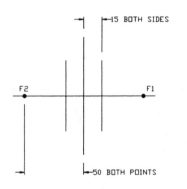

EX3-16 MILLIMETERS

Given the concentric circles and rays shown below, construct a spiral.

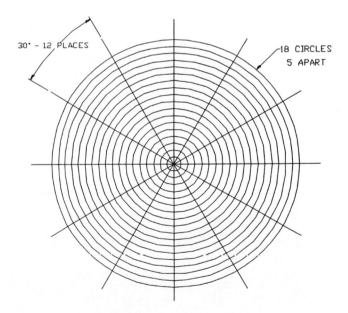

EX3-17 INCHES

Given the setup shown below, construct a helix.

Redraw the following objects based on the given dimensions.

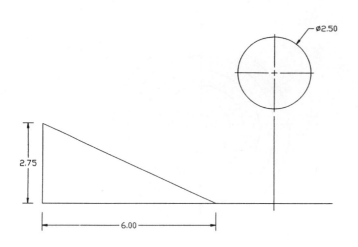

EX3-19 MILLIMETERS

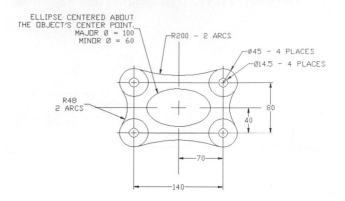

EX3-18 MILLIMETERS

Given the setup shown below, construct a helix.

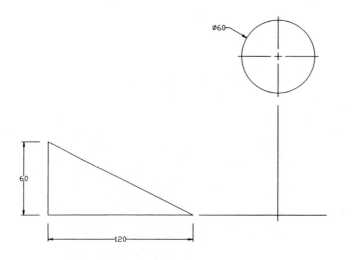

EX3-20 MILLIMETERS

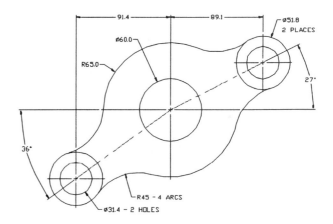

EX3-21 INCHES

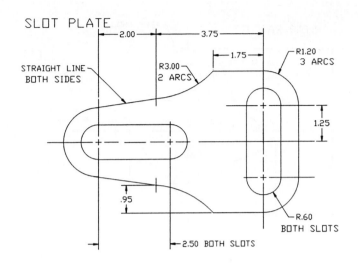

SLOT PLATE

EX3-22 INCHES

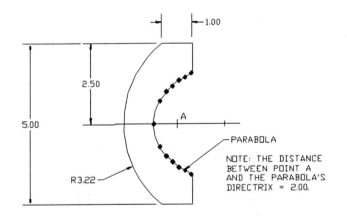

PARABOLA

NOTE: THE DISTANCE
BETWEEN POINT A
AND THE PARABOLA'S
DIRECTRIX = 2.00.

EX3-23 MILLIMETERS

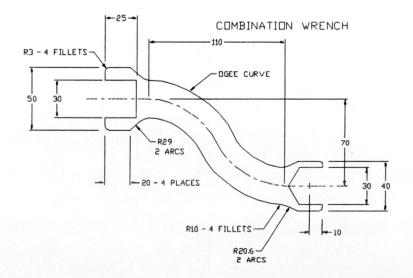

COMBINATION WRENCH

EX3-24 MILLIMETERS

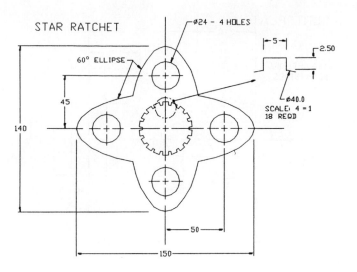

EX3-25 MILLIMETERS

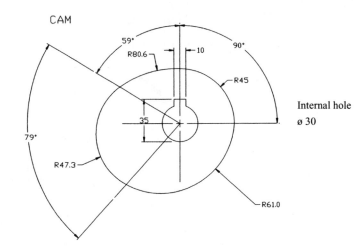

EX3-26 MILLIMETERS

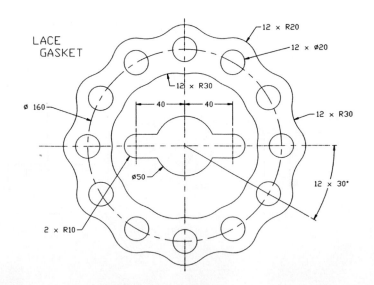

EX3-27 MILLIMETERS

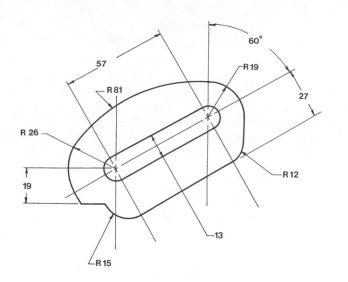

EX3-29 INCHES

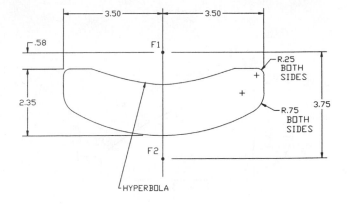

EX3-28 MILLIMETERS

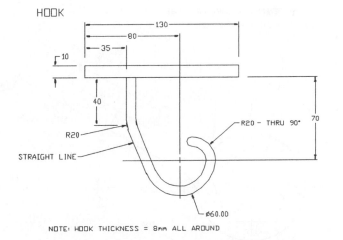

EX3-30 INCHES

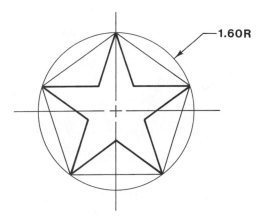

EX3-31 MILLIMETERS

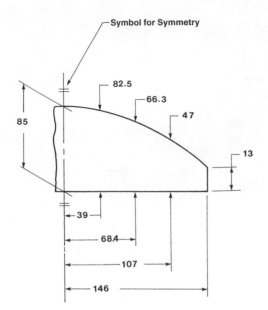

EX3-33 MILLIMETERS

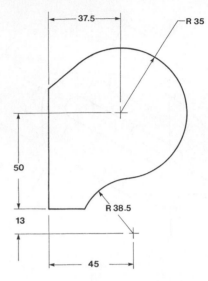

EX3-32 MILLIMETERS

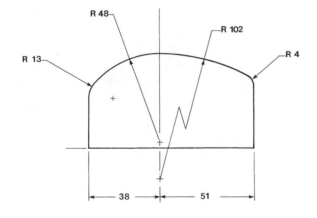

EX3-34 MILLIMETERS

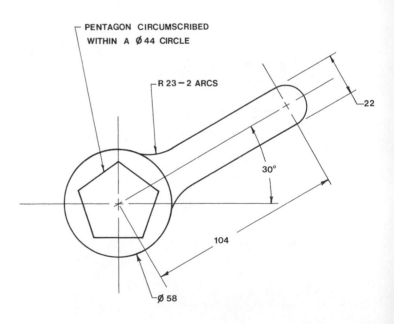

EX3-35 INCHES

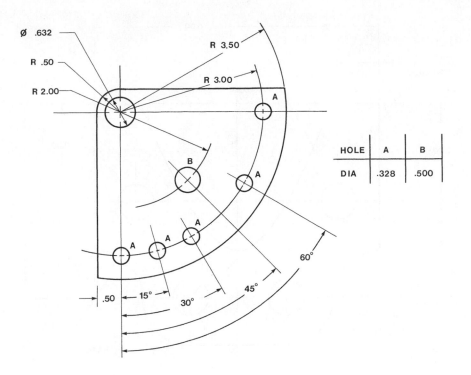

HOLE	A	B
DIA	.328	.500

EX3-36 INCHES

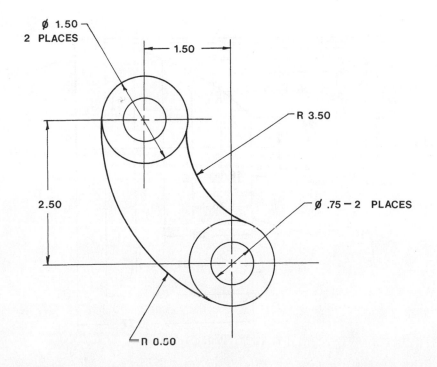

EX3-37 INCHES

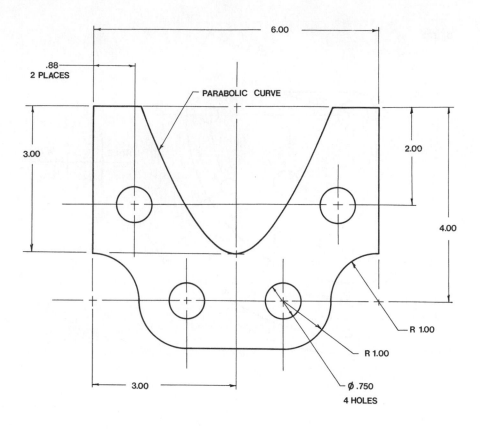

EX3-38 MILLIMETERS

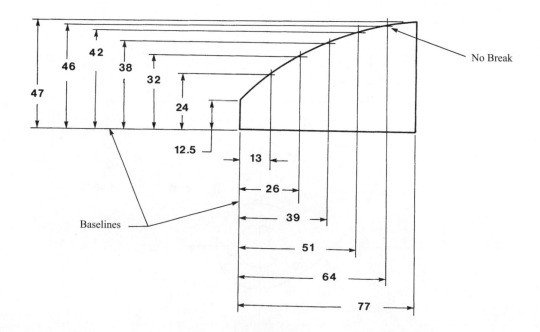

EX3-39 CENTIMETERS

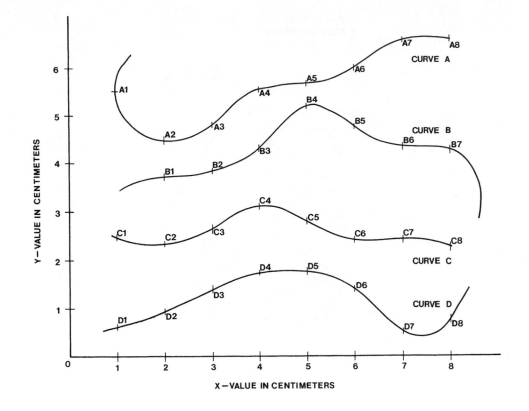

EX3-40 INCHES

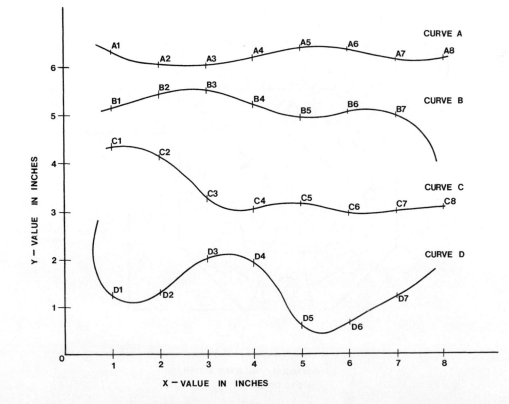

EX3-41 MILLIMETERS

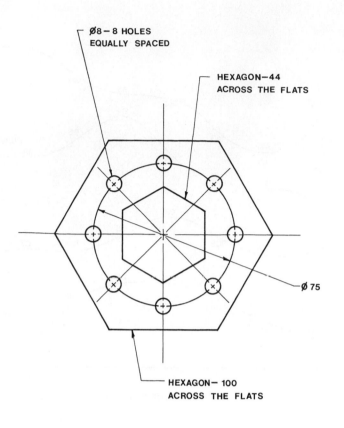

Ø8 – 8 HOLES
EQUALLY SPACED

HEXAGON–44
ACROSS THE FLATS

Ø 75

HEXAGON– 100
ACROSS THE FLATS

EX3-42 MILLIMETERS

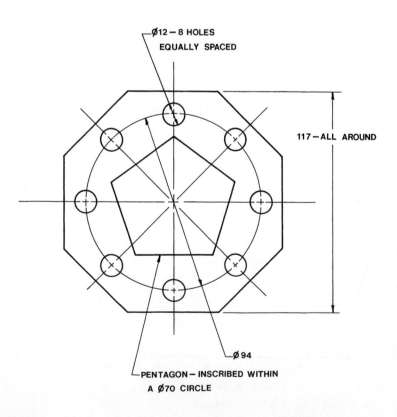

Ø12 – 8 HOLES
EQUALLY SPACED

117 – ALL AROUND

Ø 94

PENTAGON – INSCRIBED WITHIN
A Ø70 CIRCLE

EX3-43

Draw a circle, mark off 24 equally spaced points, then connect each point with every other point using only straight lines.

EX3-44 MILLIMETERS

A. Create a block of the following drawing layout. All dimensions are in millimeters.
B. Add the following attributes:

Tag = DATE
Prompt = Enter today's date
Value = (leave blank)

Tag = DRAWING
Prompt = Enter the drawing number
Value = (leave blank)

Tag = NAME
Prompt = Enter your name
Value = (leave blank)

Align the attribute tags on the drawing.

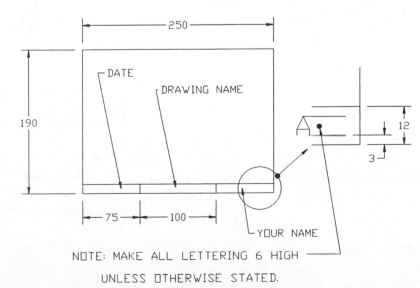

DRAWING LAYOUT — 2(MILLIMETERS)

EX3-45 MILLIMETERS

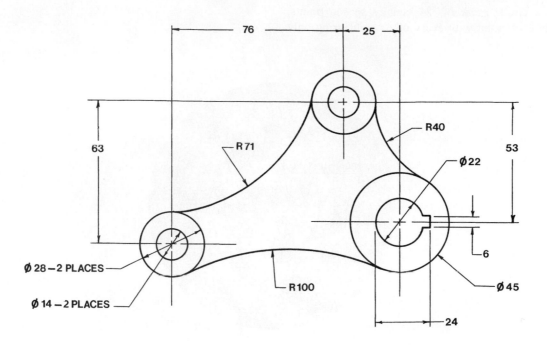

EX3-46 MILLIMETERS

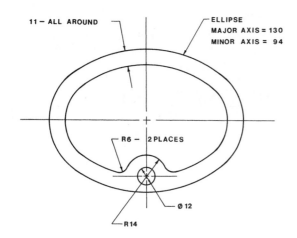

EX3-47 MILLIMETERS

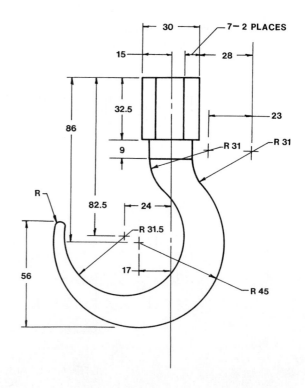

EX3-48 INCHES

Create a block for the following standard tolerance blocks shown.

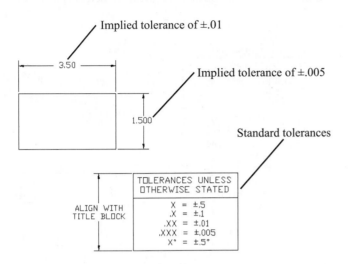

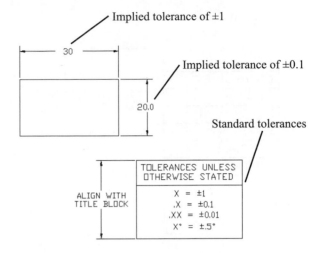

EX3-49 INCHES

A. Create a block for one of the title blocks shown. A sample completed title block is also shown.
B. Add attributes that will help the user complete the title block satisfactorily.

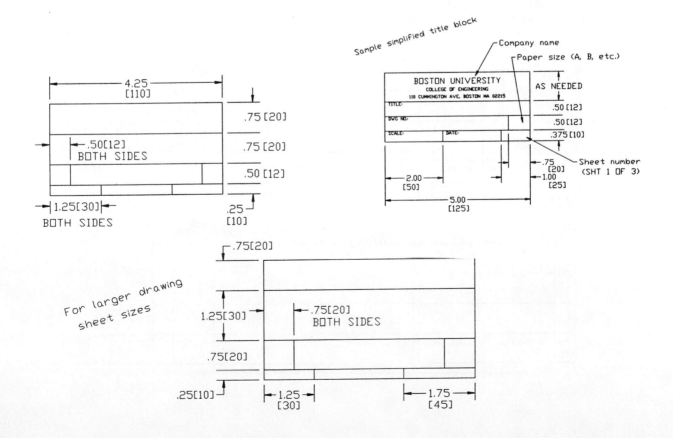

EX3-50

Create a block of the following release block. All dimensions are in inches.

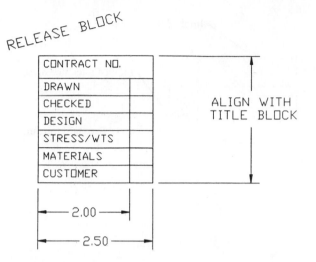

EX3-52

Create blocks for the following revision block formats. The dimensions within the dimension lines are inches; the dimensions within the brackets are millimeters.

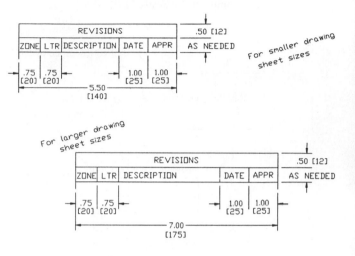

EX3-51

Create blocks for the following parts list formats. All dimensions are in inches. Add the appropriate attributes.

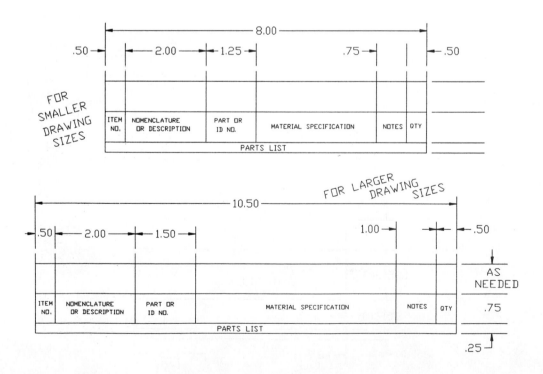

Design a shape that will support the hole patterns shown in Exercises EX3-53 through EX3-57. Design for a minimum amount of material and to satisfy the following parameters.

1. The edge distance of the material may be no closer to the center of a hole than a distance equal to the radius of the hole.
2. The minimum edge distance from any internal cutout to the edge of the part may be no less than the distance calculated in step 1.
3. The minimum inside radius for a cutout is 5 millimeters, or .125 inch.

EX3-53 INCHES

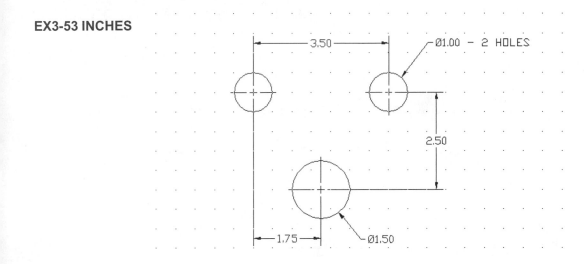

EX3-54 MILLIMETERS

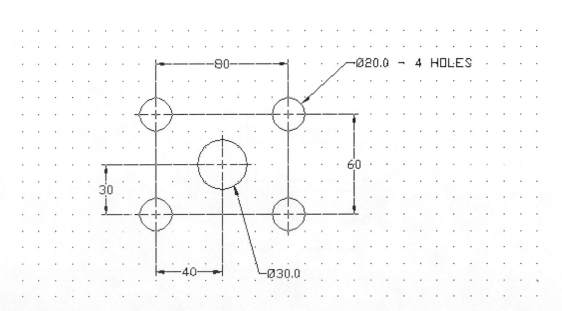

EX3-55 MILLIMETERS

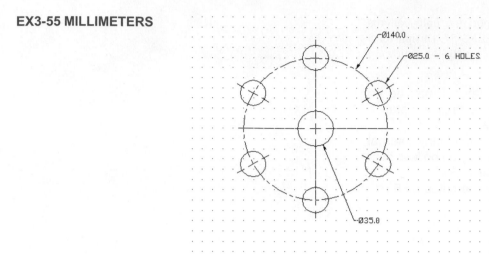

EX3-56 MILLIMETERS

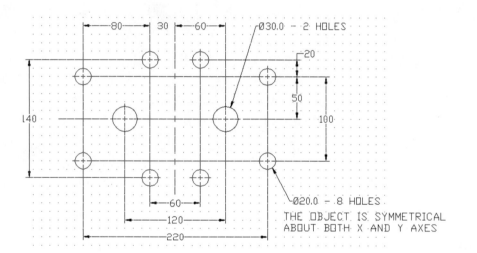

EX3-57 MILLIMETERS

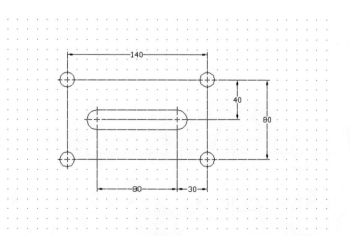

C H A P T E R 4

Sketching

4-1 INTRODUCTION

The ability to create freehand sketches is an important skill for engineers and designers to acquire. The old joke about engineers and designers not being able to talk without a pencil in their hands is not far from the truth. Many design concepts and ideas are very difficult to express verbally, so they must be expressed visually. Sketches can be created quickly and used as a powerful aid in communicating technical ideas.

This chapter presents the fundamentals of freehand sketching as applied to technical situations. It includes both two-dimensional and three-dimensional sketching. Like any skill, freehand sketching is best learned by lots of practice.

4-2 ESTABLISHING YOUR OWN STYLE

As you learn and practice how to sketch, you will find that you develop your own way of doing things, your own style. This is very acceptable, as there is no absolutely correct method for sketching, but only recommendations.

The most important requirement of freehand sketching is that you be comfortable. As you practice and experiment

you may find that you prefer a certain pencil lead hardness, a certain angle for your paper, and a certain way to make your lines, both straight and curved. You may find that you prefer to use oblique sketches rather than isometric sketches when sketching three-dimensional objects. Eventually you will develop a style that is comfortable for you and that you can consistently use to create good-quality sketches.

It is recommended that you try all the types of sketches presented in the chapter. Only after you have tried and practiced all the different types will you be able to settle on a technique and style that works best for you.

4-3 GRAPH PAPER

Graph paper is very helpful when preparing freehand sketches. It helps you sketch straight lines, allows you to set up guide points for curved lines, and can be used to establish proportions. It is recommended that you start by doing all two-dimensional sketches on graph paper. As your sketching becomes more proficient you may sketch on plain paper, but because most technical sketching requires some attention to correct proportions, grid paper will always be helpful.

Graph paper is available in many different scales and in both inch and metric units. Some graph paper is printed

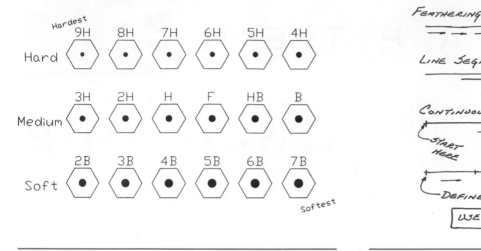

Figure 4-1

Figure 4-2

using light blue lines, since many copying machines cannot easily reproduce blue lines. This means that copies of the sketches done on light-blue guidelines will appear to have been drawn on plain paper and will include only the sketch.

4-4 PENCILS

Pencils are made with many grades of lead hardness. See Figure 4-1. Hard leads can produce thin, light lines and are well suited for the accuracy requirements of board-type drawings, but are usually too light for sketching. The soft leads produce broad, dark lines but tend to smudge easily if handled too much.

The choice of lead hardness is a personal one. Some designers use 3H leads very successfully; others use HB leads with equal results. You are probably used to a 2H lead, as this is the most commonly available. Start sketching with a 2H lead, and if the lines are too light, try a softer lead; if they are too dark, try a harder lead until you are satisfied with your work. Pencils with different grades of lead hardness are available at most stationery and art supply stores.

Most sketching is done in pencil because it can easily be erased and modified. If you want the very dark lines that inked lines produce, it is recommended that you first prepare the drawing in pencil, then use a pen to trace over the lines you want to emphasize.

4-5 LINES

Straight lines are sketched using one of two methods: by drawing a series of short lines—called *feathering*—or by a series of line segments. The line segments should be about 1 to 2 inches or 25 to 50 millimeters long. It is very difficult to keep the line segments reasonably straight if they are much longer. As you practice you will develop a comfortable line segment distance. See Figure 4-2.

Lines may also be sketched as long, continuous lines. First locate the pencil at the line's starting point, then look at the endpoint as you sketch. This will help develop straighter long lines. Continuous lines are best sketched on graph paper because the graph lines will serve as an additional guide for keeping the line straight.

Long lines can be created by a series of shorter lines, and very long lines can be sketched by first defining a series of points, then using short segments to connect the points.

It is usually more comfortable to turn the paper slightly when sketching, as shown in Figure 4-3. Right-handers turn the paper counterclockwise and left-handers, clockwise.

It is also easier to sketch all lines in the same direction, that is, with your hand motion always the same. Rather than change your hand position for lines of different angles, simply change the position of the paper and sketch the lines as before. Horizontal and vertical lines are sketched using exactly the same motion; the paper is just turned 90°. Straight lines of any angle can be sketched in the same manner.

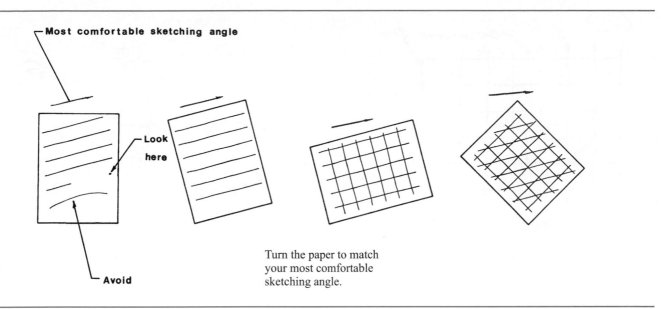

Turn the paper to match
your most comfortable
sketching angle.

Figure 4-3

Sketch lines using a gentle, easy motion. Don't squeeze the pencil too hard. Sketch lines with more of an arm motion than a wrist motion. If too much wrist motion is used, the lines will tend to curve down at the ends and look more like arcs than straight lines.

4-6 PROPORTIONS

Sketches should be proportional. A square should look like a square, and a rectangle like a rectangle. Graph paper is very helpful in sketching proportionally, but it is still sometimes difficult to be accurate even with graph paper. Start by first sketching very lightly and then checking the proportionality of the work. See Figure 4-4. Go back over the lines, making corrections if necessary, then darken in the lines. The technique of first sketching lightly, checking the proportions, making corrections, and then going over the lines is useful regardless of the type of paper used.

It is often helpful to sketch a light grid background based on the unit values of the object being sketched. This is true even if you are working on graph paper because it helps to emphasize the unit values you need. See Figure 4-5.

The exact proportions of an object are not always known. A simple technique to approximately measure an object is to use the sketching pencil. See Figure 4-6. Hold the pencil at arm's length and sight the object. Move your thumb up the pencil so that the distance between the end of

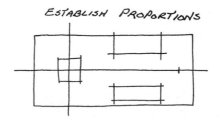

ESTABLISH PROPORTIONS

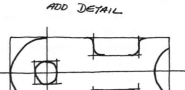

ADD DETAIL

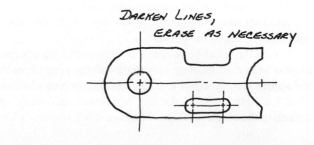

DARKEN LINES,
ERASE AS NECESSARY

Figure 4-4

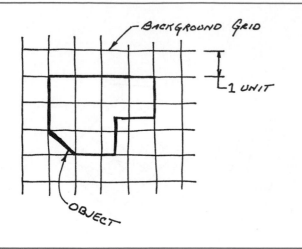

Figure 4-5

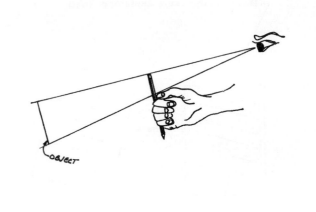

Figure 4-6

the pencil and your thumb represents a distance on the object. Transfer the distance to the sketch. Continue taking measurements and transferring them to the sketch until reasonable proportions have been created.

4-7 CURVES

Curved shapes are best sketched by first defining points along the curve, then lightly sketching the curve between the points. See Figure 4-7. Evaluate the accuracy and smoothness of the curve, make any corrections necessary, then darken in the curve.

Circles can be sketched by sketching perpendicular centerlines and marking off four points equally spaced from the center point along the centerlines. The distance between the center point and the points on the centerlines should be approximately equal to the circle's radius. See Figure 4-8.

Draw a second set of perpendicular centerlines approximately 45° to the first. Again mark four points approximately equal to the radius of the circle. Sketch a light curve through the eight points, and check the curve for accuracy and smoothness. Make any corrections necessary and darken in the circle.

An ellipse can be sketched by first sketching a perpendicular axis, then locating four marks on the centerlines that are approximately equal to major and minor axis distances. Sketch a light curve, make any corrections necessary, and darken in the elliptical shape. See Figure 4-9.

Figure 4-10 shows how to sketch a slot. To sketch the slot, centerlines for the two end semicircles are located and sketched. Two additional radius points are added, then the overall shape of the slot is lightly sketched. Corrections are made and the final lines are darkened.

The triangular object is first sketched as a triangle, guidelines and axis are added for the curved sections, and the final shape of the object is sketched.

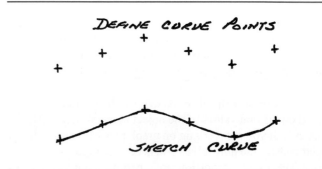

Figure 4-7

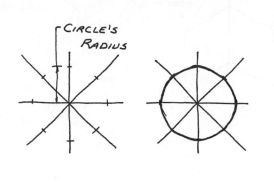

Figure 4-8

Figure 4-9

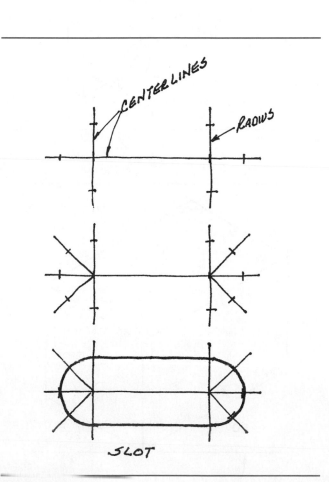

Figure 4-10

4-8 SAMPLE PROBLEM SP4-1

Sketch the object shown in Figure 4-11.

1. Sketch the overall rectangular shape of the object. See Figure 4-12.
2. Add proportional guidelines for the outside shape of the object and lightly sketch the outside shape.
3. Locate and sketch guidelines and axis lines for the other features.
4. Lightly sketch the object and make any corrections necessary.
5. Darken the final lines.

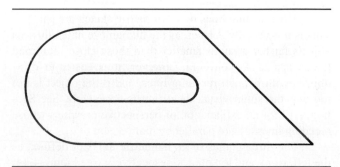

Figure 4-11

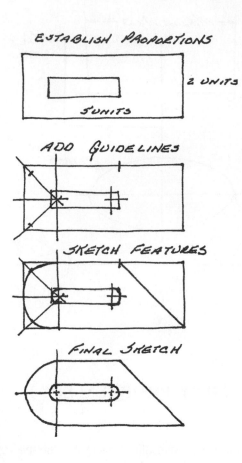

Figure 4-12

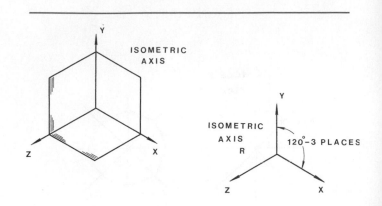

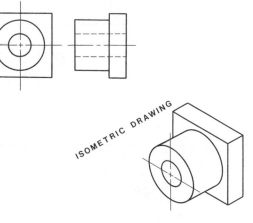

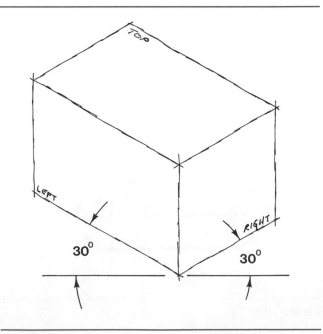

Figure 4-13

4-9 ISOMETRIC SKETCHES

Isometric sketches are based on an isometric axis that contains three lines, 120° apart. See Figure 4-13. The isometric axis can also be drawn in a modified form that contains a vertical line and two 30° lines. The modified axis is more convenient for sketching and is the more commonly used form. See Figure 4-14.

The receding lines of an isometric sketch are parallel. This is not visually correct, as the human eye naturally sees objects farther away as smaller than those closer. Railroad tracks appear to converge; however, it is easier to draw objects with parallel receding lines, and if the object is not too big, the slight visual distortion is acceptable. See Section 4-12 for an explanation of perspective drawings whose receding lines are not parallel but convergent.

The three planes of an isometric axis are defined as the left, right, and top planes, respectively. See Figure 4-14. When creating isometric sketches, it is best to start with the three planes drawn as if the object were a rectangular prism or cube. Think of creating the sketch from these

Figure 4-14

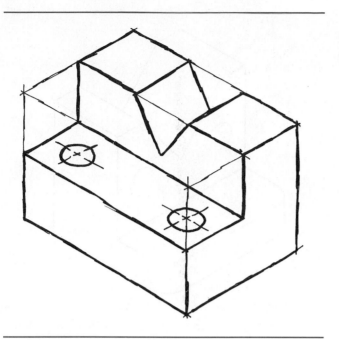

Figure 4-15

planes as working with a piece of wood, and trim away the unnecessary areas. Figure 4-15 shows an example of an isometric sketch. In the example, the overall proportions of the object were used to define the boundaries of the object, then other surfaces were added as necessary.

Isometric sketches may be sketched in different orientations. Figure 4-16 shows six possible orientations. Note how orientation 1 makes the object look as if it is below you, and orientation 6 makes it look as if it is above you. Orientation can also serve to show features that would otherwise be hidden from view. The small cutout in the rear of the object is visible in only four of the orientations. Always try to orient the isometric sketch so that it shows as many of the object's features as possible.

Figure 4-17 shows an isometric drawing of a cube that has a hole in the top plane. The hole must be sketched as an ellipse to appear visually correct in the isometric drawing. The axis lines for the ellipse are parallel to the edge lines of the plane. The proportions of the ellipse are defined by four points equidistant from the ellipse center point along the axis lines. The ellipse is then sketched lightly, checked for accuracy, then darkened.

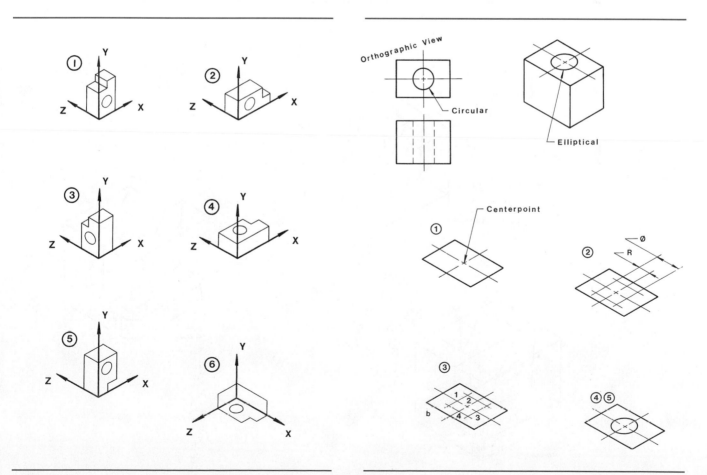

Figure 4-16

Figure 4-17

4-10 SAMPLE PROBLEM SP4-2

Sketch the object shown in Figure 4-18. Do not include dimensions, but keep the object proportional.

1. Use the overall dimensions of the object to sketch a rectangular prism of the correct proportions. This is a critical step. If the first attempt is not proportionally correct, erase it, and sketch again until a satisfactory result is achieved. See Figure 4-19.
2. Sketch the cutout.
3. Sketch the rounded surfaces. Note how axis lines are sketched and the elliptical shape added. Tangency lines are added, and visually incorrect lines are erased.
4. Sketch the holes. The holes are located by locating their centerlines and then sketching the required ellipses.

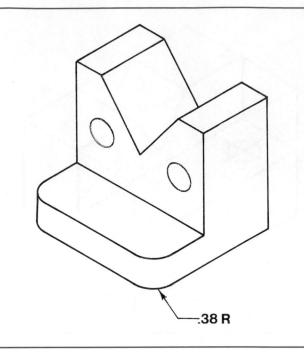

Figure 4-18

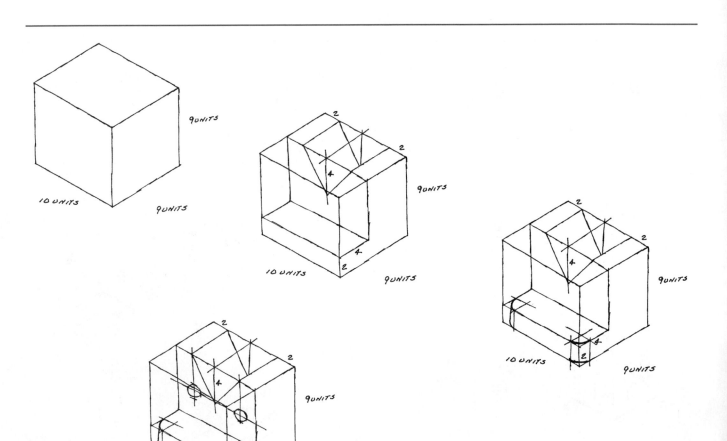

Figure 4-19

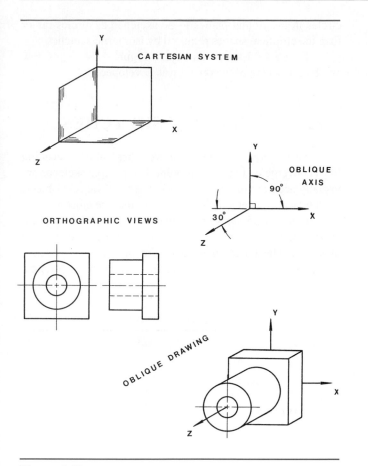

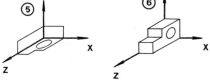

Figure 4-20

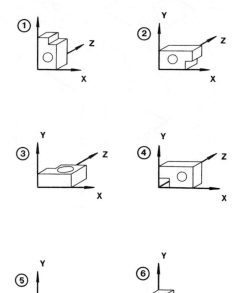

Figure 4-21

4-11 OBLIQUE SKETCHES

Oblique sketches are based on an axis system that contains one perpendicular set of axis lines and one receding line. See Figure 4-20. The front plane of an oblique axis is perpendicular, so the front face of a cube will appear as a square and the front face of a cylinder as a circle. The receding lines can be at any angle, but 30° is most common.

The receding lines of oblique sketches are parallel. As with isometric sketches this causes some visual distortions, but unless the object is very large, these distortions are acceptable.

Holes in the front plane of an oblique sketch may be sketched as circles, but holes in the other two planes are sketched as ellipses. See Figure 4-21. The axis lines for the ellipse are parallel to the edge lines of the plane. The proportions of the ellipses are determined by points equidistant from the center point along the axis.

Figure 4-22 shows an example of a circular object sketched as an oblique sketch. Oblique sketches are particularly useful in sketching circular objects because they allow

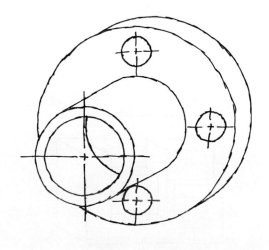

Figure 4-22

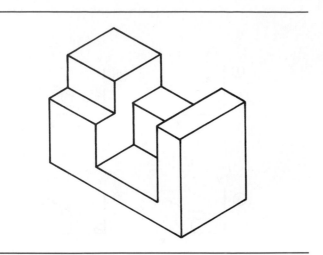

Figure 4-23

circles in the frontal planes to be sketched as circles rather than the elliptical shapes required by isometric sketches.

Figure 4-23 shows an object. Figure 4-24 shows how an oblique sketch of the object was developed.

4-12 PERSPECTIVE SKETCHES

Perspective sketches are sketches whose receding lines converge to a vanishing point. Perspective sketches are visually accurate in that they look like what we see: objects farther away appear smaller than those that are closer.

Figure 4-25 shows a comparison among the axis systems used for oblique, isometric, and two-point perspective drawings. The receding lines of the perspective drawings converge to vanishing points that are located on a theoretical

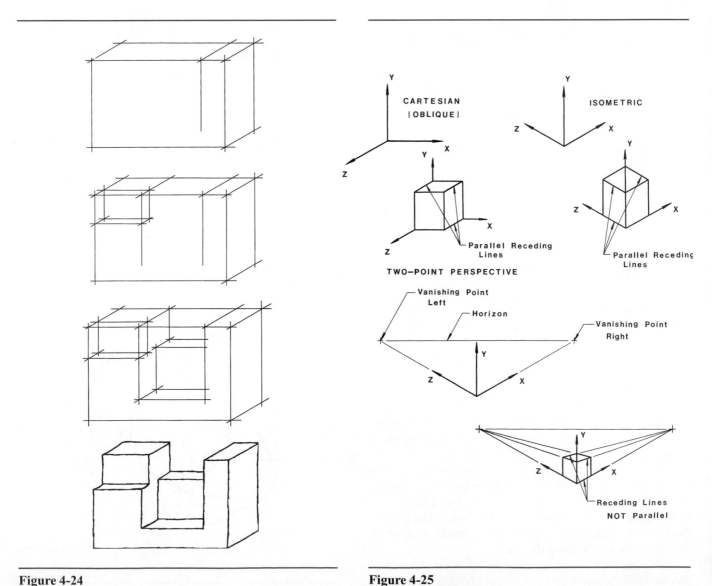

Figure 4-24

Figure 4-25

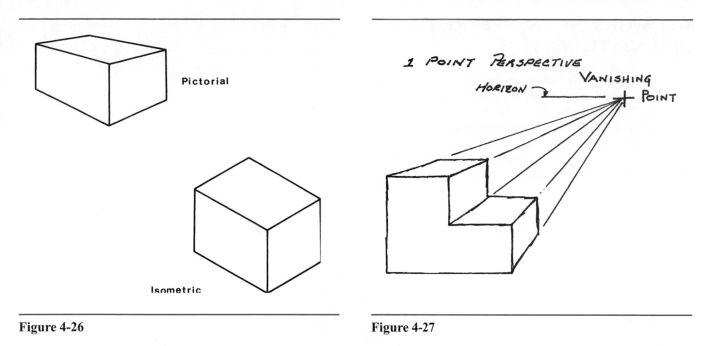

Figure 4-26

Figure 4-27

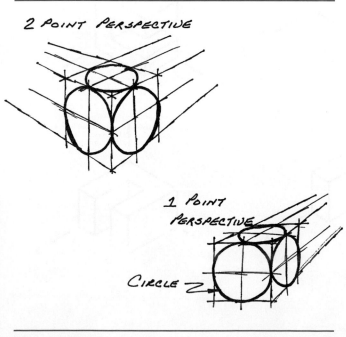

Figure 4-28

horizon. The horizon line is always located at eye level. Objects above the horizon line appear to be above, and objects below the horizon appear to be below.

Perspective drawings are often referred to as *pictorial drawings*. Figure 4-26 shows an object drawn twice: once as an isometric drawing, and again as a pictorial or two-point perspective. Note how much more lifelike the pictorial drawing looks in comparison with the isometric drawing.

Figure 4-27 shows a perspective sketch based on only one vanishing point. One-point perspective sketches are similar to oblique sketches. The front surface plane is sketched using a 90° axis, and then receding lines are sketched from the front plane to a vanishing point. As you practice sketching, eventually you will not need to include a vanishing point, but will imply its location.

Figure 4-28 shows how to sketch circular shapes in one- and two-point perspective sketches. In each style the axis lines consist of a vertical line and a line that is aligned with the receding edge lines. Circular shapes in perspective drawings are not elliptical, but are irregular splines. Each surface will require a slightly different shape depending on the angle of the receding lines to produce a visually accurate circular shape.

4-13 WORKING IN DIFFERENT ORIENTATIONS

It is important for designers to be able to sketch an object in different orientations to present a clear representation of all the design's different facets.

Figure 4-29 shows three different objects sketched in three different orientations. Each sketch was done using an isometric format. Proportions for each object were determined by using the dot grid background.

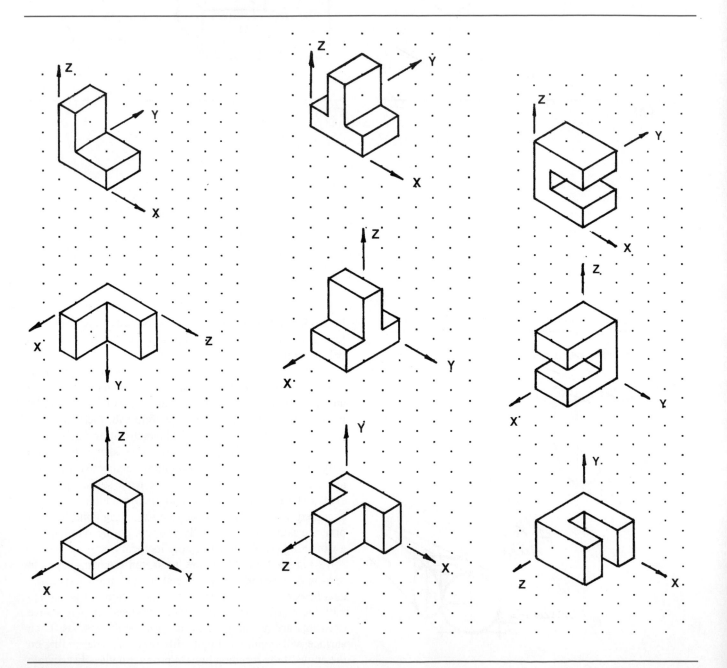

Figure 4-29

4-14 EXERCISE PROBLEMS

Sketch the shapes in Exercise Problems EX4-1 to EX4-6. Measure the shapes to determine their dimensions.

EX4-1

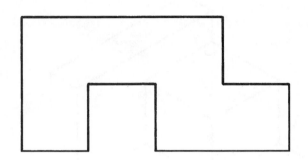

EX4-2

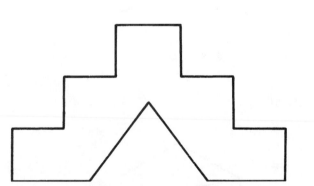

EX4-3

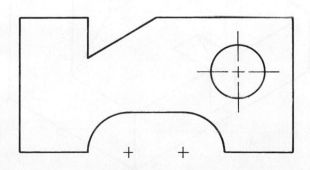

EX4-4

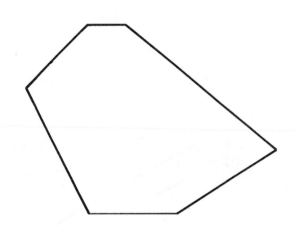

EX4-5

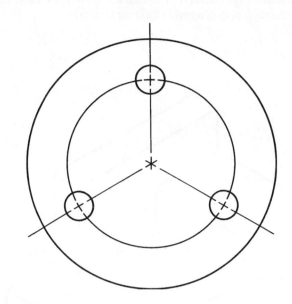

EX4-6

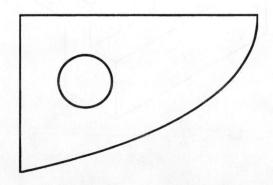

Prepare isometric or perspective sketches of the objects shown in Exercise Problems EX4-7 to EX4-18. Measure the objects to determine their dimensions.

EX4-7

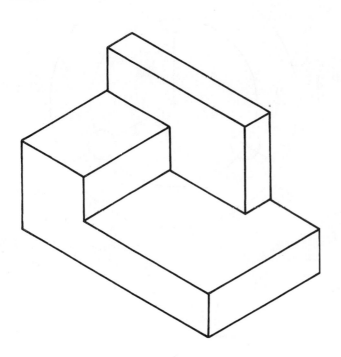

EX4-8

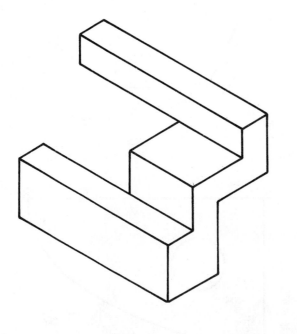

EX4-9

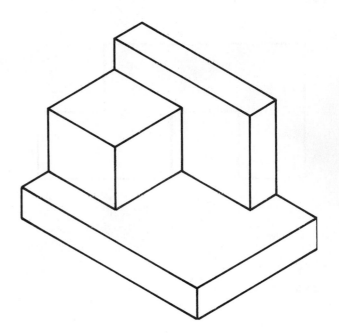

EX4-10

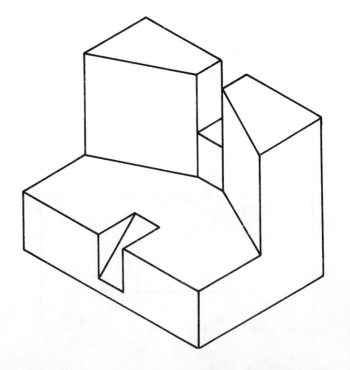

EX4-11

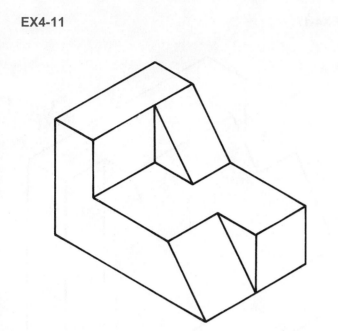

EX4-13

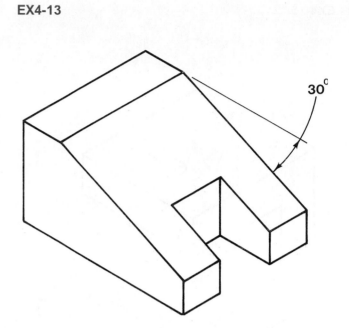

30°

EX4-12

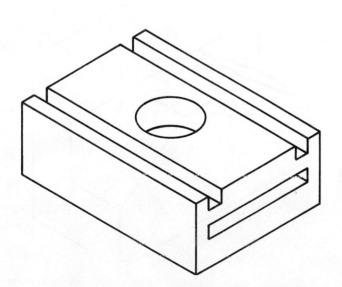

EX4-14

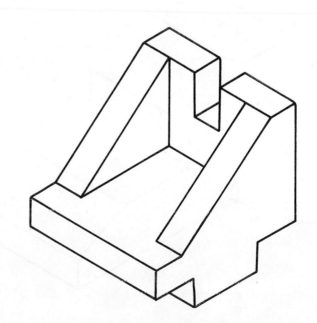

EX4-15

EX4-17

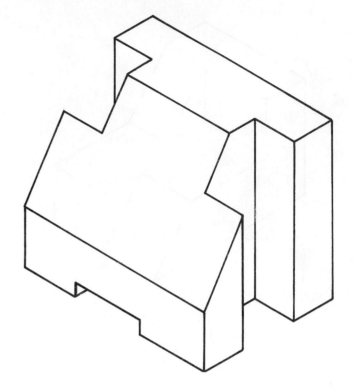

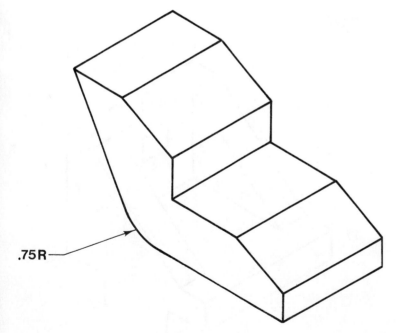

EX4-16

EX4-18

.75R

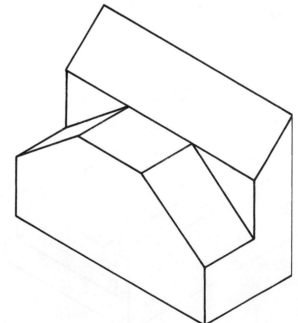

Prepare oblique or perspective sketches of the objects shown in Exercise Problems EX4-19 to EX4-22. Measure the objects to determine their dimensions.

EX4-19

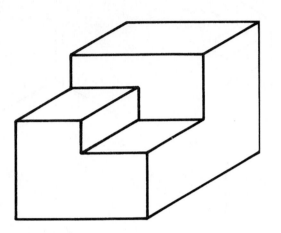

EX4-21

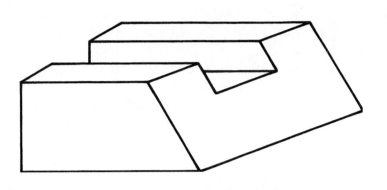

EX4-20

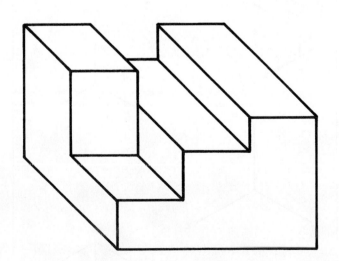

EX4-22

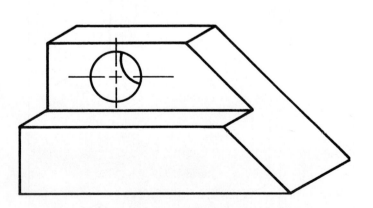

Prepare isometric sketches of the following shapes in the given orientation, then sketch the objects in the three different orientations defined by the X,Y,Z axes. Assume that the spacing between dots in the background grid is approximately .25 inch, or 10 millimeters.

EX4-23

EX4-25

EX4-24

EX4-26

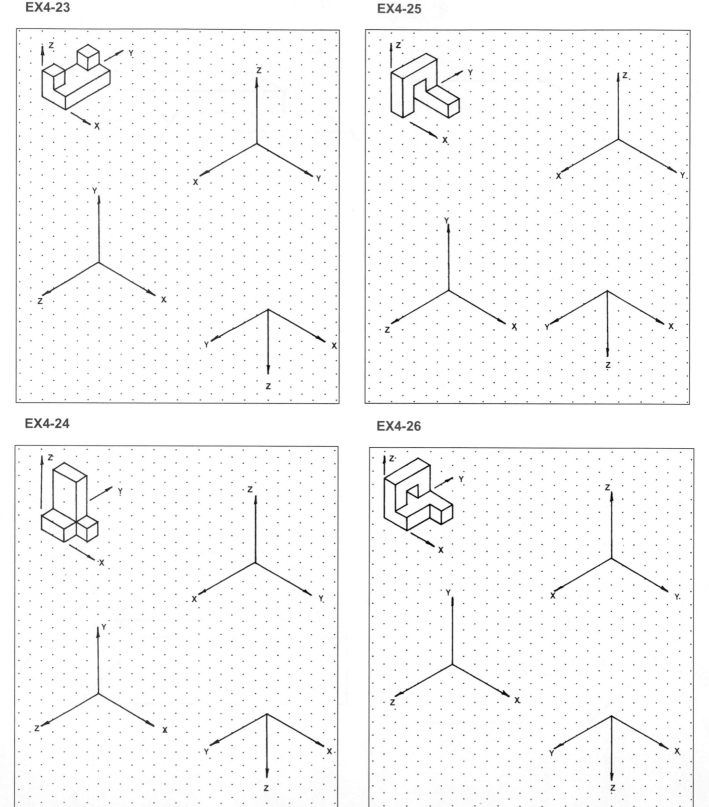

EX4-27

EX4-29

EX4-28

EX4-30

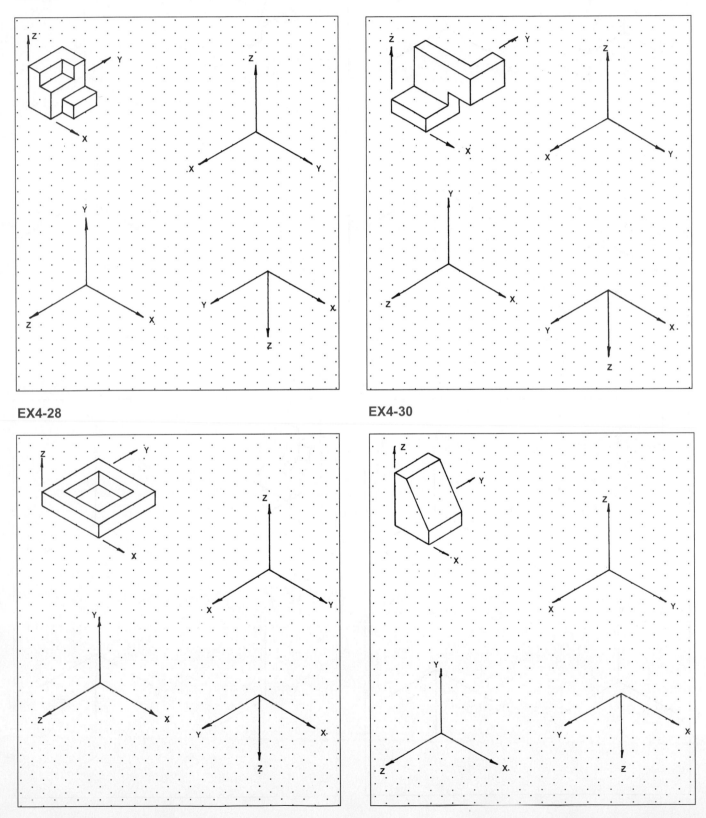

EX4-31

EX4-33

EX4-32

EX4-34

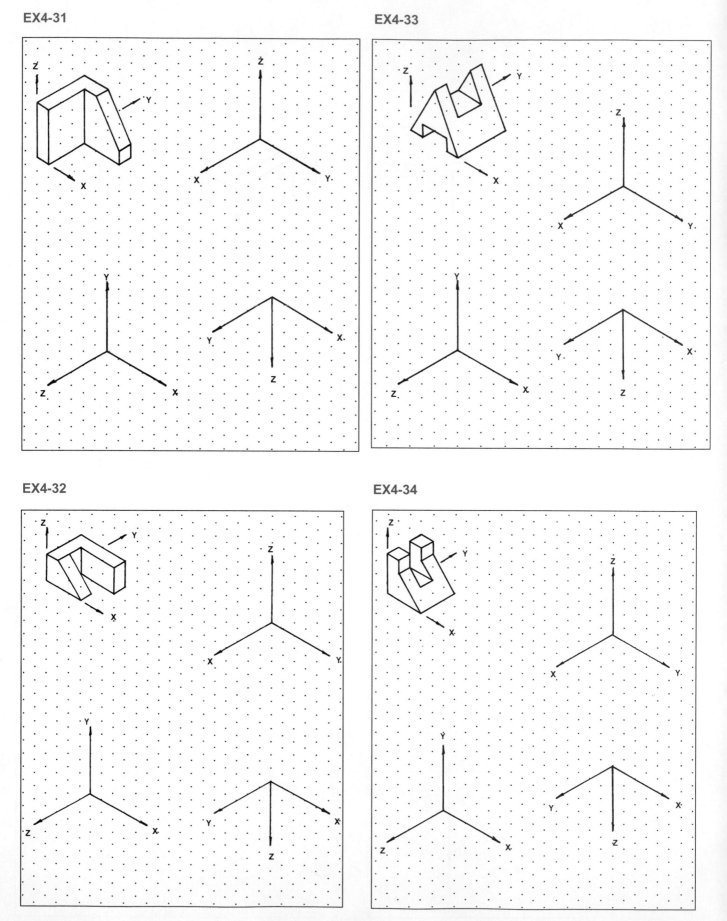

EX4-35

EX4-37

EX4-36

EX4-38

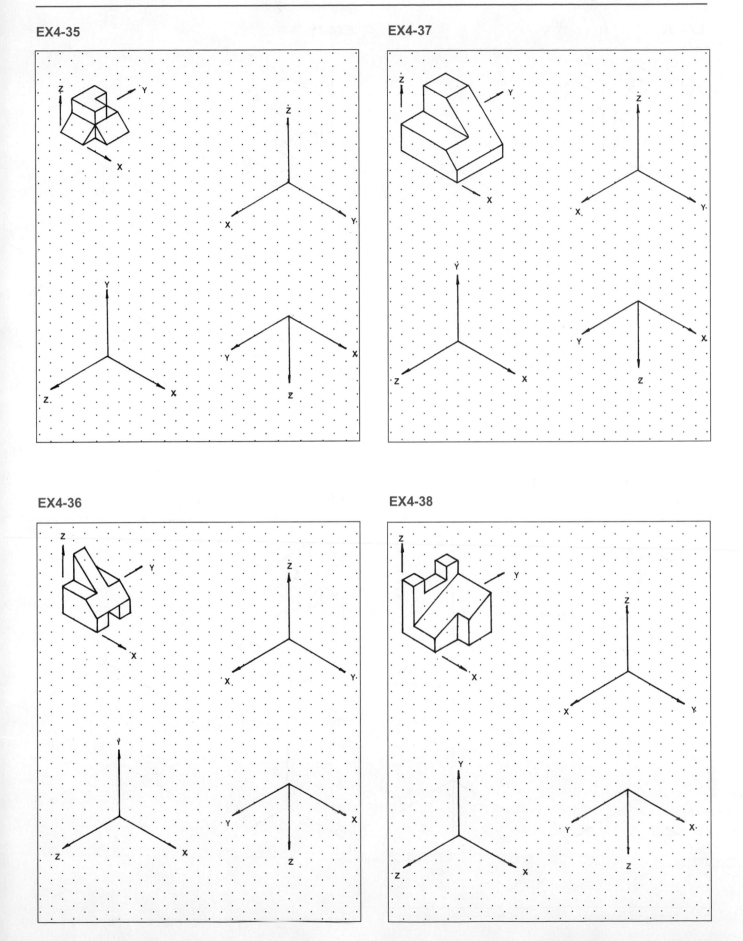

EX4-39

EX4-41

EX4-40

EX4-42

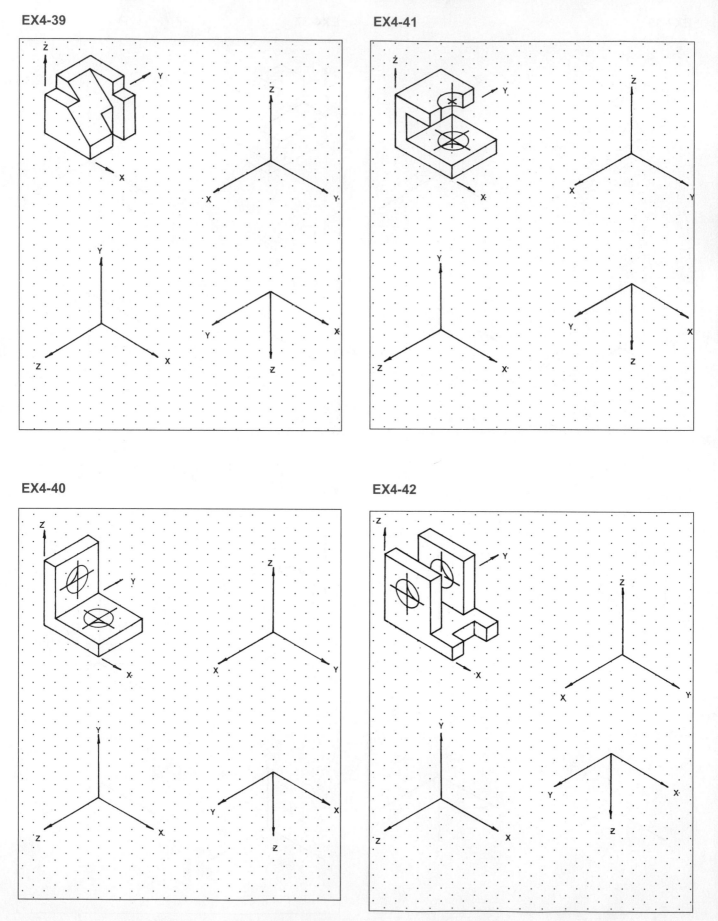

EX4-43

EX4-45

EX4-44

EX4-46

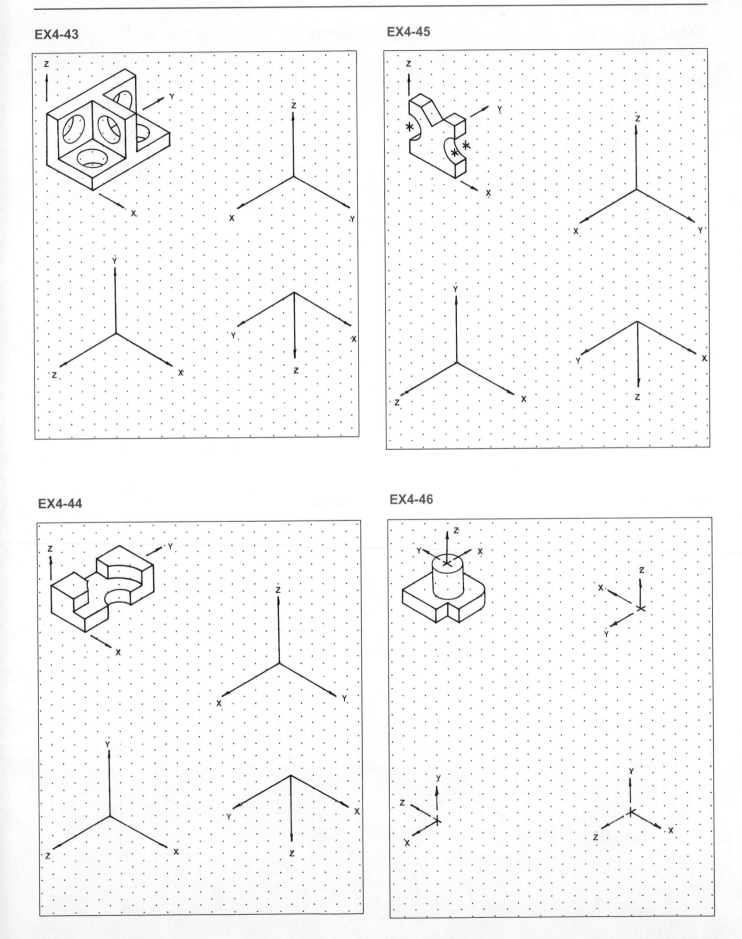

EX4-47

EX4-49

EX4-48

EX4-50

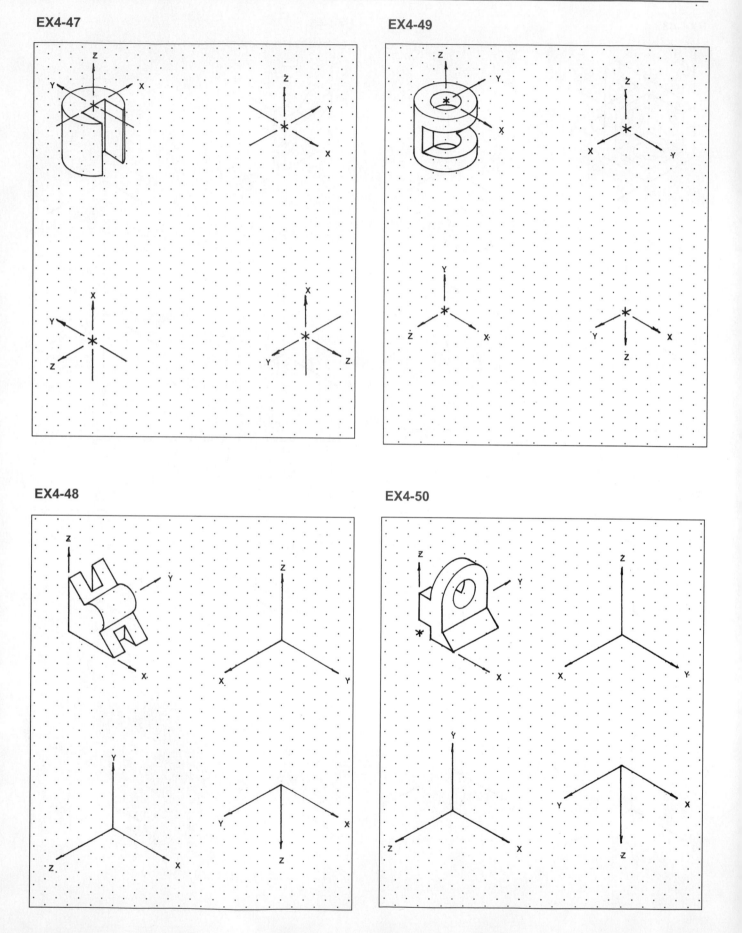

Orthographic Views

5-1 INTRODUCTION

This chapter introduces orthographic views. *Orthographic views* are two-dimensional views of three-dimensional objects. Orthographic views are created by projecting a view of an object onto a plane that is usually positioned so it is parallel to one of the planes of the object. See Figure 5-1.

Orthographic projection planes may be positioned at any angle relative to an object, but in general the six views parallel to the six sides of a rectangular object are used. See Figure 5-2. Technical drawings usually include only the front, top, and right-side orthographic views because together they are considered sufficient to completely define an object's shape.

5-2 THREE VIEWS OF AN OBJECT

Orthographic views are positioned on a technical drawing so that the top view is located directly over the front view, and the side view is directly to the right side of the front view. See Figure 5-2.

The location of the front, top, and right-side views relative to each other is critical. Correct relative positioning of views allows information to be projected between the views. Vertical lines are used to project information between the front and top views, and horizontal lines are used to project information between the front and right-side views.

Orthographic views are two-dimensional, so they cannot show depth. In Figure 5-2 plane A-B-C-D appears as a straight line in both the front and top orthographic views. Edge A-B appears as a vertical line in the top view, as a horizontal line in the side view, and as point AB in the front view. The end view of a plane is a straight line, and the end view of a line is a point. See Figure 5-3.

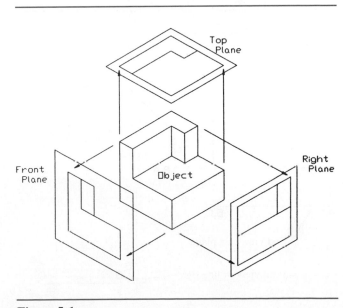

Figure 5-1

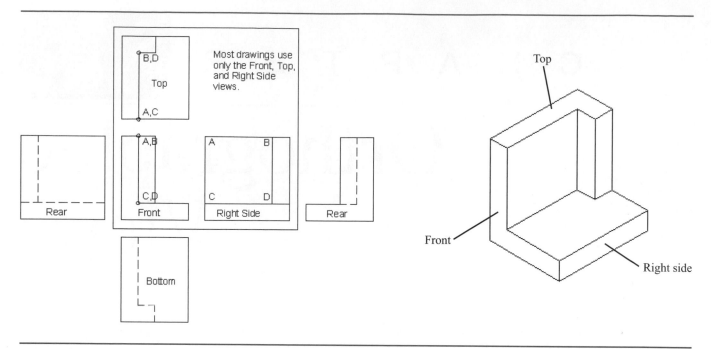

Figure 5-2

Figure 5-4 shows an object and the front, top, and right-side orthographic views of the object. Plane A-B-C-D appears as a rectangle in the front view and as a line in both the top and right-side views.

To help you better understand the relationship between orthographic views and in particular the front, top, and side views of an object, place a book on a table as shown in Figure 5-5. Note how the front, top, and side views were projected and then arranged to create a technical drawing of the book.

5-3 VISUALIZATION

Visualization is the ability to look at a three-dimensional object and mentally see the appropriate three orthographic views.

It is an important skill for a designer and an engineer to develop. Visualization also includes the ability to look at two-dimensional orthographic views and mentally picture the three-dimensional object the views represent.

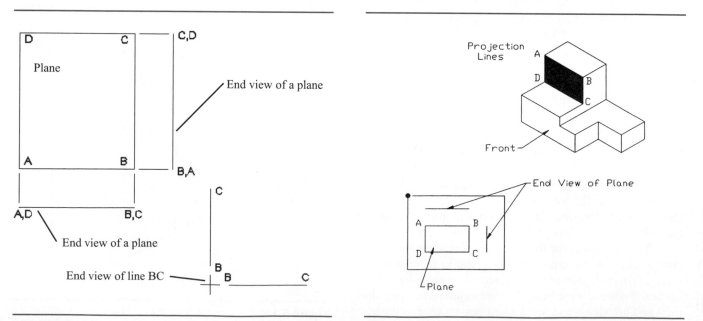

Figure 5-3

Figure 5-4

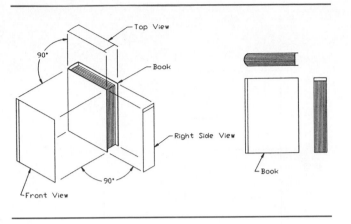

Figure 5-5

Figure 5-6 shows a 5 × 3.5 × 2-inch box and three orthographic views of the box. The creation of three-dimensional objects is explained in detail in Chapters 14, 15, and 16. This section presents a short exercise to help you better understand how orthographic views are related to three-dimensional objects.

To draw a three-dimensional box

1. Set the screen for **inches,** then select the **3D View** command from the **View** pull-down menu, then **SE Isometric.**

   The origin axis reference icon will change to the 3D alignment.

2. Type the word **box** to a command prompt.

   Specify corner of box or [Center] <0,0,0>:

3. Press **Enter.**

   Specify corner or [Cube/Length]:

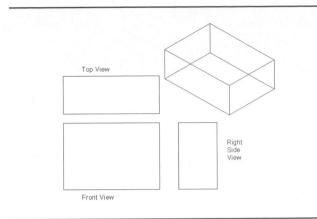

Figure 5-6

4. Type **L;** press **Enter.**

   Specify length:

5. Type **5;** press **Enter.**

   Specify width:

6. Type **3.5;** press **Enter.**

   Specify height:

7. Type **2;** press **Enter.**

   A box will appear on the screen.

8. Select the **3D Orbit** command from the **View** pull-down menu. Move the cursor around the drawing screen holding the left mouse button down. Position the box so as to create each of the three orthographic views shown in Figure 5-6.

   Figures 5-7 and 5-8 show two additional objects and their orthographic views.

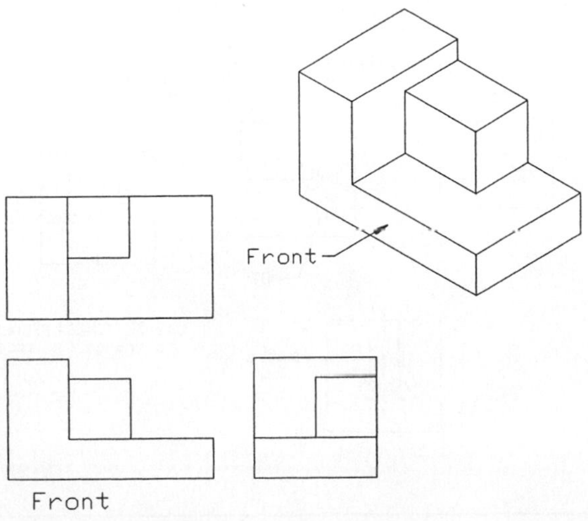

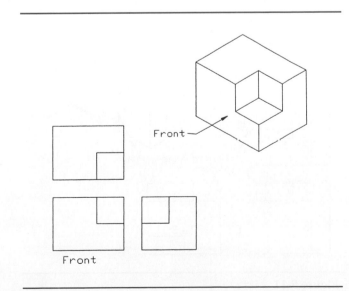

Figure 5-7

Figure 5-8

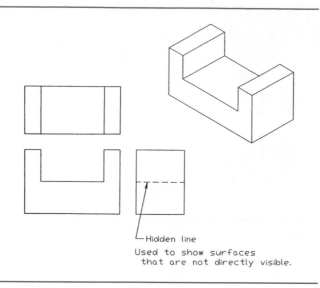

Figure 5-9

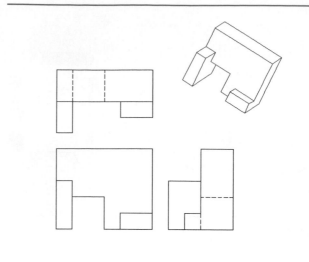

Figure 5-10

5-4 HIDDEN LINES

Hidden lines are used to represent surfaces that are not directly visible in an orthographic view. Figure 5-9 shows an object that contains a surface A-B-C-D that is not directly visible in the side view. A hidden line is used to represent the end view of plane A-B-C-D in the side view.

Figure 5-10 shows an object and three views of the object. The top and side views of the object contain hidden lines.

Figure 5-11 shows an object that contains an edge line A-B. In the top view of the object, line A-B is partially hidden and partially visible. When the line is directly visible through the square hole it is drawn as a continuous line; when it is not directly visible it is drawn as a hidden line.

5-5 HIDDEN LINE CONVENTIONS

Figure 5-12 shows several conventions associated with drawing hidden lines. Whenever possible, show intersections of hidden lines as touching lines, not as open gaps. Corners should also be shown as touching lines.

The **Linetype Scale** option on the **Properties** dialog box can be used to change the spacing of hidden lines. Figure 5-13 shows two hidden lines and a continuous line that are aligned. A small gap should be included between the two types of lines to prevent confusion as to where one type of line ends and the other begins. A visual distinction can also be created by selecting a different color for hidden lines. If continuous and hidden lines are drawn using different colors, then gaps are not necessary.

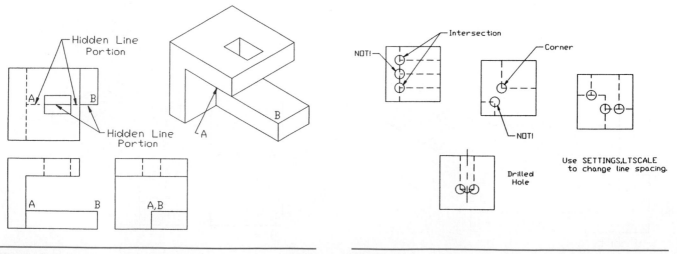

Figure 5-11

Figure 5-12

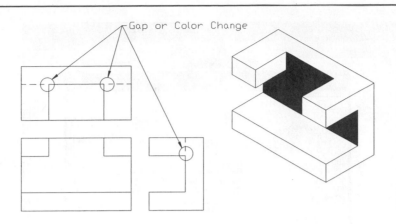

Figure 5-13

5-6 DRAWING HIDDEN LINES

There are three ways to draw hidden lines: modify the line from a continuous line to a hidden line, transfer the continuous line to a special layer created specifically for hidden lines, or use the **Properties** palette. In this example both the linetype and the line's color are changed.

To change a continuous line to a hidden line

1. Select the line.

The grip points appear.

2. Right-click the mouse and select the **Properties** option.

The **Properties** palette will appear. See Figure 5-14.

3. Select **Linetype,** then **Hidden.**

See Figure 5-15. The selected line will be changed to a hidden line. If the linetype is not listed, return to the drawing screen and click the **Format** pull-down menu. Select the **Linetype** option. The **Linetype Manager** dialog box will appear. Select the **Load** option, and the **Load or**

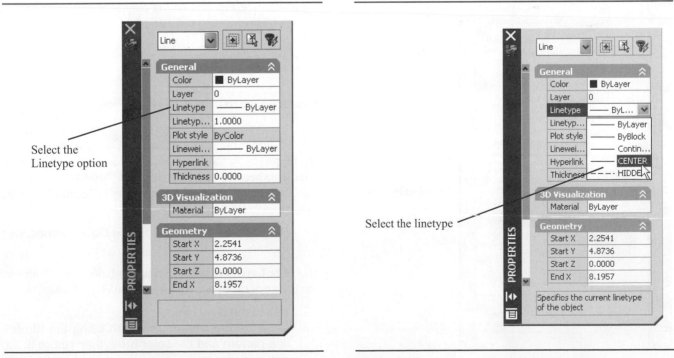

Figure 5-14

Figure 5-15

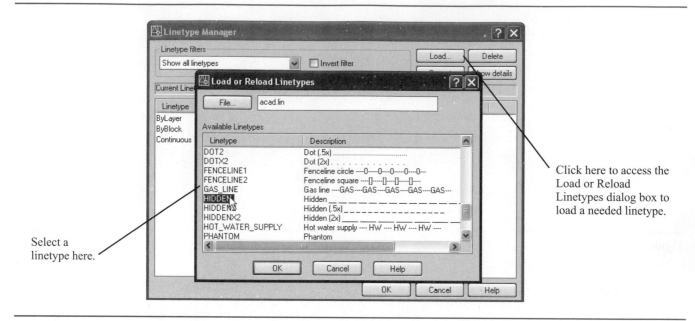

Figure 5-16

Reload Linetypes dialog box will appear. See Figure 5-16. Scroll down and select the Hidden line option. Select **OK** and return to the drawing screen. Access the **Modify Line** dialog box as explained above. This time the **Hidden** line pattern will be listed as an option.

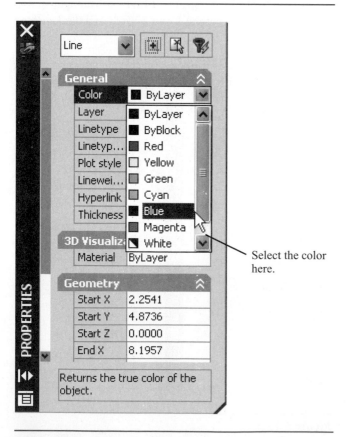

Figure 5-17

4. Select the the **Color** option in the **Properties** palette, then the scroll arrow.

   The color options will cascade down. See Figure 5-17.

5. Select **Blue** and exit the **Properties** palette.

   If the line changes color but the broken-line pattern does not appear, or if the line segments are too short or too long, use the **Linetype Scale** option found in the **Properties** palette to change the length of the hidden line segments. See Figure 5-14. It may be necessary to use a trial-and-error process to obtain the most appropriate hidden line pattern.

To use Layer to create a hidden line

   Linetype can be changed by creating a layer specifically set up for hidden lines. Both the line pattern and color can be preset on the layer. This procedure may seem a bit lengthy when compared with the **Modify Properties** procedure outlined above, but it is very efficient if there are many lines to change.

1. Select the **Layers** tool from the **Object Properties** toolbar.

   The **Layer Properties Manager** dialog box will appear. See Figure 5-18. Create a new layer called Hidden.

2. Select the **New** option.

3. Create a layer named **Hidden** using the **hidden** line pattern and the color **blue,** then return to the drawing screen.

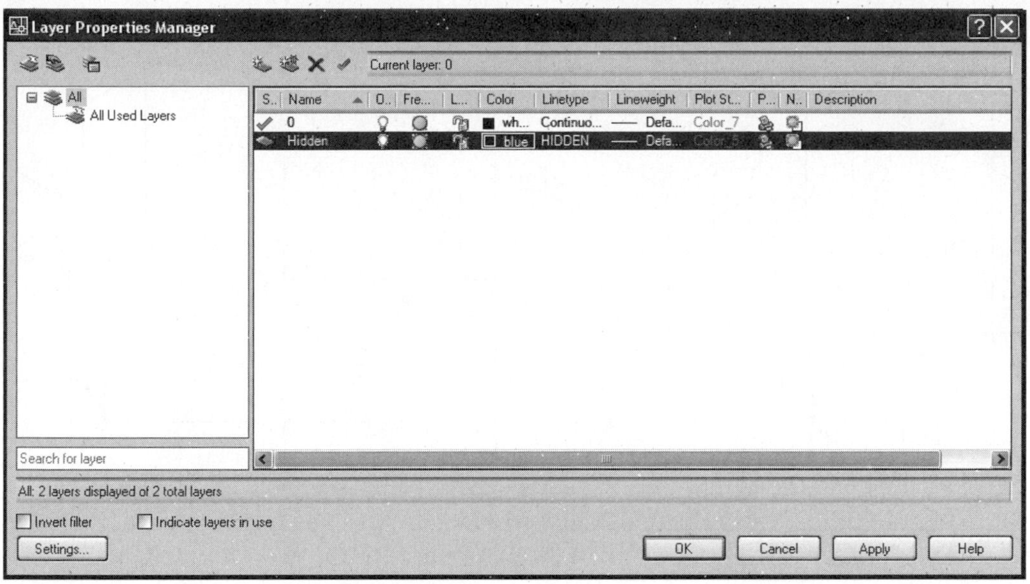

Figure 5-18

A more detailed discussion of layers is presented in Section 3-25.

4. Select the line.

   The grip points will appear.

5. Select the **Layer control** tool.

   See Figure 5-19. The available layer options will cascade down.

6. Select the **Hidden** layer, then double-click the **<Esc>** key.

To use the Properties dialog box

1. Select the line.

   The grip points will appear.

2. Right-click the mouse and select the **Properties** option.

   The **Properties** palette will appear.

3. Select the **Layer** option and change the layer to the **Hidden** layer.

See Figure 5-20.

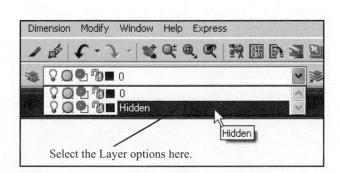

Select the Layer options here.

Figure 5-19

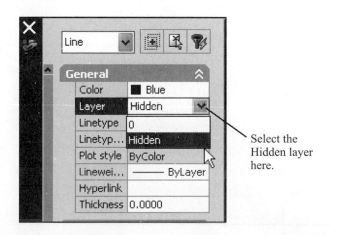

Select the Hidden layer here.

Figure 5-20

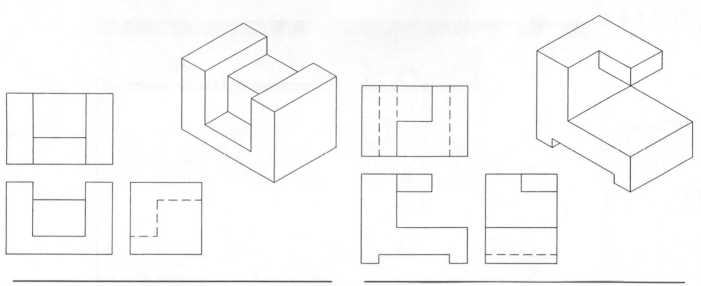

Figure 5-21

Figure 5-22

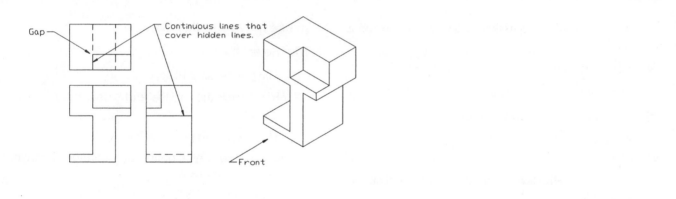

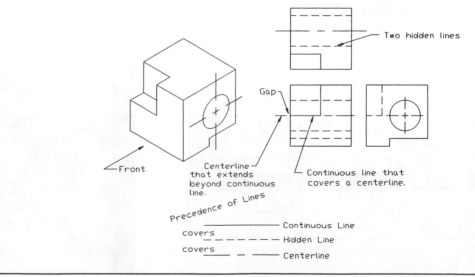

Figure 5-23

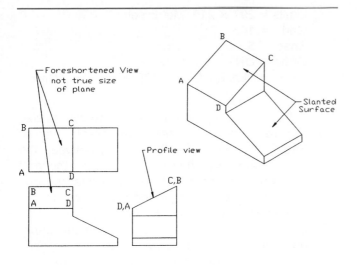

Figure 5-24

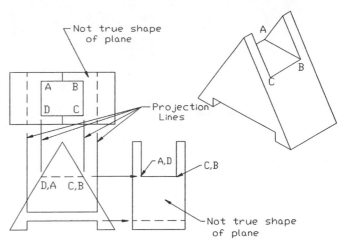

Figure 5-25

Figures 5-21 and 5-22 show an object and three views of that object. There are hidden lines in several of the orthographic views.

5-7 PRECEDENCE OF LINES

When orthographic views are prepared, it is not unusual for one type of line to be drawn over another type: a continuous line over a centerline, for example. See Figure 5-23. Drawing convention has established a precedence of lines: A continuous line takes precedence over a hidden line, and a hidden line takes precedence over a centerline.

If different linetypes are of different lengths, include a gap where the line changes from continuous to either hidden or center to create a visual distinction between the lines. Color changes may also be used to create visual distinctions between lines.

5-8 SLANTED SURFACES

Slanted surfaces are surfaces that are not parallel to either the horizontal or vertical axis. Figure 5-24 shows a slanted surface A-B-C-D. Slanted surface A-B-C-D

appears as a straight line in the side view and as a plane in both the front and top views. It is important to note that neither the front nor top view of surface A-B-C-D is a true representation of the surface. Both orthographic views are actually smaller than the actual surface. Lines B-C and A-D in the top view are true length, but lines A-B and C-D are shorter than the actual edge lengths. The same is true for the front view of the surface. The side view shows the true length of lines A-B and D-C. True lengths of lines and true shapes of planes are discussed again in Chapter 7, Auxiliary Views.

Figure 5-25 shows another object that contains slanted surfaces. Note how projection lines are used between the views. As objects become more complex the shape of a surface or the location of an edge line will not always be obvious, so projecting information between views, and therefore exact, correct relative view location, becomes critical. Information can be projected only between views that are correctly positioned and accurately drawn.

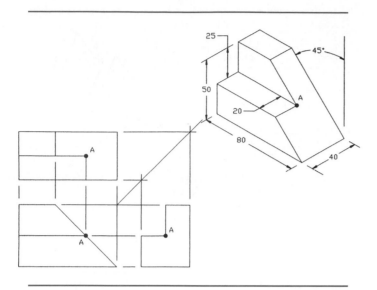

Figure 5-26

5-9 PROJECTION BETWEEN VIEWS

Information is projected between the front and side views using horizontal lines and between the front and top views using vertical lines. Information can be projected between the top and side views using a combination of horizontal and vertical lines that intersect a 45° miter line. See Figure 5-26.

A 45° miter line is constructed between the top and side views. This line allows the project lines to change direction, turn a corner, and change from horizontal to vertical lines. To go from the top view to the side view, construct horizontal projection lines from the top view so that they intersect the miter line. Construct vertical lines from the intersection points on the miter line into the side view. The reverse process is used to go from the side view to the top view.

The following sample problem shows how projection lines can be used to create orthographic views using AutoCAD.

5-10 SAMPLE PROBLEM SP5-1

Draw the front, top, and right-side orthographic views of the object shown in Figure 5-26. Set up the drawing as follows.

1. Create a separate layer for hidden lines.

Limits = **297 × 210** (Metric setup)
Grid = **10**
Snap = **5**
Ortho = **ON** (F8)
Layer = **HIDDEN** (LType = **HIDDEN**, Color = **Green**)

2. Use the overall dimensions (length = **80,** height = **50,** and depth = **40**) to define the overall size requirements of the three orthographic views. See Figure 5-27. The **20** spacing between the views is arbitrary. The distance between the front and top views does not have to equal the distance between the front and side views. In this example the two distances are equal.

3. Draw a **45°** miter line starting from the upper right corner of the front view. This step is possible only if the distances between the views are equal. If the distances are not equal, draw the front and top views, add the miter line, and project the overall size of the side view from the front and top views.

3a. (Optional) Trim away the excess lines to clearly define the areas of the front, top, and side views.

3b. (Optional) The **Layers** tool may be used to create a layer for construction lines and another layer for final drawing lines. The lines for the final drawing may then be transferred to the final drawing layer and the construction layer turned off. This method eliminates the need for extensive erasing or trimming.

4. Draw the **45°** slanted surface in the front view as shown in Figure 5-27. Project the intersection of the slanted surface and the top edge of the front view into the top view. Add a horizontal line across the front and right-side views **25** from the bottom edge line. The 25 value comes from the given dimension. Label the intersection of the slanted line and the horizontal line as point **A**.

5. Draw a horizontal line in the top view **20** from the top edge as shown in Figure 5-27. Continue the line so that it intersects the miter line. Project the line into the side view. Use **Osnap, Intersection** with **Ortho** on to ensure an accurate projection. Label point **A** in the side view.

6. Project point **A** in the front view into the top view (vertical line). Label the intersection of the projection line and the horizontal line in the top view as point **A**.

7. Use **Erase** and **Trim** to remove the excess lines.

8. Save the drawing if desired.

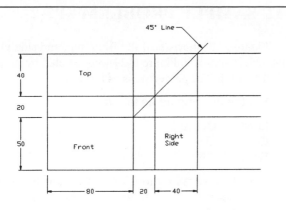

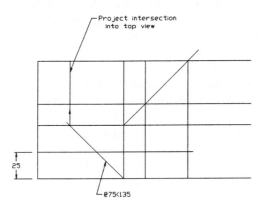

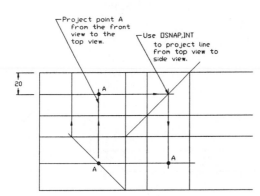

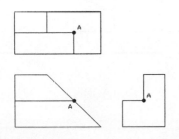

Figure 5-27

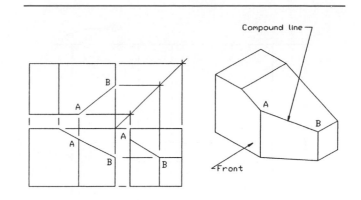

Figure 5-28

5-11 COMPOUND LINES

A *compound line* is formed when two slanted surfaces intersect. See Figure 5-28. The true length of a compound line is not shown in the front, top, or side views.

Figure 5-29 shows another object that contains compound lines. The three orthographic views of a compound line are sometimes difficult to visualize, making it more important to be able to project information accurately between views. Usually, part of an object can be visualized and part of the orthographic views created. The remainder of the drawing can be added by projecting from the known information.

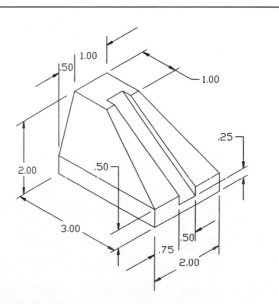

Figure 5-29

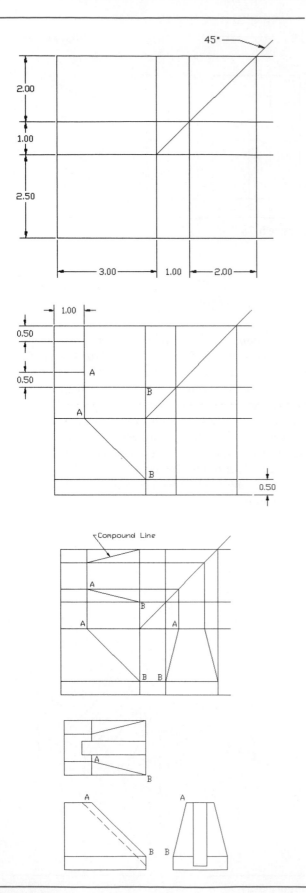

5-12 SAMPLE PROBLEM SP5-2

Figure 5-30 shows how the front, top, and side views of the object shown in Figure 5-29 were created by projecting information. The procedure is as follows.

1. Set the drawing up as follows.

   Limits = **Default**
   Grid = **.5**
   Snap = **.25**

2. Use the overall dimensions to define the space and location requirements for the views. Draw the **45°** miter line.
3. Use the given dimensions to define the starting and end points on the compound lines in the top and front views. One of the lines has been labeled **A-B**.
4. Draw the compound line in the top view and then project it into the side view using information from both the top and front views.
5. Add the dovetailed shape using the given dimensions. Draw the appropriate hidden lines for the dovetail.
6. Erase and trim the excess lines.
7. Save the drawing if desired.

Figure 5-31 shows the three orthographic views and pictorial view of another example of an object that includes compound lines. Some of the projection lines have also been included.

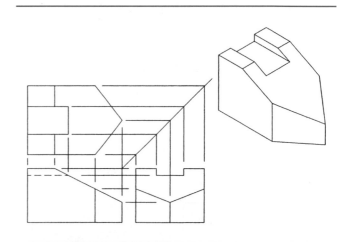

Figure 5-31

Figure 5-30

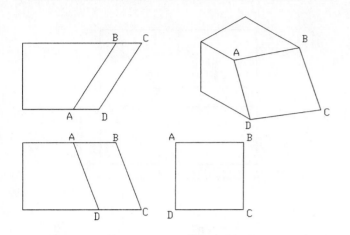

Figure 5-32

5-13 OBLIQUE SURFACES

Oblique surfaces are surfaces that do not appear correctly shaped in the front, top, or side views. Figure 5-32 shows an object that contains oblique surface A-B-C-D. Figure 5-32 also shows three views of just surface A-B-C-D. Oblique surfaces are projected by first projecting their corner points and then joining the points with straight lines. The orthographic views of oblique surfaces are sometimes visually abstract. This makes the development of their orthographic views more dependent on projection than other types of surfaces.

Figure 5-33 shows another object that contains an oblique surface. Figure 5-34 shows how the three orthographic views were developed. The oblique surface is defined using only two angles, but this is sufficient to create the three views because the surface is flat. The edge lines are parallel. Lines A-B, C-D, and E-F are parallel to each other, and lines C-B, D-E, and F-A are also parallel to each other. Figure 5-34 also shows three views of just the oblique surface along with the appropriate projection lines.

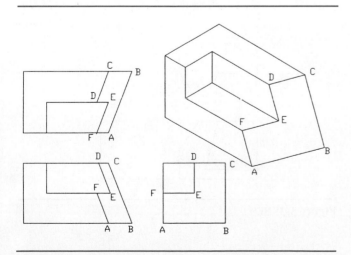

Figure 5-33

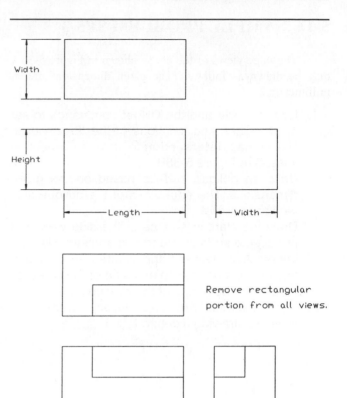

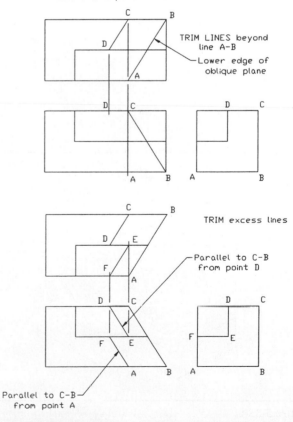

Figure 5-34

5-14 SAMPLE PROBLEM SP5-3

The three views of the object shown in Figure 5-35A may be drawn as follows. (The given dimensions are in millimeters.)

1. Use the **Line** and the **Offset** commands to set up the sizes and locations of the three views. Use **Osnap, Intersection** to draw the projection lines. See Figure 5-35B.
2. Draw an oblique surface based on the given dimensions. The edge lines of the oblique surface are parallel.
3. Draw the slanted surface in the side view and project the surface into the other views. Use the intersection between the slanted surface and flat sections on the top and side of the object to determine the shape of the oblique surface.
4. Use **Erase** and **Trim** to remove any excess lines. Save the drawing if desired.

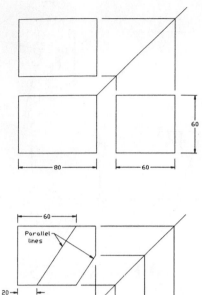

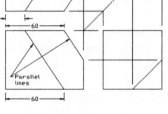

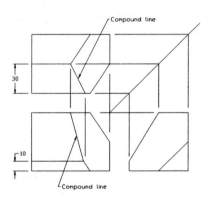

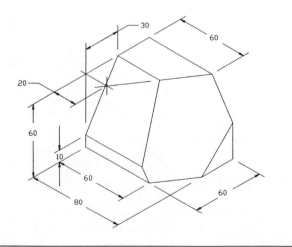

Figure 5-35A

Figure 5-35B

5-15 ROUNDED SURFACES

Rounded surfaces are surfaces that have constant radii, such as arcs or circles. Surfaces that do not have constant radii are classified as *irregular surfaces.* See Section 5-24.

Figure 5-36 shows an object with rounded surfaces. Surface A-B-C-D is tangent to both the top and side surfaces of the object, so no edge line is drawn. This means that the top and side orthographic views of the object are ambiguous. If only the top and side views are considered, then other interpretations of the views are possible. Figure 5-36 shows another object that generates the same top and side orthographic views but does not include rounded surface A-B-C-D. The front view is needed to define the rounded surfaces and limit the views to only one interpretation.

Surfaces perpendicular to an orthographic view always produce lines in that orthographic view. The vertical surface shown in the front view of Figure 5-37A requires an edge line in the top view. The vertical line in the front view is perpendicular to the top views.

The object shown in Figure 5-37B has no lines perpendicular to the top view, so no lines are drawn in the top view. Figure 5-37C shows two rounded surfaces that intersect at points tangent to their centerlines. The intersection point is considered sufficient to require an edge line in the top view. No line would be visible on the actual object, but drawing convention prescribes that a line be drawn.

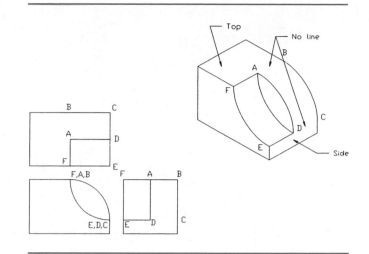

Figure 5-36

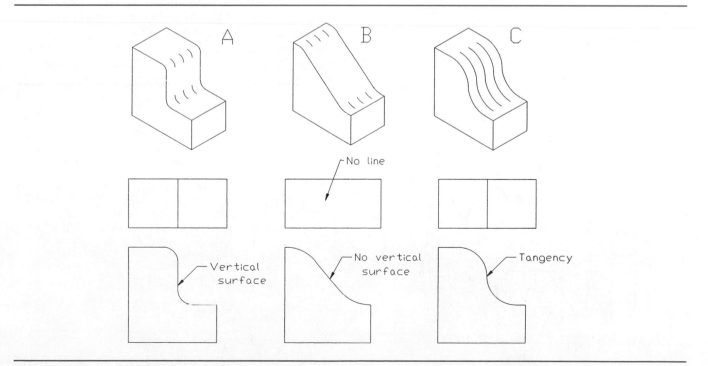

Figure 5-37

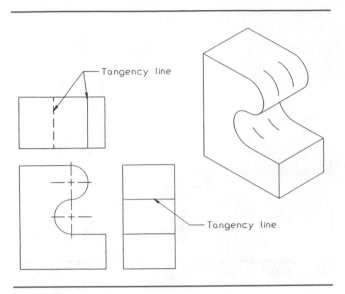

Figure 5-38

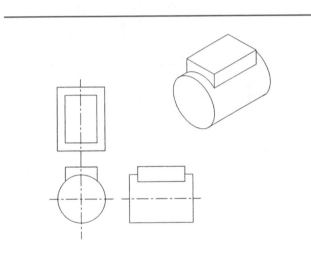

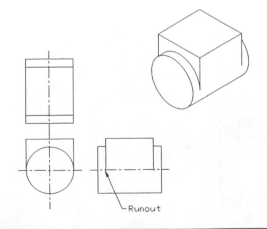

Figure 5-39

Figure 5-38 shows an object that contains two semicircular surfaces. Note how lines representing the widest and deepest point of the surfaces generate lines in the top view. No line would actually appear on the object. Figure 5-39 shows further examples of objects with curved surfaces.

5-16 SAMPLE PROBLEM SP5-4

Figure 5-40 shows an object that includes rounded surfaces, and Figure 5-41 shows how the three views of the object were developed.

The procedure is as follows.

1. Use the given overall dimensions to lay out the size and location of the three views.
2. Draw in the external details, the cutout, and the circular extension in all three views.
3. Add the appropriate hidden lines. Use either the **Modify Properties** option, or create a layer for hidden lines.
4. **Erase** and **Trim** any excess lines and save the drawing if desired.

5-17 HOLES

Holes are represented in orthographic views by circles and parallel hidden lines. All views of a hole always include centerlines. See Figure 5-42. It is suggested that a layer titled **Center** be added to any drawing being created. Specify the linetype **Center** and the color **Red.**

Holes that go completely through an object are dimensioned using only a diameter. The notation **THRU** may be added to the diameter specification for clarity.

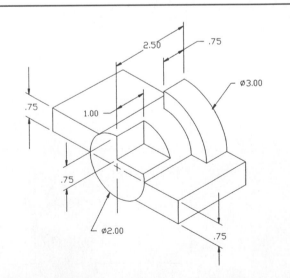

Figure 5-40

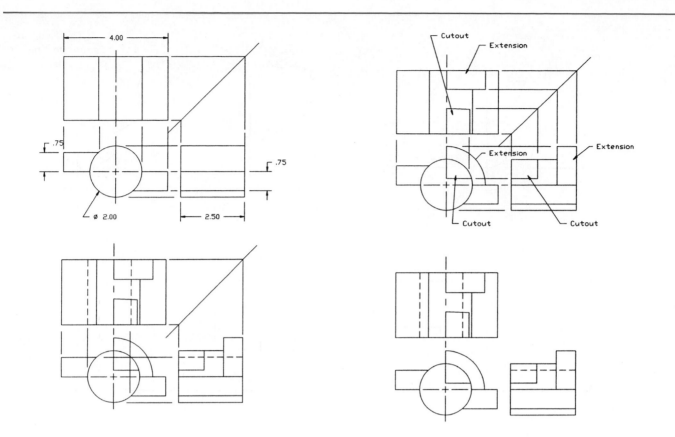

Figure 5-41

Holes that do not go completely through an object include a depth specification in their dimension. The depth specification is interpreted as shown in Figure 5-42. Holes that do not go completely through an object must include a conical point. The conical point is not included in the depth specification. Conical points must be included because most holes are produced using a twist drill with conically shaped cutting edges.

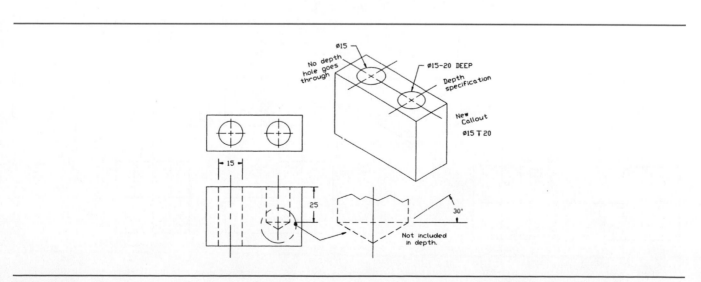

Figure 5-42

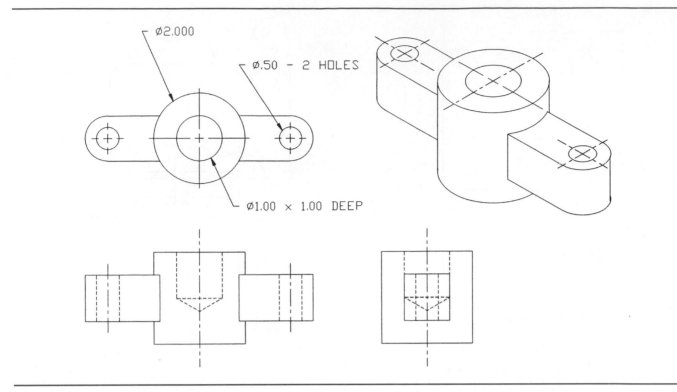

Figure 5-43

The cutting angles of twist drills vary, but they are always represented by a 30° angle as shown.

Figure 5-43 shows an object that contains two through holes and one hole with a depth specification. Note how these holes are represented in the orthographic views and how centerlines are used in the different views.

Figure 5-44 is another example of an object that includes a hole. The hole is centered about the edge line between the two normal surfaces. Figure 5-45 shows the same object with the hole's center point offset from the edge line between the two normal surfaces. Compare the differences between Figures 5-44 and 5-45.

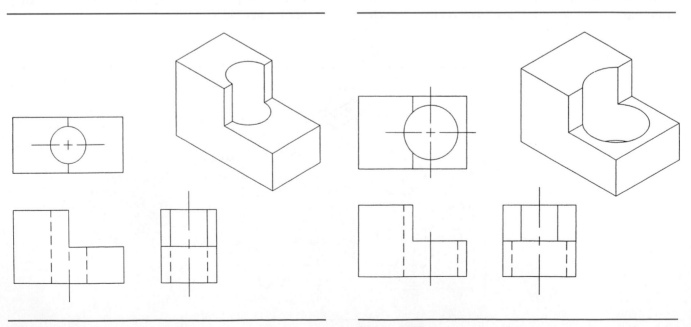

Figure 5-44

Figure 5-45

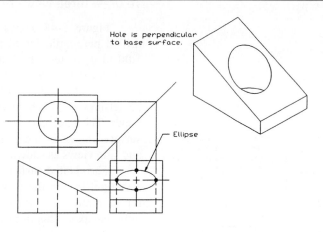

Figure 5-46

5-18 HOLES IN SLANTED SURFACES

Figure 5-46 shows a hole that penetrates a slanted surface. The hole is perpendicular to the bottom surface of the object. The top view of the hole appears as a circle because we are looking straight down into it. The side view shows a distorted view of the circle and is represented using an ellipse.

The shape of the ellipse in the side view is defined by projecting information from the front and top views. Points 1 and 2 are known to be on the vertical centerline, so their location can be projected from the front view to the side view using horizontal lines. Points 3 and 4 are located on the horizontal centerline, so their location can be projected from the top view into the side view using the 45° miter line.

The elliptical shape of the hole in the side view is drawn connecting the projected points using the Ellipse command.

To draw an ellipse representing a projected hole

See Figure 5-47.

1. Define the **first point** as one of the points where the major axis intersects one of the centerlines.
2. Define the **second point** as the other point on the major axis that intersects the same centerline.
3. Define the **third point** as one of the points on the minor axis that intersects the centerline perpendicular to the centerline used in steps 1 and 2.

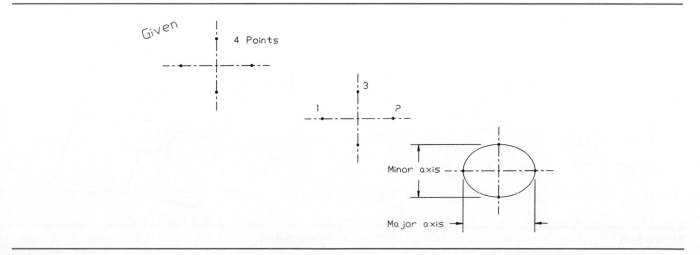

Figure 5-47

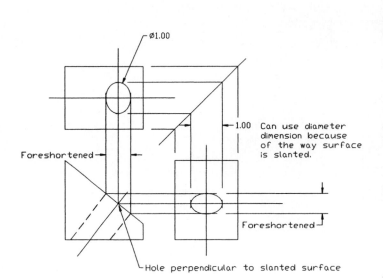

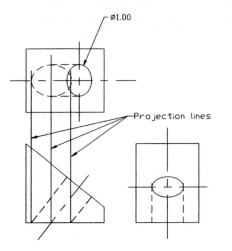

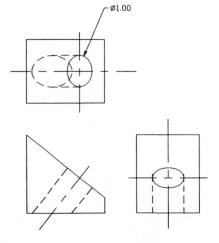

Figure 5-48

To draw three views of a hole in a slanted surface

Figure 5-48 shows an object that includes a hole drilled perpendicular to the slanted surface. The hole is 1.00 in diameter. The hidden lines in the front view that represent the edges of the hole are drawn parallel to the centerline at a distance of .50. The horizontal and vertical centerlines of the hole are located in the top and right views. The intersection of the hole edges with the slanted surface shown in the front view is projected into the top and side views as shown.

The slanted surface shown in Figure 5-48 appears as a straight line in the front view. This means that the hole representations in the top and side views are rotated about only one axis. This in turn means that the hole representation is foreshortened in only one direction. The other direction, the other axis length, remains at the original diameter distance on 1.00. The four points needed to define the projected elliptical shape can be defined by the projection lines and the measured 1.00 distance.

The hole also penetrates the bottom surface, forming a different-sized ellipse. The hole's penetration can be shown in the side view by hidden lines parallel to the given centerline. The elliptical shape in the top view is defined by drawing horizontal parallel lines from the intersection of the hole's projected shape with the vertical centerline, and by projection lines from the front view that define the hole's length.

To draw three views of a hole through an oblique surface

Figure 5-49A shows a hole drilled horizontally through the exact center of an oblique surface of an object. The oblique surface means that both axes in the top and

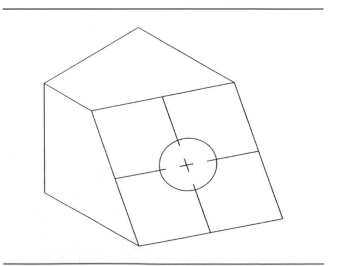

Figure 5-49A

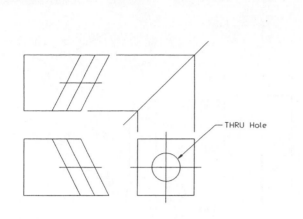

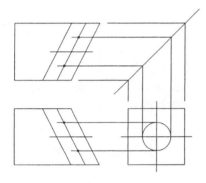

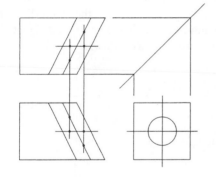

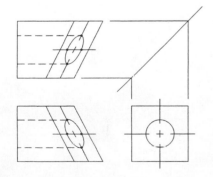

Figure 5-49B

front views will be foreshortened. The procedure used to create the hole's projection in the front and top views is as follows. (See Figure 5-49B.)

1. Draw the hole in the side view. The hole will appear as a circle because it is drilled horizontally.
2. Draw the hole's centerlines in the front and top views. The hole is located exactly in the center of the oblique surface, so lines parallel to the surface's edge lines located midway across the surface may be drawn.
3. Project the intersection of the hole and the horizontal centerline into the top view, and project the intersection of the hole with the vertical centerline into the front view.
4. Project the intersections created in step 3 with the horizontal centerline, represented by the slanted centerline, onto the horizontal centerline in the front view. Likewise, project the intersection points created in step 3 on the horizontal centerline in the front view to the slanted centerline in the top view.
5. Draw the ellipses as required, add the appropriate hidden lines, and erase and trim any excess lines. Save the drawing if desired.

5-19 CYLINDERS

Figure 5-50 shows three views of a cylinder that is cut along the centerline by surface A-B-C-D. Note that each of the three views includes centerlines. The front and

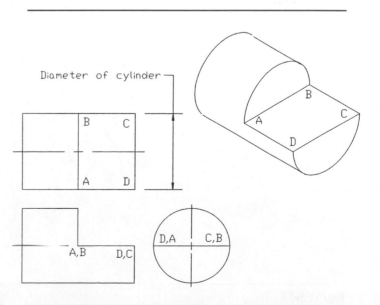

Figure 5-50

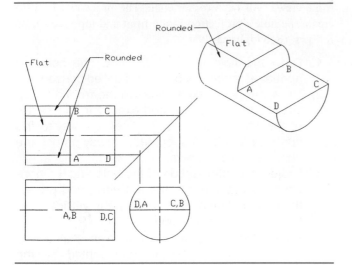

Figure 5-51

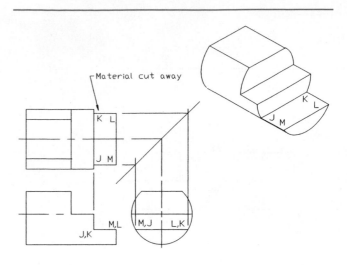

Figure 5-52

top views of a cylinder are called the *rectangular views,* and the side view (end view) is called the *circular view.* The width of the top view is equal to the diameter of the cylinder.

Figure 5-51 shows a cylinder that has a second surface located above the centerline in the front view. The top view of the cylinder shows the flat surface and the rounded portion of the cylinder. Both the flat surface and the two rounded surfaces appear as rectangles.

Figure 5-52 shows a cylinder with a surface J-K-L-M that is located below the centerline in the front view. The

top view of this surface does not include any rounded surfaces because they have been cut away.

5-20 SAMPLE PROBLEM SP5-5

Figure 5-53 shows a cylindrically shaped object that has three surface cuts: one above the centerline, one below the centerline, and one directly on the centerline. Figure 5-54 shows how the three views of the object were developed. The procedure is as follows.

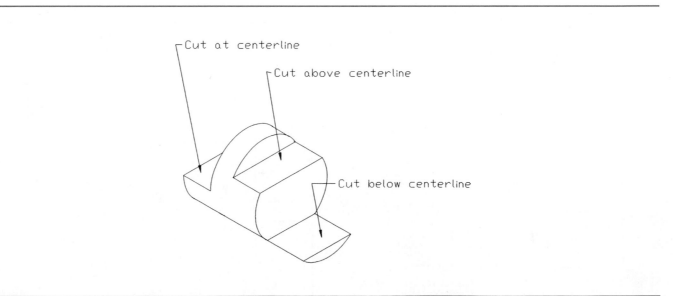

Figure 5-53

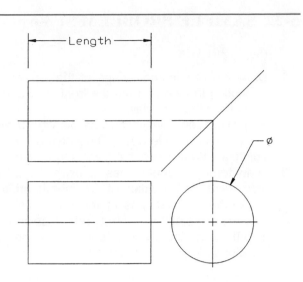

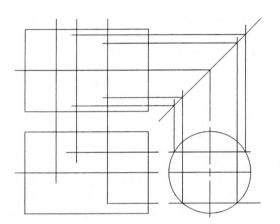

1. Use the given overall length of the cylinder and its diameter to create the two rectangular top and front views and the circular end views.
2. Draw horizontal lines in the side view that define the end views of the three surfaces. Project the size and location of the surfaces into the front and top views. Use **Osnap, Intersection,** with **Ortho** on, for accurate projection.
3. **Erase** and **Trim** the excess lines.
4. Save the drawing if desired.

5-21 CYLINDERS WITH SLANTED AND ROUNDED SURFACES

Figure 5-55 shows a front and side view of a cylindrical object that includes a slanted surface. Slanted surfaces are projected by defining points along their edges in known orthographic views and then projecting the edge points. At least two known views are needed for accurate projection. Sample Problem SP5-6 shows how to define and project a slanted cylindrical surface into the top view given the front and side views.

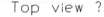

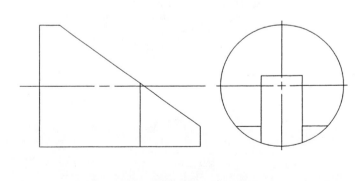

Figure 5-54

Figure 5-55

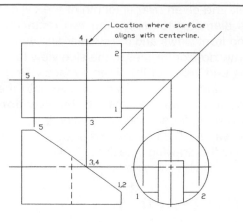

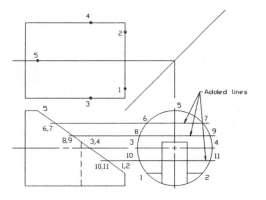

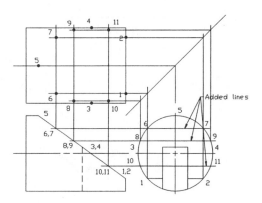

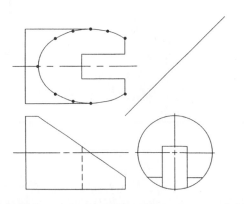

Figure 5-56

5-22 SAMPLE PROBLEM SP5-6

(See Figure 5-56.)

1. Draw the rectangular shape of the top view by projecting the length from the front view and the width from the side view.
2. Project the centerlines, and label points 1, 2, and 3 in the front and side views. The locations of these points are known from the given information.
3. Draw three horizontal lines across the front and side views. The location of the lines is arbitrary. Label the intersections of the horizontal lines with the edge of the slanted surface as shown.
4. The three horizontal lines serve to define point locations along the surface's edge.

A line contains an infinite number of points, so any six can be used.

5. Draw vertical projection lines **(Osnap, Intersection)** from the six point locations in the front view into the area of the top view.
6. Project the six point locations from the side view into the area of the top view. Use the 45° miter line to make the turn between the two views. The intersection of the vertical projection lines of step 2 and the projection lines from the side view defines the location of the points in the top view. Label the six points.
7. Use **Draw, Polyline, Edit Polyline, Fit** to create a smooth line between the six projected points.
8. Remove any excess lines and save the drawing if desired.

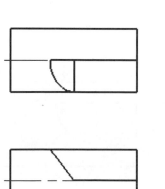

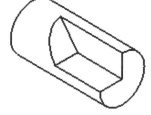

Figure 5-57

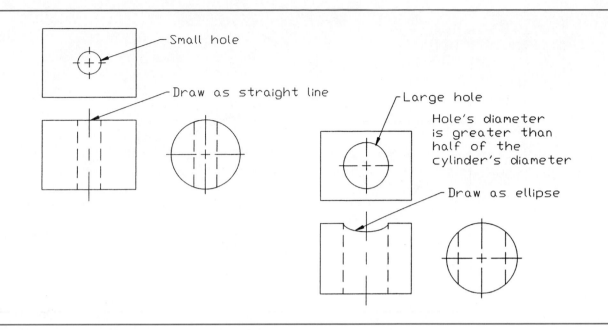

Figure 5-58

Figure 5-57 shows another cylindrical object that contains a slanted surface along with some of the projection lines.

5-23 DRAWING CONVENTIONS AND CYLINDERS

A hole drilled into a cylinder will produce an elliptically shaped edge line in a profile view of the hole. See Figure 5-58. Drawing convention allows a straight line to be drawn in place of the elliptical shape for small holes. The elliptical shape should be drawn for large holes in cylinders, that is, one whose diameter is greater than half the diameter of the cylinder.

Drawing convention also permits straight lines to be drawn for the profile view of a keyway cut into a cylinder. See Figure 5-59. As with holes, if a keyway is large relative to a cylinder, the correct offset shape should be drawn. A keyway is considered large if its width is greater than half the diameter of the cylinder.

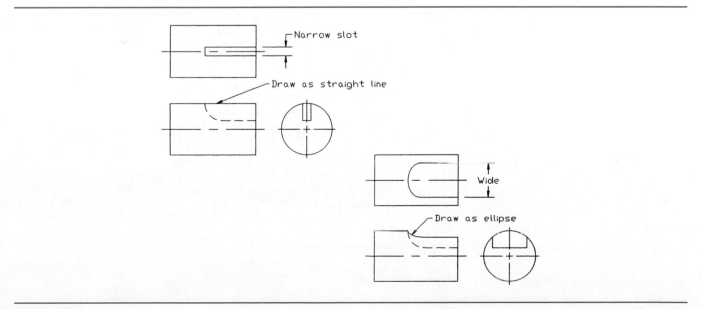

Figure 5-59

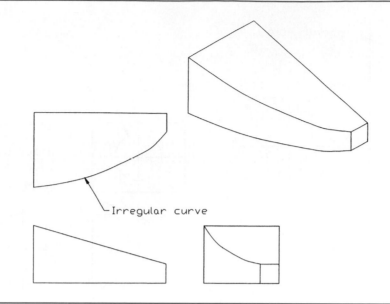

Figure 5-60

5-24 IRREGULAR SURFACES

Irregular surfaces are curved surfaces that do not have constant radii. See Figures 5-60 and 5-61. Irregular surfaces are defined using the X,Y coordinates of points located on the edge of the surface. The points may be dimensioned directly on the view but are often presented in chart form, as shown in Figure 5-61. Figure 5-61 also shows a wing surface defined relative to a given X,Y coordinate system.

Irregular surfaces are projected by defining points along their edges and then projecting the points. At least two known views are needed for accurate projection. The more points used to define the curve, the more accurate the final curve shape will be.

5-25 SAMPLE PROBLEM SP5-7

Figure 5-62 shows an object that includes an irregular surface. The irregular surface is defined by points referenced to an X,Y coordinate system. The point values are listed in a chart. Draw three views of the object shown in Figure 5-62. Your views should look similar to those in Figure 5-63.

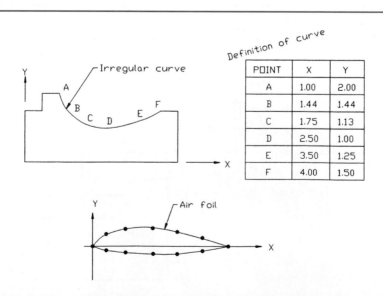

POINT	X	Y
A	1.00	2.00
B	1.44	1.44
C	1.75	1.13
D	2.50	1.00
E	3.50	1.25
F	4.00	1.50

Figure 5-61

1. Draw the front and top views using the given dimensions and chart values. Draw the outline of the side view using the given dimensions.
2. Project the points that define the irregular curve in the top view into the front view. Label the points.
3. Project the points for the curve from both the top and front views into the side view. Label the intersection points.
4. Use **Polyline, Edit Polyline,** and **Fit** to create a curve that represents the side view of the irregular surface.
5. **Erase** and **Trim** any excess lines. Save the drawing if desired.

Remember, it is possible to use **Layer** to create a layer for construction lines and for the final drawing. After the initial drawing is laid out, the final drawing lines may be copied onto another layer and the construction layer turned off. This method eliminates the need for extensive erasing and trimming.

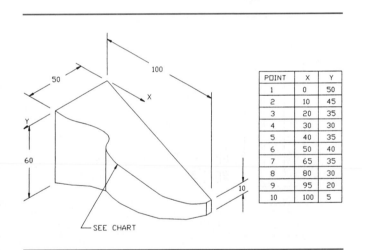

POINT	X	Y
1	0	50
2	10	45
3	20	35
4	30	30
5	40	35
6	50	40
7	65	35
8	80	30
9	95	20
10	100	5

Figure 5-62

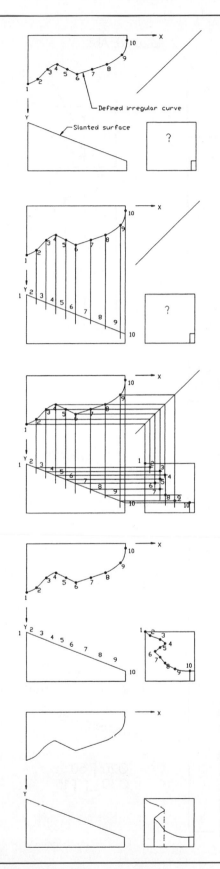

Figure 5-63

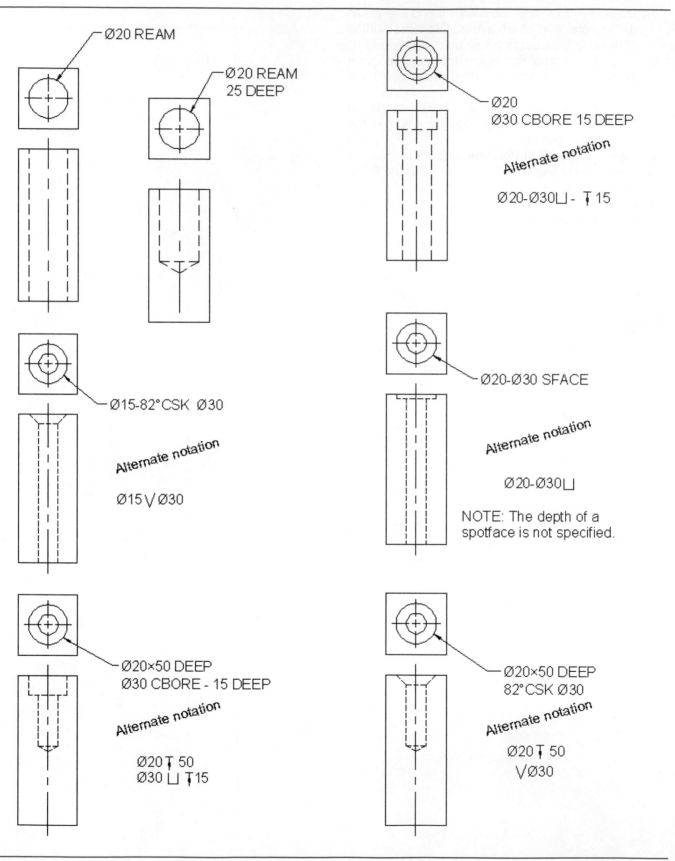

Figure 5-64

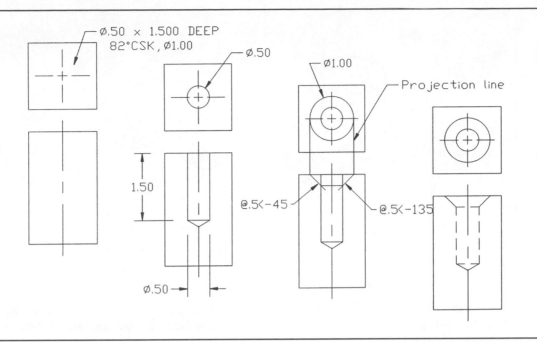

Figure 5-65

5-26 HOLE CALLOUTS

There are four hole-shape manufacturing processes that are used so often that they are defined using a standardized drawing callout: ream, counterbore, countersink, and spotface. See Figure 5-64.

A ream is a process that smooths out the inside of a drilled hole. Holes created using twist drills have spiral-shaped machine marks on their surfaces. Reaming is used to remove the spiral machine marks and to increase the roundness of the hole. Ream callouts generally include a much tighter tolerance than do drill diameter callouts.

A *counterbore* consists of two holes drilled along the same centerline. Counterbores are used to allow fasteners or other objects to be recessed, thus keeping the top surface uniform in height.

A counterbore drawing callout specifies the diameter of the small hole, the diameter of the large hole, and the depth of the large hole.

The information is given in this sequence because it is also the sequence the manufacturer uses. Note that the hidden lines used in the front view of the counterbored hole clearly show the intersection between the two holes.

Figure 5-64 shows an alternative notation for drawing callouts as defined by ISO standards (see Chapter 8) that is intended to remove language from drawings. As parts are often designed in one country and manufactured in another, it is important that drawing callouts be universally understood.

A *countersink* is a conical-shaped hole used primarily for flat-head fasteners. The drawing callout specifies the diameter of the hole, the included angle of the countersink, and the diameter of the countersink as measured on the surface of the object. Almost all countersinks are 82°, but some are drawn at 45°.

To draw a countersunk hole (See Figure 5-65.)

1. Draw a **0.50**-diameter hole in the top view. Project the hole's diameter into the front view.
2. Draw a **1.00**-diameter hole in the top view and project its diameter into the front view. In the front view draw **45°** lines from the intersection of the 1.00-diameter hole and the top surface of the front view so that the lines intersect the 0.50-diameter hole's projection lines. Use **Osnap, Intersection** to ensure accuracy.
3. Draw a horizontal line between the intersection created in step 2. Again use **Osnap, Intersection.**
4. **Erase** and **Trim** the excess lines.

A *spotface* is a very shallow counterbored hole generally used on cast surfaces. Cast surfaces are more porous than machine surfaces. Rather than machine an entire surface flat, it is cheaper to manufacture a spotface because only a small portion of the surface is machined. Spotfaces are usually used for bearing surfaces of fasteners.

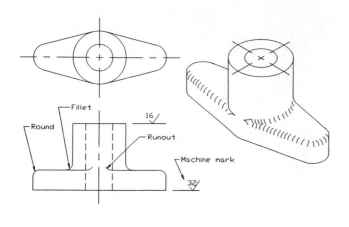

Figure 5-66

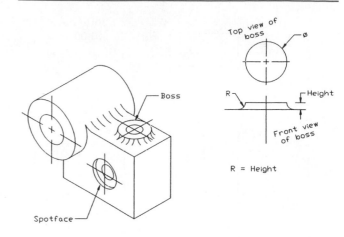

Figure 5-67

A spotface callout defines the diameter of the hole and the diameter of the spotface. A depth need not be given. When machining was done mostly by hand, the machinist would make the spotface just deep enough to produce a shiny surface (most cast surfaces are more gray in color), so no depth was needed. Automated machines require a spotface depth specification. Usually the depth is very shallow.

5-27 CASTINGS

Casting is one of the oldest manufacturing processes. Metal is heated to liquid form, then poured into molds and allowed to cool. The resulting shapes usually include many rounded edges and surface tangencies because it is very

difficult to cast square edges. Concave edges are called *rounds,* and convex edges are called *fillets.* See Figure 5-66. A runout is used to indicate that two rounded surfaces have become tangent to one another. A *runout* is a short arc of arbitrary radius, that is, arbitrary as long as the runout is visually clear.

Cast objects are often partially machined to produce flatter surfaces than can be produced by the casting process. Machined surfaces are defined using machine marks (check marks), as shown in Figure 5-66. The numbers within the machine marks specify the flatness requirements in terms of microinches or micrometers. Machine marks are explained in greater detail in Section 9-26.

A *boss* is a turretlike shape that is often included on castings to localize and minimize machining. A boss is defined by its diameter and its height. The sides of a boss

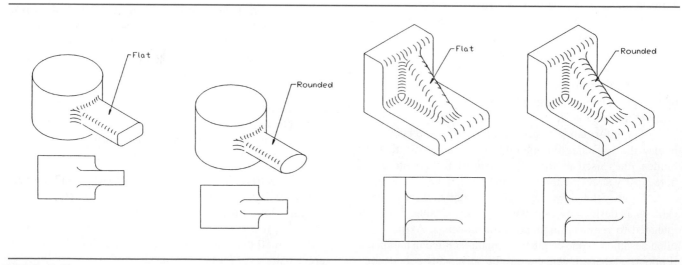

Figure 5-68

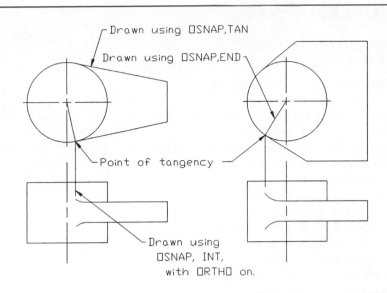

Figure 5-69

are rounded and defined by a radius usually equal to the height of the boss. See Figure 5-67.

Spotfacing is a machine process often associated with castings. As with bosses, spotfaces help localize and minimize machining requirements. Spotfaces were defined in Section 5-26.

The direction of runouts is determined by the shape of the two surfaces involved. Flat surfaces intersecting rounded surfaces generate runouts that turn out. Rounded surfaces that intersect rounded surfaces generate runouts that turn in. See Figure 5-68.

The same convention is followed for flat and rounded surfaces that intersect flat surfaces.

The location of a runout is determined by the location of the tangent it represents. See Figure 5-69. The location of a tangency point can be determined by first drawing the top view of the tangency line using **Osnap, Tangent.** A line can then be drawn from the center point of the circular top view to the end of the tangency line using **Osnap, Endpoint.** The point of tangency can then be projected from the top view to the front view using **Osnap, Intersection** with **Ortho** on.

Use **Arc** (type the word **arc** in response to a command prompt) to draw runouts. Any convenient radius may be used, providing the runout is clearly visible and visually distinct from the straight line. Figure 5-70 shows some further examples of cast surfaces that include runouts.

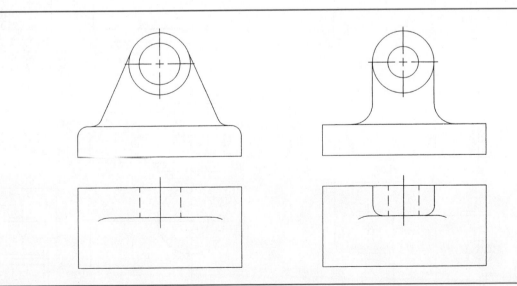

Figure 5-70

5-28 SAMPLE PROBLEM SP5-8

Draw three views of the object shown in Figure 5-71. The solution was drawn as follows:

1. Use the given overall dimensions and draw the outline of the front, top, and side views.
2. Modify the appropriate lines to hidden lines. Draw the fillets and rounds. The **Fillet** command may erase needed lines. This cannot always be avoided, so simply redraw the needed lines after the fillet or round is complete.
3. Draw the spotface and boss in each of the views. The hole in the boss is **.500** deep. Clearly show the bottom of the hole, including the conical point. Draw the hole in the front view and use **Copy** to copy the hole into the side view.
4. Use **Draw, Circle** to draw the **1.00**-diameter hole in the front view. Draw the appropriate hidden and centerlines in the other views.
5. Save the drawing if desired.

See Figure 5-72.

Figure 5-71

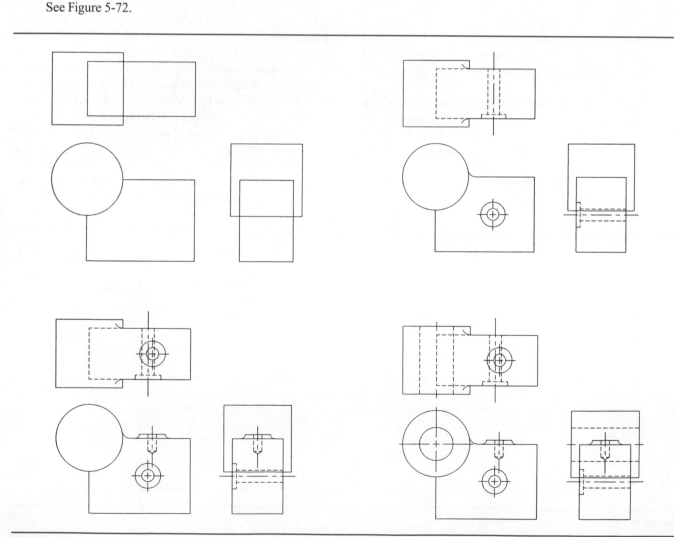

Figure 5-72

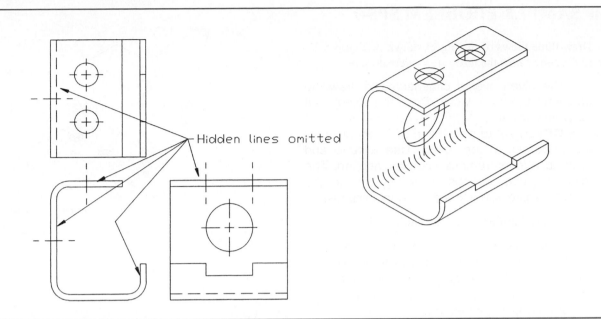

Figure 5-73

5-29 THIN-WALLED OBJECTS

Thin-walled objects such as parts made from sheet metal or tubing present some unique drawing problems. For example, the distance between surfaces is usually so small that hidden lines can't be used to define holes, and there isn't enough distance to draw a broken-line pattern. See Figure 5-73.

Hidden lines may be applied to thin-walled objects in one of three ways: draw the lines as continuous lines but use a different color from that used for the actual continuous lines; use an enlarged detail of the area including the hidden lines; or omit the hidden lines and include only a centerline.

Each of these techniques is shown in Figure 5-73. All holes must include a centerline.

Companies often have a drawing manual that states their policy on drawing hidden lines in thin-walled objects. The most important goal is to be consistent in the representation.

Many sheet-metal parts are manufactured by bending. Bending produces an inside bend radius and an outside bend radius. The inside bend radius plus the material thickness should equal the outside bend radius. See Figure 5-74. The same radius should not be used for both the inside and outside bends.

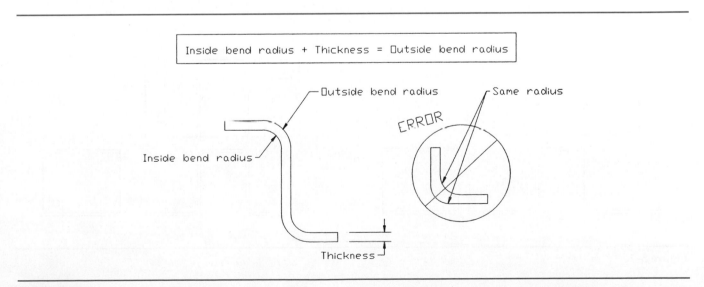

Figure 5-74

5-30 SAMPLE PROBLEM SP5-9

Draw three views of the object shown in Figure 5-75. Figure 5-76 shows how the three views were developed.

1. Use the given overall dimensions to draw the outline of the three views. Draw the object as if all bends are 90°.
2. Use **Offset,** set to a distance of **3 mm,** to draw the thickness of the object. Use **Extend** and **Trim** to add or remove lines as needed. The thickness can also be drawn by setting the **Snap** spacing equal to the material thickness.

   Change the appropriate lines to hidden lines.

3. Draw fillets using a radius of **5 mm** for the inside bend radii and **8 mm** for the outside bend radii.
4. Use **Draw, Circle** to draw the holes in the appropriate view and add centerlines as shown.
5. **Save** the drawing if desired.

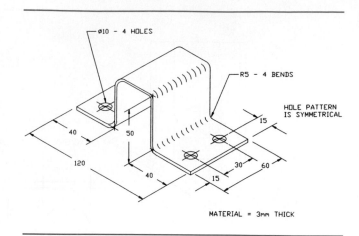

Figure 5-75

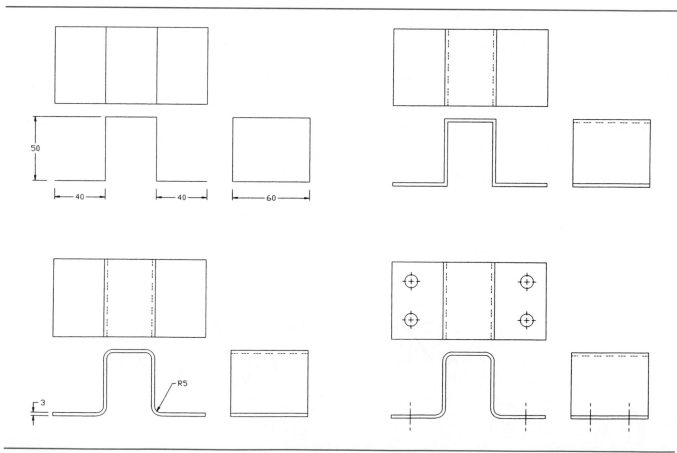

Figure 5-76

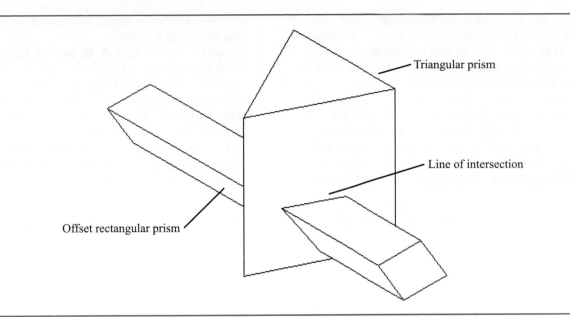

Figure 5-77

5-31 INTERSECTIONS

Intersection drawings are drawings that show the intersection of two objects. Figure 5-77 shows the intersection between two offset squares. A discussion of intersections has been included at this point in the book because they rely heavily on projection of information between views. They require not only a knowledge of the principles of projection but also an understanding about what the various lines represent. Intersections will be discussed again in Chapter 16, Solid Modeling.

Three intersection problems are presented in the following three sample problems.

5-32 SAMPLE PROBLEM SP5-10

Given the side and top views of a smaller circle intersecting a larger circle as shown in Figure 5-78, draw the front view. All dimensions are in inches. Because the object is symmetrical, the intersection for one side will be developed and then mirrored.

1. Draw the outline of the front view by projecting lines from the given top and front views. See Figure 5-79.
2. Define points on the edge of the smaller circle. In this example, 17 points were defined.

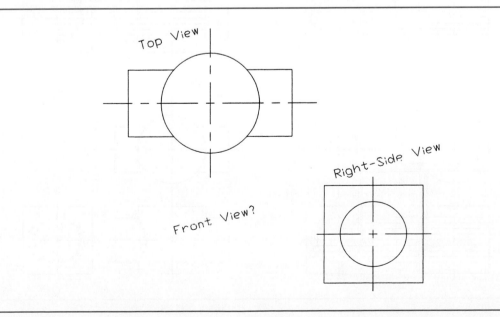

Figure 5-78

3. Extend the horizontal centerline of the smaller circle in the side view into the area of the front view. Use **Draw, Offset** and add six lines parallel to the horizontal centerline in the side view. Label the intersection of the horizontal lines with the edge of the smaller circle as shown.

4. Project points **5, 6, 7, 8, 9,** and **10** into the top view. Use **Draw, Line** along with **Osnap, Intersection** to draw vertical lines from the side view so that they intersect the 45° miter line. Then, project the intersections on the miter line into the top view using horizontal lines. Label the points as shown.

5. Project points **5, 6, 7, 8, 9,** and **10** from the top view into the area of the front view so they intersect the horizontal lines from the side view. Label the points as shown. Use **Draw, Polyline, Edit Polyline,** and **Fit** to draw the curve that represents the intersection.

6. Add the lines needed to complete the front view, remove all excess lines, and save the drawing if desired.

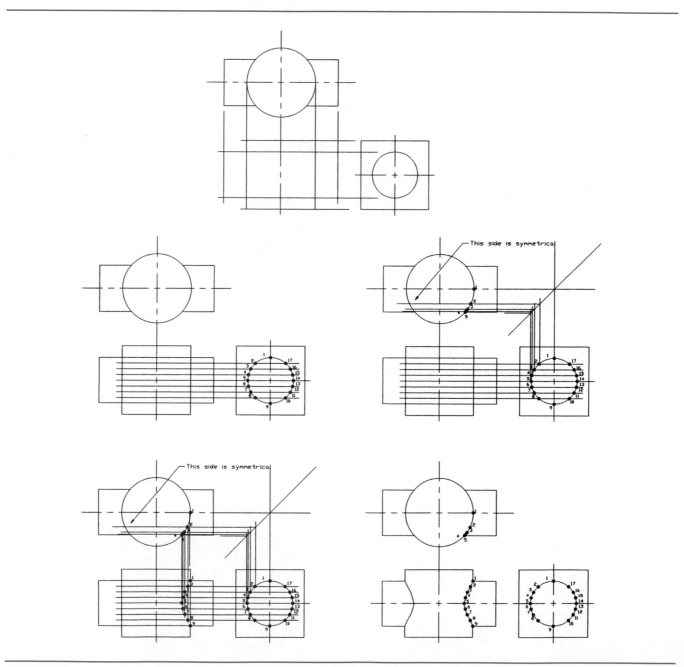

Figure 5-79

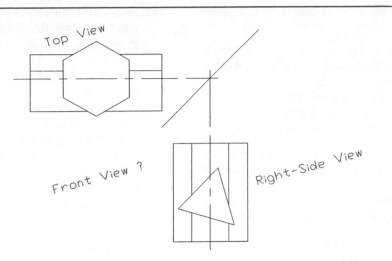

Figure 5-80

5-33 SAMPLE PROBLEM SP5-11

Figure 5-80 shows the top and right-side views of a triangular-shaped piece intersecting a hexagonal-shaped piece. What is the shape of the front view?

Figure 5-81 shows the solution obtained by projecting lines from the two given views into the front view, and shows an enlargement of the intersecting surfaces. Use a straightedge and verify the location of each labeled point by projecting horizontal lines from the front view and vertical lines from the top view into the front view.

5-34 SAMPLE PROBLEM SP5-12

Given the side and top views of a circle intersecting a cone as shown in Figure 5-82, draw the front view.

As in the previous sample problems, the problem is solved by defining intersection points in the given two views and then projecting them into the front view. The cone, however, presents a unique problem because it is both round and tapered. There are few edge lines to work with.

The solution requires that there be a more precise definition of the cone's surface than is presented by the

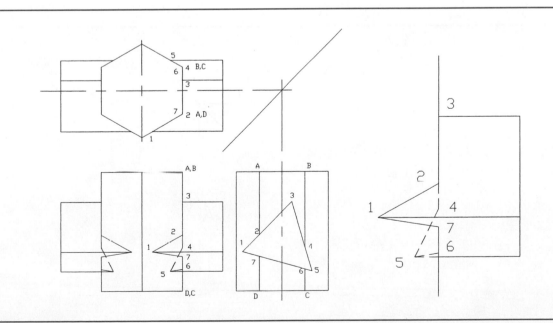

Figure 5-81

circle and centerlines in the top view and the profile front view.

1. Figure 5-83, step 1, shows how points **1** and **3**, located on the vertical centerline of the cylinder, are projected from the side view to the front view and then to the top view. The vertical centerline is aligned with the right-side profile line of the cone so that it can be used for projection.

This is not true for points 2 and 4, located on the horizontal centerline in the side view. Currently the location of points 2 and 4 is unknown in both the front and top views, so projection lines cannot be drawn. The location of points 2 and 4 can be determined in the front and top views.

2. Extend the horizontal centerline in the side view so that it intersects the edge lines of the cone. Use **Osnap, Intersection,** with **Ortho** on and project the intersection of the extended horizontal centerline and the cone's edge line so that it intersects the vertical centerline in the top view. Draw a circle in the top view using the distance between the intersection with the vertical centerline and the projection line and the center point of the cone as the radius.

The circle in the top view represents a slice of the cone located at exactly the same height as points 2 and 4. Points 2 and 4 must be located somewhere in the slice.

3. Project points **2** and **4** into the top view from the side view so that the projection line intersects the circular slice drawn in step 2. Label the intersections **2** and **4.**

4. Project the locations of points **2** and **4** in the top view into the front view using a vertical line. Project the locations of the points from the side view into the front view using a horizontal line. The intersection of the vertical and horizontal projection lines defines the locations of points 2 and 4 in the front view.

5. Expand the procedure explained in steps 2 and 3 by drawing a series of horizontal projection lines between the front and side views as shown. Use **Draw** for the first line and **Offset** for the other lines. The location of these additional lines is random.

Use **Osnap, Intersection** with **Ortho** on to project the intersections of the horizontal lines with the cone's edge lines in the top view.

6. Draw circles using the distance between the cone's center point and the projection lines' intersections with the vertical centerline as radii. Use **Osnap, Intersection** to locate the circles' center point and radii distances.

7. Use **Draw, Polyline, Edit Polyline,** and **Fit** to draw the required curves in the top and front views. Use **Osnap, Intersection** to ensure accurate curve point locations. The **Move** option located on the **Edit Polyline** command options line can be used to move vertex points on the curves to make them appear smoother and more continuous.

8. Trim and erase all excess lines and save the drawing if desired.

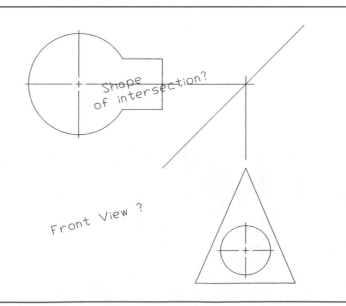

Figure 5-82

Step 1

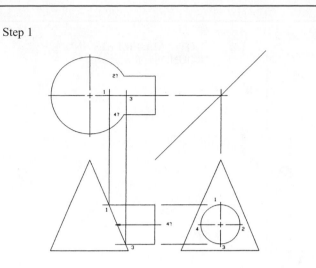

Step 2

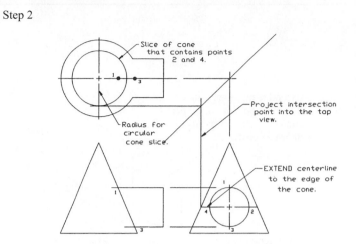

Step 3

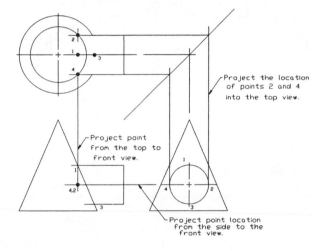

Step 4

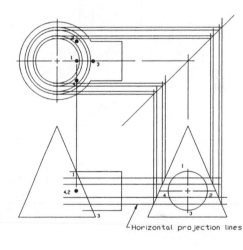

Step 5

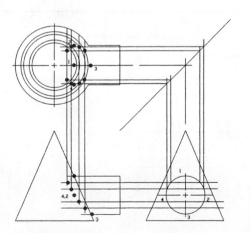

Step 6

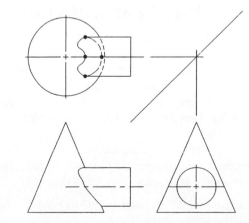

Figure 5-83

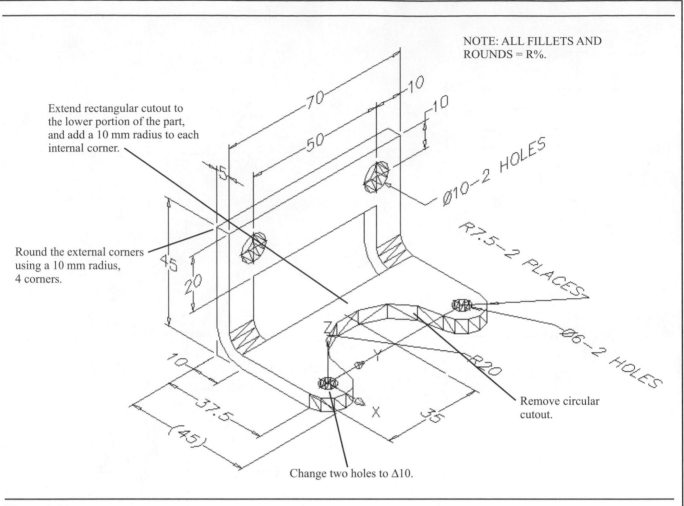

NOTE: ALL FILLETS AND
ROUNDS = R%.

Extend rectangular cutout to
the lower portion of the part,
and add a 10 mm radius to each
internal corner.

Round the external corners
using a 10 mm radius,
4 corners.

Remove circular
cutout.

Change two holes to Δ10.

Figure 5-84

5-35 DESIGNING BY MODIFYING AN EXISTING PART

Many beginning design assignments require that an existing part be modified to meet a new set of requirements. Designers must therefore create something different from the original drawing. This ability to look beyond the drawing is an important design skill.

Figure 5-84 shows a dimensioned part, and Figure 5-85 shows the three orthographic views of the part. The part is to be redesigned as follows.

1. Replace the two **Ø7.5** holes with **Ø10** holes.
2. Remove the **R20** cutout.
3. Modify the horizontal portion of the part so its overall length is **45** millimeters; make the entire part symmetrical about its **90°** axis.

Original views

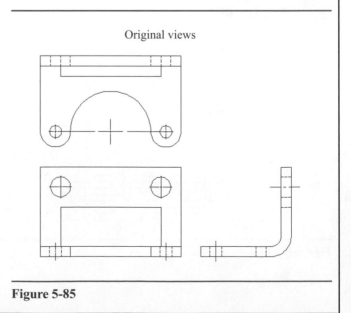

Figure 5-85

4. Round the four external and four internal corners using a **10**-millimeter radius.

Figure 5-86 shows the resulting three orthographic views.

Revised views

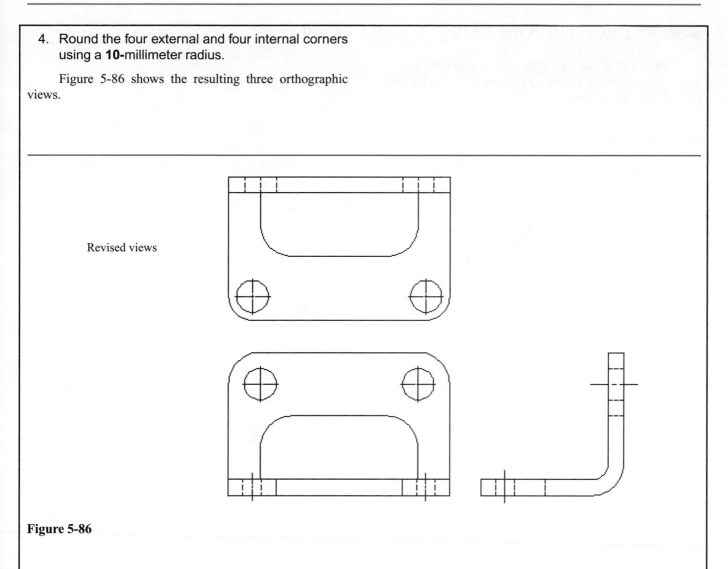

Figure 5-86

5-36 EXERCISE PROBLEMS

Draw a front, top, and right-side orthographic view of each of the objects in Exercise Problems EX5-1 to EX5-94. Do not include dimensions.

EX5-1 INCHES

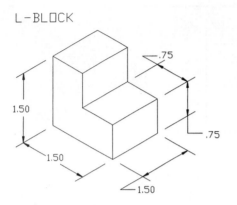

EX5-2 MILLIMETERS

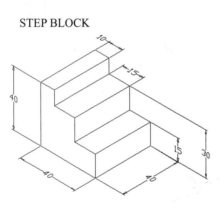

EX5-3 MILLIMETERS

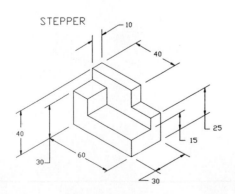

EX5-4 INCHES

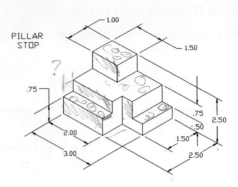

EX5-5 MILLIMETERS

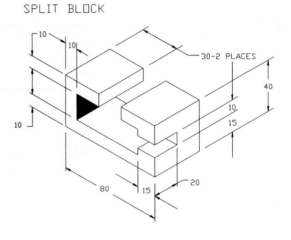

EX5-6 MILLIMETERS

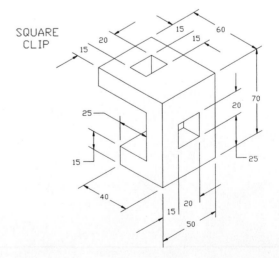

EX5-7 MILLIMETERS

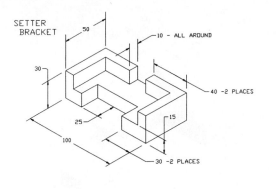

SETTER BRACKET

50

10 – ALL AROUND

30

40 –2 PLACES

25

15

100

30 –2 PLACES

EX5-10 MILLIMETERS

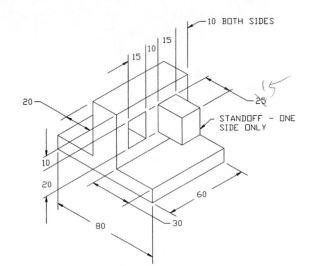

10 BOTH SIDES

15

15

10

20

15

STANDOFF – ONE SIDE ONLY

10

20

60

80

30

EX5-8 MILLIMETERS

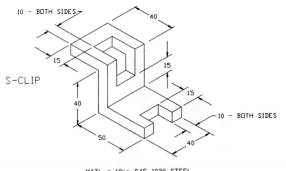

10 – BOTH SIDES

40

15

15

S-CLIP

15

40

10 – BOTH SIDES

50

40

MATL = 10mm SAE 1020 STEEL

EX5-11 MILLIMETERS

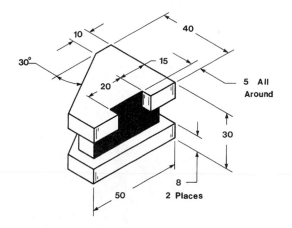

10

40

30°

15

20

5 All Around

30

50

8

2 Places

EX5-12 INCHES

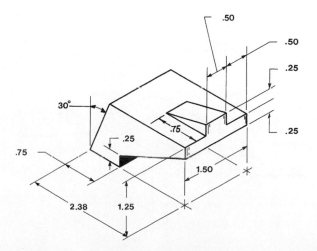

.50

.50

.25

30°

.75

.25

.25

.75

.25

1.50

2.38

1.25

EX5-9 INCHES

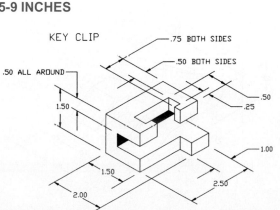

KEY CLIP

.75 BOTH SIDES

.50 BOTH SIDES

.50 ALL AROUND

1.50

.50

.25

1.00

1.50

2.50

2.00

EX5-13 MILLIMETERS

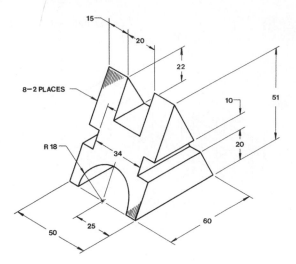

EX5-14 MILLIMETERS

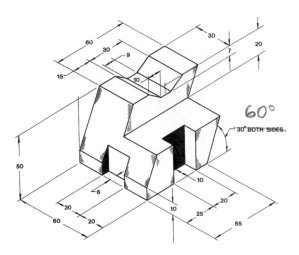

EX5-15 MILLIMETERS

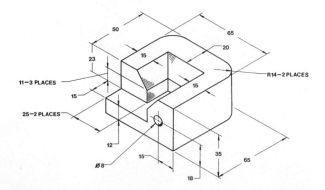

EX5-16 MILLIMETERS

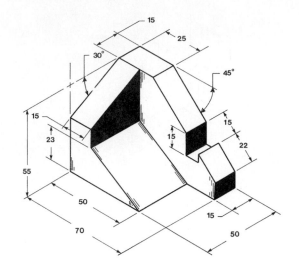

EX5-17 MILLIMETERS

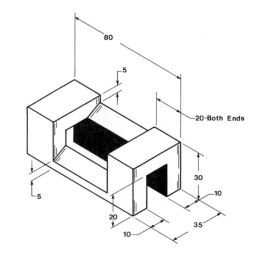

EX5-18 MILLIMETERS

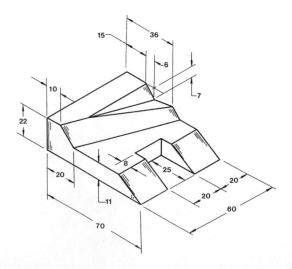

EX5-19 MILLIMETERS

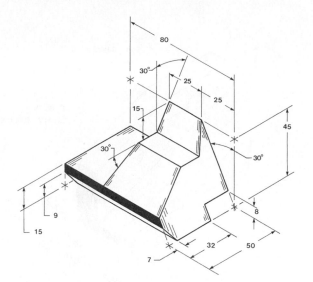

EX5-20 MILLIMETERS

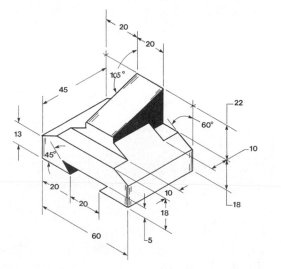

EX5-21 INCHES

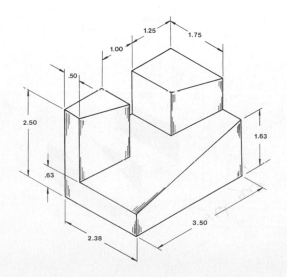

EX5-22 MILLIMETERS

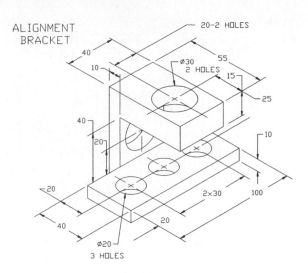

EX5-23 MILLIMETERS

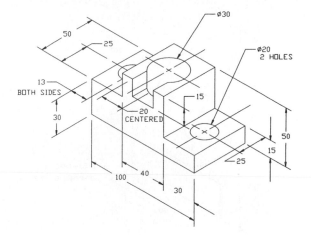

EX5-24 INCHES

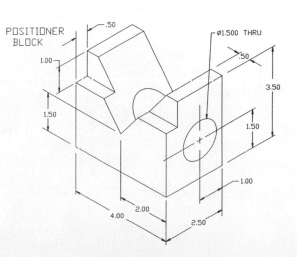

EX5-25 MILLIMETERS

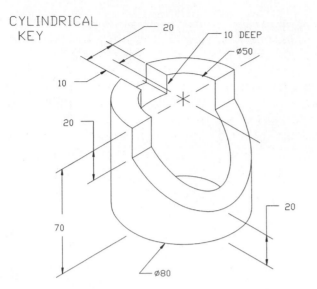

CYLINDRICAL KEY
20
10 DEEP
Ø50
10
20
20
70
Ø80

EX5-28 MILLIMETERS

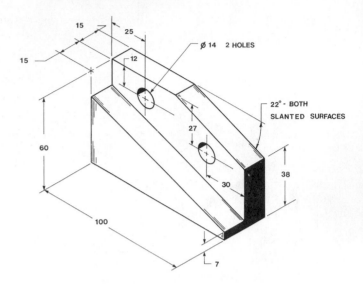

15
25
15
Ø 14   2 HOLES
12
60
27
22° - BOTH SLANTED SURFACES
30
38
100
7

EX5-26 MILLIMETERS

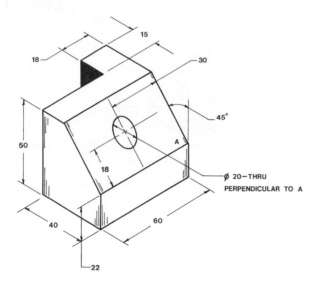

15
18
30
45°
50
18
A
Ø 20-THRU
PERPENDICULAR TO A
40
60
22

EX5-29 MILLIMETERS

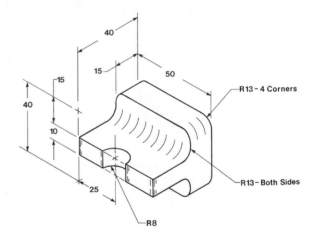

40
15
15
50
R13 - 4 Corners
40
10
R13- Both Sides
25
R8

EX5-30 MILLIMETERS

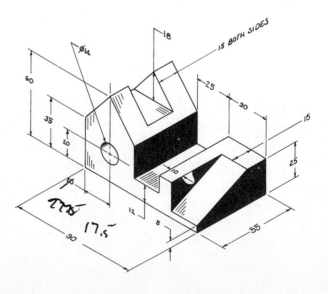

18
15 BOTH SIDES
Ø14
60
25
30
35
15
10
12
8
25
55
90
17.5

EX5-27 MILLIMETERS

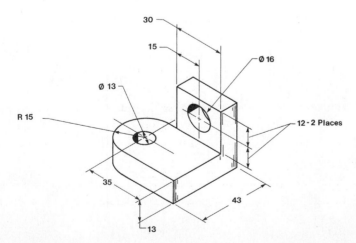

30
15
Ø 16
Ø 13
R 15
12 - 2 Places
35
43
13

EX5-31 INCHES

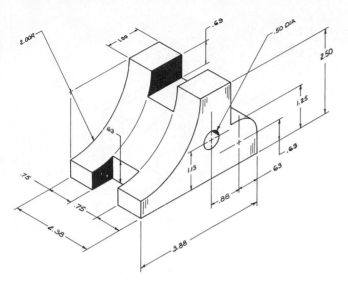

EX5-32 MILLIMETERS

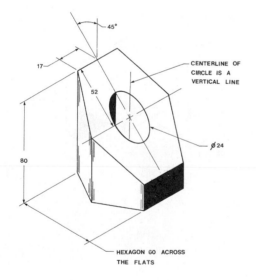

CENTERLINE OF
CIRCLE IS A
VERTICAL LINE

Ø 24

HEXAGON GO ACROSS
THE FLATS

EX5-33 INCHES

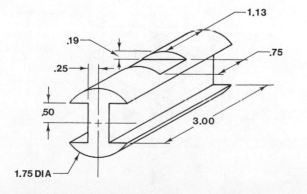

EX5-34 MILLIMETERS

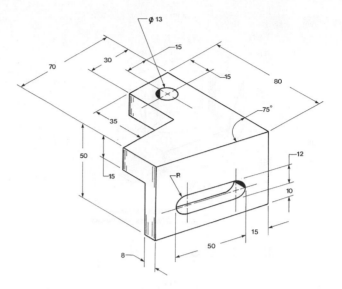

EX5-35 MILLIMETERS

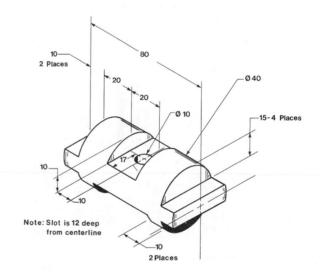

Note: Slot is 12 deep
from centerline

EX5-36 INCHES

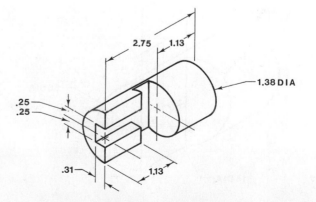

EX5-37 MILLIMETERS

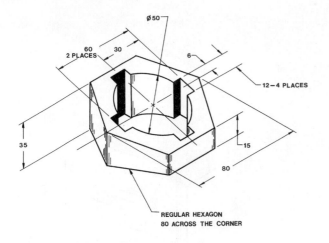

Ø50
60 2 PLACES
30
6
12–4 PLACES
35
15
80
REGULAR HEXAGON
80 ACROSS THE CORNER

EX5-40 MILLIMETERS

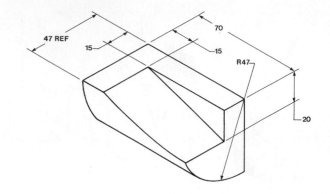

47 REF
15
70
15
R47
20

EX5-38 MILLIMETERS

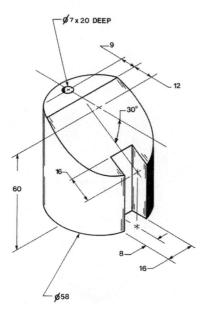

Ø7 x 20 DEEP
9
12
30°
16
60
16
8
16
Ø58

EX5-41 INCHES

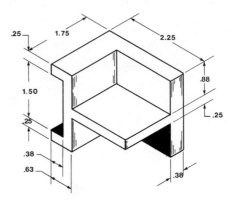

.25
1.75
2.25
.88
1.50
.25
.25
.38
.63
.38

EX5-39 MILLIMETERS

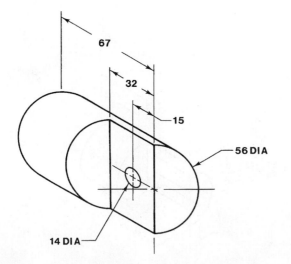

67
32
15
56 DIA
14 DIA

EX5-42 INCHES

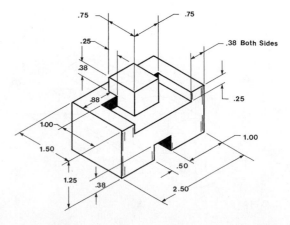

.75
.75
.25
.38 Both Sides
.38
.88
.25
1.00
.25
1.50
1.00
.50
1.25
.38
2.50

EX5-43 MILLIMETERS

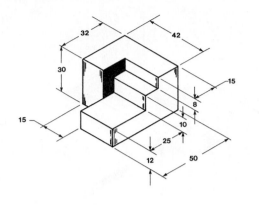

EX5-46 MILLIMETERS

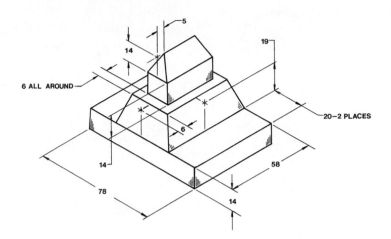

EX5-44 MILLIMETERS

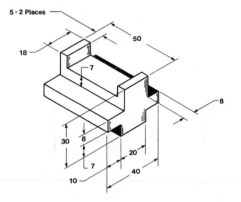

EX5-47 INCHES

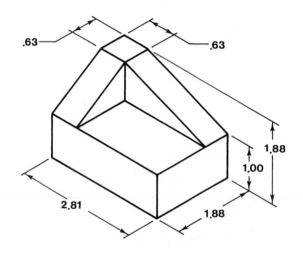

EX5-48 INCHES

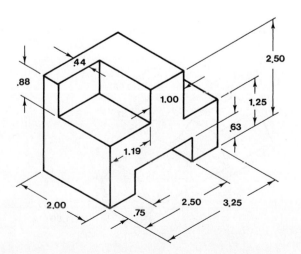

EX5-45 MILLIMETERS

NOTE: THE SLOT
IS 15 LONG

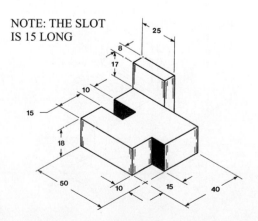

EX5-49 MILLIMETERS

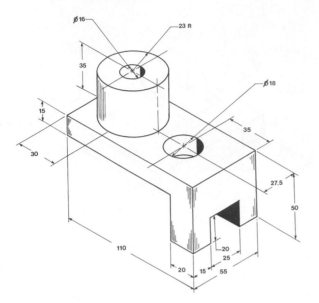

EX5-52 MILLIMETERS

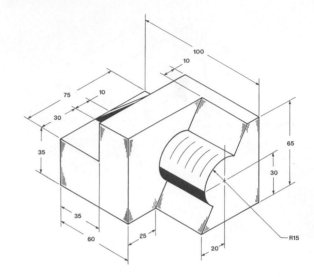

EX5-50 INCHES

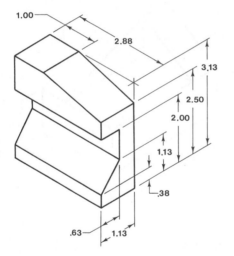

EX5-53 INCHES

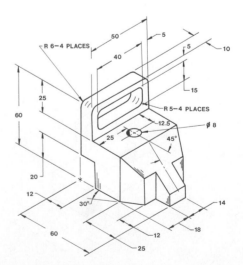

EX5-51 INCHES

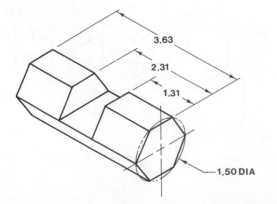

EX5-54 MILLIMETERS

EX5-55 MILLIMETERS

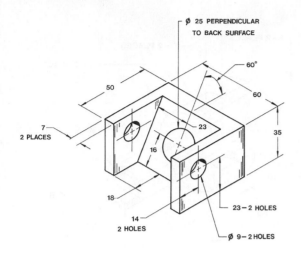

Ø 25 PERPENDICULAR
TO BACK SURFACE

60°

50

60

7
2 PLACES

23

35

16

18

23–2 HOLES

14
2 HOLES

Ø 9–2 HOLES

EX5-56 MILLIMETERS

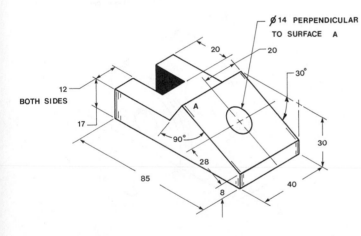

Ø 14 PERPENDICULAR
TO SURFACE A

20

20

30°

12
BOTH SIDES

17

A

90°

28

30

85

8

40

EX5-57 MILLIMETERS

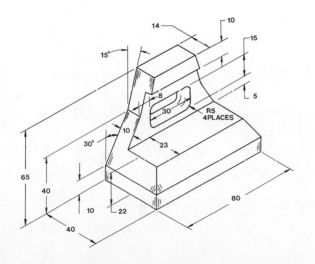

14

10

15

15°

8

5

30

R5
4 PLACES

10

30°

23

65

40

10

22

80

40

EX5-58 MILLIMETERS

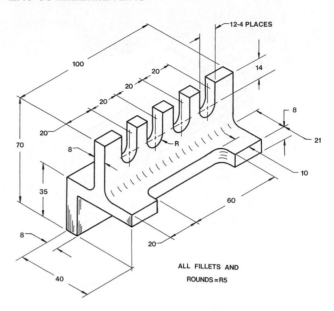

12-4 PLACES

100

20

20

14

20

20

8

70

8

21

35

R

60

10

8

40

20

ALL FILLETS AND
ROUNDS = R5

EX5-59 MILLIMETERS

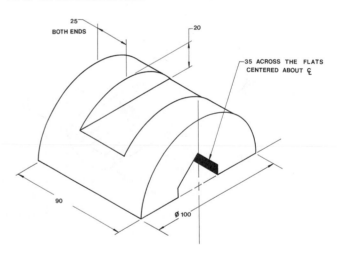

25
BOTH ENDS

20

35 ACROSS THE FLATS
CENTERED ABOUT ₵

90

Ø 100

EX5-60 INCHES

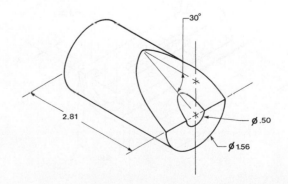

30°

2.81

Ø .50

Ø 1.56

EX5-61 MILLIMETERS

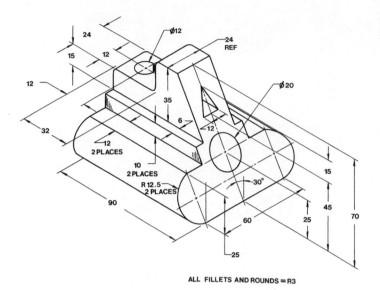

ALL FILLETS AND ROUNDS = R3

EX5-62 MILLIMETERS

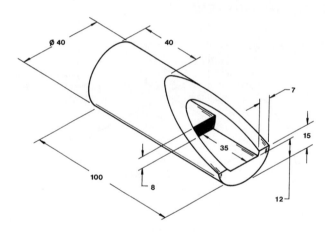

EX5-63 MILLIMETERS

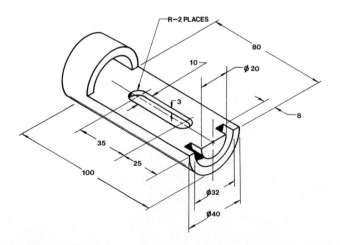

EX5-64 MILLIMETERS

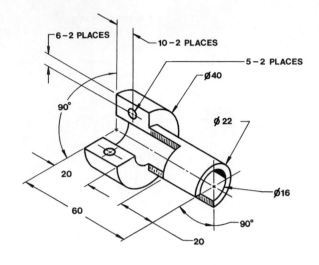

EX5-65 MILLIMETERS

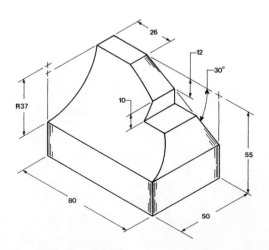

EX5-66 MILLIMETERS

EX5-67 MILLIMETERS

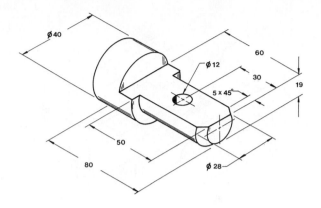

EX5-68 MILLIMETERS

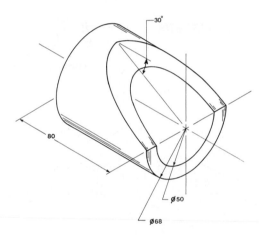

EX5-69 MILLIMETERS

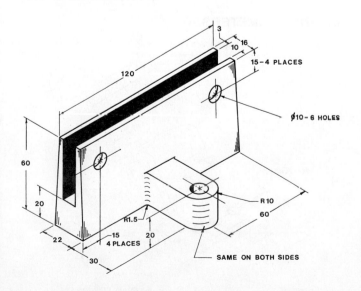

EX5-70 MILLIMETERS

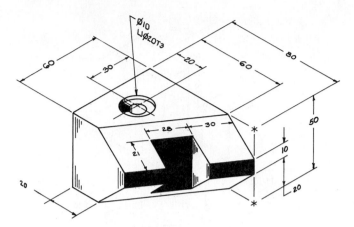

EX5-71 MILLIMETERS

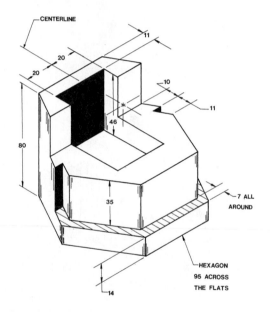

EX5-72 MILLIMETERS

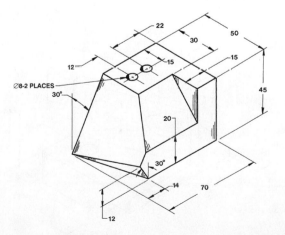

EX5-73 MILLIMETERS

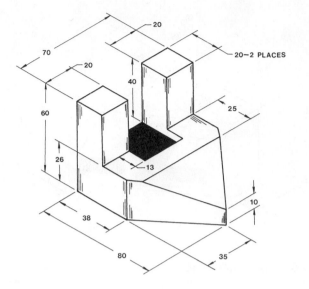

EX5-76 MILLIMETERS

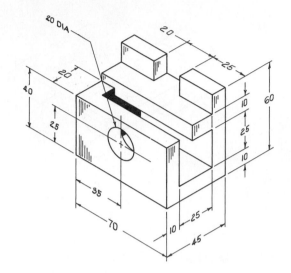

EX5-74 INCHES

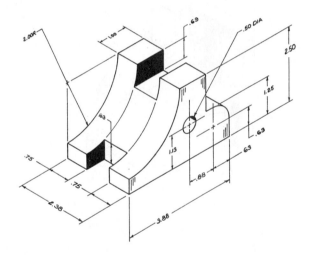

EX5-77 MILLIMETERS

NOTE: ALL FILLET AND ROUNDS=R3

EX5-78 MILLIMETERS

EX5-75 INCHES

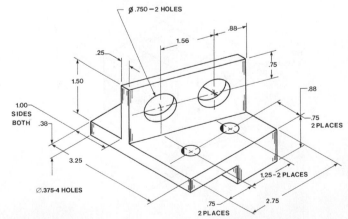

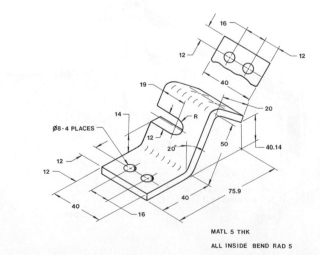

MATL 5 THK

ALL INSIDE BEND RAD 5

EX5-79 MILLIMETERS

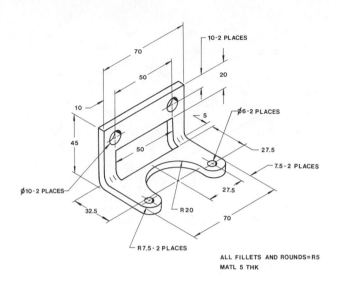

10-2 PLACES
70
50
20
10
5
Ø6-2 PLACES
45
50
27.5
7.5-2 PLACES
Ø10-2 PLACES
27.5
32.5
R 20
70
R7.5-2 PLACES

ALL FILLETS AND ROUNDS=R5
MATL 5 THK

EX5-80 MILLIMETERS

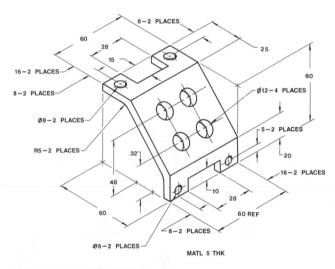

8-2 PLACES
60
28
25
15
16-2 PLACES
60
8-2 PLACES
Ø12-4 PLACES
Ø8-2 PLACES
5-2 PLACES
R5-2 PLACES
32
20
16-2 PLACES
48
10
28
60
60 REF
8-2 PLACES
Ø6-2 PLACES

MATL 5 THK

EX5-81 MILLIMETERS

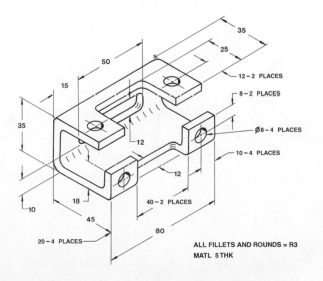

35
25
50
12-2 PLACES
15
8-2 PLACES
35
Ø8-4 PLACES
12
10-4 PLACES
12
18
40-2 PLACES
10
45
80
20-4 PLACES

ALL FILLETS AND ROUNDS = R3
MATL 5 THK

EX5-82 MILLIMETERS

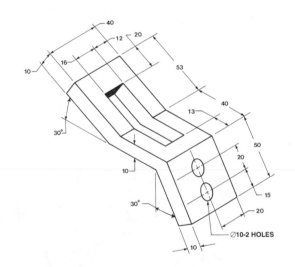

50
10
20
15
Ø10-2 PLACES
28
15
R30-2 PLACES
45
15
25
12
20
15
25
70
R5-2 PLACES

ALL FILLETS AND ROUNDS = R3
MATL 12 THK

EX5-83 MILLIMETERS

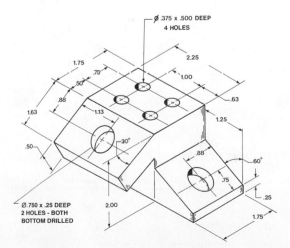

40
12
20
10
16
53
40
30°
13
50
20
10
15
30°
20
Ø10-2 HOLES
10

EX5-84 INCHES

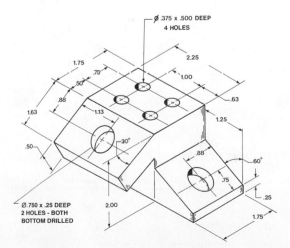

Ø .375 x .500 DEEP
4 HOLES
1.75
2.25
.70
.50
1.00
.88
.63
1.63
1.13
1.25
.50
30°
.88
60°
.75
.25
Ø .750 x .25 DEEP
2 HOLES - BOTH
BOTTOM DRILLED
2.00
1.75

EX5-85 MILLIMETERS

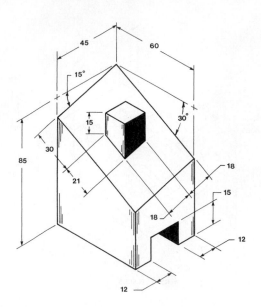

EX5-87 INCHES

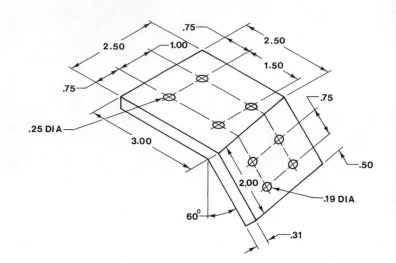

EX5-86 MILLIMETERS

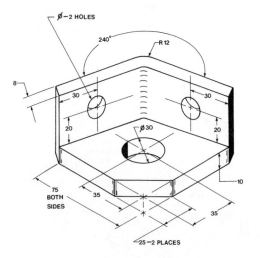

EX5-88 MILLIMETERS

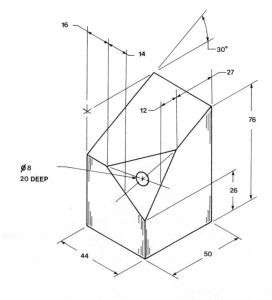

EX5-89 MILLIMETERS

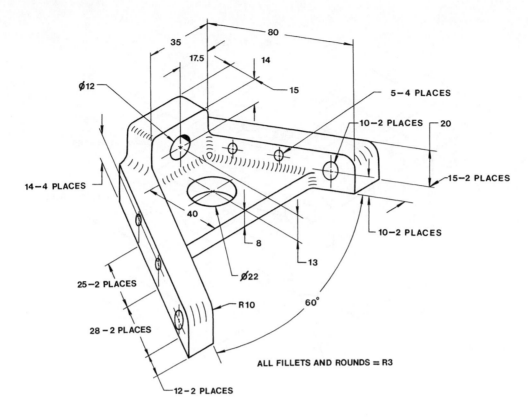

EX5-90 MILLIMETERS

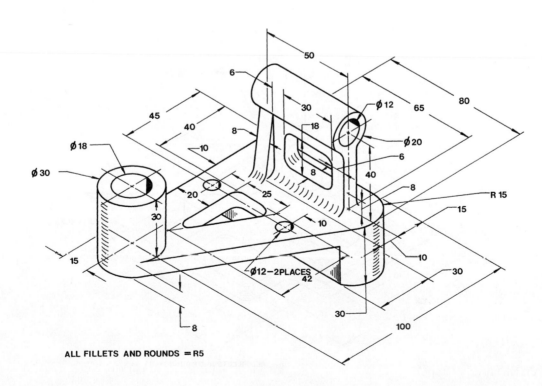

EX5-91 MILLIMETERS

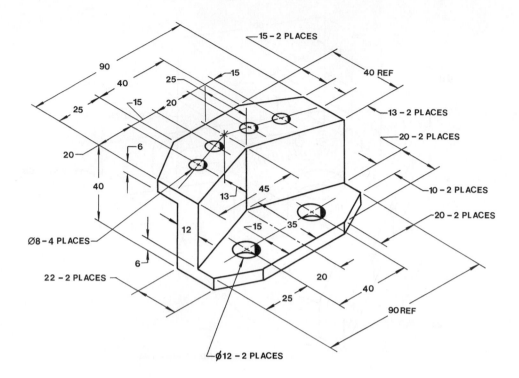

EX5-92 MILLIMETERS

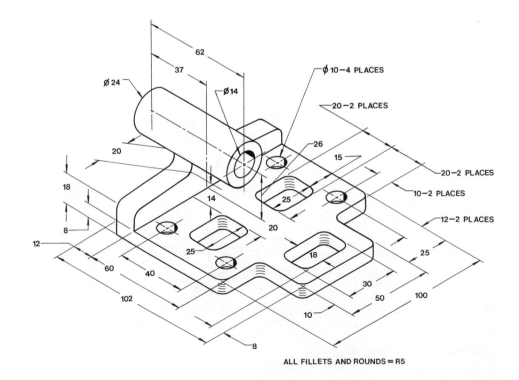

ALL FILLETS AND ROUNDS = R5

EX5-93 MILLIMETERS

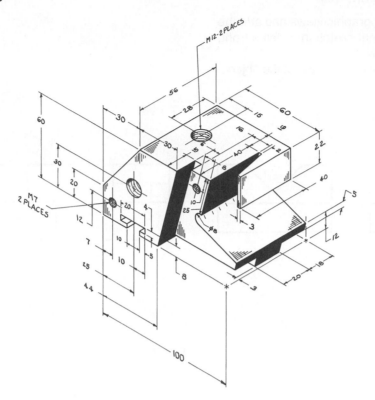

EX5-94 MILLIMETERS

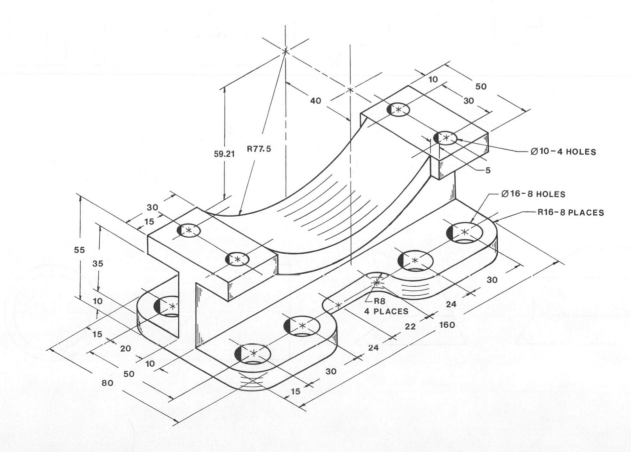

For Exercises EX5-95 to EX5-100:

A. Sketch the given orthographic views and add the top view so that the final sketch includes a front, top, and right-side view.

B. Prepare a three-dimensional sketch of the object.

EX5-95

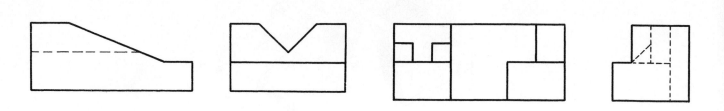

EX5-96

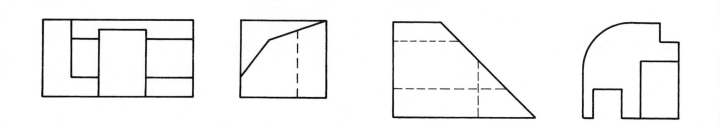

EX5-97

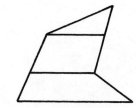

EX5-98

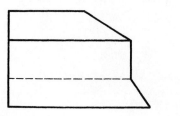

EX5-99

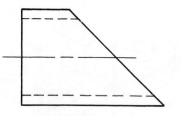

EX5-100

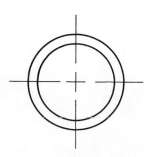

For Exercise Problems EX5-101 to EX5-128:

A. Redraw the given views and draw the third view.
B. Prepare a three-dimensional sketch of the object.

EX5-101 INCHES

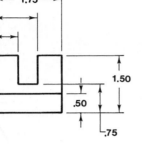
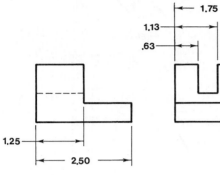

EX5-104 INCHES

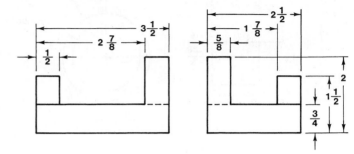

EX5-102 INCHES

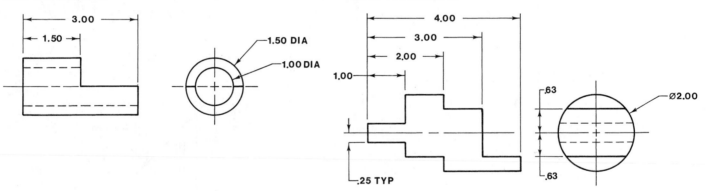

EX5-105 INCHES

EX5-103 INCHES

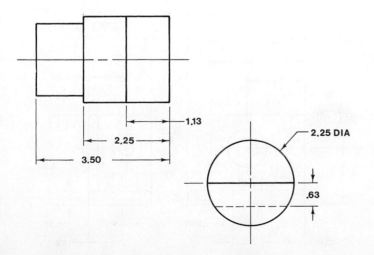

EX5-106 INCHES

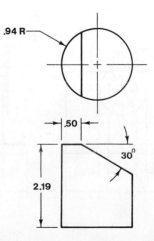

EX5-107 INCHES

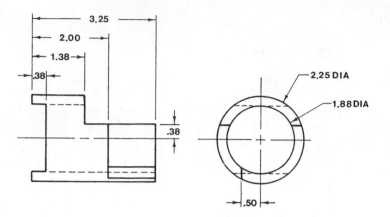

EX5-108 INCHES

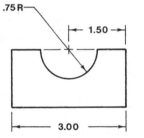

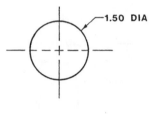

EX5-109 INCHES

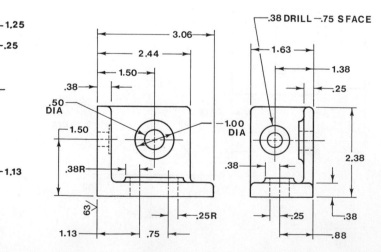

EX5-110 INCHES

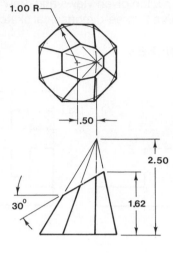

EX5-111 INCHES

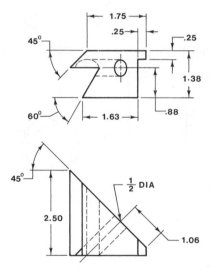

EX5-112 INCHES

All fillets and rounds = $\frac{1}{8}$R

Each exercise on this page is presented on a 10×10–mm grid.

EX5-115

EX5-113

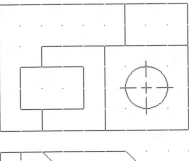
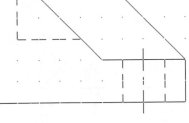

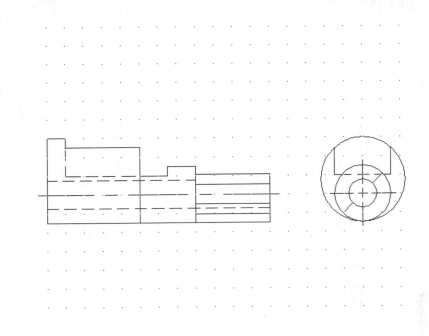

EX5-114

EX5-116

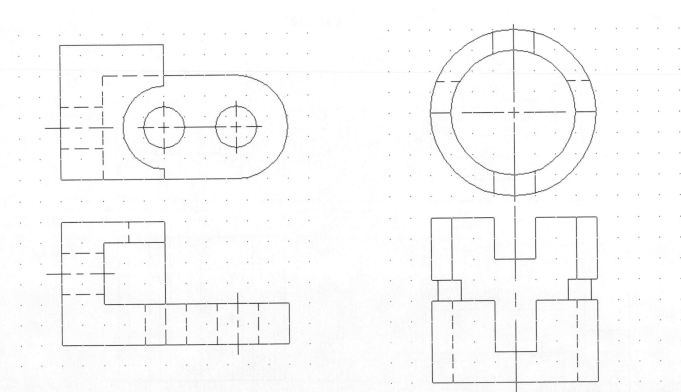

Each exercise on this page is presented on a .50″ × .50″ grid.

EX5-117

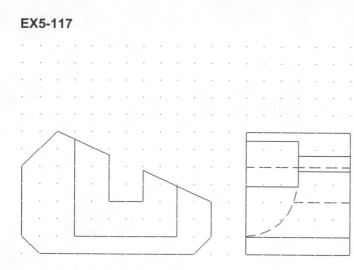

EX5-118

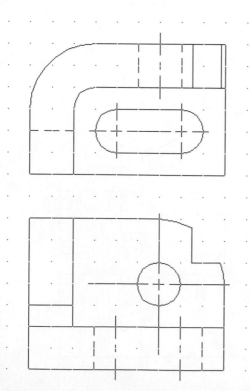

EX5-119

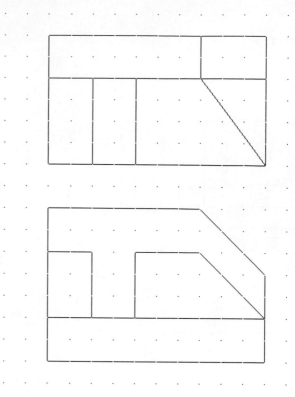

EX5-120

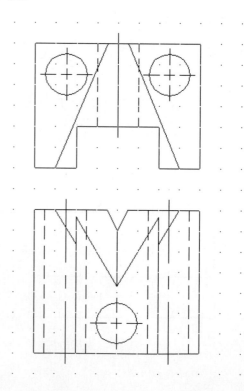

Each exercise on this page is presented on a .50″ × .50″ grid.

EX5-121

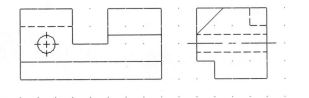

EX5-123

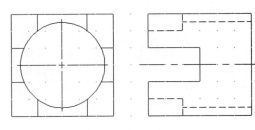

EX5-122

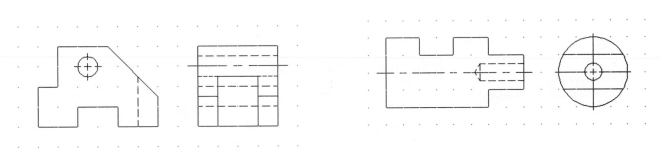

EX5-124

Each exercise on this page is presented on a 10 ×
10–mm grid.

EX5-125

EX5-127

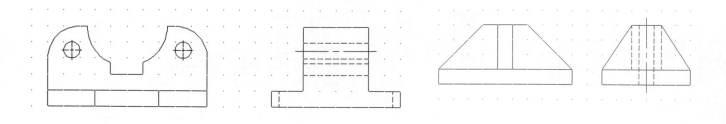

EX5-126

EX5-128

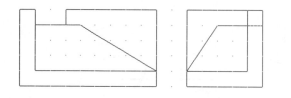

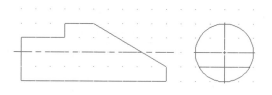

Draw the complete front, top, and side views of the two intersecting objects given in Exercise Problems EX5-129 to EX5-134 based on the given complete and partially complete orthographic views.

EX5-129 MILLIMETERS

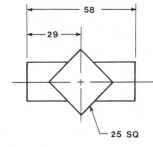

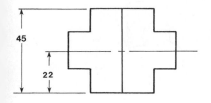

EX5-131 INCHES

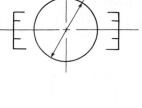

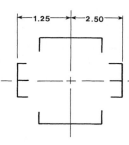

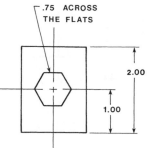

EX5-130 MILLIMETERS

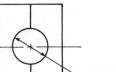

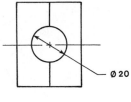

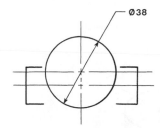

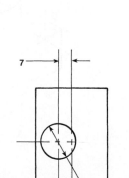

EX5-132 INCHES

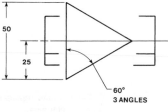

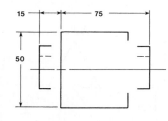

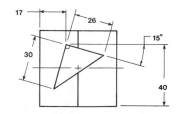

EX5-133 MILLIMETERS

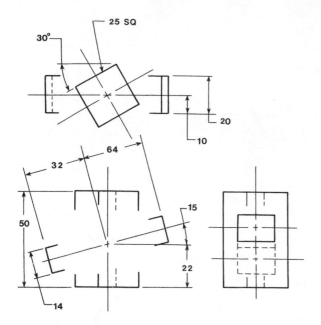

Draw the front, top, and side orthographic views of the objects given in Exercise Problems EX5-135 to EX5-138 based on the partially complete isometric drawings.

EX5-135 MILLIMETERS

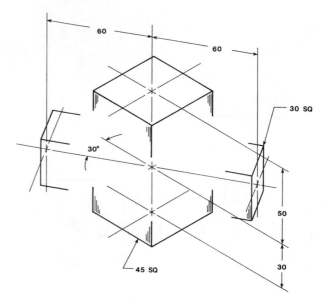

EX5-134 MILLIMETERS

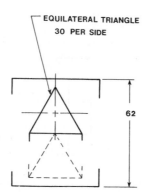

EX5-136 MILLIMETERS

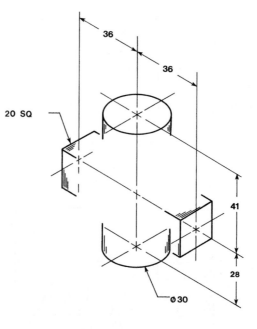

EX5-137 INCHES

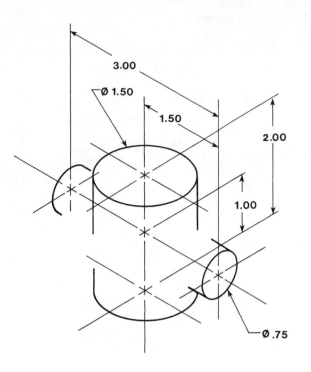

EX5-138 INCHES (SCALE: 5=1)

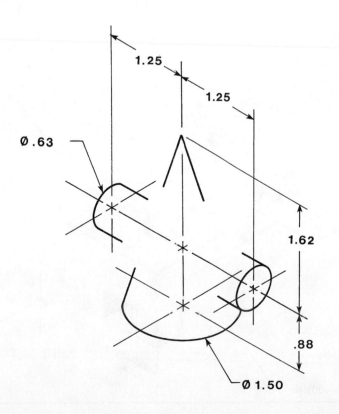

Redesign the existing objects as indicated in Exercise Problems EX5-139 to EX5-142, then prepare a front, top, and right-side view of the object.

EX5-139 MILLIMETERS

1. Replace the four existing 12-wide slots with seven slots 16 wide.
2. Increase the distance between the slots from 20 to 24.
3. Modify the overall length as needed.
4. Increase the size of the 60 slot so that it is 20 from both ends.

Change to 16-7 PLACES.

Modify as needed.

12—4 PLACES

Modify as needed.

ALL FILLETS AND ROUNDS = R5

Maintain at both ends.

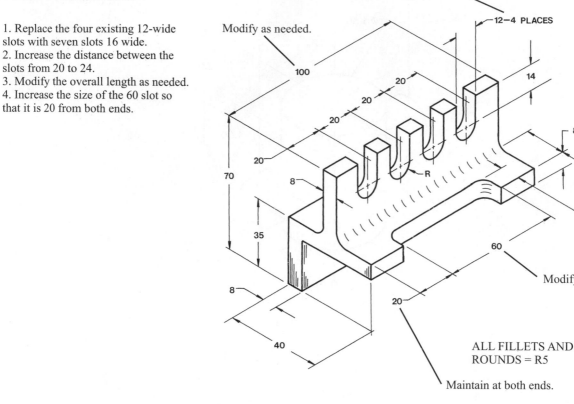

EX5-140 MILLIMETERS

Modify to
Ø16-23 DEEP
Ø30-82°CSK.

Modify to
Ø10
Ø16CBORE - 8 DEEP.

Modify to 80.

Modify to 130.

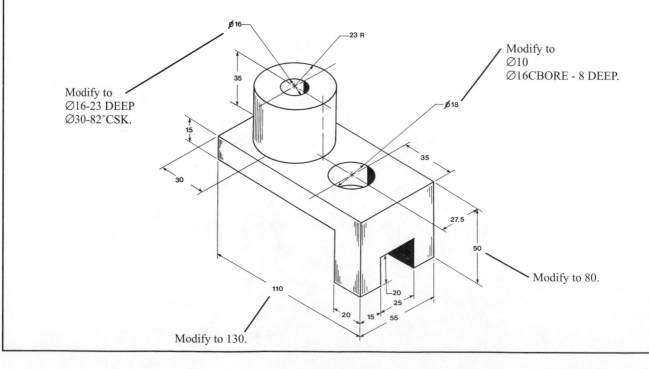

EX5-141 MILLIMETERS

Ø 25 PERPENDICULAR
TO BACK SURFACE

Modify to a ∅20 hole perpendicular
to the slanted surface.

60°

50

60

7
2 PLACES

23

35

16

Round the four front
corners using R10.

18

23 – 2 HOLES

14
2 HOLES

Ø 9 – 2 HOLES

EX5-142 MILLIMETERS

Ø 16 THRU

Ø 30 CBORE
× 3 DEEP

12 R
TYP

Ø 12 THRU
2 HOLES

130

65

10 TYP

Ø 100

Ø 75 × 5 DEEP

Modify to 15 DEEP.

55

Modify so that there are 6 standoffs
equally spaced around the top
surface of the object.

Ø 30 CBORE
× 10 DEEP

Modify to ∅30-82° CSK.

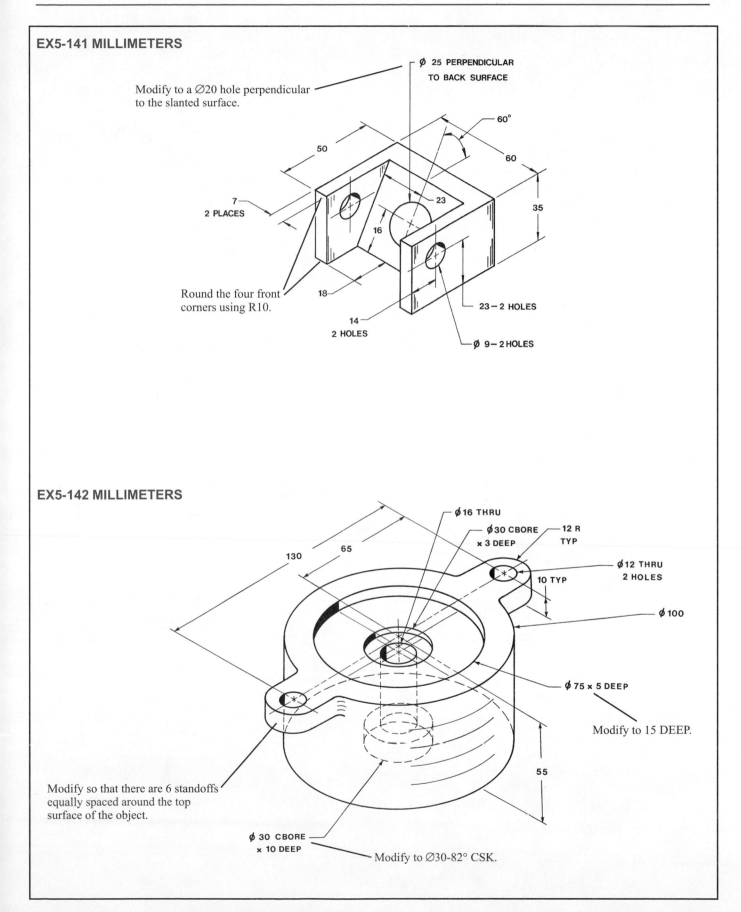

Sectional Views

6-1 INTRODUCTION

Sectional views are used in technical drawing to expose internal surfaces. They serve to present additional orthographic views of surfaces that appear as hidden lines in the standard front, top, and side orthographic views.

Figure 6-1A shows an object intersected by a cutting plane. Figure 6-1B shows the same object with its right side and the cutting plane removed. Hatch lines are drawn on the surfaces that represent where the cutting plane passed through solid material. Also shown are the front and right-side orthographic views and a sectional view. Note the

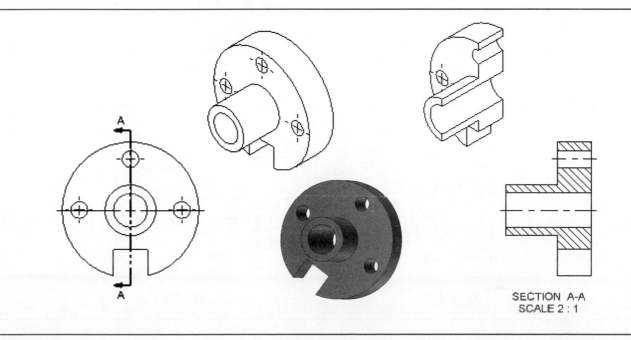

SECTION A-A
SCALE 2 : 1

Figure 6-1A

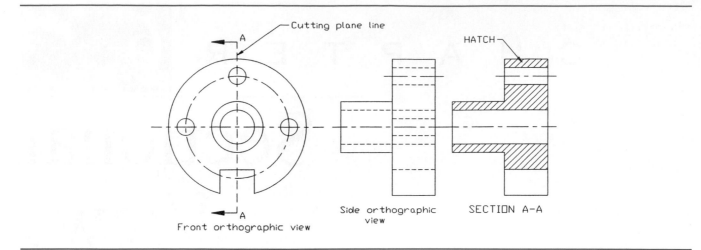

Figure 6-1B

similarity between the sectional view and the cut pictorial view.

Sectional views do not contain hidden lines, so they are used to clarify orthographic views that are difficult to understand because of excessive hidden lines. Figure 6-2 shows an object with a complex internal shape. The standard right view contains many hidden lines and is difficult to follow. Note how much easier it is to understand the object's internal shape when it is presented as a sectional view.

Sectional views *do* include all lines that are directly visible. Figure 6-3 shows an object, a cutting plane line, and a sectional view taken along the cutting plane line. Surfaces that are directly visible are shown in the sectional view. For example, part of the large hole in the back left surface is, from the given sectional view's orientation, blocked by the shorter rectangular surface. The part of the hole that appears above the blocking surface is shown; the part behind the surface is not.

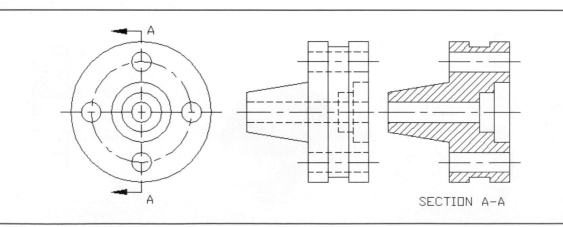

Figure 6-2

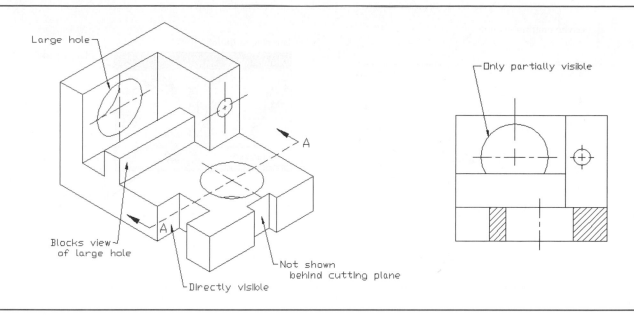

Figure 6-3

Sectional views are always viewed in the direction defined by the cutting plane arrows. Any surface that is behind the cutting plane is not included in the sectional view. Sectional views are aligned and oriented relative to the cutting plane lines as a side orthographic view is to the front view. The orientation of a sectional view may be better understood by placing your right hand on the cutting plane line so that your thumb is pointing up. Move your hand to the right and place it palm down. Your thumb should now be pointing to the left. Your thumb indicates the top of the view.

Drafters and designers often refer to sectional views as "sectional cuts" or simply "cuts." The terminology is helpful in understanding how sectional views are defined and created.

6-2 CUTTING PLANE LINES

Cutting plane lines are used to define the location for the sectional view's cutting plane. An object is "cut" along a cutting plane line.

Figure 6-4 shows two linetype patterns for cutting plane lines. Either pattern is acceptable, although some companies prefer to use only one linetype for all drawings to ensure a uniform appearance in all their drawings. The dashed line pattern will be used throughout this book.

The two patterns shown in Figure 6-4 are included in AutoCAD's linetype library as **Dashed** and **Phantom** styles.

The arrow portion of the cutting plane line is created using the **Leader** tool listed within the **Dimension** toolbar, or by creating a wblock that includes an arrowhead and extension line.

To draw a cutting plane line—Method I

Change a given continuous line, A-A, to a cutting plane line. See Figure 6-5.

1. Select **Format** (pull-down), **Linetype.**

   The **Linetype Manager** dialog box will appear.

2. Select **Load.**

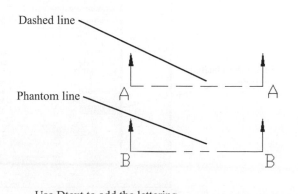

Use Dtext to add the lettering.

Figure 6-4

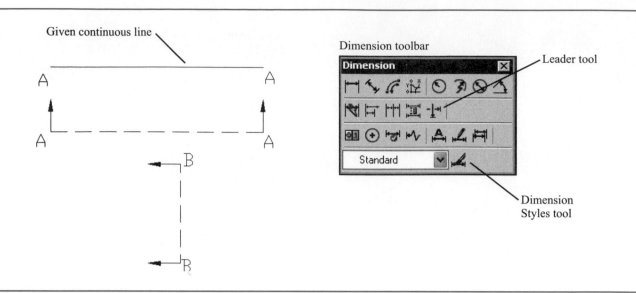

Figure 6-5

The **Load or Reload Linetypes** dialog box will appear. See Figure 6-6.

3. Select **Dashed** or **Phantom,** then return to the drawing screen.
4. Select **Modify** (pull-down), **Properties.**

   Select object(s):

5. Select line **A-A.**

   The **Change Properties** dialog box will appear.

6. Select **Linetype.**

   The **Linetype Manager** dialog box will appear. See Figure 6-7.

7. Select **Dashed** and return to the drawing screen.

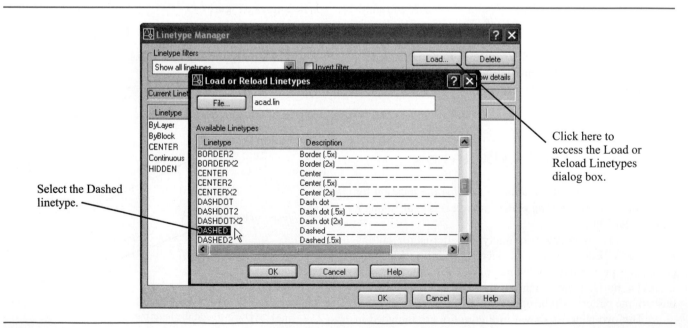

Figure 6-6

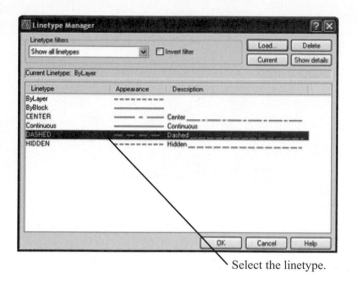

Select the linetype.

Figure 6-7

Arrowhead scale factor

Figure 6-8

To draw an arrowhead

1. Select the **Leader** tool from the **Dimension** toolbar.

   Command: _qleader
   Specify first leader point, or [Settings] <Settings>:

2. Select the location for the tip of the arrow.

   Specify next point:

3. Select the end of the line, then right-click the mouse.

   Specify text width <0.000>:

4. Press **Enter.**

   Enter first line of annotation text <Mtext>:

5. Press **Enter.**

   The **Text Formatting** dialog box will appear.

6. Click **OK.**

   If necessary, use the **Rotate** command to align the leader line with the horizontal or vertical direction.

To change the size of an arrowhead

Arrowheads used for cutting plane lines are usually drawn larger than arrowheads used for dimension lines. This serves to make them more distinctive and easier to find. The normal arrowhead scale factor is 0.19. In this example, the scale factor is 0.375.

1. Select the **Dimension Style** tool from the **Dimension** toolbar.
2. Select **Modify.**

   The **Modify Dimension Style: Standard** dialog box will appear. See Figure 6-8.

3. Select the **Symbols and Arrows** tab and use the **Arrow size** box within the **Arrowheads** portion of the dialog box to change the arrowhead's scale factor (change to **0.3750**).
4. Return to the drawing screen and use **Leader** to create the needed arrowheads.

   The arrowhead size can also be changed by typing **dimasz** in response to a command prompt and typing in the new scale factor.

To draw a cutting plane line—Method II

A cutting plane can also be created by first defining a separate layer setup for cutting plane lines. The linetype will be dashed or phantom, and a color can be assigned if desired. Section 3-25 described how to create and work with layers.

To draw cutting plane lines

Cutting plane lines should extend beyond the edges of the object. See Figure 6-9. A cutting plane line should extend far enough beyond the edges of the object so that there is a clear gap between the arrowhead and the edge of the object.

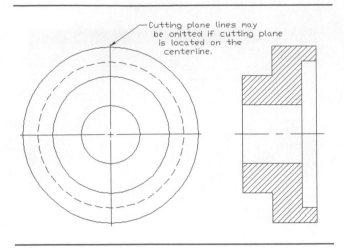

Figure 6-9

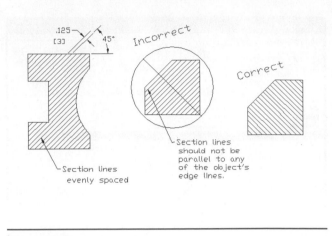

Figure 6-10

If an object is symmetrical about the centerline and only one sectional view is to be taken exactly aligned with the centerline, the cutting plane line may be omitted.

6-3 SECTION LINES

Section lines are used to define areas that represent where solid material has been cut in a sectional view. Section lines are evenly spaced at any inclined angle that is not parallel to any existing edge line and should be visually distinct from the continuous lines that define the boundary of the sectional view.

Figure 6-10 shows an area that includes uniform section lines evenly spaced at 45°. The other area shown in

Figure 6-10 includes a 45° edge line; therefore, the section lines cannot be drawn at 45°. Lines at 135° (0° is horizontal to the right) were drawn instead.

Figure 6-11 shows an object that contains edge lines at both 45° and 135°. The section lines within this area were drawn at 60°.

If two or more parts are included within the same sectional view, each part must have visually different section lines. Figure 6-12 shows a sectional view that contains two parts. Part one's section lines were spaced 3 apart at 45°, and part two's were spaced 5 apart at 135° (−45).

The recommended spacing for sectional lines is 0.125 inch or 3 millimeters, but smaller areas may use section lines spaced closer together than larger areas. See Figure 6-13. Section lines should never be spaced so close together as to

Figure 6-11

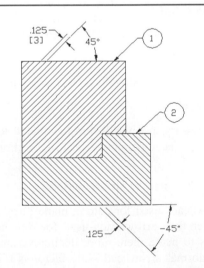

Figure 6-12

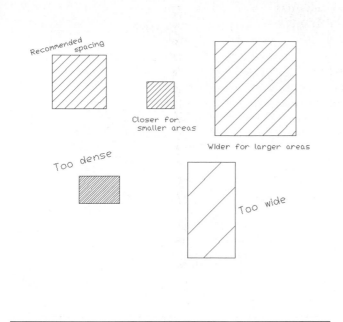

Figure 6-13

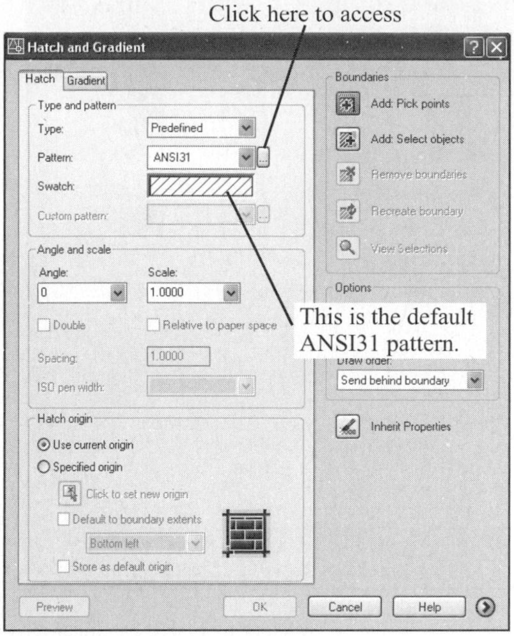

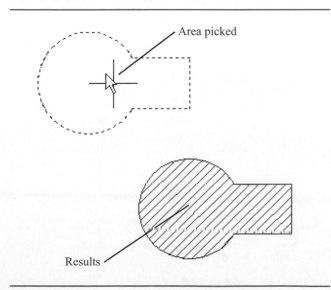

look blurry or be so far apart that they are not clearly recognizable as section lines.

Evenly spaced sectional line patterns drawn at 45° are called *general,* or *uniform,* patterns; other patterns are available. Different section line patterns are used to help distinguish different materials. The different patterns allow the drawing reader to see what materials are used in a design without having to refer to the drawing's parts list. It should be noted that not all companies use different patterns to define material differences. When you are in doubt, the general pattern is usually acceptable.

How to draw different section line patterns is explained in Section 6-4.

Figure 6-14

6-4 HATCH

Section lines are drawn in AutoCAD using the **Hatch** tool, which is located on the **Draw** toolbar. The **Hatch** tool offers many different hatch patterns and spacings. The general pattern of evenly spaced lines at 45° is defined as pattern ANSI31 and is the default setting for the **Hatch** tool.

To hatch a given area

Given an area, use **Hatch** to draw section lines.

1. Select the **Hatch** tool from the **Draw** toolbar.

   The **Hatch and Gradient** dialog box will appear. See Figure 6-14. Note that **ANSI31** is the default pattern.

2. Select the **Pick Points** option.

   Command: _bhatch
   Select internal point:

3. Select a point within the area to be hatched, then right-click the mouse, then select **Enter.**

   The **Hatch and Gradient** dialog box will reappear.

4. Select the **OK** option.

   The area will be hatched. See Figure 6-15.

To change hatch patterns

1. Select the **Hatch** tool from the **Draw** toolbar.

Figure 6-15

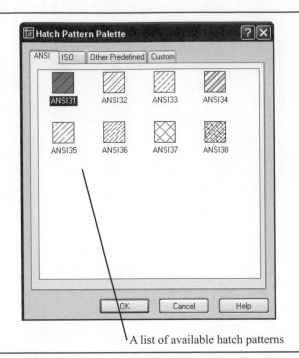

A list of available hatch patterns

Figure 6-16

The **Hatch and Gradient** dialog box will appear.

2. Click the arrow to the right of the **Pattern ANSI31** designation.

A list of available ANSI patterns will appear. See Figure 6-16.

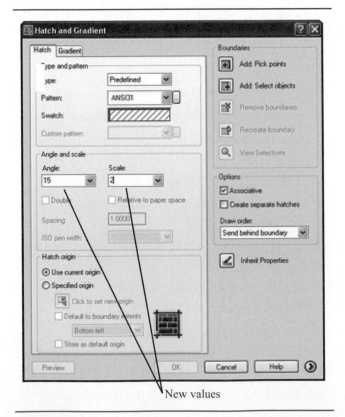

New values

Figure 6-18

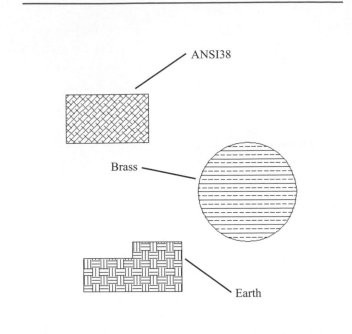

Figure 6-17

3. Select the desired pattern and apply it as described above.

A swatch of the selected pattern will appear on the **Hatch and Gradient** dialog box. Figure 6-17 shows three areas with three different hatch patterns applied.

To change the spacing and angle of a hatch pattern

1. Select the **Hatch** tool from the **Draw** toolbar.

The **Hatch and Gradient** dialog box will appear. See Figure 6-18. The **Angle** and **Scale** option boxes are located just below the **Custom pattern** box.

ANSI31 pattern

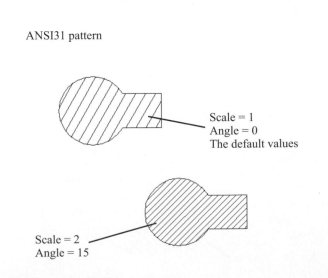

Figure 6-19

2. Change the **Angle** to **15** and the **Scale** to **2**.

Figure 6-19 shows a comparison between an ANSI31 pattern created using the default values of 1 and 0°, and the adjusted pattern of 2 and 15°.

6-5 SAMPLE PROBLEM SP6-1

Figure 6-20 shows an object with a cutting plane line. Figure 6-21 shows how a front view and a sectional view of the object are created.

1. Draw the front orthographic view of the object and lay out the height and width of the object based on the given dimensions. Sectional views must be directly aligned with the angle of the cutting plane line. Sectional views are in fact orthographic views that are based on the angle of the cutting plane line. Information is projected to sectional views as it was projected to top and side views. In this example the cutting plane line is a vertical line, so horizontal projection lines are used to draw the sectional view.
2. Draw the object features. No hidden lines are drawn.
3. Use **Hatch** to draw sectional lines within the appropriate areas.
4. Erase or trim any excess lines.
5. Modify the color of the sectional lines if desired.
6. Save the drawing if desired.

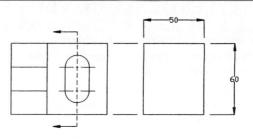

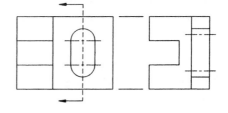

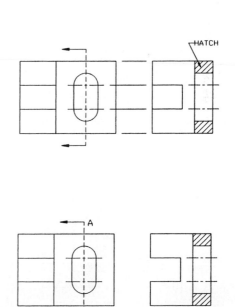

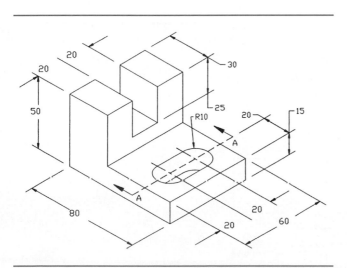

Figure 6-20

Figure 6-21

6-6 STYLES OF SECTION LINES

AutoCAD has more than 50 different hatch patterns. The different patterns can be previewed in the **Swatch** option in the **Hatch and Gradient** dialog box.

Most of the available options are for architectural use. The patterns can be used to create elevation drawings and to add texture patterns to drawings of houses, buildings, and other structures. Technical drawings usually refer to objects made from steel, aluminum, or a composite material. There are hundreds of variations of each of these materials.

In general, if you decide to assign a particular pattern to a material, clearly state which pattern has been assigned to which material on the drawing. This is best done by including a note on the drawing that includes a picture of the pattern and the material that it is to represent. Figure 6-22 shows a drawing note for a hatch pattern used to represent SAE 1040 steel. The representation is unique for the drawing shown.

6-7 SECTIONAL VIEW LOCATION

Sectional views should be located on a drawing behind the arrows. The arrows represent the viewing direction for the sectional view. See Figure 6-23. If it is impossible to locate sectional views behind the arrows, they may be located above or below, but still behind, the arrowed portion of the cutting plane line. Sectional views should never be located in front of the arrows.

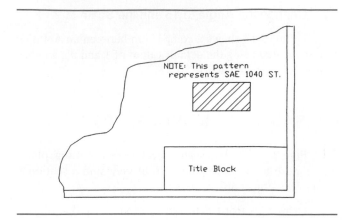

NOTE: This pattern represents SAE 1040 ST.

Title Block

Figure 6-22

Sectional views located on a different drawing sheet than the cutting plane line must be cross-referenced to the appropriate cutting plane line. See Figure 6-24. The boxed notation C/3 SHT2 next to the cutting plane line means that the sectional view may be found in zone C/3 on sheet 2 of the drawing.

The location of the sectional view must be referenced to the cutting plane line if the view's location is not near the cutting plane. The notation C/3 SHT2 defines the location of the cutting plane line for the sectional view.

The reference numbers and letters are based on a drawing area charting system similar to that used for locating

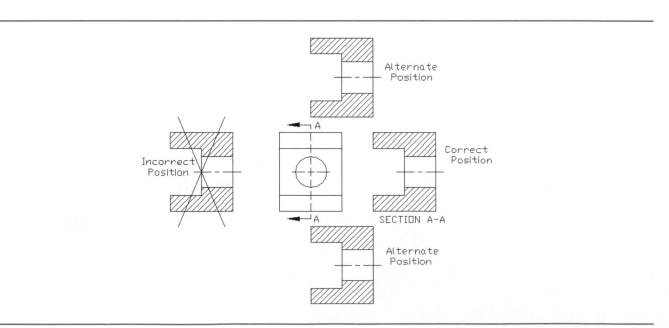

Figure 6-23

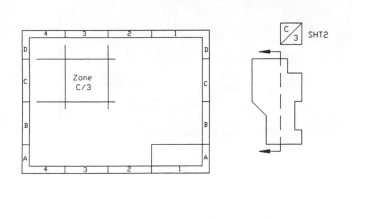

Figure 6-24

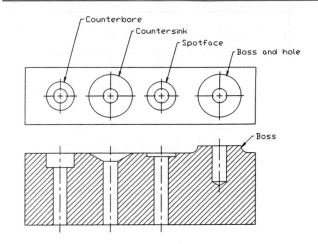

Figure 6-26

features on maps. A sectional view located at C/3 SHT2 can be found by going to sheet 2 of the drawing, then drawing a vertical line from the box marked 3 at the top or bottom of the drawing and a horizontal line from the box marked C on the left or right edge of the drawing. The sectional view should be located somewhere near the intersection of the two lines.

6-8 HOLES IN SECTIONS

Figure 6-25 shows a sectional view of an object that contains three holes. As with orthographic views, a conical

point must be included on holes that do not completely penetrate the object.

A common mistake is to omit the back edge of a hole when drawing a sectional view. Figure 6-25 shows a hole drilled through an object. Note that the sectional view includes a straight line across the top and bottom edges of the view that represents the back edges of the hole.

Figure 6-26 shows a countersunk hole, a counterbored hole, a spotface, and a hole through a boss. Again, any hole that does not completely penetrate the object must include a conical point.

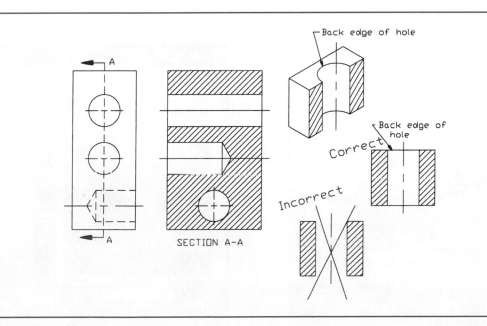

Figure 6-25

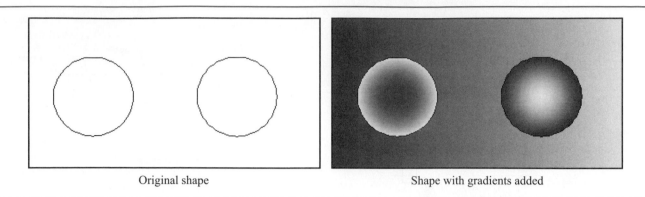

| Original shape | Shape with gradients added |

Figure 6-27

6-9 GRADIENTS

In AutoCAD a ***gradient*** is shading that varies in intensity. Figure 6-27 shows a rectangular shape with two internal circles. The figure also shows the same shape with various gradients added.

To create a gradient

1. Select the **Gradient** tool from the **Draw** toolbar.

   The **Hatch and Gradient** dialog box will appear. See Figure 6-28.

2. Select the **Add: Select objects** button.

The dialog box will disappear.

3. Select one of the circles by clicking its edge line.

4. Right-click the mouse and select the **Enter** option.

   The **Hatch and Gradient** dialog box will appear.

5. Select a gradient pattern, then click **OK.**

A gradient will appear within the circle. Use the **Send behind boundary** options to control drawings that include more than one gradient.

Click on the box to the right of the **One color** bar to change gradient colors. Select a different color by clicking on the desired color, as shown within the square field of color on the **Select Color** dialog box.

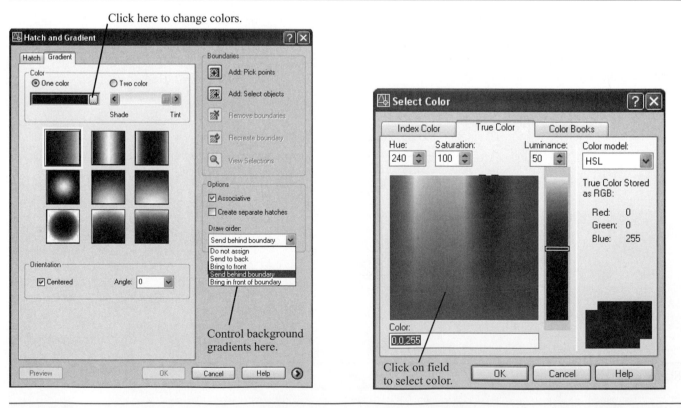

Figure 6-28

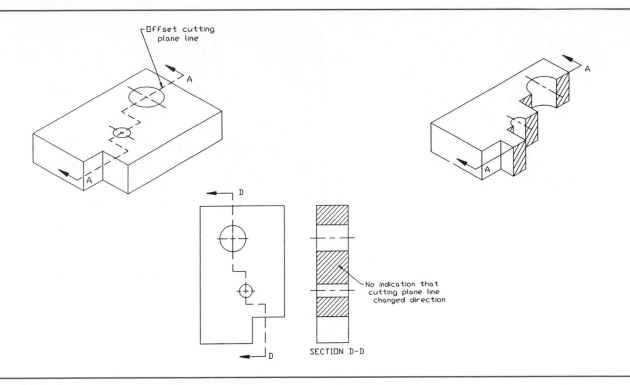

Figure 6-29

6-10 OFFSET SECTIONS

Cutting plane lines need not be drawn as straight lines across the surface of an object. They may be stepped so more features can be included in the sectional view. See Figure 6-29. In the object shown in Figure 6-29, the cutting plane line crosses two holes and a directly visible open surface that contains a hole. There is no indication in the sectional view that the cutting plane line has been offset.

Cutting plane lines should be placed to include as many features as possible without causing confusion. The intent of using sectional views is to simplify views and to clarify the drawing. Several features can be included on the same cutting plane line so that fewer views are used, giving the drawing a less cluttered look and making it easier for the reader to understand the shape of the object's features.

Figure 6-30 shows another example of an offset cutting plane line.

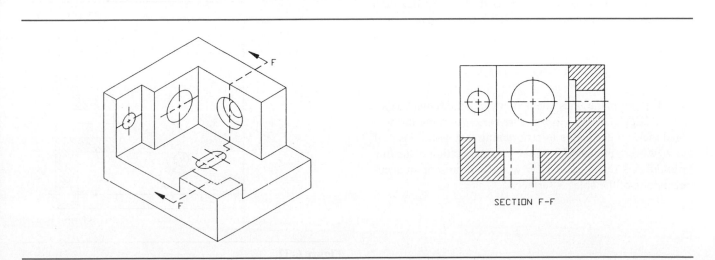

Figure 6-30

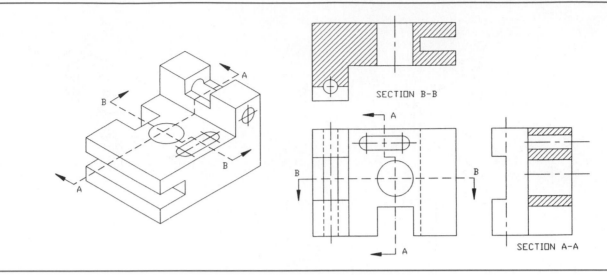

Figure 6-31

6-11 MULTIPLE SECTIONS

More than one sectional view may be taken off the same orthographic view. Figure 6-31 shows a drawing that includes three sectional views, all taken off a front view. Each cutting plane line is labeled with a letter. The letters are in turn used to identify the appropriate sectional view. Identifying letters are also placed at the ends of the cutting plane lines, behind the arrowheads. Identifying letters are also placed below the sectional view and written using the format SECTION A-A, SECTION B-B, and so on. The abbreviation SECT may also be used: SECT A-A, SECT B-B.

The letters I, O, and X are generally not used to identify sectional views because they can easily be misread. If more than 23 sectional views are used, the lettering starts again with double letters: AA-AA, BB-BB, and so on.

6-12 ALIGNED SECTIONS

Cutting plane lines taken at angles on circular shapes may be aligned as shown in Figure 6-32. Aligning the sectional views prevents the foreshortening that would result if the view were projected from the original cutting plane line location. A foreshortened view would not present an accurate picture of the object's surfaces.

6-13 DRAWING CONVENTIONS IN SECTIONS

Slots and small holes that penetrate cylindrical surfaces may be drawn as straight lines, as shown in Figure 6-33. Larger holes, that is, holes whose diameters are greater than the radii of their cylinders, should be drawn showing an elliptical curvature.

Intersecting holes are represented by crossed lines, as shown in Figure 6-34. The crossed lines are drawn from the intersecting corners of the holes.

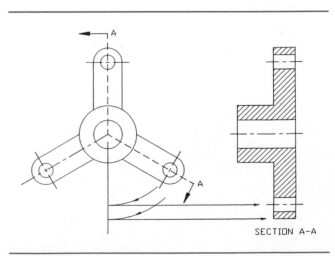

Figure 6-32

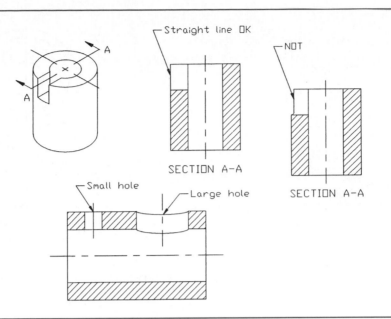

Figure 6-33

6-14 HALF, PARTIAL, AND BROKEN-OUT SECTIONAL VIEWS

Half and partial sectional views allow a designer to show an object using an orthographic view and a sectional view within one view. A half-sectional view is shown in Figure 6-35. Half the view is a sectional view; the other half is a normal orthographic view including hidden lines. The cutting plane line is drawn as shown and includes an arrowhead at only one end of the line.

Figure 6-36 shows a partial sectional view. It is similar to a half-sectional view, but the sectional view is taken at a location other than directly on a centerline or one defined by a cutting plane line. A broken line is used to separate the sectional view from the orthographic view. A broken line is a freehand line drawn using the **Sketch** command.

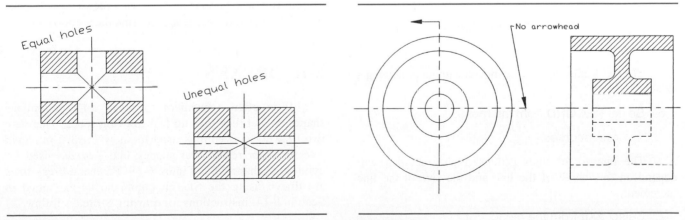

Figure 6-34

Figure 6-35

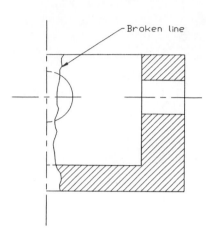

Figure 6-36

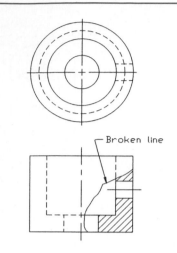

Figure 6-37

Broken-out sectional views are like small partial views. They are used to show only small internal portions of an object. Figure 6-37 shows a broken-out sectional view. Broken lines are used to separate the broken-out section from the rest of the orthographic views.

To draw a broken line

1. Type **Sketch** in response to a command prompt.

   Command:_sketch

   Record Increment <0.1000>:

2. Press **Enter.**

   Sketch. Pen/eXit/Quit/Record/Erase/Connect:

3. Press the left mouse button.

   <Pen down>

4. Move the cursor across the screen. A green freehand line will emerge from the crosshairs as it is moved.

5. Press the left mouse button again.

   <Pen up>

   The crosshairs can now be moved without producing a sketched line.

6. Select **RECORD** from the menu.

   23 lines recorded

   The number of lines generated and recorded will depend on the length of the line and the size of the line increment.

7. Select **eXit** from the menu.

   Command:

Do *not* use the right mouse button to exit the **Sketch** command. If the right mouse button is pressed, a line will be drawn from the end of the sketched line to the present location of the crosshairs. Use **eXit** to end the **Sketch** command.

6-15 REMOVED SECTIONAL VIEWS

Removed sectional views are used to show how an object's shape changes over its length. Removed sectional views are most often used with long objects whose shape changes continuously over its length. See Figure 6-38. The sectional views are not positioned behind the arrowheads but are positioned across the drawing as shown; however, the view orientation is the same as it would be if the view were projected from the arrowheads; it is simply located in a different position in the drawing.

It is good practice to identify the cutting plane lines and the sectional views in alphabetical order. This will make it easer for the drawing's readers to find the sectional views.

6-16 BREAKS

It is often convenient to break long continuous shapes so that they take up less drawing space. There are two drawing conventions used to show breaks: *freehand lines* used for rectangular shapes, and *S-breaks* used for cylindrical shapes. See Figure 6-39. Freehand break lines are drawn using the **Sketch** command as explained in Section 6-14. Instructions for drawing S-breaks follow.

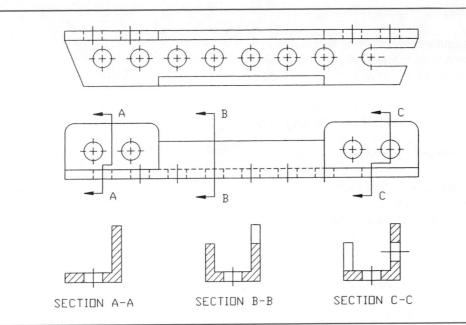

SECTION A-A   SECTION B-B   SECTION C-C

Figure 6-38

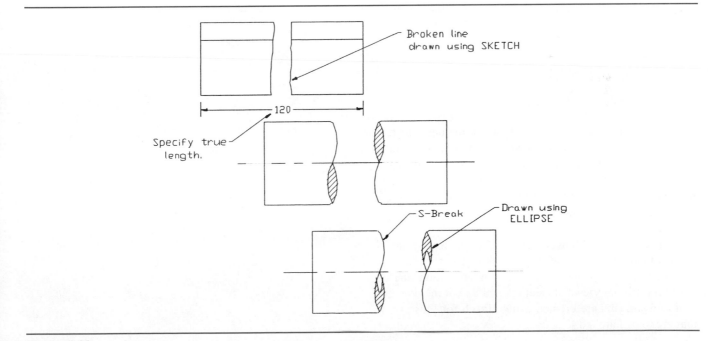

Broken line
drawn using SKETCH

120

Specify true
length.

S-Break

Drawn using
ELLIPSE

Figure 6-39

To draw an S-break (See Figure 6-40.)

1. Draw a rectangular view of the cylindrical object and draw a construction line where the break is to be located. The rectangular view should include a centerline.
2. Draw two **30°** lines: one from the intersection of the construction line and the outside edge line of the view, and the other from the intersection of the construction line and the centerline as shown.
3. Draw an arc using the intersection created in step 2 as the center point. Mirror the arc about the centerline and then about the construction line.
4. Use **Fillet,** set to a small radius, to smooth the corners between the arc and the edge lines. Any small radius may be used for the fillet, provided it produces a smooth visual transition between the arc and the edge lines.
5. Erase and trim any excess lines.
6. Hatch the area created by the two arcs and fillet as shown.

The **Copy, Mirror,** and **Move** commands may be used to create the opposing S-break as shown in Figure 6-39. The internal shapes of the tubular S-breaks are created using **Ellipse.** Position the end of the ellipse according to the thickness specifications, and determine the elliptical shape by eye. Trim the sectional lines from the inside of the ellipse.

6-17 SECTIONAL VIEWS OF CASTINGS

Cast objects are usually designed to include a feature called a *rib.* See Figure 6-41. Ribs add strength and rigidity to an object. Sectional views of ribs do not include complete section lines because this is considered misleading to the reader. Ribs are usually narrow, and a large sectioned area gives the impression of a denser and stronger area than is actually on the casting.

There are two conventions used to present sectional views of cast ribs: one that does not draw any section lines on the ribs, and one that puts every other section line on the rib. Sectional views of castings that do not include section lines on ribs are created using **Hatch** for those areas that are to be hatched.

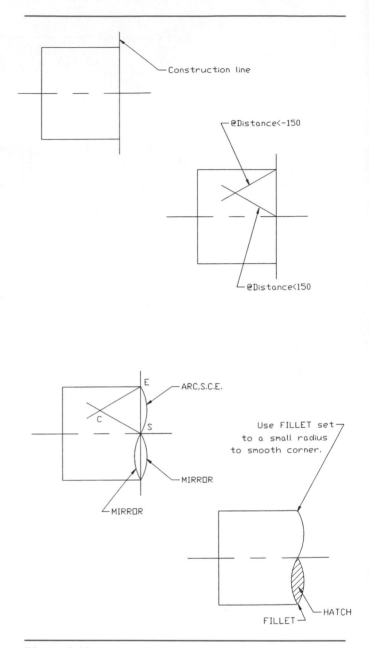

Figure 6-40

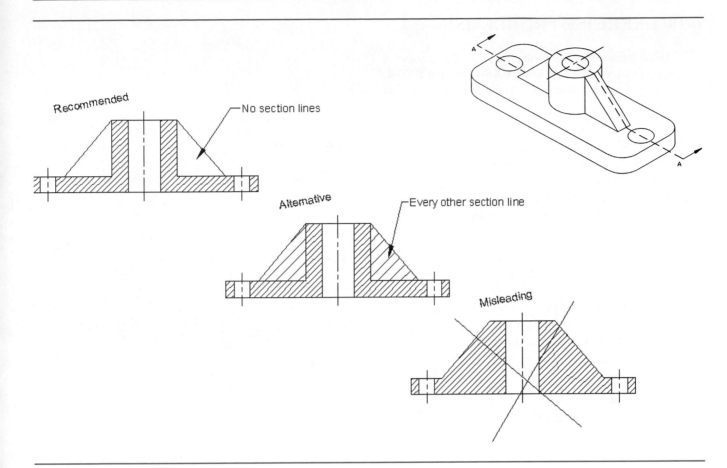

Figure 6-41

6-18 EXERCISE PROBLEMS

Draw a complete front view and a sectional view of the objects in Exercise Problems EX6-1 through EX6-4. The cutting plane line is located on the vertical centerline of the object.

EX6-1 MILLIMETERS

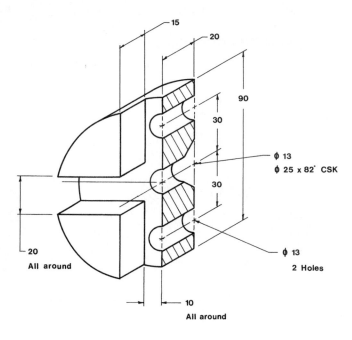

EX6-3 INCHES

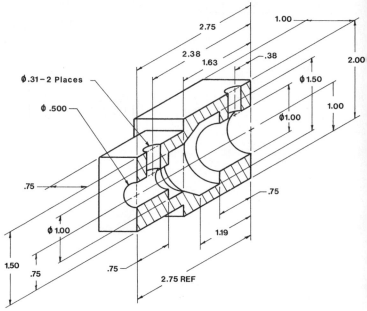

EX6-2 MILLIMETERS

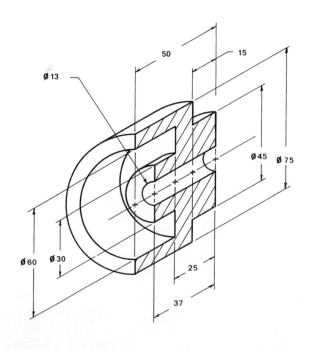

EX6-4 MILLIMETERS

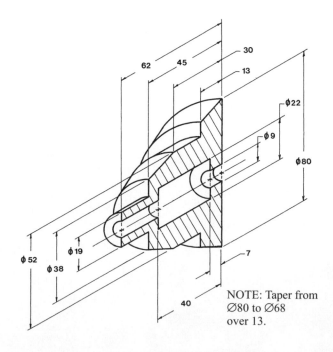

NOTE: Taper from ⌀80 to ⌀68 over 13.

Draw the following sectional views using the given dimensions.

EX6-5

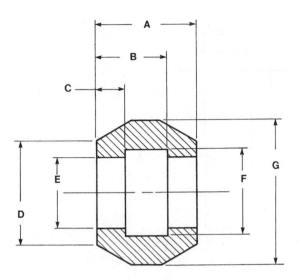

DIMENSIONS	INCHES	mm
A	.50	~~12~~ 44
B	1.25	32
C	1.75	~~44~~ 12
D	Ø1.75	44
E	Ø1.25	32
F	Ø1.50	38
G	Ø2.50	64

EX6-6

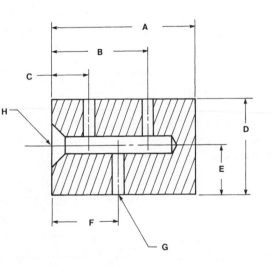

DIMENSIONS	INCHES	mm
A	3.00	72
B	2.00	48
C	.75	18
D	2.00	48
E	1.00	24
F	1.38	33
G	Ø.25	6
H	Ø.375 X 2.50 DEEP Ø.875 X 82° CSINK	Ø10 X 60 DEEP Ø24 X 82° CSINK

Resketch the given top view and Section A-A, then sketch Sections B-B, C-C, and D-D.

EX6-7

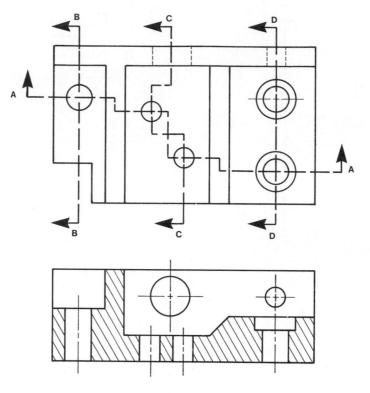

Section A-A

Resketch the given front views in Exercise Problems EX6-8 through EX6-11 and replace the given side orthographic view with the appropriate sectional view.

EX6-9

EX6-8

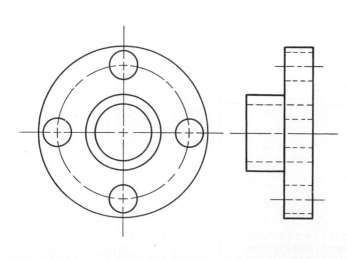

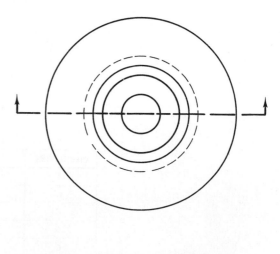

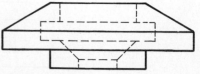

EX6-10 INCHES

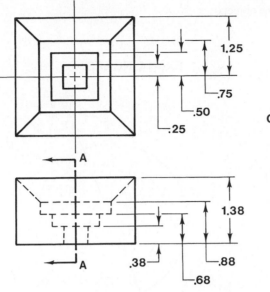

1.25
.75
.50
.25

Object is symmetrical
about both
centerlines

A

A

1.38

.38
.88
.68

EX6-11

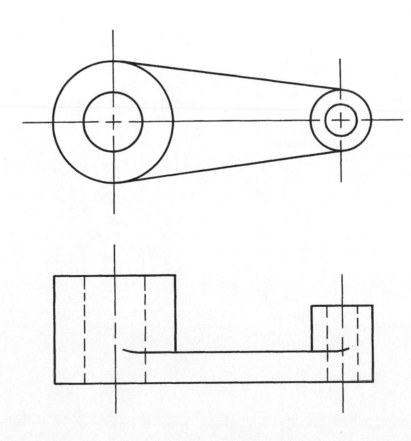

Redraw the given front views in Exercise Problems EX6-12 through EX6-15 and replace the given side orthographic view with the appropriate sectional view. The cutting plane is located on the vertical centerlines of the objects.

EX6-12 INCHES

EX6-14 INCHES

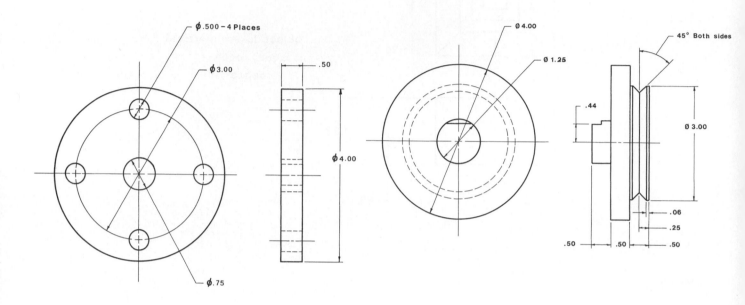

EX6-13 INCHES

EX6-15 MILLIMETERS

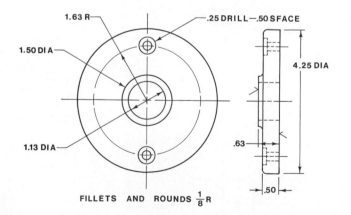

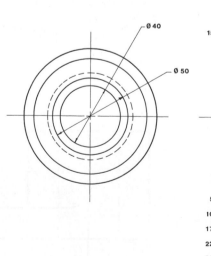

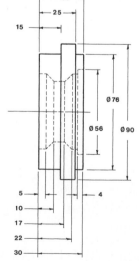

Draw the top orthographic view and the indicated sectional view.

EX6-16 MILLIMETERS

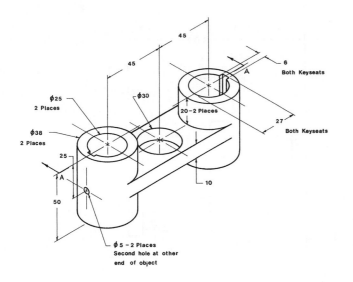

EX6-18 MILLIMETERS

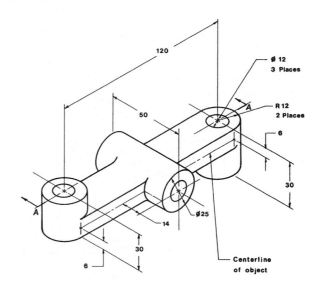

EX6-17 INCHES

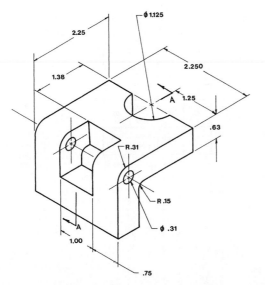

EX6-19 MILLIMETERS

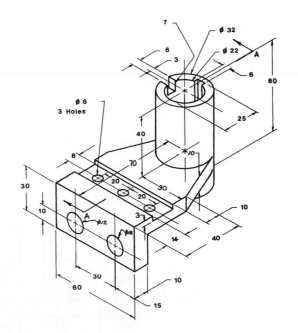

Redraw the given views and add the specified sectional views in Exercise Problems EX6-20 through EX6-34.

EX6-20 INCHES

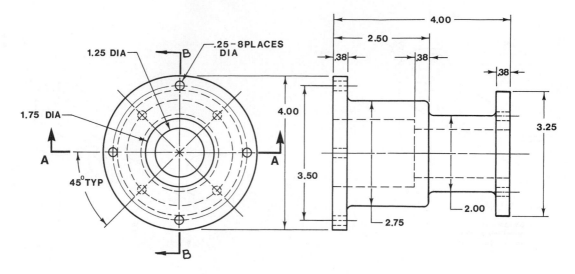

ALL FILLETS AND ROUNDS = R.125

EX6-21 INCHES

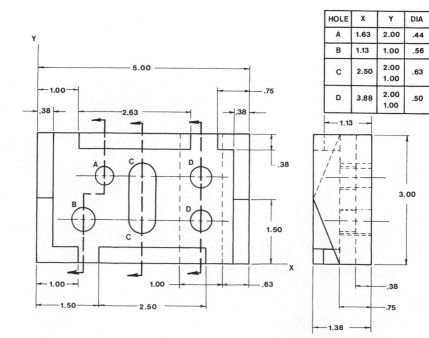

HOLE	X	Y	DIA
A	1.63	2.00	.44
B	1.13	1.00	.56
C	2.50	2.00 1.00	.63
D	3.88	2.00 1.00	.50

EX6-22 INCHES

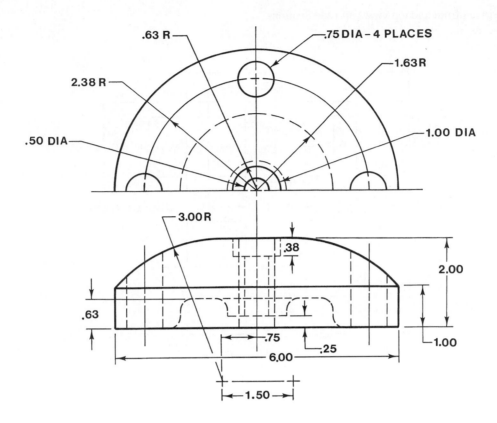

EX6-23 INCHES

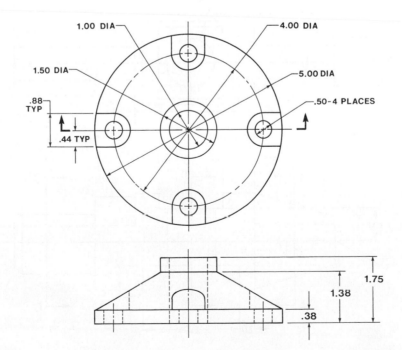

EX6-24 INCHES

Redraw the given front and top views and add Sections
A-A, B-B, and C-C as indicated.

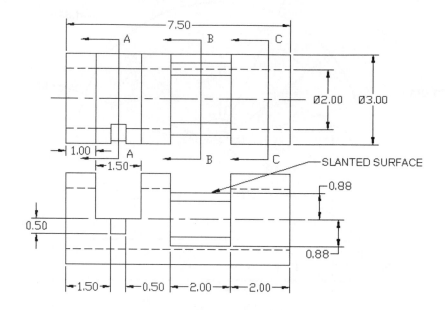

EX6-25 MILLIMETERS

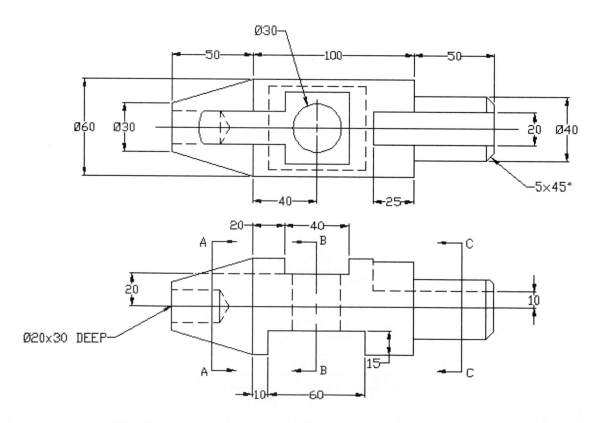

EX6-26 INCHES

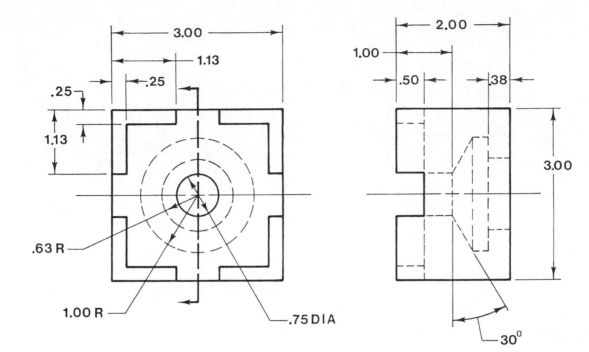

EX6-27 INCHES

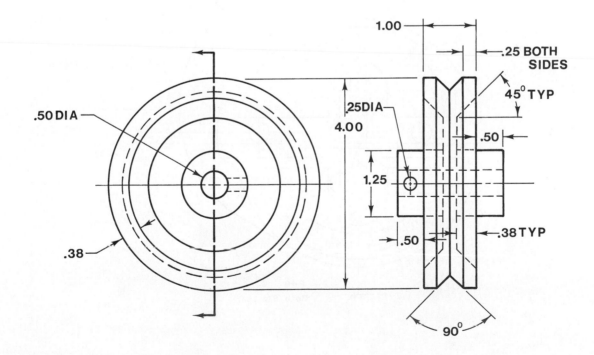

EX6-28 INCHES

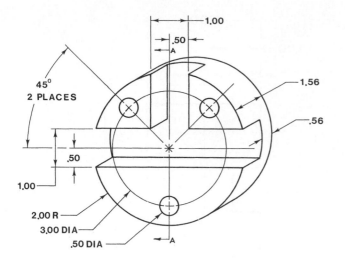

EX6-30 MILLIMETERS

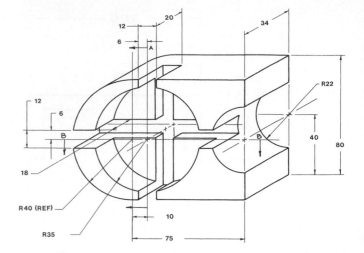

EX6-29 MILLIMETERS

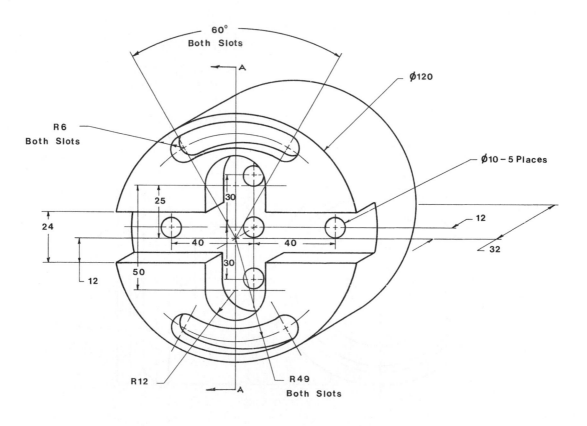

EX6-31 MILLIMETERS

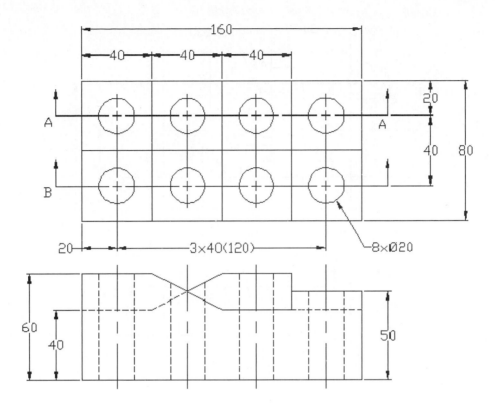

EX6-32 MILLIMETERS

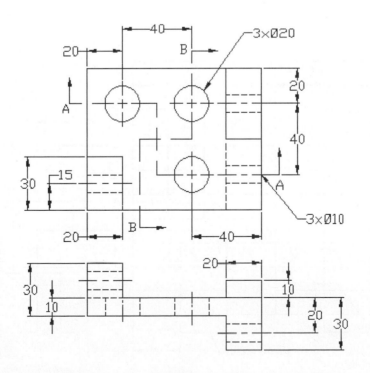

EX6-33 MILLIMETERS

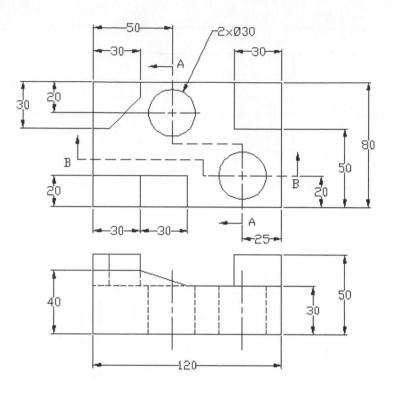

EX6-34 INCHES

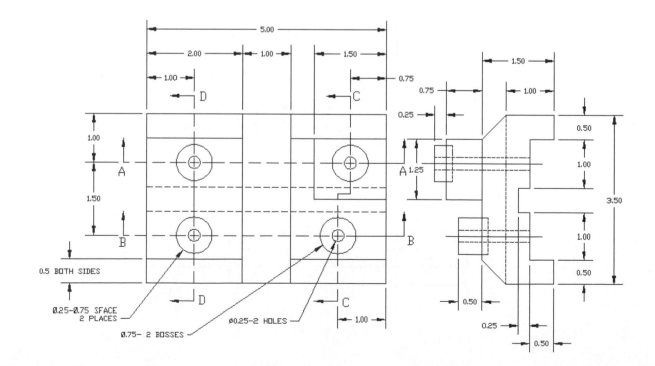

EX6-35

New customer requirements require that the following part be redesigned as follows. Draw a front and a sectional view of the redesigned part.

1. The diameters of the Ø9 internal access holes are to be increased to Ø12.
2. The diameter of the internal cavity is to be increased to Ø30.
3. The length of the internal cavity is to be increased from 33 to 50.
4. All other sizes and distances are to be increased to maintain the same wall thickness.

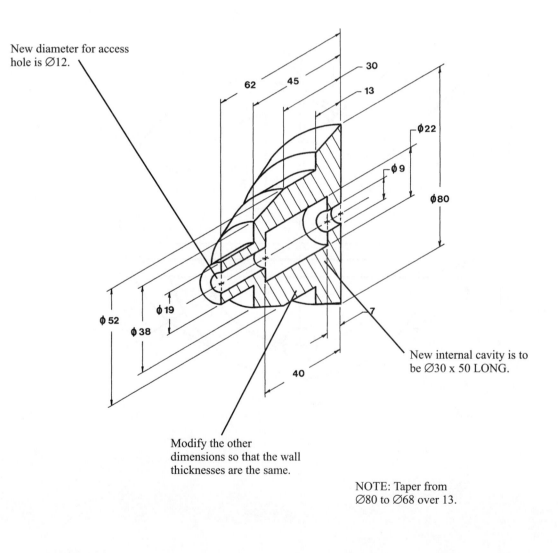

New diameter for access hole is Ø12.

New internal cavity is to be Ø30 x 50 LONG.

Modify the other dimensions so that the wall thicknesses are the same.

NOTE: Taper from Ø80 to Ø68 over 13.

EX6-36

The following object is to be redesigned as follows.

1. Extend the 1.00-inch vertical slot all the way through the object, creating four corner sections.
2. Locate countersunk holes in each of the four corner sections. Note that two of the corner sections presently have Ø.50 holes. The countersink specifications are Ø.50 - 82°, Ø.75.
3. Locate a Ø.625 hole in the center of the object.

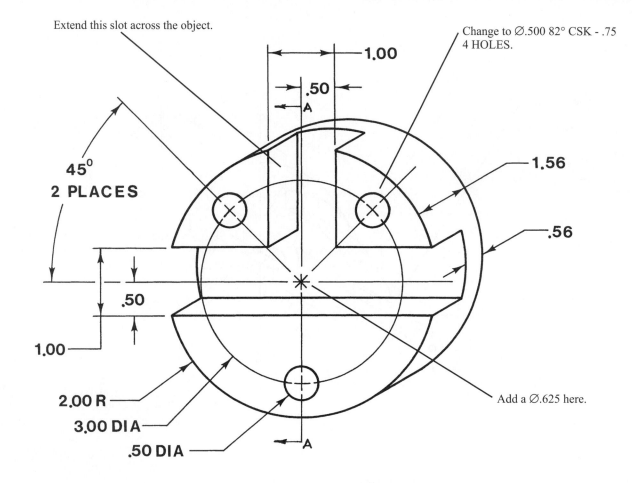

CHAPTER 7

Auxiliary Views

7-1 INTRODUCTION

Auxiliary views are orthographic views used to present true-shaped views of slanted and oblique surfaces. Slanted and oblique surfaces appear foreshortened or as edge views in normal orthographic views. Holes in the surfaces are elliptical, and other features are also distorted.

Figure 7-1 shows an object with a slanted surface that includes a hole drilled perpendicular to that surface. Note how the slanted surface is foreshortened in both the

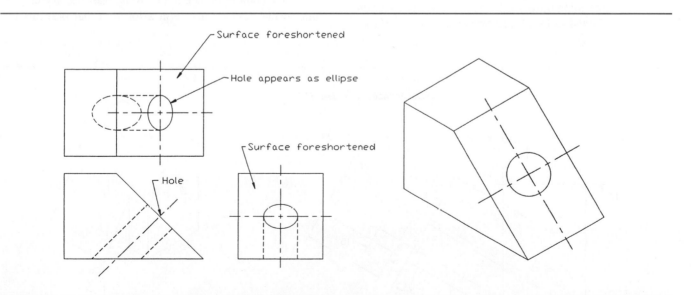

Figure 7-1

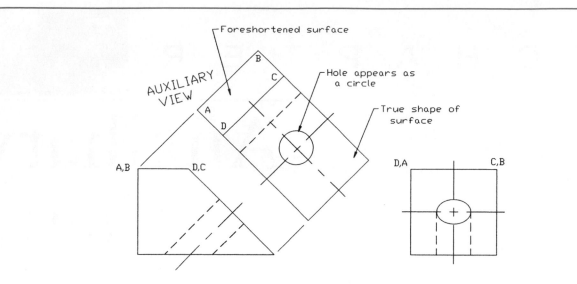

Figure 7-2

normal top and side views and that the hole appears as an ellipse in both of these views. The hole appears as an edge view with hidden lines in the front view, so none of the normal views show the hole as a circle.

Figure 7-2 shows the same object shown in Figure 7-1. The top and side views are the same, but the top view is replaced with an auxiliary view. The auxiliary view shows the true shape of the slanted surface, and the hole appears as a circle.

The auxiliary view in Figure 7-2 shows the true shape of the slanted surface but a foreshortened view of surface A-B-C-D. Positioning the auxiliary view to generate the true shape of the slanted surface foreshortened the other surfaces.

7-2 PROJECTION BETWEEN NORMAL AND AUXILIARY VIEWS

Information about an object's features can be projected between the normal views and any additional auxiliary views by establishing reference planes. A reference plane RPT is located between the top and front views of Figure 7-3. The plane appears as a horizontal line, and its location is arbitrary. In the example shown, the plane was located 10 millimeters from the top view.

A second reference plane, RPA, was established parallel to the slanted surface at a location that prevents the

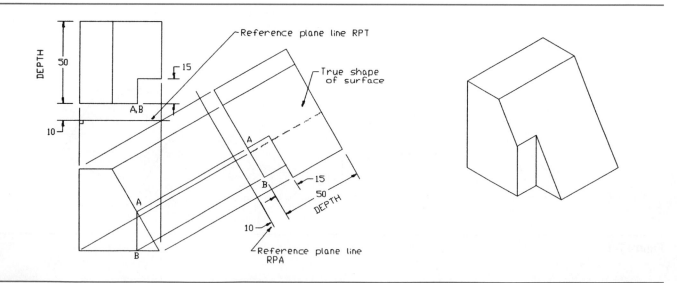

Figure 7-3

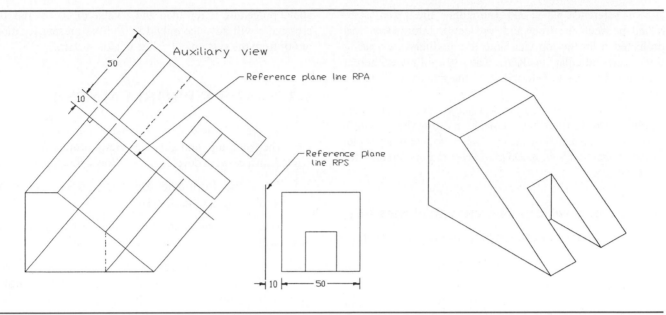

Figure 7-4

auxiliary view from interfering with the top view. The **Move** command can be used to move the top view farther away from the front view if necessary.

Information was projected from the front view into the auxiliary view by using projection lines perpendicular to reference plane RPA.

The depth of the object was transferred from the top view to the auxiliary view. Figure 7-3 shows the 50-millimeter depth and 15-millimeter slot depth dimensioned in both the top and auxiliary views.

Figure 7-4 shows information projected from given front and side views into an auxiliary view. Reference planes RPS and RPA were located 10 millimeters from and parallel to the side and auxiliary views, and information was projected from the front view into the auxiliary view using lines perpendicular to plane RPA. The depth information was transferred from the side view to the auxiliary view.

Figure 7-5 shows an object that has a slanted surface in the top view. The reference plane line, RPA, for the auxiliary view was located parallel to the slanted surface and

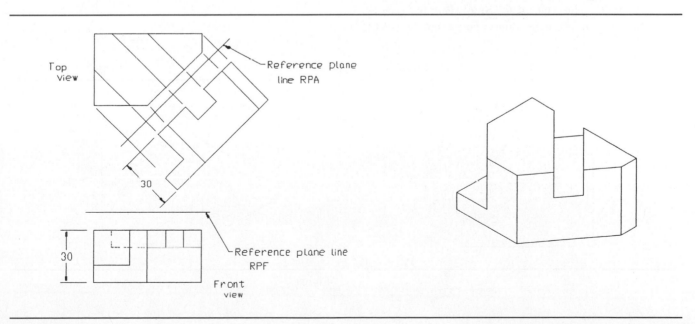

Figure 7-5

another reference plane, RPF (horizontal line), was established between the front and top views. Information was projected from the top view into the auxiliary view using lines perpendicular to RPA. The 30-millimeter height measurement was transferred from the front view into the auxiliary view.

Reference plane lines and projection lines drawn perpendicular to them are best drawn by rotating the drawing's axis system, indicated by the crosshairs, so that it is parallel to the slanted surface. A separate layer may be created for projection lines.

To rotate the drawing's axis system (See Figure 7-6.)

1. Type **Snap** in response to a command prompt.

   The command prompts are as follows.

   Command: _snap
   Specify snap spacing or [ON/OFF/Aspect/Rotate/Style/Type]<0.5000>:

2. Type **r**; press **Enter**.

   Specify base point <0.0000,0.0000>:

3. Accept the default value by pressing **Enter**.

   Specify rotation angle <0>:

4. Type **45**; press **Enter**.

The cursor rotates 45° counterclockwise. Any angle value, including negative values, may be used.

The grid also is rotated, and the **Snap** command will limit the crosshairs to spacing aligned with the rotated axis. **Ortho** will limit lines to horizontal and vertical relative to the rotated axis.

The angle value entered for **Snap, Rotate** is always interpreted as an absolute value. For example, if the

above procedure is repeated and a value of 60 entered, the crosshairs will advance only 15° from its present location, or 60° from the system's absolute 0° axis system.

7-3 SAMPLE PROBLEM SP7-1

Figure 7-7 shows an object that includes a slanted surface. The front, top, and auxiliary views projected from the slanted surface were developed as follows. See Figure 7-8.

1. Draw the front and top orthographic views as explained in Chapter 5.

   Use the **Move** command to position the top view so that it will not interfere with the auxiliary view.

2. Establish two reference plane lines: RPT parallel to the top view (horizontal line) and RPA parallel to the slanted surface (45° line).

   In this example the reference plane line for the top view is located along the lower edge of the view.
   Use **Snap, Rotate** to establish an axis system parallel to the slanted surface. The slanted surface is 45° to the horizontal. Turn **Ortho** on and draw a line parallel to the slanted surface.
   Option: Create a layer for projection lines.

3. Project lines perpendicular to RPA from the drawing's feature presented in the front view into the area of the auxiliary view. Use **Osnap, Intersection** to ensure accuracy.

4. Type **Dist** in response to a command prompt to determine the distance from the reference plane line RPT to the object's features.

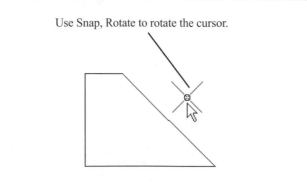

Use Snap, Rotate to rotate the cursor.

Figure 7-6

Figure 7-7

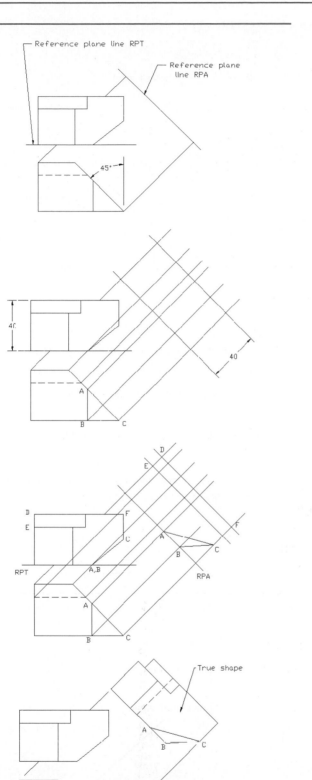

Figure 7-8

5. Use **Offset** to draw a line parallel to RPA at a distance equal to the distances of the object's features.
6. Erase and trim any excess lines and save the drawing, if desired.

7-4 TRANSFERRING LINES BETWEEN VIEWS

Objects are often dimensioned so that only some of their edge lines are dimensioned. This means that **Offset** may not be used to transfer distances to auxiliary views until the line length is known. There are three possible methods that can be used to transfer an edge line of unknown length: measure the line length using **Dist** and then use the determined distance with the **Offset** command; use **Grips,** and **Rotate** and **Move** the line; or use **Copy, Modify, Rotate** and relocate the line. The procedures are as follows.

To measure the length of a line

1. Type **Dist** in response to a command prompt.

   The command prompts are as follows:

   Command: Dist
   Specify first point:

2. Select one end of the line. The **Osnap, Endpoint** option will help identify the endpoint of the line. The intersection of one of the projection lines with a reference plane is usually used.

   Specify second point:

3. Select the other end of the line.

   The following style display will appear in the screen's prompt area.

   Distance=40, Angle in X-Y Plane=45, Angle from the X-Y Plane=0

   Delta X=28.2885, Delta Y=28.2885, Delta Z=0.0000

   Figure 7-9 shows the meaning of the displayed distance information.

   Record the distance and use it with the **Offset** command to locate the line's endpoints relative to a reference plane. See Figure 7-9.

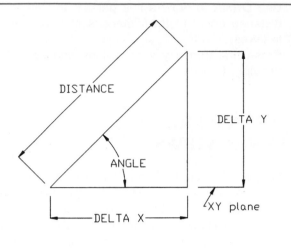

Figure 7-9

To Grip and Move a line (See Figure 7-10.)

1. Copy the line and move the copy to an open area of the drawing.
2. Click the line to highlight its grip points. See Section 3-15.
3. Select the line's lower endpoint; press the right mouse button.
4. Select the **Rotate** option from the menu.

   ROTATE
   Specify rotation angle or [Base point/Copy/Undo/Reference/eXit]:

5. Select the endpoint as the base point.
6. Type **45**; press **Enter.**

   The **Dist** command can also be used to determine the angle of a slanted surface if it is not known.

7. Select the line's lower endpoint again.
8. Select **Move** from the menu.
9. Select the lower line point as the base point.
10. Move the line to a new location or use dynamic input to define a new location.

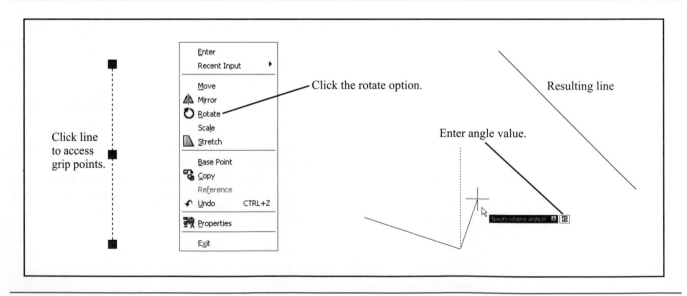

Figure 7-10

To rotate and move a line

1. Copy the line and move the line to an open area of the drawing.
2. Select **Modify** (pull-down menu), **Rotate** and rotate the line **45°**. See Section 3-15.
3. Select **Modify** (pull-down menu), **Move** and relocate the line to the auxiliary view. Use **Osnap** if necessary to ensure accuracy.

7-5 SAMPLE PROBLEM SP7-2

Figure 7-11 shows an object that contains two slanted surfaces and two cutouts. Draw a front and right-side orthographic view, and an auxiliary view. See Figure 7-12.

1. Draw the front and side orthographic views as described in Chapter 5.
2. Define a reference plane, **RPS,** between the front and side views.
3. Use **Snap, Rotate (37.5)** to align the crosshairs with the slanted surface. The 37.5° value was determined using the **Dist** command.
4. Draw a reference plane line, **RPA,** parallel to the slanted surface and project the features of the object into the auxiliary view area.
5. Transfer the distance measurements from RPS and the side view into the auxiliary view.

Label the various points as needed.

6. Erase and trim any excess lines and save the drawing, if desired.

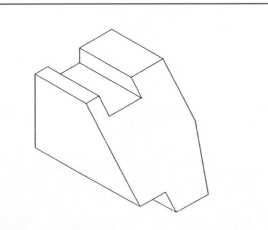

Figure 7-11

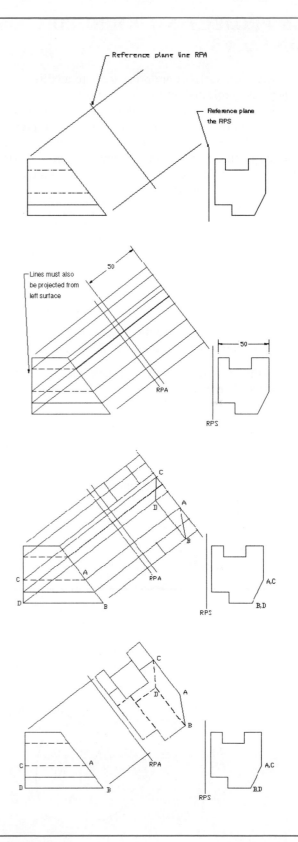

Figure 7-12

7-6 PROJECTING ROUNDED SURFACES

Rounded surfaces are projected into auxiliary views as they were projected into the normal orthographic views, as described in Section 5-15. Additional lines are added to one of the normal orthographic views and then projected into the other normal view. The additional lines define a series of points that define the outside shape of the surface. The two views with the added lines are used to project and transfer the points into the auxiliary view.

7-7 SAMPLE PROBLEM SP7-3

Figure 7-13 shows a cylindrical object that includes a slanted surface. Draw the front, side, and auxiliary views.

The procedure is as follows. See Figure 7-14.

1. Draw the front and side orthographic views.
2. Draw a reference plane line, **RPS.**

In this example the reference plane line was located on the side view's vertical centerline.

3. Draw a reference plane line, **RPA,** parallel to the slanted surface. The reference plane line will also be used as the centerline for the auxiliary view.

In this example the endpoints of both the major and minor axes of the projected elliptical auxiliary view are

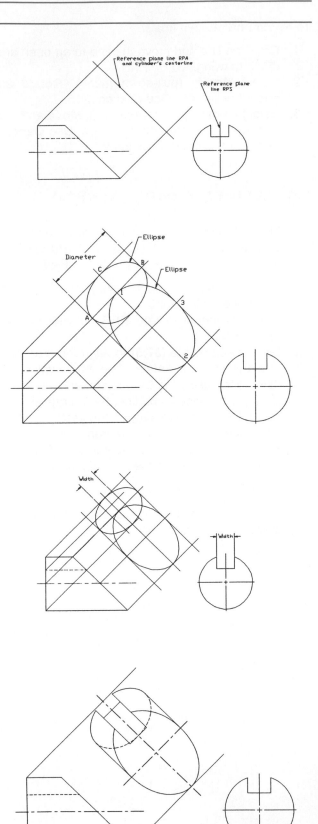

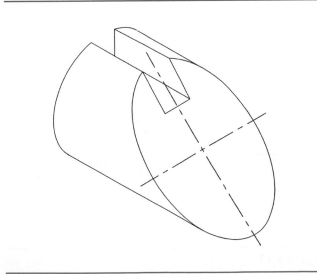

Figure 7-13

Figure 7-14

known, as well as the angle of the auxiliary view. It is therefore not necessary to define points in the circular side view, as was done in Section 5-21. Enough information is present to draw the elliptical shape directly in the auxiliary view.

4. Draw the elliptical surfaces in the auxiliary view as shown.
5. Transfer the width and location of the slot from the side view to the auxiliary view.
6. Erase and trim any excess lines and save the drawing, if desired.

7-8 PROJECTING IRREGULAR SURFACES

Auxiliary views of irregular surfaces are created by projecting information from given normal orthographic views into the auxiliary views in a manner similar to the way information was projected between orthographic views. See Section 5-28. The irregular surface is defined by a series of points along its edge line. The location of the points is random, although more points should be used when the curve's shape is changing sharply than when the curve tends to be smoother.

7-9 SAMPLE PROBLEM SP7-4

Figure 7-15 shows an object that includes an irregular surface. Draw front and side orthographic and auxiliary views of the object. The procedure is described below. See Figure 7-16.

1. Draw the front and side views as defined in Chapter 5.

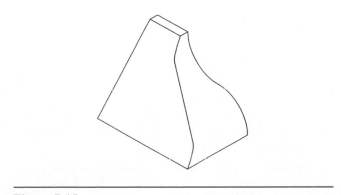

Figure 7-15

2. Draw two reference lines: RPF between the front and side views, and RPA parallel to the slanted surface in the side view.

Use **Snap, Rotate** to align the crosshairs with the slanted surface.

3. Define points along the irregular surface edge line in the front view and project the points into the side view using horizontal lines.
4. Project the points into the auxiliary view using lines perpendicular to line RPA. Label the points and their projection lines.
5. Transfer the depth measurements from RPS and the object's features and the points defining the irregular curve to the auxiliary view.

Type **Dist** in response to a command prompt to determine the distance from RPF to the points. Record the distances.

A = 3.00
B = 2.91
C = 2.62
D = 1.50
E = 0.30
F = 0.09
G = 0

Use **Offset** to draw lines parallel to RPA.

6. Use **Polyline, Edit Polyline, Fit** as explained in Section 5-26 to draw the irregular curve required in the auxiliary view. Use **Edit Vertex** to smooth the curve, if necessary.
7. Project the points defined in the front view to the back surface (far right vertical line) in the side view, and project the points into the auxiliary view.
8. Use **Polyline, Edit Polyline, Fit** to draw the required irregular curve.
9. Erase and trim any excess lines and save the drawing, if desired.

7-10 SAMPLE PROBLEM SP7-5

Figure 7-17 shows a top view and an auxiliary view of an object. Redraw the given views and add the front and right-side views. The procedure is described below.

1. Draw a reference plane line, **RPA,** parallel to the auxiliary view and a reference plane line, **RPT,** a horizontal line.

The object contains a dihedral angle, and the auxiliary view was aligned with the vertex line of the angle.

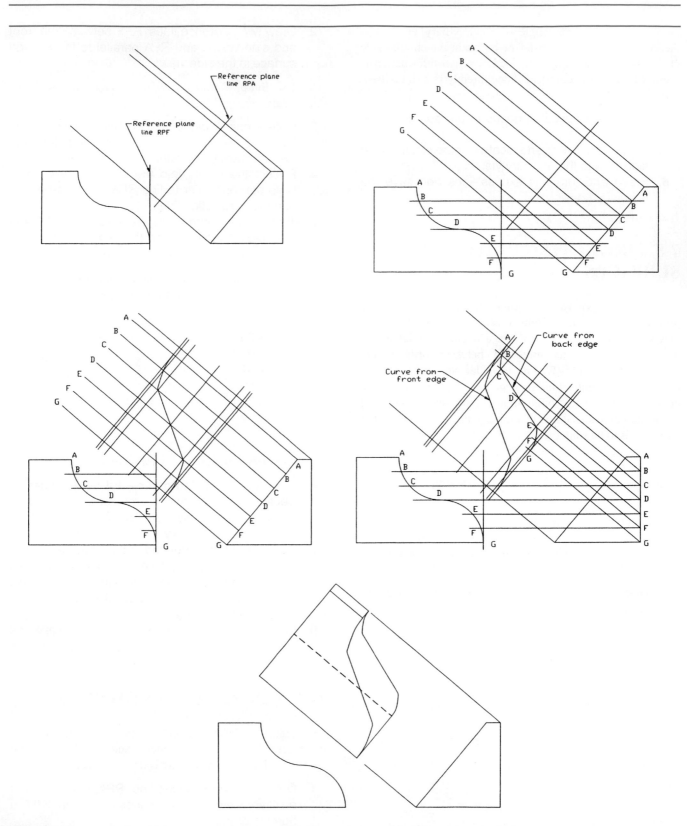

Figure 7-16

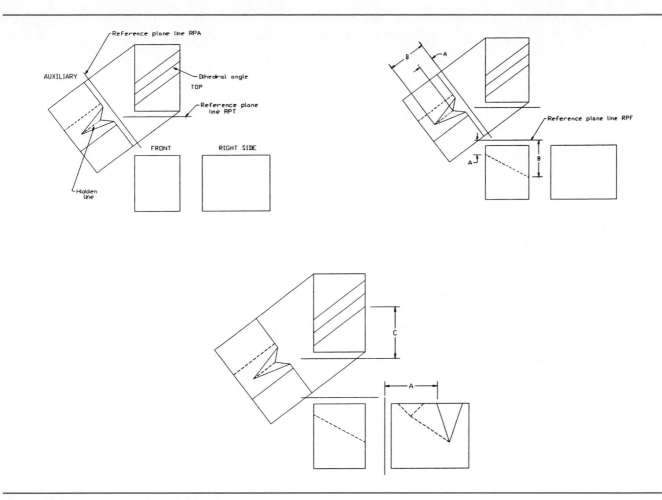

Figure 7-17

Reference plane RPA is perpendicular to the vertex of the dihedral angle's vertex.

2. Project information from the top view and transfer information from the auxiliary view into the front view.
3. Project information from the front view and transfer information from the top view into the side view.
4. Erase and trim any excess lines and save the drawing, if desired.

7-11 PARTIAL AUXILIARY VIEWS

Auxiliary views help present true-shaped views of slanted and oblique surfaces, but in doing so generate foreshortened views of other surfaces. It is often clearer to create an auxiliary view of just the slanted surface and omit the surfaces that would be foreshortened. Auxiliary views that show only one surface of an object are called *partial auxiliary views.*

Figure 7-18 shows a front view and three partial views of the object: a partial top view and two partial auxiliary views. A broken line (see Section 6-14) may be used to show that the partial auxiliary view is part of a larger view that has been omitted. Likewise, hidden lines may or may not be included. Figure 7-19 shows a front, side, and two partial auxiliary views of an object. Both the hidden lines and broken lines were omitted. If you are unsure about the interpretation of a view with hidden lines omitted, add a note to the drawing next to the partial views: ALL HIDDEN LINES OMITTED FOR CLARITY.

7-12 SECTIONAL AUXILIARY VIEWS

Sectional views may also be drawn as auxiliary views. Figure 7-20 shows an auxiliary sectional view. The cutting plane line is positioned across the object. A reference plane line is drawn parallel to the cutting plane line, and information is projected into the auxiliary sectional view, using lines perpendicular to the cutting plane line.

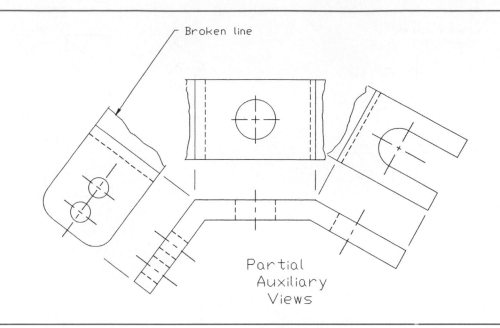

Broken line

Partial
Auxiliary
Views

Figure 7-18

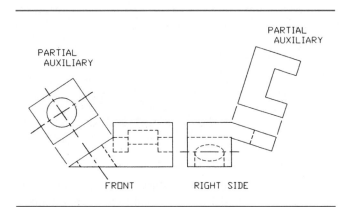

PARTIAL
AUXILIARY

PARTIAL
AUXILIARY

FRONT   RIGHT SIDE

Figure 7-19

Figure 7-21 shows front, partial auxiliary, and auxiliary sectional views of an object. In this example the partial sectional view was used to help clarify the shape of the object's feature that would appear vague in the normal top or side views.

7-13 AUXILIARY VIEWS OF OBLIQUE SURFACES

The true shape of an oblique surface cannot be determined by a single auxiliary view taken directly from the oblique surface. An auxiliary view shows the true shape of

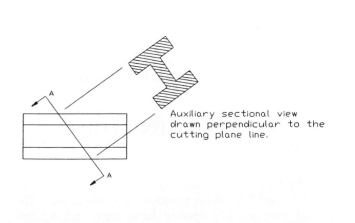

A

A

Auxiliary sectional view
drawn perpendicular to the
cutting plane line.

Figure 7-20

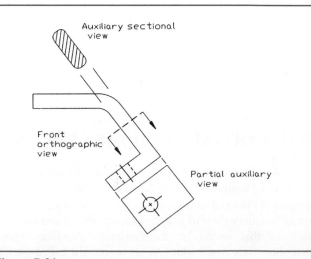

Auxiliary sectional
view

Front
orthographic
view

Partial auxiliary
view

Figure 7-21

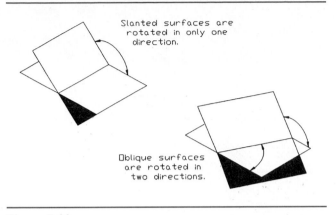

Figure 7-22

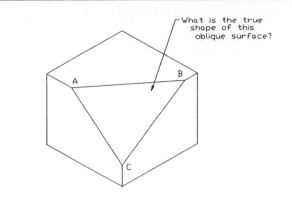

Figure 7-24

a surface only when it is taken at exactly 90° to the surface. The auxiliary views taken for Sample Problems 1 through 5 were taken off slanted surfaces.

Slanted surfaces are surfaces that are rotated about only one axis. See Figure 7-22. One of the orthographic views must be an edge view of the surface (the surface appears as a line) for an auxiliary view created by projecting lines perpendicular to that view to show the true shape of the surface.

Oblique surfaces are surfaces rotated about two axes. See Figures 7-22 and 7-23. This means that none of the normal orthographic views will show the oblique surface as an edge view; there is no given end view of the surface. This in turn means that an auxiliary view taken directly from one of the given views will not be perpendicular to the surface and therefore will not show the true shape of the surface. An auxiliary view that shows an edge view of the surface must first be created, then a second auxiliary taken perpendicular to one of the other normal orthographic views is needed to show the true shape of the surface.

7-14 SECONDARY AUXILIARY VIEWS

Consider the object shown in Figure 7-24. What is the true shape of surface A-B-C? An auxiliary view taken perpendicular to the surface will show its true shape, but what is the angle for a plane perpendicular to the surface?

Figure 7-25 shows the normal orthographic views of the object.

Start by choosing an edge line in the plane that is perpendicular to either the X or Y axis. Edge line A-B in the front view is a horizontal line, so it is parallel to the X axis. Take an auxiliary view aligned with the top view of the edge line (you're looking straight down the line). This auxiliary view will be perpendicular to the edge line and will generate an end view of the plane. A second auxiliary view can then be taken perpendicular to the end view that will show the true shape of the surface.

Figure 7-26 shows how a secondary auxiliary view is created for the object shown in Figure 7-24. The specific procedure is as follows.

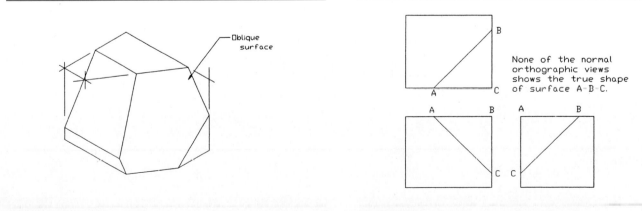

Figure 7-23

Figure 7-25

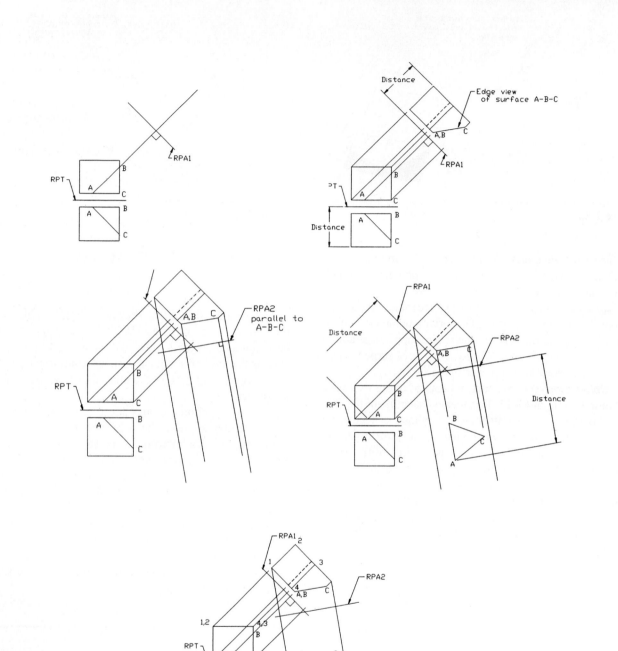

Figure 7-26

To draw the first auxiliary view

1. Draw the normal orthographic views of the object.
2. Draw a reference plane, **RPT,** between the front and top views.
3. Extend a line from the top view of line A-B in the area for the first auxiliary view.

The **Extend** command can be used by drawing a construction line slightly beyond the expected area of the first auxiliary view, then using the construction line as an **Extend** boundary line (see Section 2-24), and extending line A-B to the boundary. The construction line can then be erased.

4. Draw a reference plane line, **RPA1,** perpendicular to the line A-B extension.

Use **Snap, Rotate** to align the crosshairs with the extension line. The **Dist** command can be used to determine the line's angle if it is not known.

5. Project the feature of the object into the auxiliary view using lines perpendicular to RPA1.
6. Transfer the distance measurements from RPT and the front view as shown.
7. Erase and trim any excess lines and create the first auxiliary view.

The surface A-B-C should appear as a straight line. This is an edge view of surface A-B-C.

To draw the secondary auxiliary view

1. Use **Snap, Rotate** and align the crosshairs with the end view of surface A-B-C. **Dist** can be used to determine the angle of the edge view line.
2. Draw a reference plane line, **RPA2,** parallel to the edge view of the surface.
3. Project the features of the object into the second auxiliary view using lines perpendicular to RPA2.
4. Transfer the distance measurements from RPA1 and the top view as shown.
5. Erase and trim any excess lines and save the drawing, if desired.

The secondary auxiliary view shows the true shape of surface A-B-C.

Figure 7-27 shows another example of a secondary auxiliary view used to show the true shape of an oblique surface. In this example the first auxiliary view was taken from line F-A in the front view because it is a vertical line in the side view. The distance measurements came from the top view. The procedure used is the same as described above.

The auxiliary view shown in Figure 7-27 includes only the oblique surface. This is a partial auxiliary view. The purpose of the auxiliary view is to determine the true shape of the oblique surface. All the other surfaces would be foreshortened, not their true shape, if included in the auxiliary views.

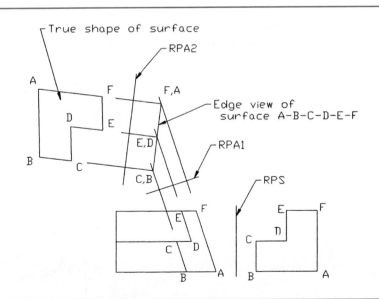

Figure 7-27

7-15 SAMPLE PROBLEM SP7-6

What is the true shape of the plane shown in Figure 7-28?

None of the edge lines are horizontal or vertical lines, so a line must be defined within the plane that is either horizontal or vertical. The added line can then be used to generate the two auxiliary views needed to define the true shape of the plane.

The procedure is as follows. See Figure 7-29.

1. Draw a horizontal line from point A in the front view.

Use **Osnap, Intersection** with **Ortho** on and draw the line across the view. Use **Trim** to remove the excess portion of the line.

2. Define the intersection of the new line with one of the plane's edge lines (D-C) as x.
3. Project line A-x into the top view.

Option: Create a layer for projection lines.

The location of point A is already known in the top view. The location of point x in the top view can be found by drawing a vertical line from point x in the front view so that the line intersects line D-C in the top view. Use **Osnap, Intersection, Ortho** on to ensure accuracy.

4. Extend line A-x into the area for the first auxiliary view.

Draw a construction line to use as a boundary line for the **Extend** command.

5. Draw two reference plane lines: **RPT** between the front and top views, and **RPA1** perpendicular to the extension of line A-x.

Use **Snap, Rotate** to align the crosshairs with the extension of line A-x.

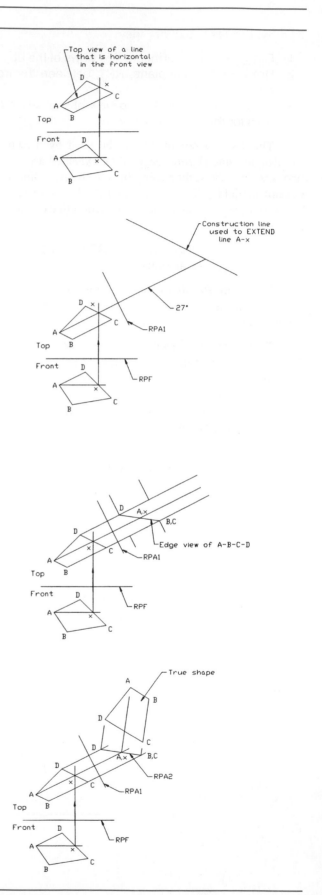

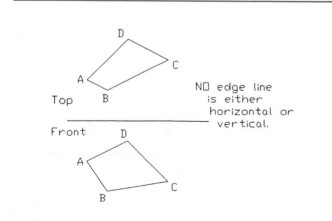

Figure 7-28

Figure 7-29

6. Project the plane's corner points into the area of the first auxiliary view from the top view. Transfer the distance measurements from RPT and the front view.

Use **Dist** to determine the distance from RPF and the corner points of the surface. Use **Offset** to transfer the point distance from RPF to RPA1. Use **Trim** to remove the internal portion of the offset lines. This will help clarify the drawing by removing excess lines from the auxiliary view but will retain intersection points needed to draw lines using **Osnap, Intersection.**

The plane should appear as a straight line, the end view of the plane.

7. Draw a third reference plane line, **RPA2,** parallel to the end view of the plane and project the plane's corner points into the secondary auxiliary view area.
8. Transfer the depth distances from RPA1 and the top view to RPA2 and the secondary auxiliary view.
9. Erase and trim any excess lines and save the drawing, if desired.

The secondary auxiliary view shows the true shape of surface A-B-C-D.

7-16 SECONDARY AUXILIARY VIEW OF AN ELLIPSE

Figure 7-30 shows the front and side views of an oblique surface and includes a foreshortened view of a hole, that is, an ellipse. Also included are two auxiliary views, the second of which shows the hole as a circle.

The ellipse is projected by first determining the correct projection angle that will produce an edge view of surface A-B-C-D. In this example, line B-C appears as a vertical line in the side view, so its front view can be used to project the required edge view.

It is known that the secondary view of the surface will show the hole as a circle, so only a radius value need be carried between the views. In this example, point 1 in the front view was projected into the first auxiliary view and then into the second auxiliary view, thereby defining the radius of the circle.

The hole is located at the center of the surface, which means that its centerlines are located on the midpoints of the edge lines in the secondary auxiliary views. If the hole's center point was not in the center of the surface, the intersections of the hole's centerlines with the surface's edge lines would also have to be projected into the secondary auxiliary view to accurately locate the hole.

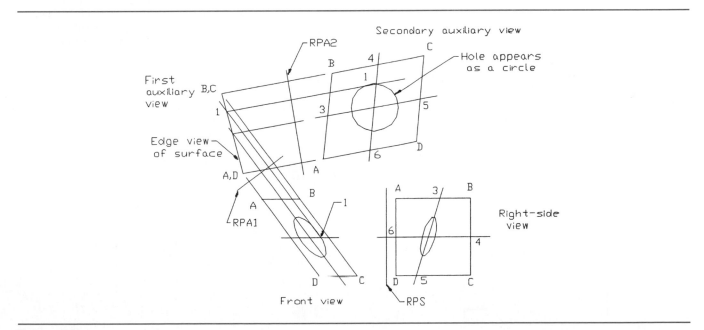

Figure 7-30

7-17 EXERCISE PROBLEMS

Given the outlines for front, side, and auxiliary views as shown, substitute one of the side views shown in Exercise Problems EX7-1A through EX7-1D and complete the front and auxiliary views. All dimensions are in millimeters.

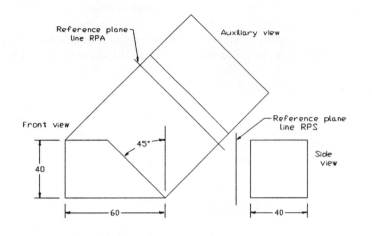

EX7-1A

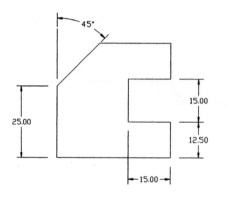

EX7-1B

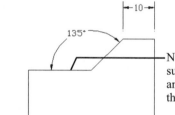

NOTE: This surface is 10 from and parallel to the top surface.

EX7-1C

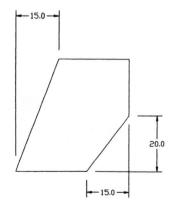

EX7-1D

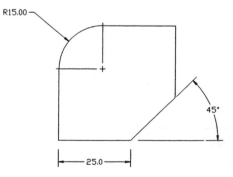

Given the outlines for front, side, and auxiliary views as shown, substitute one of the side views shown in Exercise Problems EX7-2A through EX7-2D and complete the front and auxiliary views. All dimensions are in millimeters.

Draw two orthographic views and an auxiliary view for each of the objects shown.

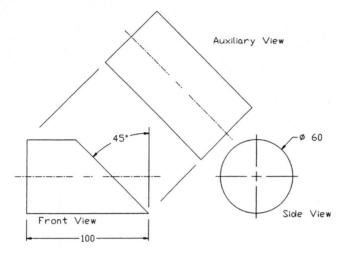

EX7-2A

EX7-2B

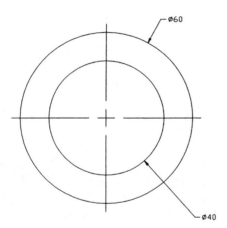

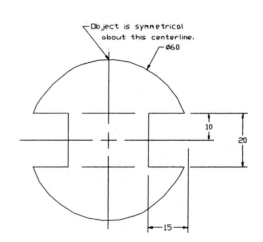

EX7-2C

EX7-2D

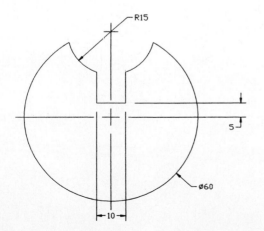

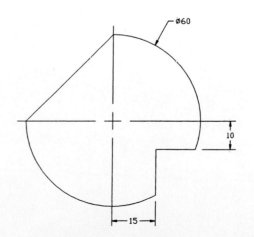

Draw two orthographic views and an auxiliary view for each of the objects in Exercise Problems EX7-3 through EX7-6.

EX7-3 MILLIMETERS

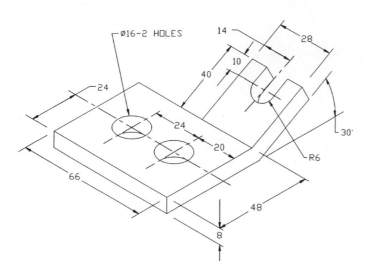

EX7-5 INCHES

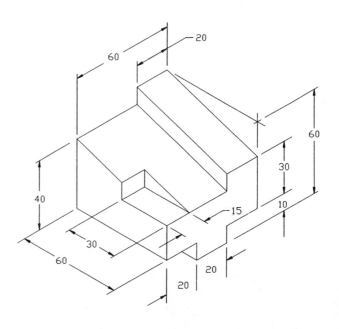

EX7-4 MILLIMETERS

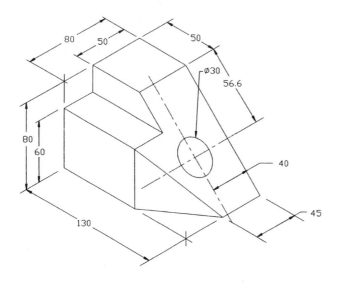

EX7-6 MILLIMETERS

Redraw the given orthographic views in Exercise Problems EX7-7 through EX7-10 and add the appropriate auxiliary view.

EX7-7 MILLIMETERS

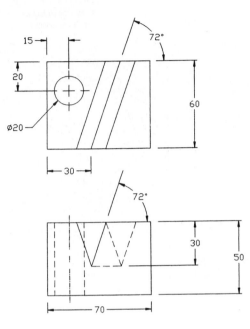

EX7-9 INCHES

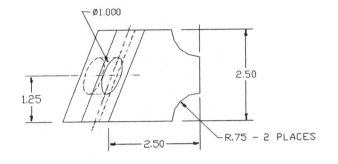

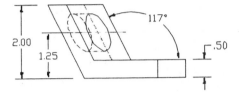

EX7-8 MILLIMETERS

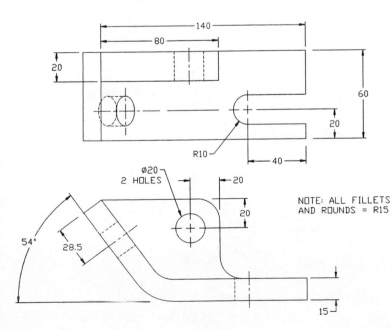

NOTE: ALL FILLETS
AND ROUNDS = R15

EX7-10 MILLIMETERS

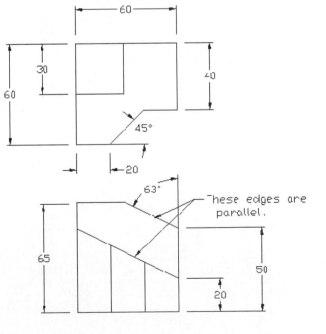

Draw two orthographic views and an auxiliary view for each of the objects in Exercise Problems EX7-11 through EX7-40.

EX7-11 MILLIMETERS

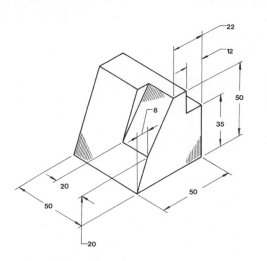

EX7-12 MILLIMETERS

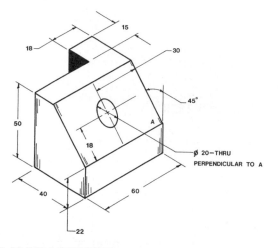

EX7-13 MILLIMETERS

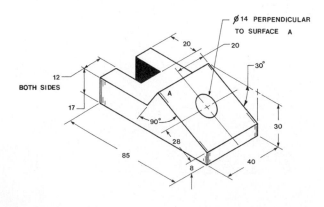

EX7-14 MILLIMETERS

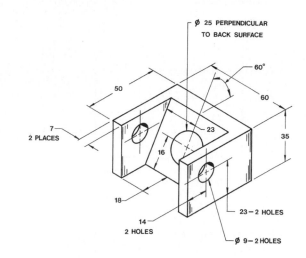

EX7-15 INCHES

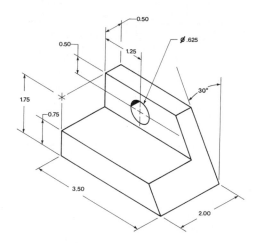

EX7-16 INCHES

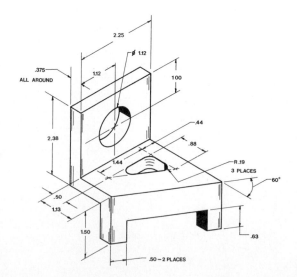

EX7-17 MILLIMETERS

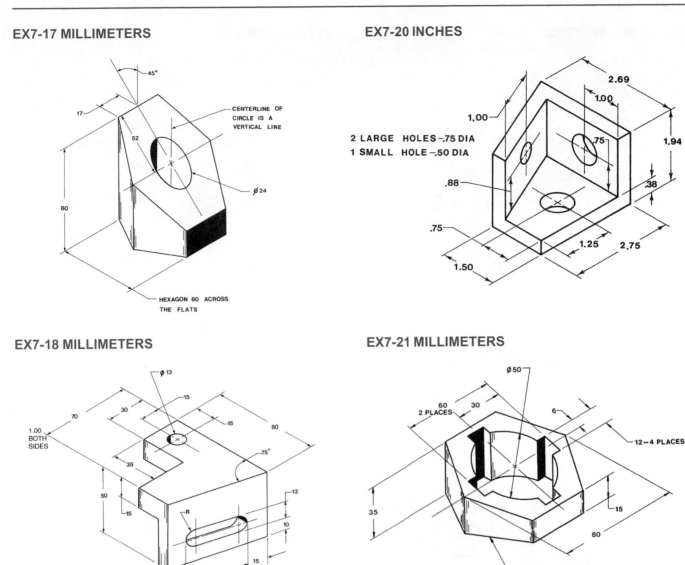

45°

17

52

80

CENTERLINE OF
CIRCLE IS A
VERTICAL LINE

Ø 24

HEXAGON 60 ACROSS
THE FLATS

EX7-20 INCHES

2.69

1.00

1.00

.75

1.94

.88

.38

.75

1.25   2.75

1.50

2 LARGE HOLES -.75 DIA
1 SMALL HOLE -.50 DIA

EX7-18 MILLIMETERS

Ø 13

15

70

30

15

80

1.00
BOTH
SIDES

35

75°

50

15

12

R

10

8

50

15

EX7-21 MILLIMETERS

Ø 50

60
2 PLACES

30

6

12 – 4 PLACES

35

15

80

REGULAR HEXAGON
80 ACROSS THE CORNER

EX7-19 MILLIMETERS

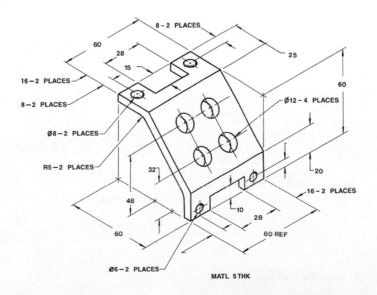

8 – 2 PLACES

60

28

25

15

16 – 2 PLACES

60

8 – 2 PLACES

Ø12 – 4 PLACES

Ø8 – 2 PLACES

R5 – 2 PLACES

32

20

16 – 2 PLACES

48

10

28

60 REF

60

Ø6 – 2 PLACES

MATL 5 THK

EX7-22 MILLIMETERS

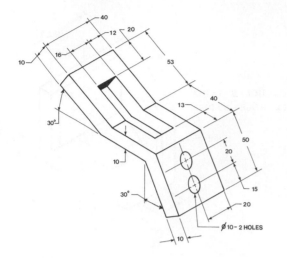

EX7-25 INCHES

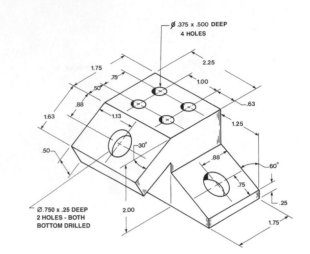

EX7-23 INCHES

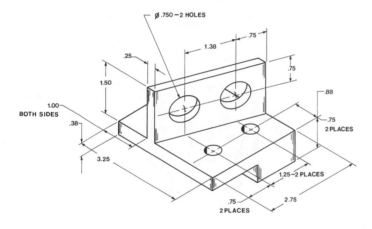

EX7-26 MILLIMETERS

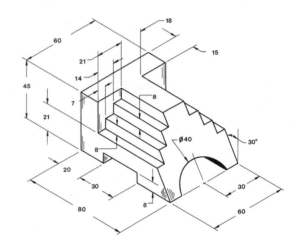

EX7-24 MILLIMETERS

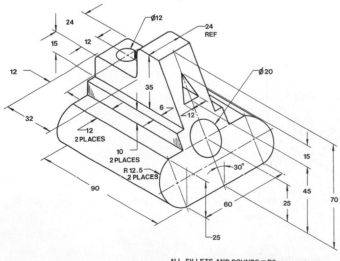

ALL FILLETS AND ROUNDS = R3

EX7-27 MILLIMETERS

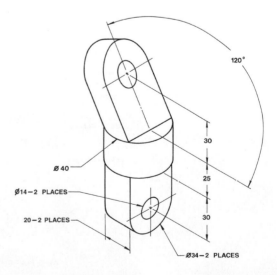

EX7-28 MILLIMETERS

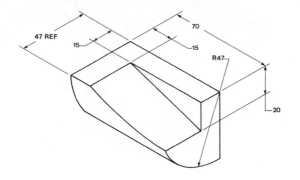

EX7-31 INCHES

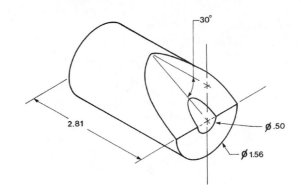

EX7-29 MILLIMETERS

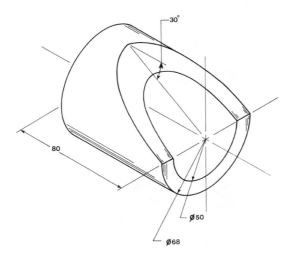

EX7-32 MILLIMETERS

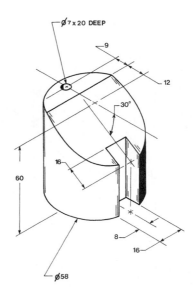

EX7-30 MILLIMETERS

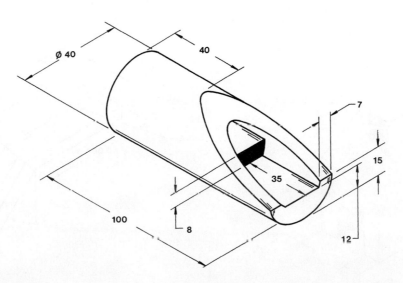

EX7-33 MILLIMETERS

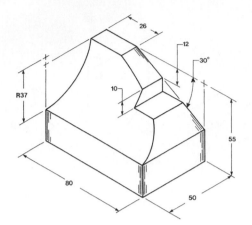

EX7-36 INCHES

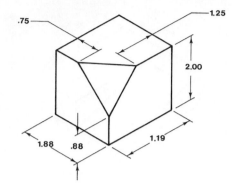

EX7-34 MILLIMETERS

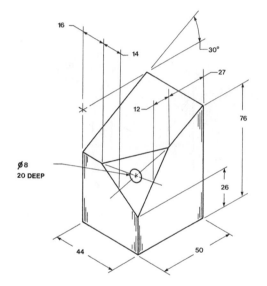

EX7-37 INCHES

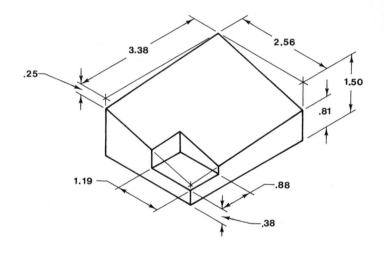

EX7-35 MILLIMETERS

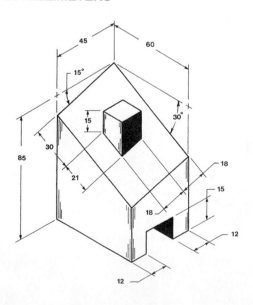

EX7-38 INCHES

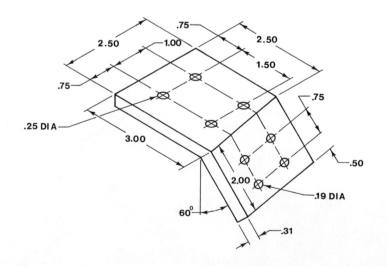

EX7-39 MILLIMETERS

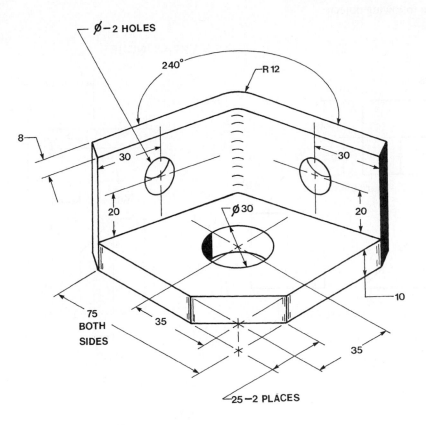

EX7-40 MILLIMETERS

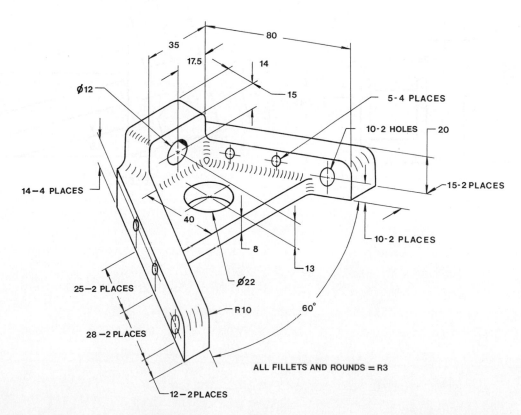

Draw at least two orthographic views and one auxiliary view for each of the following objects.

EX7-41 INCHES

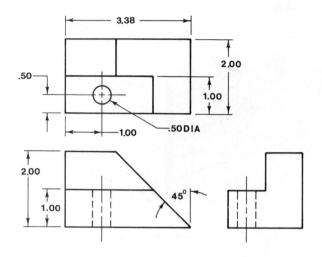

EX7-43 INCHES

EX7-42 INCHES

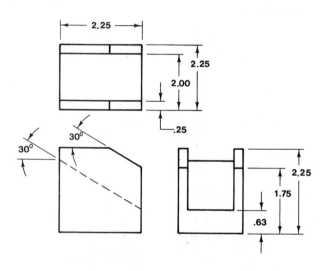

EX7-44 INCHES

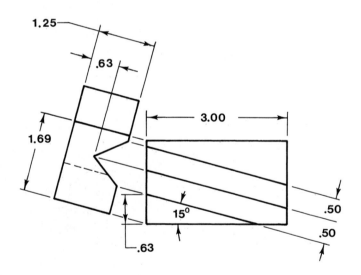

Use a secondary auxiliary view to find the true shape of the planes in Exercise Problems EX7-45 through EX7-50.

EX7-45 MILLIMETERS

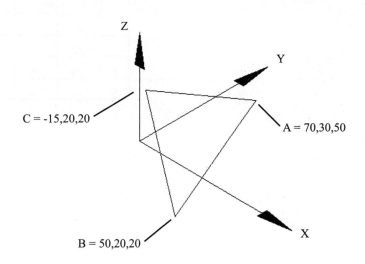

EX7-46 MILLIMETERS

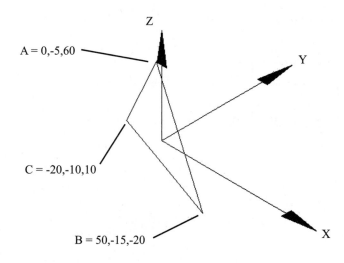

EX7-47 MILLIMETERS

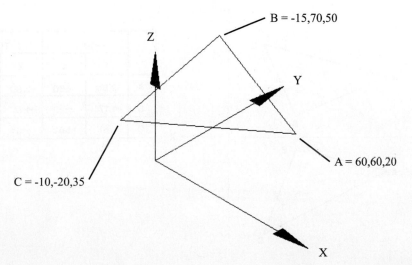

EX7-48 INCHES (SCALE: 4:1)

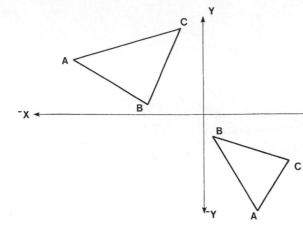

	TOP		SIDE	
	X	Y	X	Y
A	-2.30	.96	.96	-1.68
B	-.98	.15	.15	-.40
C	-.40	1.50	1 50	-.80

EX7-49 MILLIMETERS (SCALE: 2:1)

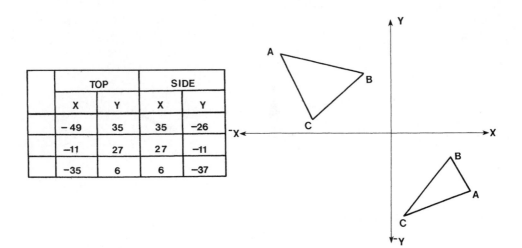

	TOP		SIDE	
	X	Y	X	Y
	-49	35	35	-26
	-11	27	27	-11
	-35	6	6	-37

EX7-50 INCHES (SCALE: 3:1)

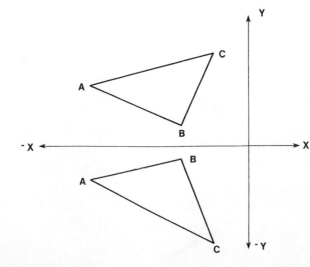

	FRONT		TOP	
	X	Y	X	Y
A	-2.80	-.60	-2.80	1.05
B	-1.17	-.20	-1.17	.35
C	-.61	-1.62	-.61	1.62

EX7-51 MILLIMETERS

Redesign the following object to include two Ø10 holes in the slanted surface. The holes should be centered along the longitudinal axis, and spaced so that the distance between the holes' centers equals the distance from the holes' centers to the upper and lower edges of the slanted surface. The holes should be perpendicular to the slanted surface.

Draw the front, top, and side views of the object plus an auxiliary view of the slanted surface.

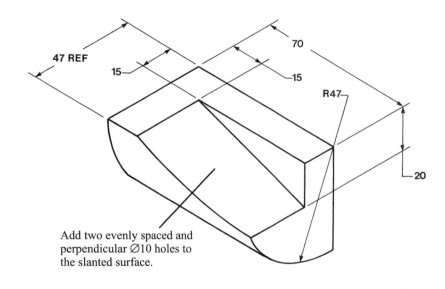

Add two evenly spaced and perpendicular Ø10 holes to the slanted surface.

EX7-52 MILLIMETERS

Redesign the following object so that the outside shape is a regular heptagon (seven-sided polygon) 80 across the flats, and the inside Ø50 hole is intersected by six evenly spaced slots each 12 wide. Draw front and top orthographic views and an auxiliary view of the object.

EX7-53 MILLIMETERS

Calculate the area of the following plane. Redesign the plane so that the new plane is congruent to the original but has an area equal to 1.5 times the original area. Draw and label the true shape of the new plane relative to the X, Y, Z axes.

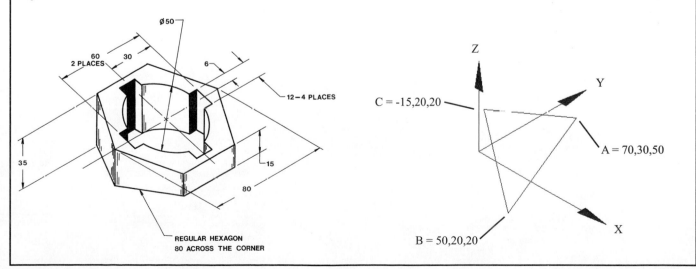

REGULAR HEXAGON
80 ACROSS THE CORNER

C = -15,20,20

A = 70,30,50

B = 50,20,20

Dimensioning

Figure 8-1

8-1   INTRODUCTION

This chapter explains the **Dimension** toolbar. See Figure 8-1. The chapter first explains dimensioning terminology and conventions then presents an explanation of each tool within the **Dimension** toolbar. The chapter also demonstrates how dimensions are applied to drawings and gives examples of standard drawing conventions and practices.

8-2   TERMINOLOGY AND CONVENTIONS

Some common terms (See Figure 8-2.)

Dimension lines: Mechanical drawings contain lines between extension lines that end with arrowheads and include a numerical dimensional value located within the line; architectural drawings contain lines between extension lines that end with tick marks and include a numerical dimensional value above the line.

Extension lines: Lines that extend away from an object and allow dimensions to be located off the surface of an object.

Leader lines: Lines drawn at an angle, not horizontal or vertical, that are used to dimension specific shapes such as holes. The start point of a leader line includes an arrowhead. Numerical values are drawn at the end opposite the arrowhead.

Linear dimensions: Dimensions that define the straight-line distance between two points.

Angular dimensions: Dimensions that define the angular value, measured in degrees, between two straight lines.

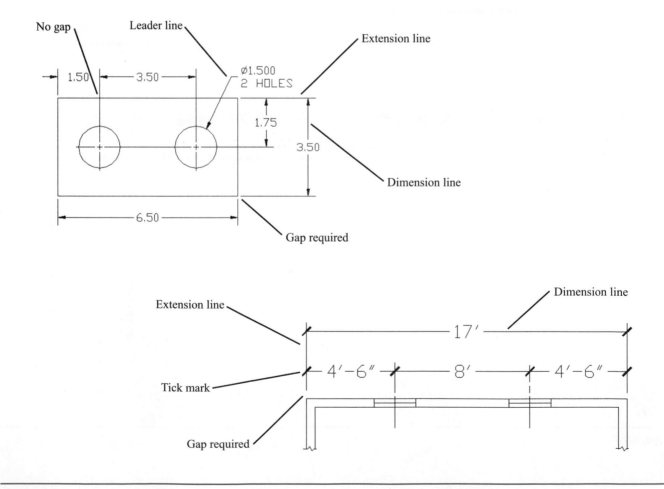

Figure 8-2

Some dimensioning conventions (See Figure 8-3.)

1. Dimension lines are drawn evenly spaced; that is, the distance between dimension lines is uniform. A general rule of thumb is to locate dimension lines about 1/2 inch, or 15 millimeters, apart.
2. There should be a noticeable gap between the edge of a part and the beginning of an extension line. This serves as a visual break between the object and the extension line. The visual difference between the linetypes can be emphasized by using different colors for the two types of lines.
3. Leader lines are used to define the size of holes and should be positioned so that the arrowhead points at the center of the hole.
4. Centerlines may be used as extension lines. No gap is used when a centerline is extended beyond the edge lines of an object.
5. Dimension lines should be aligned whenever possible to give the drawing a neat, organized appearance.

Some common errors to avoid (See Figure 8-4.)

1. Avoid crossing extension lines. Place longer dimensions farther away from the object than shorter dimensions.
2. Do not locate dimensions within cutouts; always use extension lines.
3. Do not locate any dimension too close to the object. Dimension lines should be at least 1/2 inch, or 15 millimeters, from the edge of the object.
4. Avoid long extension lines. Locate dimensions in the same general area as the feature being defined.

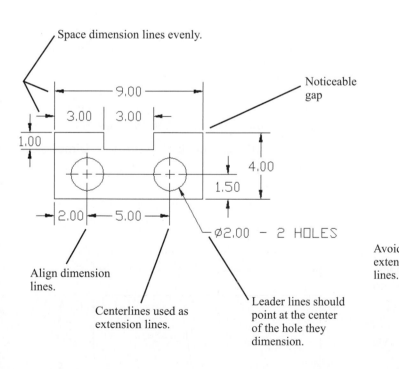

Space dimension lines evenly.

Noticeable gap

Align dimension lines.

Centerlines used as extension lines.

Leader lines should point at the center of the hole they dimension.

Figure 8-3

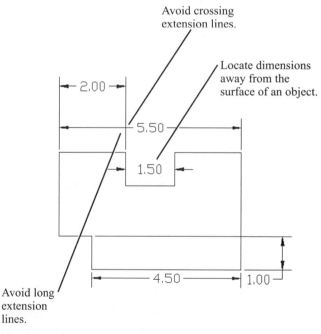

Avoid crossing extension lines.

Locate dimensions away from the surface of an object.

Avoid long extension lines.

Figure 8-4

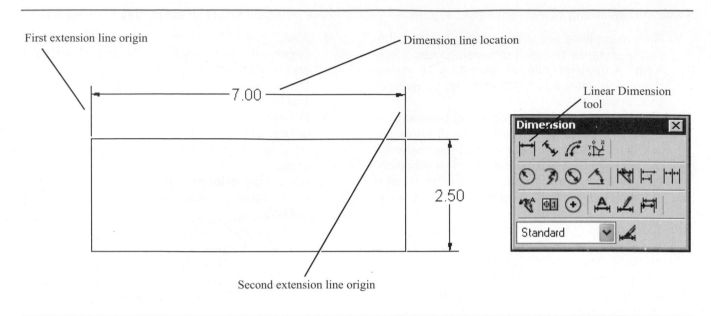

First extension line origin

Dimension line location

Linear Dimension tool

7.00

2.50

Standard

Second extension line origin

Figure 8-5

8-3 LINEAR DIMENSION

The **Linear Dimension** command is used to create horizontal and vertical dimensions.

To create a horizontal dimension by selecting extension lines (See Figure 8-5.)

1. Select the **Linear** tool from the **Dimension** toolbar.

   Command: _dimlinear
   Specify first extension line origin or <select object>:

2. Select the starting point for the first extension line.

   Specify second extension line origin:

3. Select the starting point for the second extension line.

   [MText/Text/Angle/Horizontal/Vertical/Rotated]:

4. Locate the dimension line by moving the crosshairs.
5. Press the left mouse button when the desired dimension line location has been selected.

   The dimensional value locations shown in Figure 8-5 are the default settings locations. The location and style may be changed using the **Dimension Style** command discussed in Section 8-4.

To create a vertical dimension

   The vertical dimension shown in Figure 8-5 was created using the same procedure demonstrated for the hori-

zontal dimension, except different extension line origin points were selected. AutoCAD will automatically switch from horizontal to vertical dimension lines as you move the cursor around the object.

   If there is confusion between horizontal and vertical lines when adding dimensions, that is, you don't seem to be able to generate a vertical line, type **V** and press **Enter** in response to the following prompt:

   [MText/Text/Angle/Horizontal/Vertical/Rotated]:

   The system will now draw vertical dimension lines.

To create a horizontal dimension by selecting the object to be dimensioned (See Figure 8-6.)

1. Select the **Linear** tool from the **Dimension** toolbar.

   Command: _dimlinear
   Specify first extension line origin or <select object>:

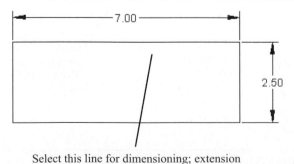

7.00

2.50

Select this line for dimensioning; extension lines will be added automatically.

Figure 8-6

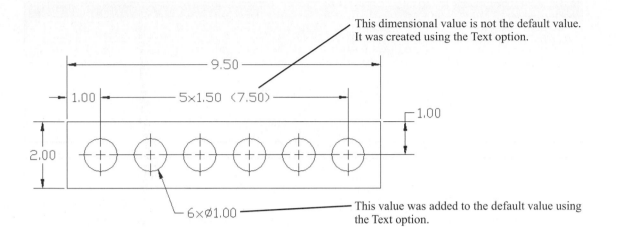

This dimensional value is not the default value.
It was created using the Text option.

This value was added to the default value using
the Text option.

Figure 8-7

2. Press the right mouse button.

   Select object to Dimension:

3. Select the line to be dimensioned.

   This option allows you to select the distance to be dimensioned directly. The option applies only to horizontal and vertical lengths. Aligned dimensions, although linear, are created using the **Aligned** tool.

**To change the default dimension text—
Text option**

   AutoCAD will automatically create a text value for a given linear distance. A different value or additional information may be added as follows. See Figure 8-7.

1. Select the **Linear** tool from the **Dimension** toolbar.

   *Command: _dimlinear
   Specify first extension line origin or <select object>:*

2. Select the starting point for the first extension line.

   Specify second extension line origin:

3. Select the starting point for the second extension line.

   [MText/ Text/Angle/Horizontal/ Vertical/Rotated]:

4. Type **t**; press **Enter.**

   Enter dimension text <7.50>:

The value given will be the linear value of the distance selected. In this example more information is required, so the default distance value must be modified.

5. Type **5 × 1.50 (7.50)**; press **Enter.**

The typed dimension will appear on the screen and can be located by moving the cursor.

**To change the default dimension text—
Mtext option**

1. Select the **Linear** tool from the **Dimension** toolbar.

   *Command: _dimlinear
   Specify first extension line origin or <select object>:*

2. Select the starting point for the first extension line.

   Specify second extension line origin:

3. Select the starting point for the second extension line.

   [MText/ Text/Angle/Horizontal/ Vertical/Rotated]:

4. Type **m**; press **Enter.**

The **Text Formatting** dialog box will appear. See Figure 8-8. The text value will appear in a box below the **Text Formatting** dialog box. To remove the existing text, highlight the text and press the **<Del>** key. New text can then be typed into the box. In the example shown, the default value of 7.0000 was replaced with a value of 7.00.

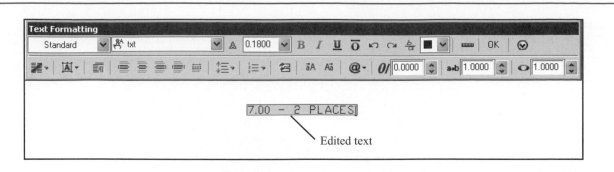

Edited text

Figure 8-8

Information can be added before or after the default text by placing the cursor in the appropriate place and typing the additional information. For example, placing the cursor to the right of the text value box and typing **-2 PLACES** would produce the text 7.00-2 PLACES on the drawing.

The **Text Formatting** dialog box can be used to edit the dimension text in the same way as it was used to edit drawing screen text (Section 2-16). Figure 8-9 shows the **Font** pull-down menu.

Figure 8-10 shows the **Color** pull-down menu that may be used to change the color of a dimension.

To edit an existing dimension

The **Dimension Text Edit** tool, located on the **Dimension** toolbar, is used to change the text justification or style. See Figure 8-11. The **New** option of the **Dimension Edit** tool located on the **Dimension** toolbar may be used to edit an existing dimensional value. See Figure 8-12. Figure 8-13 shows a revised dimensional text value.

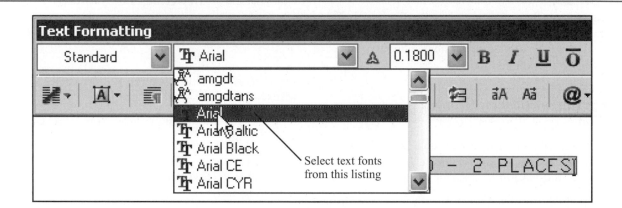

Select text fonts from this listing

Figure 8-9

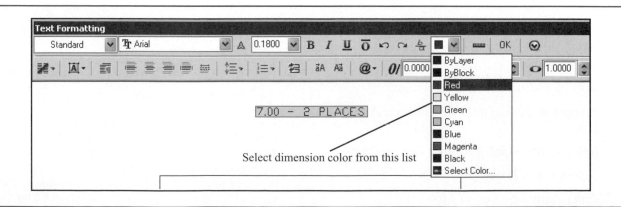

Select dimension color from this list

Figure 8-10

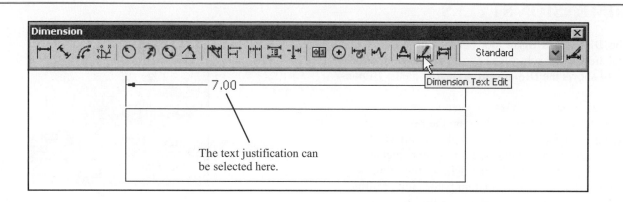

Figure 8-11

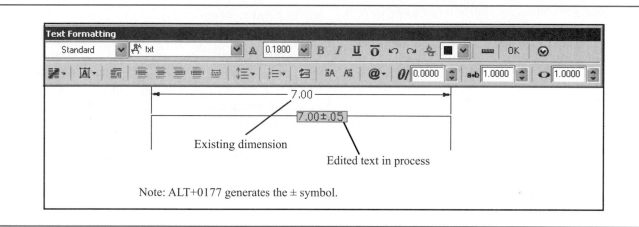

Figure 8-12

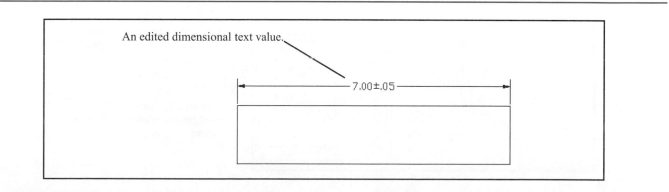

Figure 8-13

8-4 DIMENSION STYLES

The **Dimension Style** tool opens a group of dialog boxes that are used to control the appearance of dimensions. Figure 8-14 shows the **Dimension Style** tool on the **Dimension** toolbar.

A great variety of styles are used to create technical drawings. The style difference may be the result of different drawing conventions. For example, architects locate dimensions above the dimension lines, and mechanical engineers locate the dimensions within the dimension lines. AutoCAD works in decimal units for either millimeters or inches, so parameters set for inches would not be usable for millimeter drawings. The **Dimension Style** tool allows you to conveniently choose and set dimension parameters that suit your particular drawing requirements.

Figure 8-15 shows the **Dimension Style Manager** dialog box. This section will explain how to use the **Modify** option to change the **Standard** style settings to suit a specific drawing requirement. The **Set Current, New, Override,** and **Compare** options are used to create a new custom dimension style designed to meet specific applications. Figure 8-16 shows the **Primary Units** option of the **Modify Dimension Style: Standard** dialog box.

To change the scale of a drawing

Drawings are often drawn to scale because the actual part is either too big to fit on a sheet of drawing paper or too small to be seen. For example, a microchip circuit must be drawn at several thousand times its actual size to be seen.

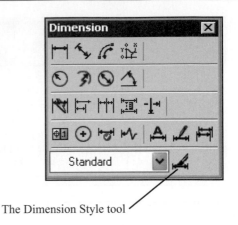

The Dimension Style tool

Figure 8-14

Drawing scales are written using the following formats:

SCALE: 1=1
SCALE: FULL
SCALE: 1000=1
SCALE: .25=1

In each example the value on the left indicates the scale factor. A value greater than 1 indicates that the drawing is larger than actual size. A value less than 1 indicates that the drawing is smaller than actual size.

Regardless of the drawing scale selected, the dimension values must be true size. Figure 8-17 shows the same rectangle drawn at two different scales. The top rectangle is drawn at a scale of 1=1, or its true size. The bottom

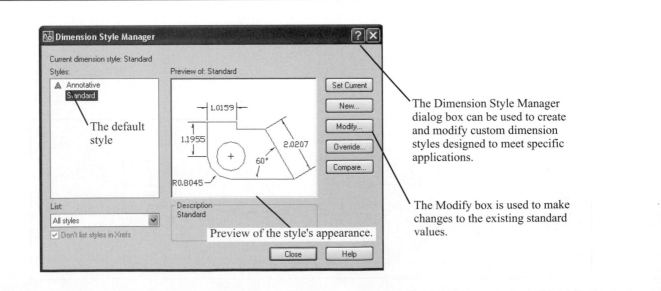

The Dimension Style Manager dialog box can be used to create and modify custom dimension styles designed to meet specific applications.

The Modify box is used to make changes to the existing standard values.

Figure 8-15

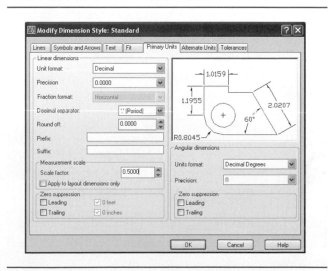

Change this value to change the scale of the drawing.

Figure 8-16

rectangle is drawn at a scale of 2=1, or twice its true size. In both examples the 3.00 dimension remains the same.

The **Measurement scale** box on the **Primary Units** option on the **Modify Dimension Style: Standard** dialog box is used to change the dimension values to match different drawing scales. Figure 8-18 shows the measurement scale set to a factor of 0.5000. If the drawing scale is 2=1,

as shown in Figure 8-17, then the scale factor for the **Measurement scale** must be 0.5000. Compare the preview in Figure 8-16 with the preview in Figure 8-18, where the scale factor has been changed.

To use the Text option

Figure 8-19 shows the **Text** option on the **Modify Dimension Style: Standard** dialog box. This option can be used to

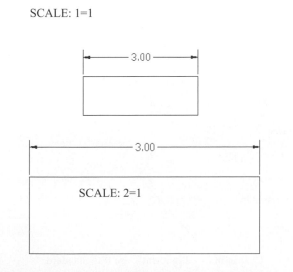

Figure 8-17

Figure 8-18

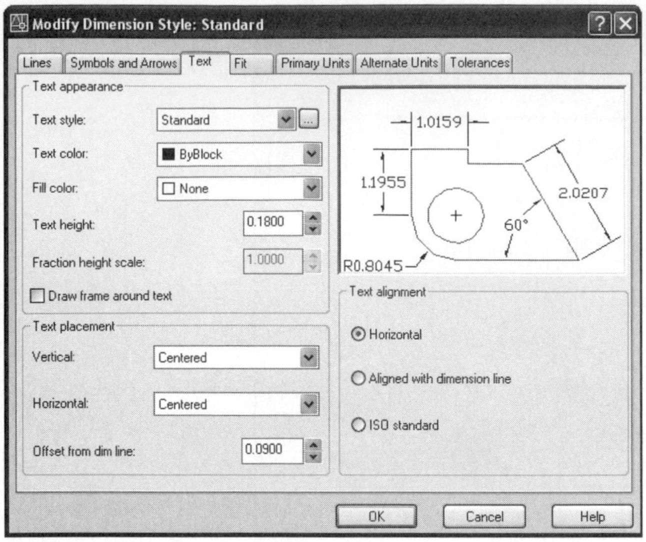

Figure 8-19

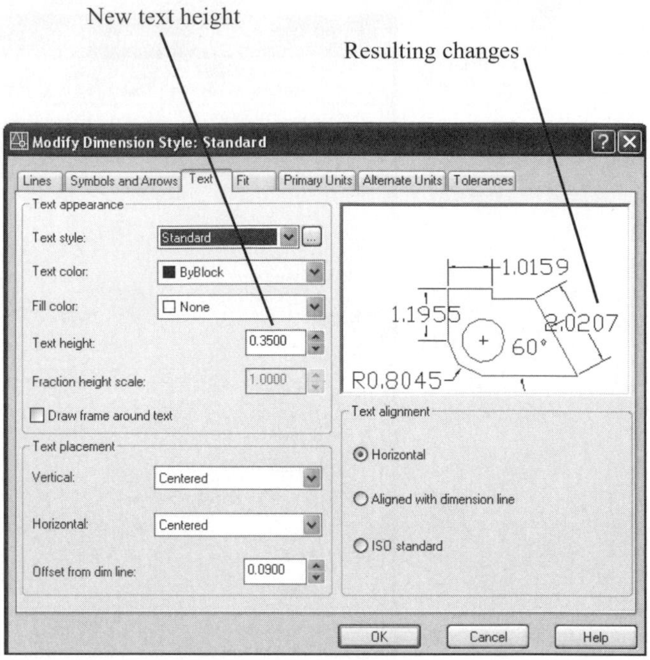

New text height

Resulting changes

Figure 8-20

change the height of dimension text or the text placement. In Figure 8-20 the text height was changed from the default value of 0.1800 to a value of 0.3500. The preview box shows the resulting changes in both the text and how it will be positioned on the drawing.

Figure 8-21 shows text located above the dimension lines and positioned nearer the first extension line. These changes were created using the **Vertical** and **Horizontal** options within the **Text placement** box.

Figure 8-22 shows text aligned with the direction of the dimension lines in accordance with ISO standards. This change was created using the ISO standard radio button within the **Text alignment** box.

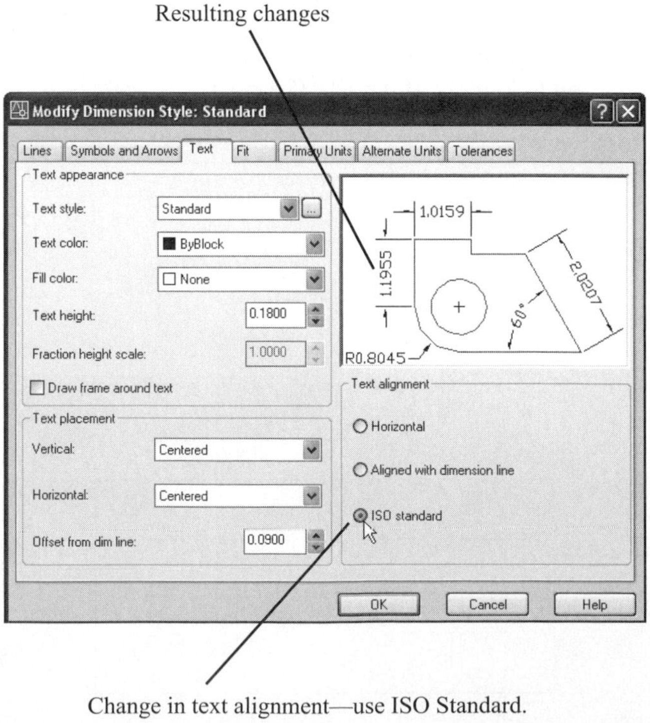

Resulting changes

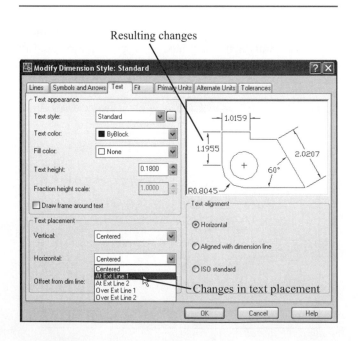

Changes in text placement

Figure 8-21

Resulting changes

Change in text alignment—use ISO Standard.

Figure 8-22

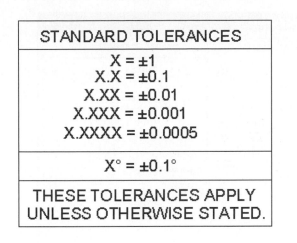

Figure 8-23

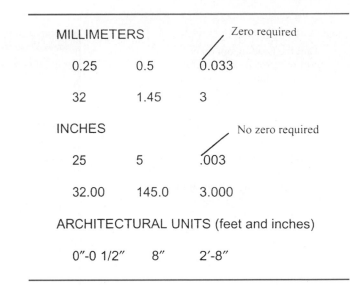

Figure 8-25

8-5 UNITS

It is important to understand that dimensional values are not the same as mathematical units. Dimensional values are manufacturing instructions and always include a tolerance, even if the tolerance value is not stated. Manufacturers use a predefined set of standard dimensions that are applied to any dimensional value that does not include a written tolerance. Standard tolerance values differ from organization to organization. Figure 8-23 shows a chart of standard tolerances.

In Figure 8-24 a distance is dimensioned twice: once as 5.50 and a second time as 5.5000. Mathematically these two values are equal, but they are not the same manufacturing instruction. The 5.50 value could, for example, have a standard tolerance of ±0.01, whereas the 5.5000 value could have a standard tolerance of ±0.0005. A tolerance of ±0.0005 is more difficult and therefore more expensive to manufacture than a tolerance of ±0.01.

Figure 8-25 shows examples of units expressed in millimeters, decimal inches, and architectural units. A zero is not required to the left of the decimal point for decimal inch values less than one. Millimeter values do not require zeros to the right of the decimal point. Architectural units should always include the feet (′) and inch (″) symbols. Millimeter and decimal inch values never include symbols; the units will be defined in the title block of the drawing.

To prevent a 0 from appearing to the left of the decimal point

1. Select the **Dimension Style** tool.

   The **Dimension Style** dialog box will appear.

2. Select **Modify.**

   The **Modify Dimension Style: Standard** dialog box will appear.

3. Select the **Primary Units** option.

   The **Primary Units** dialog box will appear. See Figure 8-26.

4. Click the box to the left of the word **Leading** within the **Zero suppression** box.

   A check mark will appear in the box indicating that the function is on.

5. Select **OK** to return to the drawing.

   Save the change, if desired. You can now dimension using any of the dimension commands; no zeros will appear to the left of the decimal point. Figure 8-27 shows the results.

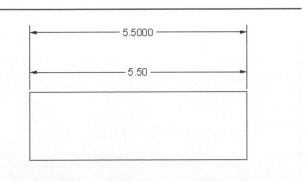

Figure 8-24

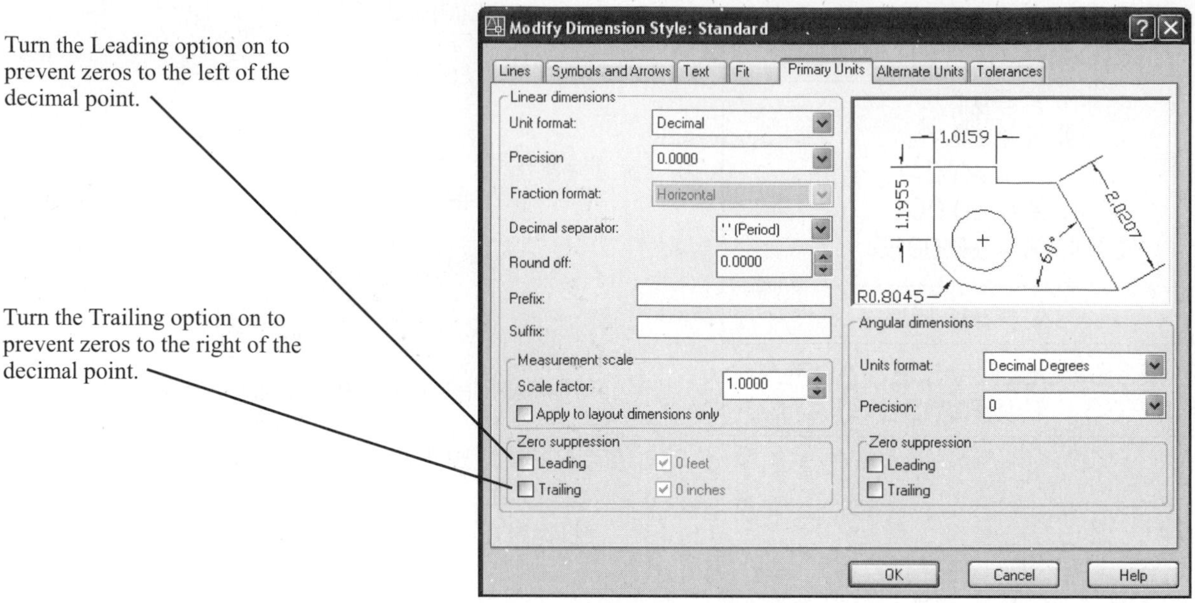

Turn the Leading option on to prevent zeros to the left of the decimal point.

Turn the Trailing option on to prevent zeros to the right of the decimal point.

Figure 8-26

To change the number of decimal places in a dimension value

1. Select the **Dimension Style** tool.

   The **Dimension Style Manager** dialog box will appear.

2. Select **Modify.**

   The **Modify Dimension Style: Standard** dialog box will appear.

3. Select the **Primary Units** option.
4. Select the arrow to the right of the **Precision** box.

   A list of precision options will cascade down. See Figure 8-28.

5. Select the desired value.

   Save the changes, if desired. You can now dimension using any of the dimension commands, and the resulting values will be expressed using the selected precision.

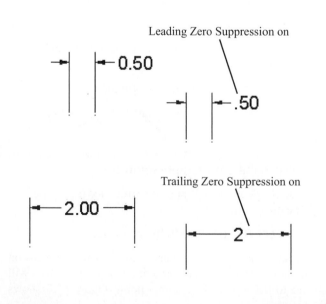

Leading Zero Suppression on

0.50

.50

Trailing Zero Suppression on

2.00

2

Figure 8-27

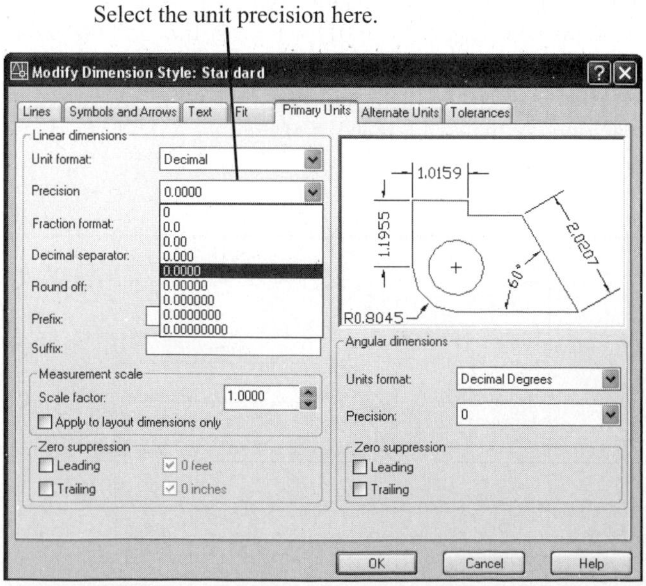

Select the unit precision here.

Figure 8-28

8-6 ALIGNED DIMENSIONS (See Figures 8-29 and 8-30.)

To create an aligned dimension

1. Select the **Aligned** tool from the **Dimension** toolbar.

   Command: _dimaligned
   Specify first extension line origin or <select object>:

2. Select the first extension line origin point.

   Specify second extension line origin:

3. Select the second extension line origin point.

   [Mtext/Text/Angle]:

4. Select the location for the dimension line.

The Select Object option

1. Select the **Aligned** tool from the **Dimension** toolbar.

   Command: _dimaligned
   Specify first extension line origin or <select object>:

2. Press the **Enter** key.

   Select object to dimension:

3. Select the line.

   [Mtext/Text/Angle]:

4. Select the dimension line location.

A response of **M** to the last prompt line will activate the **Multiline Text** option. The **Text Formatting** dialog box will appear. The **Text** option can be used to replace or supplement the default text generated by AutoCAD. The **Multiline Text** option is discussed in Section 2-16.

A dimension aligned with the Horizontal option on the Modify Dimension Styles on the aligned dimension tool.

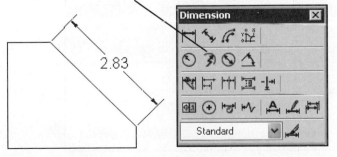

Figure 8-30

A dimension aligned using the Aligned with Dimension line option on the Modify Dimension Style: Standard dialog box.

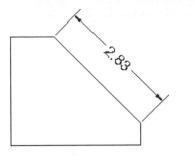

This format can also be achieved using the Angle option within the Aligned Dimension command.

Figure 8-29

A response of **A** to the prompt will activate the **Angle** option. The **Angle** option allows you to change the angle of the text within the dimension line. See Figure 8-30. The default angle value is **0°**, or horizontal. The example shown in Figure 8-29 used an angle of −45°. The prompt responses are as follows.

Specify angle of dimension text:
[MTExt/Text/Angle]:

1. Type **-45;** press **Enter.**
2. Select the location for the dimension.

8-7 RADIUS AND DIAMETER DIMENSIONS

Figure 8-31 shows an object that includes both arcs and circles. The general rule is to dimension arcs using a radius dimension, and circles using diameter dimensions.

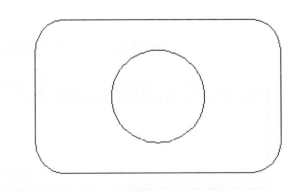

Figure 8-31

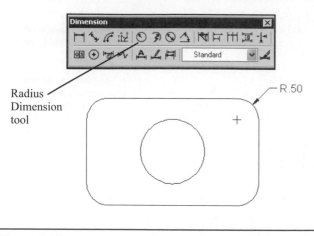

Radius
Dimension
tool

R.50

Figure 8-32

This convention is consistent with the tooling required to produce the feature shape. Any arc greater than 180° is considered a circle and is dimensioned using a diameter.

To create a radius dimension

1. Select the **Radius** tool from the **Dimension** toolbar.

   Command: _dimradius
   Select arc or circle:

2. Select the arc to be dimensioned.

   Specify dimension line location or [Mtext/Text/ Angle]:

3. Position the radius dimension so that its leader line is not horizontal or vertical.

   Figure 8-32 shows the resulting dimension. The dimension text and the angle of the text can be altered using the **[MText/Text/Angle]** options in the last prompt. In the example shown, it would be better to add the words **4 PLACES** to the radius dimension than to include four radius dimensions.

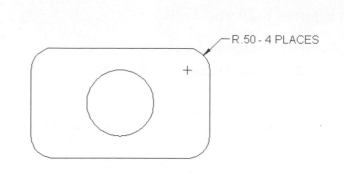

R.50 - 4 PLACES

Figure 8-34

To alter the default dimensions

1. Select the **Radius** tool from the **Dimension** toolbar.

   Command: _dimradius
   Select arc or circle:

2. Select the arc to be dimensioned.

   Specify dimension line location or [Mtext/Text/ Angle]:

3. Type **M**; press **Enter.**

   The **Text Formatting** dialog box will appear.

4. Place the flashing cursor just to the right of the dimension value box and type **- 4 PLACES.**

   See Figure 8-33.

5. Select **OK.**

   Figure 8-34 shows the resulting dimension. The **Radius** dimension command will automatically include a center point with the dimension. The center point can be excluded from the dimension as follows.

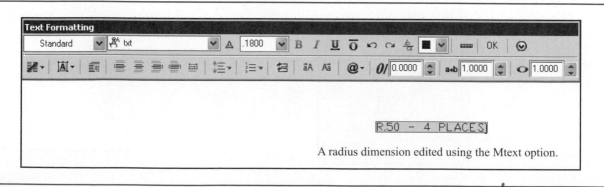

R.50 - 4 PLACES

A radius dimension edited using the Mtext option.

Figure 8-33

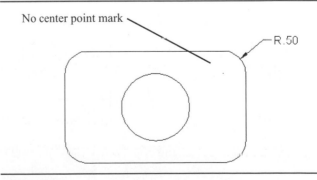

Figure 8-35

To remove the center mark from a radius dimension

1. Select the **Dimension Style** tool.

   The **Dimension Style Manager** dialog box will appear.

2. Select **Modify.**

   The **Modify Dimension Style: Standard** dialog box will appear.

3. Select the **Symbols and Arrows** option.
4. Select the **None** radio button in the **Center marks** area.

   A solid circle will appear in the preview box, indicating that it is on. See Figure 8-35.

5. Select **OK** to return to the drawing.

   You will have to redimension the arc, including the text alteration. Figure 8-36 shows the results.

To create a diameter dimension

   Circles require three dimensions: a diameter value plus two linear dimensions used to locate the circle's center point. AutoCAD can be configured to automatically add horizontal and vertical centerlines as follows.

1. Select the **Dimension Style** tool.

   The **Dimension Style Manager** dialog box will appear.

2. Select **Modify.**

   The **Modify Dimension Style: Standard** dialog box will appear.

3. Select the **Symbols and Arrows** option.
4. Select the **Line** option from the **Center marks** box.
5. Select **OK** to return to the drawing.

   Centerlines will be added to the existing radius dimension.

6. Explode the radius dimension, then erase the radius centerlines.
7. Select the **Diameter** tool from the **Dimension** toolbar.

   Command: _dimdiameter
   Select arc or circle:

Figure 8-36

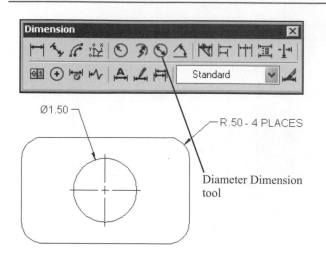

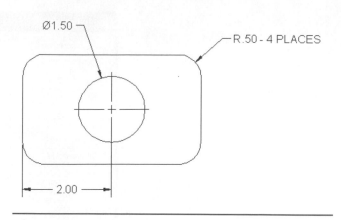

Figure 8-38

Figure 8-37

8. Select the circle.

   Specify dimension line location or [MText/Text/Angle]:

9. Locate the dimension away from the object so that the leader line is neither horizontal nor vertical.

   Figure 8-37 shows the results.

To add linear dimensions to given centerlines

1. Select the **Linear** tool on the **Dimension** toolbar.

   Command: _dimlinear
   Specify first extension line origin or <select object>:

2. Select the lower endpoint of the circle's vertical centerline.

   Specify second extension line origin:

3. Select the endpoint of the vertical edge line (the endpoint that joins with the corner arc).

   Figure 8-38 shows the results.

4. Repeat the above procedure to add the vertical dimension needed to locate the circle's center point.

5. Add the overall dimensions using the **Linear** dimension tool.

Figure 8-39 shows the results. Radius and diameter dimensions are usually added to a drawing after the linear dimensions because they are less restricted in their locations. Linear dimensions are located close to the distance they are defining, whereas radius and diameter dimensions can be located farther away and use leader lines to identify the appropriate arc or circle.

Avoid crossing extension and dimension lines with leader lines. See Figure 8-40.

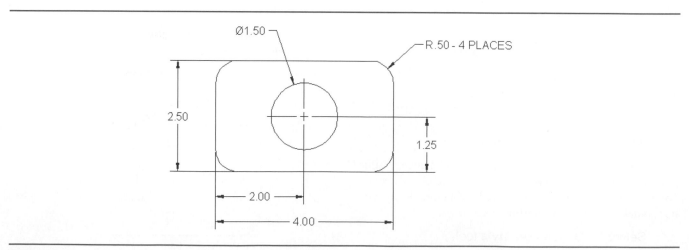

Figure 8-39

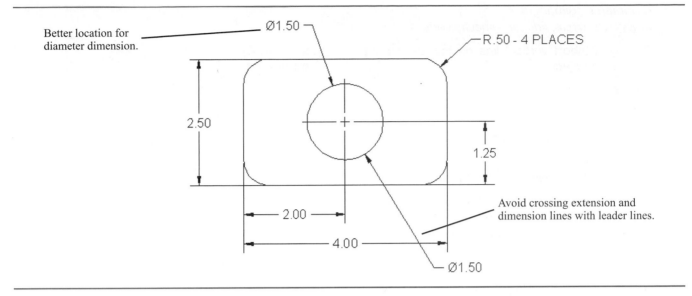

Better location for diameter dimension.

Ø1.50

R.50 - 4 PLACES

2.50

1.25

Avoid crossing extension and dimension lines with leader lines.

2.00

4.00

Ø1.50

Figure 8-40

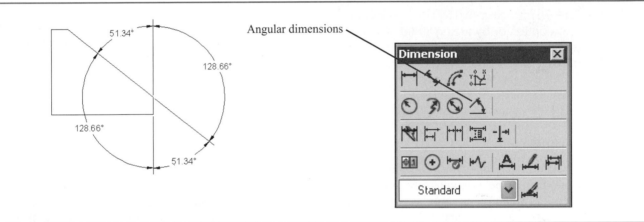

51.34°

128.66°

128.66°

51.34°

Angular dimensions

Figure 8-41

NOTE:

The diameter symbol can be added when using the **Text Formatting** dialog box by typing **%%c** or by using the **Symbols** option. See Figure 8-40. The characters **%%c** will appear on the **Text Formatting** screen but will be converted to the diameter symbol Ø when the text is applied to the drawing.

8-8 ANGULAR DIMENSIONS

Figure 8-41 shows four possible angular dimensions that can be created using the **Angular** tool on the **Dimension** toolbar. The extension lines and degree symbol will be added automatically.

To create an angular dimension (See Figure 8-42.)

1. Select the **Angular** tool on the **Dimension** toolbar.

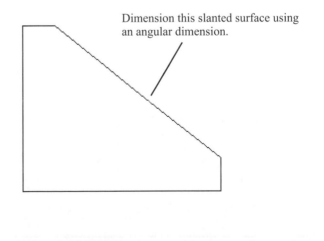

Dimension this slanted surface using an angular dimension.

Figure 8-42

Command: _dimangular
Select arc, circle, line, or <specify vertex>:

2. Select the short vertical line on the lower right side of the object.

   Select second line:

3. Select the slanted line.

   Specify dimension arc line location [MText/Text/ Angle]:

4. Locate the text away from the object.

Figure 8-43 shows the results. It is considered better to use two extension lines for angular dimensions and to not have an arrowhead touch the surface of the part.

> ## NOTE:
>
> The degree symbol can be added when using the **Text Formatting** dialog box by using the **Symbols** option or by typing **%%d**. The characters **%%d** will appear on the **Text Formatting** screen but will be converted to ° when the text is applied to the drawing.

Avoid overdimensioning

Figure 8-44 shows a shape dimensioned using an angular dimension. The shape is completely defined. Any additional dimension would be an error. It is tempting, in an effort to make sure a shape is completely defined, to add more dimensions, such as a horizontal dimension for the short horizontal edge at the top of the shape. This dimension is not needed and is considered double dimensioning.

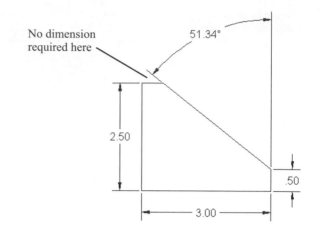

Figure 8-44

8-9 ORDINATE DIMENSIONS

Ordinate dimensions are dimensions based on an X,Y coordinate system. Ordinate dimensions do not include extension, or dimension lines, or arrowheads but simply horizontal and vertical leader lines drawn directly from the features of the object. Ordinate dimensions are particularly useful when dimensioning an object that includes many small holes.

Figure 8-45 shows an object that is to be dimensioned using ordinate dimensions. Ordinate dimensions are automatically calculated from the X, Y origin, or in this example, the lower left corner of the screen. If the object had been drawn with its lower left corner on the origin, you could proceed directly to the **Ordinate** tool on the **Dimension** toolbar; however, the lower left corner of the object is currently located at X=3, Y=4. First move the origin to the corner of the object, then use the **Ordinate** dimension tool.

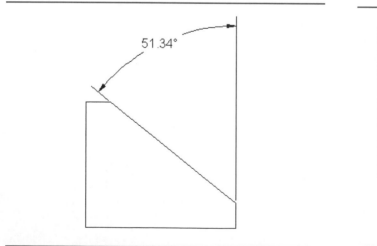

Figure 8-43

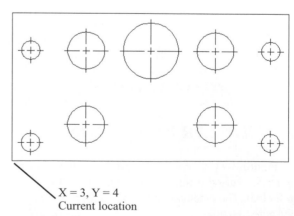

Figure 8-45

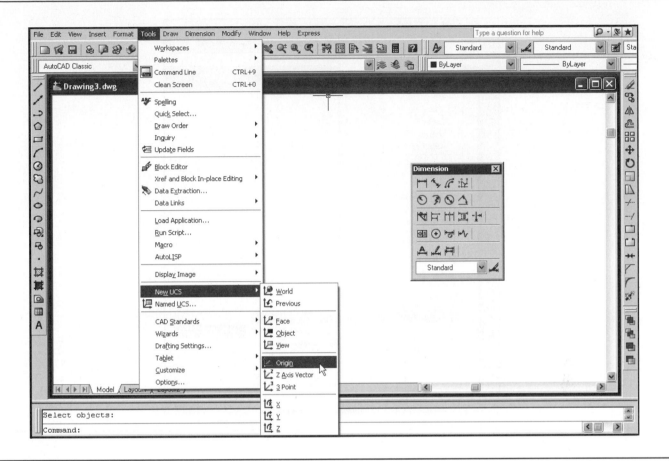

Figure 8-46

To move the origin

1. Select the **Tools** pull-down menu.

   The **Tools** pull-down menu will cascade.

2. Select **New UCS.**

   The **New UCS** menu will appear next to the **Tools** pull-down menu. See Figure 8-46.

3. Select **Origin.**

   Specify new origin point <0,0,0>:

4. Select the lower left corner of the object.

   The origin (0,0) is now located at the lower left corner of the object. This can be verified by looking at the coordinate display at the lower left corner of the screen.

   The origin icon may move to the lower left corner as shown in Figure 8-47, depending on your computer's settings. The tool can be moved back to the original screen location as follows.

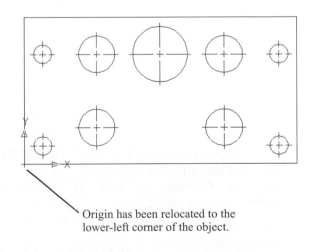

Origin has been relocated to the lower-left corner of the object.

Figure 8-47

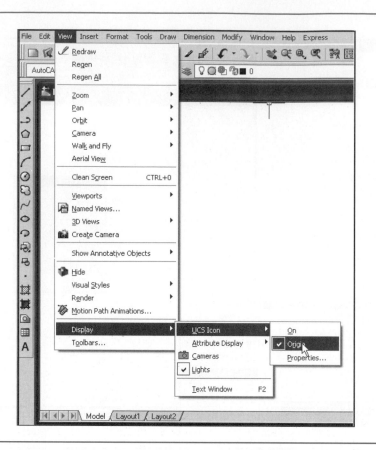

Figure 8-48

To move the Origin tool

1. Select the **View** pull-down menu, then **Display.**

The **Display** menu will appear next to the **View** pull-down menu. See Figure 8-48.

2. Select **UCS Icon,** then **Origin.**

The **Origin** tool will go to the current origin.

To add ordinate dimensions to an object

The following procedure assumes that you have already used the **Dimension Style** command (Section 8-4) to set the desired dimension style and that you have moved the origin to the lower left corner of the object as shown.

1. Turn the **Ortho** command on (click the **ORTHO** box at the bottom of the screen).
2. Select the **Ordinate** tool from the **Dimension** toolbar.

   Command: _dimordinate
   Select feature location:

3. Select the lower endpoint of the first circle's vertical centerline.

Specify leader endpoint or [Xdatum/Ydatum/ MText/Text/Angle]:

4. Select a point along the X axis directly below the vertical centerline of the circle.

The ordinate value of the point will be added to the drawing. This point should have a **0.50** value. The text value may be modified using either the **MText** or **Text** option, or by using the **Dimension Style Manager** dialog box to define the precision of the text.

5. Press the right mouse button to restart the command and dimension the object's other features.
6. Extend the centerlines across the object and add the diameter dimensions for the holes.

Figure 8-49 shows the completed drawing. The **Text** option of the prompt shown in step 3 can be used to modify or remove the default text value.

8-10 BASELINE DIMENSIONS

Baseline dimensions are a series of dimensions that originate from a common baseline or datum line. Baseline dimensions are very useful because they help eliminate tolerance buildup associated with chain-type dimensions.

Ordinate dimensions

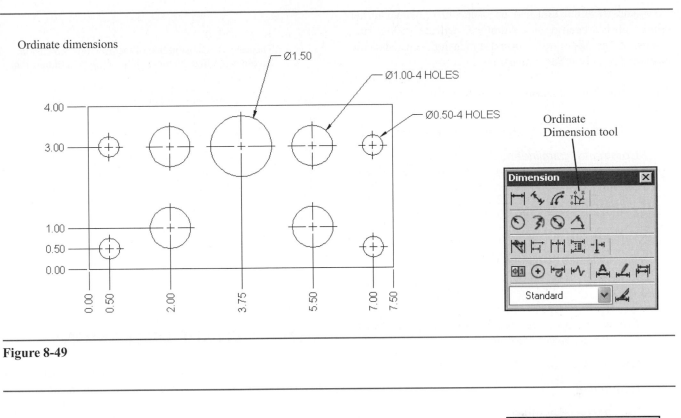

Figure 8-49

Baseline dimensions

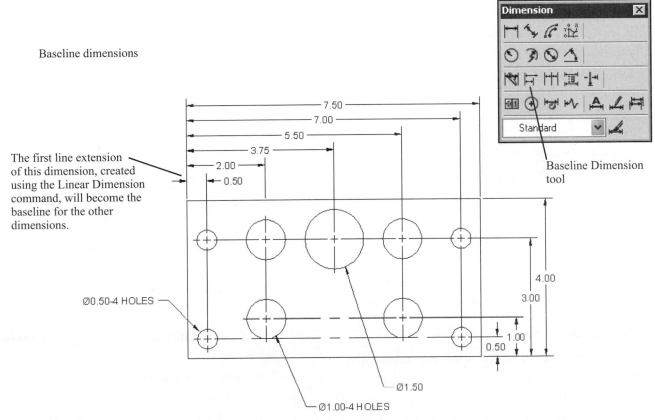

The first line extension of this dimension, created using the Linear Dimension command, will become the baseline for the other dimensions.

Baseline Dimension tool

Figure 8-50

The **Baseline** tool can be used only after an initial dimension has been drawn. AutoCAD will define the first extension line origin of the initial dimension selected as the baseline for all baseline dimensions.

To use the Baseline dimension tool (See Figure 8-50.)

1. Select the **Linear** tool on the **Dimension** toolbar.

    Command: _dimlinear
    Specify first extension line origin or <select object>:

2. Select the upper left corner of the object.

    This selection determines the baseline.

    Specify second extension line origin:

3. Select the endpoint of the first circle's vertical centerline.

    Specify dimension line location or [Text/Angle/ Horizontal/Vertical/Rotated]:

4. Select a location for the dimension line.

    Command:

5. Select the **Baseline** tool on the **Dimension** toolbar.

    Specify a second extension line origin or [Undo/ Select] <Select>:

6. Select the endpoint of the next circle's vertical centerline.

    Specify a second extension line origin or [Undo/ Select] <Select>:

7. Continue to select the circle centerlines until all circles are located.

    Specify a second extension line origin or [Undo/ Select] <Select>:

8. Select the upper right corner of the object.
9. Press the right mouse button, then select **<Enter>**.

    This will end the **Baseline** dimension command.

10. Repeat the preceding procedure for the vertical baseline dimensions.
11. Add the circles' diameter values.

    The **Baseline** dimension option can also be used with the **Angular** dimension option.

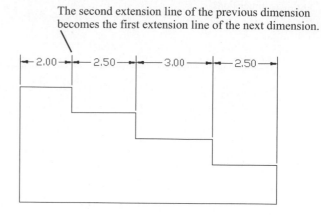

The second extension line of the previous dimension becomes the first extension line of the next dimension.

Figure 8-51

8-11 CONTINUE DIMENSION

The **Continue** tool on the **Dimension** toolbar is used to create chain dimensions based on an initial linear, angular, or ordinate dimension. The second extension line's origin becomes the first extension line origin for the continued dimension.

To use the Continue dimension command (See Figure 8-51.)

1. Select the **Linear** tool on the **Dimension** toolbar.

    Command: dimcontinue
    Specify first extension line origin or <select object>:

2. Select the upper left corner of the object.

    Specify second extension line origin:

3. Select the right endpoint of the uppermost horizontal line.

    Dimension line location (Text/Angle/Horizontal/ Vertical/Rotated):

4. Select a dimension line location.

    Command:

5. Select the **Continue** tool on the **Dimension** toolbar.

    Command: _dimcontinue

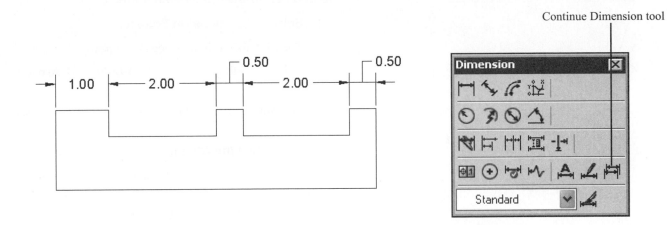

Continue Dimension tool

Figure 8-52

*Specify a second extension line origin or [Undo/
Select] <Select>:*

6. Select the next linear distance to be dimensioned.

*Specify a second extension line origin or [Undo/
Select] <Select>:*

7. Continue until the object's horizontal edges are
completely dimensioned.

AutoCAD will automatically align the dimensions.
Figure 8-52 shows how the **Continue** tool dimensions
distances that are too small for both the arrowhead and
dimension value to fit within the extension lines.

8-12 QUICK DIMENSION

The **Quick Dimension** tool is used to add a series of
dimensions. See Figure 8-53.

To use the Quick Dimension command

1. Select the **Quick Dimension** tool from the
Dimension toolbar.

   *Command: _qdim
   Select geometry to dimension:*

2. Select the left vertical edge line of the object to
be dimensioned; press **Enter.**

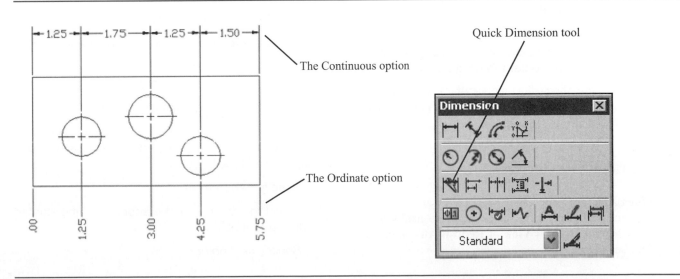

The Continuous option

The Ordinate option

Quick Dimension tool

Figure 8-53

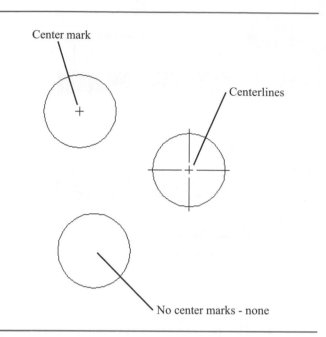

Center mark

Centerlines

No center marks - none

Figure 8-54

Specify dimension line position, or [Continuous/ Staggered/Baseline/Ordinate/Radius/Diameter/ datumPoint/Edit]<Baseline>:

3. Select the vertical centerline of the first hole.

    Select geometry to dimension:

4. Select the vertical centerline of the second hole.

    Select geometry to dimension:

5. Select the vertical centerline of the third hole.
6. Press the right mouse button.

    Specify dimension line position, or [Continuous/ Staggered/Baseline/Ordinate/Radius/Diameter/ datumPoint/Edit]<Baseline>:

7. Type **c**; press **Enter.**
8. Position the dimension lines, press the right mouse button, and enter the position.

The ordinate dimensions along the bottom of the object in Figure 8-53 were created by typing **o** rather than **c** in step 7.

8-13 CENTER MARK

When AutoCAD first draws a circle or arc, a center mark appears on the drawing; however, these marks will disappear when the **Redraw View** or **Redraw All** command is applied. See Figure 8-54.

To add centerlines to a given circle

1. Select the **Dimension Style** tool.

    The **Dimension Style** dialog box appears.

2. Select **Modify,** then the **Symbols and Arrows** option.

    The **Modify Dimension Style: Standard** dialog box will appear.

3. Select the **Line** option.

    The preview display will show a horizontal and a vertical centerline.

4. Select **OK** to return to the drawing.
5. Select the **Center Mark** tool.

    Select arc or circle:

6. Select the circle.

Horizontal and vertical centerlines will appear. The size of the center mark can be controlled using the **Size** box in the **Center marks** box. If the centerline's size appears unacceptable, try different sizes until you get an acceptable size.

8-14 MLEADER AND QLEADER

Leader lines are slanted lines that extend from notes or dimensions to a specific feature or location on the surface of a drawing. They usually end with an arrowhead or dot. The **Radius** and **Diameter** tools on the **Dimension** toolbar automatically create a leader line. The **MLEADER** and **QLEADER** commands can be used to add leader lines not associated with radius and diameter dimensions.

To create a leader line with text

1. Type **qleader** in response to a command prompt.

    Command: _qleader
    Specify first leader point, or [Settings]<Settings>:

2. Select the starting point for the leader line.

    This is the point where the arrowhead will appear. In the example shown in Figure 8-55 the upper right corner of the object was selected.

    Specify next point:

3. Select the location of the endpoint of the slanted line segment.

    Specify next point:

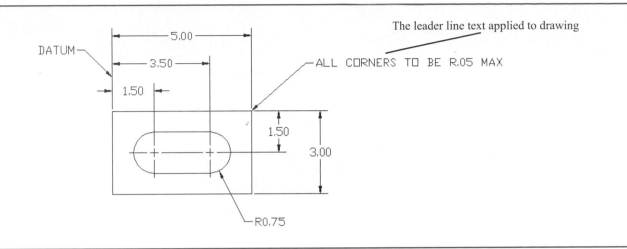

Figure 8-55

4. Draw a short horizontal line segment; press **Enter.**

   Specify text width <0.0000>:

5. Press **Enter.**

   Enter first line of annotation text <MText>:

6. Press **Enter.**

   The **Text Formatting** dialog box will appear.

7. Use the cursor and extend the text box as needed.
8. Type the desired text. See Figure 8–56.
9. Click the **OK** button.

The text will appear next to the horizontal line segment of the leader line.

To draw a curved leader line

The **Leader** command can be used to draw curved leader lines and leader lines that end with dots. See Figure 8-57.

1. Type **qleader** in response to a command prompt.

   Command: _qleader
   Specify first leader point, or [Settings] <Settings>:

2. Type **s;** press **Enter.**
3. Select the **Leader Line & Arrow** option.

The **Leader Settings** dialog box will appear. See Figure 8-58.

4. Select the **Spline** option in the **Leader Line** box, then **OK.**

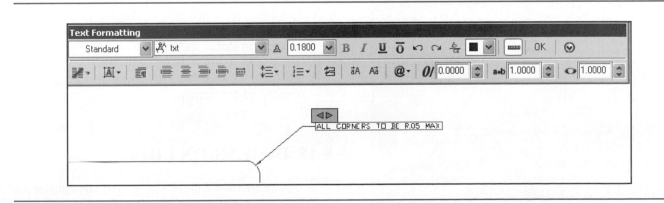

Figure 8-56

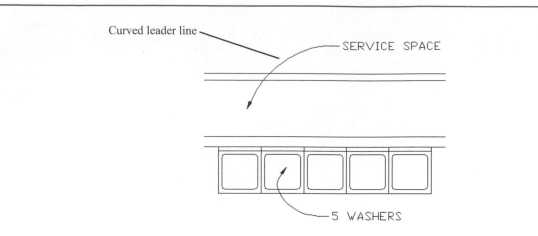

Figure 8-57

Specify first leader point, or [Settings] <Settings>:

5. Select the starting point for the leader line.

   This is the point where the arrowhead will appear.

   Specify next point:

6. Select the next point.

   Select next point:

AutoCAD will shift to the **Drag** mode, which allows you to move the cursor around and watch the changes in

shapes of the leader line. More than one point may be selected to define the shape.

7. Complete the leader line as explained previously.

To draw a leader line with a dot at its end

1. Select the **Leader** tool.

   Command: _qleader
   Specify first leader point, or [Settings] <Settings>:

2. Type **s;** press **Enter.**
3. Select the **Leader Line & Arrow** option.

   The **Leader Settings** dialog box will appear. See Figure 8-59.

4. Select the arrow to the right of the **Arrowhead** box.

   A list of arrowhead options will appear.

5. Select the **Dot small** option.
6. Select **OK** to return to the drawing.
7. Use the **Leader** tool to create leader lines as described previously.

8-15 DIMENSION EDIT

The **Dimension Edit** tool is used to edit existing dimensioning text. Existing text can be moved or rotated, or the text value can be changed. See Figure 8-60.

Figure 8-58

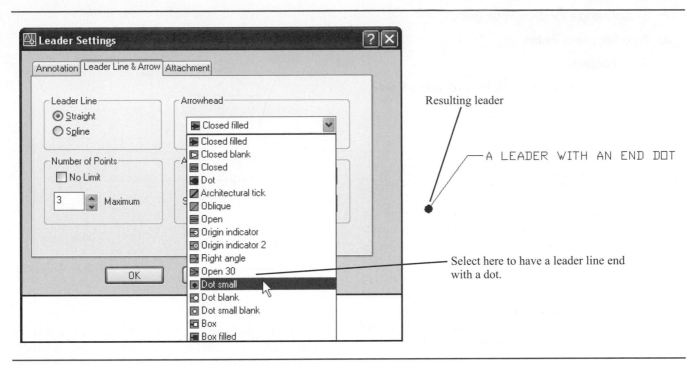

Resulting leader

A LEADER WITH AN END DOT

Select here to have a leader line end with a dot.

Figure 8-59

To use the Dimension Edit option to change a text value

1. Select the **Dimension Edit** tool.

   Command: _dimtedit
   Enter type of dimension editing [Home/New/Rotate/Oblique] <Home>:

2. Select the **New** option or type **n**; press **Enter.**

   The **Text Formatting** dialog box will appear. See Section 2-16 for a more complete explanation of the **Text Formatting** dialog box.

3. Delete the existing text value and type in a new value; click **OK.**
4. Select the dimension to be edited.
5. Press the right mouse button to enter the new text value.

To change the angle of existing dimension text

1. Select the **Dimension Edit** tool.

   Command: _dimtedit
   Enter type of dimension editing [Home/New/Rotate/Oblique] <Home>:

2. Select the **Rotate** option or type **r**; press **Enter.**

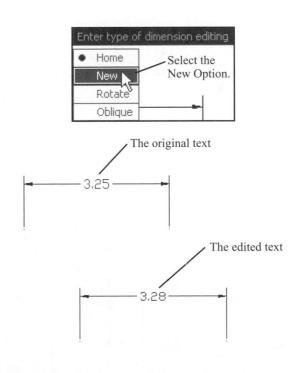

Select the New Option.

The original text

3.25

The edited text

3.28

Figure 8-60

Specify angle for dimension text:

3. Type **90;** press **Enter.**

Select objects:

4. Select the dimension to be rotated; press **Enter.**

Figure 8-61 shows the results.

8-16 TOLERANCES

Tolerances are numerical values assigned with the dimensions that define the limits of manufacturing acceptability for a distance. AutoCAD can create four types of tolerances: symmetrical, deviation, limits, and basic. See Figure 8-62. Many companies also use a group of standard tolerances that are applied to any dimensional value that is not assigned a specific tolerance. See Figure 8-23. Tolerances for numerical values expressed in millimeters are applied using a different convention than that used for inches. Tolerances are discussed in Chapters 9 and 10. The discussion of the appropriate dimensioning tools will be covered in those chapters.

8-17 DIMENSIONING HOLES

Holes are dimensioned by stating their diameter and depth, if any. The symbol **Ø** is used to represent diameter. It is considered good practice to dimension a hole using a diameter value because the tooling used to produce the

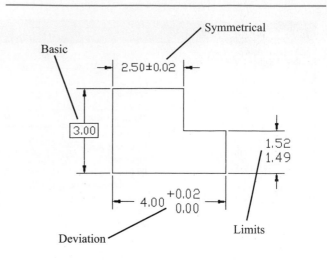

Figure 8-62

hole is also defined in terms of diameter values. A notation like 12 DRILL is considered less desirable because it specifies a machining process. Manufacturing processes should be left, whenever possible, to the discretion of the shop.

To dimension individual holes

Figure 8-63 shows three different methods that can be used to dimension a hole that does not go completely through an object. If the hole goes completely through, only the diameter need be specified. The **Radius** and **Diameter** dimension tools were covered in Section 8-7. Depth values may be added using the **Text Formatting**

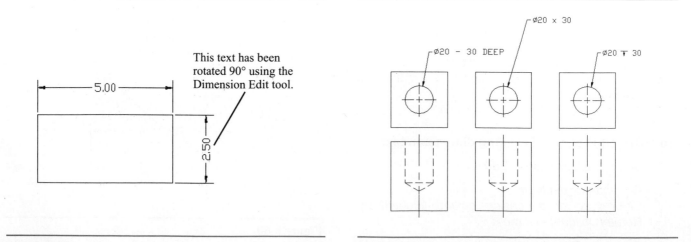

Figure 8-61

Figure 8-63

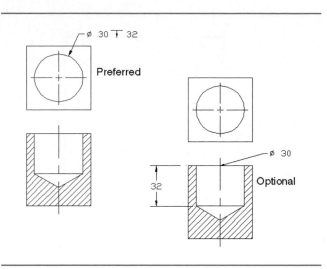

Figure 8-64

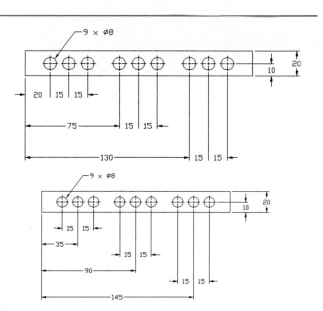

Figure 8-66

dialog box or by using the **Text** option of any of the dimension commands.

Figure 8-64 shows two methods of dimensioning holes in sectional views. The single-line note version is the preferred method.

To dimension hole patterns

Figure 8-65 shows two different hole patterns dimensioned. The circular pattern includes the note **Ø10-4 HOLES.** This note serves to define all four holes within the object.

Figure 8-65 also shows a rectangular object that contains five holes of equal diameter, equally spaced from one another. The notation **5 × Ø10** specifies five holes of 10 diameter. The notation **4 × 20 (=80)** means 4 equal spaces of 20. The notation **(=80)** is a reference dimension and is included for convenience. Reference dimensions are explained in Chapter 9.

Figure 8-66 shows two additional methods for dimensioning repeating hole patterns. Figure 8-67 shows a circular hole pattern that includes two different hole diameters. The hole diameters are not noticeably different and could be confused. One group is defined by indicating letter (A); the other is dimensioned in a normal manner.

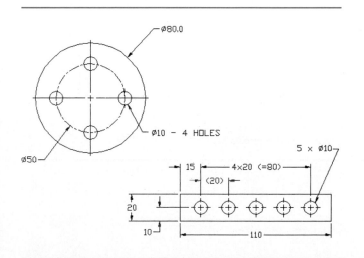

Figure 8-65

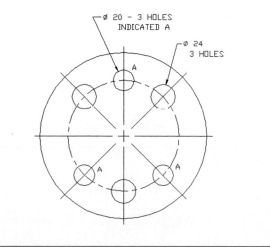

Figure 8-67

Place shorter dimensions closer to the object than longer ones.

Place dimensions near the features they are defining.

DO NOT PLACE DIMENSIONS ON THE SURFACE OF THE OBJECT.

Use the Explode, Erase, and Move commands to reconstruct and relocate inappropriate dimensions.

Align groups of dimensions.

Place overall dimensions the farthest away from the object.

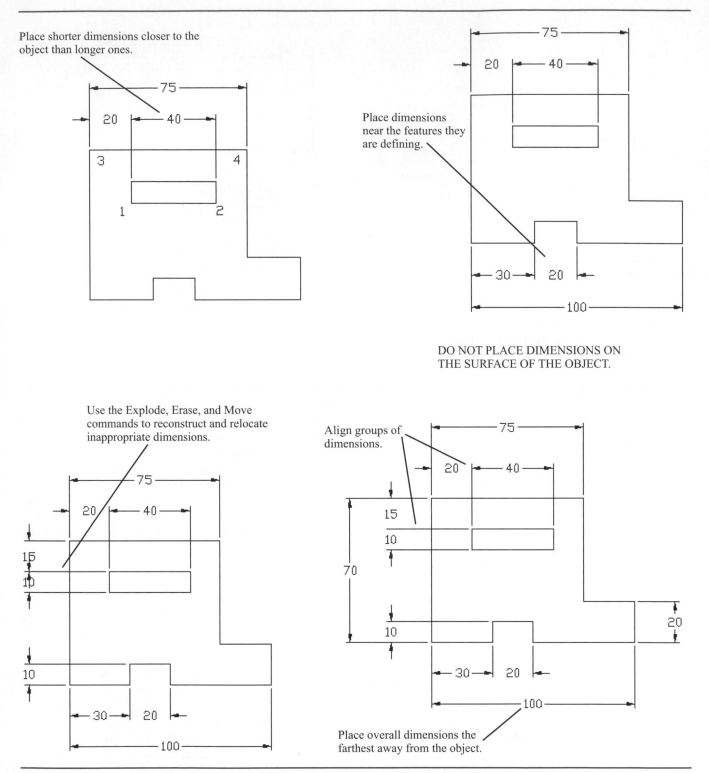

Figure 8-68

8-18 PLACING DIMENSIONS

There are several general rules concerning the placement of dimensions. See Figure 8-68.

1. Place dimensions near the features they are defining.
2. Do not place dimensions on the surface of the object.
3. Align and group dimensions so that they are neat and easy to understand.
4. Avoid crossing extension lines.

Sometimes it is impossible not to cross extension lines because of the complex shape of the object, but whenever possible, avoid crossing extension lines.

5. Place shorter dimensions closer to the object than longer ones.
6. Always place overall dimensions the farthest away from the object.
7. Do not dimension the same distance twice. This is called *double dimensioning* and will be discussed in Chapter 9.

8-19 FILLETS AND ROUNDS

Fillets and rounds may be dimensioned individually or by a note. In many design situations all the fillets and rounds are the same size, so a note as shown in Figure 8-69 is used. Any fillets or rounds that have a different radius from that specified by the note are dimensioned individually.

See Chapter 2 for an explanation of how to draw fillets and rounds using the **Fillet** command.

8-20 ROUNDED SHAPES (INTERNAL)

Internal rounded shapes are called *slots.* Figure 8-70 shows three different methods for dimensioning slots. The end radii are indicated by the note **R - 2 PLACES,** but no numerical value is given. The width of the slot is dimensioned, and it is assumed that the radius of the rounded ends is exactly half of the stated width.

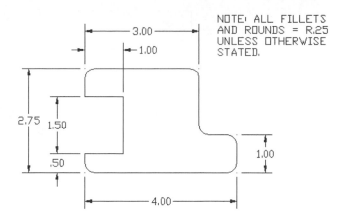

Figure 8-69

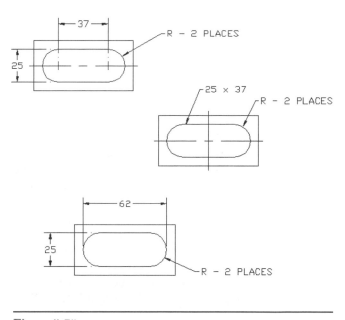

Figure 8-70

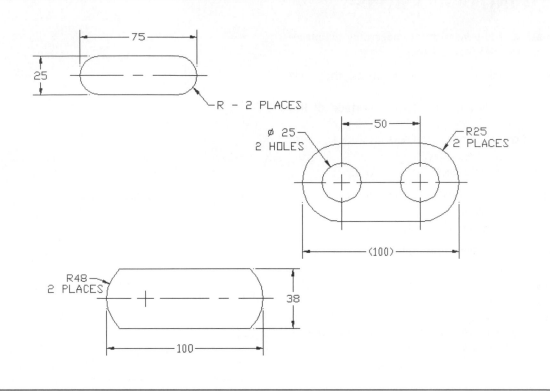

Figure 8-71

8-21 ROUNDED SHAPES (EXTERNAL)

Figure 8-71 shows two example shapes with external rounded ends. As with internal rounded shapes, the end radii are indicated but no value is given. The width of the object is given, and the radius of the rounded end is assumed to be exactly half of the stated width.

The second example shown in Figure 8-72 shows an object dimensioned using the object's centerline. This type of dimensioning is used when the distance between the hole is more important than the overall length of the object; that is, the tolerance for the distance between the holes is more exact than the tolerance for the overall length of the object.

The overall length of the object is given as a reference dimension (100). This means the object will be manufactured based on the other dimensions, and the 100 value will be used only for reference.

Objects with partially rounded edges should be dimensioned as shown in Figure 8-72. The radii of the end features are dimensioned. The center point of the radii is implied to be on the object centerline. The overall dimension is given; it is not referenced unless specific radii values are included.

8-22 IRREGULAR SURFACES

There are three different methods for dimensioning irregular surfaces: tabular, baseline, and baseline with oblique extension lines. Figure 8-72 shows an irregular surface dimensioned using the tabular method. The X,Y axes are defined using the edges of the object. Points are then defined relative to the X,Y axes. The points are assigned reference numbers, and the reference numbers and X,Y coordinate values are listed in chart form as shown.

Figure 8-73 shows an irregular curve dimensioned using baseline dimensions. The baseline method references all dimensions back to specified baselines. Usually there are two baselines, one horizontal and one vertical.

It is considered poor practice to use a centerline as a baseline. Centerlines are imaginary lines that do not exist on the object and would make it more difficult to manufacture and inspect the finished objects.

Baseline dimensioning is very common because it helps eliminate tolerance buildup (see Section 9-12) and is easily adaptable to many manufacturing processes. AutoCAD has a special **Baseline** dimension tool for creating baseline dimensions.

Station	1	2	3	4	5	6
X	0	20	40	55	62	70
Y	40	38	30	16	10	0

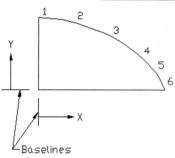

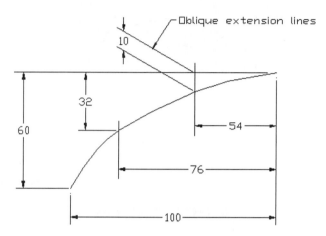

Figure 8-72

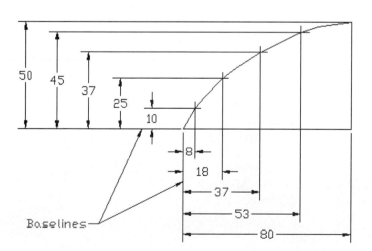

Figure 8-73

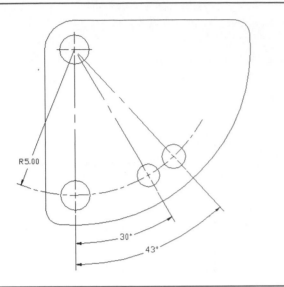

Figure 8-74

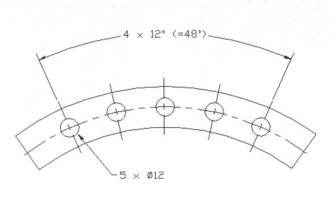

Figure 8-75

8-23 POLAR DIMENSIONS

Polar dimensions are similar to polar coordinates. A location is defined by a radius (distance) and an angle. Figure 8-74 shows an object that includes polar dimensions. The holes are located on a circular centerline, and their positions from the vertical centerline are specified using angles.

Figure 8-75 shows an example of a hole pattern dimensioned using polar dimensions.

8-24 CHAMFERS

Chamfers are angular cuts made on the edges of objects. They are usually used to make it easier to fit two parts together. They are most often made at 45° angles but may be made at any angle. Figure 8-76 shows two objects with chamfers between surfaces 90° apart and two examples between surfaces that are not 90° apart. Either of the two types of dimensions shown for the 45° dimension may be used. If an angle other than 45° is used, the angle and setback distance must be specified.

Figure 8-77 shows two examples of internal chamfers. Both define the chamfers using an angle and diameter. Internal chamfers are very similar to countersunk holes. See Section 5-26.

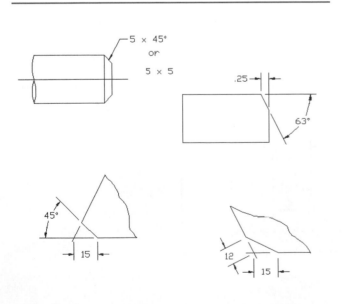

Figure 8-76

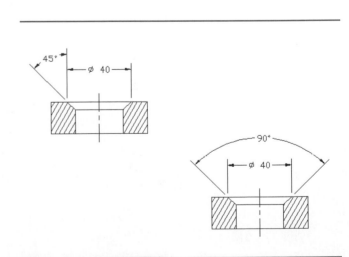

Figure 8-77

8-25 KNURLING

Knurls are used to make it easier to grip a shaft, or to roughen a surface before it is used in a press fit. There are two types of knurls: diamond and straight.

Knurls are defined by their pitch and diameter. See Figure 8-78. The *pitch* of a knurl is the ratio of the number of grooves on the circumference to the diameter. Standard knurling tools sized to a variety of pitch sizes are used to manufacture knurls for both English and metric units.

Diamond knurls may be represented by a double hatched pattern or by an open area with notes. The **Hatch** command is used to draw the double-hatched lines. See Section 6-4.

Straight knurls may be represented by straight lines in the pattern shown or by an open area with notes. The straight-line pattern is created by projecting lines from a construction circle. The construction points are evenly spaced on the circle. Once drawn, the straight-line knurl pattern can be saved as a wblock for use on other drawings. See Section 3-23 for an explanation of wblock.

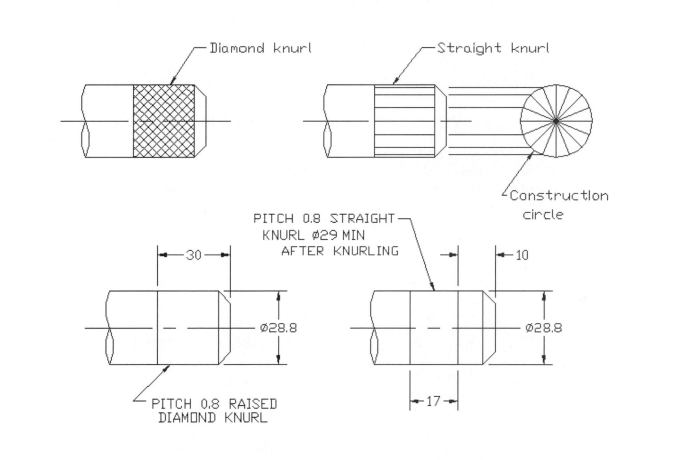

Figure 8-78

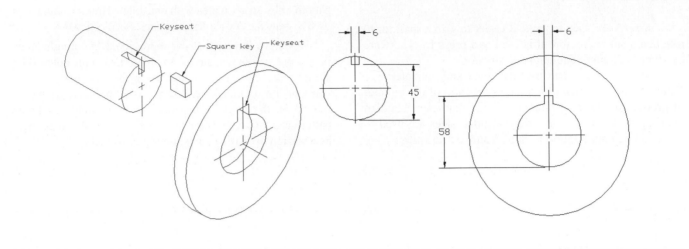

Figure 8-79

8-26 KEYS AND KEYSEATS

Keys are small pieces of material used to transmit power. For example, Figure 8-79 shows how a key can be fitted between a shaft and a gear so that the rotary motion of the shaft can be transmitted to the gear.

There are many different styles of keys. The key shown in Figure 8-79 has a rectangular cross section and is called a square key. Keys fit into grooves called *keyseats,* or *keyways.*

Keyways are dimensioned from the bottom of the shaft or hole as shown.

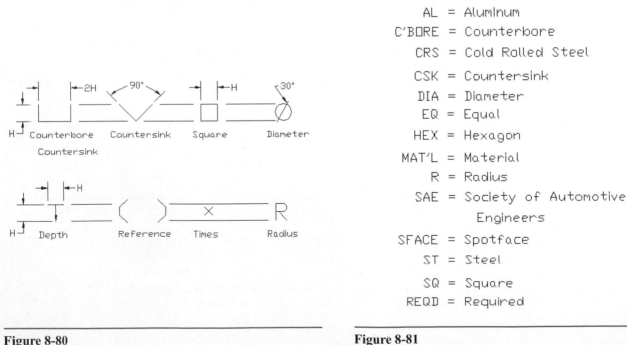

Figure 8-80

Figure 8-81

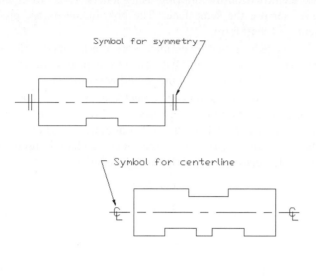

Figure 8-82

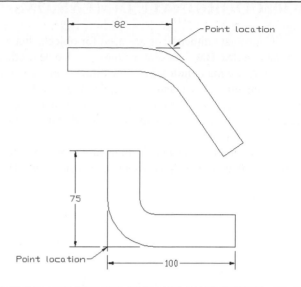

Figure 8-83

8-27 SYMBOLS AND ABBREVIATIONS

Symbols are used in dimensioning to help accurately display the meaning of the dimension. Symbols also help eliminate language barriers when reading drawings. Figure 8-80 shows a list of dimensioning symbols and their meanings. The height of a symbol should be the same as the text height.

Abbreviations should be used very carefully on drawings. Whenever possible, write out the full word including

correct punctuation. Figure 8-81 shows several standard abbreviations used on technical drawings.

8-28 SYMMETRY AND CENTERLINE

An object is symmetrical about an axis when one side is an exact mirror image of the other. Figure 8-82 shows a symmetrical object. The two short parallel lines symbol or the note **OBJECT IS SYMMETRICAL ABOUT THIS AXIS** (centerline) may be used to designate symmetry.

If an object is symmetrical, only half the object need be dimensioned. The other dimensions are implied by the symmetry note or symbol.

Centerlines are slightly different from the axis of symmetry. An object may or may not be symmetrical about its centerline. See Figure 8-82. Centerlines are used to define the center of both individual features and entire objects. Use the centerline symbol when a line is a centerline, but do not use it in place of the symmetry symbol.

8-29 DIMENSIONING TO POINTS

Curved surfaces can be dimensioned using theoretical points. See Figure 8-83. There should be a small gap between the surface of the object and the lines used to define the theoretical point. The point should be defined by the intersection of at least two lines.

There should also be a small gap between the extension lines and the theoretical point used to locate the point.

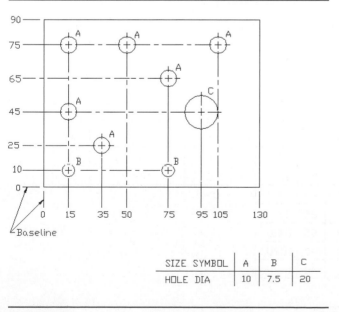

SIZE SYMBOL	A	B	C
HOLE DIA	10	7.5	20

Figure 8-84

8-30 COORDINATE DIMENSIONS

Coordinate dimensions are used for objects that contain many holes. Baseline dimensions can also be used, but when there are many holes, baseline dimensions can create a confusing appearance and will require a large area on the drawing. Coordinate dimensions use charts that simplify the appearance, use far less space on the drawing, and are easy to understand.

Figure 8-84 shows an object that has been dimensioned using coordinate dimensions without dimension lines. Holes are identified on the drawing using letters. Holes of equal diameter use the same letter. The hole diameters are presented in chart form.

Hole locations are defined using a series of centerlines referenced to baselines. The distance from the baseline to the centerline is written below the centerline as shown.

Figure 8-85 shows an object that has been dimensioned using coordinate dimensions in tabular form. Each hole is assigned both a letter and a number. Holes of equal diameter are assigned the same letter. A chart is used to define the diameter values for each hole letter.

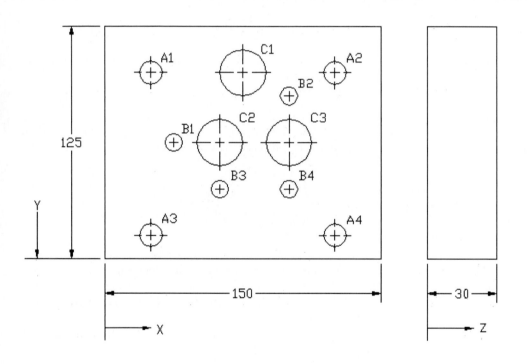

HOLE	FROM	X	Y	Z
A1	XY	15	65	THRU
A2	XY	80	65	THRU
A3	XY	15	10	THRU
A4	XY	80	10	THRU
B1	XY	25	40	12
B2	XY	65	56	12
B3	XY	40	25	12
B4	XY	65	25	12
C1	XY	48	65	THRU
C2	XY	40	40	THRU
C3	XY	65	40	THRU

HOLE	DESCRIPTION	QTY
A	⌀8	4
B	⌀5	4
C	⌀16	3

Figure 8-85

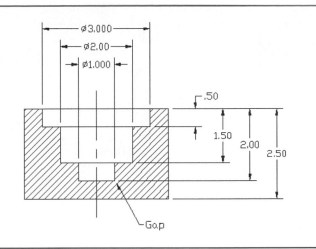

Figure 8-86

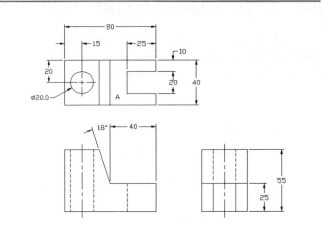

Figure 8-87

Hole locations are defined relative to X,Y axes. A Z axis is used for depth dimensions. A chart lists each hole by its letter–number designation and specifies its distance from the X, Y, or Z axis. The overall dimensions are given using extension and dimension lines.

The side view does not show any hidden lines because if all the lines were shown, it would be too confusing to understand. A note **THIS VIEW LEFT BLANK FOR CLARITY** may be added to the drawing.

8-31 SECTIONAL VIEWS

Sectional views are dimensioned, as are orthographic views. See Figure 8-86. The sectional lines should be drawn at an angle that allows the viewer to clearly distinguish between the sectional lines and the extension lines.

8-32 ORTHOGRAPHIC VIEWS

Dimensions should be added to orthographic views where the features appear in contour. Holes should be dimensioned in their circular views. Figure 8-87 shows three views of an object that has been dimensioned.

The hole dimensions are added to the top view where the hole appears circular. The slot is also dimensioned in the top view because it appears in contour. The slanted surface is dimensioned in the front view.

The height of surface A is given in the side view rather than run along extension lines across the front view. The length of surface A is given in the front view. This is a contour view of the surface.

It is considered good practice to keep dimensions in groups. This makes it easier for the viewer to find dimensions.

Be careful not to double-dimension a distance. A distance should be dimensioned only once per view. If a 30 dimension were added above the 25 dimension on the right-side view, it would be an error. The distance would be double-dimensioned: once with the 25 + 30 dimension and again with the 55 overall dimension. The 25 + 30 dimensions are mathematically equal to the 55 overall dimension, but there is a distinct difference in how they affect the manufacturing tolerances. Double dimensions are explained more fully in Chapter 9.

Figure 8-88 shows an object dimensioned from its centerline. This type of dimensioning is used when the distance between the holes relative to each other are critical.

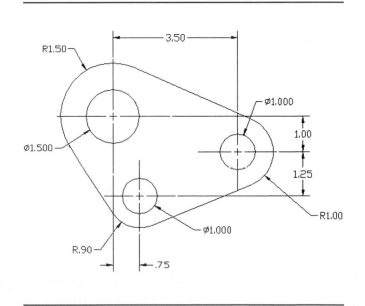

Figure 8-88

8-33 VERY LARGE RADII

Some radii are so large that it is not practical to draw the leader for the radius dimension at full size. Figure 8-89 shows an example of an object that uses foreshortened leader lines.

To create a radius dimension for large radii

1. Type **qleader** in response to a command prompt.

   Command: _leader
   Specify first leader point, or [Settings] <Settings>:

2. Type **s**; press **Enter.**

   The **Leader Settings** dialog box will appear. See Figure 8-90.

3. Select the **Leader Line and Arrow** option.
4. Select the upper arrow to the right of the **Maximum** box in the **Number of Points** box and set the maximum number of points to **4.** Select **OK.**
5. Select a point on the large arc.

   Specify next point:

6. Specify points **1, 2,** and **3** as shown in Figure 8-89.
7. Complete the leader dimension by typing in the appropriate text value in the **Text Formatting** dialog box.
8. **Explode** the leader line.
9. Use the **Break** command and create an opening in the leader line for the text value.
10. Use the **Move** command and locate the text value within the opening.
11. Erase any excess lines.

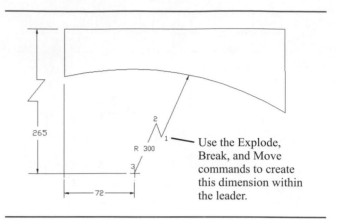

Use the Explode, Break, and Move commands to create this dimension within the leader.

Figure 8-89

Figure 8-90

8-34 EXERCISE PROBLEMS

Redraw the shapes shown in Exercise Problems EX8-1 through EX8-6. Locate the dimensions and tolerances as shown.

EX8-1 INCHES

1. 3.00
2. 1.56
3. 46°
4. .750
5. 2.75
6. 3.625
7. 45°
8. 2.250

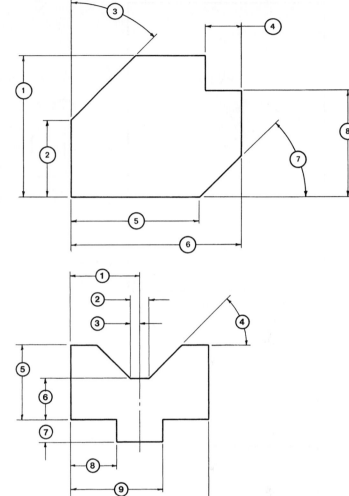

EX8-2 MILLIMETERS

1. 38
2. 10
3. 5
4. 45°
5. 40
6. 22
7. 12
8. 25
9. 51
10. 76

EX8-3 MILLIMETERS

1. 34.0
2. 17.0
3. 25.0
4. 15.00
5. 50.0
6. 80.0
7. R5 - 8 PLACES
8. 45
9. 60
10. Ø14 - 3 HOLES
11. 15.00
12. 30.00

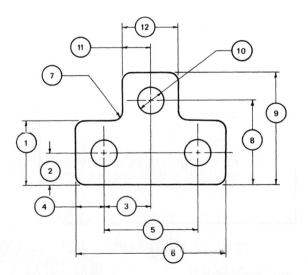

EX8-4 INCHES

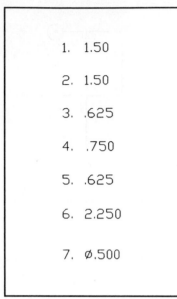

1. 1.50

2. 1.50

3. .625

4. .750

5. .625

6. 2.250

7. ⌀.500

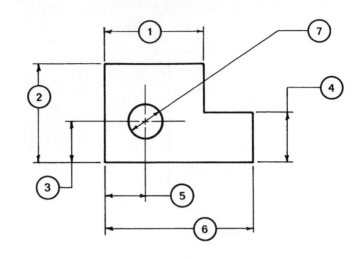

EX8-5 MILLIMETERS

Note : ⑨

1. ⌀30.0

2. ⌀15.00

3. 10.0

4. 20.0

5. 66.2

6. 15.1

7. 35.02

8. 70.00

NOTE: ALL FILLETS AND
ROUNDS = R5.0 UNLESS
OTHERWISE STATED.

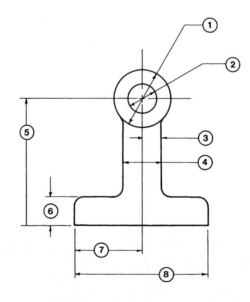

EX8-6 MILLIMETERS

1. 184.5	7. 28.0	13. 83.2	19. 120.0
2. 91.5	8. 16.00	14. 63.00	20. ⟨184.0⟩
3. 44.2	9. 16.00	15. 50.00	21. 12 × 31
4. 22.00	10. 28.0	16. 28.5	R – 3 SLOTS
5. 13.00	11. 12.5	17. 32.0	22. 6.00
6. 6.51	12. ⌀6.00	18. 76.0	23. 6.00

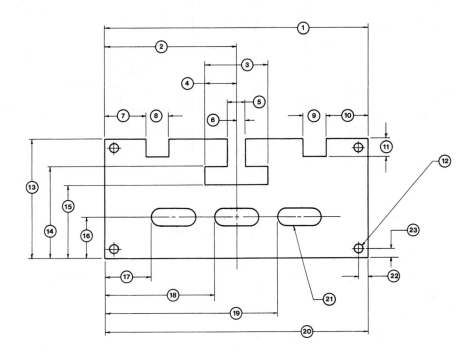

Measure and redraw the shapes in Exercise Problems EX8-7 through EX8-49. Add the appropriate dimensions. Specify the units and scale of the drawing. The dotted grid background has either .50-inch or 10-millimeter spacing.

EX8-7

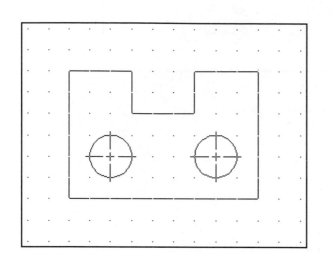

EX8-9

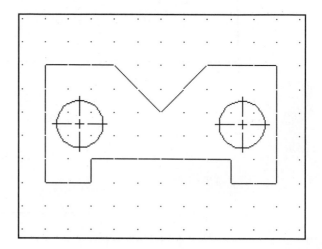

EX8-8

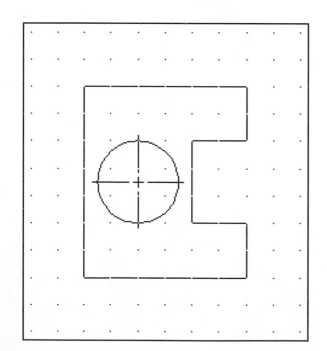

EX8-10

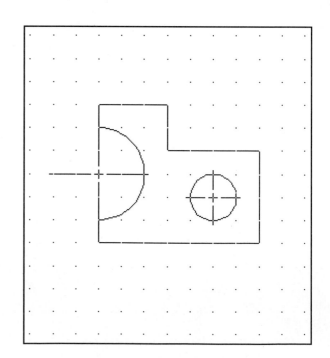

EX8-11

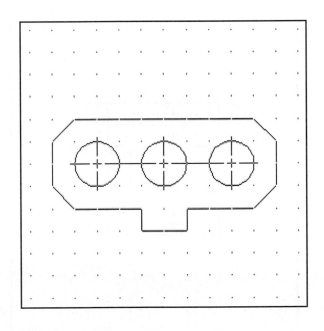

EX8-13

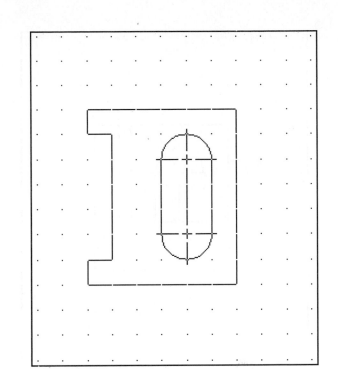

EX8-12

EX8-14

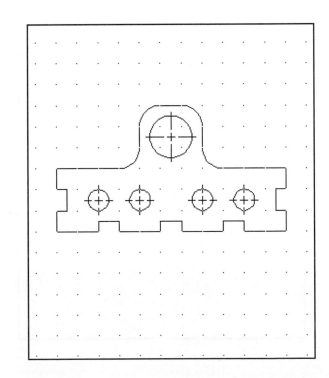

EX8-15

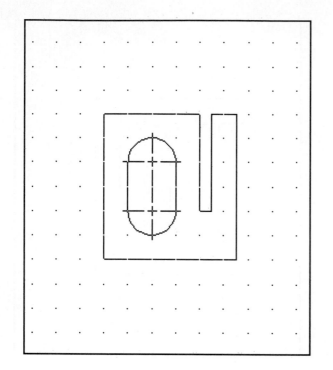

EX8-17

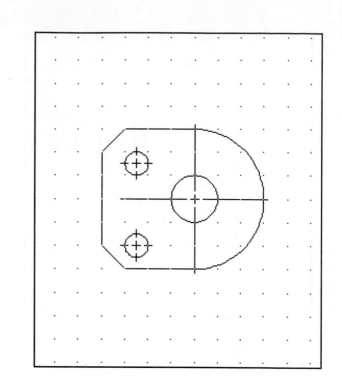

EX8-16

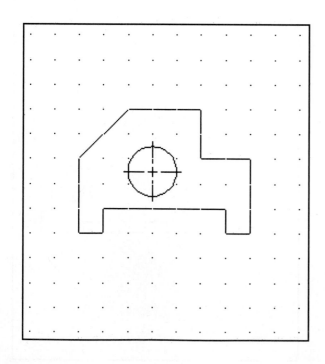

EX8-18

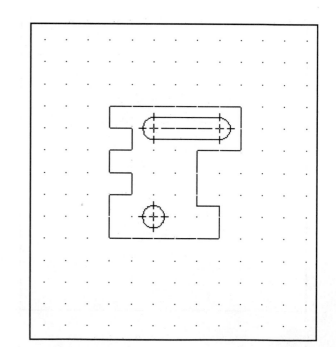

EX8-19

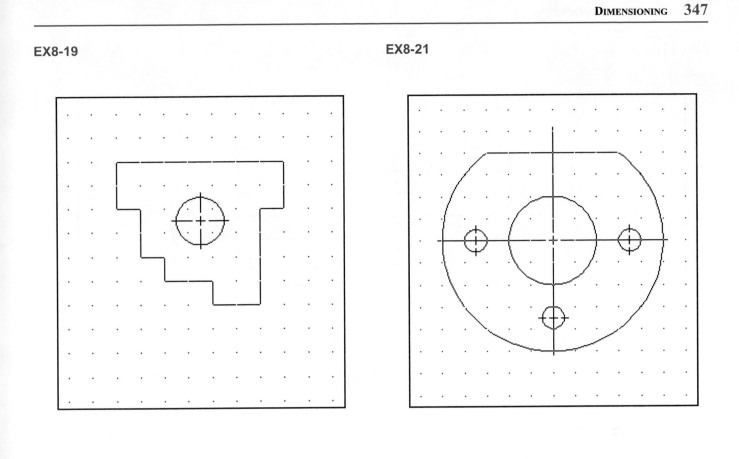

EX8-21

EX8-20

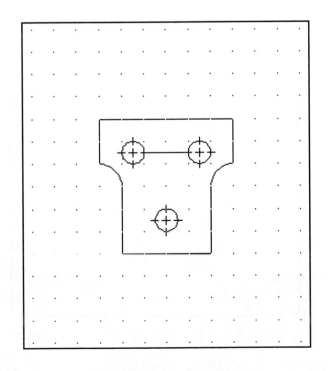

EX8-22

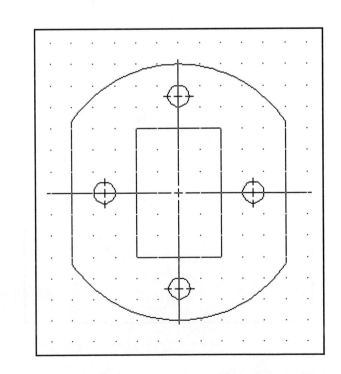

EX8-23

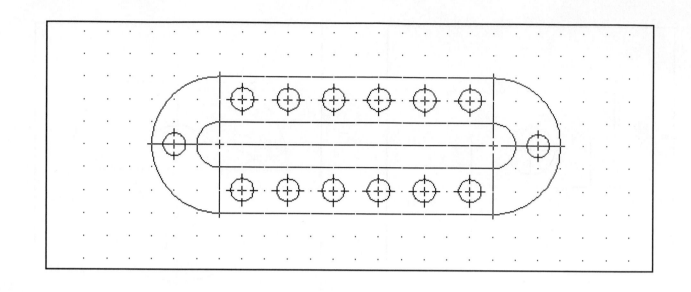

EX8-24

EX8-25

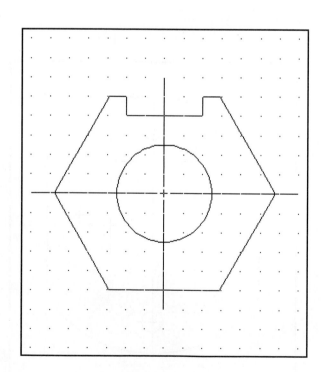

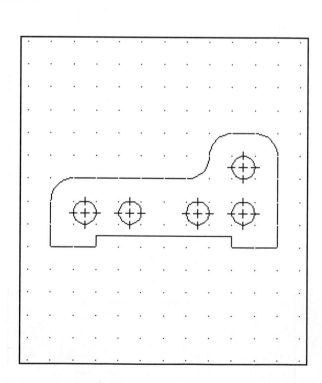

EX8-26

EX8-28

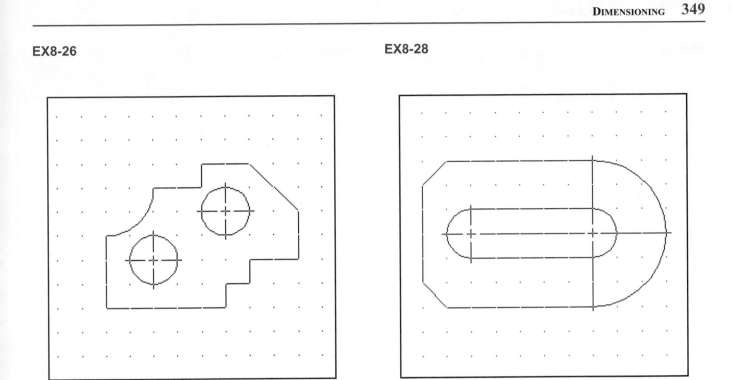

EX8-27

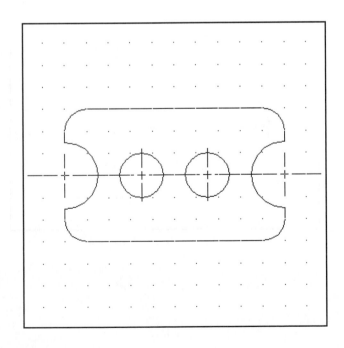

EX8-29

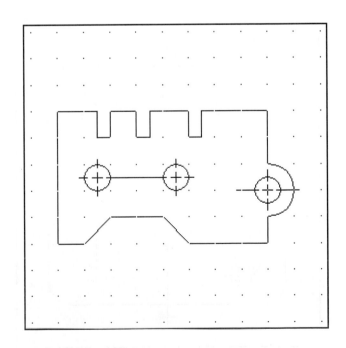

EX8-30

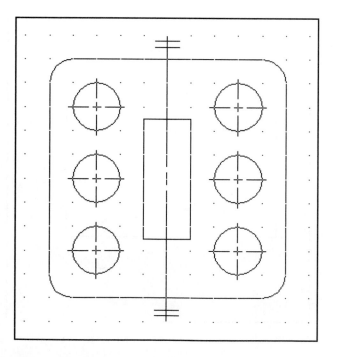

EX8-31

EX8-32

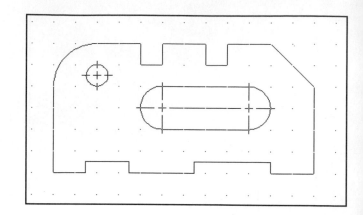

EX8-33

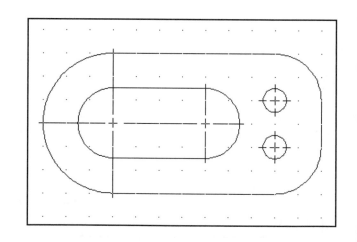

EX8-34

EX8-36

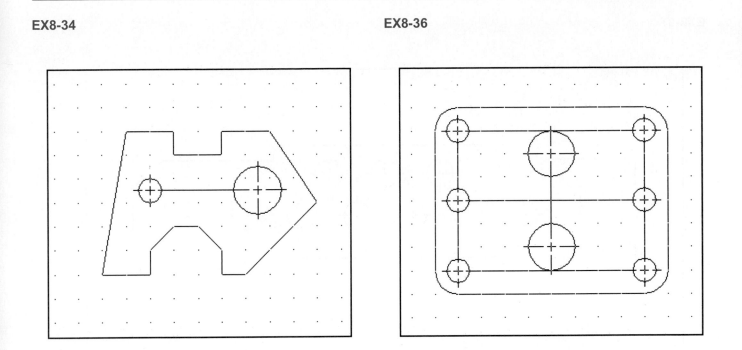

EX8-35

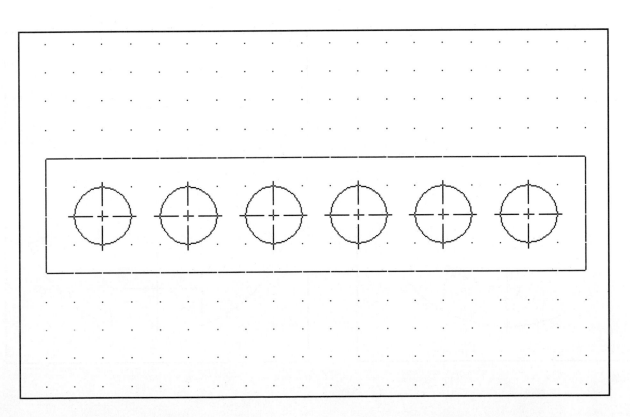

EX8-37

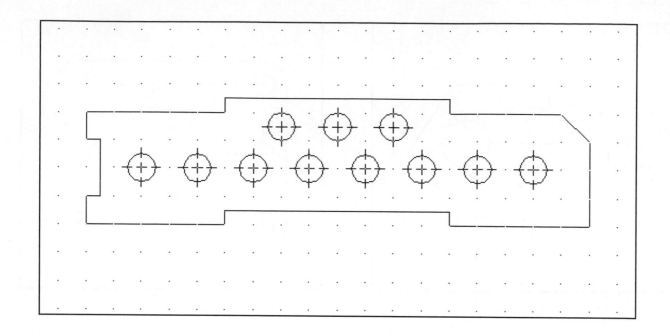

EX8-38

EX8-39

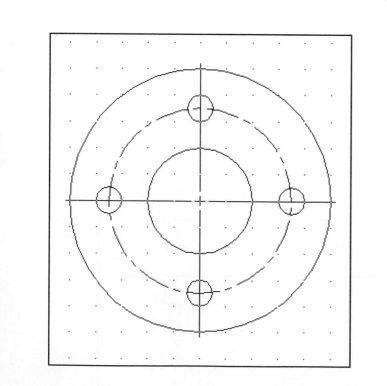

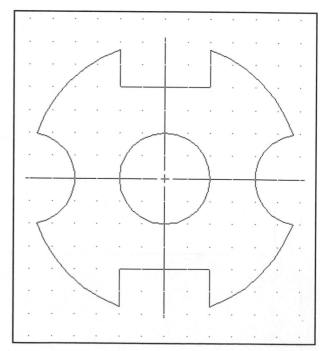

EX8-40

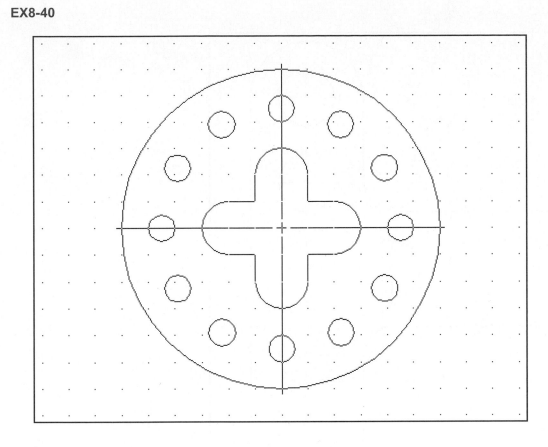

EX8-41

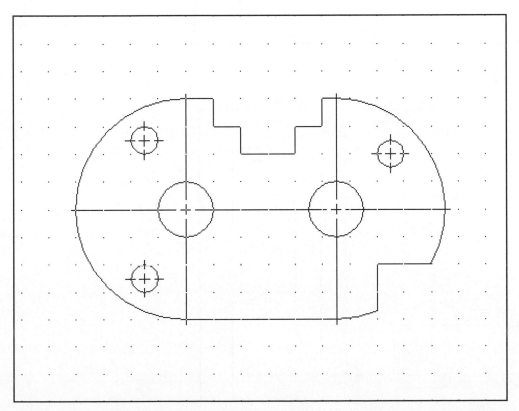

EX8-42A

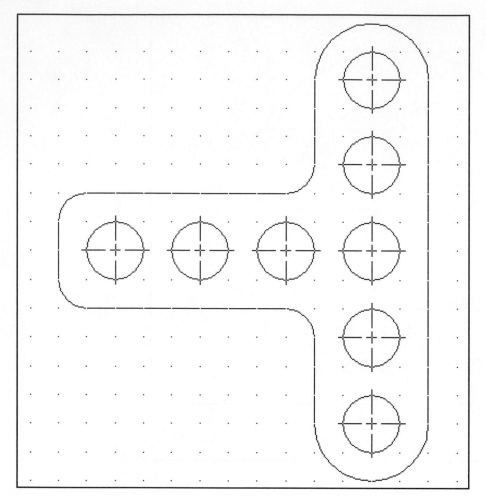

EX8-43A

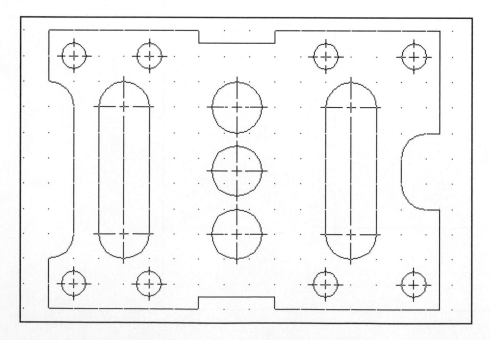

EX8-42B

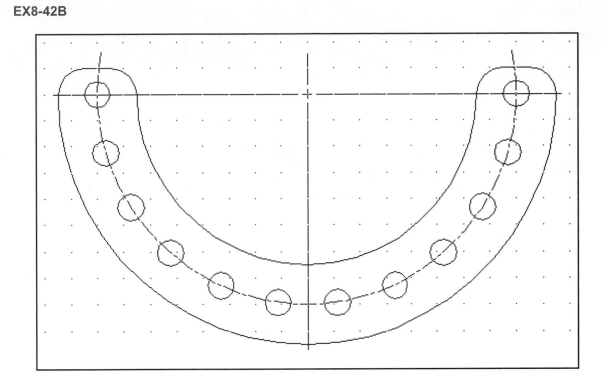

EX8-43B

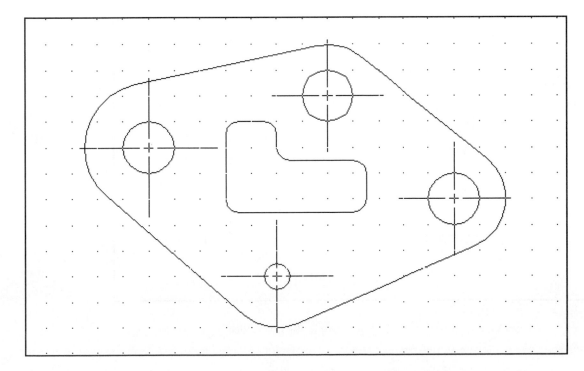

EX8-42C

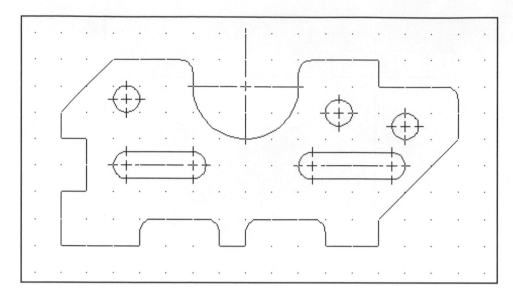

EX8-43C

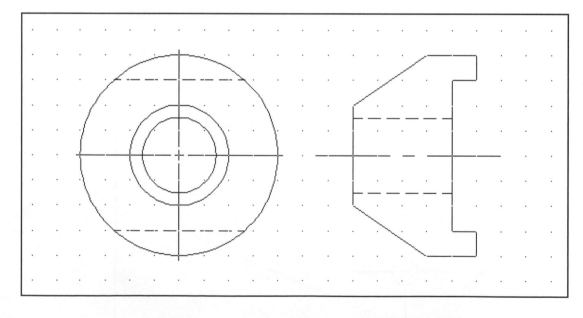

Front View   Right Side View

EX8-44

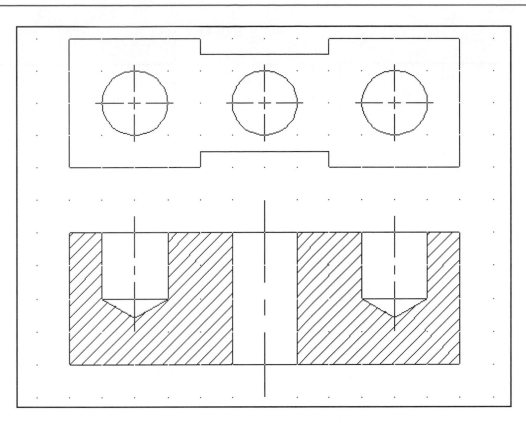

EX8-45

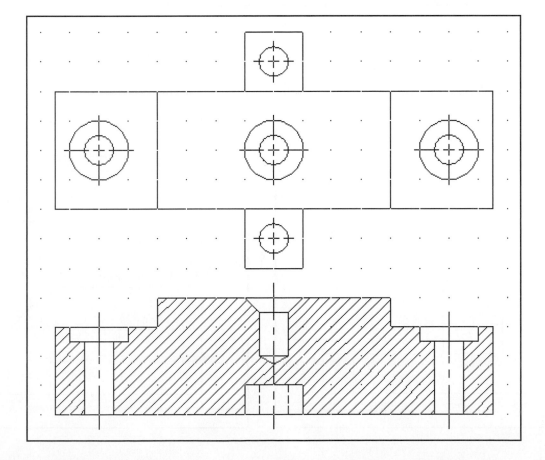

EX8-46

EX8-48

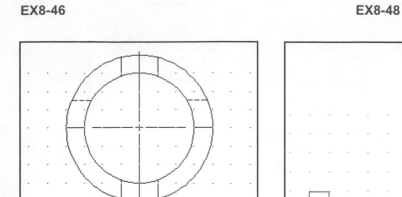

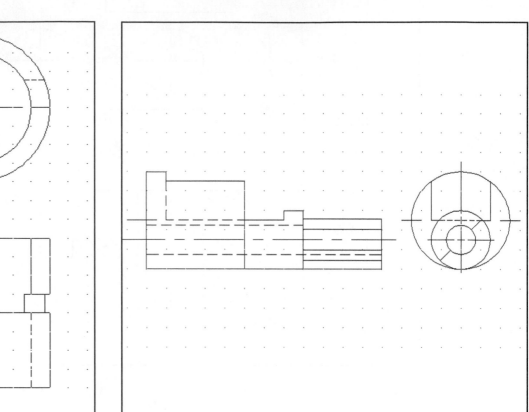

EX8-47

EX8-49

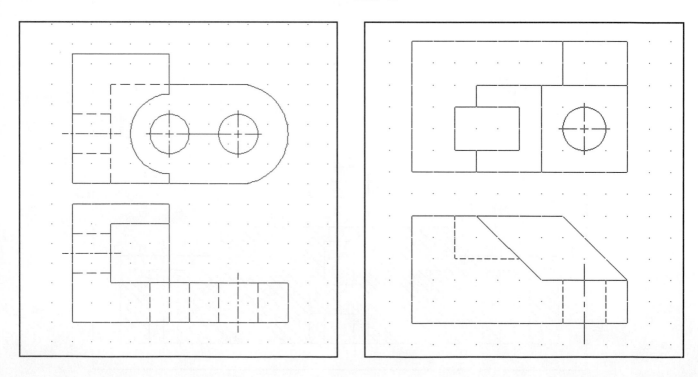

9

Tolerancing

9-1 INTRODUCTION

Tolerances define the manufacturing limits for dimensions. All dimensions have tolerances either written directly on the drawing as part of the dimension or implied by a predefined set of standard tolerances that apply to any dimension that does not have a stated tolerance.

This chapter explains general tolerance conventions and how they are applied using AutoCAD. It includes a sample tolerance study and an explanation of standard fits and surface finishes.

Chapter 10 explains geometric tolerances.

9-2 DIRECT TOLERANCE METHODS

There are two methods used to include tolerances as part of a dimension: plus and minus, and limits. Plus and minus tolerances can be expressed in either bilateral or unilateral forms.

A bilateral tolerance has both a plus and a minus value. A unilateral tolerance has either the plus or minus value equal to 0. Figure 9-1 shows a horizontal dimension of 60 millimeters that includes a bilateral tolerance of plus or minus 0.1 and another dimension of 60 millimeters that

includes a bilateral tolerance of plus 0.20 or minus 0.10. Figure 9-1 also shows a dimension of 65 millimeters that includes a unilateral tolerance of plus 1 or minus 0.

Plus or minus tolerances define a range for manufacturing. If inspection shows that all dimensioned distances on an object fall within their specified tolerance range, the object is considered acceptable; that is, it has been manufactured correctly.

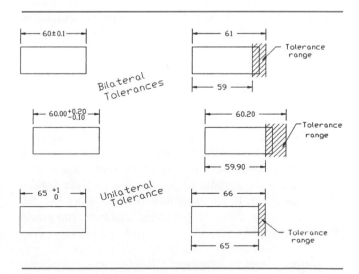

Figure 9-1

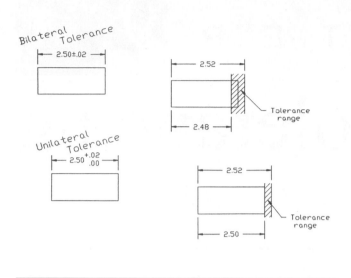

Figure 9-2

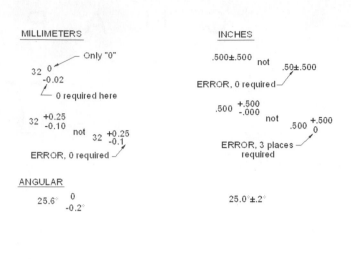

Figure 9-3

The dimension and tolerance 60 ± 0.1 means that the distance must be manufactured within a range no greater than 60.1 nor less than 59.9. The dimension and tolerance 65 + 1, −0 defines the tolerance range as 65 to 66.

Figure 9-2 shows some bilateral and unilateral tolerances applied using decimal inch values. Inch dimensions and tolerances are written using a slightly different format from that used for millimeter dimensions and tolerances, but they also define manufacturing ranges for dimension values. The horizontal bilateral dimension and tolerance 2.50 ± .02 defines the longest acceptable distance as 2.52 inches and the shortest as 2.48. The unilateral dimension 2.50 + .02, −.00 defines the longest acceptable distance as 2.52 and the shortest as 2.50.

9-3 TOLERANCE EXPRESSIONS

Dimension and tolerance values are written differently for inch and millimeter values. See Figure 9-3. Unilateral dimensions for millimeter values specify a zero limit with a single 0. A zero limit for inch values must include the same number of decimal places given for the dimension value. In the example shown in Figure 9-3, the dimension value .500 has a unilateral tolerance with minus zero tolerance. The zero limit is written as .000, three decimal places for both the dimension and the tolerance.

Both values in a bilateral tolerance must contain the same number of decimal places, although for millimeter values the tolerance values need not include the same number of decimal places as the dimension value. In Figure 9-3

the dimension value 32 is accompanied by tolerances of +0.25 and −0.10. This form is not acceptable for inch dimensions and tolerances. An equivalent inch dimension and tolerance would be written 32.00 +.25/−.10.

Degree values must include the same number of decimal places in both the dimension value and the tolerance values for bilateral tolerances. A single 0 may be used for a unilateral tolerance.

9-4 UNDERSTANDING PLUS AND MINUS TOLERANCES

A millimeter dimension and tolerance of 12.0 +0.2/−0.1 means the longest acceptable distance is 12.2000...0, and the shortest, 11.9000...0. The total range is .3000...0.

After an object is manufactured, it is inspected to ensure that the object has been manufactured correctly. Each dimensioned distance is measured and, if it is within the specified tolerance, is accepted. If the measured distance is not within the specified tolerance, the part is rejected. Some rejected objects may be reworked to bring them into the specified tolerance range, whereas others are simply scrapped.

Figure 9-4 shows a dimension with a tolerance. Assume that five objects were manufactured using the same 12 +0.2/−0.1 dimension and tolerance. The objects were then inspected and the results were as listed. Inspected measurements are usually at least one more decimal place than that specified in the tolerance. Which

GIVEN (mm)
12 +0.2
  −0.1

MEANS

TOL MAX = 12.2
TOL MIN = 11.9
TOTAL TOL = 0.3

OBJECT	AS MEASURED	ACCEPTABLE?
1	12.160	OK
2	12.020	OK
3	12.203	TOO LONG
4	11.920	OK
5	11.895	TOO SHORT

Figure 9-4

GIVEN (inches)
3.50±.02

MEANS

TOL MAX = 3.52
TOL MIN = 3.48
TOTAL TOL = .04

OBJECT	AS MEASURED	ACCEPTABLE?
1	3.520	OK
2	3.486	OK
3	3.470	TOO SHORT
4	3.521	TOO LONG
5	3.515	OK

Figure 9-5

objects are acceptable and which are not? Object 3 is too long, and object 5 is too short because their measured distances are not within the specified tolerances.

Figure 9-5 shows a dimension and tolerance of 3.50 ±.02 inches. Object 3 is not acceptable because it is too short, and object 4 is too long.

9-5 CREATING PLUS AND MINUS TOLERANCES USING AUTOCAD

Plus and minus tolerances may be created using AutoCAD in four ways: using the **Text** option, the **Mtext** option, typing the tolerances directly using **Dtext**, or by setting the plus and minus values using the **Dimension Style** tool.

To create plus and minus tolerances using the Text option

The example given is for a horizontal dimension, but the procedure is the same for any linear or radial dimension.

1. Select the **Linear** tool from the **Dimension** toolbar.
2. Select the extension line origins as explained in Section 8-3 and locate the dimension line.

Specify dimension line location or [Mtext/Text/Angle/Horizontal/Vertical/Rotated]:

3. Type **t**; press **Enter**.

Dimension text <x.xxxx>:

4. Type the appropriate text value; press **Enter**.

Type the dimension value, then **%%p**, then the tolerance value.

Dimension text <x.xxxx>:5.00%%p.02.

In this example, the resulting dimension will be 5.00±.02.

To create plus and minus tolerances using the Mtext option

1. Select the **Linear** tool from the **Dimension** toolbar.
2. Select the extension line origins as explained in Section 8-3 and locate the dimension line.

Specify dimension line location or [Mtext/Text/Angle/Horizontal/Vertical/Rotated]:

3. Type **m**; press **Enter**.

The **Text Formatting** dialog box will appear. See Figure 9-6. The existing dimension value will appear in a box. The value may be changed by deleting the existing values, and then typing in a new value.

4. Place the flashing cursor to the right of the dimension value.
5. Type **%%p** or press the **<Alt>** key and type **0177**.
6. Type in the tolerance value; select **OK**.

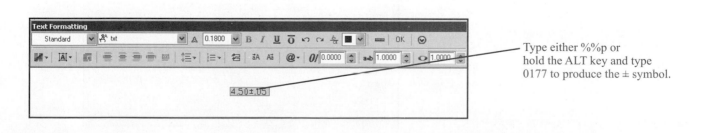

Type either %%p or hold the ALT key and type 0177 to produce the ± symbol.

Figure 9-6

The modified dimension will appear on the drawing screen and can be located by moving the mouse.

The **Dimension Style** tool on the **Dimension** toolbar can be used to change the precision of the inital AutoCAD select dimension values. (The default value is four places; 1.0000.) If, for example, two decimal places are desired, use the **Precision** box on the **Primary Units** tab on the **Modify Dimension Style: Standard** dialog box to reset the system for two decimal places. See Section 8-5.

To use Dtext to create a plus and minus tolerance

1. Type **Dtext** in response to a command prompt.
2. Place the starting point for the text, then define the appropriate height and angle.
3. Type the desired dimension.

   Text: 5.00%%p.02

4. Press **Enter.**

In the example shown, the resulting dimension will be **5.00±.02.** The symbol will initially appear on the screen as %%p but will change to ± when the last text line is entered. The symbol ± can also be created using the Windows character map by holding down the <**Alt**> key and typing **0177.**

To use the Dimension Style tool

1. Select the **Dimension Style** tool from the **Dimension** toolbar.
2. Select the **Modify** option, then the **Tolerances** tab.

3. Select the arrow to the right of the **Method box.**

A list of available tolerancing methods will cascade. See Figure 9-7. AutoCAD offers two options for plus and minus tolerancing: **Symmetrical** and **Deviation.**

The symmetrical method

4a. Select the **Symmetrical** method.

The **Tolerance format** box within the **Modify Dimension Style: Standard** dialog box is used to create a symmetrical tolerance by entering a value in the **Upper value** box. See Figure 9-8.

5a. Type a value.

In this example, a value of **0.0300** was entered. Only an upper value needed to be entered, as the tolerance value is symmetrical.

6a. Return to the drawing screen.

Dimensions created using the **Dimension** toolbar will now automatically include a ±0.03000 tolerance.

The deviation method

4b. Select the **Deviation** method.

The **Tolerance Format** dialog box is used to create a deviation tolerance by entering values in the **Upper value** and **Lower value** boxes. See Figure 9-9.

5b. Type in values.

Select the tolerance method here.

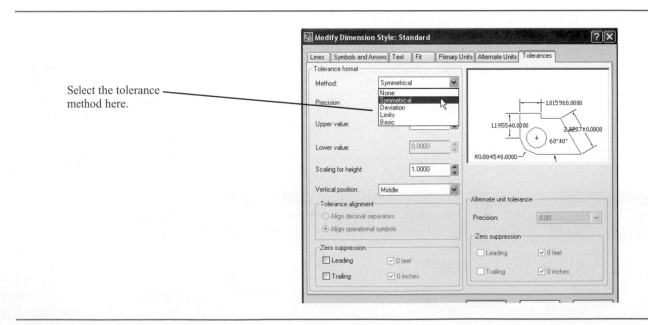

Figure 9-7

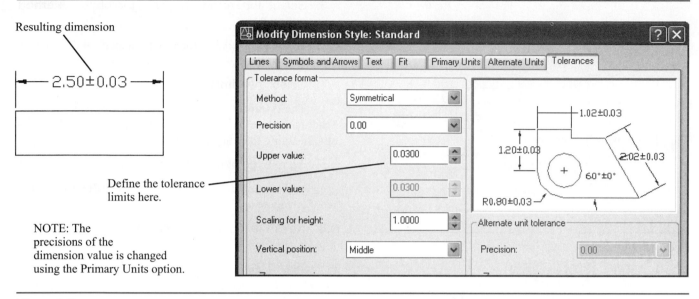

Resulting dimension

Define the tolerance limits here.

NOTE: The precisions of the dimension value is changed using the Primary Units option.

Figure 9-8

In this example values of **0.0100** and **0.0200** were entered.

6b. Return to the drawing screen.

Dimensions created using the **Dimension** toolbar will now automatically include a +0.0100, −0.0200 tolerance.

The **Dimension Style** tool on the **Dimension** toolbar can also be used to change the precision of the initial AutoCAD selected dimension values. (The default value is four places: 1.0000.) If, for example, two decimal places are desired, use the **Precision** box on the **Primary Units** tab on the **Dimension Style Manager** dialog box to reset the system for two decimal places. See Section 8-5.

9-6 LIMIT TOLERANCES

Figure 9-10 shows examples of limit tolerances. Limit tolerances replace dimension values. Two values are given: the upper and lower limits for the dimension value. The limit tolerance of 62.1 and 59.9 is mathematically equal to 62 ± 0.1, but the stated limit tolerance is considered easier to read and understand.

Limit tolerances define a range for manufacture. Final distances on an object must fall within the specified range to be acceptable.

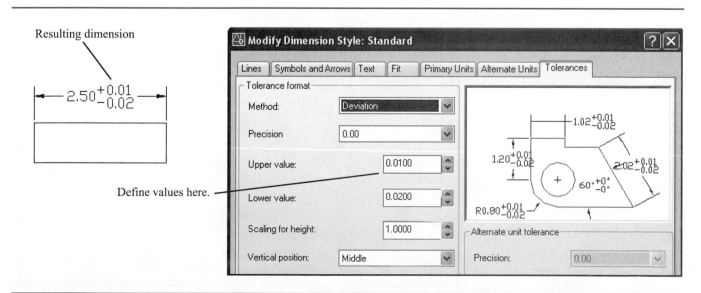

Resulting dimension

Define values here.

Figure 9-9

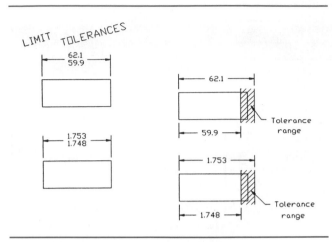

Figure 9-10

9-7 CREATING LIMIT TOLERANCES USING AUTOCAD

Limit tolerances may be created using AutoCAD in two ways: by using the **Dimension Style** tool, or by modifying a given dimension using the **Dimension Edit** tool.

To create a limit tolerance using the Dimension Style tool

1. Select the **Dimension Style** tool from the **Dimension** toolbar.

   The **Dimension Style Manager** dialog box will appear.

2. Select the **Modify** option, then the **Tolerances** tab.

   The **Modify Dimension Style: Standard** dialog box will appear.

3. Select the arrow to the right of the **Method** option box.

   A list of available tolerance methods will cascade. See Figure 9-7.

4. Select the **Limits** option.

   The **Annotation** dialog box will reappear with the headings **Upper value** and **Lower value** now in black letters, meaning that they can be accessed.

5. Type in the upper and lower values.

   In this example values of **0.1000** and **0.3000** were chosen. See Figure 9-11.

6. Return to the drawing screen.

   Every dimension created using the **Linear** dimension tool will now automatically create a limit tolerance based on the selected distance and the selected upper and lower values. Figure 9-11 shows the results of the **Linear** dimension tool applied to a distance of 3.2500. The upper limit is 3.2500 + 0.1000 and the lower limit is 3.2500 − 0.3000.

   If only a few limit-type tolerances are to be used on a drawing, it is sometimes easier simply to modify an existing dimension rather than to change the **Tolerance** settings.

To modify an existing dimension into a limit tolerance

1. Select the **Dimension Edit** tool from the **Dimension** toolbar.

   Enter type of dimension editing [Home/New/Rotate/Oblique] <Home>:

2. Select the **New** option or type **n; press Enter.**

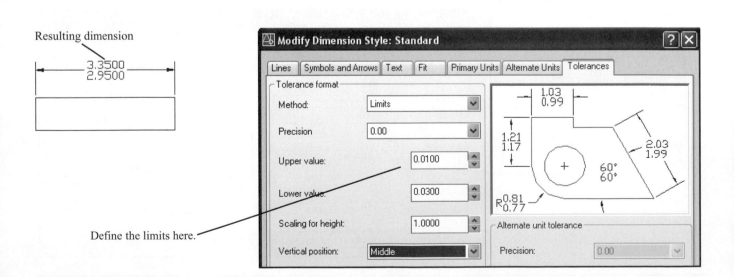

Resulting dimension

Define the limits here.

Figure 9-11

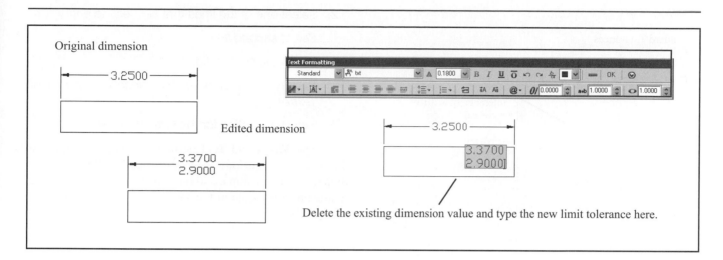

Figure 9-12

The **Text Formatting dialog** box will appear. See Figure 9-12.

3. Delete the existing dimension value and type in the new limit tolerance value.
4. Type **3.3700,** press **Enter,** then type **2.9000;** select **OK.**

Select objects:

5. Select the dimension to be changed; press **Enter.**

Figure 9-12 shows the resulting changed dimension.

9-8 ANGULAR TOLERANCES

Figure 9-13 shows an example of an angular dimension with a tolerance. The procedure explained for plus and minus tolerances applies to angular as well as to linear dimensions and tolerances.

The precision of angular dimensions is set using the **Primary Units** dialog box. In the following example, a deviation tolerance of $+0.10, -0.30$ was also specified.

To set the precision for angular dimensions and tolerances

1. Select the **Dimension Style** tool, then the **Modify** option, then the **Primary Units** option.
2. Select the arrow to the right of the **Precision** box.

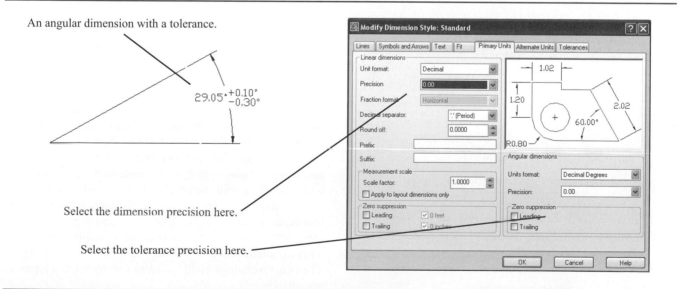

Figure 9-13

Standard Tolerances

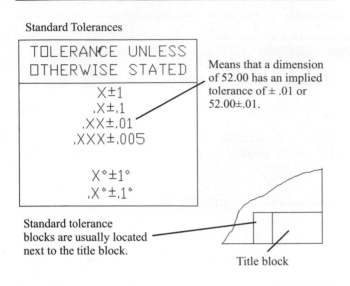

Means that a dimension of 52.00 has an implied tolerance of ± .01 or 52.00±.01.

Standard tolerance blocks are usually located next to the title block.

Title block

Figure 9-14

A list of available precision factors will cascade. See Figure 9-13.

3. Select two significant figures for both **Dimension** and **Tolerance** and return to the drawing screen.

The precision for the angular value can be changed using the **Precision** option under **Angular Dimensions.**

To create an angular dimension and tolerance

1. Select the **Angular** tool from the **Dimension** toolbar.

   Select arc, circle, line, or <specify vertex>:

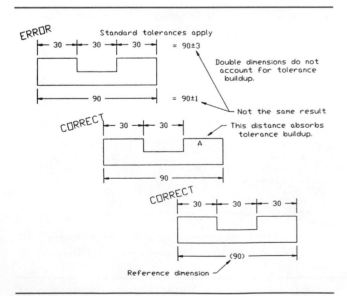

Figure 9-15

2. Select one of the lines that defines the angle.

   Select second line:

3. Select the second line.

   Specify dimension arc line location or [Mtext/Text/Angle]:

4. Select a location for the dimension; press **Enter.**

The **Mtext** and **Text** options are used as explained for linear dimensions and tolerance in Section 9-5. Symmetrical, deviation, and limit tolerances can be applied to angular tolerances in the same way they were to linear tolerances.

9-9 STANDARD TOLERANCES

Most manufacturers establish a set of standard tolerances that are applied to any dimension that does not include a specific tolerance. Figure 9-14 shows some possible standard tolerances. Standard tolerances vary from company to company. Standard tolerances are usually listed on the first page of a drawing to the left of the title block, but this location may vary.

The X value used when specifying standard tolerances means any X stated in that format. A dimension value of 52.00 will have an implied tolerance of ±.01 because the stated standard tolerance is .XX ±.01, so any dimension value with two decimal places will have a standard implied tolerance of ±.01. A dimension value of 52.000 will have an implied tolerance of ±.005.

9-10 DOUBLE DIMENSIONING

It is an error, called *double dimensioning,* to dimension the same distance twice. Double dimensioning is an error because it does not allow for tolerance buildup across a distance.

Figure 9-15 shows an object that has been dimensioned twice across its horizontal length; once using three 30-millimeter dimensions and a second time using the 90-millimeter overall dimension. The two dimensions are mathematically equal but are not equal when tolerances are considered. Assume that each dimension has a standard tolerance of ±1 millimeter. The three 30-millimeter dimensions could create an acceptable distance of 90±3 millimeters, or a maximum distance of 93 and a minimum distance of 87. The overall dimension of 90 millimeters allows a maximum distance of 91 and a minimum distance of 89. The two dimensions yield different results when tolerances are considered.

The size and location of a tolerance depends on the design objectives of the object, how it will be manufactured, and how it will be inspected. Even objects that have similar shapes may be dimensioned and toleranced very differently.

One possible solution to the double dimensioning shown in Figure 9-15 is to remove one of the 30-millimeter dimensions and to allow that distance to "float," that is, absorb the cumulated tolerances. The choice of which 30-millimeter dimension to eliminate depends on the design objectives of the part. For this example the far-right dimension was eliminated to remove the double-dimensioning error.

Another possible solution to the double-dimensioning error is to retain the three 30-millimeter dimensions and to change the 90-millimeter overall dimension to a reference dimension. A reference dimension is used only for mathematical convenience. It is not used during the manufacturing or inspection process. A reference dimension is designated on a drawing with parentheses (90).

If the 90-millimeter dimension were referenced, then only the three 30-millimeter dimensions would be used to manufacture and inspect the object. This would eliminate the double-dimensioning error.

9-11 CHAIN DIMENSIONS AND BASELINE DIMENSIONS

There are two systems used to apply dimensions and tolerances to a drawing: *chain* and *baseline*. Figure 9-16 shows examples of both systems. Chain dimensions relate each feature to the feature next to it; baseline dimensions relate all features to a single baseline or datum

Chain and baseline dimensions may be used together. Figure 9-16 also shows two objects with repetitive features: one object includes two slots, and the other three sets of three holes. In each example, the center of the repetitive feature is dimensioned to the left side of the object, which serves as a baseline. The sizes of the individual features are dimensioned using chain dimensions referenced to centerlines.

Baseline dimensions eliminate tolerance buildup and can be related directly to the reference axis of many machines. They tend to take up much more area on a drawing than do chain dimensions.

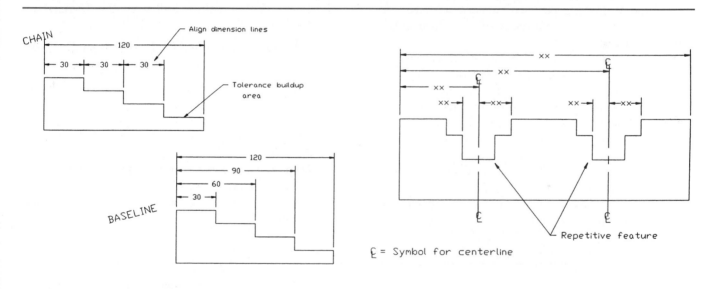

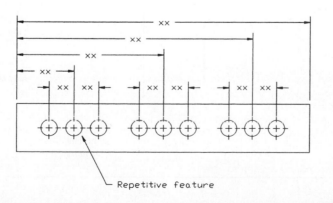

Figure 9-16

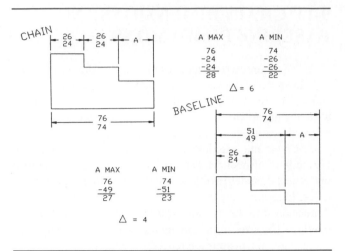

Figure 9-17

reduces the tolerance variations for the object simply because it applies the tolerances and dimensions differently. So why not always use baseline dimensions? For most applications, the baseline system is probably better, but if the distance between the individual features is more critical than the distance from the feature to the baseline, use the chain system.

To create baseline dimensions using AutoCAD

The **Baseline** tool can be applied only after a linear dimension has been created. The object shown in Figure 9-18 was dimensioned by first creating the 1.00 dimension using the **Linear** dimension tool, then using the **Baseline** tool as follows.

1. Create a linear dimension using the **Linear** tool from the **Dimension** toolbar.

The point selected as the origin for the first extension line will become the origin of the baseline.

2. Select the **Baseline** tool from the **Dimension** toolbar.

   Specify a second extension line origin or [Undo/ <Select>]:

3. Select the origin for the second extension line of the baseline dimension.

   Specify a second extension line origin or [Undo/ Select] <Select>:

4. Repeat the process until the baseline dimensioning is complete.

5. Press the right mouse button and **Enter** the dimensions.

Chain dimensions are useful in relating one feature to another, such as the repetitive hole pattern shown in Figure 9-16. In this example the distance between the holes is more important than the distance of the individual hole from the baseline.

Figure 9-17 shows the same object dimensioned twice, once using chain dimensions and once using baseline dimensions. All distances are assigned a tolerance range of 2 millimeters stated using limit tolerances. The maximum distance for surface A is 28 millimeters using the chain system and 27 millimeters using the baseline system. The 1-millimeter difference comes from the elimination of the first 26–24 limit dimension found on the chain example but not on the baseline.

The total tolerance difference is 6 millimeters for the chain and 4 millimeters for the baseline. The baseline

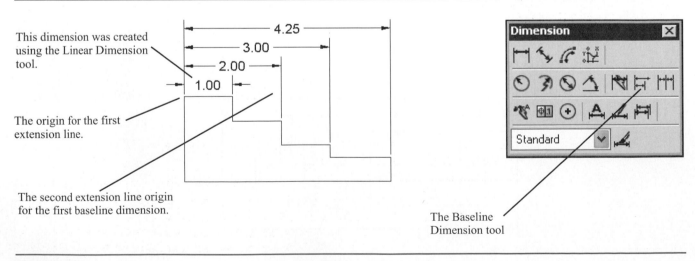

Figure 9-18

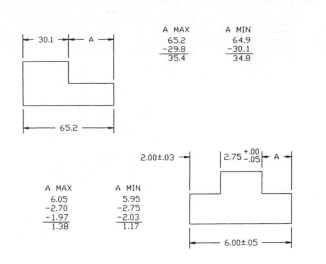

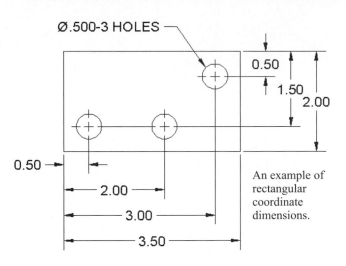

Figure 9-19

Figure 9-20

9-12 TOLERANCE STUDIES

The term *tolerance study* is used when analyzing the effects of a group of tolerances on each other and on an object. Figure 9-19 shows an object with two horizontal dimensions. The horizontal distance A is not dimensioned. Its length depends on the tolerances of the two horizontal dimensions.

To calculate A's maximum length

Distance A will be longest when the overall distance is at its longest and the other distance is at its shortest:

$$\begin{array}{r} 65.2 \\ -29.8 \\ \hline 35.4 \end{array}$$

To calculate A's minimum length

Distance A will be shortest when the overall length is at its shortest and the other length is at its longest:

$$\begin{array}{r} 64.9 \\ -30.1 \\ \hline 34.8 \end{array}$$

Figure 9-19 also shows an object that includes three horizontal dimensions. Surface A is at its maximum length when the overall dimension is at its longest and the other dimensions are at their shortest. Surface A is at its minimum when the overall length is at its shortest and the other dimensions are at their longest.

9-13 RECTANGULAR DIMENSIONS

Figure 9-20 shows an example of rectangular dimensions referenced to baselines. Figure 9-21 shows a circular object for which dimensions are referenced to a circle's centerlines. Dimensioning to a circle's centerline is critical to accurate hole location.

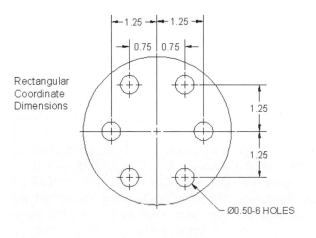

Figure 9-21

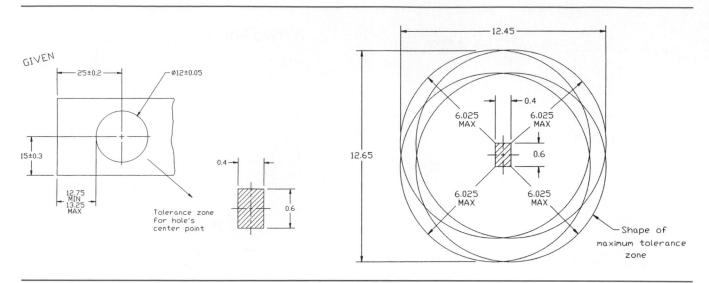

Figure 9-22

9-14 HOLE LOCATIONS

When rectangular dimensions are used, the location of a hole's center point is defined by two linear dimensions. The result is a rectangular tolerance zone whose size is based on the linear dimension's tolerances. The shape of the center point's tolerance zone may be changed to circular using positional tolerancing, as described in Section 10-16.

Figure 9-22 shows the location and size dimensions for a hole. Also shown is the resulting tolerance zone and the overall possible hole shape. The center point's tolerance is 0.2 by 0.3, based on the given linear locating tolerances.

The hole diameter has a tolerance of ±0.05. This value must be added to the center point location tolerances to define the maximum overall possible shape of the hole. The maximum possible hole shape is determined by drawing the maximum radius from the four corner points of the tolerance zone.

This means that the left edge of the hole could be as close to the vertical baseline as 12.75 or as far as 13.25. The 12.75 value was derived by subtracting the maximum hole diameter value, 12.05, from the minimum linear distance, 24.80 (24.80 − 12.05 = 12.75). The 13.25 value was derived by subtracting the minimum hole diameter, 11.95, from the maximum linear distance, 25.20 (25.20 − 11.95 = 13.25).

Figure 9-23 shows a hole's tolerance zone based on polar dimensions. The zone has a sector shape, and the possible hole shape is determined by locating the maximum radius at the four corner points of the tolerance zone.

9-15 CHOOSING A SHAFT FOR A TOLERANCED HOLE

Given the hole location and size shown in Figure 9-23, what is the largest diameter shaft that will always fit into the hole?

Figure 9-24 shows the hole's center point tolerance zone based on the given linear locating tolerances. Four circles have been drawn centered at the four corners of the linear tolerance zone that represent the smallest possible hole diameter. The circles define an area that represents the maximum shaft size that will always fit into the hole, regardless of how the given dimensions are applied.

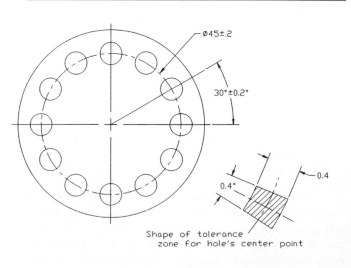

Figure 9-23

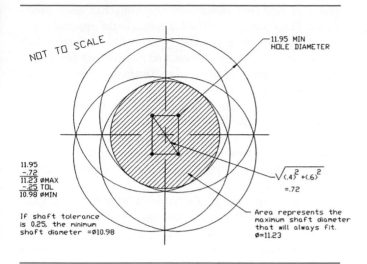

Figure 9-24

Figure 9-25

The diameter of this circular area can be calculated by subtracting the maximum diagonal distance across the linear tolerance zone (corner to corner) from the minimum hole diameter.

The results can be expressed as a formula.

For linear dimensions and tolerances

$$S_{max} = H_{min} - DTZ$$

where

S_{max} = Maximum shaft diameter

H_{min} = Minimum hole diameter

DTZ = Diagonal distance across the tolerance zone

In the example shown the diagonal distance is determined using the Pythagorean theorem:

$$DTZ = \sqrt{(.4)^2 + (.6)^2}$$
$$= \sqrt{.16 + .36}$$
$$DTZ = .72$$

This means that the maximum shaft diameter that will always fit into the given hole is 11.23.

$$S_{max} = H_{min} - DTZ$$
$$= 11.95 - .72$$
$$S_{max} = 11.23$$

This procedure represents a restricted application of the general formula presented in Chapter 10 for positional tolerances. For a more complete discussion see Section 10-16.

Once the maximum shaft size has been established, a tolerance can be applied to the shaft. If the shaft had a total tolerance of .25, the minimum shaft diameter would be 11.23 − .25, or 10.98. Figure 9-24 shows a shaft dimensioned and toleranced using these values.

The formula presented is based on the assumption that the shaft is perfectly placed on the hole's center point. This assumption is reasonable if two objects are joined by a fastener and both objects are free to move. When both objects are free to move about a common fastener, they are called *floating objects*.

9-16 SAMPLE PROBLEM SP9-1

Parts A and B in Figure 9-25 are to be joined by a common shaft. The total tolerance for the shaft is to be 0.05. What are the maximum and minimum shaft diameters?

Both objects have the same dimensions and tolerances and are floating relative to each other.

$$S_{max} = H_{min} - DTZ$$
$$= 15.93 - .85$$
$$S_{max} = 15.08$$

The shaft's minimum diameter is found by subtracting the total tolerance requirement from the calculated maximum diameter.

$$15.08 - .05 = 15.03$$

Therefore,

Shaft max = 15.08
Shaft min = 15.03

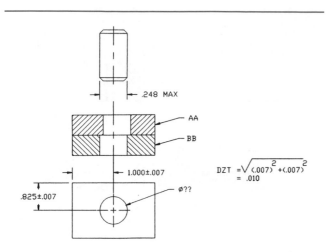

Figure 9-26

9-17 SAMPLE PROBLEM SP9-2

The procedure presented in Sample Problem SP9-1 can be worked in reverse to determine the maximum and minimum hole size based on a given shaft size.

Objects AA and BB as shown in Figure 9-26 are to be joined using a bolt whose maximum diameter is .248. What is the minimum hole size for the objects that will always accept the bolt? What is the maximum hole size if the total hole tolerance is .007?

$$S_{max} = H_{min} - DZT$$

In this example the H_{min} is the unknown factor, so the equation is rewritten:

$$H_{min} = S_{max} + DZT$$
$$= .248 + .010$$
$$H_{min} = .258$$

This is the minimum hole diameter, so the total tolerance requirement is added to this value:

$$.258 + .007 = .265$$

Therefore,

Hole max = .265
Hole min = .258

9-18 STANDARD FITS (METRIC VALUES)

Calculating tolerances between holes and shafts that fit together is so common in engineering design that a group of standard values and notations has been established. These values are listed in tables in the appendix.

There are three possible types of fits between a shaft and a hole: clearance, transition, and interference. See Figure 9-27. There are several subclassifications within each of these categories.

A *clearance fit* always defines the maximum shaft diameter as smaller than the minimum hole diameter. The difference between the two diameters is the amount of clearance. It is possible for a clearance fit to be defined with zero clearance; that is, the maximum shaft diameter is equal to the minimum hole diameter.

An *interference fit* always defines the minimum shaft diameter as larger than the maximum hole diameter, or more simply said, the shaft is always bigger than the hole. This definition means that an interference fit is the converse of a clearance fit. The difference between the diameter of the shaft and the hole is the amount of interference.

An interference fit is primarily used to assemble objects together. Interference fits eliminate the need for threads, welds, or other joining methods. Using an interference for joining two objects is generally limited to light-load applications.

It is sometimes difficult to visualize how a shaft can be assembled into a hole with a diameter smaller than that of the shaft. It is sometimes done using a hydraulic press that slowly forces the two parts together. The joining process can be augmented by the use of lubricants or heat. The hole is heated, causing it to expand, the shaft is inserted, and the hole is allowed to cool and shrink around the shaft.

A *transition fit* may be either a clearance or an interference fit. It may have a clearance between the shaft and the hole or an interference.

Figure 9-27 shows two graphic representations of 20 different standard hole/shaft tolerance ranges. The figure shows ranges for hole tolerances, shaft tolerances, and the amount of clearance or interference for each classification. The notations are based on Standard International Tolerance values. A specific description for each category of fit is as follows.

Clearance Fits

H11/c11 or C11/h11 = Loose running fit
H9/d9 or D9/h9 = Free running fit
H8/f7 or F8/h7 = Close running fit
H7/g6 or G7/h6 = Sliding fit
H7/h6 = Locational clearance fit

Transition Fits

H7/k6 or K7/h6 = Locational transition fit
H7/n6 or N7/h6 = Locational transition fit

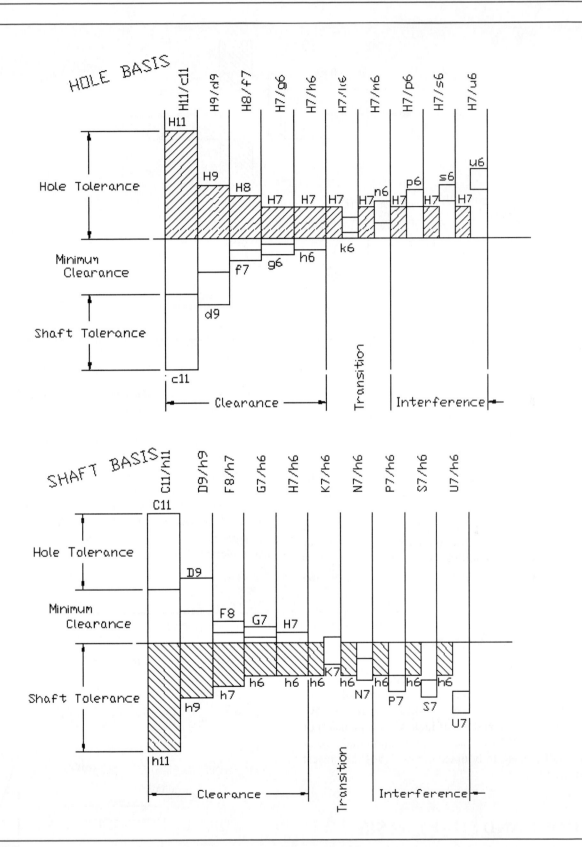

Figure 9-27

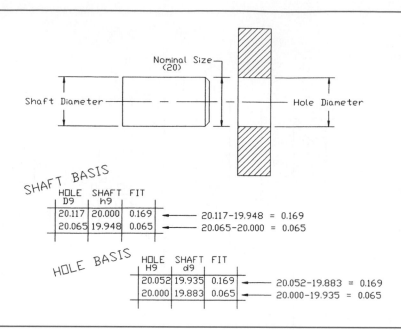

Figure 9-28

Interference Fits

H7/p6 or P7/h6 = Locational transition fit
H7/s6 or S7/h6 = Medium drive fit
H7/u6 or U7/h6 = Force fit

Not all possible sizes are listed in the tables in the appendix. Only preferred sizes are listed. Tolerances for sizes between the stated sizes are derived by going to the next nearest given size. The values are not interpolated. A basic size of 27 would use the tolerance values listed for 25. Sizes that are exactly halfway between two stated sizes may use either set of values depending on the design requirements.

9-19 NOMINAL SIZES

The term *nominal* refers to the approximate size of an object that matches a common fraction or whole number. A shaft with a dimension of 1.500±.003 is said to have a nominal size of "one and a half inches." A dimension of 1.500 +.000/−0.005 is still said to have a nominal size of one and a half inches. In both examples, 1.5 is the closest common fraction.

9-20 HOLE AND SHAFT BASIS

One of the charts shown in Figure 9-27 applies tolerances starting with the nominal hole sizes, called *hole basis tolerances*; the other applies tolerances starting with the shaft

nominal sizes, called *shaft basis tolerances.* The choice of which set of values to use depends on the design application. In general, hole basis numbers are used more often because it is more difficult to vary hole diameters manufactured using specific drill sizes than shaft sizes manufactured using a lathe. Shaft sizes may be used when a specific fastener diameter is used to assemble several objects.

Figure 9-28 shows a hole, a shaft, and a set of sample values taken from the tables found in the appendix. One set of values is for hole basis tolerance and the other for shaft basis tolerance. The fit values are the same for both sets of

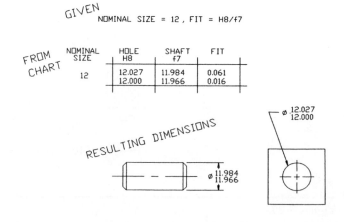

Figure 9-29

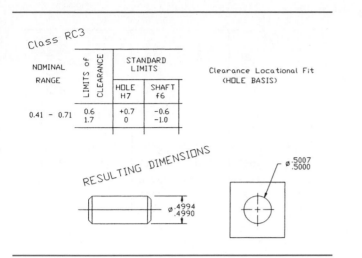

Figure 9-30

Figure 9-31

values. The hole basis values were derived starting with a nominal hole size of 20.000, whereas the shaft basis values were derived starting with a shaft nominal size of 20.000. The letters used to identify holes are always written using capital letters, and the letters for shaft values use lowercase.

Additional fit tolerances may be found on the Web. Search for fits and tolerances.

9-21 SAMPLE PROBLEM SP9-3

Dimension a hole and a shaft that are to fit together using a preferred clearance fit. Use hole basis values based on a nominal size of 12 mm.

Figure 9-29 shows values taken from the appropriate table in the appendix. The values may be applied directly to the shaft and hole as shown.

9-22 STANDARD FITS (INCH VALUES)

The appendix also includes tables of standard fit tolerances for inch values. The tables for inches are presented for a range of nominal values and are not for specific values, as are the metric value tables. The values may be hole or shaft basis.

Fits defined using inch values are classified as follows:

RC = Running and sliding fits
LC = Clearance locational fits

LT = Transitional locational fits
LN = Interference fits
FN = Force fits

Each of these general categories has several subclassifications within it defined by a number, for example, Class RC1, Class RC2, through Class RC9. The letter designations are based on International Tolerance Standards, as are metric designations.

The values are listed in thousandths of an inch. A table value of 1.1 means .0011 inch. A table value of .5 means .0005 inch.

Figure 9-30 shows a set of values for a Class RC3 clearance fit hole basis taken from the appendix. If the values are applied to a nominal size of .5 inch, the resulting hole and shaft sizes will be as shown. Plus table values are added to the nominal value; minus values are subtracted from the nominal value.

Nominal values that are common to two nominal ranges (0.71) may use values from either range.

9-23 SAMPLE PROBLEM SP9-4

Dimension a hole and shaft for a Class LN1 Interference fit based on a nominal diameter of .25 in. Use hole basis values.

Figure 9-31 shows the values for the 0.24 – 0.40 nominal range as listed in the appendix. The values are in thousandths of an inch. Plus values are added to the nominal size. The resulting shaft and hole dimensions are as shown. The diameter of the shaft is larger than that of the hole because this example calls for an interference fit.

PREFERRED SIZES (mm)			
First Choice	Second Choice	First Choice	Second Choice
1	1.1	12	14
1.2	1.4	16	18
1.6	1.8	20	22
2	2.2	25	28
2.5	2.8	30	35
3	3.5	40	45
4	4.5	50	55
5	5.5	60	70
6	7	80	90
8	9	100	110
10	11	120	140

Figure 9-32

Fraction	Decimal Equivalent	Fraction	Decimal Equivalent	Fraction	Decimal Equivalent
7/64	.1094	21/64	.3281	11/16	.6875
1/8	.1250	11/32	.3438	3/4	.7500
9/64	.1406	23/64	.3594	13/16	.8125
5/32	.1562	3/8	.3750	7/8	.8750
11/64	.1719	25/64	.3906	15/16	.9375
3/16	.1875	13/32	.4062	1	1.0000
13/64	.2031	27/64	.4219		
7/32	.2188	7/16	.4375	PARTIAL LIST	
1/4	.2500	29/64	.4531	of standard	
17/64	.2656	15/32	.4688	Twist Drill Sizes	
9/32	.2812	1/2	.5000	(Fractional	
19/64	.2969	9/16	.5625	Sizes)	
5/16	.3125	5/8	.6250		

Figure 9-33

9-24 PREFERRED AND STANDARD SIZES

It is important that designers always consider preferred and standard sizes when selecting sizes for designs. Most tooling is set up to match these sizes, so manufacturing is greatly simplified when preferred and standard sizes are specified. Figure 9-32 shows a list of preferred sizes for metric values.

Consider the case of design calculations that call for a 42-mm diameter hole. A 42-mm diameter hole is not a preferred size. A diameter of 40 mm is the closest preferred size, and a 45-mm diameter is a second choice. A 42-mm hole could be manufactured but would require an unusual drill size that may not be available. It would be wise to reconsider the design to see if a 40-mm diameter hole could be used, and if not, possibly a 45-mm diameter hole.

A very large quantity production run could possibly justify the cost of special tooling, but for smaller runs it is probably better to use preferred sizes. Machinists will have the required drills, and maintenance people will have the appropriate tools for these sizes.

Figure 9-33 shows a list of standard fractional drill sizes. Most companies now specify metric units or decimal inches; however, many standard items are still available in fractional sizes, and many older objects may still require fractional-sized tools and replacement parts. A more complete listing is available in the appendix.

9-25 SURFACE FINISHES

The term *surface finish* refers to the accuracy (flatness) of a surface. Metric values are measured using micrometers (μm), and inch values are measured in microinches (μin.).

The accuracy of a surface depends on the manufacturing process used to produce the surface. Figure 9-34 shows a list of several manufacturing processes and the quality of the surface finish they can be expected to produce.

Surface finishes have several design applications. *Datum surfaces,* or surfaces used for baseline dimensioning, should have fairly accurate surface finishes to help assure accurate measurements; bearing surfaces should have good-quality surface finishes for better load distribution, and parts that operate at high speeds should have smooth finishes to help reduce friction.

Figure 9-35 shows a screw head sitting on a very wavy surface. Note that the head of the screw is in contact with only two wave peaks, meaning all the bearing load is concentrated on the two peaks. This situation could cause stress

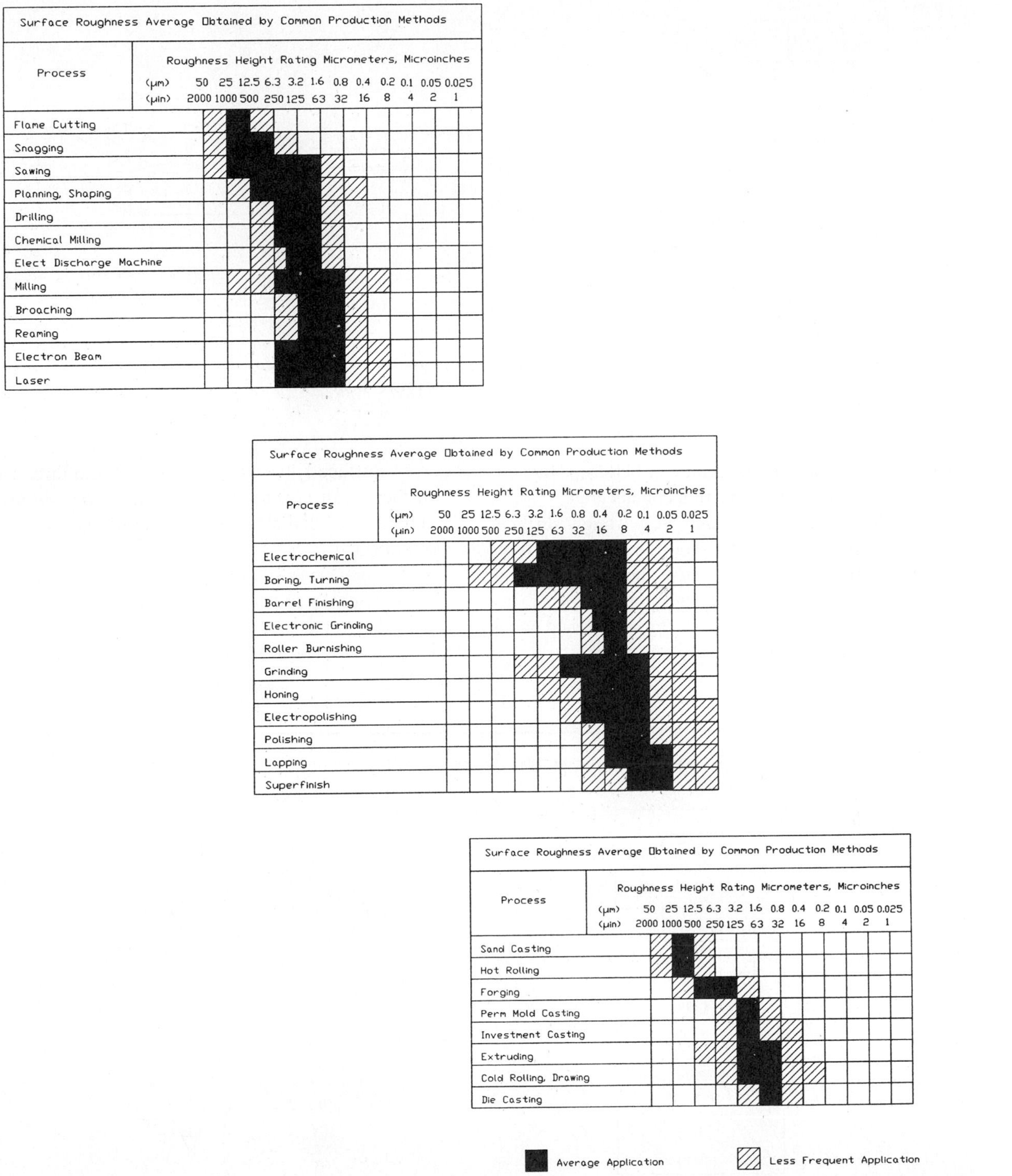
Surface Roughness Average Obtained by Common Production Methods
Process
Roughness Height Rating Micrometers, Microinches
(μm) 50 25 12.5 6.3 3.2 1.6 0.8 0.4 0.2 0.1 0.05 0.025
(μin) 2000 1000 500 250 125 63 32 16 8 4 2 1
Flame Cutting
Snagging
Sawing
Planning, Shaping
Drilling
Chemical Milling
Elect Discharge Machine
Milling
Broaching
Reaming
Electron Beam
Laser
Surface Roughness Average Obtained by Common Production Methods
Process
Roughness Height Rating Micrometers, Microinches
(μm) 50 25 12.5 6.3 3.2 1.6 0.8 0.4 0.2 0.1 0.05 0.025
(μin) 2000 1000 500 250 125 63 32 16 8 4 2 1
Electrochemical
Boring, Turning
Barrel Finishing
Electronic Grinding
Roller Burnishing
Grinding
Honing
Electropolishing
Polishing
Lapping
Superfinish
Surface Roughness Average Obtained by Common Production Methods
Process
Roughness Height Rating Micrometers, Microinches
(μm) 50 25 12.5 6.3 3.2 1.6 0.8 0.4 0.2 0.1 0.05 0.025
(μin) 2000 1000 500 250 125 63 32 16 8 4 2 1
Sand Casting
Hot Rolling
Forging
Perm Mold Casting
Investment Casting
Extruding
Cold Rolling, Drawing
Die Casting
Average Application
Less Frequent Application

Figure 9-34

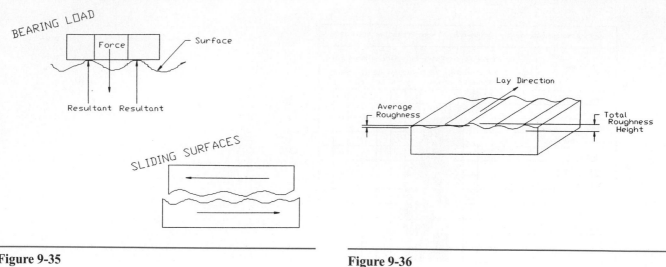

Figure 9-35

Figure 9-36

cracks and greatly weaken the surface. A better-quality surface finish would increase the bearing contact area.

Figure 9-35 also shows two very rough surfaces moving in contact with each other. The result will be excess wear to both surfaces because the surfaces touch only on the peaks, and these peaks will tend to wear faster than flatter areas. Excess vibration can also result when interfacing surfaces are too rough.

Surface finishes are classified into three categories: surface texture, roughness, and lay. ***Surface texture*** is a general term that refers to the overall quality and accuracy of a surface.

Roughness is a measure of the average deviation of a surface's peaks and valleys. See Figure 9-36.

Lay refers to the direction of machine marks on a surface. See Figure 9-37. The lay of a surface is particularly

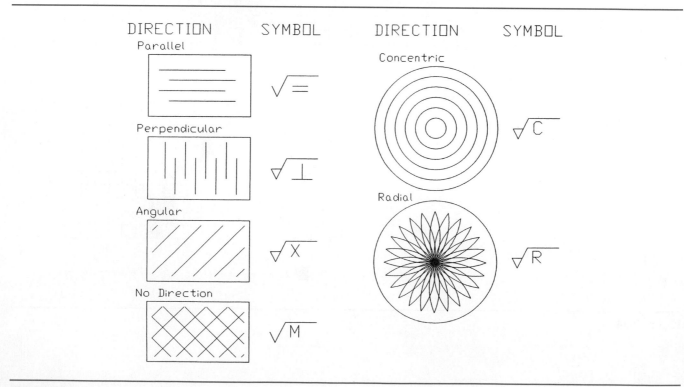

Figure 9-37

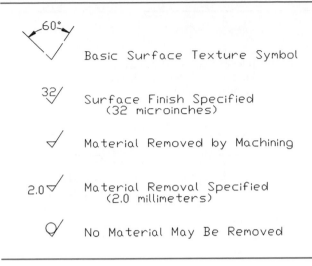

Figure 9-38

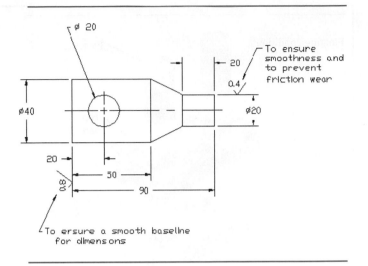

Figure 9-39

important when two moving objects are in contact with each other, especially at high speeds.

9-26 SURFACE CONTROL SYMBOLS

Surface finishes are indicated on a drawing with surface control symbols. See Figure 9-38. The general surface control symbol looks like a check mark. Roughness values may be included with the symbol to specify the required accuracy. Surface control symbols can also be used to specify the manufacturing process that may or may not be used to produce a surface.

Figure 9-39 shows two applications of surface control symbols. In the first example, a 0.8-μm (32-μin.) surface finish is specified on the surface that serves as a datum for several horizontal dimensions. A 0.8-μm surface finish is generally considered the minimum acceptable finish for datums.

A second finish mark with a value of 0.4 μm is located on an extension line that refers to a surface that will be in contact with a moving object. The extra flatness will help prevent wear between the two surfaces.

It is suggested that a general finish mark be drawn and saved as a wblock so it can be inserted as needed on future drawings. Add the machine mark wblock to any prototype drawings created.

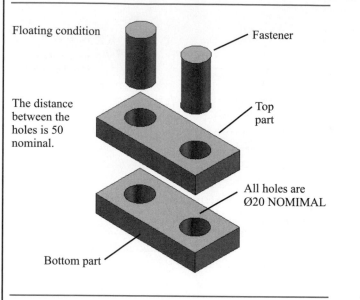

Floating condition

Fastener

The distance between the holes is 50 nominal.

Top part

All holes are Ø20 NOMIMAL

Bottom part

Figure 9-40

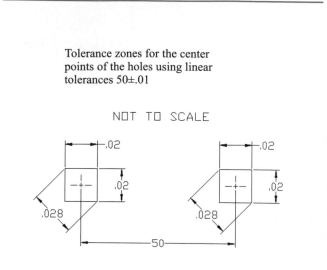

Tolerance zones for the center points of the holes using linear tolerances 50±.01

NOT TO SCALE

Figure 9-41

9-27 DESIGN PROBLEMS

Figure 9-40 shows two objects that are to be fitted together using a fastener such as a screw-and-nut combination. For this example a cylinder will be used to represent a fastener. Only two nominal dimensions are given. The dimensions and tolerances were derived as follows.

The distance between the centers of the holes is given as 50 NOMINAL. The term *nominal* means that the stated value is only a starting point. The final dimensions will be close to the given value, but do not have to equal it.

Assigning tolerances is an iteration process; that is, a tolerance is selected and other tolerance values are calculated from the selected initial values. If the results are not satisfactory, go back and modify the initial value and calculate the other values again. As your experience grows, you will become better at selecting realistic initial values.

In the example shown in Figure 9-40, start by assigning a tolerance of ±0.01 to both the TOP and BOTTOM parts for both the horizontal and vertical dimensions used to locate the holes. This means that there is a possible center point variation of 0.02 for both parts. The parts must always fit together, so tolerances must be assigned based on the worst-case condition, or when the parts are made at the extreme ends of the assigned tolerances.

Figure 9-41 shows a greatly enlarged picture of the worst-case condition created by a tolerance of ±0.01. The center points of the two holes could be as much as 0.028 apart if they were located at opposite corners of the tolerance zones. This means that the minimum hole diameter must always be at least 0.028 larger than the maximum stud diameter. See Section 9-15. In addition, there should be a clearance tolerance assigned so that the hole and stud are

never exactly the same size. Figure 9-42 shows the resulting tolerances.

Floating condition

The TOP and BOTTOM parts shown in Figure 9-40 are to be joined by two independent fasteners; that is, the location of one fastener does not depend on the location of the other. This is called a ***floating condition.*** This means that the tolerance zones for both the TOP and BOTTOM

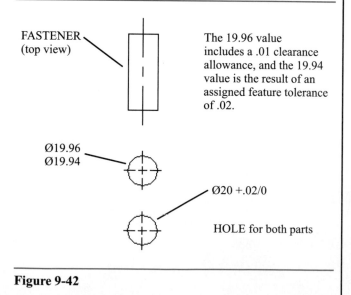

FASTENER (top view)

The 19.96 value includes a .01 clearance allowance, and the 19.94 value is the result of an assigned feature tolerance of .02.

Ø19.96
Ø19.94

Ø20 +.02/0

HOLE for both parts

Figure 9-42

parts can be assigned the same values, and that a fastener diameter selected to fit one part will also fit the other part.

The final tolerances were developed by first defining a minimum hole size of 20.00. An arbitrary tolerance of .02 was assigned to the hole and was expressed as 20.00 +.02/−0, so that the hole can never be any smaller than 20.00.

The 20.00 minimum hole diameter dictates that the maximum fastener diameter can be no greater than 19.97, or .03 (the rounded-off diagonal distance across the tolerance zone −.028) less than the minimum hole diameter. A .01 clearance was assigned. The clearance ensures that the hole and fastener are never exactly the same diameter. The resulting maximum allowable diameter for the fastener is 19.96. Again, an arbitrary tolerance of .02 was assigned to the fastener. The final fastener dimensions are therefore 19.96 to 19.94.

The assigned tolerances ensure that there will always be at least .01 clearance between the fastener and the hole. The other extreme condition occurs when the hole is at its largest possible size (10.02) and the fastener is at its smallest (19.94). This means that there could be as much as .08 clearance between the parts. If this much clearance is not acceptable, then the assigned tolerances will have to be re-evaluated.

Figure 9-43 shows the TOP and BOTTOM parts dimensioned and toleranced. Any dimensions that do not have assigned tolerances are assumed to have standard tolerances. See Figure 9-14.

Note, in Figure 9-43, that the top edge of each part was assigned a surface finish. This was done to help ensure the accuracy of the 20±.01 dimension. If this edge surface were rough, it could affect the tolerance measurements.

This example will be repeated in Chapter 10 using geometric tolerances. Geometric tolerance zones are circular rather than rectangular.

Fixed condition

Figure 9-44 shows the same nominal conditions presented in Figure 9-40, but the fasteners are now fixed to the TOP part. This is called the *fixed condition.* In analyzing the tolerance zones for the fixed condition, one must consider two positional tolerances: the positional tolerances for the holes in the BOTTOM part and the positional tolerances for the fixed fasteners in the TOP part. This may be expressed in an equation as follows:

$$S_{max} + DTSZ = H_{min} - DTZ$$

where

S_{max} = Maximum shaft (fastener) diameter

H_{min} = Minimum hole diameter

DTSZ = Diagonal distance across the shaft's center point tolerance zone

DTZ = Diagonal distance across the hole's center point tolerance zone

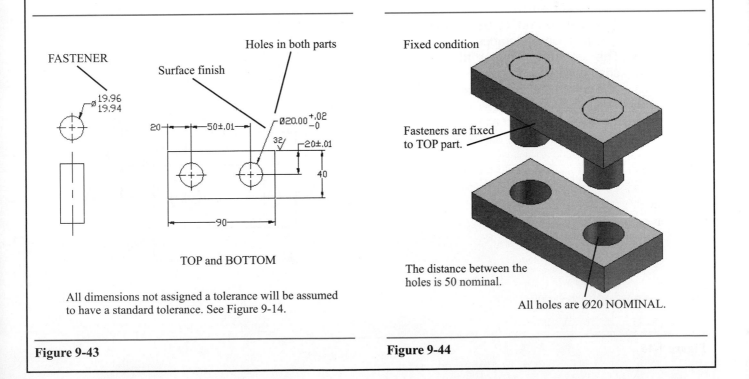

TOP and BOTTOM

All dimensions not assigned a tolerance will be assumed to have a standard tolerance. See Figure 9-14.

Figure 9-43

Figure 9-44

If a dimension and tolerance of 50±.01 and 20±.01 are assigned to both the center distance between the holes and the center distance between the fixed fasteners, the values for DTSZ and DTZ will be equal. The formula can then be simplified as follows:

$$S_{max} = H_{min} - 2(DTZ)$$

where DTZ equals the diagonal distance across the tolerance zone. If a hole tolerance of 20.00 +.02/−0 is also defined, the resulting maximum shaft size can be determined, assuming that the calculated distance of .028 is rounded off to .03. See Figure 9-45.

$$\begin{aligned} S_{max} &= 20.00 - 2(.03) \\ &= 19.94 \end{aligned}$$

This means that the largest possible shaft diameter that will just fit equals 19.94. If a clearance tolerance of .01 is assumed to ensure that the shaft and hole are never exactly the same size, the maximum shaft diameter becomes 19.93. See Figure 9-46.

A feature tolerance of .02 on the shaft will result in a minimum shaft diameter of 19.91. Note that the .01 clearance tolerance and the .02 feature tolerance were arbitrarily chosen. Other possible values could have been used.

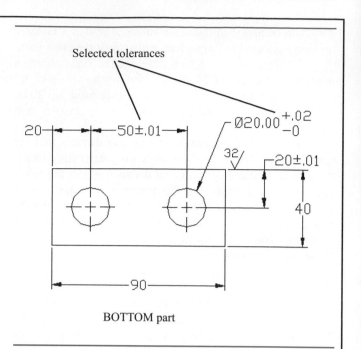

BOTTOM part

Figure 9-45

The shaft values were derived as follows.

20.00	The selected value for the minimum hole diameter
−.03	The rounded-off value for the hole positional tolerance
−.03	The rounded-off value for the shaft positional tolerance
−.01	The selected clearance value
19.93	The maximum shaft value
−.02	The selected tolerance value
19.91	The minimum shaft diameter

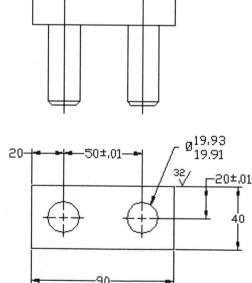

TOP part

Figure 9-46

To design a hole given a fastener size

The previous two examples started by selecting a minimum hole diameter and then calculating the resulting fastener size. Figure 9-47 shows a situation in which the fastener size is defined, and the problem is to determine the appropriate hole sizes. Figure 9-48 shows the dimensions and tolerances for both top and bottom parts.

Requirements:
Clearance, minimum = .003
Hole tolerances = .005
Positional tolerance = .002

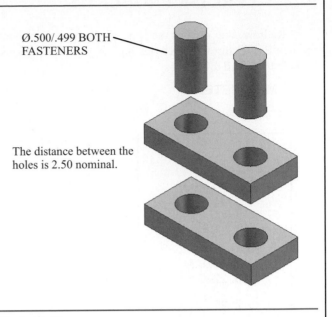

Ø.500/.499 BOTH FASTENERS

The distance between the holes is 2.50 nominal.

Figure 9-47

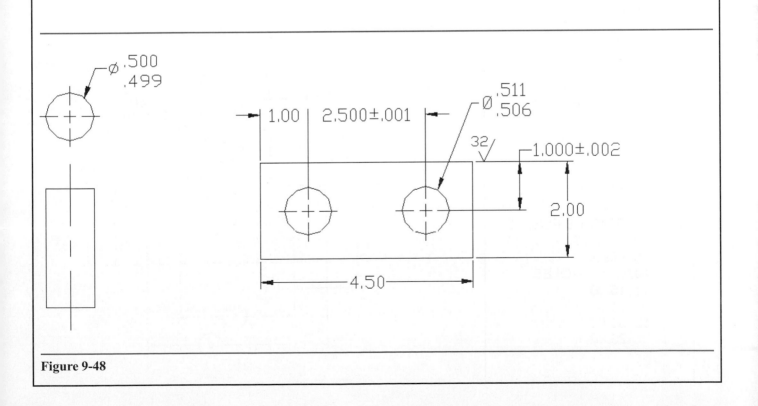

Figure 9-48

9-28 EXERCISE PROBLEMS

Redraw the objects shown in Exercise Problems through EX9-1 through EX9-4 using the given dimensions. Include the listed tolerances.

EX9-1 MILLIMETERS

1. 38±0.05
2. 10±0.1
3. 5±0.05
4. 45.50°
   44.50°
5. 40±0.1
6. 22±0.1
7. 25 $^{+0}_{-0.1}$
8. 25 $^{+0.05}_{-0}$
9. 51.50
   50.75
10. 76±0.1

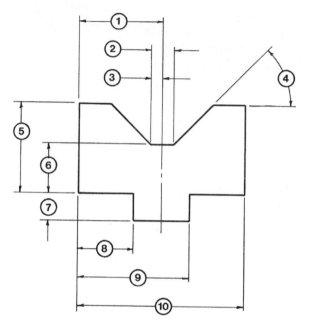

EX9-2 MILLIMETERS

1. 34±0.25
2. 17±0.25
3. 25±0.05
4. 15.00
   14.80
5. 50±0.05
6. 80±0.1
7. R5±0.1-8 PLACES
8. 45±0.25
9. 60±0.1
10. Ø14-3 HOLES
11. 15.00
    14.80
12. 30.00
    29.80

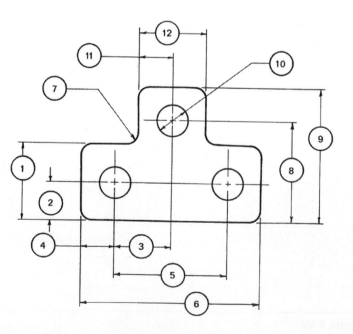

EX9-3 INCHES

1. 3.00±.01
2. 1.56±.01
3. 46.50 / 45.50
4. .750±.005
5. 2.75 / 2.70
6. 3.625±.010
7. 45°±.5°
8. 2.250±.005

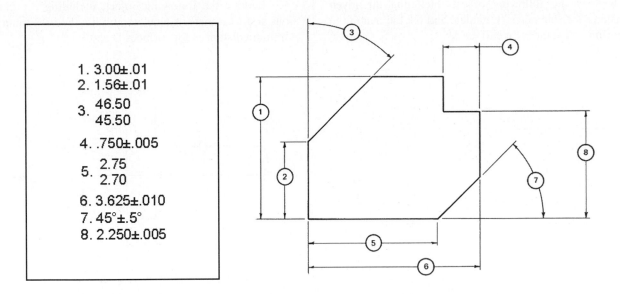

EX9-4 INCHES

1. $50^{+.2}_{0}$
2. R45±.1-2 PLACES
3. $63.5^{0}_{0.2}$
4. 76±.1
5. 38±.1
6. Ø12.00$^{+.05}_{0}$ -3 HOLES
7. 30±.03
8. 30±.03
9. $100^{+.4}_{0}$

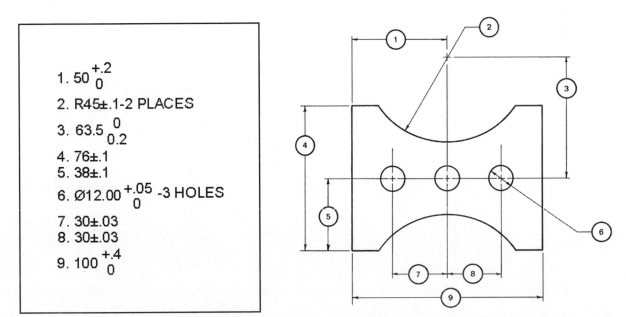

EX9-5 MILLIMETERS

Redraw the following object, including the given dimensions and tolerances. Calculate and list the maximum and minimum distances for surface A.

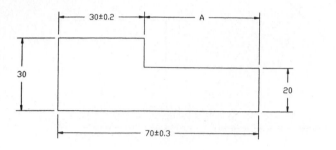

EX9-6 INCHES

A. Redraw the following object, including the dimensions and tolerances. Calculate and list the maximum and minimum distances for surface A.
B. Redraw the given object and dimension it using baseline dimensions. Calculate and list the maximum and minimum distances for surface A.

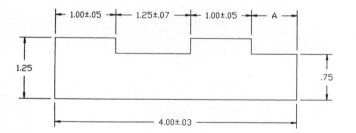

EX9-7 MILLIMETERS

Redraw the following object, including the dimensions and tolerances. Calculate and list the maximum and minimum distances for surfaces D and E.

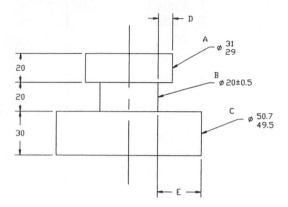

EX9-8 MILLIMETERS

Dimension the following object twice: once using chain dimensions and once using baseline dimensions. Calculate and list the maximum and minimum distances for surface D for both chain and baseline dimensions. Compare the results.

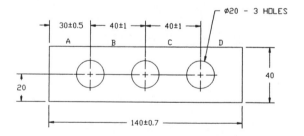

EX9-9 INCHES

Redraw the following shapes, including the dimensions and tolerances. Also list the required minimum and maximum values for the specified distances.

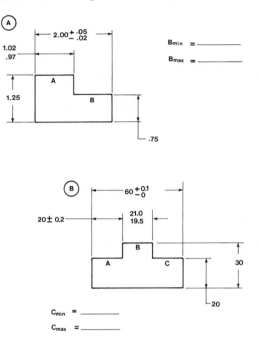

B_{min} = _____

B_{max} = _____

C_{min} = _____

C_{max} = _____

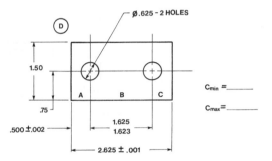

C_{min} = _____

C_{max} = _____

C_{min} = _____

C_{max} = _____

EX9-10

Redraw and complete the following inspection report. Under the Results column classify each "AS MEASURED" value as OK if the value is within the stated tolerances, REWORK if the value indicates that the measured value is beyond the stated tolerance but can be reworked to bring it into the acceptable range, or SCRAP if the value is not within the tolerance range and cannot be reworked to make it acceptable.

INSPECTION REPORT

PART NAME AND NO: 1075500 2

INSPECTOR:

DATE:

BASE DIMENSION	TOLERANCES		AS MEASURED	RESULTS
	MAX	MIN		
(1) 100 ± 0.5			99.8	
(2) ϕ^{57}_{56}			57.01	
(3) 22 ± 0.3			21.72	
(4) $^{40.05}_{39.95}$			39.98	
(5) 22 ± 0.3			21.68	
(6) $R52^{+0}_{-0.2}$			51.99	
(7) $35^{+0.2}_{-0.3}$			35.20	
(8) $30^{+0.4}_{0}$			30.27	
(9) $6.0^{+.1}_{-.2}$			5.85	
(10) 12.0 ± 0.2			11.90	

1.00 3 PLACES

.50 — 10 PLACES

EX9-11 MILLIMETERS

Redraw the following charts and complete them based on the following information. All values are in millimeters.

A. Nominal=16, Fit=H9/d9
B. Nominal=30, Fit=H11/c11
C. Nominal=22, Fit=H7/g6
D. Nominal=10, Fit=C11/h11
E. Nominal=25, Fit=F8/h7
F. Nominal=12, Fit=H7/k6
G. Nominal=3, Fit=H7/p6
H. Nominal=19, Fit=H7/s6
I. Nominal=27, Fit=H7/u6
J. Nominal=30, Fit=N7/h6

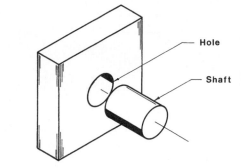

half space

NOMINAL	HOLE		SHAFT		CLEARANCE	
	MAX	MIN	MAX	MIN	MAX	MIN
A						
B						
C						
D						
E						

3.75
6 equal spaces

1.5   6.0 – 6 equal spaces

NOMINAL	HOLE		SHAFT		INTERFERENCE	
	MAX	MIN	MAX	MIN	MAX	MIN
F						
G						
H						
I						
J						

Use the same dimensions given above

EX9-12

Redraw the following charts and complete them based
on the following information. All values are in inches.

A. Nominal=0.25, Fit=Class LC5
B. Nominal=1.00, Fit=Class LC7
C. Nominal=1.50, Fit=Class LC10
D. Nominal=0.75, Fit=Class RC3
E. Nominal=2.50, Fit=Class RC6
F. Nominal=.500, Fit=Class LT2
G. Nominal=1.25, Fit=Class LT5
H. Nominal=3.00, Fit=Class LN3
I. Nominal=1.625, Fit=Class FN1
J. Nominal=2.00, Fit=Class FN4

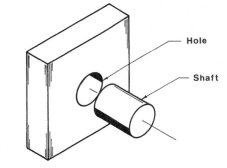

half space

3.75
6 equal
spaces

NOMINAL	HOLE		SHAFT		CLEARANCE	
	MAX	MIN	MAX	MIN	MAX	MIN
A						
B						
C						
D						
E						

|← 1.5 →|← 6.0 – 6 equal spaces →|

NOMINAL	HOLE		SHAFT		INTERFERENCE	
	MAX	MIN	MAX	MIN	MAX	MIN
F						
G						
H						
I						
J						

Use the same dimensions given above

Draw the chart shown and add the appropriate values based on the dimensions and tolerances given in Exercise Problems EX9-13 through EX9-16.

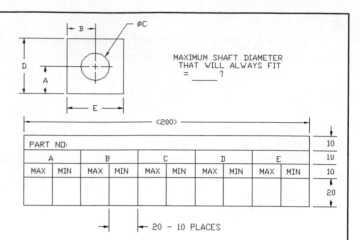

MAXIMUM SHAFT DIAMETER
THAT WILL ALWAYS FIT
= _____ ?

PART NO:									
A		B		C		D		E	
MAX	MIN	MAX	MIN	MAX	MIN	MAX	MIN	MAX	MIN

20 – 10 PLACES

EX9-13 MILLIMETERS

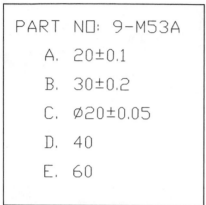

PART NO: 9-M53A

A. 20±0.1

B. 30±0.2

C. Ø20±0.05

D. 40

E. 60

EX9-15 MILLIMETERS

PART NO: 9-M53B

A. 32.02 / 31.97

B. 47.52 / 47.50

C. Ø18 +0.05 / 0

D. 64±0.05

E. 100±0.05

EX9-14 MILLIMETERS

PART NO: 9-M53B

A. 32.02 / 31.97

B. 47.52 / 47.50

C. Ø18 +0.05 / 0

D. 64±0.05

E. 100±0.05

EX9-16 MILLIMETERS

PART NO: 9-E47B

A. 18 +0 / -0.02

B. 26 +0 / -0.04

C. Ø 24.03 / 23.99

D. 52±0.04

E. 36±0.02

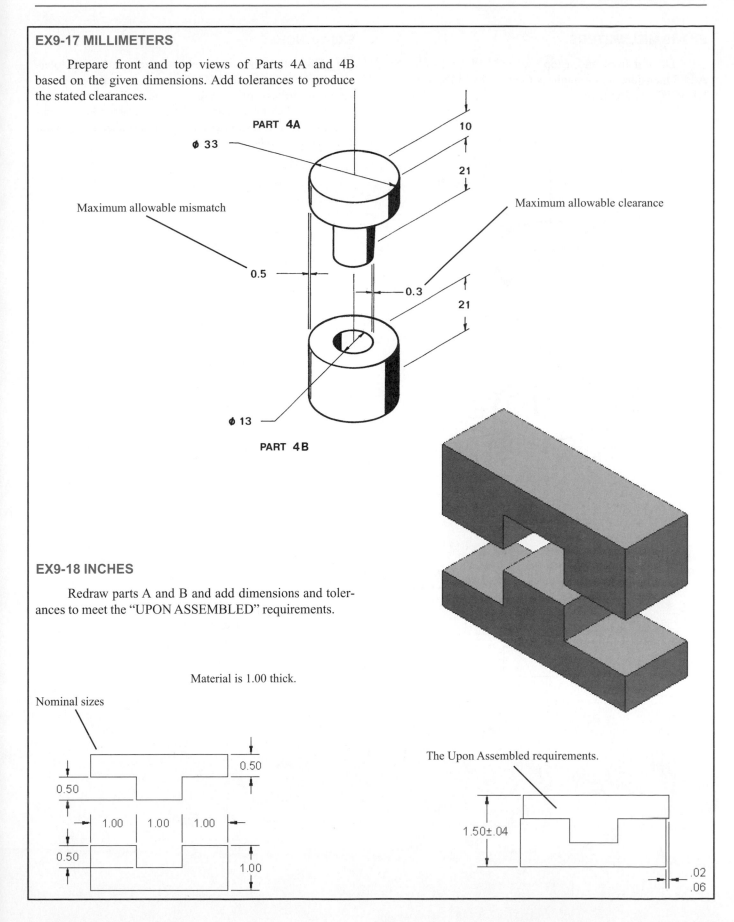

EX9-17 MILLIMETERS

Prepare front and top views of Parts 4A and 4B based on the given dimensions. Add tolerances to produce the stated clearances.

PART 4A

φ 33

10

21

Maximum allowable mismatch

Maximum allowable clearance

0.5

0.3

21

φ 13

PART 4B

EX9-18 INCHES

Redraw parts A and B and add dimensions and tolerances to meet the "UPON ASSEMBLED" requirements.

Material is 1.00 thick.

Nominal sizes

0.50

0.50

1.00 1.00 1.00

0.50

1.00

The Upon Assembled requirements.

1.50±.04

.02
.06

EX9-19 MILLIMETERS

Draw a front and a top view of both given objects. Add dimensions and tolerances to meet the "FINAL CONDITION" requirements.

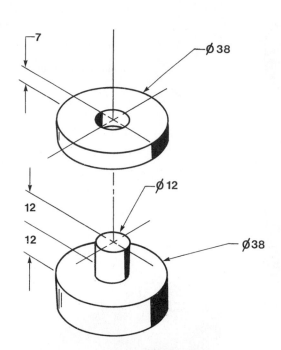

FINAL CONDITION

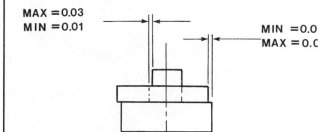

EX9-20 INCHES

Given the following nominal sizes, dimension tolerance parts AM311 and AM312 so that they always fit together regardless of orientation. Further, dimension the overall lengths of each part so that, in the assembled condition, they will always pass through a clearance gauge with an opening of 90.00±.02.

In the assembled condition, both parts must always pass through the clearance gauge.

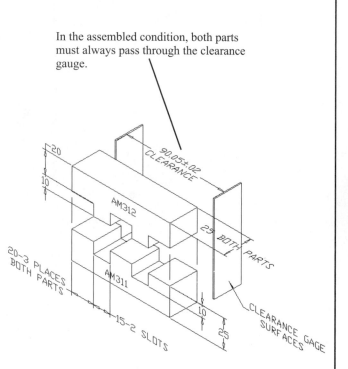

All given dimensions, except for the clearance gauge, are nominal.

EX9-21 MILLIMETERS

Design a bracket that will support the three Ø100 wheels shown. The wheels will utilize three Ø5.00±0.01 shafts attached to the bracket. The bottom of the bracket must have a minimum of 10 millimeters from the ground. The wall thickness of the bracket must always be at least 5 millimeters, and the minimum bracket opening must be at least 15 millimeters.

1. Prepare a front and a side view of the bracket.
2. Draw the wheels in their relative positions using phantom lines.
3. Add all appropriate dimensions and tolerances.

ALL SIZES ARE NOMINAL, UNLESS
OTHERWISE STATED.

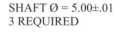

SHAFT Ø = 5.00±.01
3 REQUIRED

ROLLER BLADE ASSEMBLY
PART NUMBER BU110-44

Given a TOP and BOTTOM part in the floating condition as shown in Figure EX9-22 add dimensions and tolerances to satisfy the following conditions. Size the TOP and BOTTOM parts and FASTENER length as needed.

EX9-22 INCHES–CLEARANCE FIT

A. The distance between the holes is 2.00 nominal.
B. The diameter of the fasteners is Ø.375 nominal.
C. The fasteners have a total tolerance of .001.
D. The holes have a tolerance of .002.
E. The minimum allowable clearance between the fastener and the holes is .003.
F. The material is .375 inch thick.

EX9-23 MILLIMETERS–CLEARANCE FIT

A. The distance between the holes is 80 nominal.
B. The nominal diameter of the fasteners is Ø12.
C. The fasteners have a total tolerance of 0.05.
D. The holes have a tolerance of 0.03.
E. The minimum allowable clearance between the fastener and the holes is 0.02.
F. The material is 12 millimeters thick.

EX9-24 INCHES–CLEARANCE FIT

A. The distance between the holes is 3.50 nominal.
B. The diameter of the fasteners is Ø.625.
C. The fasteners have a total tolerance of .005.
D. The holes have a tolerance of .003.
E. The minimum allowable clearance between the fastener and the holes is .002.
F. The material is .500 inch thick.

EX9-25 MILLIMETERS–CLEARANCE FIT

A. The distance between the holes is 120 nominal.
B. The diameter of the fasteners is Ø24 nominal.
C. The fasteners have a total tolerance of 0.01.
D. The holes have a tolerance of 0.02.
E. The minimum allowable clearance between the fastener and the holes is 0.04.
F. The material is 20 millimeters thick.

EX9-26 INCHES–INTERFERENCE FIT

A. The distance between the holes is 2.00 nominal.
B. The diameter of the fasteners is Ø.250 nominal.
C. The fasteners have a total tolerance of .001.
D. The holes have a tolerance of .002.
E. The maximum allowable interference between the fastener and the holes is .0065.
F. The material is .438 inch thick.

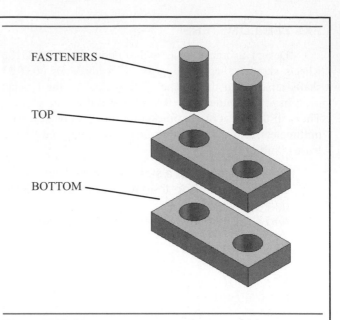

Figure EX9-22

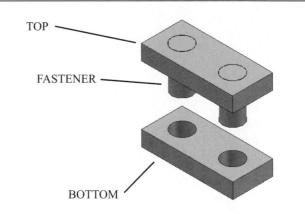

Figure EX9-23

EX9-27 MILLIMETERS–INTERFERENCE FIT

A. The distance between the holes is 80 nominal.
B. The diameter of the fasteners is Ø10 nominal.
C. The fasteners have a total tolerance of 0.01.
D. The holes have a tolerance of 0.02.
E. The maximum allowable interference between the fastener and the holes is 0.032.
F. The material is 14 millimeters thick.

EX9-28 INCHES–LOCATIONAL FIT

A. The distance between the holes is 2.25 nominal.
B. The diameter of the fasteners is Ø.50 nominal.

C. The fasteners have a total tolerance of .001.
D. The holes have a tolerance of .002.
E. The minimum allowable clearance between the fastener and the holes is .0010.
F. The material is .370 inch thick.

EX9-29 MILLIMETERS–TRANSITIONAL FIT

A. The distance between the holes is 100 nominal.
B. The diameter of the fasteners is Ø16 nominal.
C. The fasteners have a total tolerance of 0.01.
D. The holes are to have a tolerance of 0.02.
E. The minimum allowable clearance between the fastener and the holes is 0.01.
F. The material is 20 millimeters thick.

Given a TOP and BOTTOM part in the fixed condition as shown in Figure EX9-23 add dimensions and tolerances to satisfy the following conditions. Size the TOP and BOTTOM parts and FASTENER length as needed.

EX9-30 INCHES–CLEARANCE FIT

A. The distance between the holes is 2.00 nominal.
B. The diameter of the fasteners is Ø.375 nominal.
C. The fasteners have a total tolerance of .001.
D. The holes have a tolerance of .002.
E. The minimum allowable clearance between the fastener and the holes is .003.
F. The material is .375 inch thick.

EX9-31 MILLIMETERS–CLEARANCE FIT

A. The distance between the holes is 80 nominal.
B. The nominal diameter of the fasteners is Ø12.
C. The fasteners have a total tolerance of 0.05.
D. The holes have a tolerance of 0.03.
E. The minimum allowable clearance between the fastener and the holes is 0.02.
F. The material is 12 millimeters thick.

EX9-32 INCHES–CLEARANCE FIT

A. The distance between the holes is 3.50 nominal.
B. The diameter of the fasteners is Ø.625.
C. The fasteners have a total tolerance of .005.
D. The holes have a tolerance of .003.
E. The minimum allowable clearance between the fastener and the holes is .002.
F. The material is .500 inch thick.

EX9-33 MILLIMETERS–CLEARANCE FIT

A. The distance between the holes is 120 nominal.
B. The diameter of the fasteners is Ø24 nominal.
C. The fasteners have a total tolerance of 0.01.
D. The holes have a tolerance of 0.02.
E. The minimum allowable clearance between the fastener and the holes is 0.04.
F. The material is 20 millimeters thick.

EX9-34 INCHES–INTERFERENCE FIT

A. The distance between the holes is 2.00 nominal.
B. The diameter of the fasteners is Ø.250 nominal.
C. The fasteners have a total tolerance of .001.
D. The holes have a tolerance of .002.
E. The maximum allowable interference between the fastener and the holes is .0065.
F. The material is .438 inch thick.

EX9-35 MILLIMETERS–INTERFERENCE FIT

A. The distance between the holes is 80 nominal.
B. The diameter of the fasteners is Ø10 nominal.
C. The fasteners have a total tolerance of 0.01.
D. The holes have a tolerance of 0.02.
E. The maximum allowable interference between the fastener and the holes is 0.032.
F. The material is 14 millimeters thick.

EX9-36 INCHES–LOCATIONAL FIT

A. The distance between the holes is 2.25 nominal.
B. The diameter of the fasteners is Ø.50 nominal.
C. The fasteners have a total tolerance of .001.
D. The holes have a tolerance of .002.
E. The minimum allowable clearance between the fastener and the holes is .0010.
F. The material is .370 inch thick.

EX9-37 MILLIMETERS–TRANSITIONAL FIT

A. The distance between the holes is 100 nominal.
B. The diameter of the fasteners is Ø16 nominal.
C. The fasteners have a total tolerance of 0.01.
D. The holes are to have a tolerance of 0.02.
E. The minimum allowable clearance between the fastener and the holes is 0.01.
F. The material is 20 millimeters thick.

EX9-38 MILLIMETERS

Given the following two assemblies, size the individual parts so that they always fit together. Create a drawing for each part including dimensions and tolerances.

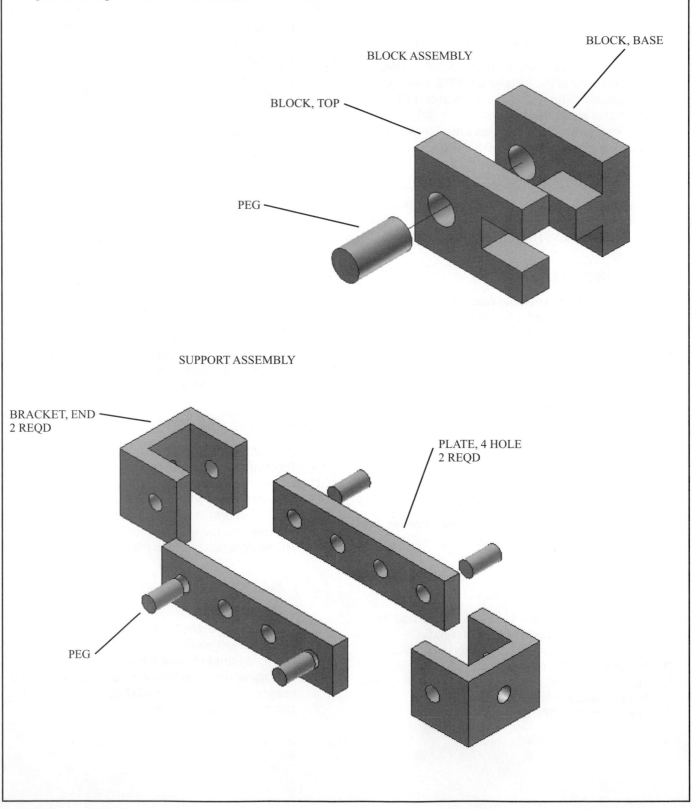

BLOCK ASSEMBLY

BLOCK, TOP

BLOCK, BASE

PEG

SUPPORT ASSEMBLY

BRACKET, END
2 REQD

PLATE, 4 HOLE
2 REQD

PEG

C H A P T E R 10

Geometric Tolerances

10-1 INTRODUCTION

Geometric tolerancing is a dimensioning and tolerancing system based on the geometric shape of an object. Surfaces may be defined in terms of their flatness or roundness, or in terms of how perpendicular or parallel they are to other surfaces.

Geometric tolerances allow a more exact definition of the shape of an object than do conventional coordinate-type tolerances. Objects can be toleranced in a manner more closely related to their design function, or so that their features and surfaces are more directly related to each other.

Figure 10-1 shows a square shape dimensioned and toleranced using plus and minus tolerances. The resulting tolerance zone has an outside length of 51 and an inside length of 49 square. The defined tolerance zone allows any shape that falls within in it to be deemed acceptable, or correctly manufactured. Figure 10-1 shows an exaggerated shape that fits within the defined tolerance zone and is not square yet would be acceptable under the specified dimensions and tolerances. Geometric tolerancing can be used to more precisely define the tolerance zone when a more nearly square shape is required.

It should be pointed out that geometric tolerancing is not a panacea for all dimensioning and tolerancing problems. In many cases coordinate tolerancing, as presented in

Chapter 9, is sufficient to accurately define an object. Unnecessary or excessive use of geometric tolerances can increase production costs. Most objects are toleranced using a combination of coordinate and geometric tolerances, depending on the design function of the object.

The key to using tolerances and types of tolerances may be simply stated as "decimal points cost money." Every tolerance should be made as loose as possible while

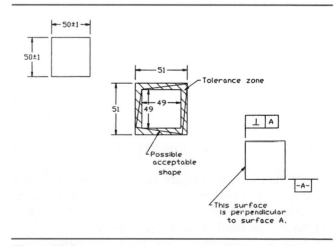

Figure 10-1

397

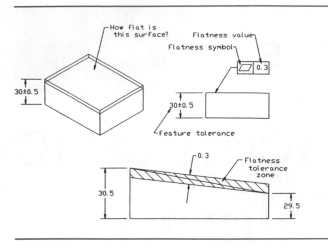

Figure 10-2

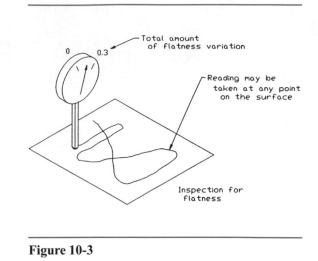

Figure 10-3

still maintaining the design integrity of the object. If a surface flatness is critical to the correct functioning of the object, then, of course, it will require a very close tolerance. But every tolerance should be considered individually and loosened wherever possible to make the object's manufacture easier and therefore less expensive.

10-2 TOLERANCES OF FORM

Tolerances of form are used to define the shape of a surface relative to itself. There are four classifications: flatness, straightness, roundness and cylindricity. Tolerances of form are not related to other surfaces but apply only to an individual surface.

10-3 FLATNESS

Flatness tolerances are used to define the amount of variation permitted in an individual surface. The surface is thought of as a plane not related to the rest of the object.

Figure 10-2 shows a rectangular object. How flat is the top surface? The given plus or minus tolerances allow a variation of ±0.5 across the surface. Without additional tolerances the surface could look like a series of waves that vary between 30.5 and 29.5.

If the example in Figure 10-2 is assigned a flatness tolerance of 0.3, the height of the object, the feature tolerance could continue to vary based on the 30 ± 0.5 tolerance, but the surface itself could not vary by more than 0.3. In the most extreme condition, one end of the surface could be 30.5 above the bottom surface and the other end 29.5, but the surface would still be limited to within two parallel planes 0.3 apart as shown.

To better understand the meaning of flatness, consider how the surface would be inspected. The surface would be acceptable if a gauge could be moved all around the surface and never varied by more than 0.3. See Figure 10-3. Every point in the plane must be within the specified tolerance.

10-4 STRAIGHTNESS

Straightness tolerances are used to measure the variation of an individual feature along a straight line in a specified direction. Figure 10-4 shows an object with a straightness tolerance applied to its top surface. Straightness differs from flatness because straightness measurements are checked by moving a gauge directly across the surface in a single direction. The gauge is not moved randomly about the surface, as is required by flatness.

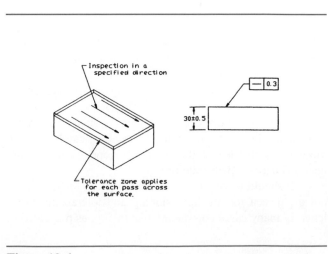

Figure 10-4

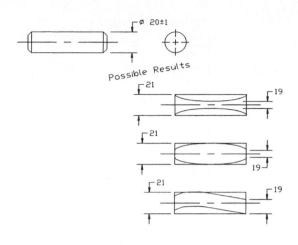

Figure 10-5

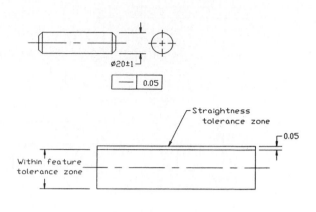

Figure 10-6

Straightness tolerances are most often applied to circular or matching objects to help ensure that the parts are not barreled or warped within the given feature tolerance range and therefore not fit together well. Figure 10-5 shows a cylindrical object dimensioned and toleranced using a standard feature tolerance. The surface of the cylinder may vary within the specified tolerance range as shown.

Figure 10-6 shows the same object shown in Figure 10-5 dimensioned and toleranced using the same feature tolerance, but also including a 0.05 straightness tolerance. The straightness tolerance limits the surface variation to 0.05 as shown.

10-5 STRAIGHTNESS (RFS AND MMC)

Figure 10-7 again shows the same cylinder shown in Figures 10-5 and 10-6. This time the straightness tolerance is applied about the cylinder's centerline. This type of tolerance permits the feature tolerance and geometric tolerance to be used together to define a virtual condition. A *virtual condition* is used to determine the maximum possible size variation of the cylinder or the smallest diameter hole that will always accept the cylinder. See Section 10-17.

The geometric tolerance specified in Figure 10-7 is applied to any circular segment along the cylinder, regardless of the cylinder's diameter. This means that the 0.05 tolerance is applied equally when the cylinder's diameter measures 19 or when it measures 21. This application is called *RFS, regardless of feature size.* RFS conditions are specified in a tolerance either by an S with a circle around

it or implied tacitly when no other symbol is used. In Figure 10-7 no symbol is listed after the 0.05 value, so it is assumed to be applied RFS.

Figure 10-8 shows the cylinder dimensioned with an *MMC* condition applied to the straightness tolerance. MMC stands for *maximum material condition* and means that the specified straightness tolerance (0.05) is applied only at the MMC condition or when the cylinder is at its maximum diameter size (21).

A shaft is an external feature, so its largest possible size or MMC occurs when it is at its maximum diameter. A hole is an internal feature. A hole's MMC condition occurs when it is at its smallest diameter. The MMC condition for holes will be discussed later in the chapter along with positional tolerances.

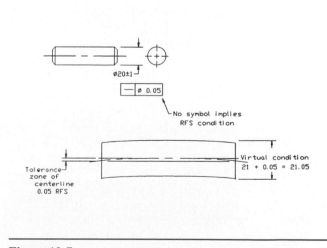

Figure 10-7

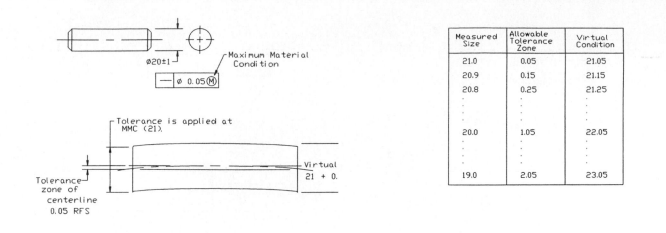

Measured Size	Allowable Tolerance Zone	Virtual Condition
21.0	0.05	21.05
20.9	0.15	21.15
20.8	0.25	21.25
.	.	.
.	.	.
.	.	.
20.0	1.05	22.05
.	.	.
.	.	.
19.0	2.05	23.05

Figure 10-8

Applying a straightness tolerance at MMC allows for a variation in the resulting tolerance zone. Because the 0.05 flatness tolerance is applied at MMC, the virtual condition is still 21.05, the same as with the RFS condition; however, the tolerance is applied only at MMC. As the cylinder's diameter varies within the specified feature tolerance range the acceptable tolerance zone may vary to maintain the same virtual condition.

Figure 10-8 lists how the tolerance zone varies as the cylinder's diameter varies. When the cylinder is at its largest size, or MMC, the tolerance zone equals 0.05, or the specified flatness variation. When the cylinder is at its smallest diameter the tolerance zone equals 2.05, or the

total feature size plus the total flatness size. In all variations the virtual size remains the same, so at any given cylinder diameter value, the size of the tolerance zone can be determined by subtracting the cylinder's diameter value from the virtual condition.

Figure 10-9 shows a comparison between different methods used to dimension and tolerance a .750 shaft. The first example uses only a feature tolerance. This tolerance sets an upper limit of .755 and a lower limit of .745. Any variations within that range are acceptable.

The second example in Figure 10-9 sets a straightness tolerance of .003 about the cylinder's centerline. No conditions are defined, so the tolerance is applied RFS.

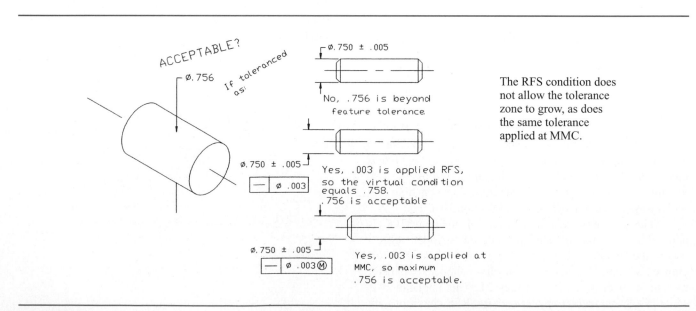

The RFS condition does not allow the tolerance zone to grow, as does the same tolerance applied at MMC.

Figure 10-9

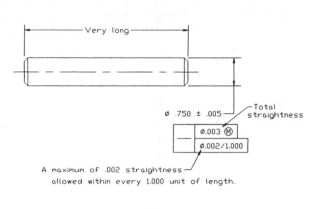

Figure 10-10

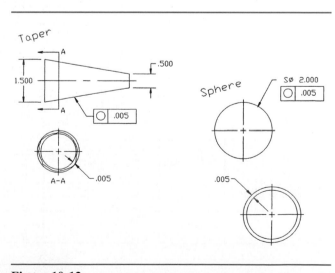

Figure 10-11

This limits the variations in straightness to .003 at all feature sizes. For example, when the shaft is at its smallest possible feature size, .745, the .003 still applies. This means that a shaft measuring .745 that had a straightness variation greater than .003 would be rejected. If the tolerance has been applied at MMC, the part would be accepted. This does not mean that straightness tolerances should always be applied at MMC. If straightness is critical to the design integrity or function of the part, then straightness should be applied in the RFS condition.

The third example in Figure 10-9 applies the straightness tolerance about the centerline at MMC. This tolerance creates a virtual condition of .758. The MMC condition allows the tolerance to vary as the feature tolerance varies, so when the shaft is at its smallest feature size, .745, a straightness tolerance of .013 is acceptable (.010 feature tolerance +.003 straightness tolerance).

If the tolerance specification for the cylinder shown in Figure 10-9 were to have a 0.000 tolerance applied at MMC, it would mean that the shaft would have to be perfectly straight at MMC, or when the shaft was at its maximum value (.755); however, the straightness tolerance could vary as the feature size varied, as discussed for the other tolerance conditions. A 0.000 tolerance means that the MMC and the virtual conditions are equal.

Figure 10-10 shows a very long .750-diameter shaft. Its straightness tolerance includes a length qualifier that serves to limit the straightness variations over each inch of the shaft length and prevents excess waviness over the full length. The tolerance Ø.002/1.000 means that the total straightness may vary over the entire length of the shaft by .003 but that the variation is limited to .002 per 1.000 of shaft length.

10-6 CIRCULARITY

Circularity tolerances are used to limit the amount of variation in the roundness of a surface of revolution and are measured at individual cross sections along the length of the object. The measurements are limited to the individual cross sections and are not related to other cross sections. This means that in extreme conditions the shaft shown in Figure 10-11 could actually taper from a diameter of 21 to a diameter of 19 and never violate the circularity requirement. It also means that qualifications such as MMC cannot be applied.

Figure 10-11 shows a shaft that includes a feature tolerance and a circularity tolerance of 0.07. To understand circularity tolerances, consider an individual cross section, or

Figure 10-12

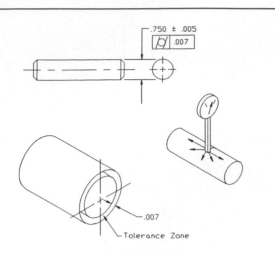

Figure 10-13

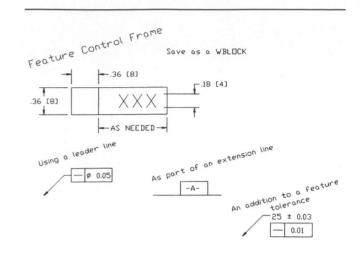

Figure 10-14

slice, of the cylinder. The shape of the outside edge of the slice varies around the slice. The difference between the maximum diameter and the minimum diameter of the slice can never exceed the stated circularity tolerance.

Circularity tolerances can be applied to tapered sections and spheres, as shown in Figure 10-12. In both applications, circularity is measured around individual cross sections, as it was for the shaft shown in Figure 10-11.

10-7 CYLINDRICITY

Cylindricity tolerances are used to define a tolerance zone both around individual circular cross sections of an object and also along its length. The resulting tolerance zone looks like two concentric cylinders.

Figure 10-13 shows a shaft that includes a cylindricity tolerance that establishes a tolerance zone of .007. This means that if the maximum measured diameter is determined to be .755, the minimum diameter cannot be less than .748 anywhere on the cylindrical surface. Figure 10-14 shows how to draw the feature control frames that surround the tolerance symbol and size specifications.

Cylindricity and circularity are somewhat analogous to flatness and straightness. Flatness and cylindricity are concerned with variations across an entire surface or plane. In the case of cylindricity, the plane is shaped like a cylinder. Straightness and circularity are concerned with variations of a single element of a surface, that is, a straight line across the plane in a specified direction for straightness, and a path around a single cross section for circularity.

	TYPE OF TOLERANCE	CHARACTERISTIC	SYMBOL
FOR INDIVIDUAL FEATURES	FORM	STRAIGHTNESS	—
		FLATNESS	▱
		CIRCULARITY	○
		CYLINDRICITY	⌀
INDIVIDUAL OR RELATED FEATURES	PROFILE	PROFILE OF A LINE	⌒
		PROFILE OF A SURFACE	⌓
RELATED FEATURES	ORIENTATION	ANGULARITY	∠
		PERPENDICULARITY	⊥
		PARALLELISM	//
	LOCATION	POSITION	⊕
		CONCENTRICITY	◎
	RUNOUT	CIRCULAR RUNOUT	↗
		TOTAL RUNOUT	↗↗

TERM	SYMBOL
AT MAXIMUM MATERIAL CONDITION	Ⓜ
REGARDLESS OF FEATURE SIZE	Ⓢ
AT LEAST MATERIAL CONDITION	Ⓛ
PROJECTED TOLERANCE ZONE	Ⓟ
DIAMETER	⌀
SPHERICAL DIAMETER	S⌀
RADIUS	R
SPHERICAL RADIUS	SR
REFERENCE	( )
ARC LENGTH	⌒

Figure 10-15

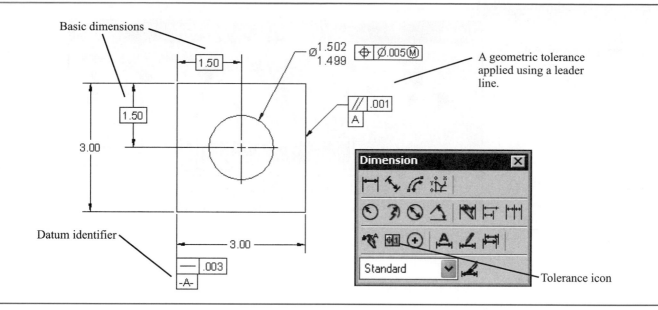

Figure 10-16

10-8 GEOMETRIC TOLERANCES USING AUTOCAD

Geometric tolerances are tolerances that limit dimensional variations based on the geometric properties of an object. Figure 10-15 shows a list of geometric tolerance symbols. Figure 10-16 shows an object dimensioned using geometric tolerances. The geometric tolerances were created as follows.

To define a datum

1. Select the **Tolerance** tool from the **Dimension** toolbar.

The **Geometric Tolerance** dialog box will appear. See Figure 10-17.

2. Click the **Datum Identifier** box and type **-A-,** then click **OK.**

   Command: _tolerance
   Enter tolerance location:

3. Position the datum identifier and press the left mouse button. See Figure 10-16.

To define a straightness value

1. Select the **Tolerance** tool from the **Dimension** toolbar.

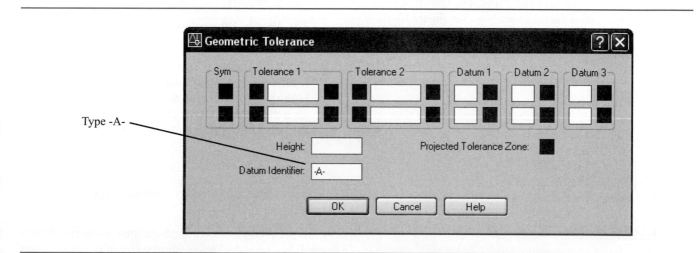

Figure 10-17

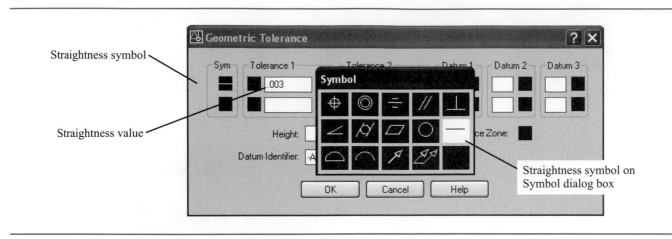

Straightness symbol

Straightness value

Straightness symbol on Symbol dialog box

Figure 10-18

The **Geometric Tolerance** dialog box will appear.

2. Select the top open box under the heading **Sym.**

The **Symbol** dialog box will reappear. See Figure 10-18.

3. Select the straightness symbol, then **OK.**

The **Geometric Tolerance** dialog box will reappear with the straightness symbol in the first box under the heading **Sym.**

4. Click the open box in the **Tolerance 1** box and type **.003;** click **OK.**

Command: _tolerance
Enter tolerance location:

5. Position the straightness tolerance and press the left mouse button

See Figure 10-19. Use the **Move** and **Osnap** tools if necessary to reposition the tolerance box. See Figure 10-16.

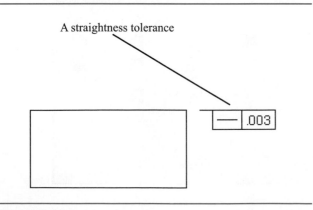

A straightness tolerance

Figure 10-19

To create a positional tolerance

A *positional tolerance* is used to locate and tolerance a hole in an object. Positional tolerances require base locating dimensions for the hole's center point. Positional tolerances also require a feature tolerance to define the diameter tolerances of the hole, and a geometric tolerance to define the position tolerance for the hole's center point.

To create a basic dimension

See the two 1.50 dimensions in Figure 10-16 used to locate the center position of the hole.

1. Select the **Dimension Style** tool or type **DDIM** in response to a command prompt.

The **Dimension Style Manager** dialog box will appear.

2. Select **Modify.**

The **Modify Dimension Style: Standard** dialog box will appear. See Figure 9-7.

3. Select the **Basic** option next to the heading **Method** in the **Tolerance** box.
4. Return to the drawing screen.
5. Use the **Linear** tool from the **Dimension** toolbar and add the appropriate dimensions.

See Figure 10-16.

To create basic dimensions from existing dimensions

Figure 10-20 shows a shape that includes dimensions. It has been decided to change two of the dimensions to base dimensions. The procedure is as follows.

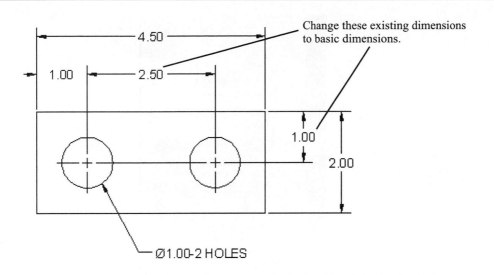

Figure 10-20

1. Click on the **2.50** dimension.

   Blue squares will appear on the dimension.

2. Right-click the mouse.

A dialog box will appear.

3. Select the **Properties** option.

   The **Properties** dialog box will appear. See Figure 10-21.

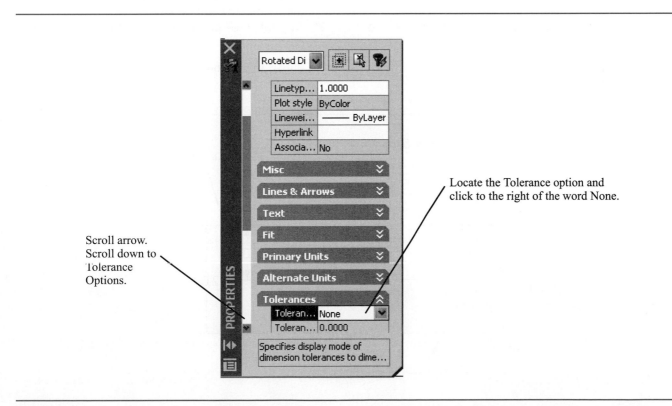

Figure 10-21

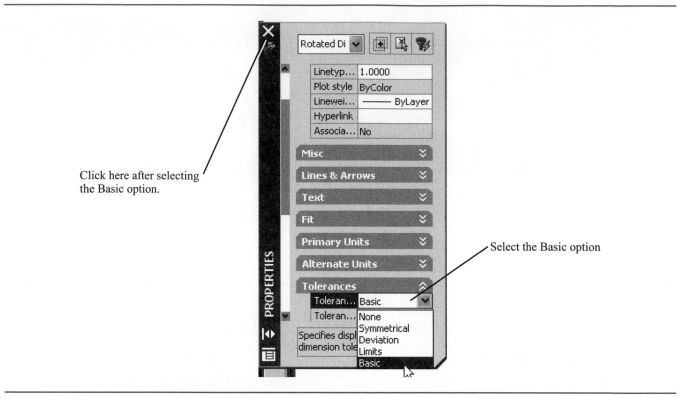

Click here after selecting the Basic option.

Select the Basic option

Figure 10-22

4. Use the scroll arrow next to the **P** in **Properties** and locate the **Tolerances** option.
5. Click the open space next to the word **None** and scroll down to the **Basic** option.

   See Figure 10-22.

6. Select the **Basic** option, then click the return **X,** then press the **<Esc>** key.
7. Repeat the procedure for the vertical 1.00 dimension.

   Figure 10-23 shows the final results.

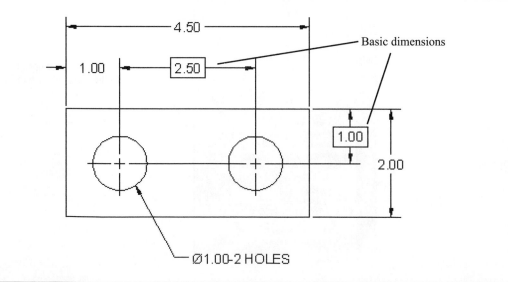

Figure 10-23

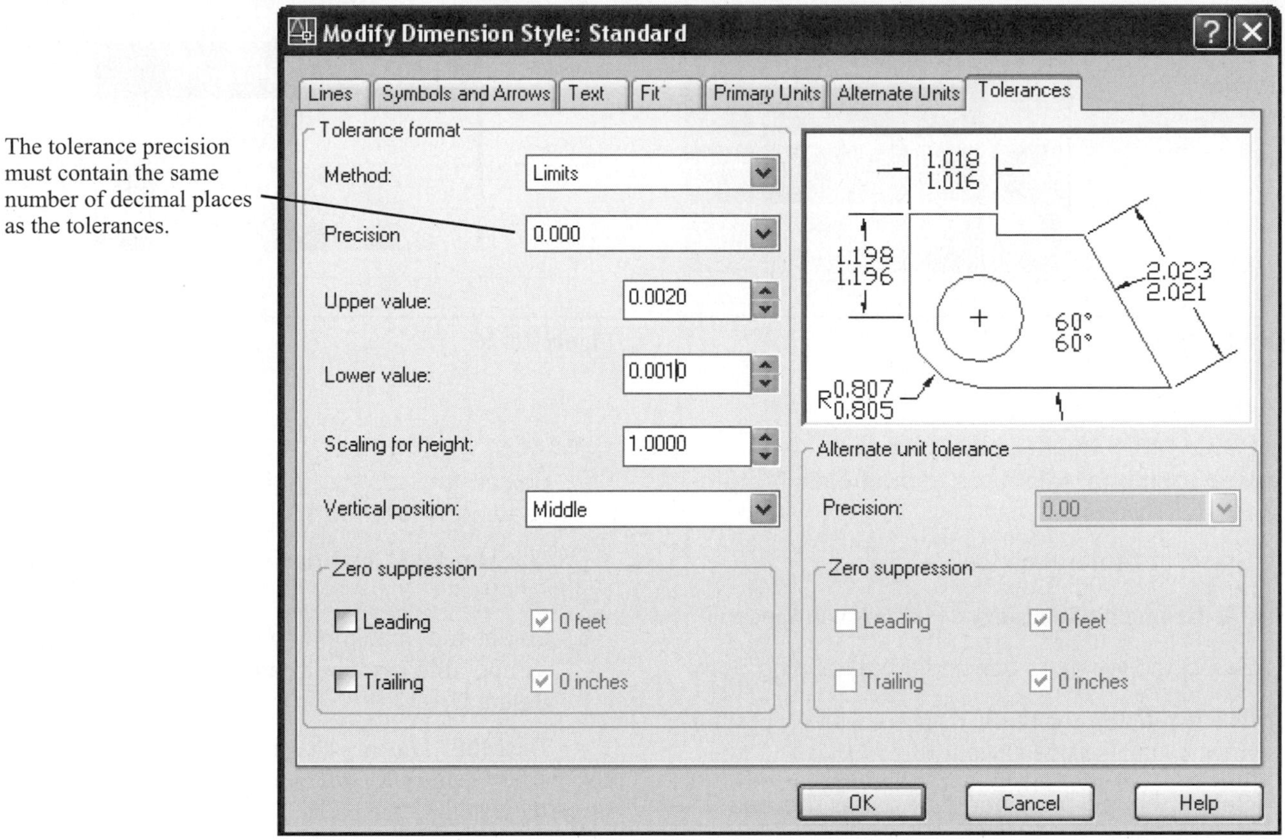

The tolerance precision must contain the same number of decimal places as the tolerances.

Figure 10-24

To add a limit feature tolerance to a hole

1. Select the **Dimension Style** tool or type **DDIM** in response to a command prompt.

The **Dimension Style Manager** dialog box will appear.

2. Select **Modify,** then the **Tolerance** tab.

The **Modify Dimension Style: Standard** dialog box will appear.

3. Select the **Limits** option in the **Method** box located in the **Tolerance format** box.

See Figure 10-24.

4. Change the upper value to **0.0020** by placing the cursor within the **Upper value** box, backspacing out the existing value, and typing in the new value.

5. Change the lower value to **0.0010** by placing the cursor within the **Lower value** box, backspacing out the existing value, and typing in the new value.

6. Select the arrow to the right of the **Precision** box.

The **Precision** options will cascade down.

7. Change the precision for both **Dimension** and **Tolerance** boxes to three decimal places (**0.000**).

See Figure 10-24. AutoCAD will truncate any input according to the number of decimal places allowed by the precision settings. If the precision settings had been two decimal places (0.00), the resulting limit dimensions would have both been 1.50. The values defined in the third decimal place would have been ignored.

8. Return to the drawing screen.
9. Select the **Diameter** tool on the **Dimension** toolbar.

Select arc or circle:

10. Select the hole.

Dimension line location (Text/Angle):

11. Locate the diameter dimension.

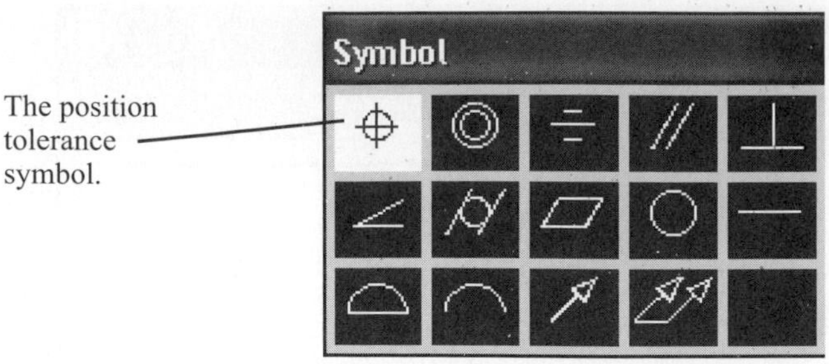

Figure 10-25

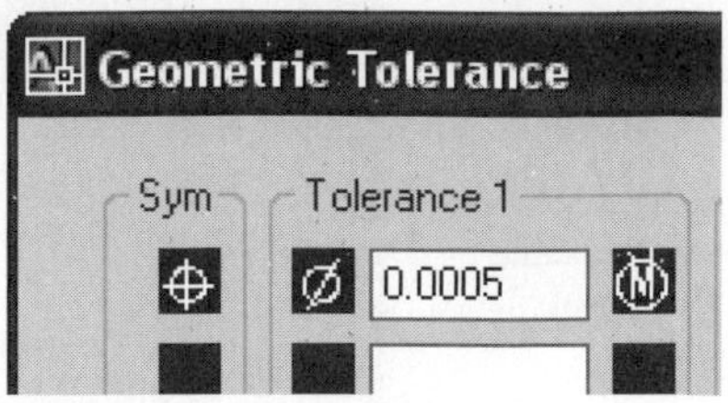

The position tolerance symbol.

Figure 10-26

To add a positional tolerance to the hole's feature tolerance

1. Select the **Tolerance** tool.

   The **Geometric Tolerance** dialog box will appear.

2. Select the top open box under the heading **Sym.**

   The **Symbol** dialog box will appear with the positioning symbol highlighted. See Figure 10-25.

3. Select the top left open box under the heading **Tolerance 1.**

   A diameter symbol will appear.

4. Select the **Value** box and type **0.0005.**

   The numbers will appear in the box.

5. Select the top far right open box under the heading **Tolerance 1.**

   The **Material Condition** dialog box will appear. See Figure 10-25.

6. Select the maximum material condition (MMC) symbol (the circle with an M in it).

7. Select **OK.**

   The MMC symbol will appear in the material condition box in the **Geometric Tolerance** dialog box. Figure 10-26 shows the resulting **Geometric Tolerance** dialog box.

8. Select **OK.**

   Enter tolerance location:

9. Locate the tolerance box.

   Use the **Move** and **Osnap** commands to position the box, if necessary.

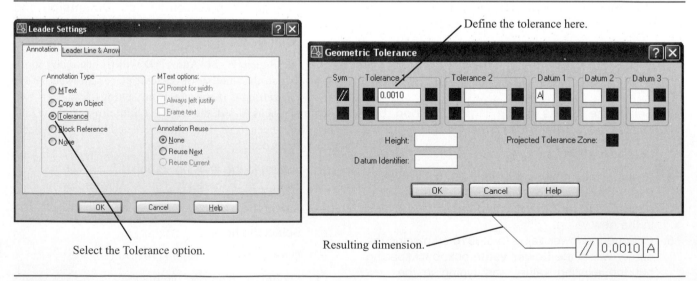

Select the Tolerance option.

Define the tolerance here.

Resulting dimension.

Figure 10-27

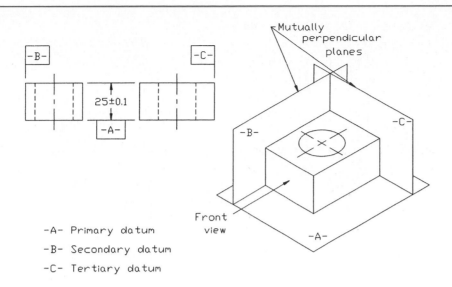

-A- Primary datum

-B- Secondary datum

-C- Tertiary datum

Figure 10-28

To add a geometric tolerance with a leader line

1. Type **qleader** in response to a command prompt.

   Specify first leader point, or [Settings] <Settings>:

2. Select **Settings** by pressing the **Enter** key.

   The **Leader Settings** dialog box will appear. See Figure 10-27.

3. Select the **Tolerance** option in the **Annotation Type** box, then **OK**.

   Specify first leader point, or [Settings] <Settings>:

4. Select a starting point for the leader line.

   Specify next point:

5. Draw a short horizontal segment.

   The **Geometric Tolerance** dialog box will appear. See Figure 10-27.

6. Select the top **Sym** box.

   The **Symbol** dialog box will appear.

7. Select the parallel symbol.
8. Select the open box under the heading **Tolerance 1** and type **0.0010**.
9. Select the open box under the heading **Datum 1**, type **A**, then **OK**.

   See Figure 10-27.

10-9 TOLERANCES OF ORIENTATION

Tolerances of orientation are used to relate a feature or surface to another feature or surface. Tolerances of orientation include perpendicularity, parallelism, and angularity. They may be applied using RFS or MMC conditions, but they cannot be applied to individual features by themselves. To define a surface as parallel to another surface is very much like assigning a flatness value to the surface. The difference is that flatness applies only within the surface; every point on the surface is related to a defined set of limiting parallel planes. Parallelism defines every point in the surface relative to another surface. The two surfaces are therefore directly related to each other, and the condition of one affects the other.

Orientation tolerances are used with locational tolerances. A feature is first located, then it is oriented within the locational tolerances. This means that the orientation tolerance must always be less than the locational tolerances. The next four sections will further explain this requirement.

10-10 DATUMS

A datum is a point, axis, or surface used as a starting reference point for dimensions and tolerances. Figure 10-28 shows a rectangular object with three datum planes labeled –A–, –B–, and –C–. The three datum planes are called the primary, secondary, and tertiary datums, respectively. The three datum planes are, by definition, oriented exactly 90° to one another.

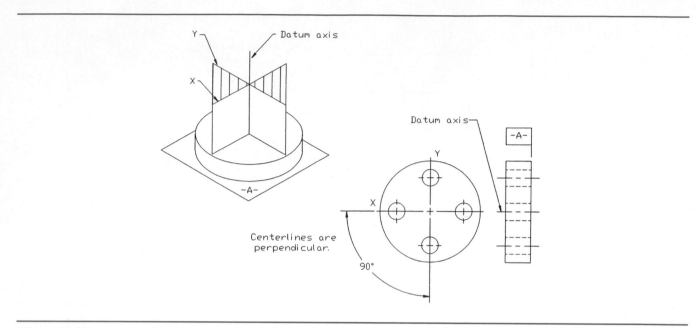

Figure 10-29

Figure 10-29 shows a cylindrical datum frame that includes three datum planes. The X and Y planes are perpendicular to each other, and the base A plane is perpendicular to the datum axis between the X and Y planes.

Datums are defined on a drawing by using letters enclosed in rectangular boxes, as shown. The defining letters are written between dash lines: –A– , –B– , and –C–.

Datum planes are assumed to be perfectly flat. When assigning a datum status to a surface, be sure that the surface is reasonably flat. This means that datum surfaces should be toleranced using surface finishes, or created using machine techniques that produce flat surfaces.

10-11 PERPENDICULARITY

Perpendicularity tolerances are used to limit the amount of variation for a surface or feature within two planes perpendicular to a specified datum. Figure 10-30 shows a rectangular object. The bottom surface is assigned as datum A, and the right vertical edge is toleranced so that it must be perpendicular within a limit of 0.05 to datum A. The perpendicularity tolerance defines a tolerance zone 0.05 wide between two parallel planes that are perpendicular to datum A.

The object also includes a horizontal dimension and tolerance of 40 ± 1. This tolerance is called a ***locational tol-***

erance because it serves to locate the right edge of the object. As with rectangular coordinate tolerances discussed in Chapter 9, the 40±1 controls the location of the edge, how far away or how close it can be to the left edge, but does not directly control the shape of the edge. Any shape that falls within the specified tolerance range is acceptable. This may, in fact, be sufficient for a given design, but if a more controlled shape is required, a perpendicularity tolerance must be added. The perpendicularity tolerance works

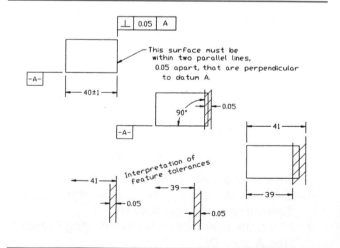

Figure 10-30

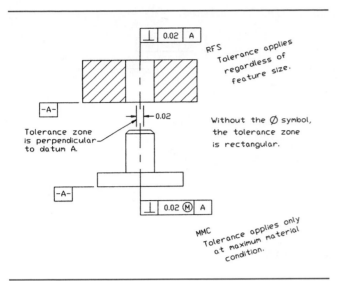

RFS
Tolerance applies regardless of feature size.

Tolerance zone is perpendicular to datum A.

Without the Ø symbol, the tolerance zone is rectangular.

MMC
Tolerance applies only at maximum material condition.

Figure 10-31

If tolerance is

Ø20±0.03

⊥ 0.02 A

Feature Tolerance	Allowable Tolerance
20.03	.02
20.02	.02
20.01	.02
20.00	.02
19.99	.02
19.98	.02
19.97	.02

If tolerance is

Ø20±0.03

⊥ Ø0.02 Ⓜ A

Feature Tolerance	Allowable Tolerance
20.03	.02
20.02	.03
20.01	.04
20.00	.05
19.99	.06
19.98	.07
19.97	.08

Tolerance zone shape

0.02 SQUARE

Tolerance zone shape

Ø0.02

Figure 10-32

within the locational tolerance to ensure that the edge is not only within the locational tolerance but is also perpendicular to datum A.

Figure 10-30 shows the two extreme conditions for the 40±1 locational tolerance. The perpendicularity tolerance is applied by first measuring the surface and determining its maximum and minimum lengths. The difference between these two measurements must be less than 0.05. Thus, if the measured maximum distance is 41, then no other part of the surface may be less than 41 − 0.05 = 40.95.

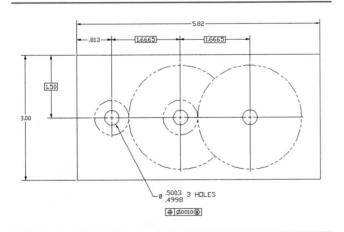

Ø .5003 3 HOLES
.4998

⊕ Ø.0010 Ⓜ

Figure 10-33

Tolerances of perpendicularity serve to complement locational tolerances, to make the shape more exact, so tolerances of perpendicularity must always be smaller than tolerances of location. It would be of little use, for example, to assign a perpendicularity tolerance of 1.5 for the object shown in Figure 10-30. The locational tolerance would prevent the variation from ever reaching the limits specified by such a large perpendicularity tolerance.

Figure 10-31 shows a perpendicularity tolerance applied to cylindrical features: a shaft and a hole. The figure includes examples of both RFS and MMC applications. As with straightness tolerances applied at MMC, perpendicularity tolerances applied about a hole or shaft's centerline allow the tolerance zone to vary as the feature size varies.

The inclusion of the Ø symbol in a geometric tolerance is critical to its interpretation. See Figure 10-32. If the Ø symbol is not included, the tolerance applies only to the view in which it is written. This means that the tolerance zone is shaped like a rectangular slice, not a cylinder, as would be the case if the Ø symbol were included. In general, it is better always to include the Ø symbol for cylindrical features because it generates a tolerance zone more like that used in positional tolerancing.

Figure 10-33 shows a perpendicularity tolerance applied to a slot, a noncylindrical feature. Again, the MMC specification is always for variations in the tolerance zone.

10-12 PARALLELISM

Parallelism is used to ensure that all points within a plane are within two parallel planes that are parallel to a referenced datum plane. Figure 10-34 shows a rectangular object that is toleranced so that its top surface is parallel to the bottom surface within 0.02. This means that every point on the top surface must be within a set of parallel planes 0.02 apart. These parallel tolerancing planes are located by determining the maximum and minimum distances from the datum surface. The difference between the maximum and minimum values may not exceed the stated 0.02 tolerance.

In the extreme condition of maximum feature size, the top surface is located 40.5 above the datum plane. The parallelism tolerance is then applied, meaning that no point on the surface may be closer than 40.3 to the datum. This is an RFS condition. The MMC condition may also be applied, thereby allowing the tolerance zone to vary as the feature size varies.

10-13 ANGULARISM

Angularism tolerances are used to limit the variance of surfaces and axes that are at an angle relative to a datum. Angularism tolerances are applied like perpendicularity and parallelism tolerances as a way to better control the shape of locational tolerances.

Figure 10-35 shows an angularism tolerance and several ways in which it is interpreted at extreme conditions.

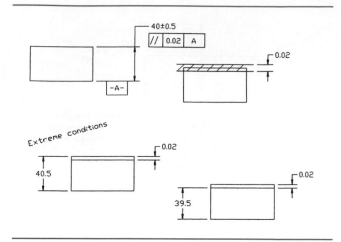

Figure 10-34

10-14 PROFILES

Profile tolerances are used to limit the variations of irregular surfaces. They may be assigned as either bilateral or unilateral tolerances. There are two types of profile tolerances: surface and line. **Surface profile tolerances** limit the variation of an entire surface, whereas a **line profile tolerance** limits the variations along a single line across a surface.

Figure 10-36 shows an object that includes a surface profile tolerance referenced to an irregular surface. The tolerance is considered a bilateral tolerance because no other specification is given. This means that all points on the surface must be located between two parallel planes 0.08 apart that are centered about the irregular surface. The measurements are taken perpendicular to the surface.

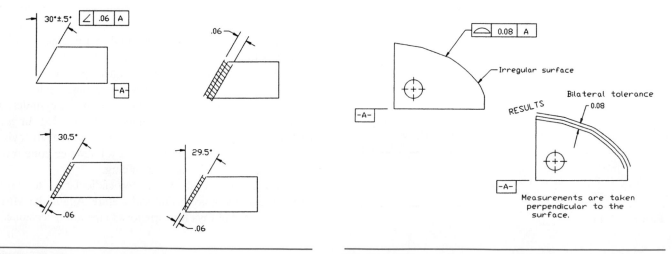

Figure 10-35

Figure 10-36

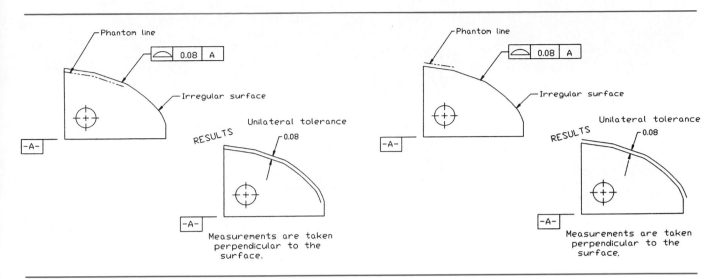

Figure 10-37

Unilateral applications of surface profile tolerances must be indicated on the drawing using phantom lines. The phantom line indicates on which side of the true profile line of the irregular surface the tolerance is to be applied. A phantom line above the irregular surface indicates that the tolerance is to be applied using the true profile line as 0 and then adding a specified tolerance range above that line. See Figures 10-37 and 10-38.

Profiles of line tolerances are applied to irregular surfaces, as shown in Figure 10-37. Profiles of line toler-

ances are particularly helpful for tolerancing an irregular surface that is constantly changing, such as the surface of an airplane wing.

Surface and line profile tolerances are somewhat analogous to flatness and straightness tolerances. Flatness and surface profile tolerances are applied across an entire surface; straightness and line profile tolerances are applied only along a single line across the surface.

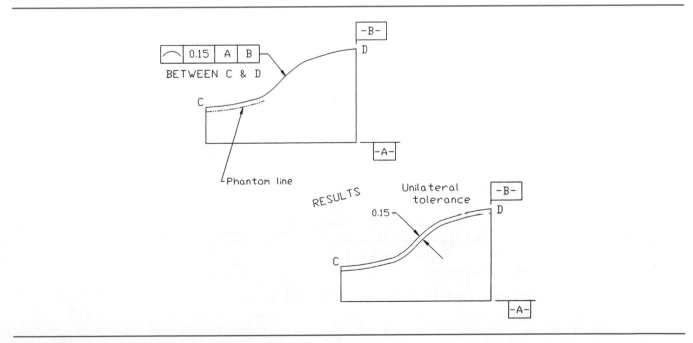

Figure 10-38

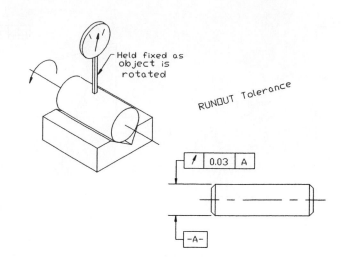

Figure 10-39

10-15 RUNOUTS

A *runout tolerance* is used to limit the variations between features of an object and a datum. More specifically, runout tolerances are applied to surfaces around a datum axis such as a cylinder or to a surface constructed perpendicular to a datum axis. There are two types of runout tolerances: circular and total.

Figure 10-39 shows a cylinder that includes a circular runout tolerance. The runout requirements are checked by rotating the object about its longitudinal axis or datum axis while holding an indicator gauge in a fixed position on the surface of the object.

Runout tolerances may be either bilateral or unilateral. A runout tolerance is assumed to be bilateral unless otherwise indicated. If a runout tolerance is to be unilateral, a phantom line is used to indicate to which side of the object's true surface the tolerance is to be applied. See Figure 10-40.

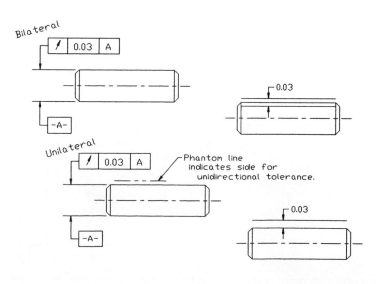

Figure 10-40

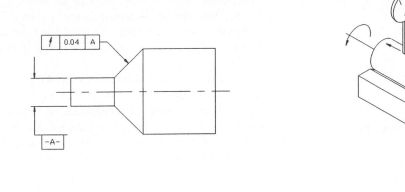

Figure 10-41

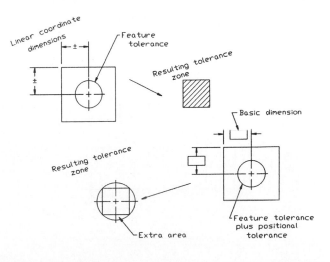

Gage is moved over entire surface.

Total Runout Tolerance

Figure 10-42

Runout tolerances may be applied to tapered areas of cylindrical objects, as shown in Figure 10-41. The tolerance is checked by rotating the object about a datum axis while holding an indicator gauge in place.

A total runout tolerance limits the variation across an entire surface. See Figure 10-42. An indicator gauge is not held in place while the object is rotated, as it is for circular runout tolerances, but is moved about the rotating surface.

Figure 10-43 shows a circular runout tolerance that references two datums. The two datums serve as one datum. The object can then be rotated about both datums simultaneously as the runout tolerances are checked.

10-16 POSITIONAL TOLERANCES

Positional tolerances are used to locate and tolerance holes. Positional tolerances create a circular tolerance zone for hole center point locations that differs from the rectangular-shaped tolerance zone created by linear coordinate dimensions. See Figure 10-44. The circular tolerance zone allows for an increase in acceptable tolerance variation without compromising the design integrity of the object. Note that some of the possible hole center points fall in an area outside the rectangular tolerance zone but are still within the circular tolerance zone. If the hole had been located using linear coordinate dimensions, center points located beyond

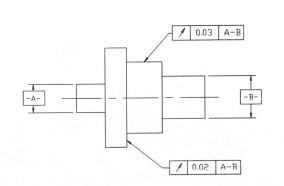

Figure 10-43

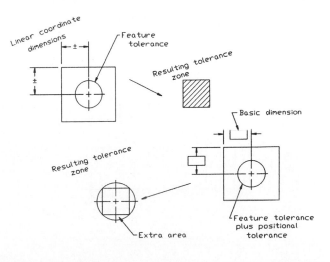

Figure 10-44

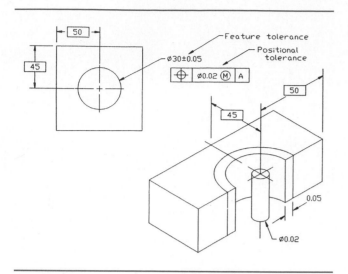

Figure 10-45

the rectangular tolerance zone would have been rejected as beyond tolerance, and yet holes produced using these locations would function correctly from a design standpoint. The center point locations would be acceptable if positional tolerances had been specified. The finished hole is round, so a round tolerance zone is appropriate. The rectangular tolerance zone rejects some holes unnecessarily.

Holes are dimensioned and toleranced using geometric tolerances by a combination of locating dimensions, feature dimensions and tolerances, and positional tolerances. See Figure 10-45. The locating dimensions are enclosed in rectangular boxes and are called **basic dimensions.** Basic dimensions are assumed to be exact.

The feature tolerances for the hole are as presented in Chapter 9. They can be presented using plus or minus or limit-type tolerances. In the example shown in Figure 10-45 the diameter of the hole is toleranced using a plus and minus 0.05 tolerance.

The basic locating dimensions of 45 and 50 are assumed to be exact. The tolerances that would normally accompany linear locational dimensions are replaced by the positional tolerance. The positional tolerance also specifies that the tolerance be applied at the centerline at maximum material condition. The resulting tolerance zones are as shown in Figure 10-45.

Figure 10-46 shows an object containing two holes that are dimensioned and toleranced using positional tolerances. There are two consecutive horizontal basic dimensions. Because basic dimensions are exact, they do not have tolerances that accumulate; that is, there is no tolerance buildup.

10-17 VIRTUAL CONDITION

Virtual condition is a combination of a feature's MMC and its geometric tolerance. For external features (shafts) it is the MMC plus the geometric tolerance; for internal features (holes) it is the MMC minus the geometric tolerance.

The following calculations are based on the dimensions shown in Figure 10-47.

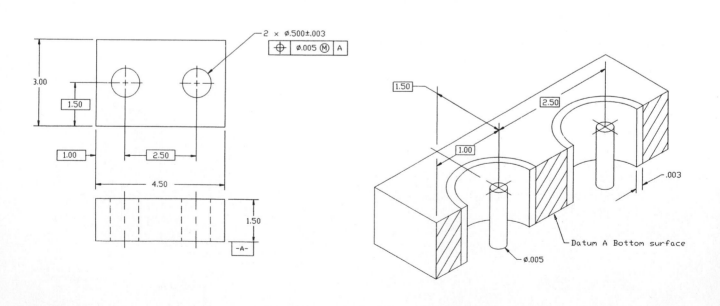

Figure 10-46

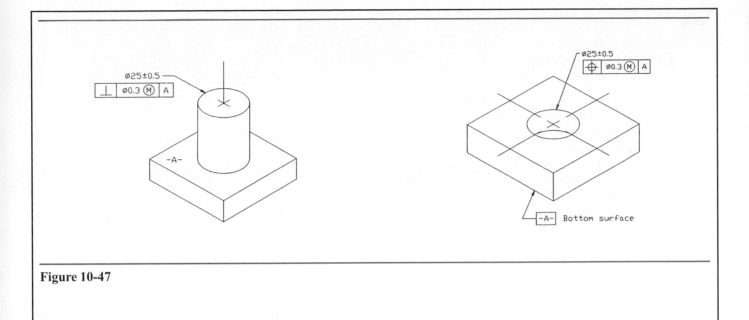

Figure 10-47

To calculate the virtual condition for a shaft

25.5 MMC for shaft—maximum diameter
+0.3 Geometric tolerance
25.8 Virtual condition

To calculate the virtual condition for a hole

24.5 MMC for hole—minimum diameter
−0.3 Geometric tolerance
24.2 Virtual condition

10-18 FLOATING FASTENERS

Positional tolerances are particularly helpful when dimensioning matching parts. Because basic locating dimensions are considered exact, the sizing of mating parts is dependent only on the MMC of the hole and shaft and the geometric tolerance between them.

The relationship for floating fasteners and holes in objects may be expressed as a formula:

$$H - T = F$$

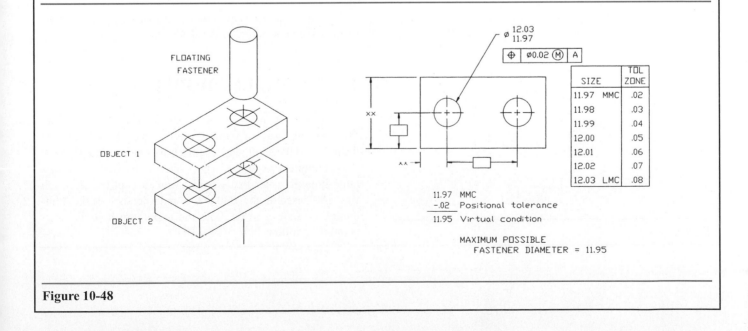

Figure 10-48

where

H = Hole at MMC
T = Geometric tolerance
F = Shaft at MMC

A *floating fastener* is one that is free to move in either object. It is not attached to either object and it does not screw into either object. Figure 10-48 shows two objects that are to be joined by a common floating shaft, such as a bolt or screw. The feature size and tolerance and the positional geometric tolerance are both given. The minimum size hole that will always just fit is determined using the preceding formula:

$$H - T = F$$
$$11.97 - .02 = 11.95$$

Therefore, the shaft's diameter at MMC, the shaft's maximum diameter, equals 11.95. Any required tolerance would have to be subtracted from this shaft size.

The 0.02 geometric tolerance is applied at the hole's MMC. Thus, as the hole's size expands within its feature tolerance, the tolerance zone for the acceptable matching parts also expands. See the table in Figure 10-48.

10-19 SAMPLE PROBLEM SP10-1

The situation presented in Figure 10-48 can be worked in reverse; that is, hole sizes can be derived from given shaft sizes.

The two objects shown in Figure 10-49 are to be joined by a .250-inch bolt. The parts are floating; that is,

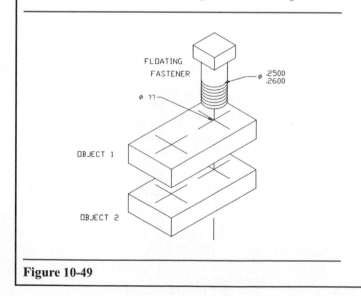

Figure 10-49

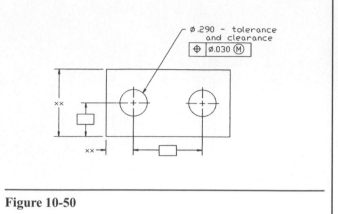

Figure 10-50

they are both free to move, and the fastener is not joined to either object. What is the MMC of the holes if the positional tolerance is to be .030?

A manufacturer's catalog specifies that the tolerance for a .250 bolt is .2500 to .2600.

Rewriting the formula

$$H - T = F$$

to isolate the H, we have

$$H = F + T$$
$$= .260 + .030$$
$$= .290$$

The .290 value represents the minimum hole diameter, MMC, for all four holes that will always accept the .250 bolt. Figure 10-50 shows the resulting drawing callout.

Any clearance requirements or tolerances for the hole would have to be added to the .290 value.

10-20 SAMPLE PROBLEM SP10-2

Repeat the problem presented in SP10-1 but be sure that there is always a minimum clearance of .002 between the hole and the shaft and assign a hole tolerance of .0010.

Sample Problem SP10-1 determined that the maximum hole diameter that would always accept the .250 bolt was .290 based on the .030 positional tolerance. If the minimum clearance is to be .002, the maximum hole diameter is found as follows:

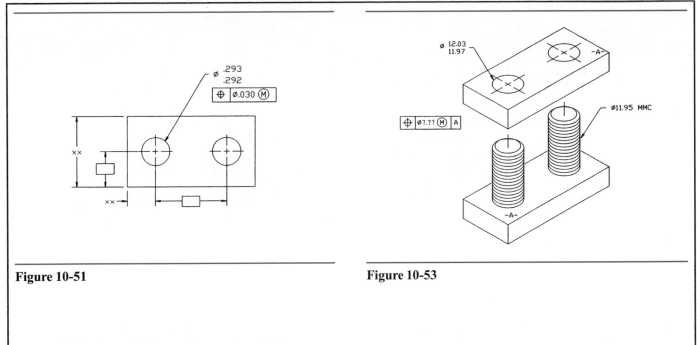

Figure 10-51

Figure 10-53

.290 Minimum hole diameter that will always accept the bolt (0 clearance at MMC)
+.002 Mininum clearance
.292 Minimum hole diameter including clearance

Now assign the tolerance to the hole.

.292 Minimum hole diameter
+.001 Tolerance
.293 Minimum hole diameter

See Figure 10-51 for the appropriate drawing callout. The choice of clearance size and hole tolerance varies with the design requirements for the objects.

10-21 FIXED FASTENERS

A *fixed fastener* is one that is attached to one of the mating objects. See Figure 10-52. Because the fastener is fixed to one of the objects, the geometric tolerance zone must be smaller than that used for floating fasteners. The fixed fastener cannot move without moving the object it is attached to. The relationship between fixed fasteners and holes in mating objects is defined by the formula

$$H - 2T = F$$

The tolerance zone is cut in half, as can be demonstrated by the objects shown in Figure 10-53. The same feature sizes that were used in Figure 10-48 are assigned, but in this example the fasteners are fixed. Solving for the geometric tolerance, we obtain the following value:

$$H - F = 2T$$
$$11.97 - 11.95 = 2T$$
$$.02 = 2T$$
$$.01 = T$$

The resulting positional tolerance is half of that obtained for floating fasteners.

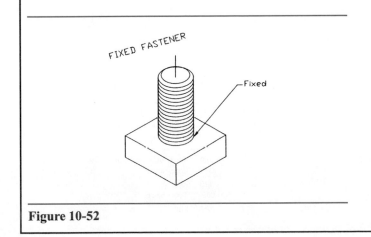

Figure 10-52

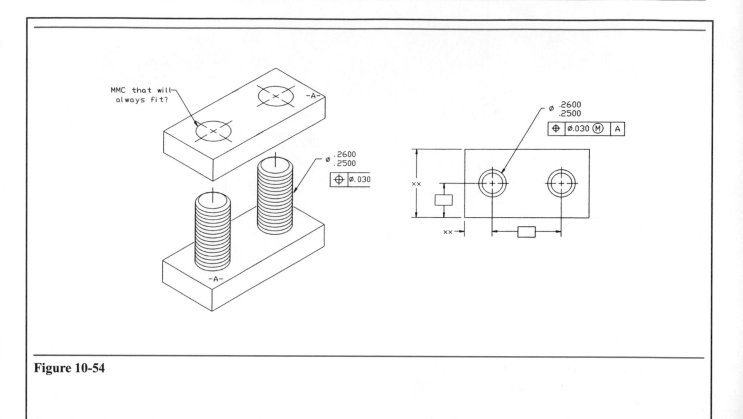

Figure 10-54

10-22 SAMPLE PROBLEM SP10-3

This problem is similar to Sample Problem SP10-1, but the given conditions are applied to fixed fasteners rather than to floating fasteners. Compare the resulting shaft diameters for the two problems. See Figure 10-54.

A. What is the minimum-diameter hole that will always accept the fixed fasteners?
B. If the minimum clearance is .005 and the hole is to have a tolerance of .002, what are the maximum and minimum diameters of the hole?

$$H - 2T = F$$
$$H = F + 2T$$
$$= .260 + 2(.030)$$
$$= .260 + .060$$
$$= .320 \quad \text{Minimum diameter that will always accept the fixed fastener}$$

If the minimum clearance = .005 and the hole tolerance is .002,

```
   .320  Virtual condition
+ .005  Clearance
   .325  Minimum hole diameter
```

```
   .325  Minimum hole diameter
+ .002  Tolerance
   .327  Maximum hole diameter
```

The maximum and minimum values for the hole's diameter can then be added to the drawing of the object that fits over the fixed fasteners. See Figure 10-55.

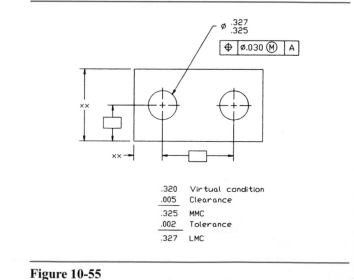

Figure 10-55

10-23 DESIGN PROBLEMS

This problem was originally done in Section 9-27 using rectangular tolerances. It is done in this section using positional geometric tolerances so that the two systems can be compared. It is suggested that you review Section 9-27 before reading this section.

Figure 10-56 shows top and bottom parts that are to be joined in the floating condition. A nominal distance of 50 between hole centers and 20 for the holes has been assigned. In Section 9-27 a rectangular tolerance of ±.01 was selected and there was a minimum hole diameter of 20.00. Figure 10-57 shows the resulting tolerance zones.

The diagonal distance across the rectangular tolerance zone is .028 and was rounded off to .03 to yield a maximum possible fastener diameter of 19.97. If the same .03 value is used to calculate the fastener diameter using positional tolerance, the results will be as follows:

$$H - T = F$$
$$20.00 - .03 = 19.97$$

The results seem to be the same, but because of the circular shape of the positional tolerance zone, the manufactured results are not the same. The minimum distance between the inside edges of the rectangular zones is 49.98, or .01 from the center point of each hole. The minimum distance from the innermost points of the circular tolerance zones is 49.97, or .015 (half of the rounded-off .03 value)

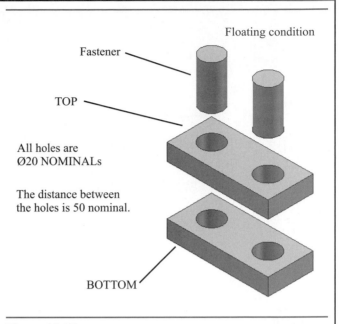

Floating condition

Fastener

TOP

All holes are
Ø20 NOMINALs

The distance between
the holes is 50 nominal.

BOTTOM

Figure 10-56

from the center point of each hole. The same value difference also occurs for the maximum distance between center points, where 50.02 is the maximum distance for the rectangular tolerances, and 50.03 is the maximum distance for the circular tolerances. The size of the circular tolerance zone increased more because the hole tolerances are assigned at

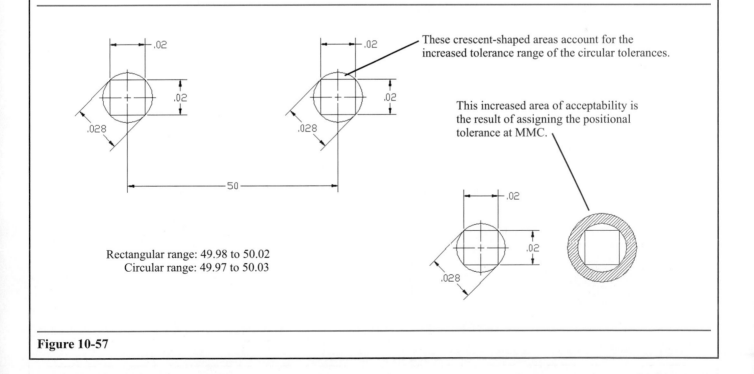

These crescent-shaped areas account for the increased tolerance range of the circular tolerances.

This increased area of acceptability is the result of assigning the positional tolerance at MMC.

Rectangular range: 49.98 to 50.02
Circular range: 49.97 to 50.03

Figure 10-57

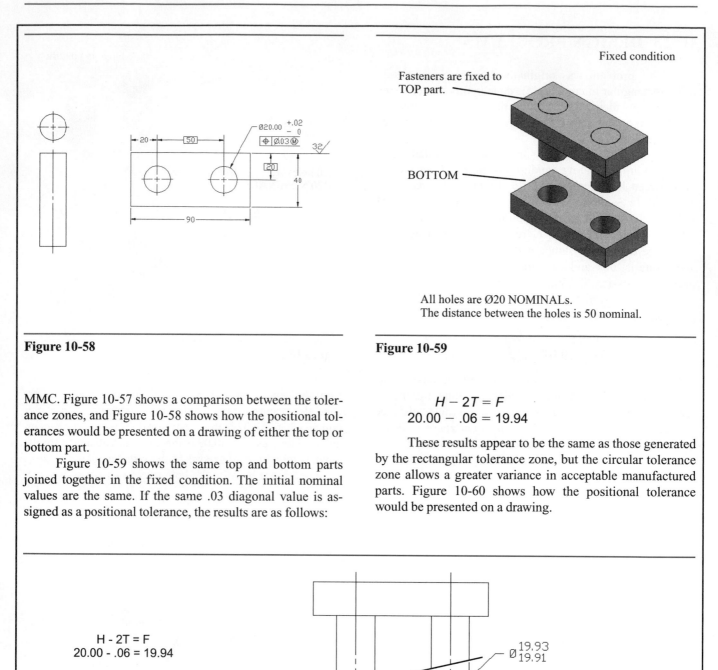

Figure 10-58

Figure 10-59

MMC. Figure 10-57 shows a comparison between the tolerance zones, and Figure 10-58 shows how the positional tolerances would be presented on a drawing of either the top or bottom part.

Figure 10-59 shows the same top and bottom parts joined together in the fixed condition. The initial nominal values are the same. If the same .03 diagonal value is assigned as a positional tolerance, the results are as follows:

$$H - 2T = F$$
$$20.00 - .06 = 19.94$$

These results appear to be the same as those generated by the rectangular tolerance zone, but the circular tolerance zone allows a greater variance in acceptable manufactured parts. Figure 10-60 shows how the positional tolerance would be presented on a drawing.

$$H - 2T = F$$
$$20.00 - .06 = 19.94$$

Subtracting .01 for clearance results in a maximum shaft diameter of 19.93.

Assigning a shaft tolerance of .02 results in a minimum shaft diameter of 19.91.

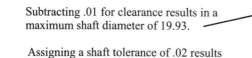

Figure 10-60

10-24 EXERCISE PROBLEMS

EX10-1

Redraw the object shown. Include all dimensions and tolerances.

EX10-3

A. Given the shaft shown, what is the minimum hole diameter that will always accept the shaft?

B. If the minimum clearance between the shaft and a hole is equal to 0.02, and the tolerance on the hole is to be 0.6, what are the maximum and minimum diameters for the hole?

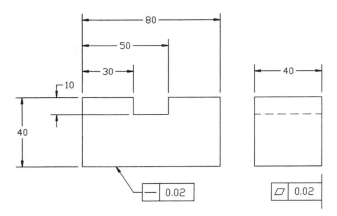

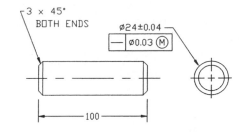

EX10-2

Redraw the following shaft and add a feature dimension and tolerance of 36 ± 0.1 and a straightness tolerance of 0.07 about the centerline at MMC.

EX10-4

A. Given the shaft shown, what is the minimum hole diameter that will always accept the shaft?

B. If the minimum clearance between the shaft and a hole is equal to .005, and the tolerance on the hole is to be .007, what are the maximum and minimum diameters for the hole?

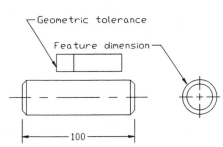

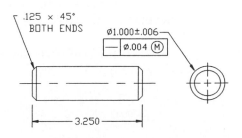

EX10-5

Draw a front and a right-side view of the object shown in Figure EX10-5 and add the appropriate dimensions and tolerances based on the following information. Numbers located next to an edge line indicate the length of the edge.

A. Define surfaces A, B, and C as primary, secondary, and tertiary datums, respectively.
B. Assign a tolerance of ±0.5 to all linear dimensions.
C. Assign a feature tolerance of 12.07−12.00 to the protruding shaft.
D. Assign a flatness tolerance of 0.01 to surface A.
E. Assign a straightness tolerance of 0.03 to the protruding shaft.
F. Assign a perpendicularity tolerance to the centerline of the protruding shaft of 0.02 at MMC relative to datum A.

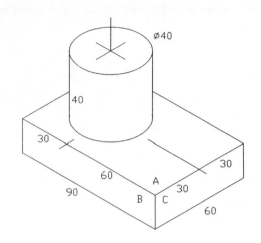

Figure EX10-5

EX10-6

Draw a front and right-side view of the object shown in Figure EX10-6 and add the following dimensions and tolerances.

A. Define the bottom surface as datum A.
B. Assign a perpendicularity tolerance of 0.4 to both sides of the slot relative to datum A.
C. Assign a perpendicularity tolerance of 0.2 to the centerline of the 30-diameter hole centerline at MMC relative to datum A.
D. Assign a feature tolerance of ±0.8 to all three holes.
E. Assign a parallelism tolerance of 0.2 to the common centerline between the two 20-diameter holes relative to datum A.
F. Assign a tolerance of ±0.5 to all linear dimensions.

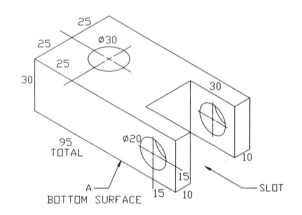

Figure EX10-6

EX10-7

Draw a circular front and the appropriate right-side view of the object shown in Figure EX10-7 and add the following dimensions and tolerances.

A. Assign datum A as indicated.
B. Assign the object's longitudinal axis as datum B.
C. Assign the object's centerline through the slot as datum C.
D. Assign a tolerance of ±0.5 to all linear tolerances.
E. Assign a tolerance of ±0.5 to all circular-shaped features.
F Assign a parallelism tolerance of 0.01 to both edges of the slot.
G. Assign a perpendicularity tolerance of 0.01 to the outside edge of the protruding shaft.

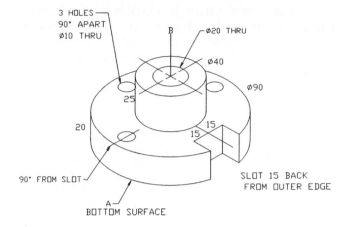

Figure EX10-7

EX10-8

Given the two objects shown in Figure EX10-8, draw a front and a side view of each. Assign a tolerance of ±0.5 to all linear dimensions. Assign a feature tolerance of ±0.4 to the shaft, and also assign a straightness tolerance of 0.2 to the shaft's centerline at MMC.

Tolerance the hole so that it will always accept the shaft with a minimum clearance of 0.1 and a feature tolerance of 0.2. Assign a perpendicularity tolerance of 0.05 to the centerline of the hole at MMC.

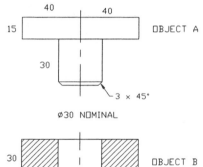

Figure EX10-8

EX10-9

Given the two objects shown in Figure EX10-9, draw a front and a side view of each. Assign a tolerance of ±.005 to all linear dimensions. Assign a feature tolerance of ±.004 to the shaft, and also assign a straightness tolerance of .002 to the shaft's centerline at MMC.

Tolerance the hole so that it will always accept the shaft with a minimum clearance of .001 and a feature tolerance of .002.

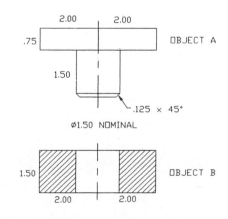

Figure EX10-9

EX10-10

Refer to parts A through G for this exercise problem. Use the format shown in Figure EX10-10 and redraw the given geometric tolerance symbols and frame as shown in the sample. Express in words (**Dtext**) the meaning of each tolerance callout.

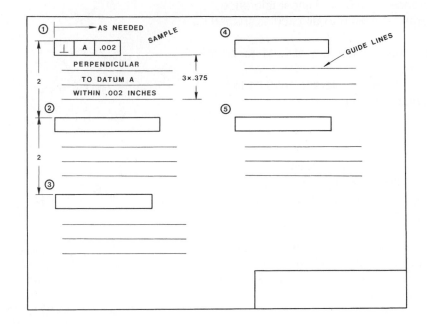

Figure EX10-10

A.

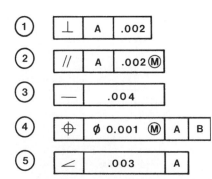

B.

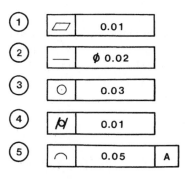

EX10-10, continued

C.

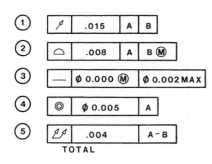

① | ↗ | .015 | A | B |

② | ⌒ | .008 | A | B Ⓜ |

③ | — | ⌀ 0.000 Ⓜ | ⌀ 0.002 MAX |

④ | ◎ | ⌀ 0.005 | A |

⑤ | ⌀↗ | .004 | A – B |
TOTAL

F.

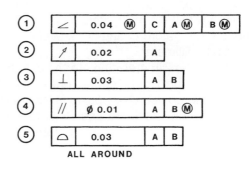

① | ∠ | 0.04 Ⓜ | C | A Ⓜ | B Ⓜ |

② | ↗ | 0.02 | A |

③ | ⊥ | 0.03 | A | B |

④ | // | ⌀ 0.01 | A | B Ⓜ |

⑤ | ⌒ | 0.03 | A | B |
ALL AROUND

D.

① | ⊕ | 0.25 Ⓜ | A | B | C |

② | ⊕ | 0.8 Ⓜ | A | B Ⓜ |

③ | ⊕ | ⌀ 0.4 Ⓜ | A | B | C |

④ | ⊕ | ⌀ 0.3 Ⓜ | A | B Ⓜ | C Ⓜ |

⑤ | ⊕ | ⌀ 0.1 Ⓜ |

G.

① | ⊕ | ⌀ .002 Ⓜ |

② | ⊕ | .005 Ⓜ | A | B | C |

③ | ⊕ | .002 | A | B Ⓜ |

④ | ⊕ | ⌀ .003 Ⓜ | A | B Ⓜ | C Ⓜ |

⑤ | ⊕ | ⌀ .015 Ⓜ | A | B |

E.

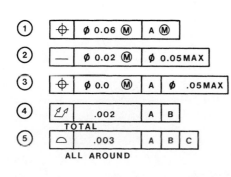

① | ⊕ | ⌀ 0.06 Ⓜ | A Ⓜ |

② | — | ⌀ 0.02 Ⓜ | ⌀ 0.05 MAX |

③ | ⊕ | ⌀ 0.0 Ⓜ | A | ⌀ .05 MAX |

④ | ⌀↗ | .002 | A | B |
TOTAL

⑤ | ⌒ | .003 | A | B | C |
ALL AROUND

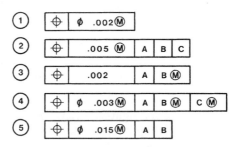

EX10-11 MILLIMETERS

Draw front, top, and right-side views of the object in Figure EX10-11, including dimensions. Add the following tolerances and specifications to the drawing.

A. Surface 1 is datum A.
B. Surface 2 is datum B and is perpendicular to datum A within 0.1 millimeter.
C. Surface 3 is datum C and is parallel to datum A within 0.3 millimeter.
D. Locate a 16-millimeter-diameter hole in the center of the front surface that goes completely through the object. Use positional tolerances to locate the hole. Assign a positional tolerance of 0.02 at MMC perpendicular to datum A.

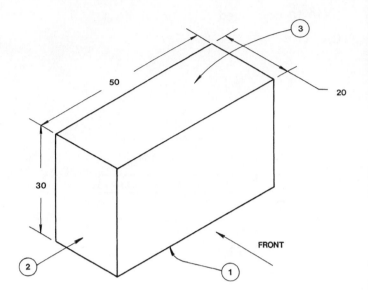

Figure EX10-11

EX10-12

Draw front, top, and right-side views of the object in Figure EX10-12, including dimensions. Add the following tolerances and specifications to the drawing.

A. Surface 1 is datum A.
B. Surface 2 is datum B and is perpendicular to datum A within .003 inch.
C. Surface 3 is parallel to datum A within .005 inch.
D. The cylinder's longitudinal centerline is to be straight within .001 inch at MMC.
E. Surface 2 is to have circular accuracy within .002 inch.

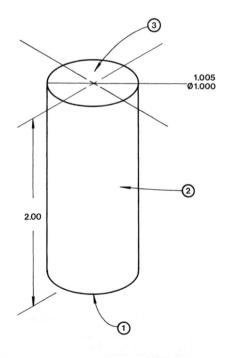

Figure EX10-12

EX10-13

Draw front, top, and right-side views of the object in Figure EX10-13, including dimensions. Add the following tolerances and specifications to the drawing.

A. Surface 1 is datum A.
B. Surface 4 is datum B and is perpendicular to datum A within 0.08 millimeter.
C. Surface 3 is flat within 0.03 millimeter.
D. Surface 5 is parallel to datum A within 0.01 millimeter.
E. Surface 2 has a runout tolerance of 0.2 millimeter relative to surface 4.
F. Surface 1 is flat within 0.02 millimeter.
G. The longitudinal centerline is to be straight within 0.02 millimeter at MMC and perpendicular to datum A.

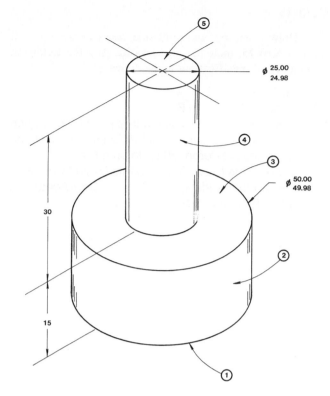

Figure EX10-13

EX10-14

Draw front, top, and right-side views of the object in Figure EX10-14, including dimensions. Add the following tolerances and specifications to the drawing.

A. Surface 2 is datum A.
B. Surface 6 is perpendicular to datum A with 0.000 allowable variance at MMC but with a .002 inch MAX variance limit beyond MMC.
C. Surface 1 is parallel to datum A within .005 inch.
D. Surface 4 is perpendicular to datum A within .004 inch.

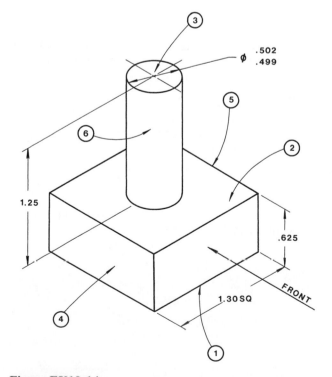

Figure EX10-14

EX10-15

Draw front, top, and right-side views of the object in Figure EX10-15, including dimensions. Add the following tolerances and specifications to the drawing.

A. Surface 1 is datum A.
B. Surface 2 is datum B.
C. The hole is located using a true position tolerance value of 0.13 millimeter at MMC. The true position tolerance is referenced to datums A and B.
D. Surface 1 is to be straight within 0.02 millimeter.
E. The bottom surface is to be parallel to datum A within 0.03 millimeter.

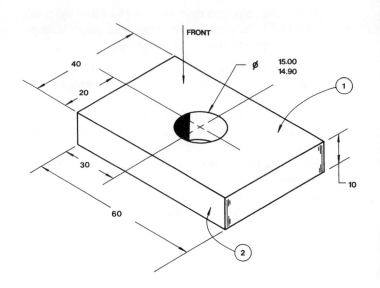

Figure EX10-15

EX10-16

Draw front, top, and right-side views of the object in Figure EX10-16, including dimensions. Add the following tolerances and specifications to the drawing.

A. Surface 1 is datum A.
B. Surface 2 is datum B.
C. Surface 3 is perpendicular to surface 2 within 0.02 millimeter.
D. The four holes are to be located using a positional tolerance of 0.07 millimeter at MMC referenced to datums A and B.
E. The centerlines of the holes are to be straight within 0.01 millimeter at MMC.

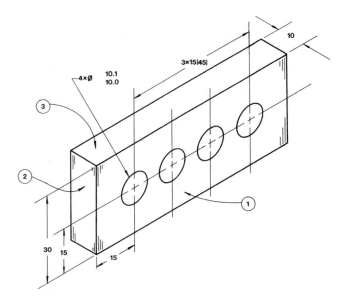

Figure EX10-16

EX10-17

Draw front, top, and right-side views of the object in Figure EX10-17, including dimensions. Add the following tolerances and specifications to the drawing.

A. Surface 1 has a dimension of .378−.375 inch and is datum A. The surface has a dual primary runout with datum B to within .005 inch. The runout is total.

B. Surface 2 has a dimension of 1.505−1.495 inch. Its runout relative to the dual primary datums A and B is .008 inch. The runout is total.

C. Surface 3 has a dimension of 1.000 ±.005 and has no geometric tolerance.

D. Surface 4 has no circular dimension but has a total runout tolerance of .006 inch relative to the dual datums A and B.

E. Surface 5 has a dimension of .500−.495 inch and is datum B. It has a dual primary runout with datum A within .005 inch. The runout is total.

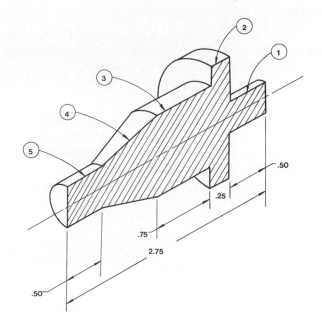

Figure EX10-17

EX10-18

Draw front, top, and right-side views of the object in Figure EX10-18, including dimensions. Add the following tolerances and specifications to the drawing.

A. Hole 1 is datum A.

B. Hole 2 is to have its circular centerline parallel to datum A within 0.2 millimeter at MMC when datum A is at MMC.

C. Assign a positional tolerance of 0.01 to each hole's centerline at MMC.

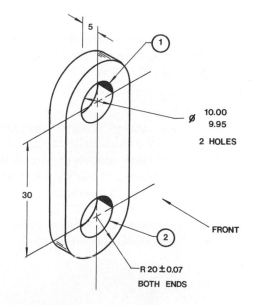

Figure EX10-18

EX10-19

Draw front, top, and right-side views of the object in Figure EX10-19, including dimensions. Add the following tolerances and specifications to the drawing.

A. Surface 1 is datum A.
B. Surface 2 is datum B.
C. The six holes have a diameter range of .502–.499 inch and are to be located using positional tolerances so that their centerlines are within .005 inch at MMC relative to datums A and B.
D. The back surface is to be parallel to datum A within .002 inch.

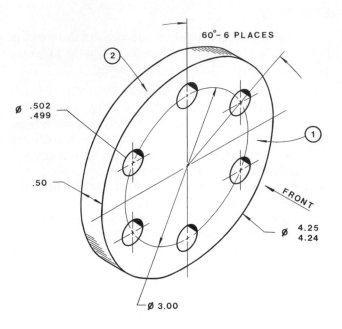

Figure EX10-19

EX10-20

Draw front, top, and right-side views of the object in Figure EX10-20, including dimensions. Add the following tolerances and specifications to the drawing.

A. Surface 1 is datum A.
B. Hole 2 is datum B.
C. The eight holes labeled 3 have diameters of 8.4–8.3 millimeters with a positional tolerance of 0.15 millimeter at MMC relative to datums A and B. Also, the eight holes are to be counterbored to a diameter of 14.6–14.4 millimeters and to a depth of 5.0 millimeters.
D. The large center hole is to have a straightness tolerance of 0.2 at MMC about its centerline.

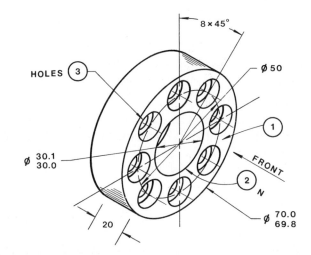

Figure EX10-20

EX10-21

Draw front, top, and right-side views of the object in Figure EX10-21, including dimensions. Add the following tolerances and specifications to the drawing.

A. Surface 1 is datum A.
B. Surface 2 is datum B.
C. Surface 3 is datum C.
D. The four holes labeled 4 have a dimension and tolerance of 8 +0.3, −0 millimeters. The holes are to be located using a positional tolerance of 0.05 millimeter at MMC relative to datums A, B, and C.
E. The six holes labeled 5 have a dimension and tolerance of 6 +0.2, −0 millimeters. The holes are to be located using a positional tolerance of 0.01 millimeter at MMC relative to datums A, B, and C.

EX10-22

The objects on page 000 labeled A and B, are to be toleranced using four different tolerances as shown. Redraw the charts shown in Figure EX10-22 and list the appropriate allowable tolerance for "as measured" increments of 0.1 millimeter or .001 inch. Also include the appropriate geometric tolerance drawing called out above each chart.

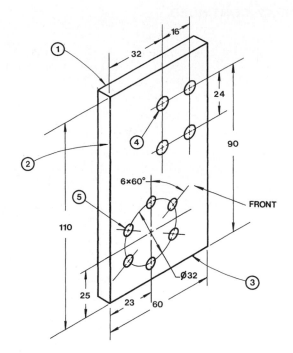

Figure EX10-21

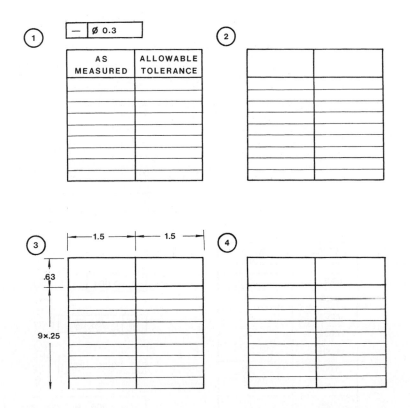

Figure EX10-22

EX10-22, CONTINUED

A.

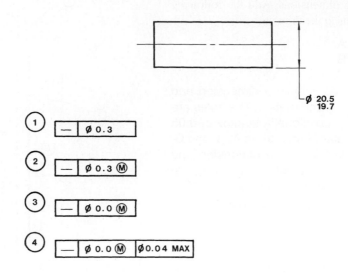

B.

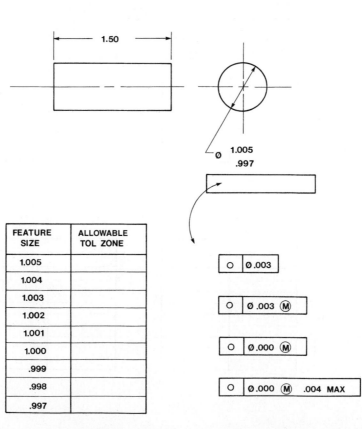

FEATURE SIZE	ALLOWABLE TOL ZONE
1.005	
1.004	
1.003	
1.002	
1.001	
1.000	
.999	
.998	
.997	

EX10-23

Dimension and tolerance parts 1 and 2 of Figure EX10-23 so that part 1 always fits into part 2 with a minimum clearance of .005 inch. The tolerance for part 1's outer matching surface is .006 inch.

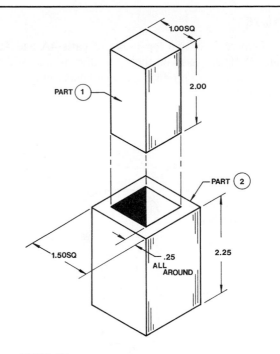

Figure EX10-23

EX10-24

Dimension and tolerance parts 1 and 2 of Figure EX10-24 so that part 1 always fits into part 2 with a minimum clearance of 0.03 millimeter. The tolerance for part 1's diameter is 0.05 millimeter. Take into account the fact that the interface is long relative to the diameters.

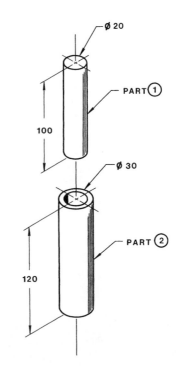

Figure EX10-24

EX10-25

Prepare front and top views of parts 4A and 4B of Figure EX10-25 based on the given dimensions. Add geometric tolerances to produce the stated maximum clearance and mismatch.

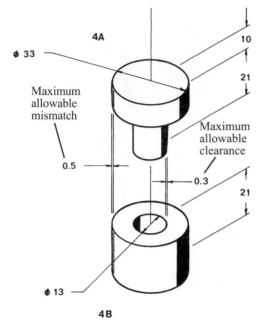

Figure EX10-25

EX10-26

Redraw parts A and B of Figure EX10-26 and add dimensions and tolerances to meet the "UPON ASSEMBLY" requirements.

The UPON ASSEMBLY condition.

Nominal sizes

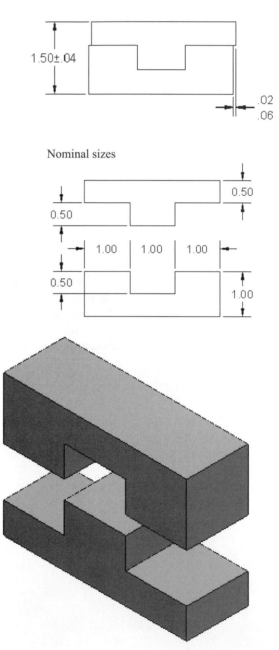

Figure EX10-26

EX10-27

Draw front and top views of both objects in Figure EX10-27. Add dimensions and geometric tolerances to meet the "FINAL CONDITION" requirements.

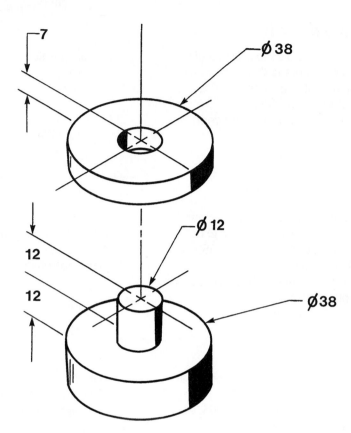

FINAL CONDITION

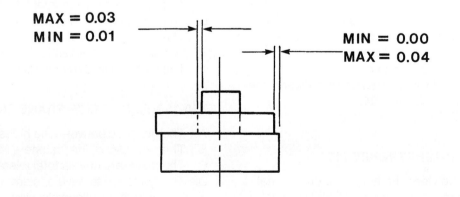

MAX = 0.03
MIN = 0.01

MIN = 0.00
MAX = 0.04

Figure EX10-27

Given a TOP and BOTTOM part in the floating condition as shown in Figure EX10-28 add dimensions and tolerances to satisfy the following conditions. Size the TOP and BOTTOM parts and FASTENER length as needed. Use geometric and positional tolerances.

EX10-28 INCHES–CLEARANCE FIT

A. The distance between the holes is 2.00 nominal.
B. The diameter of the fasteners is Ø.375 nominal.
C. The fasteners have a total tolerance of .001.
D. The holes have a tolerance of .002.
E. The minimum allowable clearance between the fastener and the holes is .003.
F. The material is .375 inch thick.

EX10-29 MILLIMETERS–CLEARANCE FIT

A. The distance between the holes is 80 nominal.
B. The nominal diameter of the fasteners is Ø12.
C. The fasteners have a total tolerance of 0.05.
D. The holes have a tolerance of 0.03.
E. The minimum allowable clearance between the fastener and the holes is 0.02.
F. The material is 12 millimeters thick.

EX10-30 INCHES–CLEARANCE FIT

A. The distance between the holes is 3.50 nominal.
B. The diameter of the fasteners is Ø.625.
C. The fasteners have a total tolerance of .005.
D. The holes have a tolerance of .003.
E. The minimum allowable clearance between the fastener and the holes is .002.
F. The material is .500 inch thick.

EX10-31 MILLIMETERS–CLEARANCE FIT

A. The distance between the holes is 120 nominal.
B. The diameter of the fasteners is Ø24 nominal.
C. The fasteners have a total tolerance of 0.01.
D. The holes have a tolerance of 0.02.
E. The minimum allowable clearance between the fastener and the holes is 0.04.
F. The material is 20 millimeters thick.

EX10-32 INCHES–INTERFERENCE FIT

A. The distance between the holes is 2.00 nominal.
B. The diameter of the fasteners is Ø.250 nominal.
C. The fasteners have a total tolerance of .001.
D. The holes have a tolerance of .002.

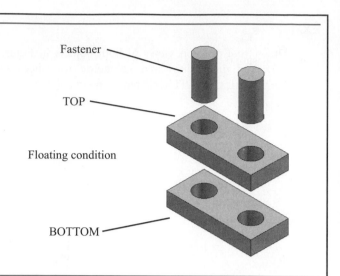

Figure EX10-28

E. The maximum allowable interference between the fastener and the holes is .0065.
F. The material is .438 inch thick.

EX10-33 MILLIMETERS–INTERFERENCE FIT

A. The distance between the holes is 80 nominal.
B. The diameter of the fasteners is Ø10 nominal.
C. The fasteners have a total tolerance of 0.01.
D. The holes have a tolerance of 0.02.
E. The maximum allowable interference between the fastener and the holes is 0.032.
F. The material is 14 millimeters thick.

EX10-34 INCHES–LOCATIONAL FIT

A. The distance between the holes is 2.25 nominal.
B. The diameter of the fasteners is Ø.50 nominal.
C. The fasteners have a total tolerance of .001.
D. The holes have a tolerance of .002.
E. The minimum allowable clearance between the fastener and the holes is .0010.
F. The material is .370 inch thick.

EX10-35 MILLIMETERS–TRANSITIONAL FIT

A. The distance between the holes is 100 nominal.
B. The diameter of the fasteners is Ø16 nominal.
C. The fasteners have a total tolerance of 0.01.
D. The holes are to have a tolerance of 0.02.
E. The minimum allowable clearance between the fastener and the holes is 0.01.
F. The material is 20 millimeters thick.

Given a TOP and BOTTOM part in the fixed condition as shown in Figure EX10-36 add dimensions and tolerances to satisfy the following conditions. Size the TOP and BOTTOM parts and FASTENERlength as needed. Use geometric and positional tolerances.

EX10-36 INCHES–CLEARANCE FIT

A. The distance between the holes is 2.00 nominal.
B. The diameter of the fasteners is Ø.375 nominal.
C. The fasteners have a total tolerance of .001.
D. The holes have a tolerance of .002.
E. The minimum allowable clearance between the fastener and the holes is .003.
F. The material is .375 inch thick.

EX10-37 MILLIMETERS–CLEARANCE FIT

A. The distance between the holes is 80 nominal.
B. The nominal diameter of the fasteners is Ø12.
C. The fasteners have a total tolerance of 0.05.
D. The holes have a tolerance of 0.03.
E. The minimum allowable clearance between the fastener and the holes is 0.02.
F. The material is 12 millimeters thick.

EX10-38 INCHES–CLEARANCE FIT

A. The distance between the holes is 3.50 nominal.
B. The diameter of the fasteners is Ø.625.
C. The fasteners have a total tolerance of .005.
D. The holes have a tolerance of .003.
E. The minimum allowable clearance between the fastener and the holes is .002.
F. The material is .500 inch thick.

EX10-39 MILLIMETERS–CLEARANCE FIT

A. The distance between the holes is 120 nominal.
B. The diameter of the fasteners is Ø24 nominal.
C. The fasteners have a total tolerance of 0.01.
D. The holes have a tolerance of 0.02.
E. The minimum allowable clearance between the fastener and the holes is 0.04.
F. The material is 20 millimeters thick.

EX10-40 INCHES–INTERFERENCE FIT

A. The distance between the holes is 2.00 nominal.
B. The diameter of the fasteners is Ø.250 nominal.
C. The fasteners have a total tolerance of .001.
D. The holes have a tolerance of .002.

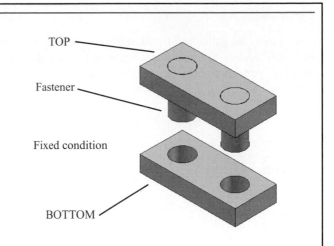

Figure EX10-36

E. The maximum allowable interference between the fastener and the holes is .0065.
F. The material is .438 inch thick.

EX10-41 MILLIMETERS–INTERFERENCE FIT

A. The distance between the holes is 80 nominal.
B. The diameter of the fasteners is Ø10 nominal.
C. The fasteners have a total tolerance of 0.01.
D. The holes have a tolerance of 0.02.
E. The maximum allowable interference between the fastener and the holes is 0.032.
F. The material is 14 millimeters thick.

EX10-42 INCHES–LOCATIONAL FIT

A. The distance between the holes is 2.25 nominal.
B. The diameter of the fasteners is Ø.50 nominal.
C. The fasteners have a total tolerance of .001.
D. The holes have a tolerance of .002.
E. The minimum allowable clearance between the fastener and the holes is .0010.
F. The material is .370 inch thick.

EX10-43 MILLIMETERS–TRANSITIONAL FIT

A. The distance between the holes is 100 nominal.
B. The diameter of the fasteners is Ø16 nominal.
C. The fasteners have a total tolerance of 0.01.
D. The holes are to have a tolerance of 0.02.
E. The minimum allowable clearance between the fastener and the holes is 0.001.
F. The material is 20 millimeters thick.

EX10-44

Given the following two assemblies, size the parts so that they always fit together. Create individual drawings of each part including dimensions and tolerances. Use geometric and positional tolerances.

SUPPORT ASSEMBLY

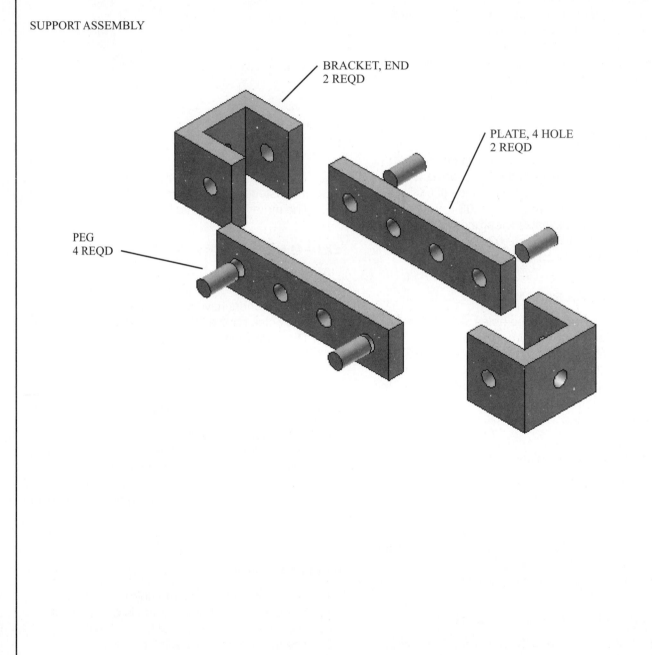

BRACKET, END
2 REQD

PLATE, 4 HOLE
2 REQD

PEG
4 REQD

EX10-45

   Assume that there are two copies of the part in Figure EX10-45 and that these parts are to be joined together using four fasteners in the floating condition. Draw front and top views of the object, including dimensions and tolerances. Add the following tolerances and specifications to the drawing, then draw front and top views of a shaft that can be used to join the two objects. The shaft should be able to fit into any of the four holes.

A. Surface 1 is datum A.
B. Surface 2 is datum B.
C. Surface 3 is perpendicular to surface 2 within 0.02 millimeter.
D. Specify the positional tolerance for the four holes applied at MMC.
E. The centerlines of the holes are to be straight within 0.01 millimeter at MMC.
F. The clearance between the shafts and the holes is to be 0.05 minimum and 0.10 maximum.

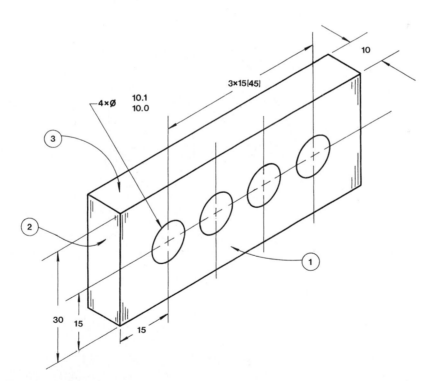

Figure EX10-45

EX10-46

Assume that there are two copies of the part in Figure EX10-46 and that these parts are to be joined together using six fasteners in the floating condition. Draw front and top views of the object, including dimensions and tolerances. Add the following tolerances and specifications to the drawing, then draw front and top views of a shaft that can be used to join the two objects. The shaft should be able to fit into any of the six holes.

A. Surface 1 is datum A.
B. Surface 2 is round within .003.
C. Specify the positional tolerance for the six holes applied at MMC.
D. The clearance between the shafts and the holes is to be .001 minimum and .003 maximum.

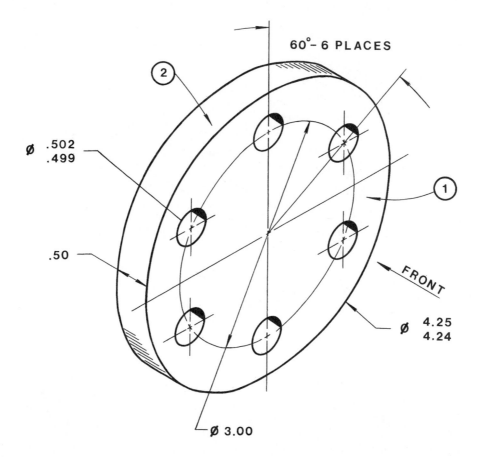

Figure EX10-46

C H A P T E R 11

Threads and Fasteners

11-1 INTRODUCTION

This chapter explains how to draw threads, washers, keys, and springs. It explains how to use fasteners to join parts together and design uses for washers, keys, and springs.

Throughout the chapter it will be suggested that blocks and wblocks be created of the various thread and fastener shapes. Thread representations, fastener head shapes, setscrews, and both internal and external thread representations for orthographic views and sectional views are so common in technical drawings that it is good practice to create a set of wblocks that can be used on future drawings to prevent having to redraw a thread shape every time it is needed.

See Chapter 3 for an explanation of the **Block** command.

11-2 THREAD TERMINOLOGY

Figure 11-1 shows a thread. The peak of a thread is called the *crest*, and the valley portion is called the *root*. The *major diameter* of a thread is the distance across the thread from crest to crest. The *minor diameter* is the distance across the thread from root to root.

The *pitch* of a thread is the linear distance along the thread from crest to crest. Thread pitch is usually referred

Detailed representation

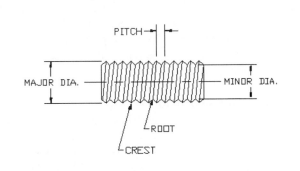

Figure 11-1

to in terms of a unit of length such as 20 threads per inch or 1.5 threads per millimeter.

11-3 THREAD CALLOUTS (METRIC UNITS)

Threads are specified on a drawing using drawing callouts. See Figure 11-2. The M preceding a drawing callout specifies that the callout is for a metric thread. Holes that are not threaded use the Ø symbol.

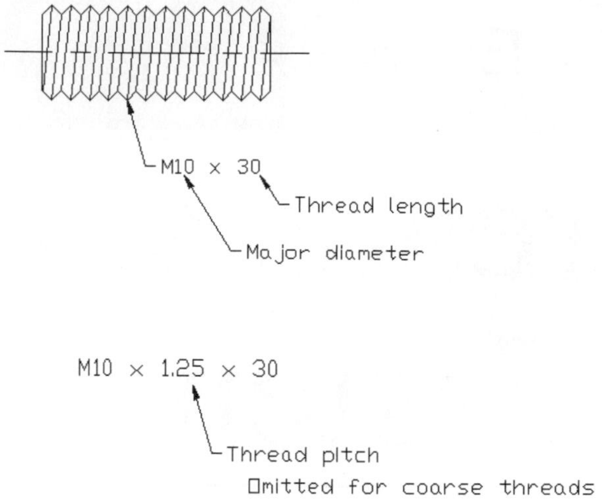

Figure 11-2

Major Dia	Coarse		Fine	
	Pitch	Tap Drill Dia	Pitch	Tap Drill Dia
1.6	0.35	1.25		
2	0.4	1.6		
2.5	0.45	2.05		
3	0.5	2.5		
4	.7	3.3		
5	0.8	4.2		
6	1	5.0		
8	1.25	6.7	1	7.0
10	1.5	8.5	1.25	8.7
12	1.75	10.2	1.25	10.8
16	2	14	1.5	14.5
20	2.5	17.5	1.5	18.5
24	3	21	2	22
30	3.5	26.5	2	28
36	4	32	3	33
42	4.5	37.5	3	39
48	5	43	3	45

Figure 11-3

The number following the M is the major diameter of the thread; for example, an M10 thread has a major diameter of 10 millimeters. The pitch of a metric thread is assumed to be a coarse thread unless otherwise stated. The callout M10 × 30 assumes a coarse thread, or 1.5 threads per millimeter. The number 30 is the thread length in millimeters. The "×" is read as "by," so the thread is called a "ten by thirty."

The callout M10 × 1.25 × 30 specifies a pitch of 1.25 threads per millimeter. This is not a standard coarse thread size, so the pitch must be specified.

Figure 11-3 shows a list of preferred thread sizes. These sizes are similar to the standard sizes shown in Figure 9-32. A list of other metric thread sizes is included in the appendix.

Whenever possible use preferred thread sizes for designing. Preferred thread sizes are readily available and are usually cheaper than nonstandard sizes. In addition, tooling such as wrenches is also readily available for preferred sizes.

11-4 THREAD CALLOUTS (ENGLISH UNITS)

English unit threads always include a thread form specification. Thread form specifications are designated by capital letters, as shown in Figure 11-4, and are defined as follows:

UNC—Unified National Coarse
UNF—Unified National Fine
UNEF—Unified National Extra Fine
UN—Unified National, or constant pitch threads

An English unit thread callout starts by defining the major diameter of the thread followed by the pitch specification. The callout .500 –13 UNC means a thread whose major diameter is .500 inch with 13 threads per inch and is manufactured to the Unified National Coarse standards.

There are three possible classes of fit for a thread: 1, 2, and 3. The different class specifications specify a set of manufacturing tolerances. A class 1 thread is the loosest and a class 3, the most exact. A class 2 fit is the most common.

The letter A designates an external thread, B an internal thread. The symbol × means "by" as in 2 × 4, "two by four." The thread length (3.00) may be followed by the word LONG to prevent confusion about which value represents the length.

Drawing callouts for English unit threads are sometimes shortened, as in Figure 11-4. The callout .500–13UNC–2A × 3.00 LONG is shortened to .500–13 × 3.00. Only a coarse thread has 13 threads per inch, and it should be obvious whether a thread is internal or external, so these specifications may be dropped. Most threads are class 2, so it is tacitly accepted that all threads are class 2 unless otherwise specified. The shortened callout form is not universally accepted. When in doubt, use a complete thread callout.

A partial list of standard English unit threads is shown in Figure 11-5. A more complete list is included in the appendix. Some of the drill sizes listed use numbers and letters. The decimal equivalents to the numbers and letters are listed in the appendix.

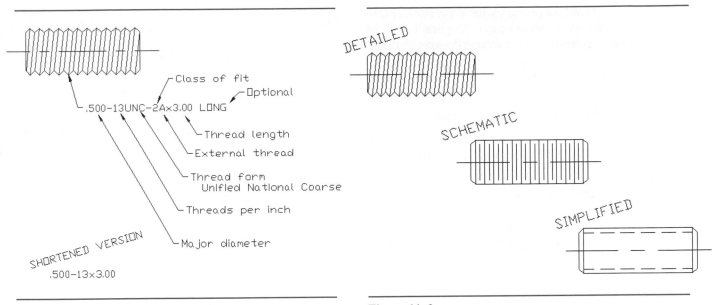

Class of fit
Optional

.500-13UNC-2A×3.00 LONG

Thread length

External thread

Thread form
Unifed National Coarse

Threads per inch

Major diameter

SHORTENED VERSION
.500-13×3.00

Figure 11-4

DETAILED

SCHEMATIC

SIMPLIFIED

Figure 11-6

11-5 THREAD REPRESENTATIONS

There are three ways to graphically represent threads on a drawing: detailed, schematic, and simplified. Figure 11-6 shows the three representations.

Detailed representations look the most like actual threads but are time-consuming to draw. Creating a wblock of a detailed shape will help eliminate this time constraint.

Schematic and simplified thread representations are created using a series of straight lines. The simplified representation uses only two hidden lines and can be mistaken for an internal hole if it is not accompanied by a thread specification callout. The choice of which representation to use depends on individual preferences. The resulting drawing should be clear and easy to understand. All three representations may be used on the same drawing, but in gen- eral, only very large threads (those over 1.00 in., or 25 mm) are drawn using the detailed representation.

Ideally, thread representations should be drawn with each thread equal to the actual pitch size. This is not practical

Major Dia	Decimal	UNC		UNF		UNEF	
		Thread/in	Tap drill Dia.	Thread/in	Tap drill Dia.	Thread/in	Tap drill Dia.
#6	.138	40	#38	44	#37		
#8	.164	32	#29	36	#29		
#10	.190	24	#25	32	#21		
1/4	.250	20	7	28	3	32	.219
5/16	.312	18	F	24	1	32	.281
3/8	.375	16	.312	24	Q	32	.344
7/16	.438	14	U	20	.391	28	Y
1/2	.500	13	.422	20	.453	28	.469
9/16	.562	12	.484	18	.516	24	.516
5/8	.625	11	.531	18	.578	24	.578
3/4	.750	10	.656	16	.688	20	.703
7/8	.875	9	.766	14	.812	20	.828
1	1.000	8	.875	12	.922	20	.953
1 1/4	1.250	7	1.109	12	1.172	18	1.188
1 1/2	1.500	6	1.344	12	1.422	18	1.438

UNC = Unified National Coarse

UNF = Unified National Fine

UNEF = Unified National Extra Fine

Figure 11-5

for smaller threads and not necessary for larger ones. Thread representations are not meant to be exact duplications of the threads but representations, so convenient drawing distances are acceptable.

To draw a detailed thread representation

Draw a detailed thread representation for a 1.00-inch-diameter thread that is 3.00 inches long. See Figure 11-7.

1. Set **GRID = .5** and **Snap = .125.**
2. Draw a **4.00-inch** centerline near the center of the screen.
3. Zoom the area around the centerline.
4. Draw a zigzag pattern **.375** above the centerline using the .125 snap points. Start the zigzag line **.50** from the left end of the centerline.
5. Access the **Array** command.

The **Array** dialog box will appear.

6. Select the **Rectangular** option.
7. Click the **Select objects** button, then select the zigzag pattern.
8. Set the **Row value** for **1** and the **Column value** for **12.**
9. Set the **Row offset** for **0.0000**
10. Set the **Column offset** for **0.25,** then select the **OK** button.
11. Mirror the arrayed zigzag line about the center-line.
12. Draw vertical lines at both ends of the thread and two slanted lines between the thread's roots and crests as shown.
13. Array both slanted lines using the same array parameters used for the zigzag line: **12** columns **.25** apart.
14. Save the thread representation as a block and wblock named **DETLIN.** Define the insertion point as shown.

Creating blocks is explained in Chapter 3.

Another technique for drawing a detailed thread representation is to draw a single thread completely and then use the **Copy** or **Array** command to generate as many additional threads as is necessary.

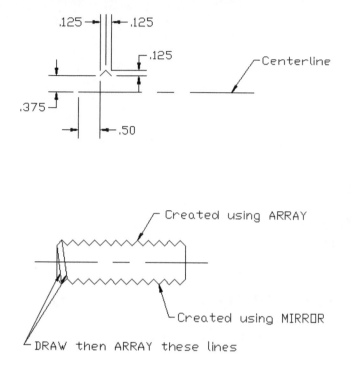

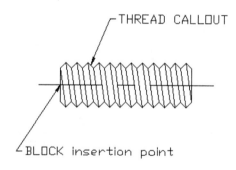

Figure 11-7

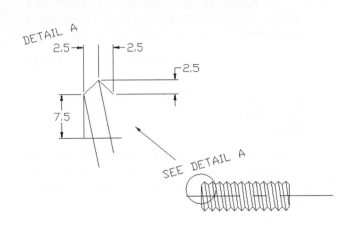

Figure 11-8

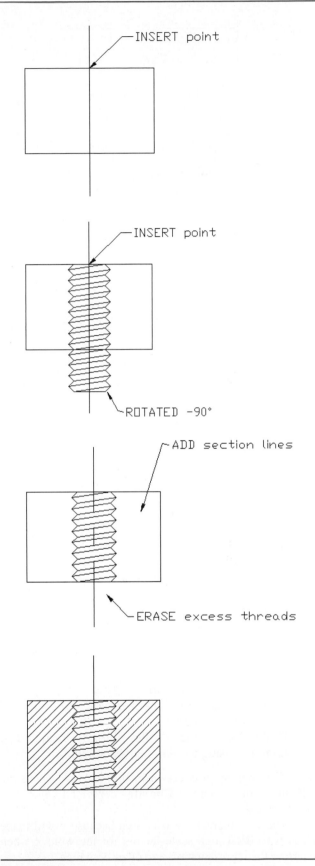

It is recommended that you save all thread wblocks on a separate disk. This disk will become a reference disk that you can use when creating other drawings that require threads.

Figure 11-8 shows a metric unit detailed thread representation. It was created using the procedure outlined. **Grid** was set at **10, Snap** was set at **2.5,** and the distance from the centerline to the zigzag pattern was **10.** Draw and save the metric detailed thread representation shown in Figure 11-8 as a wblock named **DETLMM.**

To create an internal detailed thread representation in a sectional view

Figure 11-9 shows how to create a 1.00-inch internal detailed thread representation from the wblock DETLIN of the external detailed thread created above.

1. Use **Block, Insert** and locate the detailed wblock on the drawing screen at the indicated insert point. In this example the thread is to be drawn in a vertical orientation, so the block is rotated **90°** when it is inserted. The same scale size is used for the wblock as was drawn for a 1.00-inch diameter thread.
2. The thread created from wblock DETLIN is longer than needed, so explode the block and then erase the excess lines.
3. Define the hatch pattern as **ANSI31** and apply hatching to the areas outside the thread as shown.

Figure 11-9

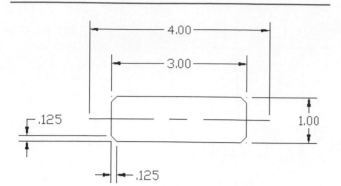

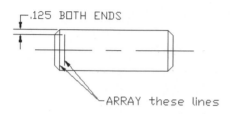

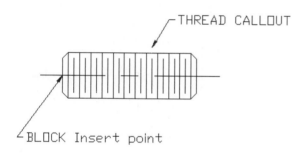

ARRAY these lines

THREAD CALLOUT

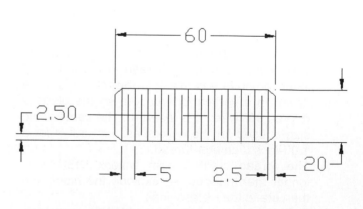

BLOCK Insert point

Figure 11-10

4. Draw three vertical lines as shown.
5. Use the **Array, Rectangular** command to draw **11** vertical lines across the thread. The distance between the 11 lines (COLUMNS) is **.25**.

A distance of −.25 would create lines to the left of the original line.

6. Save the representation as a wblock named **SCHMINCH**. Define the insertion point as shown.

Figure 11-11 shows a metric unit version of a schematic thread representation in a sectional view. The procedure used to create the representation is the same as explained previously but with different drawing limits and different values. The major diameter is **20**, and the spacing between lines is **2.5** and **5** as shown. Draw and save the representation as a wblock named **SCHMMM.**

To create an internal schematic thread representation

Figure 11-12 shows a 36-millimeter-diameter internal schematic thread representation. It was developed from the wblock SCHMMM created previously.

1. Set **Grid** = **10**
   Snap = **5**
   Limits = **297,210**
   Zoom = **ALL**

2. Insert the wblock SCHMMM at the indicated insert point.

The required thread diameter is **36** millimeters. The wblock SCHMMM was drawn using a diameter of **20** mil-

To create a schematic thread representation

Draw a schematic representation of a 1.00-inch-diameter thread. See Figure 11-10.

1. Set Grid to **.50** and **Snap** to **.125.**
2. Draw a **4.00-inch** centerline near the middle of the drawing screen.
3. Draw the outline of the thread, including a chamfer, using the dimensions shown.

The chamfer was drawn in this example using the .125 snap points, but the **Chamfer** command could also have been used.

The 1.00 diameter was chosen because it will make it easier to determine scale factors for the wblock when inserting the representation into other drawings.

Figure 11-11

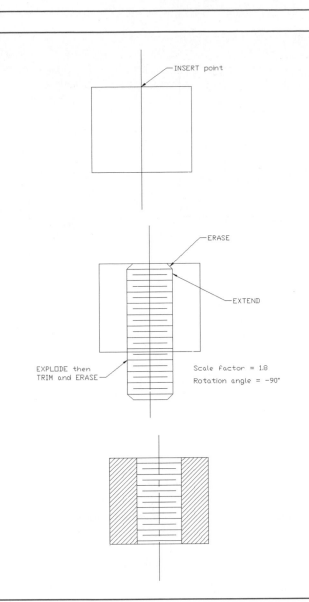

Figure 11-12

limeters. This means that the block must be enlarged by using a scale factor. The scale factor is determined by dividing the desired diameter by the wblock's diameter.

$$36/20 = 1.8$$

The wblock must also be rotated **90°** to give it the correct orientation.

3. The inserted thread shape is longer than desired, so first explode the wblock and then use the **Erase, Trim,** and **Extend** commands as needed.
4. Draw the sectional lines using **Hatch ANSI31.**
5. Save the drawing as a wblock if desired.

To create a simplified thread representation

The simplified representation looks very similar to the orthographic view of an internal hole, so it is important to always include a thread callout with the representation. In the example shown, a leader line was included with the representation. See Figure 11-13. The leader serves as a reminder to add the appropriate drawing callout. If the leader line is in an inconvenient location when the wblock is inserted into a drawing, the leader line can be moved or simply erased.

1. Set **Grid** = .5
   Snap = .125
2. Draw a **4.00** centerline near the center of the screen.
3. Draw the thread outline using the given dimensions.
4. Draw the hidden lines using the given dimensions.
5. Save the thread representation as a wblock named **SIMPIN.** Define the insertion point as shown.

Figure 11-14 shows an internal simplified thread representation in a sectional view. Note how hidden lines that cross over the sectional lines are used. The hatch must be drawn first and the hidden lines added over the pattern. If the hidden lines are drawn first, the **Hatch** command may not add section lines to the portion between the hidden line and the solid line that represents the edge of the threaded hole.

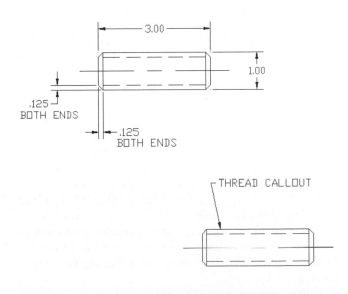

Figure 11-13

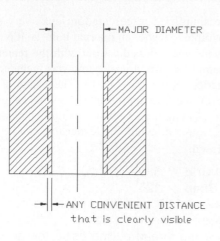

Figure 11-14

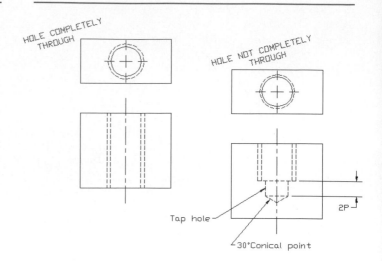

Figure 11-15

11-6 ORTHOGRAPHIC VIEWS OF INTERNAL THREADS

Figure 11-15 shows top and front orthographic views of internal threads. One thread goes completely through the object, the other only partially.

Internal threads are represented in orthographic views using parallel hidden lines. The distance between the lines should be large enough so that there is a clear distinction between the lines; that is, the lines should not appear to blend together or become a single, very thick line.

Circular orthographic views of threaded holes are represented by two circles: one drawn using a continuous line and the other drawn using a hidden line. The distance between the circles should be large enough to be visually distinctive. The two circles shown should both be clearly visible.

Threaded holes are created by first drilling a tap hole and then tapping (cutting) the threads using a tapping bit. Tapping bits have cutting surfaces on their side surfaces, not on the bottom. This means that if the tapping bit were forced all the way to the bottom of the tap hole, the bit could be damaged or broken. It is good design practice to make the tap hole deep enough so that a distance equivalent to at least two thread lengths $(2P)$ extends beyond the tapped portion of the hole.

Threaded holes that do not go completely through an object must always show the unused portion of the tap hole. The unused portion should also include the conical point. See Chapter 5.

If an internal .500–13 UNC thread does not go completely through an object, the length of the unused portion of the tap hole is determined as follows:

Inches/Thread
= Pitch length = 1.00/13
= .077 inch

so

$$2P = 2(.077) = .15 \text{ inch}$$

The distance .15 represents a minimum. It would be acceptable to specify a pilot hole depth greater than .15 depending on the specific design requirements.

If an internal M12 × 1.75 thread does not go completely through an object, the length of the unused portion of the tap hole is determined as follows:

$$2(\text{Pitch length}) = 2(1.75) = 3.50 \text{ mm}$$

11-7 SECTIONAL VIEWS OF INTERNAL THREAD REPRESENTATIONS

Figure 11-16 shows sectional views of internal threads that do not go completely through an object. Each example was created from wblocks of the representations. The hatch pattern used was ANSI32. Each example includes both a threaded portion and an untapped pilot hole that extends approximately $2P$ beyond the end of the tapped portion of the hole.

When drawing a simplified representation in a sectional view, draw the hidden lines that represent the outside edges of the threads after the section lines have been added. If the lines are drawn before the section lines are added, the section lines will stop at the outside line. Section lines should be drawn up to the solid line, as shown.

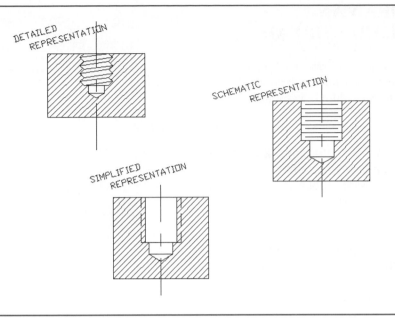

Figure 11-16

11-8 TYPES OF THREADS

Figure 11-17 shows the profiles of four different types of thread: American National, square, acme, and knuckle. There are many other types of threads. In general, square and acme threads are used when heavy loading is involved. A knuckle thread can be manufactured from sheet metal and is most commonly found on a lightbulb.

The American National thread is the thread shape most often used in mechanical design work. All threads in this chapter are assumed to be American National threads unless otherwise stated.

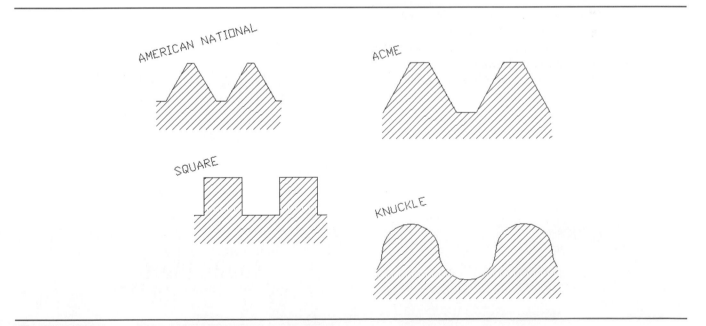

Figure 11-17

11-9 HOW TO DRAW AN EXTERNAL SQUARE THREAD

Figure 11-18 shows how to draw a 4.00-inch-long external square thread that has a major diameter of 5.25 and 2 threads per inch. The procedure is as follows.

1. Set **Grid** = **.50**
   Snap = **.125**
2. Draw two **5.00**-long horizontal lines **2.00** apart and a **4.50** centerline between them.
3. Draw a rhomboid center about the centerline using the given dimensions.

   In this example $P = .5$, so $.5P = .25$.

4. Use the **Array, Rectangular** command to create eight columns (2 threads per inch) **.50** apart.
5. Draw slanted lines 1-2 and 3-4.
6. Zoom the upper right portion of the thread as needed.
7. Draw a horizontal line **.25** (.5*P*) from the outside edge of the thread as shown.
8. Draw line 5-6 from the intersection of the horizontal line drawn in step 7 with line 1-2, labeled point 5, to the intersection of the toothline 1-6 and the thread's centerline.
9. Trim the horizontal line to create line 5-7, and trim line 1-2 below point 5.
10. Array lines 1-5, 5-6, and 5-7 using **Array, Rectangular** with eight columns −**.25** apart.

    The minus sign will generate a right-to-left array.

11. Repeat steps 7 through 10 for the lower portion of the thread. Array eight columns, +**.25** apart. The plus sign generates a left-to-right array.
12. Erase any excess lines and add any necessary shaft information to the drawing.
13. Save the drawing as a wblock named **SQIN** using the indicated insertion point.

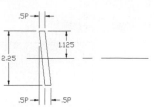

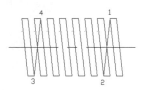

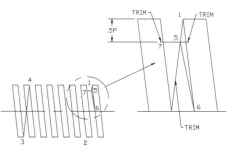

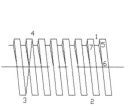

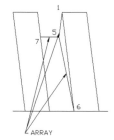

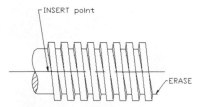

Figure 11-18

11-10 HOW TO DRAW AN INTERNAL SQUARE THREAD

Figure 11-19 shows an external 2.25 × 2 square thread. The drawing was developed from wblock SQIN created in Section 11-9. The wblock was rotated to the correct orientation. **Erase** and **Trim** were used to fit the thread within the required depth. No scale factor was needed.

11-11 HOW TO DRAW AN EXTERNAL ACME THREAD

Draw a 2.25 × 2 × 4.00–long external acme thread. The procedure is as follows. See Figure 11-20.

1. Set **Grid** = **.5**
   Snap = **.125**
2. Draw a **5.00**-long horizontal centerline.
3. Draw two **.5**-long horizontal lines **1.125** above and below.

These lines establish the major diameter of the thread.

4. Draw a single acme thread using the given dimensions. Use **Zoom** to help create an enlarged working area. The first thread should start at the right end of the short horizontal line above the centerline.

The width dimensions are taken along the horizontal centerline of the individual thread. Each side of the thread is slanted at **14.5°** [.5(29)]. In this example $P = .5$, so $.5P = .25$, and $.25P = .125$.

5. Use the **Array** command and draw eight columns (two threads per inch) **.5** apart to develop the top portion of the thread.
6. Copy and move the top portion of the thread to create the lower portion.

Do not use **Mirror.** The lower portion of the thread is not a mirror image of the upper portion.

7. Erase and extend lines as necessary along the lower left portion of the thread to blend the thread into the shaft.

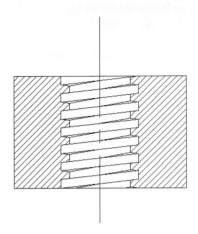

Figure 11-19

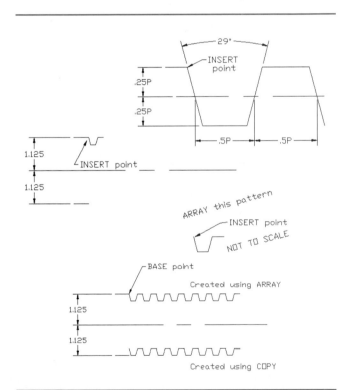

Figure 11-20, Part 1

8. Draw two slanted lines between the thread's root lines at the left end of the thread.
9. Array the lines drawn in step 8 so that there are eight columns, **.5** apart.
10. Draw two slanted lines across the thread's crest lines at the left end of the thread.
11. Array the lines drawn in step 10 so that there are eight columns, **.5** apart.

12. Draw a vertical line at each end of the thread, establishing the thread's 4.00-inch length.
13. Erase and trim the excess lines from the left end of the thread.
14. Copy and trim a crest-to-crest line to complete the left end of the thread.
15. Save the drawing as a wblock named **ACMEIN** using the indicated insertion point.

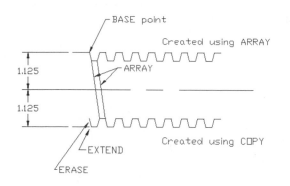

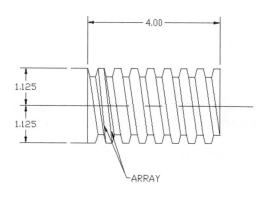

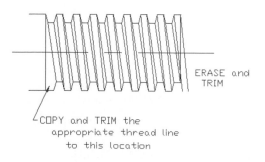

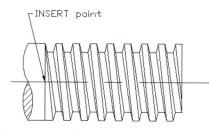

Figure 11-20, Part 2

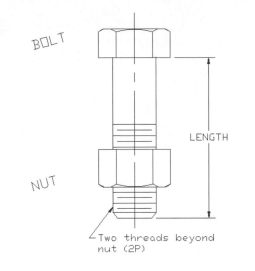

Figure 11-21

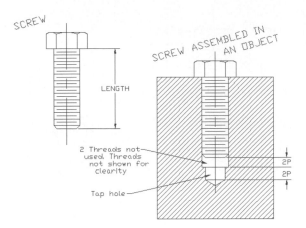

Figure 11-22

11-12 BOLTS AND NUTS

A *bolt* is a fastener that passes through a clearance hole in an object and is joined to a nut. There are no threads in the object. See Figure 11-21. Note that there are no hidden lines within the nut to indicate that the bolt is passing through. Drawing convention allows for nuts to be drawn without hidden lines.

Threads on a bolt are usually made just long enough to correctly accept a nut. This is done to minimize the amount of contact between the edges of threads and the inside surfaces of the clearance holes. The sharp, knife-like thread edges could cut into the object, particularly if the application involves vibrations.

It is considered good design practice to specify a bolt length long enough to allow at least two threads to extend beyond the end of the nut. This ensures that the nut is fully attached to the bolt.

Bolt drawings can be created from wblocks of threads. Remember that a block must be exploded before it can be edited.

11-13 SCREWS

A *screw* is a fastener that assembles into an object. It does not use a nut. The joining threads are cut into the assembling object. See Figure 11-22.

Screws may or may not be threaded over their entire length. If a screw passes through a clearance hole in an object before it assembles into another object, it is good design practice to minimize the number of threads that contact the sides of the clearance hole.

It is also considered good design practice to allow a few unused threads in the threaded hole beyond the end of an assembled screw. If a screw was forced to the bottom of a tapped hole, it might not assemble correctly or could possibly be damaged.

It is good design practice to allow at least two unused threads beyond the end of the screw. Figure 11-22 shows a schematic thread representation of a screw mounted in a threaded hole.

There is a distance of $2P$ (two threads) between the end of the screw and the end of the threaded portion of the hole. There should also be a $2P$ distance between the end of the threaded hole and the end of the pilot hole, plus the conical point of the tap hole, as described in Section 11-6.

Threads are usually not drawn in a threaded hole beyond the end of an assembling screw. This makes it easier to visually distinguish the end of the screw.

Figure 11-23 shows a screw assembled into a hole drawn using the detailed, schematic, and simplified representations in a sectional view. An orthographic view of a screw in a threaded hole is also shown.

The top view shown in Figure 11-23 applies to all three representations and the orthographic view.

11-14 STUDS

A *stud* is a threaded fastener that both screws into an object and accepts a nut. See Figure 11-24. The thread callouts and representations for studs are the same as they are for bolts and screws.

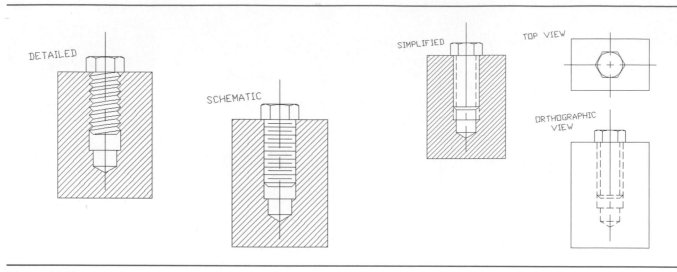

Figure 11-23

11-15 HEAD SHAPES

Bolts and screws are manufactured with a variety of different head shapes, but hexagon (hex) and square head shapes are the most common. There are many different head sizes available for different applications. Extra-thick heads are used for heavy-load applications, and very thin heads are used for applications where space is limited. The exact head size specifications are available from fastener manufacturers.

This section shows how to draw hex and square heads based on accepted average sizes that are functions of the major diameters of both the bolt and the screw. It is suggested that hexagon and square head drawings be saved as wblocks for both inch and millimeter values so that they can be combined with the thread wblocks to form fasteners.

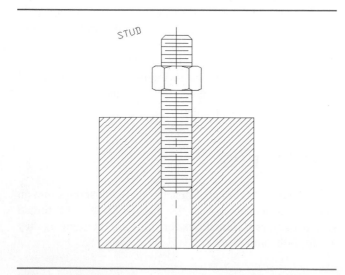

Figure 11-24

To draw a hexagon (hex) shaped head

Draw front and top orthographic views of a hex head based on a thread with an M24 major diameter. See Figure 11-25.

1. Set **Limits** = **297,210**
    Grid = **10**
    Snap = **5**
2. Draw a vertical line and two horizontal lines using the given dimensions.

The two horizontal lines are used to locate the center of the head in the top view and the bottom of the head in the front view.

3. Use **Polygon** and draw a hexagon distance across the flats equal to **1.5D**, where **D** is the major diameter of the thread.

In this example a radius of **18** was used to draw the hexagon. $1.5D = 1.5(24) = 36$ is the distance across the flats of the hexagon. Use the **Polygon** command to circumscribe a six-sided polygon around a circle of radius 18.

4. Offset a line **.67D** from the lower horizontal.

This line defines the thickness of the head. The head thickness is .67D, where D is the major diameter of the thread. In this example $D = 24$, so $.67D = .67(24) = 16.08$, which can be rounded off to 16.

5. Draw projection lines from the corners of the hexagon in the top view into the front view.
6. **Zoom** the front view portion of the drawing if necessary. Draw two **60°** lines from the corners of the front view as shown so that they intersect on the vertical centerline.

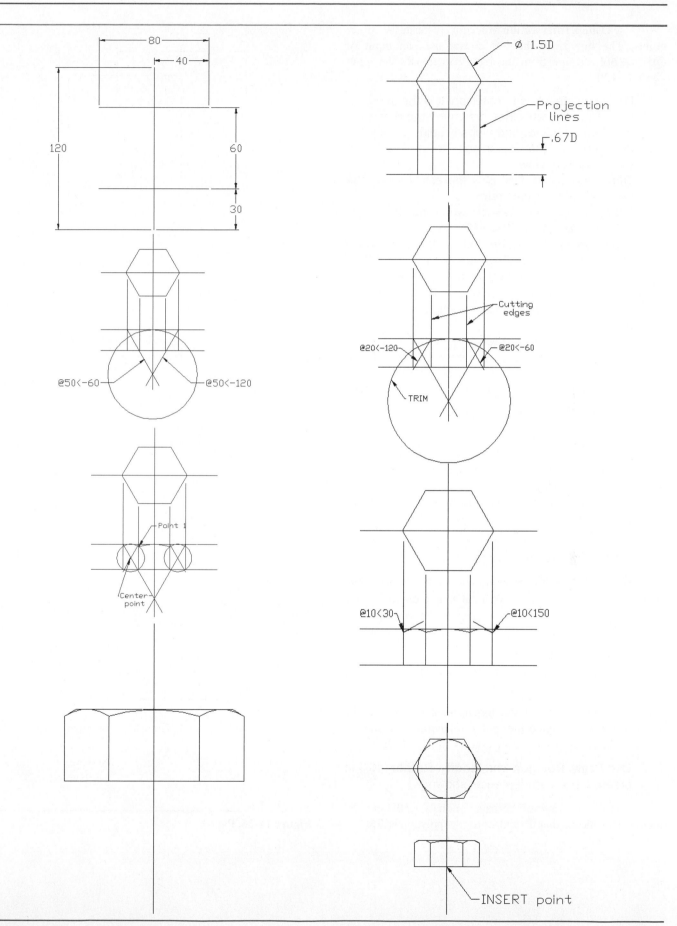

Figure 11-25

Use **Osnap, Intersection** to accurately locate the corner points. The line from the left corner uses an input of @50<−60; the line from the right corner uses the input @50<−120.

7. Draw a circle whose center point is the intersection of the two 60° lines drawn in step 6 and the vertical centerline, and whose radius equals the distance from the center point to the line at the top of the front view.
8. Trim the circle so that only the arc between the inside projection lines remains.
9. Draw two **60°** lines as shown using the inputs @**20**<−**120** and @**20**<−**60.**
10. Draw two circles centered about the intersection of the slanted lines drawn in steps 6 and 8. The radius of each circle equals the distance from the center point to the intersection of the circle drawn in step 7 and the inside projection lines labeled point 1.
11. Trim and erase as needed.
12. Draw two lines from the intersections of the smaller arcs with the outside edge of the head. The input for the lines is @**10**<**30** and @**10**<**150.**

These lines could have been generated using the **Chamfer** command.

13. Trim the chamfer lines to the top of the head.
14. **Zoom All** and draw a circle that is circumscribed within the hexagon as shown.
15. Save the drawing as a wblock named **HEXHEAD,** using the indicated insertion point.

To draw a square-shaped head

Draw front and top orthographic views of a hex head based on a thread with a 1.00-inch major diameter. See Figure 11-26.

1. Set **Grid** = **.50**
   Snap = **.25**
2. Draw two horizontal lines and a vertical line using the given dimensions.

The intersection of the top horizontal line and the vertical line will be the center point of the top view, and the lower horizontal line will be the bottom edge of the head.

3. Use **Draw, Polygon** and **Modify, Rotate (45°)** to create a square oriented as shown.

The distance across the square equals 1.50D or 1.50 inches. This means that the radius for the polygon is **.75.**

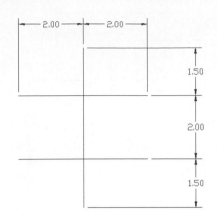

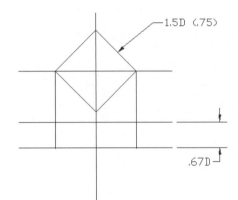

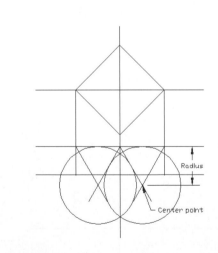

Figure 11-26, Part 1

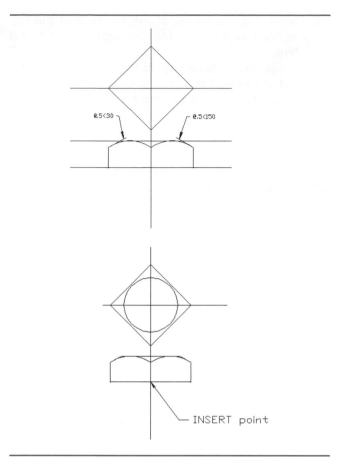

Figure 11-26, Part 2

4. Offset a line **.67** (.67*D*) from the lower horizontal line.

The offset distance is equal to the thickness of the head.

5. Project the corners of the square in the top view into the front view.
6. Zoom the front view and draw four **60°** lines from the head's upper corners and intersection of the centerline and the top surface of the head as shown. The inputs for the lines are **@1.5<−120** and **@1.5<−60**. Use **Osnap, Intersection** to ensure accuracy.
7. Use **Draw, Circle, Center, Radius** and draw two circles about the center points created in step 6. Trim the excess portions of the circle.
8. Trim and erase any excess lines.
9. Add the **30°** chamfer lines as shown. The line inputs are **@.5<150** and **@.5<30**. Use **Osnap, Intersection** to accurately locate the intersection

between the arc and the vertical side line of the head.
10. Trim any excess lines.
11. Zoom the drawing back to its original size and draw a circle in the top view tangent to the inside edges of the square.
12. Save the drawing as a wblock named **SQHEAD** using the insertion point indicated.

11-16 NUTS

This section explains how to draw hexagon- and square-shaped nuts. Both construction methods are based on the head shape wblocks created in Section 11-15.

There are many different styles of nuts. A *finished nut* has a flat surface on one side that acts as a bearing surface when the nut is tightened against an object. A finished nut has a thickness equal to .88*D*, where *D* is the major diameter of the nut's thread size.

A *locknut* is symmetrical, with the top and bottom surfaces identical. A locknut has a thickness equal to .5*D*, where *D* is the major diameter of the nut's thread size.

To draw a hexagon-shaped finished nut

Draw a hexagon-shaped finished nut for an M36 thread. See Figure 11-27.

1. Set **Limits = 297,210**
   Grid = 10
   Snap = 2.5
2. Construct a vertical line that is intersected by two horizontal lines **32** millimeters apart.

The thickness of a finished nut is .88*D*. In this example, .88(36) = 31.68, or 32.

3. Use **Block, Insert** and insert the HEXHEAD wblock created in Section 11-15.

The wblock HEXHEAD was created for an M24 thread, so the X and Y scale factors must be increased to accommodate the larger thread size. The scale factor is determined by dividing the desired size by the wblock size. In this example, 36/24 = 1.5. The scale factor is **1.5.** Respond to the command prompts as follows:

Command: INSERT

The **Insert** dialog box will appear.

4. Type or select **HEXHEAD;** press **Enter.**

5. Select the indicated insertion point.
6. Specify an X and Y scale of **1.5** and a **0** rotation angle; select **OK.**

   Specify insertion point or [Scale/X/Y/Z/Rotate/ PScale/PX/PY/PZ/PRotate]:

7. Press **Enter.**

   If you have not created a wblock, refer to Section 11-15 and draw a hexagon-shaped head.

8. Explode the wblock.

9. Move the front view of the nut so that the top surface aligns with the top parallel horizontal line.
10. Erase the bottom line of the front view of the nut and extend the vertical line to the lower horizontal line. Erase the top horizontal line in the front view.
11. Use **Offset** to draw a horizontal line **2** millimeters below the bottom of the nut.

    This line defines the shoulder surface of the nut. Any offset distance may be used as long as the line is clearly

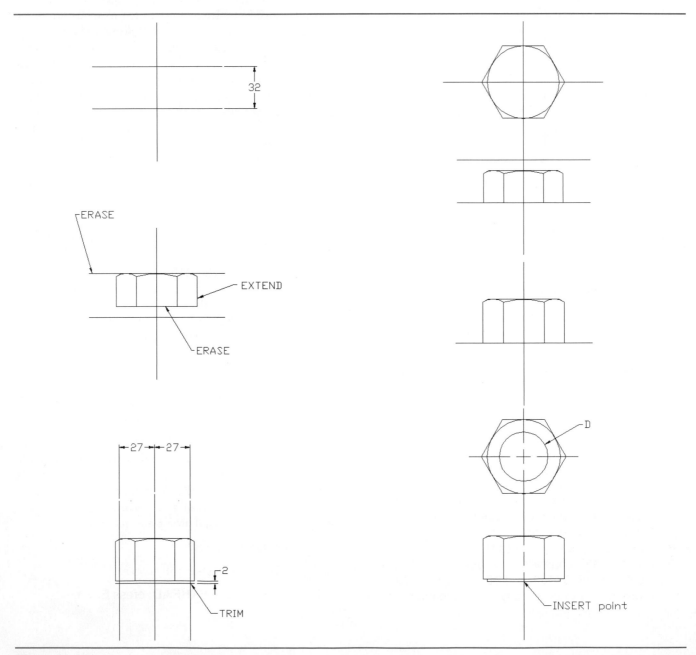

Figure 11-27

visible. The actual shoulder surface is less than 1 millimeter deep and would not appear clearly on the drawing.

The shoulder surface has a diameter equal to 1.5*D* [1.5(36) = 54] or, in this example, 54.

12. Offset the centerline **27** (half of 54) to each side and trim the excess lines.
13. Save the drawing as a wblock named **FNUTHEX**. Define the insertion point as shown.

To draw a locking nut

Draw a locking nut for an M24 thread. See Figure 11-28. The HEXHEAD wblock was originally drawn for an M24 thread, so no scale factor is needed.

1. Set **Limits = 297,210**
   Grid = 10
   Snap = 5
2. Draw a vertical line and two horizontal lines using the given dimensions.

The 6 distance is half the .5*D* [.5(24) = 12] thickness distance recommended for locknuts. Locknuts are symmetrical, so half the nut will be drawn and then mirrored.

3. Insert the HEXHEAD wblock at the indicated insert point.
4. Explode the wblock.
5. Move the front view of the hex head so that it aligns with the horizontal line as shown.
6. Trim and erase the lines that extend beyond the lower horizontal line.
7. Mirror the remaining portion of the front view about the lower horizontal line, then erase the horizontal line.
8. Draw a circle of diameter **D** in the top view as shown.
9. Save the drawing as a wblock named **LNUTHEX** using the indicated insertion point.

The procedure explained previously for hexagon-shaped finish nuts and locknuts is the same for square-shaped nuts. Use the wblock SQHEAD in place of the HEXHEAD wblock.

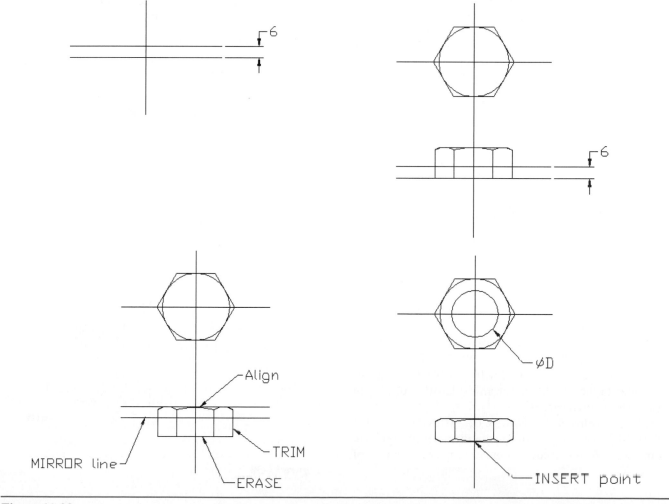

Figure 11-28

11-17 SAMPLE PROBLEM SP11-1

Draw and specify the mimimum threaded hole depth and pilot hole depth for an M12 × 1.75 × 50 hex head screw. Assemble the screw into the object shown in Figure 11-29. Use the schematic thread representation and a sectional view.

The SCHMMM wblock was originally drawn for an M20 thread, so a scale factor is needed. The scale factor to reduce an M20 diameter to an M12 is 12/20 = .6.

1. Insert the SCHMMM wblock and insert it at the indicated point. Use a **.6** X and Y scale factor.
2. Explode the block and trim any excess threads.
3. Insert the HEXHEAD wblock as indicated. The HEXHEAD wblock was also created for an M20 thread, so the scale factor is again **.6.**

The thread pitch equals 1.75, so $2P = 3.5$. This means that the threaded hole should be at least $50 + 3.5$, or 53.5, deep to allow for two unused threads beyond the end of the screw.

The unused portion of the pilot hole should also extend $2P$ beyond the end of the threaded portion of the hole, so the minimum pilot hole depth equals $53.5 + 3.5 = 57$.

The diameter of the tap hole for the thread is given in a table in the appendix as 10.3. For this example a diameter of 10 was used for drawing purposes. If required, the 10.3 value would be given in the hole's drawing callout.

4. Draw the unused portion of the thread hole and pilot hole as shown. Omit the thread representation in the portion of the threaded hole beyond the end of the screw for clarity.

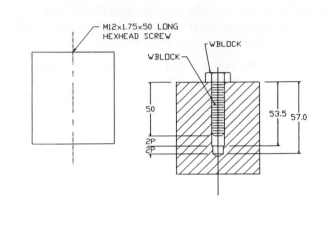

Figure 11-29

1. Insert the SCHMINCH wblock at the indicated point. Use an X and Y scale factor of **.75.** Rotate the block **−90°** to the correct orientation.

The wblock SCHMINCH is not long enough to go completely through the objects, so a second wblock is inserted below the first.

2. Insert the SQHEAD wblock at the indicated point. Use an X and Y scale factor of **.75.**

11-18 SAMPLE PROBLEM SP11-2

Determine the length of a .750–10UNC square head bolt needed to pass through objects 1 and 2 shown in Figure 11-30. Include a hexagon-shaped locknut on the end of the bolt. Allow at least two threads beyond the end of the nut. Use the closest standard bolt length and draw a schematic representation as a sectional view.

This construction is based on the wblocks SCHMINCH, HEXHEAD, and LNUTHEX created in Sections 11-5, 11-15, and 11-16. If no wblocks are available, refer to these sections for an explanation of how to create the required shapes.

The wblocks SCHMINCH, HEXHEAD, and LNUTHEX were created for a major diameter of 1.00 inch, so the scale factor needed to create a major diameter of .750 is .75.

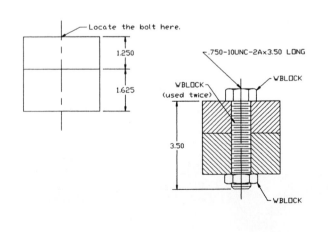

Figure 11-30

3. Insert the wblock LNUTHEX at the indicated point. Use an X and Y scale factor of **.75.**
4. Explode the wblocks.
5. Trim the excess lines within the nut.
6. Calculate the minimum length for the bolt.

The total depth of objects 1 and 2 equals 1.250 + 1.625 = 2.875.

The thickness of the nut equals .88D or .88(.75) = .375.

The pitch length of the thread equals 1.00/10 = .1, so 2P = .2.

The minimum bolt length equals 2.875 + .375 + .200 = 3.450.

From the table of standard bolt lengths in the appendix, the standard bolt length that is greater than 3.45 is 3.50, so the bolt callout can now be completed:

$$.750 - 10\text{UNC} - 2\text{A} \times 3.50 \text{ LONG}$$

The completed drawing with the appropriate bolt callout is shown in Figure 11-30.

11-19 STANDARD SCREWS

Figure 11-31 shows a group of standard screw shapes. The proportions given in Figure 11-30 are acceptable for general drawing purposes and represent average values. The exact dimensions for specific screws are available from manufacturers' catalogs. A partial listing of standard screw sizes is included in the appendix.

The given head shape dimensions are all in terms of D, the major diameter of the screw's thread. Information about the available standard major diameters and lengths is included in the appendix.

The choice of head shape is determined by the specific design requirements. For example, a flat head mounted flush with the top surface is a good choice when space is critical, when two parts butt against each other, or when aerodynamic considerations are involved. A round head can be assembled using a common blade screwdriver, but it is more susceptible to damage than a hex head. The hex head, however, requires a specific wrench for assembly.

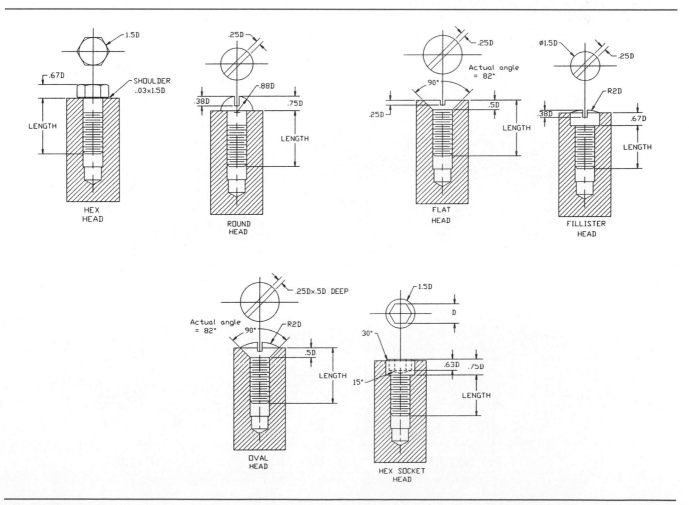

Figure 11-31

11-20 SETSCREWS

Setscrews are fasteners used to hold parts like gears and pulleys to rotating shafts or other objects to prevent slippage between the two objects. See Figure 11-32.

Most setscrews have recessed heads to help prevent interference with other parts. Many different head styles and point styles are available. See Figure 11-33.

Setscrews are referenced on a drawing using the following format.

THREAD SPECIFICATION
HEAD SPECIFICATION POINT SPECIFICA-
   TION
SET SCREW

.250 − 20UNC − 2A 1.00 LONG
SLOT HEAD FLAT POINT
SET SCREW

The words LONG, HEAD, and POINT are optional.

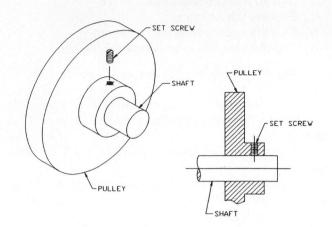

Figure 11-32

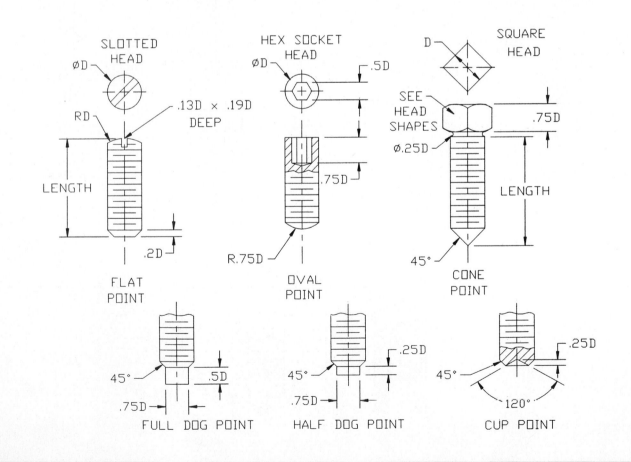

Figure 11-33

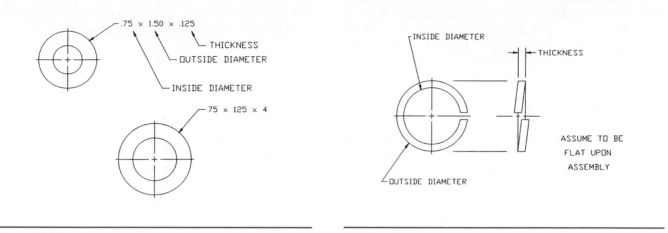

Figure 11-34

Figure 11-36

11-21 WASHERS

There are many different styles of washers available for different design applications. The three most common types of washers are *plain, lock,* and *star.* Plain washers can be used to help distribute the bearing load of a fastener or used as a spacer to help align and assemble objects. Lock and star washers help absorb vibrations and prevent fasteners from loosening prematurely. All washers are identified as follows:

Inside diameter × Outside diameter × Thickness

Examples of plain washers and their callouts are shown in Figure 11-34. A listing of standard washer sizes is included in the appendix.

To draw a plain washer (See Figure 11-35.)

1. Draw concentric circles for the inside and outside diameters.
2. Project lines from the circular view and draw a line that defines the width of the washer.
3. Use **Offset** to define the thickness.

Figure 11-36 shows two views of a lock washer. As the lock washer is compressed during assembly it tends to flatten, so the slanted end portions are not usually included on the drawing. The drawing callout should include the words LOCK WASHER.

Figure 11-37 shows an internal and an external tooth lock-type washer. They may also be called *star washers.*

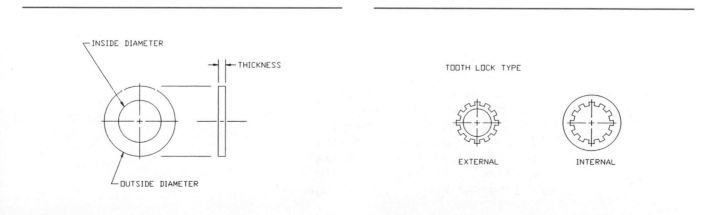

Figure 11-35

Figure 11-37

They are best drawn by first drawing an individual tooth, then using **Array** to draw a total of 12 teeth. Because there are 12 teeth, each tooth and the space between the tooth and the next tooth require a total of $30° - 15°$ for the tooth and 15° for the space between.

11-22 KEYS

Keys are used to help prevent slippage in power transmission between parts, for example, a gear and a drive shaft. Grooves called *keyways* are cut into both the gear and the drive shaft and a key is inserted between, as shown in Figure 11-38.

There are four common types of keys: *square, Pratt & Whitney, Woodruff,* and *gib head*. Each has design advantages and disadvantages. See Figure 11-39. A list of standard key sizes is included in the appendix.

Square keys are called out on a drawing by specifying the length of one side of the square cross section and the overall length; Pratt & Whitney and Woodruff keys are specified by numbers; and gib head keys are defined by a group of dimensions. See the appendix for the appropriate tables and charts of standard key sizes.

Keyways are dimensioned as shown in Figure 8-46. Note that the depth of a keyway in a shaft is dimensioned from the bottom of the shaft. Because material has been cut away, the intersection between the shaft's centerline and top outside edge does not exist. It is better to dimension from a real surface than from a theoretical one. The same dimensioning technique also applies to the gear.

11-23 RIVETS

Rivets are fasteners that hold adjoining or overlapping objects together. A rivet starts with a head at one end and a straight shaft at the other. The rivet is then inserted into the object, and the headless end is "bucked" or otherwise forced into place. A force is applied to the headless end that changes its shape so that another head is formed holding the objects together.

There are many different shapes and styles of rivets. Figure 11-40 shows five common head shapes for rivets. Aircraft use hollow rivets because they are very lightweight. A design advantage for using rivets is that they can be drilled out and removed or replaced without damaging the objects they hold together.

Rivet types are represented on technical drawings using a coding system. See Figure 11-41. The lines used to code a rivet must be clearly visible on the drawing. Because rivets are sometimes so small and the material they hold together so thin that it is difficult to clearly draw the rivets, companies draw only the rivet's centerline in the side view and identify the rivet using a drawing callout.

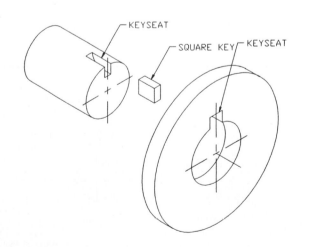

Figure 11-38

SQUARE

GIB HEAD

PRATT & WHITNEY

WOODRUFF

Figure 11-39

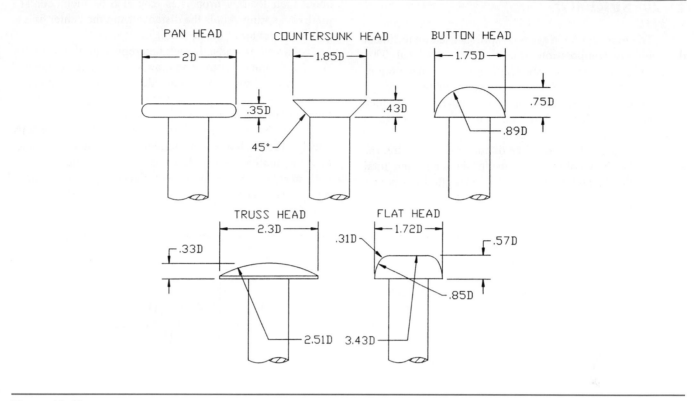

Figure 11-40

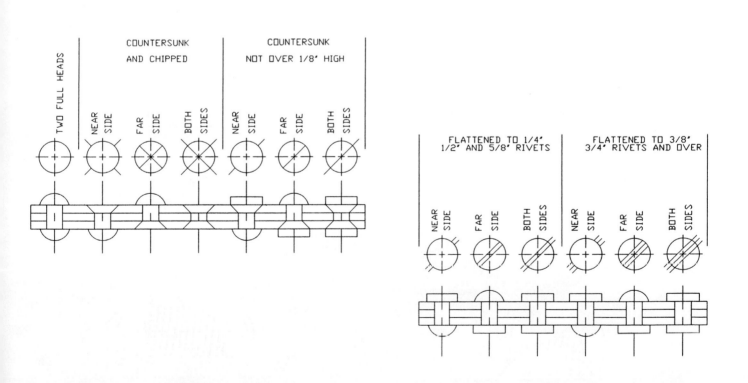

Figure 11-41

11-24 SPRINGS

The most common types of springs used on technical drawings are compression, extension, and torsional. This section explains how to draw detailed and schematic representations of springs. Springs are drawn in their relaxed position (not expanded or compressed). Phantom lines may be used to show several different positions for springs as they are expanded or compressed.

Springs are defined by the diameter of their wire, the direction of their coils, their outside diameter, their total number of coils, and their total relaxed length. Information about their loading properties may also be included. The pitch of a spring equals the distance from the center of one coil to the center of the next coil.

As with threads, the detailed representations are difficult and time consuming to draw, but the Array command shortens the drawing time considerably. Saving the representation as a wblock also helps avoid having to redraw the representation for each new drawing.

Ideally, the distance between coils should be exactly equal to the pitch of the spring, but this is not always practical for smaller springs. Any convenient distance between coils may be used that presents a clear, easy-to-understand representation of the spring.

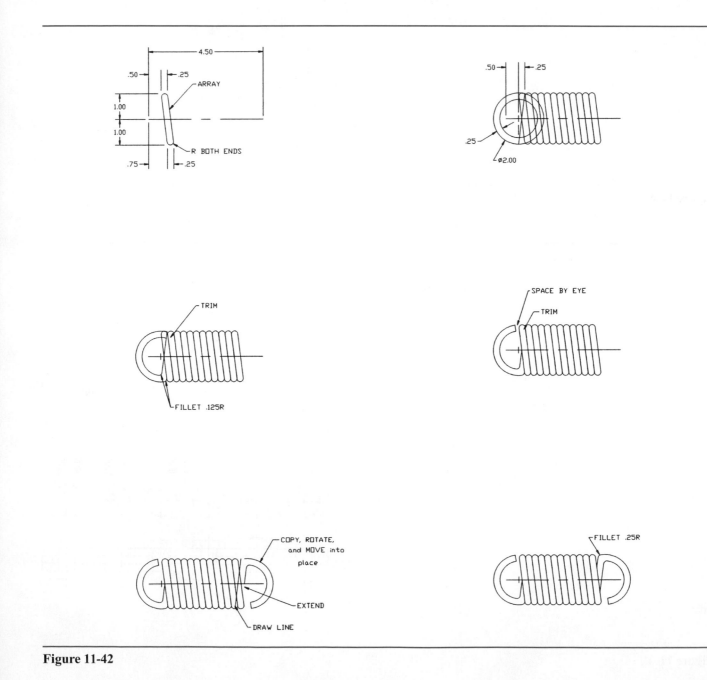

Figure 11-42

To draw a detailed representation of an extension spring

Draw an extension spring 2.00 inches in diameter with right-hand coils made from .250-inch wire. The spring's pitch equals .25, and the total length of the coils is 3.00 inches. See Figure 11-42.

Because the pitch of the spring equals the wire diameter, the coils will touch each other.

1. Set **Grid** = **.5**
   Snap = **.25**
2. Draw the rounded shape of one of the coils.

The distance between the lines equals the diameter of the spring wire. The diameter of the rounded ends also equals the diameter of the wire. The coil's offset equals the spring's pitch.

3. Array the coil shape.

A rectangular array was used with one row, and 12 columns **.25** apart.

4. Draw two concentric circles at the left end of the spring so that the larger circle's diameter equals the spring's diameter. The vertical centerline of the circles is tangent to the left edge of the first coil.
5. Draw a line across the thread offset a distance equal to one pitch. The endpoints of the line are on snap points.
6. Trim the excess portion of the circles and fillet the circles as shown.
7. Trim the excess portion of the slanted line and cut back the end of the circles so that there is a noticeable gap between the first coil and the end of the circles. The gap distance is determined by eye.
8. **Copy, Rotate,** and **Move** the circular end portion to the right end of the spring.
9. Draw a slanted line across the farthest right coil as shown.
10. Trim and extend the lines as necessary and add the **FILLET .25R** as shown.
11. Save the drawing as a wblock named **EXTSPRNG.**

To draw a detailed representation of a compression spring

Draw a compression spring with six right-hand coils made from .25-inch-diameter wire. The pitch of the spring equals .50 inch. The diameter of the spring is 2.00 inches. See Figure 11-43.

1. Set **Grid** = **.50**
   Snap = **.25**

2. Draw the rounded shape of the first left coil as shown. The diameter of the rounded ends of the coil equals the diameter of the spring's wire.
3. Array the coil shape.

In this example a rectangular array was used to draw one row, and six columns **.50** inch apart.

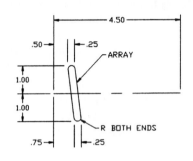

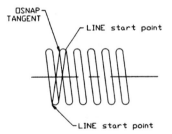

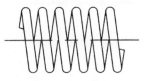

Figure 11-43

4. Draw two slanted lines as shown.

These lines represent the back portion of the coil. They are most easily drawn from a snap point next to the rounded portion of the coil shape (this point would be the corner point of the coil if the coil were not rounded) to a point tangent to the opposite rounded end of the coil.

5. Trim and array the slanted lines as shown.

Use a rectangular array with one row, and six columns **.50** inch apart.

6. Draw the right end of the spring so that it appears to end just short of the spring's centerline. Any convenient distance may be used.

7. Draw the left end of the spring using **Copy,** and copy one of the existing slanted lines. Use **Zoom,** if necessary, to align the copied line with the coil. Draw the left end of the spring so that it appears to end just short of the centerline.

8. Save the drawing as a wblock named **COMPSPNG.**

Figure 11-44 shows schematic representations of a compression and an extension spring. The distance between the peaks of the slanted lines should equal the pitch of the spring.

11-25 DESIGNCENTER

AutoCAD includes a **DesignCenter** that includes, among other things, a set of fastener blocks. (See Section 3-21 for an explanation of blocks.) The **DesignCenter** is accessed using the **DesignCenter** tool located on the **Standard** toolbar.

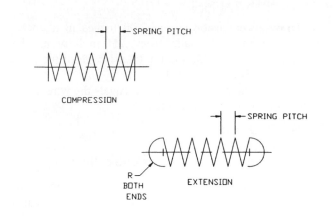

Figure 11-44

To access U.S. fastener blocks

1. Click the **DesignCenter** tool on the **Standard** toolbar.

The **DesignCenter** dialog box will appear. See Figure 11-45.

2. Select the **Blocks** tool (double-click) under the **Fasteners - US.dwg** heading.

A grouping of available blocks will appear in the dialog box.

3. Double-click the **Hex Bolt 1/2 in. -side** option in the display.

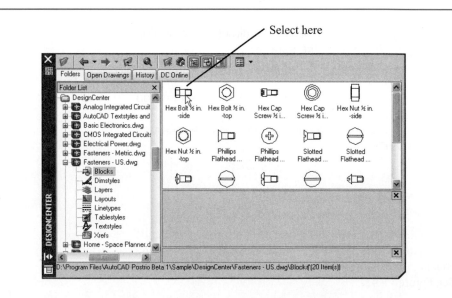

Figure 11-45

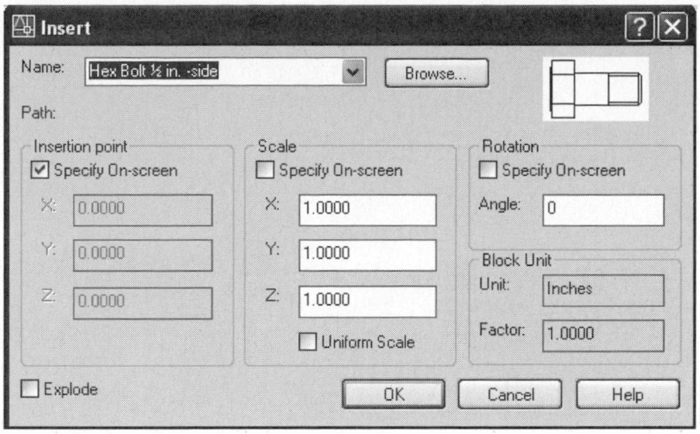

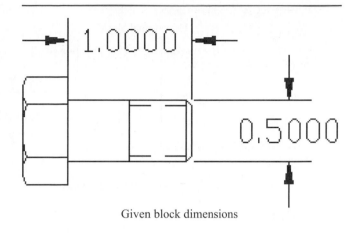

The **Insert** dialog box will appear. See Figure 11-46.

4. Accept the default scale values and click **OK**.

A side view of a hex bolt will appear on the screen. See Figure 11-47.

To change the size of a fastener block

The available blocks are common sizes and can easily be modified to different sizes using the **Scale** options on the **Insert** dialog box. Figure 11-48 shows the dimension values for the **Hex Bolt** block. The dimensions were created using the **Linear** dimension tool. The dimensions for any block can be determined by first inserting a drawing of the block and then using the **Dimension** tool.

To create a hex bolt that has a .75-inch diameter and a length of 2.00 inches

1. Select the **Hex Bolt 1/2 in. -side** block from the **DesignCenter**.

The **Insert** dialog box will appear. See Figure 11-49.

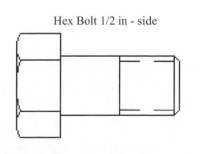

Hex Bolt 1/2 in - side

Figure 11-47

Given block dimensions

Figure 11-48

2. Set the **X** scale factor for **2.0000** and the **Y** scale factor for **1.5000,** then click **OK**.

Figure 11-50 shows the resulting fastener drawing. The dimensions were added for demonstration purposes. Dimensions will not appear on the initial drawing.

To use the Scale tool

See Figure 11-49.

1. Click the **Scale** tool located on the **Modify** toolbar.

    Command: _scale
    Select objects:

2. Window the fastener; press the right mouse button.

    Specify base point.

3. Select a base point.

    The base point need not be on the fastener.

    Specify scale factor or [Copy/Reference] <1.0000>:

4. Type **2.50**; press **Enter**.

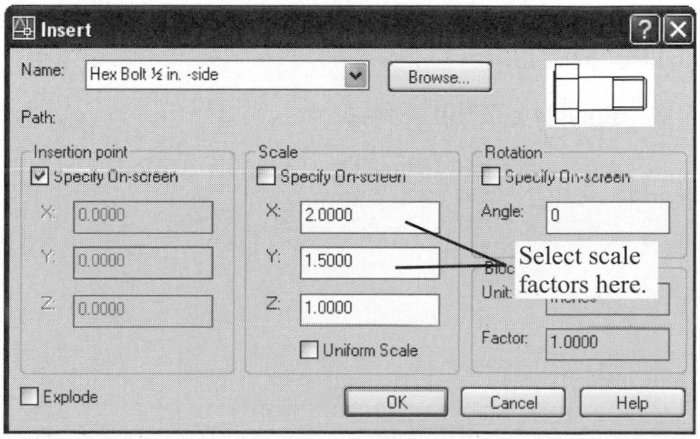

Figure 11-49

The original block

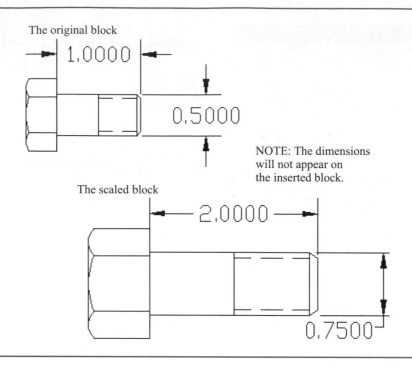

NOTE: The dimensions will not appear on the inserted block.

The scaled block

Figure 11-50

The scale tool can be used to increase the size of an object, but the object's proportions remain the same.

To change a fastener's proportions

Figure 11-51 shows a Hex Bolt ¹/₂-in. side view taken from the **DesignCenter.** The original proportions are shown in Figure 11-48. Draw a bolt that has a ⌀0.375 inch and a length of 2.00 inches.

1. Click the **Scale** tool on the **Modify** toolbar.

The **Scale** tool is used because the proportions of a fastener's head are generally related to its major diameter. The given major diameter is 0.5000 inch. Scaling this dimension down to 0.375 will also reduce the proportions of the fastener's head.

2. Specify a **0.750** scale factor.

This factor will reduce the fastener's major diameter to 0.375.

3. Use the **Explode** tool from the **Modify** toolbar and explode the fastener.

The fastener as presented from the **DesignCenter** is a block. It must be exploded so that individual lines may be modified.

4. Use the **Move** tool and move the right portion of the fastener 1.00 inch to the right as shown in Figure 11-52.

5. Use the **Extend** tool and extend lines from the fastener's head to the extended protion.

6. Use the **Move** and **Extend** tools to define any required thread length.

7. Save the fastener as a new block. See Figure 11-52.

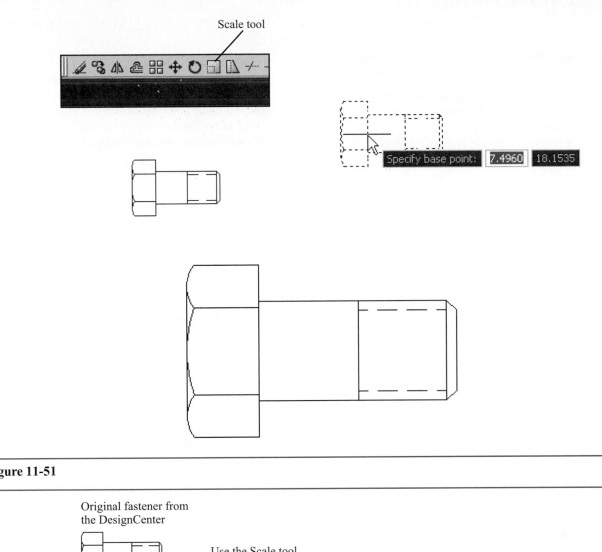

Figure 11-51

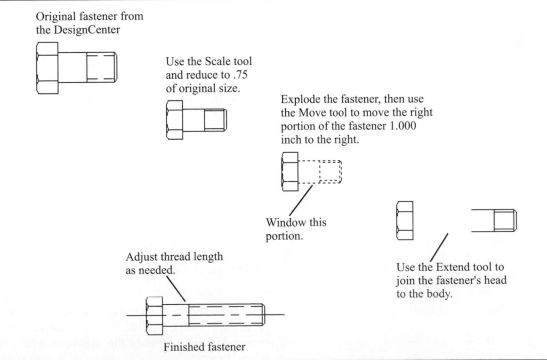

Figure 11-52

11-26 EXERCISE PROBLEMS

EX11-1

Create wblocks for the following thread major diameters.

A. 1.00-in. detailed representation
B. 1.00-in. schematic thread representation
C. 1.00-in. simplified thread representation
D. 100-mm detailed representation
E. 100-mm schematic thread representation
F. 100-mm simplified thread representation

EX11-2

Draw a .750–10UNC–2_ × ___ LONG thread. Include the thread callout.

A. External −3.00 LONG.
B. Internal to fit into the object shown in Figure EX11-2.
C. Use a detailed representation.
D. Use a schematic representation.
E. Use a simplified representation.

EX11-3

Draw a .250 − 28UNF − 2_ × ___ LONG thread. Include the thread callout.

A. External−1.50 LONG.
B. Internal to fit into the object shown in Figure EX11-2.
C. Use a detailed representation.
D. Use a schematic representation.
E. Use a simplified representation.

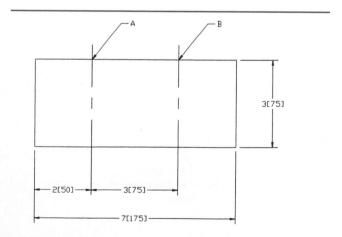

Figure EX11-2

EX11-4

Draw an M36 × 4 × ___ LONG thread. Include the thread callout.

A. External −100 LONG.
B. Internal to fit into the object shown in Figure EX11-2.
C. Use a detailed representation.
D. Use a schematic representation.
E. Use a simplified representation.

EX11-5

Draw an M12 × 1.75 × ___ LONG thread. Include the thread callout.

A. External −40 LONG.
B. Internal to fit into the object shown in Figure EX11-2.
C. Use a detailed representation.
D. Use a schematic representation.
E. Use a simplified representation.

EX11-6

Draw a 2.75 × 2 × 5.00 inch–long external square thread.

EX11-7

Draw a 50 × 2 × 100 millimeter–long external square thread.

EX11-8

Draw a 3 × 1.5 × 6 inch–long external acme thread.

EX11-9

Draw an 80 × 1.5 × 200 millimeter–long external acme thread.

Draw the bolts, nuts, and screws in Exercise Problems EX11-10 through EX11-13 assembled into the object shown in Figure EX11-2. If not given, determine the length of the fasteners using the tables given in this chapter or in the appendix. Include the appropriate drawing callout.

EX11-10

A. .500−13UNC × ____ LONG HEX HEAD BOLT. Include a finished nut on the end of the bolt.
B. 5/8(.625)−18UNF × 1.5 LONG HEX HEAD SCREW. Specify the diameter and length of the tap hole.
C. Draw a detailed representation.

D. Draw a schematic representation.
E. Draw a simplified representation.
F. Draw an orthographic view.
G. Draw a sectional view.

EX11-11

A. M24–____ LONG HEX HEAD BOLT. Include a finished nut on the end of the bolt.
B. M16–30 LONG HEX HEAD SCREW. Specify the diameter and length of the tap hole.
C. Draw a detailed representation.
D. Draw a schematic representation.
E. Draw a simplified representation.
F. Draw an orthographic view.
G. Draw a sectional view.

EX11-12

A. M30 × ____ LONG SQUARE HEAD BOLT. Include two locknuts on the end of the bolt.
B. M12 × 1.4 × 24 LONG SQUARE HEAD SCREW. Specify the diameter and length of the tap hole.
C. Draw a detailed representation.
D. Draw a schematic representation.
E. Draw a simplified representation.
F. Draw an orthographic view.
G. Draw a sectional view.

EX11-13

A. 1.25–7UNC × ____ LONG SQUARE HEAD BOLT. Include two locknuts on the end of the bolt.
B. .375–32UNEF × 1.5 LONG SQUARE HEAD SCREW. Specify the diameter and length of the tap hole.
C. Draw a detailed representation.
D. Draw a schematic representation.
E. Draw a simplified representation.
F. Draw an orthographic view.
G. Draw a sectional view.

EX11-14

Redraw the sectional view shown in Figure EX11-14. Include a drawing callout that defines the threads as M12. Specify the depth of the threaded hole and the diameter and depth of the tap hole.

EX11-15

Redraw the drawing shown in Figure EX11-15. Add a drawing callout for the bolt and nut based on the following thread major diameters. Use only standard bolt lengths as defined in the appendix.

A. .375–24UNF–2A × _____ LONG
B. M16 × 2 × _____ LONG

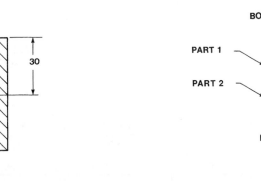

Figure EX11-14

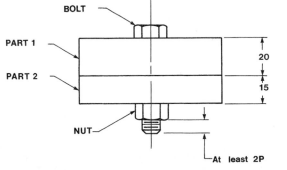

Figure EX11-15

EX11-16

Draw a front sectional view and a top orthographic view based on the following drawing and table information. Include the following setscrews in the indicated holes. Include the appropriate drawing callout for each setscrew.

FOR INCH VALUES:
1. .250–20UNC–2A × 1.00
   SLOT HEAD, FLAT POINT
   SET SCREW
2. .375–24UNF–2A × .750
   HEX SOCKET, FULL DOG
   SET SCREW
3. #10–28UNF–2A × .625
   SQUARE, OVAL
   SET SCREW
4. #6–32UNC–2A × .50
   SLOT, CONE POINT
   SET SCREW

FOR MILLIMETER VALUES:
1. M12 × 20
   SLOT HEAD, FLAT POINT
   SET SCREW
2. M16 × 2 × 30
   SQUARE HEAD, HALF DOG
   SET SCREW
3. M6 × 20
   HEX SOCKET, CONE
   SET SCREW
4. M10 × 1.5 × 20
   CUP, SLOT
   SET SCREW

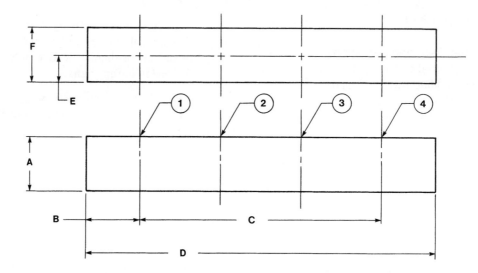

DIMENSION	INCHES	mm
A	2.00	50
B	1.00	25
C	3 X 1.00	3 X 40
D	6.5	170
E	.75	20
F	1.50	40

EX11-17

Draw a front sectional view and a top orthographic view based on the following drawing and table information.

A. Use inch values.
B. Use millimeter values.

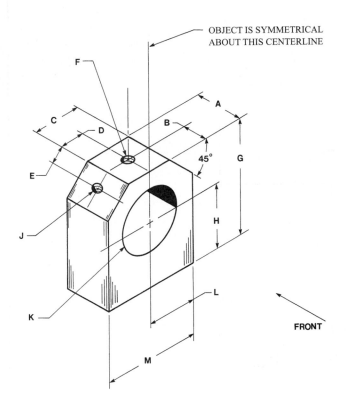

OBJECT IS SYMMETRICAL
ABOUT THIS CENTERLINE

FRONT

DIMENSION	INCHES	mm
A	1.00	26
B	.50	13
C	1.00	26
D	.50	13
E	.38	10
F	.190-32 UNF	M8 X 1
G	2.38	60
H	1.38	34
J	.164-36 UNF	M6
K	Ø1.25	Ø30
L	1.00	26
M	2.00	52

EX11-18

Redraw the following drawing as a sectional view. Select and include the drawing callouts for the nut and washer based on the following bolts. Specify the bolt lengths based on standard lengths listed in the appendix.

A. .625–11UNC–2A × _____ LONG HEX HEAD BOLT
B. M16 × 2 × _____ LONG HEX HEAD

C. .500-20 UNF × _____ LONG SQUARE HEAD
D. M20 × _____ SQUARE HEAD
E. .438-28 UNEF × _____ LONG HEX HEAD
F. M12 × 1.25 × _____ LONG HEX HEAD
G. .750-10 UNC × _____ LONG HEX HEAD
H. M24 × 3 × _____ LONG HEX HEAD

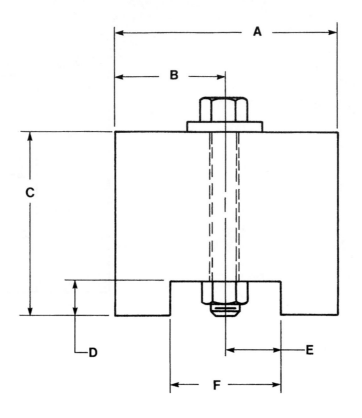

DIMENSION	INCHES	mm
A	3.00	76
B	1.50	38
C	2.50	64
D	.50	13
E	.75	19
F	1.50	38

EX11-19

Draw a front sectional view and a top orthographic view based on the following drawing and table information. Add fasteners to the labeled holes based on the following information. Include the appropriate drawing callouts. Use only standard sizes as listed in the appendix.

FOR INCH VALUES:
1. Nominal diameter = .250, UNC
   Square head bolt and nut
   A washer between the bolt head and part 23
   A washer between the nut and part 24
2. .375−24UNF−2A × 1.00
   SLOT, OVAL
   SETSCREW
3. Nominal diameter =.375, UNF
   Flat head screw, 1.25 LONG

FOR MILLIMETER VALUES:
1. Nominal diameter = 12, coarse
   Square head bolt and nut
   A washer between the bolt head and part 23
   A washer between the nut and part 24
2. M10 × .5 × 20
   SQUARE HEAD, FULL DOG
   SET SCREW
3. Nominal diameter = 12, fine
   Flat head screw, 20 LONG

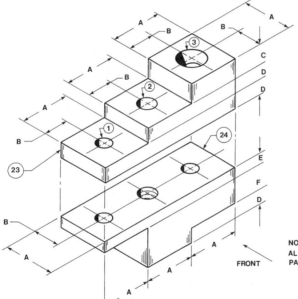

NOTE: HOLES IN PART 24 ALIGN WITH THOSE IN PART 23.

FRONT

DIMENSION	INCHES	mm
A	1.25	32
B	.63	16
C	.50	13
D	.38	10
E	.25	7
F	.63	16

EX11-20

Redraw the following sectional view and include the appropriate bolts and nuts at locations W, X, Y, and Z so that the selected bolt heads sit flat on their bearing surfaces. Select the bolts from the standard sizes listed in the appendix.

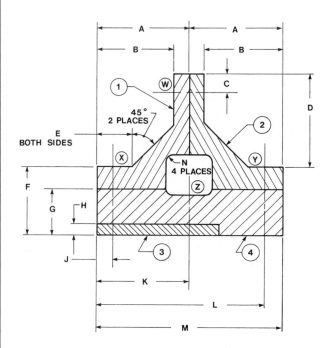

(W)(X)(Y)(Z) INDICATES FASTENER LOCATIONS

DIMENSION	INCHES	mm
A	1.00	50
B	1.63	41
C	.38	10
D	2.00	50
E	.75	19
F	1.50	38
G	1.00	25
H	.25	6
J	.38	10
K	2.00	50
L	3.63	92
M	4.00	100
N	R.13	3

EX11-21

Redraw the following drawing based on the following information. Include all bolt, nut, washer, and spring callouts. Use only standard sizes as listed in the appendix.

FOR INCH VALUES:

A. Major diameter of bolt = .375 coarse thread.
B. Add the appropriate nut.
C. Washer is .125 thick and allows a minimum clearance from the bolt of at least .125.
D. Compression springs are made from .125-diameter wire and are 1.00 long. They have an inside diameter that always clears the bolt by at least .125.
E. Parts 1 and 2 are 1.00 high and 4.00 wide.
F. Locate the bolts at least 1.00 from each end.

FOR MILLIMETER VALUES:

A. Major diameter of bolt = 12 coarse thread.
B. Add the appropriate nut.
C. Washer is 3 thick and allows a minimum clearance from the bolt of at least 2.
D. Compression springs are made from 4-diameter wire and are 24 long. They have an inside diameter that always clears the bolt by at least 3.
E. Parts 1 and 2 are 25 high and 100 wide.
F. Locate the bolts at least 25 from each end.

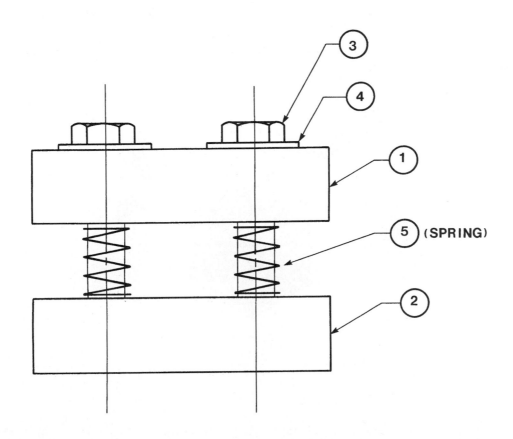

EX11-22

Figure EX11-22 shows two identical blocks that are to be held together using a bolt and a nut. Two washers are also used. Specify the appropriate fastener, nut, and washer for the following block sizes.

Block and nominal hole sizes:
A. 1.50 × 1.50 × 0.75 inches: Ø.375 inch
B. 20 × 20 × 12 mm; Ø8 mm
C. 2.00 × 2.00 × 0.625 inches; Ø.500 inch
D. 30 × 30 × 15 mm; Ø10 mm

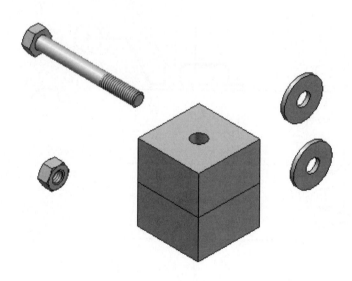

Figure EX11-22

EX11-23

Figure EX11-23 shows three parts that are to be held together using six screws. Given the part sizes, specify the six fasteners needed to assemble the parts.

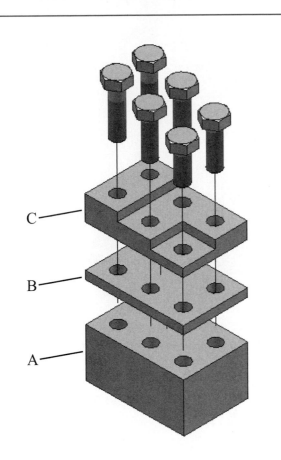

Figure EX11-23

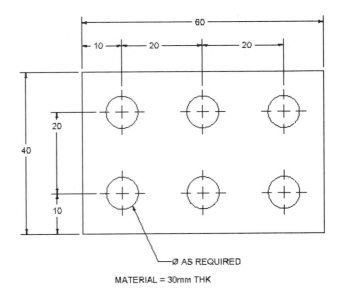

Ø AS REQUIRED

MATERIAL = 30mm THK

A.

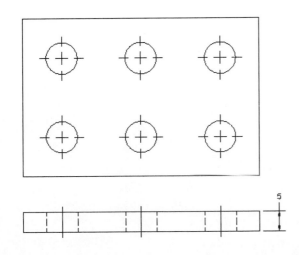

B.

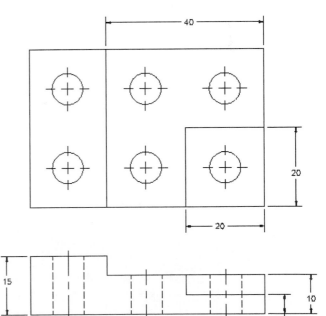

C.

CHAPTER 12

Working Drawings

12-1 INTRODUCTION

This chapter explains how to create assembly drawings, parts lists, and detail drawings. It includes guidelines for titles, revisions, tolerances, and release blocks. The chapter shows how to create a design layout, then use the layout to create assembly and detail drawings using the **Layer** command.

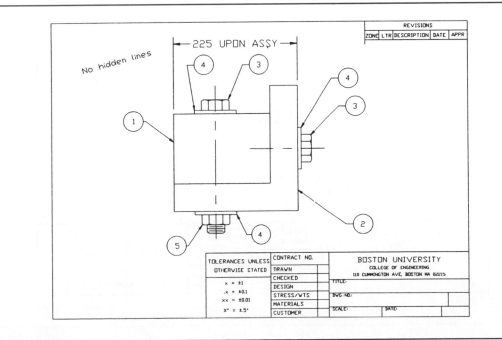

Figure 12-1

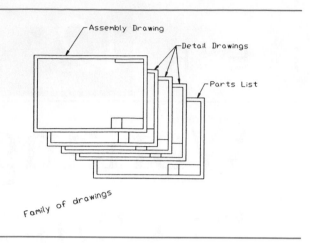

Figure 12-2

12-2 ASSEMBLY DRAWINGS

Assembly drawings show how objects fit together. See Figure 12-1. All information necessary to complete the assembly must be included on the drawing. This information may include specific assembly dimensions, torque require-

ments for bolts, finishing and shipping instructions, and any other appropriate company or customer specifications.

Assembly drawings are sometimes called *top drawings* because they are the first of a series of drawings used to define a group of parts that are to be assembled together. The group of drawings, referred to as a *family of drawings*, may include subassemblies, modification drawings, detail drawings, and a parts list. See Figure 12-2.

Assembly drawings do *not* contain hidden lines. A sectional view may be used to show internal areas critical to the assembly. Specific information about the internal surfaces of objects that make up the assembly can be found on the detail drawings of the individual objects. See Figure 12-3.

Each part of an assembly is identified by an assembly, or item, number. Assembly numbers are enclosed in a circle or ellipse, and a leader line is drawn between the assembly number and the part. Assembly numbers are unique to each assembly drawing; that is, a part that is used in several different assemblies may have a different assembly number on each assembly drawing.

If several of the same part are used in the same assembly, each part should be assigned the same assembly number. A leader line should be used to identify each part

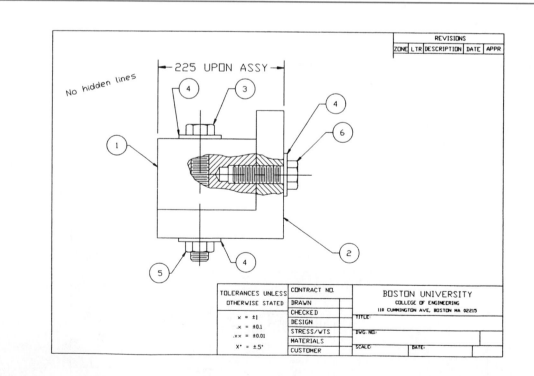

Figure 12-3

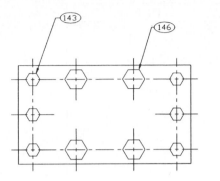

Figure 12-4

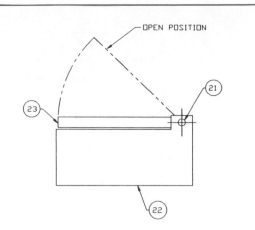

Figure 12-6

unless the differences between the parts are obvious. In Figure 12-4 the difference between head sizes for the fasteners is obvious, so all parts need not be identified.

Assemblies should be shown in their natural or neutral positions. Figure 12-5 shows a clamp in a slightly open position. It is best to show the clamp jaws partially open,

not fully closed or fully open. In general, a drawing should show an assembly in its most common position.

The range of motion for an assembly is shown using phantom lines. See Figure 12-6. Note how the range of motion for the hinged top piece is displayed using phantom lines.

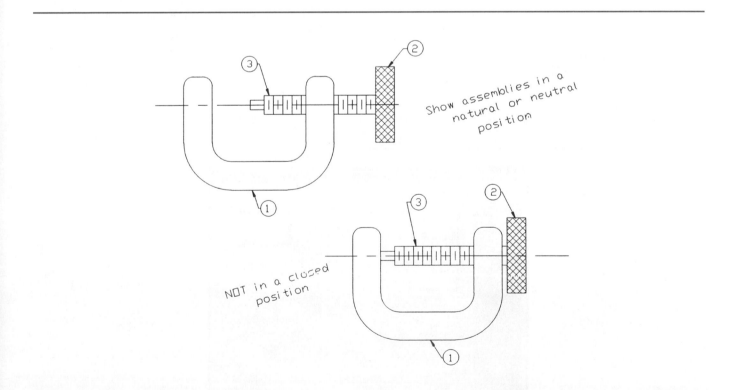

Figure 12-5

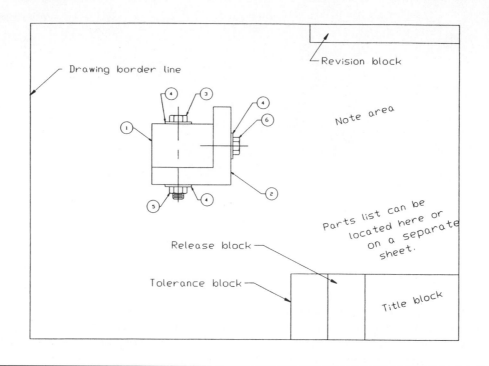

Figure 12-7

12-3 DRAWING FORMATS (TEMPLATES)

Figure 12-7 shows a general format for an assembly drawing. The format varies from company to company and with the size of the drawing paper.

AutoCAD includes many different drawing templates that conform to ANSI and ISO standards. Templates automatically include a drawing border, title block, revision block, and a partial release block.

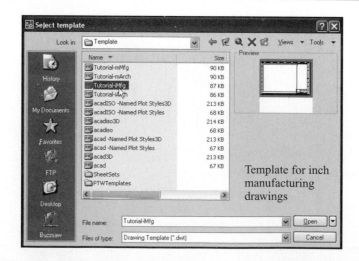

Figure 12-8

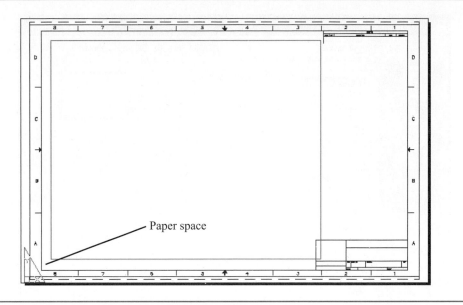

Paper space

Figure 12-9

To add a drawing template

1. Click the **New** tool.

The **Select template** dialog box will appear. See Figure 12-8.

2. Select the **Tutorial-iMfg** template; click the **Open** box.

The template will appear on the screen. See Figure 12-9. Drawings may be created on the template just as they are with a blank, no template, drawing screen.

Figure 12-10 shows a completed title block for an iMfg template. It was completed using the **Text** tool. A

listing of standard drawing sheet sizes is included in Section 1-8, Drawing Limits. An A-size drawing sheet is 8.5 × 11 inches. If an A-size template is selected, the screen units will automatically be fitted to the sheet size and will be in inches.

Standard metric drawing sheet sizes are also listed in Section 1-8. The closest equivalent to an A-size inch drawing is an A4-size drawing, which is 210 × 297 millimeters. Figure 12-11 shows the **Select template** dialog box set to select a **Tutorial-mMfg** template. Figure 12-12 shows the resulting template. The screen units will be in millimeters.

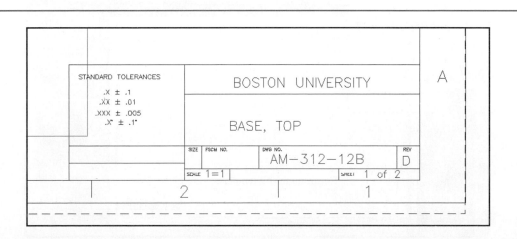

Figure 12-10

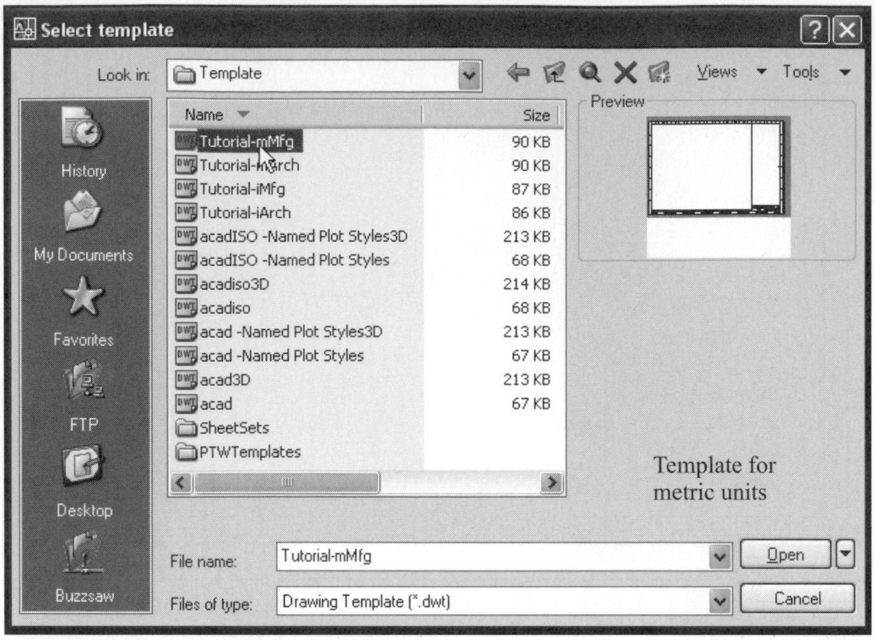

Figure 12-11

Template can also be accesed by clicking the **Insert** heading at the top of the screen and selecting **Layout,** then the **Layout from Template** option. See Figure 12-12.

12-4 TITLE BLOCK

Title blocks are located in the lower right corner of a drawing and include, at a minimum, the drawing's name and number, the company's name, the drawing scale, the release date of the drawing, and the sheet number of the drawing. Other information may be included. Figure 12-13 shows a sample title block created from an **iMfg Layout** template. The text was added using the **Multiline text** tool and positioned using the **Move** tool.

Figure 12-14 shows the title block from an mMfg template. The title block will appear with a series of XXX inputs. Use the **Explode** command and explode the title block.

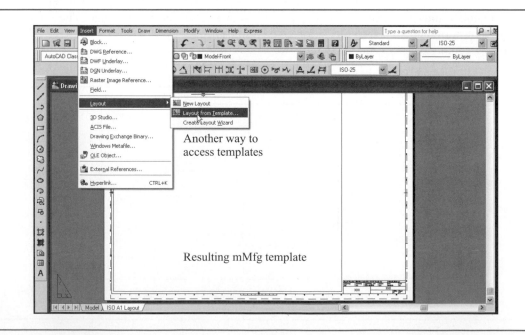

Figure 12-12

The XXX inputs will be replaced with words. Erase the unwanted words and replace them with the appropriate names and numbers.

Drawing titles (names)

Drawing titles should be chosen so that they clearly define the function of the part. They should be presented in the following word sequence:

Noun, modifier, modifying phrase

For example,

SHAFT, HIGH SPEED, LEFT HAND
GASKET, LOWER

Noun names may be two words if normal usage includes the two words.

GEAR BOX, COMPOSITE
SHOCK ABSORBER, LEFT

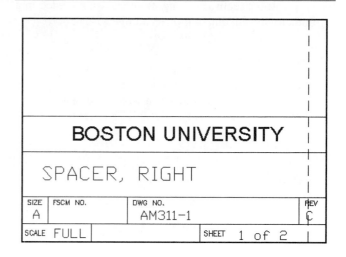

Figure 12-13

Drawing numbers

Drawing numbers are assigned by companies according to their usage requirement. The numbering system

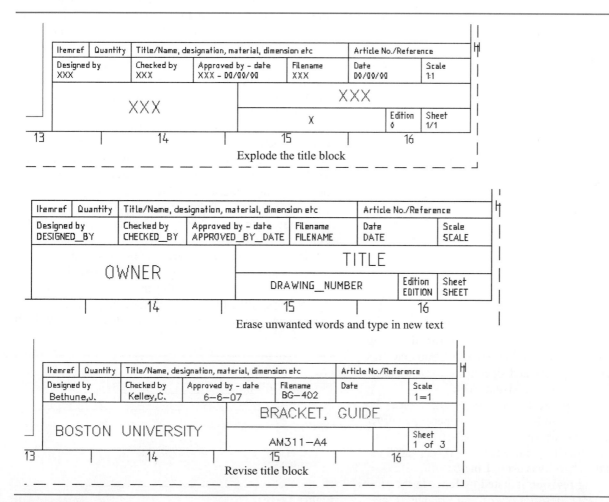

Figure 12-14

varies greatly. Usually, drawing numbers are recorded in a log book to prevent duplication of numbers.

Company name

The company's name and logo are preprinted on drawing paper or are included as a wblock so that they can be inserted on each drawing.

Scale

Define the scale of the drawing.

SCALE: FULL, or 1 = 1

The drawing size is the same as the object size.

SCALE: 2 = 1

The drawing size is twice as large as the object size.

SCALE: 1 = 2

The drawing size is half as large as the object size.

Release date

A drawing is released only after all persons required by the company's policy have reviewed it and added their signatures to the release block. Once released, a drawing becomes a legal document. Drawings that have not been officially released are often stamped with statements such as "NOT RELEASED" or "FOR REFERENCE ONLY."

Sheet

The number of the sheet relative to the total number of sheets that make up the drawing should be stated clearly.

SHEET 2 OF 3

or

SH 2 OF 3

12-5 REVISION BLOCK

Drawings used in industry are constantly being changed. Products are improved or corrected, and drawings must be changed to reflect and document these changes. Figure 12-15 shows a sample revision block.

Drawing changes are listed in the revision block by letter. Revision blocks are located in the upper right corner of the drawing. See Figure 12-7. Figure 12-16 shows a revision block from an mMfg template. The text was added using **Mtext,** and the chart lines using **Line.**

Each drawing revision is listed by letter in the revision block. A brief description of the change is also

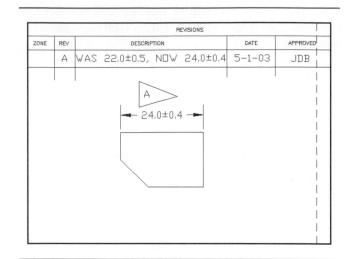

Figure 12-15

included. It is important that the description be as accurate and complete as possible. Revisions are often used to check drawing requirements on parts manufactured before the revisions were introduced.

The revision letter is also added to the field of the drawing in the area where the change was made. The letter is located within a "flag" to distinguish it from dimensions and drawing notes. The flag serves to notify anyone reading the drawing that revisions have been made.

Most companies have systems in place that allow engineers and designers to make quick changes to drawings. These change orders are called *engineering change orders* (ECOs), *change orders* (COs), or *engineering orders* (EOs), depending on the company's preference. Change orders are documented on special drawing sheets that are usually stapled to a print of the drawing. Figure 12-17 shows a change order attached to a drawing.

After a group of change orders accumulate, they are incorporated into the drawing. This process is called a *drawing revision,* which is different from a revision to the drawing. Drawing revisions are usually indicated by a letter located somewhere in the title block. The revision letters may be included as part of the drawing number or in a separate box in the title block. Whenever you are working on a drawing, make sure you have the latest revision and all

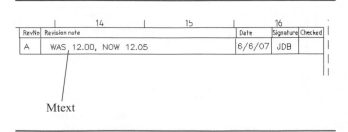

Mtext

Figure 12-16

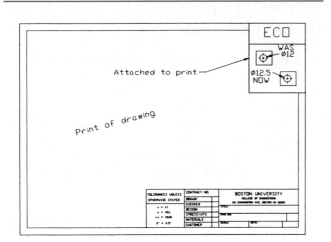

Figure 12-17

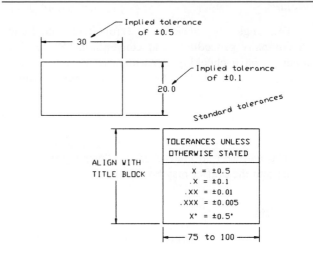

Figure 12-19

appropriate change orders. Companies have recording and referencing systems for listing all drawing revisions and drawing changes.

12-6 TOLERANCE BLOCK

Most drawings include a ***tolerance block*** next to the title block that lists the standard tolerances that apply to the dimensions on the drawing. A dimension that does not include a specific tolerance is assumed to have the appropriate standard tolerance.

Figure 12-18 shows a sample tolerance block for inch values, and Figure 12-19 shows a sample tolerance

block for millimeter values. In Figure 12-18 the dimension 2.00 has an implied tolerance of ±.01. In Figure 12-19 the 20.0 dimension has an implied tolerance of ±0.1.

12-7 RELEASE BLOCK

A ***release block*** contains a list of approval signatures or initials required before a drawing can be released for production. See Figure 12-20. The required signatures are generally as follows.

Drawn

Person that created the drawing.

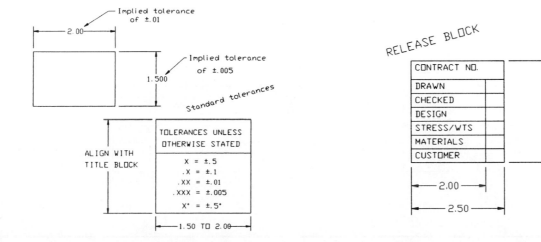

Figure 12-18

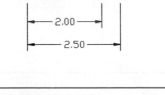

Figure 12-20

Checked

Drawings are checked for errors and compliance with company procedures and conventions. Some large companies have checking departments, whereas smaller companies have drawings checked by a senior person or the drafting supervisor.

Design

The engineer in charge of the design project. The designer and the drafter may be the same person.

Stress/Wts

The department or person responsible for the stress analysis of the design.

Materials

Usually, a person in the production department checks the design and makes sure that the necessary materials and the machine times for the design are available. This person may also schedule production time.

Customer

The customer for the design may have a representative on site at the production facility to check that its design requirements are being met. For example, it is not unusual for the Air Force to assign an officer to a plant that manufactures its fighter aircraft.

12-8 PARTS LIST (BILL OF MATERIALS)

A *parts list* is a listing of all parts used on an assembly. See Figure 12-21. The parts list was created using the **Table** tool. See Section 2-28. Parts lists may be located on an assembly drawing above the title block or on a separate sheet of paper. Assembly drawings done using AutoCAD often include the parts list on a separate layer within the drawing. Use only capital letters on a parts list.

Figure 12-22 shows two sample parts list formats, including dimensions. Parts list formats vary greatly from company to company. The dimensions are given in inches. They may be converted to millimeters using the conversion factor 1.00 inch = 25.4 millimeters.

A parts list serves as a way to cross-reference detail drawing numbers to assembly item numbers. It also provides a list of the materials needed for production and is very helpful for scheduling and materials purchasing.

Parts purchased from a vendor and used exactly as they are supplied, without any modification, will not have detail drawings but are included on the parts list. The washers, bolts, and screws listed on the parts list shown in Figure 12-21 would not have detail drawings. This means that the information on the parts list must be sufficient for a purchaser to know exactly what size washers, bolts, and screws to buy. The information used to define an object on the drawing, the drawing callout, is also used on the parts list. See Chapter 11 for an explanation of drawing callouts for fasteners.

PARTS LIST					
NO.	DESCRIPTION	PART NO.	MATL	NOTE	QTY
1	BLOCK, TOP	BU311S1	SAE 1020		1
2	BLOCK, BOTTOM	BU311S2	SAE 1020		1
3	M16×24 HEX HEAD BOLT	SPM16H	STEEL	▷2	1
4	20×40×3 PLAIN WASHER		STEEL		3
5	M16 HEX NUT	2PM16N	STEEL		1
6	M16×20 HEX HEAD SCREW		STEEL		1

Figure 12-21

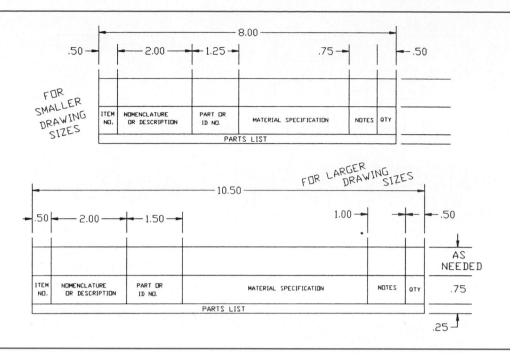

Figure 12-22

12-9 DETAIL DRAWINGS

A *detail drawing* is a drawing of a single part. The drawing should include all the information necessary to accurately manufacture the part, including orthographic views with all appropriate hidden lines, dimensions, toler-ances, material requirements, and any special manufacturing requirements. Figure 12-23 shows a sample detail drawing.

Detail drawings include title, release, tolerance, and revision blocks located on the drawing in the same places as they are found on assembly drawings.

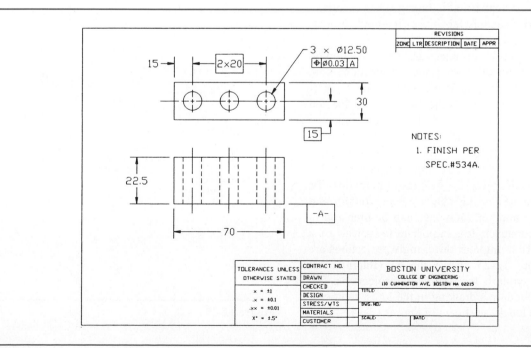

Figure 12-23

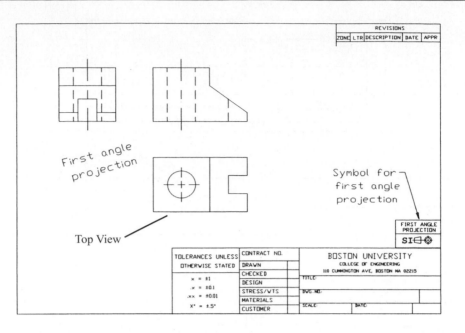

Figure 12-24

12-10 FIRST-ANGLE PROJECTION

The instructions for the creation of orthographic views as presented in this book are based on third-angle projection. See Chapter 5. Third-angle projection is used in the United States, Canada, and Great Britain, among other countries. Many other countries such as Japan use first-angle projection to create orthographic views.

Figure 12-24 shows an object and the orthographic views of the object created in first-angle projection. Note that the top view in the first-angle projection is the same as the top view in the third-angle projection, but it is located below the front view, not above the front view, as in the third-angle projection. Right-side views are also the same but are located to the left of the front view in first-angle projections and to the right in third-angle projections.

Many companies now do business internationally. They can have manufacturing plants in one country and assembly plants in another. Drawings can be prepared in several different countries. It is important to indicate on a drawing whether first-angle or third-angle projections are being used. Figure 12-25 shows the SI (International System of Units) symbols for first-angle projections. The SI symbol is included on a drawing in the lower right corner above the title block, as shown in Figure 12-24.

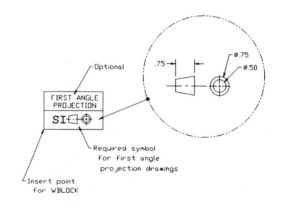

Figure 12-25

12-11 DRAWING NOTES

Drawing notes are used to provide manufacturing information that is not visual, for example, finishing instructions, torque requirements for bolts, and shipping instructions.

Drawing notes are usually located above the title block on the right side of the drawing. See Figure 12-26. Drawing notes are listed by number. If a note applies to a specific part of the drawing, the note number is enclosed in a triangle. The note numbers enclosed in triangles are also drawn next to the corresponding areas of the drawing.

12-12 DESIGN LAYOUTS

A design layout is not a drawing. It is like a visual calculation sheet used to size and locate parts as a design is developed. A design layout allows you to "build the assembly" on paper.

When drawings are created on a drawing board an initial layout is made locating and sizing the parts. Then the individual detail drawings and the assembly drawing are traced from the layout.

The same procedure can be followed using AutoCAD. First, create a design layout on one layer, then either transfer the individual parts and assembly drawing to other layers or create a new drawing from the layout drawing using the **Save As** command.

The sample problem in the next section demonstrates how to create a design layout and then use the layout to create the required detail and assembly drawings.

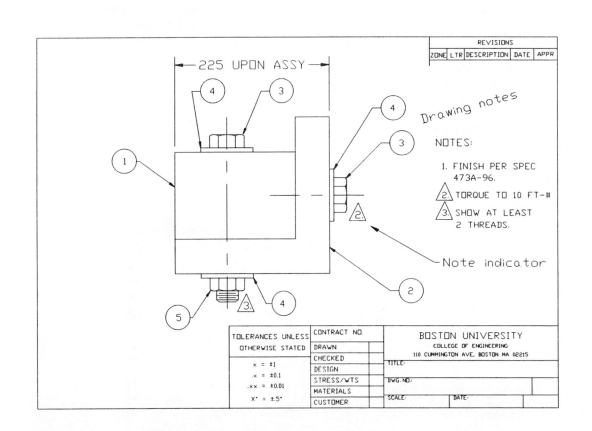

Figure 12-26

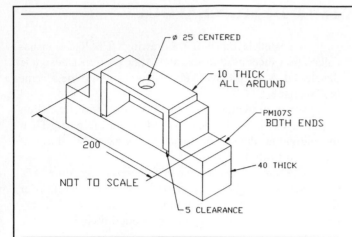

Figure 12-27

12-13 SAMPLE PROBLEM SP12-1

Figure 12-27 shows an engineer's sketch for a design problem. Prepare an assembly drawing and a detail drawing of the PN123 BASE and PN124 CENTER PLATE based on the following information.

1. Parts PM107S END BRACKET are existing parts and can be used as is. Figure 12-28 shows a detail drawing of the part.
2. Place the mounting holes for the two END BRACKETS 200 millimeters apart, as shown in the engineer's sketch.
3. Establish the size of the BASE so that it aligns with the END BRACKETS and is 40 millimeters thick.

4. Determine the sizes for the CENTER BRACKET so that it just fits between the end plates and has a 25-millimeter-diameter hole centered in its top surface. The CENTER BRACKET should extend 10 millimeters above the END BRACKETS and 5 millimeters above the BASE. The CENTER BRACKET should be 10 millimeters thick all around.
5. Mount the END BRACKETS to the BASE using M12 FLAT HEAD screws, and attach the END BRACKETS to the CENTER PLATE using M20 bolts with appropriate nuts.
6. Use only standard length fasteners.

To create the design layout (See Figure 12-29.)

1. Draw horizontal and vertical lines that define the top of the BASE and the distance between the mounting holes in the END BRACKETS.
2. Draw front and top views of the END BRACKETS using the vertical lines drawn in step 1. The vertical lines should align with the centerlines of the mounting holes.
3. Draw the BASE 40 millimeters thick.
4. Size the CENTER BRACKET.

Note that at this stage of the layout all lines are drawn using the same pattern. Lines can be changed to hidden lines or centerlines when the detail and assembly drawings are created.

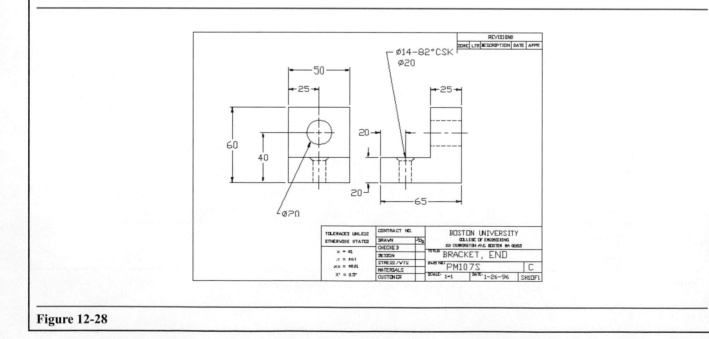

Figure 12-28

5. Add the M12 FLAT HEAD screws. A length of 20 millimeters was chosen from the tables in the appendix. The threaded holes in the BASE are M12 to match the screws. The tap drill size is included on the layout for reference.
6. Add the M20 × 60 bolts and nuts.

The bolt length is determined by adding 25 (END BRACKET) + 10 (CENTER BRACKET) + 15 (nut thickness = .75D) + 2 thread pitches = 52 millimeters. The next largest standard bolt length, as listed in the appendix, is 60, so an M20 × 60 bolt is specified.

7. The diameter of the clearance holes was chosen as 14 and 22 and is listed on the layout.

The design layout is complete. It may be used to create the required assembly drawing and detail drawings in one of two ways: using the **Layer** command and including all the drawings under one file name, or by using the **Save As** command to create separate drawings.

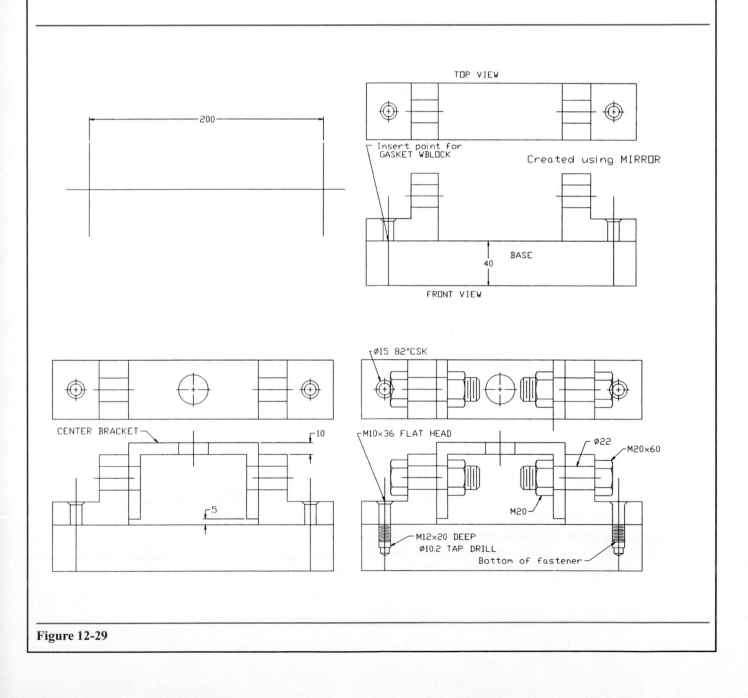

Figure 12-29

To create a drawing using layers

When lines are moved from one layer to another the lines will disappear from the original layer. This will, in essence, erase the layout drawing because all the lines will have been moved to other layers. The layout may be preserved by first making a copy of the layout, or that part of the layout you wish to transfer, then using the copy for the transfers.

To make a copy of the layout, use the **Copy** command. Select the entire screen and select a base point. Select the exact same point as the second point of displacement, and AutoCAD will copy the layout exactly over itself. You can then move the appropriate lines to different layers.

Note: Explode all blocks before transferring them between layers. A block cannot be edited once it is moved from its original layer.

Create a detail drawing of the BASE by transferring the BASE from the design layout. Respond to the prompts as follows.

1. Copy the layout onto itself.

   COMMAND:_Copy
   Select objects:

2. Window the entire layout; press **Enter.**

   <Base point or displacement>/Multiple:

3. Select a point on the screen.

   Second point of displacement:

4. Select exactly the same point as that used in the previous step.
5. Create the following layers. See Section 3-24.

   ASSEMBLY
   END
   END-DIM
   BASE
   BASE-DIM
   CENTER
   CENTER-DIM
   PARTSLIST

The "DIM" layers will be used for the dimensions for the detail drawings. This will allow the dimensions to be shut off if they are not needed.

6. Transfer the BASE to a layer called **BASE**. See Section 3-24. Turn off all the other layers. The resulting transfer should look like Figure 12-30.

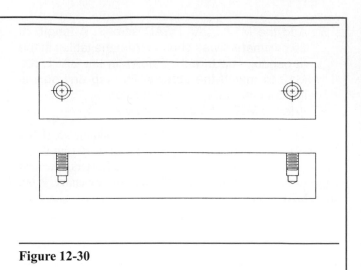

Figure 12-30

It will be difficult to make a perfectly clean transfer, that is, only the lines of the BASE. If other lines appear on the transferred BASE, erase and trim them as necessary.

7. Move the views closer together, if necessary.
8. Turn on the **BASE-DIM** layer and make it the current layer. Add the appropriate dimensions, tolerances, and callouts.
9. Create the drawing format and add the necessary information (drawing title, etc.).

The title block, and so forth, may be saved as a block or as part of a prototype drawing. Figure 12-31 shows the resulting drawing.

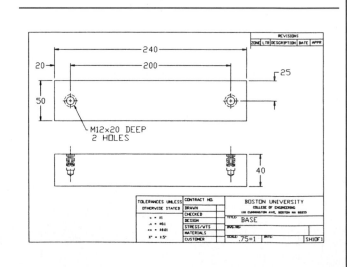

Figure 12-31

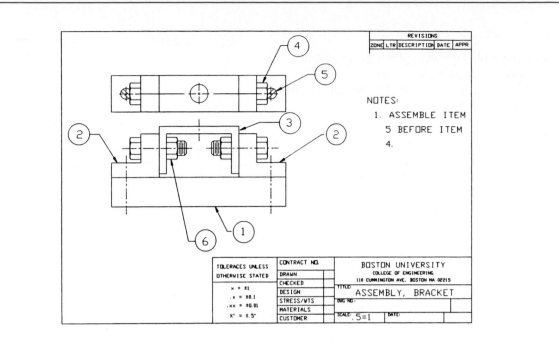

NOTES:
1. ASSEMBLE ITEM 5 BEFORE ITEM 4.

TOLERACES UNLESS OTHERWISE STATED	CONTRACT NO.		BOSTON UNIVERSITY
	DRAWN		COLLEGE OF ENGINEERING
	CHECKED		110 CUMMINGTON AVE, BOSTON MA 02215
x = ±1	DESIGN		TITLE: ASSEMBLY, BRACKET
.x = ±0.1	STRESS/VTS		DWG NO.:
.xx = ±0.01	MATERIALS		
X° = ±.5°	CUSTOMER		SCALE: .5=1 DATE:

Figure 12-32

To create a drawing from a layout

Figure 12-32 shows an assembly drawing that was created from the design layout. The procedure is as follows.

1. Save the design layout.
2. Create the assembly drawing from the layout drawing, but use the **Save As** command to save it under a different drawing name.
3. Add the appropriate drawing format and notes.
4. Save the assembly drawing using its new name.

Figure 12-33 shows a parts list for the assembly.

PARTS LIST			
ITEM NO	DESCRIPTION	MATL	QTY
1	BASE	STEEL	1
2	END BRACKET - PM107S	STEEL	2
3	CENTER PLATE	STEEL	1
4	M20x60 HEX HEAD BOLT	STEEL	2
5	M12x36 FLAT HEAD SCREW	STEEL	2
6	M20 NUT	STEEL	2

Figure 12-33

12-14 EXERCISE PROBLEMS

EX12-1

A. Draw an assembly drawing of the given objects.
B. Draw detail drawings for all nonstandard parts.
C. Prepare a parts list. Specify the length for the M10 hex head screw.

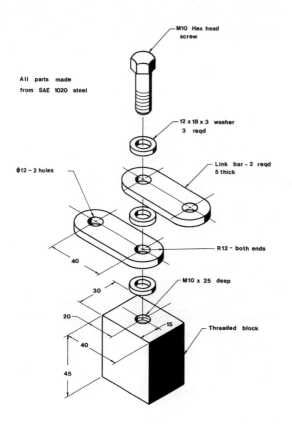

EX12-2

A. Draw an assembly drawing of the given objects.
B. Draw detail drawings for all nonstandard parts.
C. Prepare a parts list.

EX12-3

A. Draw an assembly drawing of the given objects.
B. Draw detail drawings for all nonstandard parts.
C. Prepare a parts list.

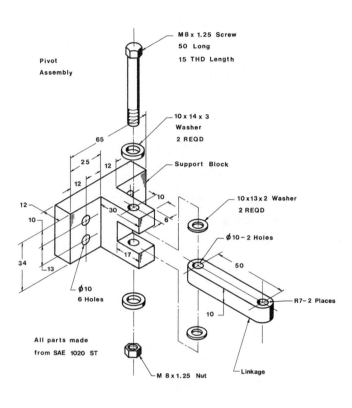

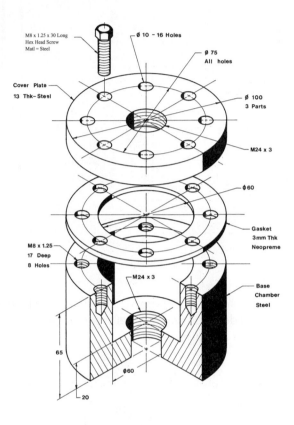

EX12-4

A. Draw an assembly drawing of the given objects.
B. Draw detail drawings for all nonstandard parts.
C. Prepare a parts list.

DESIGN

D. Replace part 3 with a bolt and appropriate nut. Add four washers.

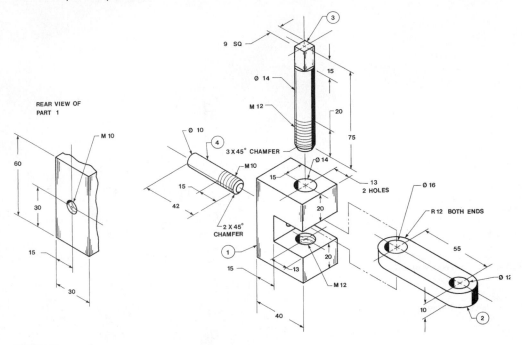

EX12-5

A. Draw an assembly drawing of the given objects.
B. Draw detail drawings for all nonstandard parts.
C. Prepare a parts list.

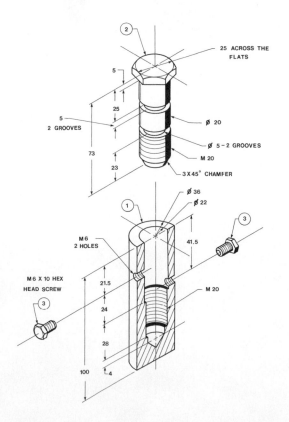

EX12-6

A. Draw an assembly drawing of the given objects.
B. Draw detail drawings for all nonstandard parts.
C. Prepare a parts list.

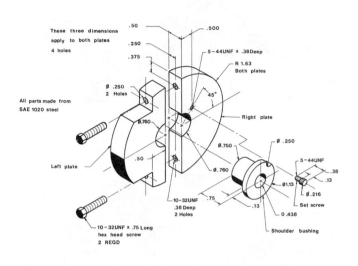

EX12-7

A. Draw an assembly drawing of the given objects.
B. Draw detail drawings for all nonstandard parts.
C. Prepare a parts list.

DESIGN

D. Replace the rivets with the appropriate screws and nuts.

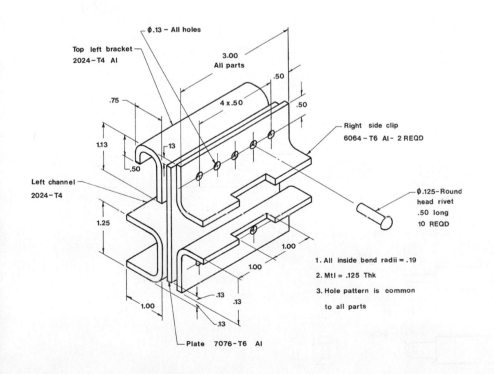

EX12-8 MILLIMETERS

A. Draw an assembly drawing of the given objects.
B. Prepare a parts list.

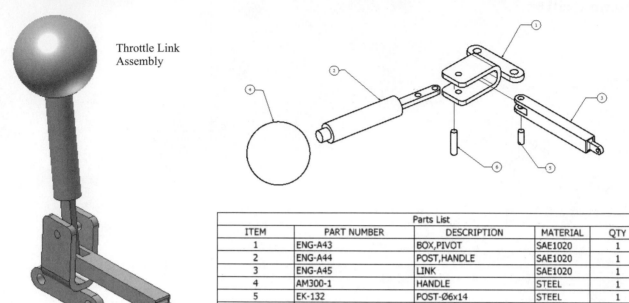

Throttle Link
Assembly

Parts List				
ITEM	PART NUMBER	DESCRIPTION	MATERIAL	QTY
1	ENG-A43	BOX,PIVOT	SAE1020	1
2	ENG-A44	POST,HANDLE	SAE1020	1
3	ENG-A45	LINK	SAE1020	1
4	AM300-1	HANDLE	STEEL	1
5	EK-132	POST-Ø6x14	STEEL	1
6	EK-131	POST-Ø6x26	STEEL	1

Pivot Box

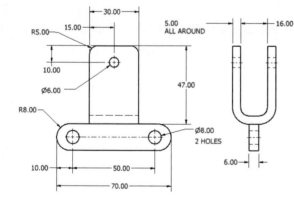

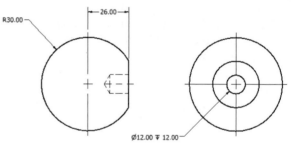

Handle Post

Link

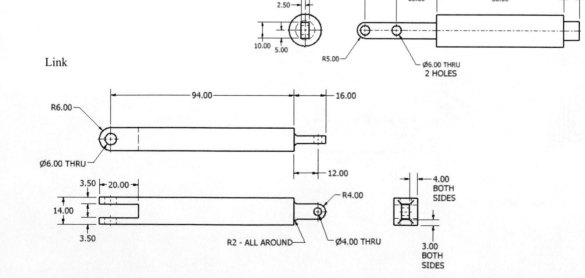

EX12-9 INCHES

A. Draw an assembly drawing and parts list for the Guide Assembly.

Adjustable Assembly

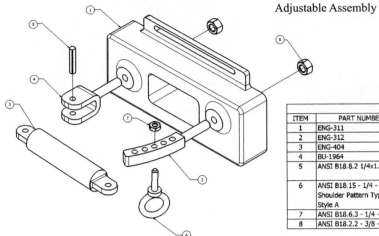

Parts List

ITEM	PART NUMBER	DESCRIPTION	MATERIAL	QTY
1	ENG-311	BASE #4, CAST	Cast Iron	1
2	ENG-312	SUPPORT, ROUNDED	SAE 1040 STEEL	1
3	ENG-404	POST, ADJUSTABLE	Steel, Mild	1
4	BU-1964	YOKE	Cast Iron	1
5	ANSI B18.8.2 1/4x1.3120	Grooved pin, Type C - 1/4x1.312 ANSI B18.8.2	Steel, Mild	1
6	ANSI B18.15 - 1/4 - 20. Shoulder Pattern Type 2 - Style A	Forged Eyebolt	Steel, Mild	1
7	ANSI B18.6.3 - 1/4 - 20	Hex Machine Screw Nut	Steel, Mild	1
8	ANSI B18.2.2 - 3/8 - 16	Hex Nut	Steel, Mild	2

Cast Base #4

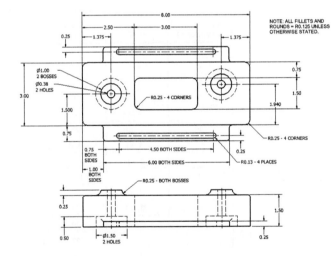

NOTE: ALL FILLETS AND ROUNDS = R0.125 UNLESS OTHERWISE STATED.

Rounded Support

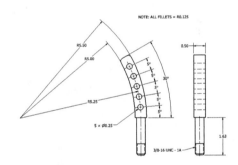

NOTE: ALL FILLETS = R0.125

Adjustable Post

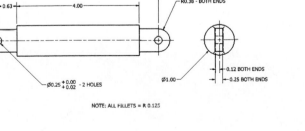

NOTE: ALL FILLETS = R 0.125

Yoke

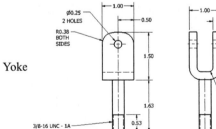

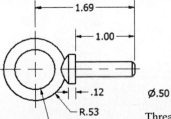

Forged Eyebolt

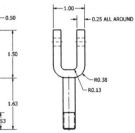

Thread = 1/4 = 20 UNC
0.03 x 45° Chamfer

EX12-10

Redraw the given objects as an assembly drawing and add bolts with the appropriate nuts at the L and H holes. Add the appropriate drawing callouts. Specify standard bolt lengths.

A. Use inch values.
B. Use millimeter values.
C. Draw the front assembly view using a sectional view.
D. Draw the fasteners using schematic representations.
E. Draw the fasteners using simplified representations.
F. Prepare a parts list.
G. Prepare detail drawings for each part.

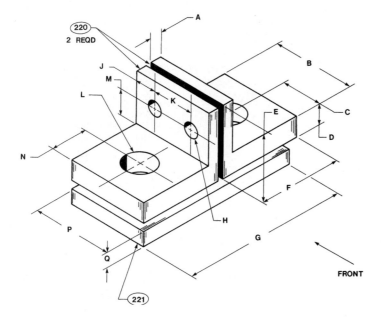

DIMENSION	INCHES	mm
A	.25	6
B	2.00	50
C	1.00	25
D	.50	13
E	1.75	45
F	2.00	50
G	4.00	100
H	Ø.438	Ø11
J	.50	12.5
K	1.00	25
L	Ø.781	Ø19
M	.63	16
N	.88	22
P	2.00	50
Q	.25	6

EX12-11

Redraw the given views and add the appropriate hex head machine screws at M and N. Use standard length screws and allow at least two unused threads at the bottom of each threaded hole. Add a bolt with the appropriate nut at hole P.

A. Use inch values.
B. Use millimeter values.
C. Draw the front assembly view using a sectional view.
D. Draw the fasteners using schematic representations.
E. Draw the fasteners using simplified representations.
F. Prepare a parts list.
G. Prepare detail drawings for each part.

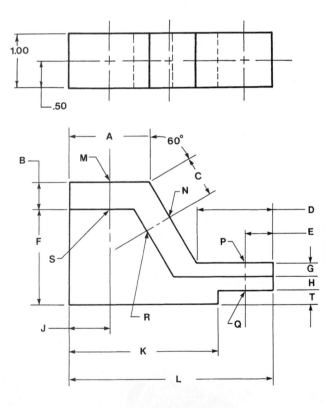

DIMENSION	INCHES	mm
A	1.50	38
B	.50	13
C	.75	19
D	1.38	35
E	.50	13
F	1.75	44
G	.25	6
H	.25	6
J	.75	19
K	2.75	70
L	3.75	96
M	Ø.31	Ø8
N	Ø.25	Ø6
P	Ø.41	Ø12
Q	Ø.41	Ø12
R	.164–32 UNF X .50 DEEP	M4 X 14 DEEP
S	.250–20 UNC X 1.63 DEEP	M6 X 14 DEEP
T	.25	6

EX12-12 MILLIMETERS

Draw an assembly drawing and parts list for the Circular Damper Assembly shown.

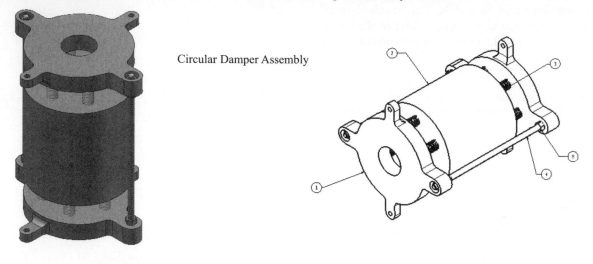

Circular Damper Assembly

Parts List

ITEM	PART NUMBER	DESCRIPTION	MATERIAL	QTY
1	BU2008-1	BASE, HOLDER	Steel	2
3	BU2008-2	SPRING, COMPRESSION	Steel, Mild	12
2	BU2008-3	COUNTERWEIGHT	Steel	1
4	AM312-12	POST, THREADED	Steel	2
5	AS 1112 - M18 Type	HEX NUT	Steel, Mild	8

Note: AM-312-12 Threaded Post is M18 x 560 long with 1 x 45° chamfers at each end

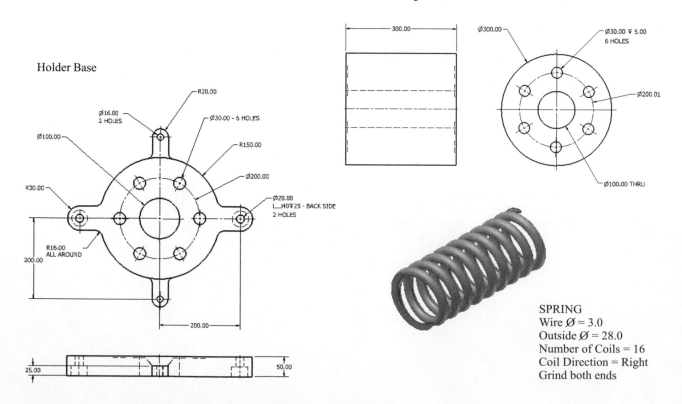

Counterweight

Holder Base

SPRING
Wire Ø = 3.0
Outside Ø = 28.0
Number of Coils = 16
Coil Direction = Right
Grind both ends

EX12-13 INCHES

The given objects are to be assembled as shown. Select sizes for the parts that make the assembly possible. (Choose dimensions for the end blocks and then determine the screw and stud lengths.) The hex head screws (5) have a major diameter of either .375 inch or M10. The studs (3) are to have the same thread sizes as the screws and are to

be screwed into the top part (2). The holes in the lower part (1) that accept the studs are to be clearance holes.

A. Draw an assembly drawing.
B. Draw detail drawings of each nonstandard part. Include positional tolerances for all holes.
C. Prepare a parts list.

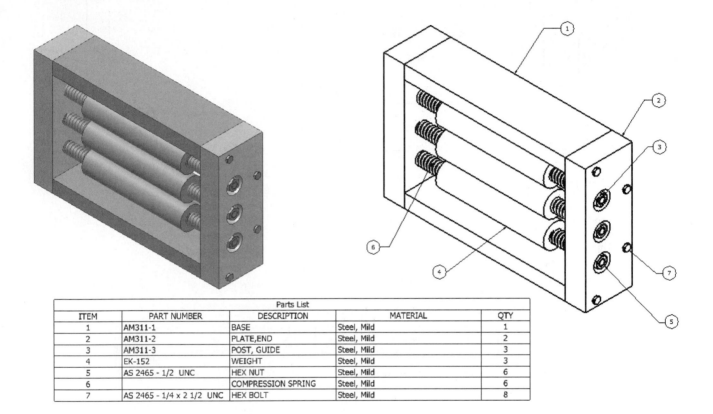

Parts List				
ITEM	PART NUMBER	DESCRIPTION	MATERIAL	QTY
1	AM311-1	BASE	Steel, Mild	1
2	AM311-2	PLATE,END	Steel, Mild	2
3	AM311-3	POST, GUIDE	Steel, Mild	3
4	EK-152	WEIGHT	Steel, Mild	3
5	AS 2465 - 1/2 UNC	HEX NUT	Steel, Mild	6
6		COMPRESSION SPRING	Steel, Mild	6
7	AS 2465 - 1/4 x 2 1/2 UNC	HEX BOLT	Steel, Mild	8

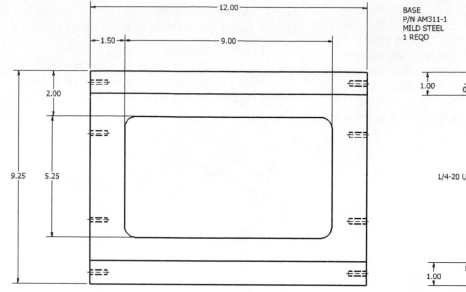

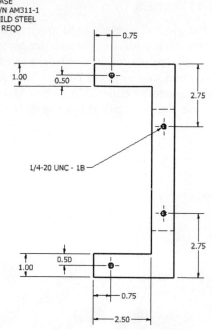

BASE
P/N AM311-1
MILD STEEL
1 REQD

1/4-20 UNC - 1B

WEIGHT
P/N EK-152
MILD STEEL
3 REQD

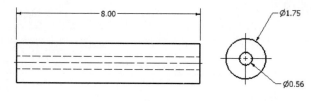

End

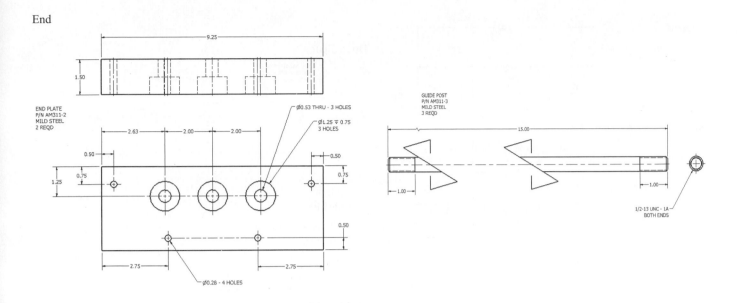

END PLATE
P/N AM311-2
MILD STEEL
2 REQD

GUIDE POST
P/N AM311-3
MILD STEEL
3 REQD

1/2-13 UNC - 1A
BOTH ENDS

SPRING
Wire Ø = 0.125
Outside Ø = 1.000
Length = 2.00
Coil direction = light
Number of coils = 10
Grind both ends.

EX12-14 INCHES

Draw an assembly drawing and parts list for the Winding Assembly.

Winding Assembly

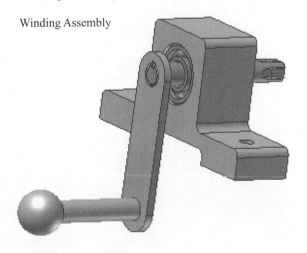

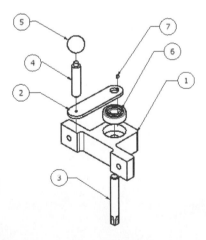

Parts List				
ITEM	PART NUMBER	DESCRIPTION	MATERIAL	QTY
1	EK131-1	SUPPORT	STEEL	1
2	EK131-2	LINK	STEEL	1
3	EK131-3	SHAFT,DRIVE	STEEL	1
4	EK131-4	POST, THREADED	STEEL	1
5	EK131-5	BALL	STEEL	1
6	BS 292 - BRM 3/4	Deep Groove Ball Bearings	STEEL,MILD	1
7	3/16x1/8x1/4	RECTANGULAR KEY	STEEL	1

Support

Drive Shaft

NOTE: ALL FILLETS AND ROUNDS R=0.250
UNLESS OTHERWISE STATED.

Threaded Post

Link

Ball

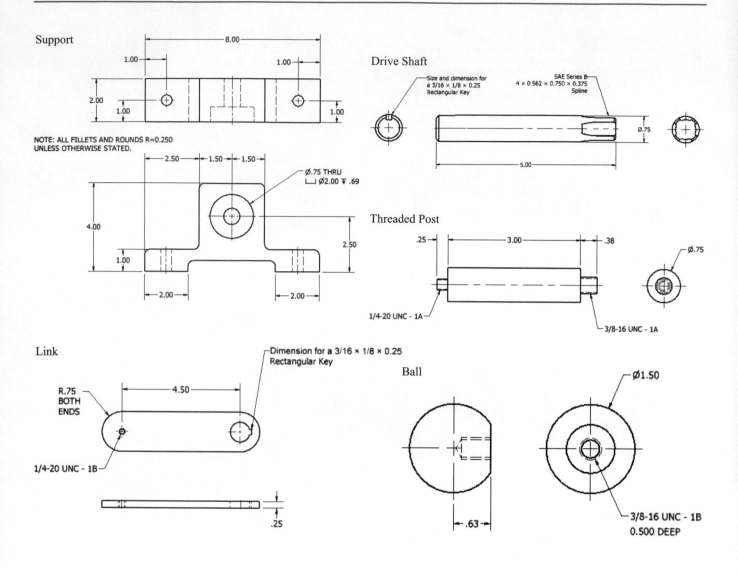

EX12-15 INCHES

Design a hand-operated grinding wheel specifically for sharpening a chisel. The chisel is to be located on an adjustable rest while it is being sharpened. The mechanism should be able to be clamped to a table during operation using two thumbscrews.

A standard grinding wheel is Ø6.00 inch and 1/2 inch thick, and has an internal mounting hole with a 50.00±.03 bore.

Prepare the following drawings.

A. Draw an assembly drawing.
B. Draw detail drawings of each nonstandard part. Include positional tolerances for all holes.
C. Prepare a parts list.

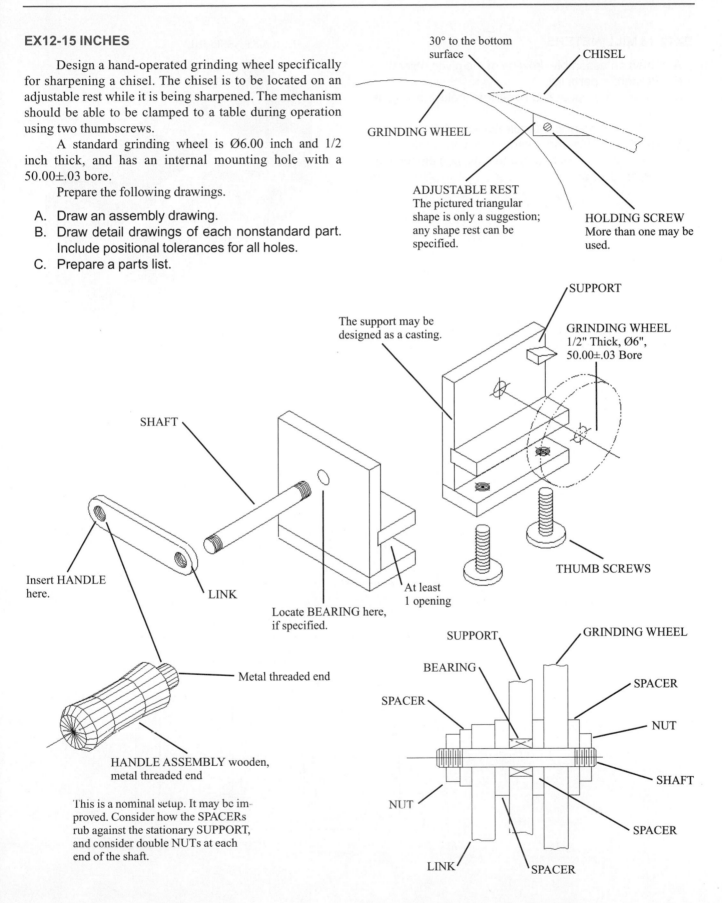

30° to the bottom surface

CHISEL

GRINDING WHEEL

ADJUSTABLE REST
The pictured triangular shape is only a suggestion; any shape rest can be specified.

HOLDING SCREW
More than one may be used.

SUPPORT

GRINDING WHEEL
1/2" Thick, Ø6", 50.00±.03 Bore

The support may be designed as a casting.

SHAFT

Insert HANDLE here.

LINK

Locate BEARING here, if specified.

At least 1 opening

THUMB SCREWS

Metal threaded end

HANDLE ASSEMBLY wooden, metal threaded end

This is a nominal setup. It may be improved. Consider how the SPACERs rub against the stationary SUPPORT, and consider double NUTs at each end of the shaft.

SUPPORT

BEARING

SPACER

NUT

SPACER

GRINDING WHEEL

SPACER

NUT

SHAFT

SPACER

LINK

SPACER

EX12-16 MILLIMETERS

A. Draw an assembly drawing of the given object.
B. Prepare a parts list.
C. Select appropriate fasteners to hold the object together.
D. Define the appropriate tolerances.
E. Define an assembly sequence considering the size of any screw heads and nuts, and the tooling required for assembly.

BRACKET ASSEMBLY

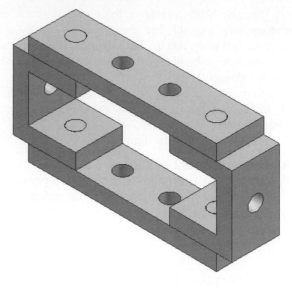

C-BRACKET, SAE1020 STEEL, 2 REQD
Part Number: AM311-1

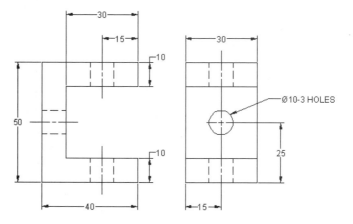

QUAD SPACER, SAE 1020 STEEL
2 REQD, Part Number: AM311-2

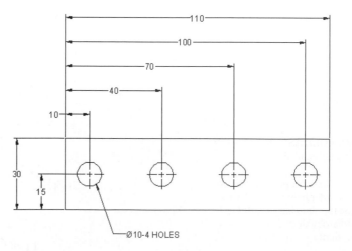

EX12-17 MILLIMETERS

A. Draw an assembly drawing of the given object.
B. Prepare a parts list.
C. Select appropriate fasteners to hold the object together.
D. Define the appropriate tolerances.
E. Define an assembly sequence considering the size of any screw heads and nuts, and the tooling required for assembly.

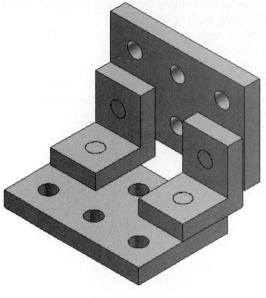

CLIP ASSEMBLY

SUPPORT PLATE
SAE 1040 STEEL, 2 REQD
Part Number: AM312-3

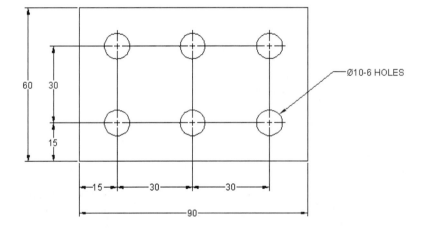

L-CLIP
SAE 1040 STEEL, 2 REQD
Part Number: AM312-4

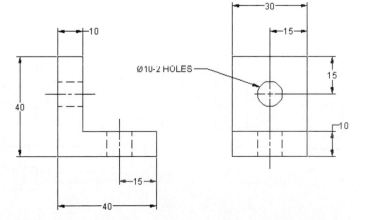

EX12-18 MILLIMETERS

A. Draw an assembly drawing of the given object.
B. Prepare a parts list.
C. Select appropriate fasteners to hold the object together.
D. Define the appropriate tolerances.
E. Define an assembly sequence considering the size of any screw heads and nuts, and the tooling required for assembly.

C-BRACKET
SAE 1020 STEEL, 4 REQD
Part Number: AM311-1

GUIDE ASSEMBLY

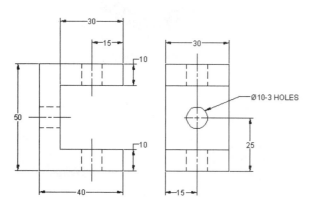

L-CLIP
SAE 1040 STEEL, 2 REQD
Part Number: AM312-4

SUPPORT PLATE
SAE 1040 STEEL, 2 REQD
Part Number: AM312-3

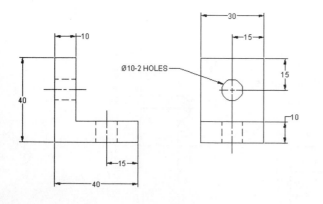

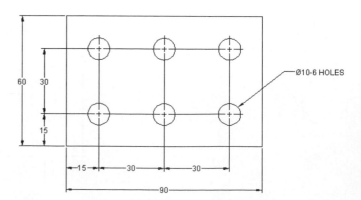

EX12-19 MILLIMETERS

A. Draw an assembly drawing of the given object.
B. Prepare a parts list.
C. Select appropriate fasteners to hold the object together. Note that the fastener used to join the center link and the drive link passes over the web plate as the drive link rotates.
D. Define the appropriate tolerances.
E. Use phantom lines and define the motion of the center link and the rocker link if the drive link rotates 360°.

ROCKER ASSEMBLY

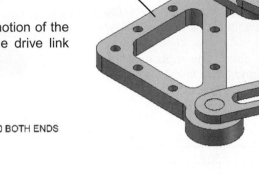

DRIVE LINK
Part Number:
AM311-22A

SAE 1040 STEEL
5mm THK

ROCKER LINK
Part Number: AM311-2C
SAE 1040 STEEL
5mm THK

ALL FILLETS AND ROUNDS = R3

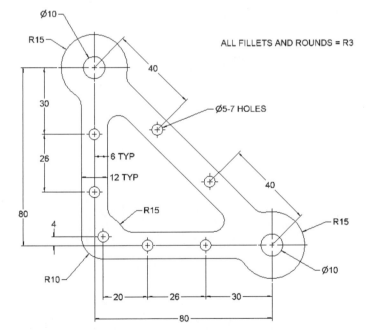

WEB PLATE
Part Number: AM311-22B
SAE 1040 STEEL
10mm THK

CENTER LINK
Part Number: AM311-22D
SAE 1040 STEEL
5mm THK

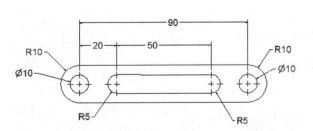

EX12-20 MILLIMETERS

A. Draw an assembly drawing of the given object.
B. Prepare a parts list.
C. Select appropriate fasteners to hold the object together. Note that the fasteners used to join the side links to the holder arm pass over the holder base.
D. Define the appropriate tolerances. Use an H7/p6 tolerance between bushing-A and the holder base.
E. Use phantom lines to define the motion of the holder arm, side links, and cross link if the holder arm rotates between +30° and −30°.

LINK ASSEMBLY

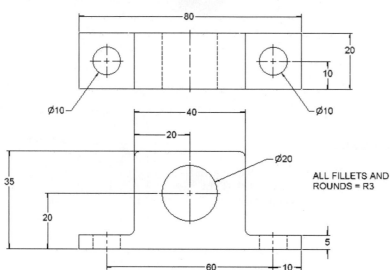

ALL FILLETS AND
ROUNDS = R3

HOLDER ARM, BU100-2, 7075-T6, AL, 5mm THK

SIDE LINK
BU100-4
7075-T6 AL
5mm THK
2 REQD

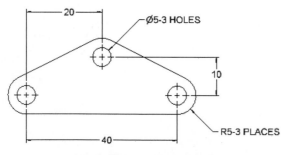

CROSS LINK, BU100-3, 7075-T6, AL, 5mm THK

BUSHING-A
A/M CORP - B20-AD
TEFLON

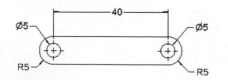

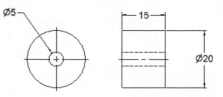

EX12-21 MILLIMETERS

A. Draw an assembly drawing of the minivise.
B. Prepare a parts list.
C. Select the appropriate fasteners to hold the vise together.

D. Redesign the interface between the drive screw and the holder plate.

MINIVISE

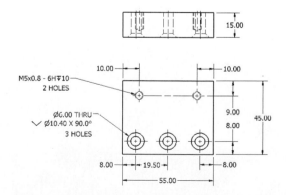

MINIVISE

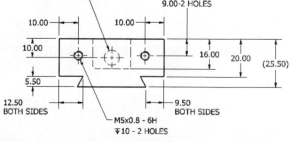

1. BASE
   SAE 1040 STEEL

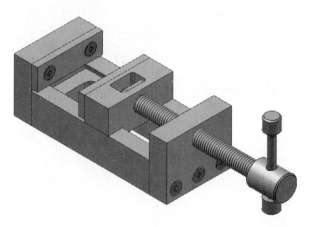

2. DRIVE SCREW
   SAE 1040 STEEL

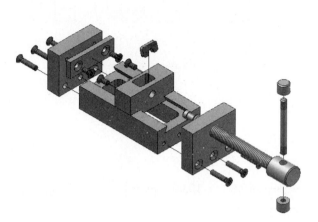

3. END PLATE
   SAE 1040 STEEL

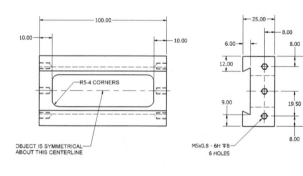

4. SLIDER
   SAE 1040
   STEEL

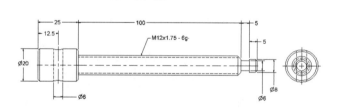

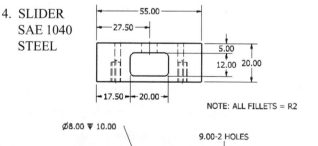

5. HOLDER PLATE
   SAE 1040 STEEL

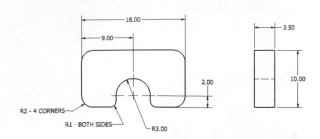

6. FACE PLATE
   SAE 1040 STEEL

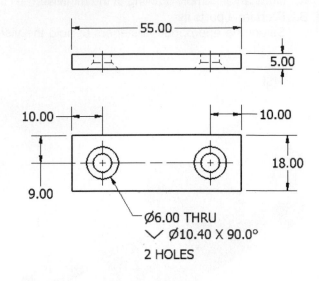

7. HANDLE
   SAE 1040 STEEL

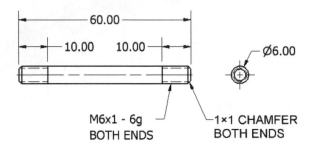

8. END CAP
   SAE 1040 STEEL

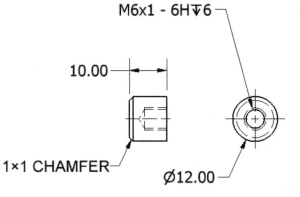

9. DRIVE PLATE
   SAE 1040 STEEL

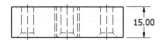

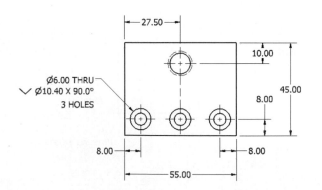

10. M5×10 RECESSED COUNTERSUNK HEAD-STEEL
11. M5×22 RECESSED COUNTERSUNK HEAD-STEEL

C H A P T E R 13

Gears, Bearings, and Cams

13-1 INTRODUCTION

This chapter explains how to draw and design with gears and bearings. The chapter does not discuss how to design specific gears and bearings but how to design using existing parts selected from manufacturers' catalogs. Various gear terms are defined, and design applications demonstrated.

The chapter also discusses how to design and draw a cam based on a displacement diagram. Different types of follower motion are explained as well as different types of followers.

13-2 TYPES OF GEARS

There are many types of gears, including spur, bevel, worm, helical, and rack. See Figure 13-1. Each type has its own terminology, drawing requirements, and design considerations.

13-3 GEAR TERMINOLOGY—SPUR

Following is a list of common spur gear terms and their meanings. Figure 13-2 illustrates the terms, and Figure 13-3 shows a listing of relative formulas.

For spur gears using English units

Pitch diameter (PD)—The diameter used to define the spacing of gears.

Diametral pitch (DP)—The number of teeth per inch or millimeter.

Circular pitch (CP)—The circular distance from a fixed point on one tooth to the same position on the next tooth as measured along the pitch circle. The circumference of the pitch circle divided by the number of teeth.

Preferred pitches—The standard sizes available from gear manufacturers. Whenever possible, use preferred gear sizes.

Center distance (CD)—The distance between the center points of two meshing gears.

Backlash—The difference between a tooth width and the engaging space on a meshing gear.

Addendum (a)—The height of a tooth above the pitch diameter.

Dedendum (d)—The depth of a tooth below the pitch diameter.

Whole depth—The total depth of a tooth. The addendum plus the dedendum.

Working depth—The depth of engagement of one gear into another. Equal to the sum of the two gear addendums.

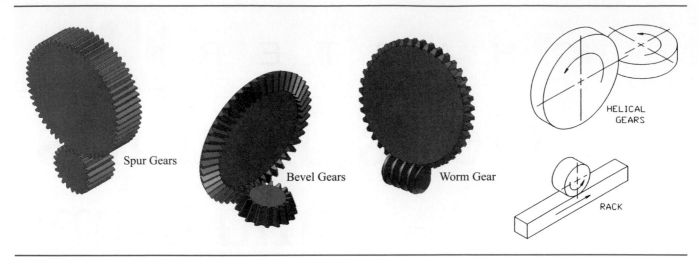

Figure 13-1

Circular thickness—The distance across a tooth as measured along the pitch circle.

Face width (FW)—The distance from front to back along a tooth as measured perpendicular to the pitch circle.

Outside diameter (OD)—The largest diameter of the gear. Equals the pitch diameter plus the addendum.

Root diameter (RD)—The diameter of the base of the teeth. The pitch circle minus the dedendum.

Clearance—The distance between the addendum of a meshing gear and the dedendum of the mating gears.

Pressure angle—The angle between the line of action and a line tangent to the pitch circle. Most gears have pressure angles of either 14.5° or 20°.

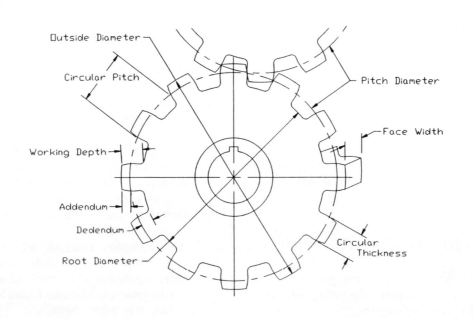

Figure 13-2

Pitch Diameter (PD)	See catalogs
Circular Pitch (CP)	$CP = \dfrac{\pi}{DP}$
Diametral Pitch (DP)	$DP = \dfrac{\pi}{CP}$
Number of Teeth (N)	$N = (PD)(DP)$
Outside Diameter (OD)	See catalogs
Addendum (a)	$a = \dfrac{1}{DP}$
Dedendum (d)	$d = a + .125$ (for drawing purposes ONLY)
Root Diameter (RD)	$RD = PD - d$
Circular Thickness (CT)	$CT = \dfrac{PD}{2N}$
Face Width (F)	See catalogs

Figure 13-3

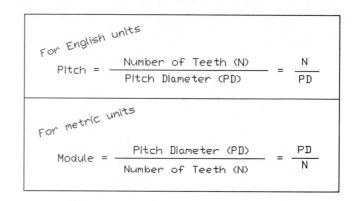

For English units

$$Pitch = \frac{Number\ of\ Teeth\ (N)}{Pitch\ Diameter\ (PD)} = \frac{N}{PD}$$

For metric units

$$Module = \frac{Pitch\ Diameter\ (PD)}{Number\ of\ Teeth\ (N)} = \frac{PD}{N}$$

Figure 13-4

For spur gears using metric units

The preceding definitions apply to both English unit and metric unit spur gears with the exception of pitch. For English unit gears, pitch is defined as the number of teeth per inch relative to the pitch diameter and is expressed by the formula shown in Figure 13-4. Gears made to metric specifications are defined in terms of the amount of pitch diameter per tooth, called the gear's *module.* Figure 13-4 shows the formula for calculating a gear's module. Metric gears also have a slightly different tooth shape that makes them incompatible with English unit gears.

13-4 SPUR GEAR DRAWINGS

Figure 13-5 shows a representation of spur gears. The individual teeth are not included in the front view but are represented by three phantom lines. The diameters of the three lines represent the outside diameter, the pitch diameter, and the root diameter.

The outside diameter and the pitch circle are usually given in manufacturers' catalogs. The root circle can be calculated from the pitch circle using the formula presented in Figure 13-3.

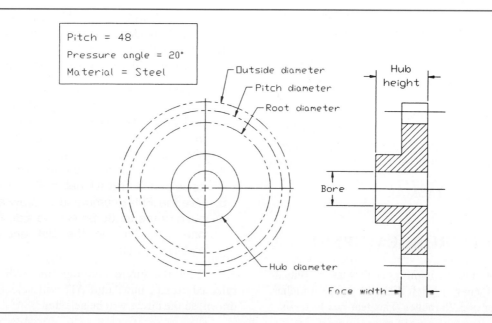

Pitch = 48
Pressure angle = 20°
Material = Steel

Outside diameter
Pitch diameter
Root diameter
Hub height
Bore
Hub diameter
Face width

Figure 13-5

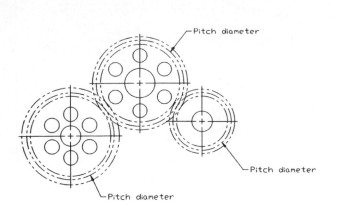

Figure 13-6

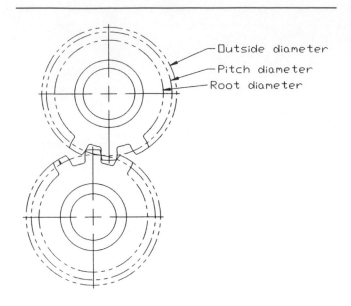

Figure 13-7

The side view of the gear is drawn as a sectional view taken along the vertical centerline. The gear size is defined using the manufacturer's stock number, dimensions, and a list of appropriate design information.

Figure 13-6 shows the front representation view of three meshing gears. The gears are positioned so that the pitch circle diameters are tangent. Ideally, mating gears always mesh exactly tangent to their pitch circles.

Gear representations were developed because it was both difficult and time consuming to accurately draw individual teeth when creating drawings on a drawing board. AutoCAD can be used to create detailed gear drawings that include all teeth by using the **Array** and **Block** commands, among others; however, it is usually sufficient to show a few meshing teeth and use the representative centerlines for the remaining portions of both gears. See Figure 13-7.

Most gear tooth shapes are based on an involute curve. Involute-based teeth fit together well, transfer forces smoothly, and can use one cutter to generate all gear tooth variations within the same pitch. Standards for tooth proportions have been established by the American National Standards Institute (ANSI) and the American Gear Manufacturers Association (AGMA).

13-5 SAMPLE PROBLEM SP13-1

Draw a front view of a spur gear that has an outside diameter of 6.50 inches, a pitch diameter of 6.00 inches, and a root diameter of 5.25 inches. The gear has 12 teeth.

The method presented is a simplified method and is an acceptable representation for most drawing applications. See Figure 13-8.

1. Draw three concentric circles of diameter **6.50, 6.00,** and **5.25.** Use the **Center Mark** command modified to draw a centerline for the circles.
2. Use the **Array** command and create **48** ray lines as shown. Label three rays on each side of the top vertical centerline as shown. Zoom the labeled area.

The number of rays should equal four times the number of teeth to be drawn. Two adjoining sectors are used to define the width of the tooth, and the next two adjacent sectors are used to define the space between teeth.

3. Draw a circle whose center point is at the intersection of the pitch circle and the ray labeled **2,** and whose radius is determined by the distance from the center point to the intersection of the pitch circle and the ray labeled **−1.**
4. Repeat step 3 using the intersection of the pitch circle and the point labeled **−2** as the center point.
5. Use the **Fillet** command to draw a radius at the base of the tooth. Select the side of the tooth as one of the lines for the fillet, and the root circle as the other line.

Both the circle and ray line will be within the selected cursor, but AutoCAD will select the last entity drawn, so the circle will be selected.

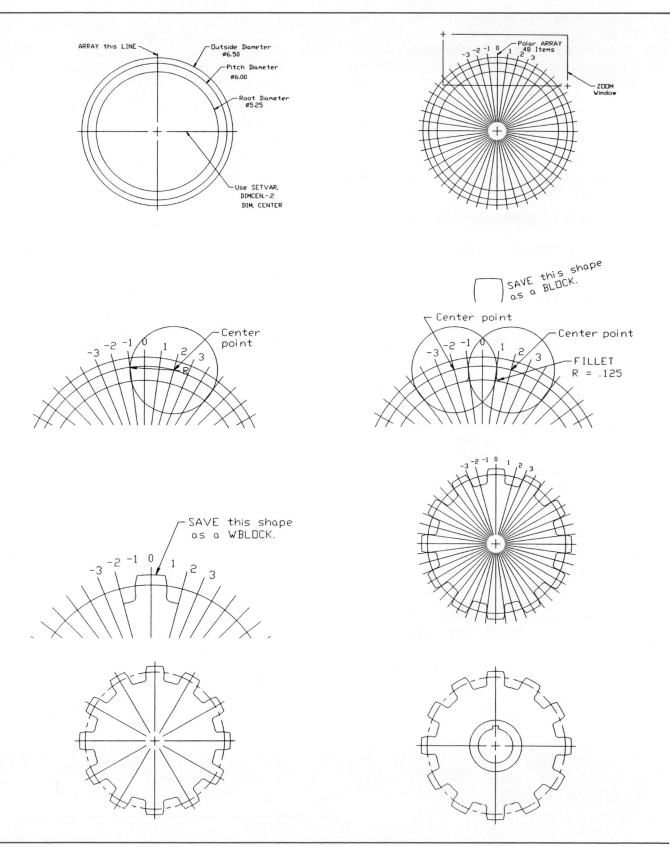

Figure 13-8

The tooth shape can be saved as a wblock named **TOOTH** and used when drawing other gears. Only the lines that represent the top of the gear and the two side sections need to be saved. A different size gear will have a different root diameter, and new fillets can be drawn that align with the new root circle.

6. Use the **Trim** and **Erase** commands to remove excess lines and create the tooth shape between rays −**2** and **2** as shown.

This tooth shape may be saved as a wblock and used when drawing other gears.

7. Array the tooth shape about the gear's center point.
8. Erase the excess ray lines and change the pitch circle to a centerline.
9. Draw the gear's bore hole or center hole and hub as required.

13-6 SAMPLE PROBLEM SP13-2

Figure 13-9A shows two meshing gears. In the example shown, the larger gear has a diameter of 6.00 inches with 24 teeth; the smaller gear has a diameter of 3.00 inches and 12 teeth. The drawing utilizes the wblock created in Sample Problem SP13-1 as follows.

The circular thickness of the wblock tooth as measured along the pitch diameter equals $\frac{1}{12}$ the circumference of the pitch diameter.

$$(\tfrac{1}{12})(\pi \, PD)$$

For the gear in SP13-1, where PD = 6

$$(\tfrac{1}{12})(\pi \, 6) = \pi/2$$
$$= 1.57 \text{ inches}$$

The large gear requires 24 teeth on a pitch diameter of 6, or twice as many teeth as the gear used to create the wblock. The teeth on the 24-tooth gear must be half the size of the tooth created in the wblock. The teeth on the smaller gear must be the same size as the teeth on the larger gear.

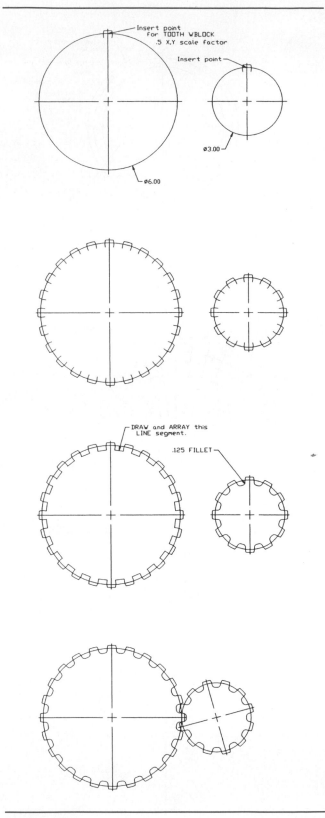

Figure 13-9A

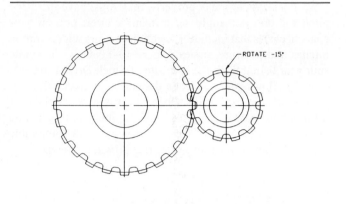

Figure 13-9B

To draw meshing spur gears (See Figure 13-9B.)

1. Draw two circles of diameter **6.00** and **3.00** inches. Include the circles' centerlines. Use **Setvar, Dimcen,** set to −.2, or use a modified **Center Mark** tool.

The gears will be drawn separately, then meshed.

2. Insert the TOOTH wblock on both gears as shown on the pitch diameter. Use an X and Y scale factor of **.5.** Explode the wblock.
3. Use the **Array** command to create the required 24 and 12 teeth.
4. Draw a line between the roots of two of the teeth as shown. Use the **Array** command to create the root circle. Add the fillet to each tooth base.

The line in this example is a straight line acceptable for smaller gears. If more accuracy of shape is required, draw a complete root circle, then trim all of it away except the portion between two tooth roots. Array this sector between all the other teeth.

5. Use the **Rotate** command to orient the small gear with the large gear.

Each tooth on the large gear requires 360/24 = 15°. Rotate the smaller gear 15°.

6. Use the **Move** command to position the small gear.

The pitch circles of the two gears should be tangent.

7. Rotate the smaller gear's centerline **−15°.**
8. Add the center hubs to both gears as shown.

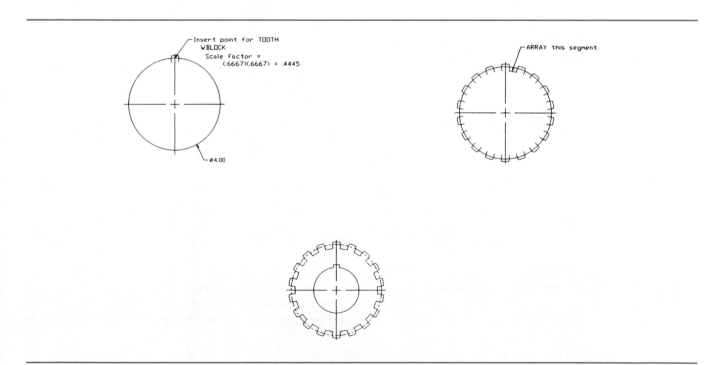

Figure 13-10

13-7 SAMPLE PROBLEM SP13-3

Figure 13-10 shows a gear that has a pitch diameter of 4.00 with 18 teeth. The TOOTH wblock can be used as follows.

The wblock is first reduced to accommodate the smaller diameter.

1. Draw a circle with a **4.00**-inch diameter. Include centerlines.
2. Insert the wblock using an X and Y scale factor of **.4445.**

The scale factor was derived by first considering the ratio between the number of teeth on the gear used to create the TOOTH wblock (12) and the number of teeth on the desired gear, 18, or 12/18 = .6667. The ratio between the diameters is also considered: 4.00/6.00 = .6667. The two ratios are multiplied together: (.6667)(.6667) = .4445.

3. Use the **Array** command to create the required 18 teeth.
4. Draw a line between the end lines of two of the teeth, then array the line **18** times around the gear.

The line may be drawn as a straight line for small gears or as an arc for larger gears. The arc may be created by drawing a circle, then using the Trim command to create the desired length.

5. Add the center hub as required.

The same wblock can be used for metric gears using the conversion factor 1.00 inch = 25.4 millimeters. It is probably easier to create a separate wblock for a metric tooth.

13-8 SELECTING SPUR GEARS

When two spur gears are engaged, the smaller gear is called the *pinion gear* and the larger gear is called simply the *gear*. The relationship between the relative speed of two mating gears is directly proportional to the gears' pitch diameters. Also, the number of teeth on a gear is proportional to the gear's pitch diameter. This means that the ratio of speed between two meshing gears is equal to the ratio of the number of teeth on the two gears. If one gear has 40 teeth and the other 20, the speed ratio between the two gears is 2:1.

For gears to mesh properly, they must have the same pitch and pressure angle. Gear manufacturers present their gears in charts that include a selection of gears with common pitches and pressure angles. Figure 13-11 shows a sample spur gear listing from the Web site of Stock Drive Products. The site allows you to select a diametral pitch and all other appropriate product details. Once a gear is selected, the gear information will be displayed. You may be able to generate an AutoCAD drawing of the gear if you have a compatible setup. Many manufacturers offer free product catalogs.

13-9 CENTER DISTANCE BETWEEN GEARS

The center distance between meshing spur gears is needed to align the gears properly. Ideally, gears mesh exactly on their pitch diameters, so the ideal center distance between two meshing gears is equal to the sum of the two pitch radii, or the sum of the two diameters divided by 2:

$$CD1 = (PD1 + PD2)/2$$

If two gears were chosen from the chart in Figure 13-11, and one had 30 teeth and a pitch diameter of .9375, and the other had 60 teeth and a pitch diameter of 1.8750, the center distance between the gears would be

$$CD1 = (.9375 + 1.8750)/2 = 2.8125$$

The center distance of gears is dependent on the tolerance of the gears' bores, the tolerance of the supporting shafts, and the feature and positional tolerance of the holes in the shaft's supporting structure. The following sample problem shows how these tolerances are considered and applied when matching two spur gears.

13-10 SAMPLE PROBLEM SP13-4

An electric motor generates power at 1750 rpm. Reduce this speed by a factor of 2 using steel metric gears with a module of 1.5 and a pressure angle of 20°. Figure 13-12 shows a list of gears. Determine the center distance between the gears, and specify the shaft sizes required for both gears.

Gear number A 1C22MYKW150 50A and number A 1C22MKYW150 100A were selected from the chart shown in Figure 13-12. The pinion gear has 50 teeth, and the large gear has 100, so if the pinion gear is mounted on

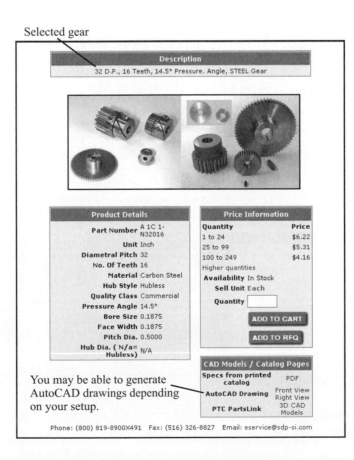

Select Diametral Pitch here

This gear selected

Selected gear

You may be able to generate AutoCAD drawings depending on your setup.

Figure 13-11
Courtesy of Stock Drive Products.

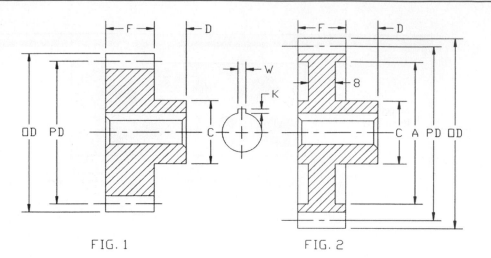

FIG. 1   FIG. 2

This information is available from the Stock Drive Products catalog or website.

Catalog Number	Fig No.	No. of Teeth	P.D.	O.D.	B* Bore (H7)	F Face Width	C Hub Dia	D Hub Proj	A Dia	W Dim	K Dim
A 1C22MYKW150 20A		20	30	33	14		25				
A 1C22MYKW150 24A		24	36	39	16		30			5	2.3
A 1C22MYKW150 25A		25	37.5	40.5	18	18	32				
A 1C22MYKW150 28A		28	42	45			36				
A 1C22MYKW150 30A		30	45	48							
A 1C22MYKW150 32A	1	32	48	51							
A 1C22MYKW150 36A		36	54	57							
A 1C22MYKW150 40A		40	60	63				14			
A 1C22MYKW150 48A		48	72	75						6	2.8
A 1C22MYKW150 50A		50	75	78	20		40				
A 1C22MYKW150 56A		56	84	87		16					
A 1C22MYKW150 60A		60	90	93					76		
A 1C22MYKW150 64A		64	96	99					82		
A 1C22MYKW150 70A		70	105	108					91		
A 1C22MYKW150 72A	2	72	108	111					94		
A 1C22MYKW150 80A		80	120	123	25		50		106	8	3.3
A 1C22MYKW150 100A		100	150	153					136		

* Gears with 14, 16, and 18mm bores have a tolerance of +.018, +.000.
20 and 25mm bores have a tolerance of +0.021/– 0

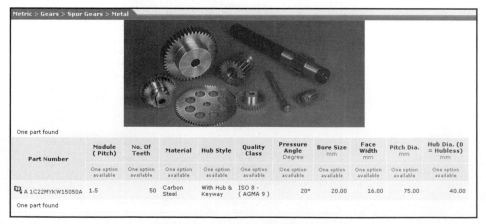

From the website www.SDP-SI.com

Figure 13-12
Courtesy of Stock Drive Products.

the motor shaft, the larger gear will turn at 875 rpm, or half the 1750 motor speed.

$$\frac{1}{2} = \frac{x}{1750}$$

$$x = \frac{1750}{2} = 875 \text{ rpm}$$

The specific design information for the selected gears is as follows:

PINION GEAR
  PD = 75
  OD = 78
  N = 50
  Bore = $20 \begin{smallmatrix} +.021 \\ -.000 \end{smallmatrix}$
  Tolerance = H7

LARGE GEAR
  PD = 150
  OD = 153
  N = 100
  Bore = $25 \begin{smallmatrix} +.021 \\ -.000 \end{smallmatrix}$
  Tolerance = H7

The center distance between the gears is equal to the sum of the two pitch diameters divided by 2:

$$\frac{PD1 + PD2}{2} =$$

$$\frac{75 + 150}{2} = 112.5$$

The manufacturer's catalog lists the bore tolerance as an H7. One gear has a nominal bore diameter of 20 and the other one of 25. Standard fit tolerances for metric values are discussed in Chapter 9, and appropriate tables are included in the appendix.

Tolerance values for a sliding fit (H7/g6) hole basis were selected for this design application. This means that the shaft tolerances are 19.993 and 19.980 for the pinion gear and 24.993 and 24.980 for the large gear.

13-11 COMBINING SPUR GEARS

Gears may be mounted on the same shaft. Gears on a common shaft have the same turning speed. Combining gears on the same shaft enables the designer to develop larger gear ratios within a smaller space.

Combining gears can also be used to help reduce the size of gear ratios between individual gears and the amount of space needed to create the reductions. Figure 13-13 shows a four-gear setup. Gears B and C are mounted on the same shaft. Gear A is the driver gear and is turning at a speed of 1750 rpm. The speed of gear D is determined as follows.

The ratio between gears A and B is

$$\frac{48}{72} = .6667$$

The speed of gear B is therefore

$$1750 \times .6667 = 1166.7 \text{ rpm}$$

Gears B and C are on the same shaft, so they have the same speed. Gear C is turning at 1166.7 rpm.

The ratio between gears C and D is

$$\frac{24}{48} = .5000$$

The speed of gear D is therefore

$$1166.7 \times .5000 = 583 \text{ rpm}$$

The ratio between gears A and D is 3:1.

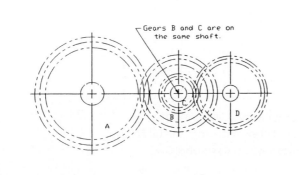

Figure 13-13

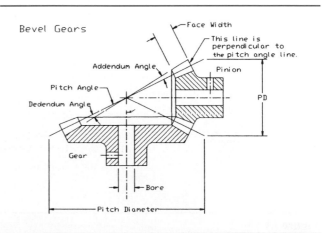
Figure 13-14
Courtesy of Stock Drive Products.

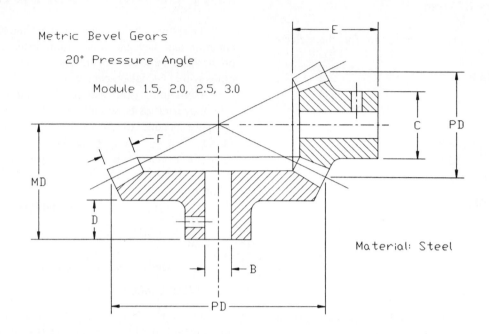

Stock Drive Products

Metric Bevel Gears

20° Pressure Angle

Module 1.5, 2.0, 2.5, 3.0

Material: Steel

Catalog Number	Module	No. of Teeth	Ratio	PD	OD	B* Bore (H8)	F Face Width	E Length	C Hub Dia	D Hub Proj	MD Dim
A 1C 3MYK 15018	1:5	18		27	29.7	8	9.8	23	22	12.5	40.74
A 1C 3MYK 15036		36		54	55.4	10		18.5	30	10	26.75
A 1C 3MYK 20018	2:0	18		36	39.6	9	12.6	29	28	15	53.12
A 1C 3MYK 20036		36	1:2	72	73.8	12		24	36	13	35.21
A 1C 3MYK 25018	2.5	18		45	49.5	12	16.7	35	36	17	64.29
A 1C 3MYK 25036H		36		90	92.2	14		29	50	15	42.55
A 1C 3MYK 30018	3.0	18		54	59.4	12	20	40	41	18	75.27
A 1C 3MYK 30036H		36		108	110.7	16		36	60	19	52.32

* Gears with: 8, 9, 10mm bores have a tolerance of +0.022/-0
        12, 14, 16mm bores have a tolerance of +0.027/-0

This information is also available on the Stock Drive Products website www.SDP-SI.com.

Figure 13-15
Courtesy of Stock Drive Products.

13-12 GEAR TERMINOLOGY— BEVEL

Bevel gears align at an angle with each other. An angle of 90° is most common. Bevel gears use much of the same terminology as spur gears but with the addition of several terms related to the angles between the gears and the shape and position of the teeth. Figure 13-14 defines the related terminology.

Bevel gears must have the same pitch or module value and have the same pressure angle for them to mesh properly. Manufacturers' catalogs usually list bevel gears in matched sets designated by ratios that have been predetermined to fit together correctly. Figure 13-15 shows a sample list of matched set bevel gears using millimeter values, and Figure 13-16 shows a list for inch values.

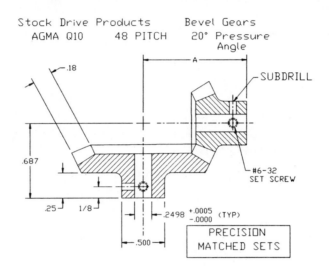

Catalog Number	Ratio	No. of Teeth	PD	A	Material
S1346Z-48S30A030	1:1	30 30	.625	.687	St Steel Aluminum
S1346Z-48S30S030	1:1	30 30	.625	.687	St Steel
S1346Z-48A30A030	1:1	30 30	.625	.687	Aluminum
S1346Z-48S30A045	1:1-1/2	30 45	.625 .937	.812	St Steel Aluminum
S1346Z-48S30S045	1:1-1/2	30 45	.625 .937	.812	St Steel
S1346Z-48A30A045	1:1-1/2	30 45	.625 .937	.812	Aluminum
S1346Z-48S30A060	1:2	30 60	.625 1.250	.937	St Steel Aluminum
S1346Z-48S30S060	1:2	30 60	.625 1.250	.937	St Steel
S1346Z-48A30A060	1:2	30 60	.625 1.250	.937	Aluminum
S1346Z-48S30A090	1:3	30 90	.625 1.875	1.250	St Steel Aluminum
S1346Z-48S30S090	1:3	30 90	.625 1.875	1.250	St Steel
S1346Z-48S30A090	1:3	30 90	.625 1.875	1.250	Aluminum
S1346Z-48S30A120	1:4	30 120	.625 2.500	1.531	St Steel Aluminum
S1346Z-48S30S120	1:4	30 120	.625 2.500	1.531	St Steel
S1346Z-48A30A120	1:4	30 120	.625 2.500	1.531	Aluminum

Figure 13-16
Courtesy of Stock Drive Products.

13-13 HOW TO DRAW BEVEL GEARS

Bevel gears are usually drawn by working from dimensions listed in manufacturers' catalogs for specific matching sets. The gears are drawn using either a sectional view or a half-sectional view that shows the profiles of the two gears. Figure 13-17 shows a matching set of bevel gears that were drawn from the information given in Figure 13-15.

The procedure used to draw the gears, based on information given in manufacturers' catalogs, is as follows.

To draw a matched set of beveled gears

1. Draw a perpendicular centerline pattern and use the **Offset** command to define the pitch diameters of the pinion and gear.
2. Extend the pitch diameter lines and draw lines from the center point to the intersections as shown.

These lines are called the *face angle lines*. The ends of beveled gears are drawn perpendicular to the face angle lines. Gear manufacturers do not always include the outside

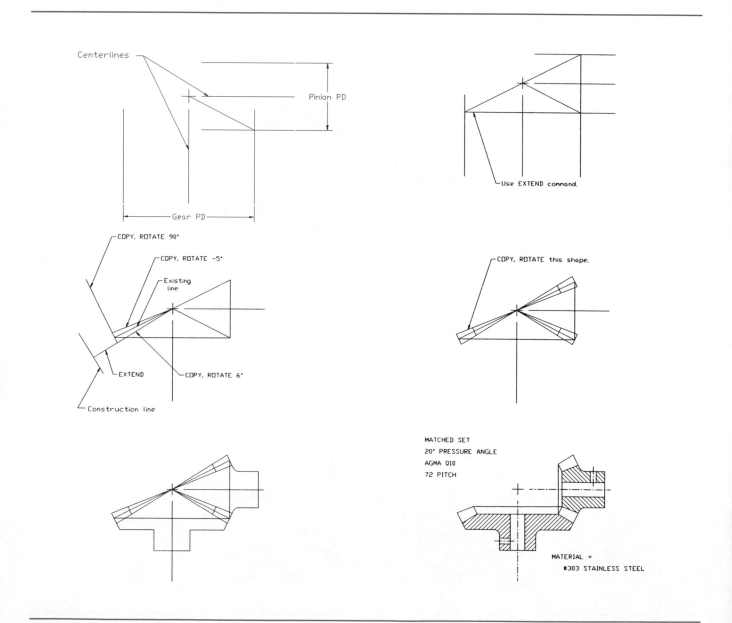

Figure 13-17

diameter values with matching sets of gears. The values are sometimes listed with data for the individual gears elsewhere in the catalog. If the outside diameter values are not given, they can be conservatively estimated by drawing the addendum angle approximately −5.0° from the face angle and the dedendum 6.0° from the face angle.

The perpendicular end lines may be constructed using the **Copy** and **Rotate** commands. Copy the existing face angle line directly over the existing line, then rotate the line 90° from the face angle line.

Use the **Extend** command to extend the rotated lines as needed. Add a construction line to help define an intersection between the dedendum ray line and the face line perpendicular to the face angle line. Trim and erase any excess lines.

3. Use the **Copy** and **Rotate** commands to copy the face shape and rotate it into the two other positions shown.
4. Use the given dimensions to complete the profiles. Erase and trim lines as necessary.
5. Draw the bore holes and holes for the setscrews based on the manufacturer's specifications.
6. Use the **Hatch** command to draw the appropriate hatch lines.

The two gears should have hatch patterns at different angles. In this example the hatch pattern on the pinion is at 90° to the pattern on the gear.

13-14 WORM GEARS

A worm gear setup is created using a cylindrical gear called a ***worm***, and a circular matching gear called a ***worm gear***. See Figure 13-18. As with other types of gears, worm gears must have the same pitch and pressure angle to mesh correctly. Manufacturers list matching worms and worm gears together in their catalogs. Figure 13-20 shows a gear manufacturer's listing for a worm gear and the appropriate worms.

Worm gears are drawn using the representation shown in Figure 13-20 or using sectional views as shown in Figure 13-19. The worm teeth shown can be drawn using the procedure explained in Chapter 11 for acme threads.

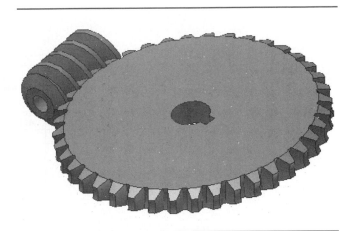

Figure 13-18

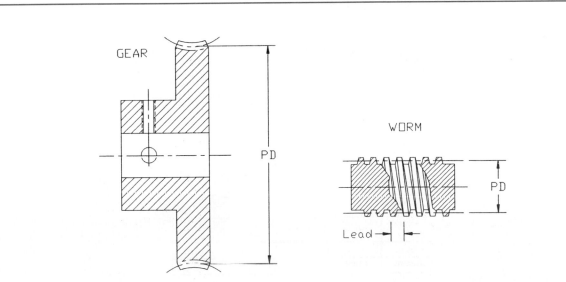

Figure 13-19

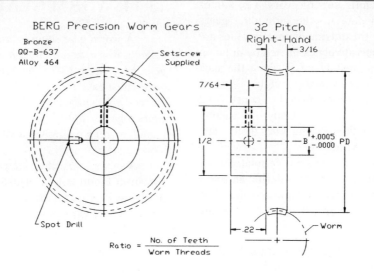

BERG Precision Worm Gears

32 Pitch
Right-Hand

$$Ratio = \frac{No.\ of\ Teeth}{Worm\ Threads}$$

FOR SINGLE THREAD WORM		FOR DOUBLE THREAD WORM		NO. OF TEETH	PITCH DIA
CIRCULAR PITCH	.0982	CIRCULAR PITCH	.1963		
HELIX ANGLE	4° - 5'	HELIX ANGLE	8° - 8'		
PRESSURE ANGLE	14-1/2°	PRESSURE ANGLE	20°		
STOCK NUMBER		STOCK NUMBER			
W32B29-S20		W32B29-D20		20	.625
W32B29-S30		W32B29-D30		30	.938
W32B29-S40		W32B29-D40		40	1.250
W32B29-S50		W32B29-D50		50	1.562
W32B29-S60		W32B29-D60		60	1.875
W32B29-S80		W32B29-D80		80	2.500
W32B29-S96		W32B29-D96		96	3.000
W32B29-S100		W32B29-D100		100	3.125
W32B29-S120		W32B29-D120		120	3.750
W32B29-S180		W32B29-D180		180	5.625

Precision Worms

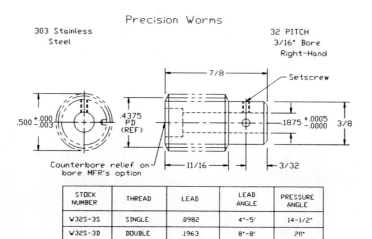

303 Stainless
Steel

32 PITCH
3/16' Bore
Right-Hand

STOCK NUMBER	THREAD	LEAD	LEAD ANGLE	PRESSURE ANGLE
W32S-3S	SINGLE	.0982	4°-5'	14-1/2°
W32S-3D	DOUBLE	.1963	8°-8'	20°

Figure 13-20
Courtesy of W. M. Berg Inc.

The relationship between worm gears is determined by the *lead* of the worm thread. The lead of a worm thread is similar to the pitch of a thread discussed in Chapter 11. Worm threads may be single, double, or quadruple. If a worm has a double thread, it will advance the gear twice as fast as a worm with a single thread.

13-15 HELICAL GEARS

Helical gears are drawn as shown in Figure 13-21. The two gears are called the *driver* and the *driven,* as indicated. Figure 13-22 shows a manufacturer's listing of compatible helical gears.

Helical gears may be manufactured with either left- or right-hand threads. Left- and right-hand threads are used to determine the relative rotation direction of the gears.

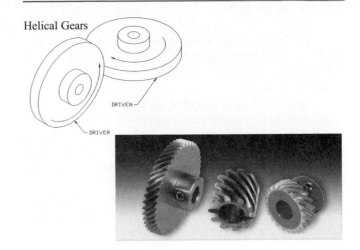

Figure 13-21
Courtesy of Stock Drive Products.

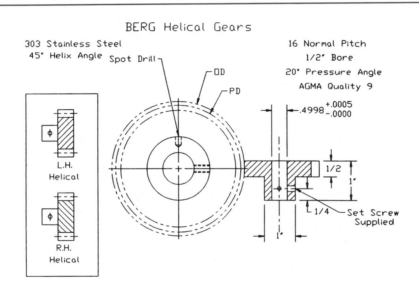

BERG Helical Gears

303 Stainless Steel
45° Helix Angle

16 Normal Pitch
1/2" Bore
20° Pressure Angle
AGMA Quality 9

STOCK NUMBER		NO. OF TEETH	PITCH DIAMETER	OUTSIDE DIAMETER
RIGHT-HAND	LEFT-HAND			
H16S38-R12	H16S38-L12	12	1.060	1.185
H16S38-R16	H16S38-L16	16	1.4142	1.539
H16S38-R20	H16S38-L20	20	1.7677	1.892
H16S38-R24	H16S38-L24	24	2.1213	2.246
H16S38-R32	H16S38-L32	32	2.8284	2.953
H16S38-R40	H16S38-L40	40	3.5355	3.660
H16S38-R48	H16S38-L48	48	4.2426	4.367

Figure 13-22
Courtesy of Stock Drive Products.

13-16 RACKS

Racks are gears whose teeth are in a straight row. Racks are used to change rotary motion into linear motion. See Figure 13-23. Racks are usually driven by a spur gear called a *pinion*.

One of the most common applications of gear racks is the steering mechanism of an automobile. Rack-and-pinion steering helps create a more positive relationship between the rotation of the steering wheel and the linear input to the car's wheel than did the mechanical linkages used on older-model cars.

Figure 13-24 shows a manufacturer's list for racks. The rack and pinions must have the same pitch and pressure angle to mesh correctly.

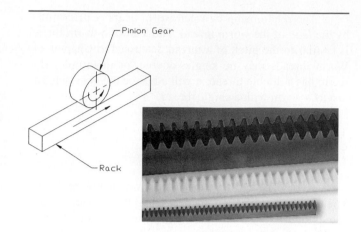

Figure 13-23
Courtesy of Stock Drive Products.

Berg Precision Racks

416 ST. Steel &
2024T4 Aluminum
Anodized

24 to 120 Pitch
20° Pressure Angle
AGMA Quality 10

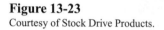

.156 (72, 80, and 120 Pitch)
.218 (24, 48, and 64 Pitch)

#26 Dr (4 Holes) on
24, 32 48, and 64 Pitch
#31 Dr (4 Holes) On
72, 80, 96, and 120 Pitch

STOCK NUMBER	MATERIAL	PITCH	A	P	W	F	C
R4-5 R4-6	STAINLESS ST ALUMINUM	24	10'	.4383	.480	.230	3.208
R4-9 R4-10	STAINLESS ST ALUMINUM	32	10'	.4487	.480	.230	3.208
R4-11 R4-12	STAINLESS ST ALUMINUM	48	9'	.4592	.480	.230	2.879
R4-15 R4-16	STAINLESS ST ALUMINUM	64	7'	.4644	.480	.230	2.208
R4-17 R4-18	STAINLESS ST ALUMINUM	72	5'	.3411	.355	.167	1.541
R4-19 R4-20	STAINLESS ST ALUMINUM	80	5'	.3425	.355	.167	1.541
R4-21 R4-22	STAINLESS ST ALUMINUM	96	3'	.3446	.355	.167	.875
R4-23 R4-24	STAINLESS ST ALUMINUM	120	3'	.3467	.355	.167	.875

Figure 13-24
Courtesy of W. M. Berg Inc.

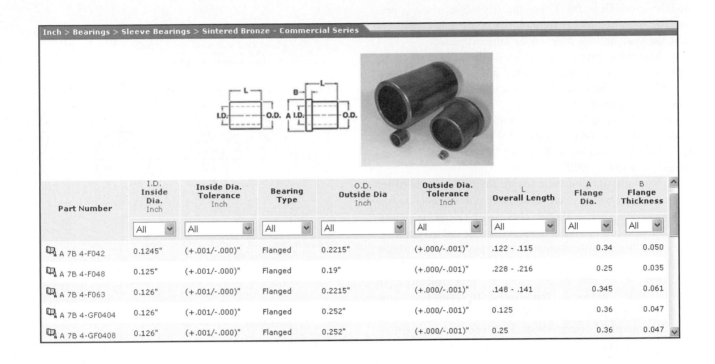

Figure 13-25
Courtesy of Stock Drive Products.

13-17 BALL BEARINGS

Ball bearings are used to help eliminate friction between moving and stationary parts. The moving and stationary parts are separated by a series of balls that ride in a *race*.

Figure 13-25 shows a listing for ball bearings that includes applicable dimensions and tolerances, which is from the Stock Drive Products website. There are many other types and sizes of ball bearings available.

Ball bearings may be drawn as shown in Figure 13-25 or by using one of the representations shown in Figure 13-26. It is recommended that the representations be drawn and saved as wblocks for use on future drawings.

The outside diameter of the bearings listed in Figure 13-25 has a tolerance of +.0000/−.0002. The same tolerance range applies to the center hole. These tight tolerances are manufactured because this type of ball bearing is usually assembled using a force fit. See Chapter 9 for an explanation of fits. Ball bearings are available that do not assemble using force fits.

The following sample problem shows how ball bearings can be used to support the gear's shafts.

13-18 SAMPLE PROBLEM SP13-5

Figure 13-27 shows two spur gears and a dimensioned drawing of a shaft used to support both gears. Design a support plate for the shafts. Use ball bearings to support the shafts. Specify dimensions and tolerances for the support plate and assume that the bearings are to be fitted into the support plate using an LN2 medium press fit. The tolerance for the center distance between the gears is to be +.001,−.000.

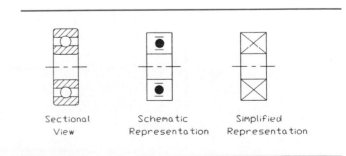

Figure 13-26

The maximum interference permitted for an LN2 medium press fit is .0011. See the fits tables in the appendix. For purposes of calculation, the bearing is considered to be the shaft and the holes in the support plate.

Maximum interference occurs when the shaft (bearing) is at its maximum diameter and the hole (support plate) is at its minimum. The maximum shaft diameter, the maximum diameter of the bearing, is .5000, as defined in the manufacturer's listing. This means that the minimum hole diameter should be .5000−.0011 = .4989. An LN2 fit has a hole tolerance of +.0007, so the maximum hole size should be .4989 + .0007 = .4996.

The minimum interference, or the difference between the minimum shaft diameter and the maximum hole diameter, is .4998−.4996 = .0002. There will always be at least .0002 interference between the ball bearing and the hole.

The bore of the selected bearing, listed in Figure 13-25, has a limit tolerance of .2500−.2498. The shafts specified in Figure 13-27 have a limit tolerance of .2497−.2495. This means that there will always be a slight clearance between the shaft and the bore hole.

The nominal center distance between the two gears is 3.000 inches. The given tolerance for the center distance is +.001, −.000. This tolerance can be ensured by assigning a positional tolerance of .0005 to each of the two holes applied at maximum material condition at the centerline. The base distance between the two holes is defined as 3.0000. The maximum center distance, including the positional tolerance, is 3.0000 + .0005 = 3.0005, and the minimum is 3.0000−.0005 = 2.9995, or a total maximum tolerance of .001.

Figure 13-28 shows a detail drawing of the support plate.

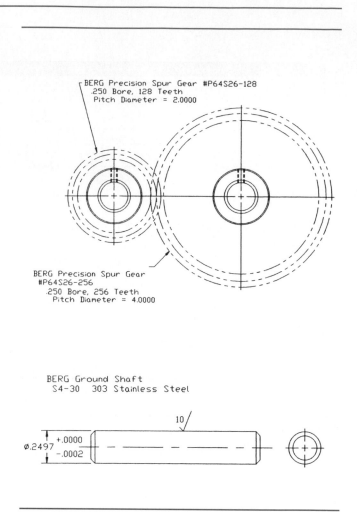

Figure 13-27
Courtesy of W. M. Berg Inc.

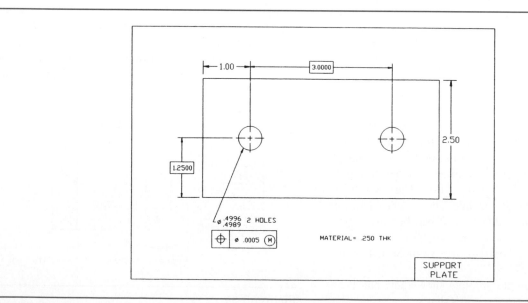

Figure 13-28

13-19 BUSHINGS

A *bushing* is a cylindrically shaped bearing that helps reduce friction between a moving part and a stationary part. Bushings, unlike ball bearings, have no moving parts. Bushings are usually made from oil-impregnated bronze or Teflon. Figure 13-29 shows a manufacturer's list of bronze bushings, and Figure 13-30 shows a list for Teflon bushings.

Bushings are cheaper than ball bearings but they wear over time, particularly if the application is high speed or one with heavy loading. Bushings are usually pressed into a supporting plate. Gear shafts must always have clearance from the inside diameters of bushings.

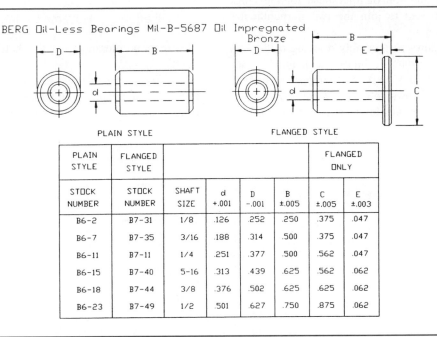

| PLAIN STYLE | FLANGED STYLE | | | | FLANGED ONLY | |
STOCK NUMBER	STOCK NUMBER	SHAFT SIZE	d +.001	D −.001	B ±.005	C ±.005	E ±.003
B6-2	B7-31	1/8	.126	.252	.250	.375	.047
B6-7	B7-35	3/16	.188	.314	.500	.375	.047
B6-11	B7-11	1/4	.251	.377	.500	.562	.047
B6-15	B7-40	5-16	.313	.439	.625	.562	.062
B6-18	B7-44	3/8	.376	.502	.625	.625	.062
B6-23	B7-49	1/2	.501	.627	.750	.875	.062

Figure 13-29
Courtesy of W. M. Berg Inc.

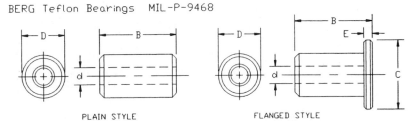

| PLAIN STYLE | FLANGED STYLE | | | | FLANGED ONLY | |
STOCK NUMBER	STOCK NUMBER	SHAFT SIZE	d +.001	D −.001	B ±.005	C ±.005	E ±.003
	B9-3	1/8	.126	.252	.250	.312	.047
B8-5	B9-6	3/16	.188	.315	.250	.375	.047
B8-11	B9-11	1/4	.251	.377	.500	.500	.047
B8-14	B9-15	5/16	.313	.439	.500	.562	.093
B8-19	B9-19	3/8	.376	.502	.625	.687	.093
B8-23	B8-29	1/2	.501	.628	.750	.875	.125

Figure 13-30
Courtesy of W. M. Berg Inc.

13-20 SAMPLE PROBLEM SP13-6

Figure 13-31 shows two support plates used to support and align a matched set of bevel gears. The gears selected are numbered S1346Z–48S30S60 in the Berg listings presented in Figure 13-16. The calculations are similar to those presented earlier for Sample Problem SP13-5 but with the addition of tolerances for the holes and machine screws used to join the two perpendicular support plates together.

Figure 13-31 shows an assembly drawing of the two gears along with appropriate bushings, stock number B6-11 from Figure 13-29, and shafts and supporting parts 1 and 2. The figure also shows detailed drawings of the two support plates with appropriate dimensions and tolerances.

The bushings are fitted into the support plates using an FN1 fit. The fit tables in the appendix define the maximum interference for an FN1 fit as .0075 and the minimum interference as .0001. The outside diameter of the bushing has a tolerance of .3770−.3760, per Figure 13-29. The feature tolerances for the holes in the support parts are found as follows:

Shaft max − Hole min = Interference max

Shaft min − Hole max = Interference min

The feature tolerance for the hole is therefore .3750−.3695.

The positional tolerance is determined as described in SP13-5 and is based on a tolerance of .001 between gear centers.

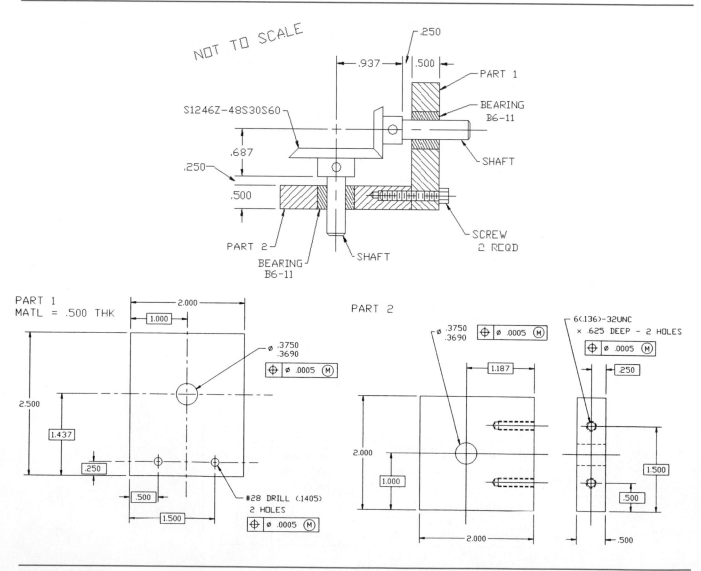

Figure 13-31

13-21 CAM DISPLACEMENT DIAGRAMS

A *displacement diagram* is used to define the motion of a cam follower. Displacement diagrams are set up as shown in Figure 13-32. The horizontal axis is marked off in 12 equal spaces that represent 30° on the cam. The vertical axis is used to define the linear displacement of the follower and is defined using either inches or millimeters.

The vertical axis of a displacement diagram must be drawn to scale because, once defined, the vertical distances are transferred to the cam's base circle to define the cam's shape. Figure 13-32 shows distances A, B, and C on both the displacement diagram and cam. The distances define the follower displacement at the 30°, 60°, and 90° marks, respectively.

The horizontal axis may use any equal spacing to indicate the angle because the vertical distances will be transferred to the cam along ray lines. The lower horizontal line represents the circumference of the base circle.

Figure 13-33 shows a second displacement diagram. Note how the distance between the 60° and 90° lines has been further subdivided. The additional lines are used to more accurately define the cam motion. Additional degree

lines are often added when the follower is undergoing a rapid change of motion.

The term *dwell* means the cam follower does not move either up or down as the cam turns. Dwells are drawn as straight horizontal lines on a displacement diagram. Note the horizontal line between the 90° and 210° lines on the displacement diagram shown in Figure 13-33. Dwells are drawn as sectors of constant radius on the cam.

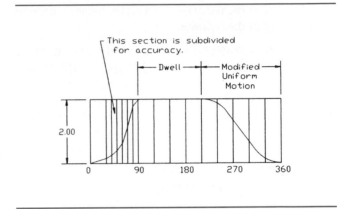

Figure 13-33

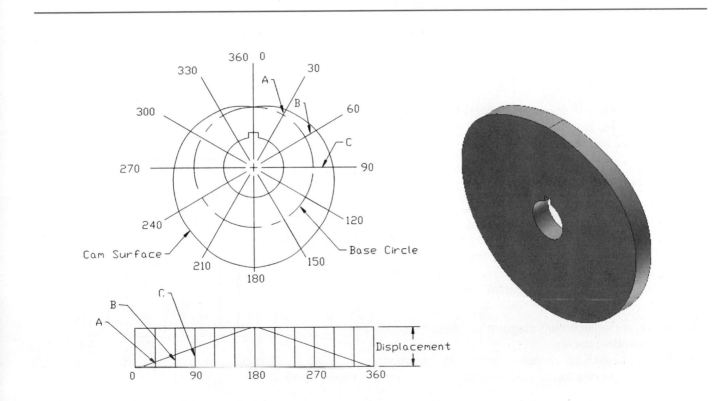

Figure 13-32

To set up a displacement diagram

See Figure 13-34. The given dimensions are in inches. The values in the brackets, [ ], are in millimeters.

1. Set **Grid** = .5 [10]
   Snap = .25 [5]
2. Draw a horizontal line **6** [120] long.
3. Draw a vertical line **2** [40] from the left end of the horizontal line as shown.

The length 2 [40] was chosen arbitrarily for this example. The vertical distance should be equal to the total displacement of the follower.

4. Use the **Array** command to create a rectangular array with **12** columns .5 [10] apart. Draw a horizontal line across the top of the diagram.
5. Label the horizontal axis in degrees, with each vertical line representing **30°**, and the vertical axis in inches [millimeters] of displacement.

13-22 CAM MOTIONS

The shape of a cam surface is designed to move a follower through a specific distance. The surface also determines the acceleration, deceleration, and smoothness of motion of the follower. It is important that a cam surface be shaped to maintain continuous contact with the follower. Several standard cam motions are defined next.

Uniform motion

Uniform motion is drawn as a straight line on a displacement diagram. See Figure 13-35. The follower rises the same distance for each degree of rotation by the cam.

Modified uniform motion

Modified uniform motion is similar to uniform motion but has a curved radius shape added to each end of the line to facilitate a smooth transition from the uniform motion to another type of motion or to a dwell section.

Figure 13-36 shows how to create a modified uniform motion on a displacement diagram. The procedure is as follows.

1. Set up a displacement diagram as presented in the preceding section.
2. Draw two arcs or circles of radius no greater than half the required displacement.

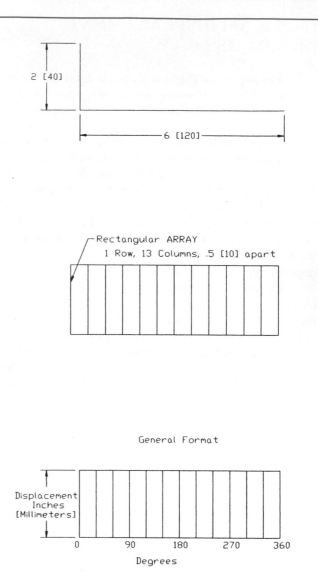

Figure 13-34

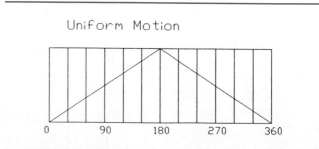

Figure 13-35

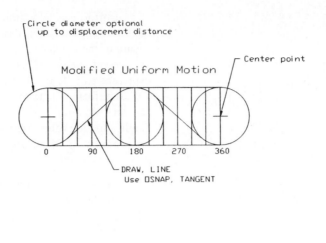

Circle diameter optional
up to displacement distance

Modified Uniform Motion

Center point

0 90 180 270 360

DRAW, LINE
Use OSNAP, TANGENT

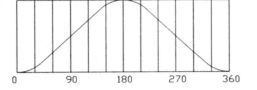

0 90 180 270 360

Figure 13-36

Any radius value can be used. In general, the larger the radius, the smoother the transition. In the example shown, a radius equal to half the total displacement was used.

3. Use **Osnap, Tangent,** and draw a line between the two arcs.
4. Erase and trim any excess lines.

Harmonic motion

Figure 13-37 shows how to create a harmonic cam motion. The procedure is as follows.

1. Set up a displacement diagram as presented in Section 13-21.
2. Draw a semicircle aligned with the left side of the displacement diagram. The diameter of the semicircle equals the total height of the displacement.

3. Use the **Array** command or the **Line** command with relative coordinate inputs, and draw rays every **30°** on the circle as shown. Label the rays from 0° to 180° in 30° segments.
4. Use **Osnap, Intersection** with **Ortho** on <F8>, and draw projection lines from the intersections of the rays with the circumference of the circle across the displacement diagram.
5. Draw a polyline starting at the lower left corner of the diagram and connecting the intersections of like angle lines.

For this example, the horizontal line from the circle's 30° increment intersects with the vertical line from the 30° mark on the displacement diagram.

6. Use **Edit Polyline, Fit** option, to change the straight polyline into a smooth curved line.
7. Project the same lines to the far side of the diagram to define the deceleration harmonic motion path.

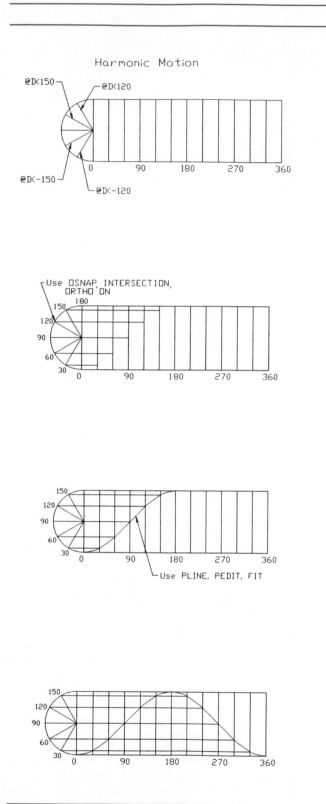

Uniform acceleration and deceleration

Uniform acceleration and deceleration are based on the knowledge that acceleration is related to distance by the square of the distance. Acceleration is measured in distance per second squared. Distances of units 1, 2, and 3 may be expressed as 1, 4, and 9, respectively.

Uniform acceleration and deceleration motions create smooth transitions between various displacement heights and are often used in high-speed applications.

Figure 13-38 shows how to create a uniform acceleration cam motion. The procedure is as follows.

1. Set up a displacement diagram as presented in Section 13-21.
2. Draw a horizontal construction line to the left and align it with the bottom horizontal line of the displacement diagram.
3. Use the **Array** command and create **1** column and **19** rows, **.1111** apart.

The 19 rows create 18 spaces. The uniform motion shape is created by combining six horizontal steps (30°, 60°, 90°, 120°, 150°, 180°) and the squares of six vertical steps (1, 4, 9, 4, 1, 0).

The vertical spacing is symmetrical about the centerline of the displacement diagram, so the original spacing is 1, 2, 3, 2, 1. The square of these values is used to create the uniform acceleration and deceleration.

The .1111 value was derived by dividing the displacement distance by the number of spaces, 2.00/18 = .1111.

4. Label the stack of vertical construction lines as shown.
5. Use the **Extend** command to extend the horizontal construction lines so that they intersect with the appropriate vertical degree line.
6. Use the **Edit Polyline** command's **Fit** option to create a smooth, continuous curve between the 0° and 180° lines.

The same line may be used to create a deceleration curve as shown.

Figure 13-37

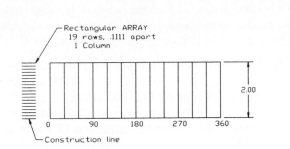

Rectangular ARRAY
19 rows, .1111 apart
1 Column

2.00

Construction line

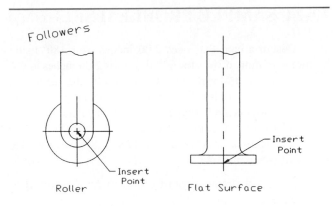

Followers

Insert
Point

Roller

Insert
Point

Flat Surface

Figure 13-39

13-23 CAM FOLLOWERS

There are two basic types of cam followers: those that roll as they follow the cam's surface, and those that have a fixed surface that slides in contact with the cam surface. Figure 13-39 shows an example of a roller follower and a fixed- or flat-surface follower. The flat-surface-type followers are limited to slow-moving cams with low force requirements.

Followers are usually spring-loaded to keep them in contact with the cam surface during operation. Springs were discussed in Section 11-24.

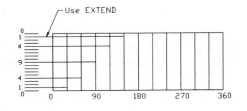

Use EXTEND

0
1
4
9
4
1
0

0   90   180   270   360

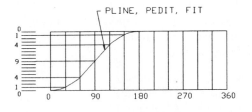

PLINE, PEDIT, FIT

0
1
4
9
4
1
0

0   90   180   270   360

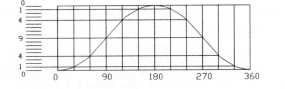

0
1
4
9
4
1
0

0   90   180   270   360

Figure 13-38

13-24 SAMPLE PROBLEM SP13-7

Design a cam that rises 2.00 inches over 180° using harmonic motion, dwells for 60°, descends 2.00 inches in 90° using modified uniform motion, and dwells the remaining 30°. The base circle for the cam is 3.00 inches in diameter, and the follower is a roller type with a 1.00-inch diameter. The cam will rotate in a counterclockwise direction. The center hole is .75 inch in diameter with a .125 × .875 keyway. See Figures 13-40 and 13-41.

1. Set up a displacement diagram as described in Section 13-21.
2. Define a path between the 0° and 180° vertical lines using the method described for harmonic motion. Draw the required circle on the left end of the diagram as shown.
3. Draw a horizontal line from the 180° to the 240° line.

This line defines the follower's dwell.

4. Draw two arcs of **.50** radius, one tangent to the top horizontal line of the displacement diagram and the second tangent to the bottom line.

Draw one arc on the 240° line and the other on the 330° line as shown. In this example the arcs have a radius equal to .25 of the total displacement. Any convenient radius value may be used.

5. Use the **Osnap, Tangent** command and draw a line tangent to the two arcs.
6. Erase and trim any excess lines and constructions.

This completes the displacement diagram. The follower distances are now transferred to the cam's base circle to define the surface shape. See Figure 13-41.

7. Draw two concentric circles of **3.00** and **8.00** diameter, respectively.

The 3.00 diameter is the base circle. The 8.00-diameter-circle value is derived from the radius of the base circle plus the maximum displacement plus the radius of the follower: 1.50 + 2 + .5 = 4.00 radius, or 8.00 diameter.

8. Use the **Setvar, Dimcen, −.2, Dim, Center** commands, or a modified **Center Mark** tool to draw the centerlines for the 8.00 circle.
9. Use the **Array** command, and polar array the top portion of the vertical centerline **12** times around the full 360°.
10. Label the ray lines as shown.

Note that the top vertical ray line is labeled both as 0 and 360.

11. Transfer the follower distances from the displacement diagram to the cam drawing.

Several different techniques can be used to transfer the distances: use the **Linear** dimension tool, the **Distance** command found under the **Tools** pull-down menu, or the **Inquiry** command. Use the **Osnap, Intersection** command to ensure accuracy. In this example the measured distance values are listed below the displacement diagram in Figure 13-40. The .50 addition to the bottom of the diagram is to account for the .50 follower radius. Note how the distance of 2.1000 was measured.

The distance values can be used to draw lines from the base circle along the appropriate ray line on the cam drawing using relative coordinate values. The values for this example are as follows:

.5000-(0)	@.5000<90
.6340-(30)	@.6340<60
1.0000-(60)	@1.0000<30
1.5000-(90)	@1.5000<0
2.0000-(120)	@2.0000<−30
2.3660-(150)	@2.3660<−60

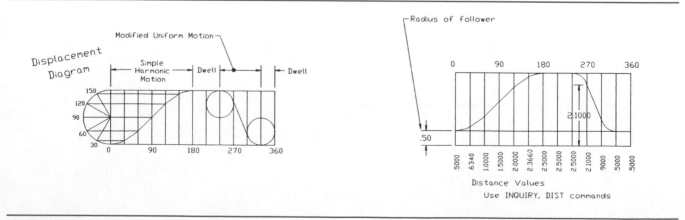

Figure 13-40

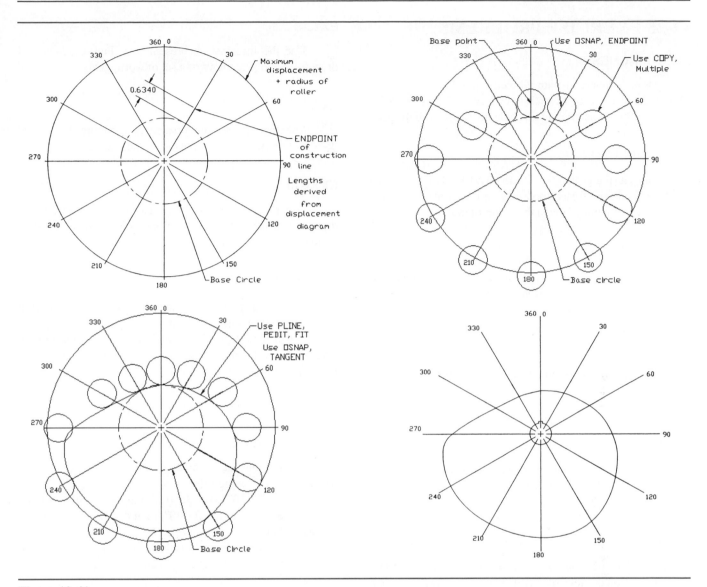

Figure 13-41

2.5000-(180) @2.5000<−90
2.5000-(210) @2.5000<−120
2.5000-(240) @2.5000<−150
2.1000-(270) @2.1000<180
.9000-(300) @.9000<150
.5000-(330) @.5000<120

Lines can be drawn over the existing vertical lines between the base line and the displacement path on the displacement diagram. The **Move, Rotate** or **Grips, Rotate** command can be used to transfer the lines to the base circle and appropriate ray line.

12. Draw circles of diameter **.50,** representing the roller follower, with their center points on the ends of the lines created in step 11.

Note that these lines and their endpoints are not visible on the screen because they are drawn directly over the existing ray lines. The **Osnap, Endpoint** command should be used to locate the circle's center point.

13. Use the **Draw, Polyline** command and draw a line tangent to the follower at each ray line.

The tangent point for each ray may not be exactly on the ray line.

14. Use the **Edit Polyline, Spline** command to create the cam surface.

15. Draw the center hole and keyway using the given dimensions.

16. Save the drawings, if desired.

13-25 EXERCISE PROBLEMS

EX13-1 INCHES

Draw a spur gear with 24 teeth on a 4.00-pitch circle. The fillets at the base of each tooth have a radius of .0625. Locate a 1.00-diameter mounting hole in the center of the gear.

EX13-2 MILLIMETERS

Draw a spur gear with 36 teeth on a 100-pitch circle. The fillets at the base of each tooth have a radius of 3. Locate a 20-diameter mounting hole in the center of the gear.

EX13-3

Use the information presented in Figure 13-12 and draw front and side sectional views of gear number A 1C22MYKW150 32A.

EX13-4

Use the information presented in Figure 13-12 and draw front and side sectional views of gear number A 1C22MYKW150 64A.

EX13-5

Use the information presented in Figure 13-15 and draw a sectional view of bevel gears A 1C 3MYK 30018 and A 1C 3MYK 30036H.

EX13-6

Use the information presented in Figure 13-15 and draw a sectional view of bevel gears A 1C 3MYK 15018 and A 1C 3MYK 15036.

EX13-7

Use the information presented in Figure 13-16 and draw a sectional view of the matched set of bevel gears S1346Z-48S30A120.

EX13-8

Use the information presented in Figure 13-20 and draw front and side views of a set of worm gears.

EX13-9

Use the information presented in Figure 13-22 and draw a front view of a matched set of helical gears.

EX13-10

Use the information presented in Figure 13-24 and draw a front view of a set of rack-and-pinion gears. Select a pinion gear from Figure 13-12.

EX13-11 INCHES

Draw a displacement diagram and appropriate cam based on the following information.

Dwell for 60°, rise 1.00 inch using harmonic motion over 180°, dwell for 30°, then descend 1.00 inch using harmonic motion over 90°.

The cam's base circle is 4.00 inches in diameter. Include a 1.25-inch center mounting hole.

EX13-12 MILLIMETERS

Draw a displacement diagram and appropriate cam based on the following information.

Dwell for 60°, rise 30 millimeters using harmonic motion over 180°, dwell for 30°, then descend 30 millimeters using harmonic motion over 90°.

The cam's base circle is 120 millimeters in diameter. Include a 20-millimeter center mounting hole.

EX13-13 INCHES

Draw a displacement diagram and appropriate cam based on the following information.

Rise 1.25 inches over 180° using uniform acceleration motion, dwell for 60°, descend 1.25 inches over 90° using modified uniform motion, dwell for 30°.

The cam's base circle is 3.25 inches in diameter. Include a .75-inch-diameter center mounting hole.

EX13-14 MILLIMETERS

Draw a displacement diagram and appropriate cam based on the following information.

Rise 20 millimeters over 180° using uniform acceleration motion, dwell for 60°, descend 20 millimeters over 90° using modified uniform motion, dwell for 30°.

The cam's base circle is 80 millimeters in diameter. Include a 16-millimeter-diameter center mounting hole.

EX13-15 INCHES

Draw a displacement diagram and appropriate cam based on the following information.

Rise .60 inch in 90° using harmonic motion, dwell for 30°, rise .60 inch in 60° using modified uniform motion, dwell 60°, descend 1.20 inches using uniform deceleration in 120°.

The cam's base circle is 3.20 inches in diameter. Include a 1.75-inch-diameter center mounting hole.

EX13-16 MILLIMETERS

Draw a displacement diagram and appropriate cam based on the following information.

Rise 16 millimeters in 90° using harmonic motion, dwell for 30°, rise 16 millimeters in 60° using modified uniform motion, dwell 60°, descend 32 millimeters using uniform deceleration in 120°.

The cam's base circle is 84 millimeters in diameter. Include a 20-millimeter-diameter center mounting hole.

EX13-17 INCHES

Draw a displacement diagram and appropriate cam based on the following information. The cam's base circle is 2.00 inches. Include a $\varnothing 0.625$ center mounting hole. Rise 0.375 inch in 90° using harmonic motion, dwell for 90°, rise 0.375 inch in 90° using harmonic motion, descend 0.750 inch using harmonic motion in 90°.

EX13-18 MILLIMETERS

Draw a displacement diagram and appropriate cam based on the following information. The cam's base circle is 98 millimeters. Include a $\varnothing 18$ center mounting hole. Rise 15 millimeters in 90° using harmonic motion, dwell for 90°, rise 15 millimeters in 90° using harmonic motion, descend 30 millimeters using harmonic motion in 90°.

EX13-19 INCHES

Draw a displacement diagram and appropriate cam based on the following information. The cam's base circle is 2.00 inches. Include a $\varnothing 0.625$ center mounting hole. Rise 0.375 inch in 90° using uniform acceleration and deceleration motion, dwell for 90°, rise 0.375 inch in 90° using uniform acceleration and deceleration motion, descend 0.750 inch using motion in 90°.

EX13-20 MILLIMETERS

Draw a displacement diagram and appropriate cam based on the following information. The cam's base circle is 98 millimeters. Include a $\varnothing 18$ millimeters center mounting hole. Rise 15 millimeters in 90° using uniform acceleration and deceleration motion, dwell for 90°, rise 15 millimeters in 90° using uniform acceleration and deceleration motion, descend 30 millimeters using motion in 90°.

EX13-21 INCHES

Draw an assembly drawing and parts list for the 2-gear assembly shown.

2-Gear Assembly

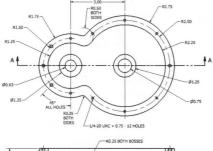

	Small Gear	Large Gear
Number of teeth	48	96
Face width	0.50	0.50
Diametral pitch	24	24
Pressure angle	20	20
Hub Ø	0.750	1.00
Base Ø	0.500	0.625

Parts List				
ITEM	PART NUMBER	DESCRIPTION	MATERIAL	QTY
1	ENG-453-A	GEAR, HOUSING	CAST IRON	1
2	BU-1123	BUSHING Ø0.75	Delrin, Black	1
3	BU-1126	BUSHING Ø0.625	Delrin, Black	1
4	ASSEMBLY-6	GEAR ASSEEMBLY	STEEL	1
5	AM-314	SHAFT, GEAR Ø.625	STEEL	1
6	AM-315	SHAFT, GEAR Ø.0.500	STEEL	1
7	ENG -566-B	COVER, GEAR	CAST IRON	1
8	ANSI B18.6.2 - 1/4-20 UNC - 0.75	Slotted Round Head Cap Screw	Steel, Mild	12

Gear Housing

Gear Cover

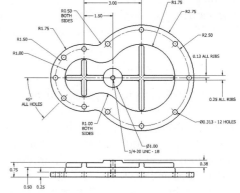

Bushing Ø0.625

Bushing Ø0.75

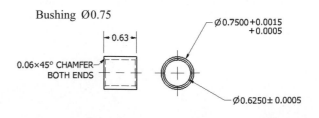

Shaft, Gear Ø0.625

Shaft, Gear Ø0.500

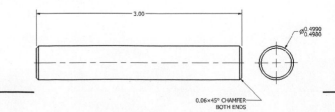

EX13-22

An electric motor operates at 3600 rpm. Design a gear that includes at least four spur gears (more may be used if needed) and reduces the motor speed to 200 rpm. Select gears and bearings from the tables in this book or from manufacturers' websites. Design each shaft as needed.

Enclose the gears in a box made from .50-inch [12-millimeter] plates, assembled using flat head screws. There should be at least three screws per edge on the box.

Extend the shafts for the input and output at least .50[12] outside the box. The other shafts should end at the edge of the box. Mount each shaft using two ball bearings, one mounted in each support plate.

Assume that the center distances have a tolerance of +.001, −.000 and the support shafts have tolerances of +.0000, −.0002, or their metric equivalent.

A. Draw an assembly drawing showing the gears in their assembled positions.

B. Support the gear shafts with ball bearings press fitted into the support plate. The support plate is to be .50 inch or 12 millimeters thick. The length and width dimensions are arbitrary, but there should be at least .25 [6] clearance between the gears and the support plate.

C. Draw detail drawings of the required support plates and shafts. Include dimensions and tolerances. Locate all support holes using positional tolerances.

Suggested websites:

www.wmberg.com
www.bostgear.com
www.newmantools.com

Do your own search for gears and bearings.

Top and end pieces omitted for clarity

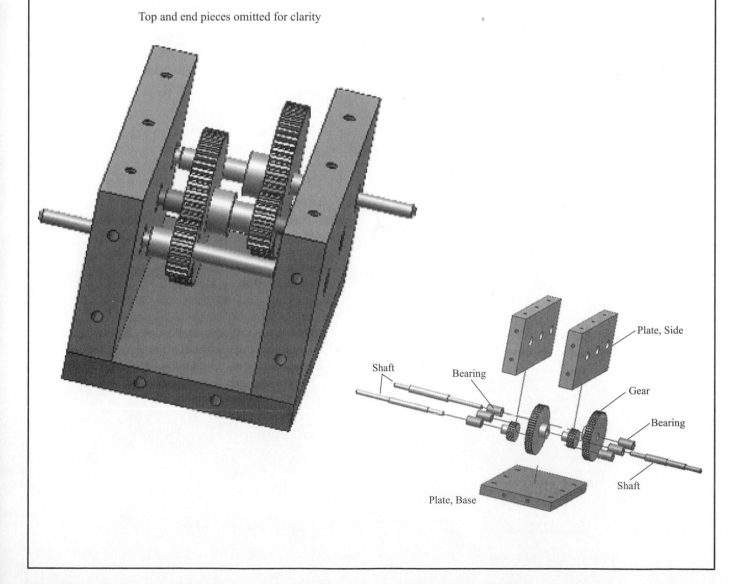

Shaft · Bearing · Plate, Side · Gear · Bearing · Shaft · Plate, Base

EX13-23

The following figure shows a general setup for matched bevel gears. Complete the design for a gear box with a ratio of 3:1 between the two gears. Select appropriate bearings and fasteners. Dimension and tolerance the shaft sizes and each of the six supporting plates. Extend the input and output shafts at least 1.00 inch [24 millimeters] beyond the surface of the box.

Assume that the center distances have a tolerance of +.001, −.000 and the support shafts have a tolerance of +.0000, −.0002, or their metric equivalent.

The figure shown represents half the gear box. The other half includes another three plates without the holes

for the shafts. There should be at least three screws in each edge of the box.

A. Draw an assembly drawing showing the gears in their assembled positions.
B. Support the gear shafts with ball bearings press fitted into the support plate. The support plate is to be .50 inch or 12 millimeters thick. The length and width dimensions are arbitrary, but there should be at least .375 [10] clearance between the gears and the support plates.
C. Draw detail drawings of the required support plates. Include dimensions and tolerances. Locate all support holes using positional tolerances.

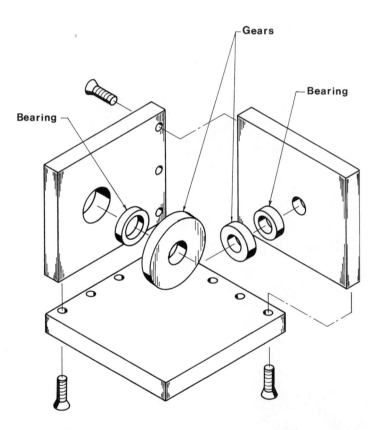

EX13-24 MILLIMETERS

Draw an assembly drawing and parts list for the 4-gear assembly shown.

4-Gear Assembly

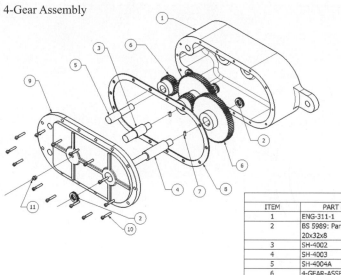

Parts List				
ITEM	PART NUMBER	DESCRIPTION	MATERIAL	QTY
1	ENG-311-1	4-GEAR HOUSING	CAST IRON	1
2	BS 5989: Part 1 - 0 10 - 20x32x8	Thrust Thrust Ball Bearing	Steel, Mild	4
3	SH-4002	SHAFT, NEUTRAL	STEEL	1
4	SH-4003	SHAFT, OUTPUT	STEEL	1
5	SH-4004A	SHAFT,INPUT	STEEL	1
6	4-GEAR-ASSEMBLY		STEEL	2
7	CSN 02 1181 - M6 x 16	Slotted Headless Set Screw - Flat Point	Steel, Mild	2
8	ENG-312-1	GASKET	Brass, Soft Yellow	1
9	COVER			1
10	CNS 4355 - M 6 x 35	Slotted Cheese Head Screw	Steel, Mild	14
11	CSN 02 7421 - M10 x 1coned short	Lubricating Nipple, coned Type A	Steel, Mild	1

4-Gear Housing

Cover

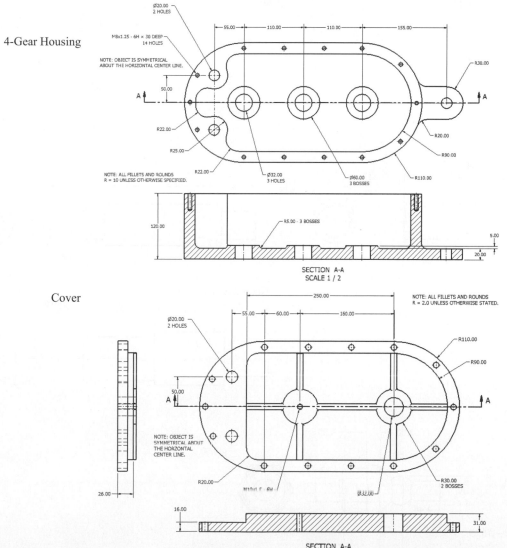

Gasket

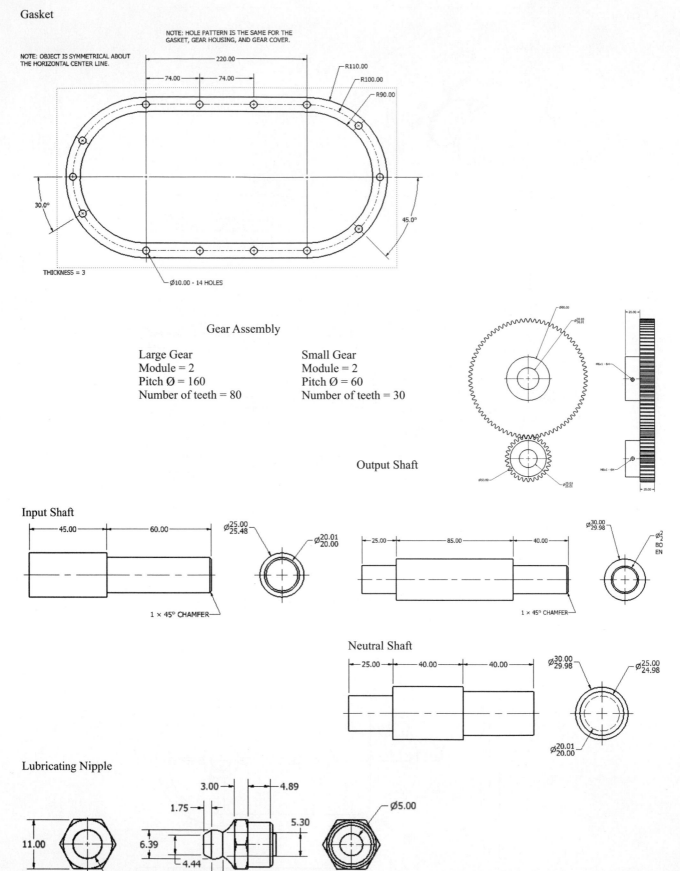

Gear Assembly

Large Gear
Module = 2
Pitch Ø = 160
Number of teeth = 80

Small Gear
Module = 2
Pitch Ø = 60
Number of teeth = 30

Output Shaft

Input Shaft

Neutral Shaft

Lubricating Nipple

EX13-25

Create the following for the slider assembly shown.

1. Select the appropriate fasteners.
2. Create an assembly drawing.
3. Create a parts list.
4. Specify the tolerance for the Guide Shaft/Bearings interface.

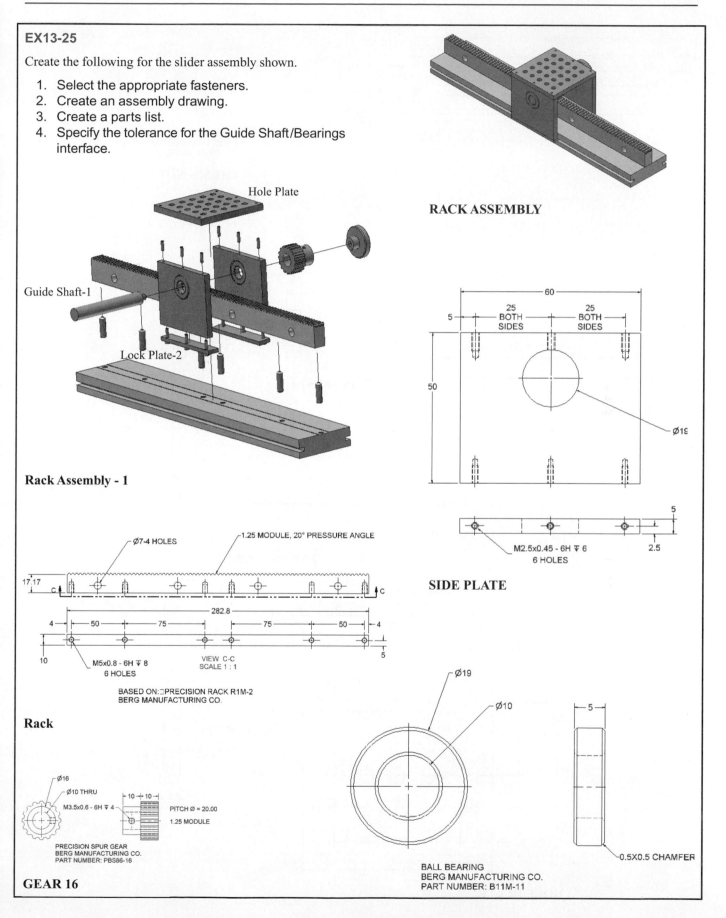

RACK ASSEMBLY

Rack Assembly - 1

Guide Shaft-1

Hole Plate

Lock Plate-2

Ø7-4 HOLES

1.25 MODULE, 20° PRESSURE ANGLE

17.17

C

282.8

4   50   75   75   50   4

10

M5x0.8 - 6H ▼ 8
6 HOLES

VIEW C-C
SCALE 1 : 1

5

BASED ON:⊐PRECISION RACK R1M-2
BERG MANUFACTURING CO.

Rack

Ø16

Ø10 THRU

M3.5x0.6 - 6H ▼ 4

10 ─ 10

PITCH Ø = 20.00

1.25 MODULE

PRECISION SPUR GEAR
BERG MANUFACTURING CO.
PART NUMBER: PBS86-16

GEAR 16

60

25
BOTH
SIDES

25
BOTH
SIDES

5

50

Ø19

5

M2.5x0.45 - 6H ▼ 6
6 HOLES

2.5

SIDE PLATE

Ø19

Ø10

5

BALL BEARING
BERG MANUFACTURING CO.
PART NUMBER: B11M-11

0.5X0.5 CHAMFER

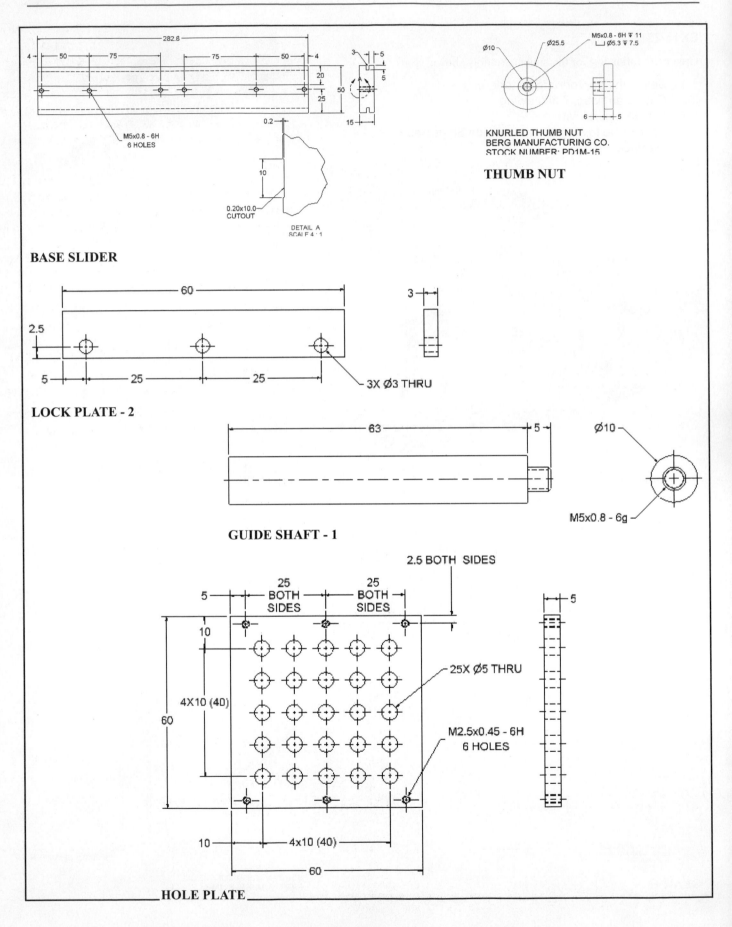

BASE SLIDER

LOCK PLATE - 2

GUIDE SHAFT - 1

THUMB NUT

KNURLED THUMB NUT
BERG MANUFACTURING CO.
STOCK NUMBER: PD1M-15

HOLE PLATE

EX13-26 MILLIMETERS

Draw an assembly drawing and parts list for the cam asembly shown.

Cam parameters

. Base circle = ⌀146
   Face width = 16
   Motion: rise 10 using harmonic motion in 90°, dwell for 180°, fall 10 in 90°
   Bore = ⌀16.0
   Keyway = 2.3 × 5 × 16

Follower ⌀ = 16
Follower width = 4
Square key = 5 × 5 × 16

Bearing overall dimensions

DIN625-SKF 6203 (ID × OD × THK) 17 × 40 × 10
DIN625-SKF 634 4 × 13 × 4
GB 2273.2-87-7/70 8 × 18 × 5

Spring parameters

Wire ⌀ = 1.5
Inside ⌀ = 9.0
Length = 20
Coil direction = Right
Active coils = 10
Grind both ends

Parts List			
ITEM	QTY	PART NUMBER	DESCRIPTION
1	1	ENG-2008-A	BASE, CAST
2	1	DIN625 - SKF 6203	Single row ball bearings
3	1	SHF-4004-16	SHAFT: Ø16×120, WITH 2.3×5×16 KEYWAY
4	1		SUB-ASSEMBLY, FOLLOWER
5	1	SPR-C22	SPRING, COMPRESSION
6	1	GB 273.2-87 - 7/70 - 8 x 18 x 5	Rolling bearings - Thrust bearings - Plan of boundary dimensions
7	1	IS 2048 - 1983 - Specification for Parallel Keys and Keyways B 5 x 5 x 16	Specification for Parallel Keys and Keyways

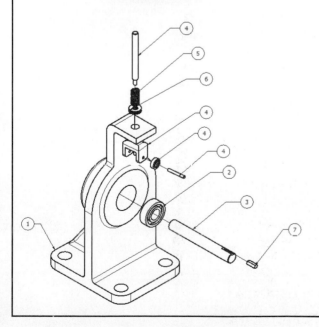

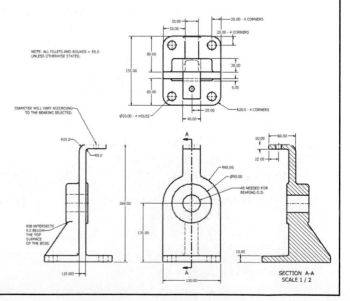

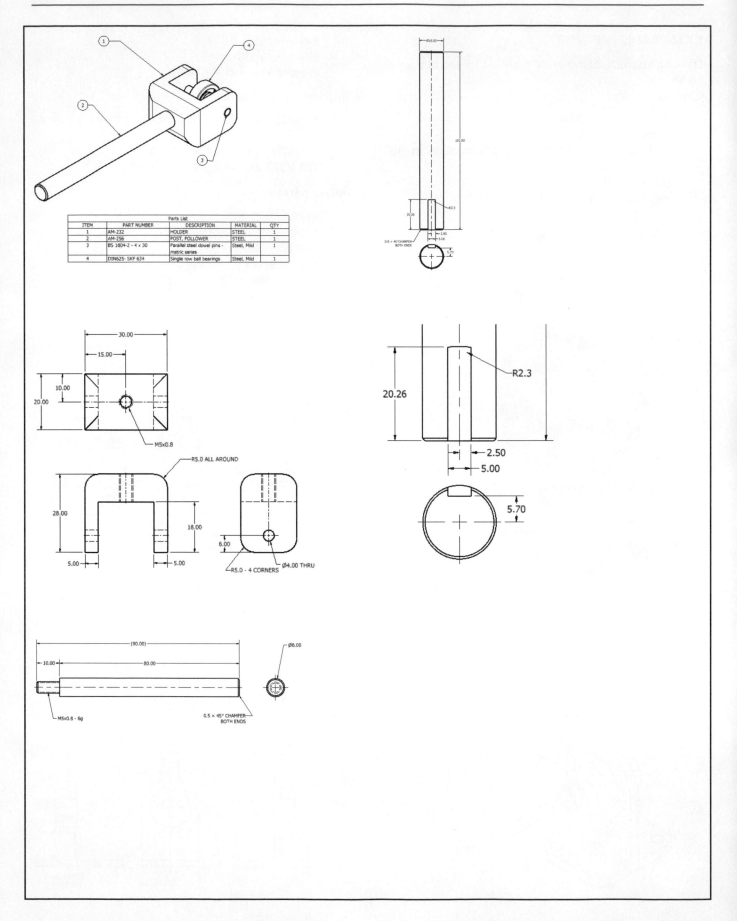

Parts List				
ITEM	PART NUMBER	DESCRIPTION	MATERIAL	QTY
1	AM-232	HOLDER	STEEL	1
2	AM-256	POST, FOLLOWER	STEEL	1
3	BS 1804-2 - 4 x 30	Parallel steel dowel pins - metric series	Steel, Mild	1
4	DIN625- SKF 634	Single row ball bearings	Steel, Mild	1

14

Fundamentals of 3D Drawing

14-1 INTRODUCTION

This chapter introduces the fundamental concepts needed to produce 3D drawings using AutoCAD using the **View** command and **acad 3D** template. The chapter shows how to change viewpoints and how to create, save, and work with user-defined coordinate systems called *user coordinate systems,* or UCSs.

The chapter demonstrates how to use both the **Viewports** and **UCS** toolbars. It also shows how to create orthographic views from given 3D objects using the **View** toolbar.

14-2 THE WORLD COORDINATE SYSTEM

AutoCAD's absolute coordinate system is called the *world coordinate system* (WCS). The default setting for the WCS is a viewing position located so that you are looking at the system 90° to its X,Y plane. See Figure 14-1. The Z axis is also perpendicular to the X,Y plane, or directly aligned with your viewpoint. This setup is ideal for 2D drawings and is called a *plan view.* All your drawings up to now have been done in this orientation.

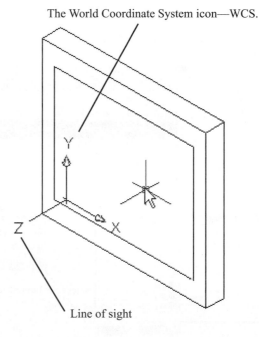

The World Coordinate System icon—WCS.

Y

Z

X

Line of sight

Figure 14-1

563

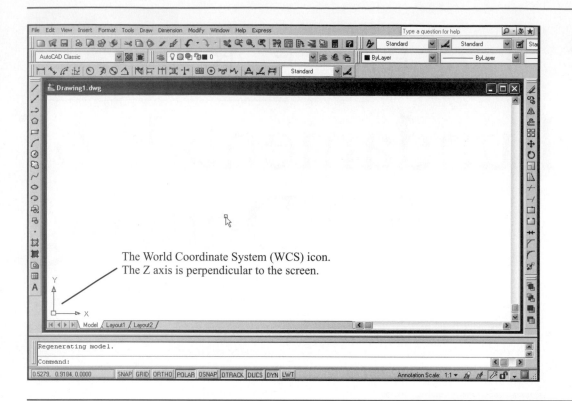

The World Coordinate System (WCS) icon.
The Z axis is perpendicular to the screen.

Figure 14-2

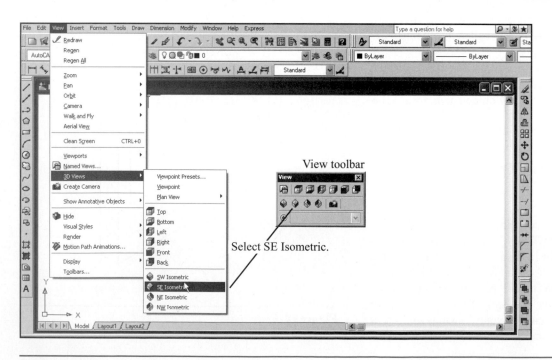

View toolbar

Select SE Isometric.

Figure 14-3

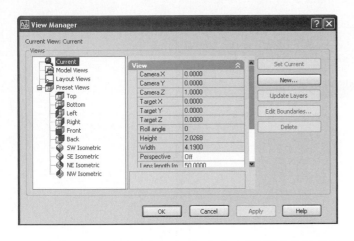

Figure 14-4

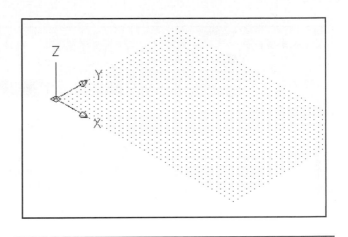

Figure 14-5

Figure 14-2 shows a standard drawing screen created using the acad template manager oriented so that you are looking directly down on the X,Y plane of the world coordinate system. Note the icon in the lower left corner of the screen. The box around the icon's origin indicates that it is the WCS, and the X and Y indicate the orientation of these axes. The WCS icon will always be present when you are in the world coordinate system.

14-3 VIEWPOINTS

The orientation of the WCS may be changed by changing the drawing's viewpoint. There are three ways to change the viewpoint: use the **View** pull-down menu, type the word **view,** or use the **View** toolbar.

To change the viewpoint using the View pull-down menu

This exercise assumes that the screen includes a grid. The grid serves to help define a visual orientation.

1. Select the **View** pull-down menu.
2. Select **3D Views.**

See Figure 14-3. The current WCS orientation is the **Plan View** setting.

3. Select **SE Isometric.**

To change the viewpoint using the View command

1. Type the word **view** in response to a command prompt.

The **View Manager** dialog box will appear. See Figure 14-4.

2. Click **Preset Views;** then select **SE Isometric,** then select **OK.**

The drawing will now be oriented as shown in Figure 14-5. Note the change in the WCS icon. It is important to remember that you are still in the WCS but are merely looking at it from a different perspective. This concept may be verified by drawing some simple shapes. Figure 14-6 shows a rectangular shape created using the **Line** command, and a circle created using the **Circle** command. These shapes are 2D shapes drawn in the WCS.

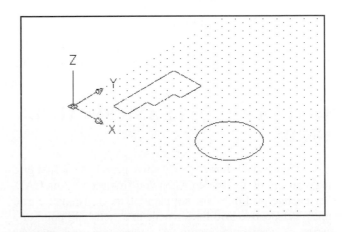

Figure 14-6

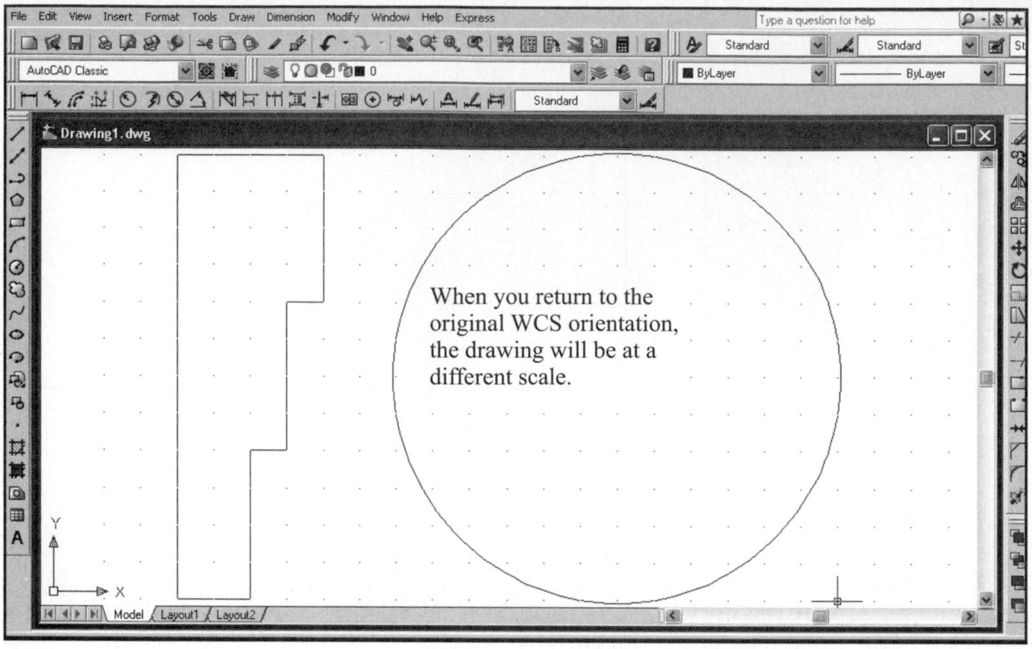

Figure 14-7

To return to the original WCS orientation

1. Select the **View** pull-down menu.
2. Select **3D Views, Plan View, World UCS.**

See Figure 14-7. The drawing will reappear on the screen at a different scale than was originally defined.

3. Type **zoom;** press **Enter.**

   [All/Center/Dynamic/Extents/Previous/Scale/Window/] <real time>:

4. Type **.75;** press **Enter.**

Figure 14-8 shows the resulting screen scale. The drawing may also be zoomed using the center mouse wheel.

14-4 PERSPECTIVE AND PARALLEL GRIDS

Figure 14-9 shows a perspective grid. It is the first grid that appears on the screen when you first access AutoCAD. Perspective grid lines are not parallel, as are isometric grid lines. Perspective grid lines recede to a vanishing point or a series of vanishing points. A perspective grid may be changed to a parallel grid.

Right-click the mouse and select the **Zoom** option, then right-click again and select the **Parallel** option. The same sequence is used to return the grid to a perspective grid. You may work directly on this initial screen or create a new drawing with a perspective grid using the **acad3D** template.

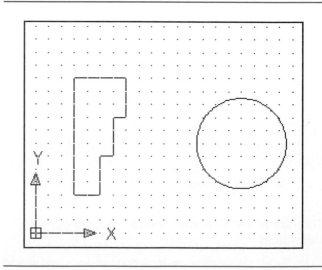

Figure 14-8

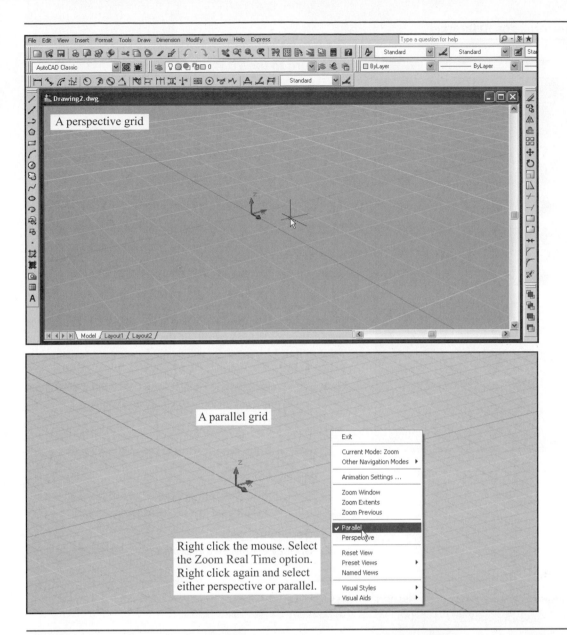

A perspective grid

A parallel grid

Right click the mouse. Select
the Zoom Real Time option.
Right click again and select
either perspective or parallel.

Figure 14-9

To create a drawing with a perspective grid

1. Click the **New** tool on the **Standard** toolbar
2. Select the **acad3D** template; click **Open**.

The perspective grid will appear. See Figure 14-10.

The two-dimensional drawing may be applied to the perspective grid background as it was to the isometric grid background. Figure 14-11 shows the same two shapes drawn in Figure 14-6. The difference in the size of the grid background was created using the center mouse wheel. The grid size may also be controlled using the **Zoom All** command. Drawings created on the perspective grid may be viewed on a two-dimensional grid.

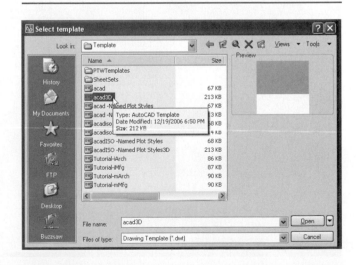

Figure 14-10

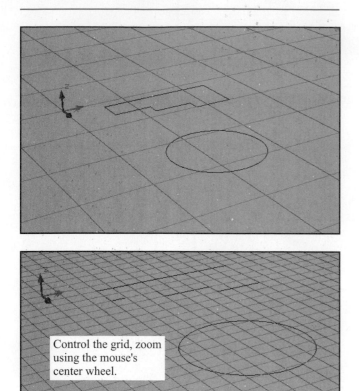

Figure 14-11

To return to the 2D WCS

1. Select the **View** pull-down menu.
2. Select **3D views, Plan View, World UCS.**

   See Figure 14-12.

3. Zoom the drawing as desired using the mouse's center wheel or the **Zoom All** command.

14-5 USER COORDINATE SYSTEM (UCS)

User coordinate systems are coordinate systems that you define relative to the WCS. Drawings often contain several UCSs. UCSs can be saved and recalled.

AutoCAD drawings may be created only on an XY plane. So, for example, if you had a box whose bottom surface was drawn on the XY plane of the WCS and you wished to draw a circle's top surface, you would have to create a new XY plane, a new UCS, on the box's top surface. This section will show how to create new UCSs. A solid box will be drawn and then used to demonstrate how to create and use new UCSs.

To draw a solid box

1. Create a **3D SE isometric** axis using WCS.

   See Figure 14-13.

2. Activate the **Modeling** and **UCS** toolbars.
3. Click the **Box** tool on the **Modeling** toolbar.

   Command:_box
   Specify corner of the box or [Center] <0,0,0>:

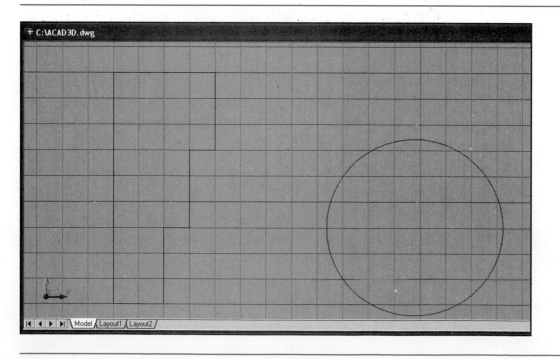

Figure 14-12

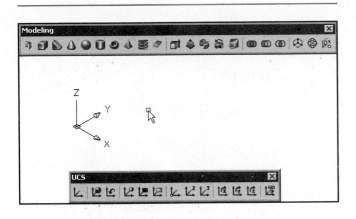

Figure 14-13

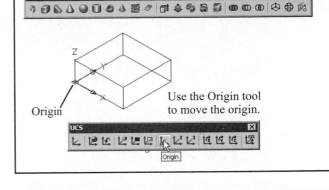

Figure 14-15

4. Type **0, 0, 0**; press **Enter**.

    Specify corner or [Cube/Length]:

5. Type **5,5,0,** or use the **Snap** tool with dynamic coordinates, press **Enter**.

    Specify height:

6. Type **2.0**; press **Enter**.

    A box will appear on the screen. See Figure 14-14. The corner of the box is located on the WCS origin.

To create a UCS on the top surface

1. Click on the **Origin** tool on the **UCS** toolbar.

    See Figure 14-15.

    Specify new origin point <0,0,0>:

2. Specify the top left corner of the box as the new origin.

See Figure 14-16. You must use the **Endpoint** option of the **Object Snap** (accessed by pressing **<Shift>**/right mouse button) commands to define the corner. The cursor moves only in the XY plane. If you locate the cursor on the corner without using the **Endpoint** option, the cursor may appear to be on the corner, but, in fact, it is located on the XY plane at a point behind the corner point.

The origin is now located on the top left corner of the box. This is a new UCS.

3. Use the **Circle** command and draw a circle on the top surface of the box.

To save a UCS

The UCS created in Figure 14-16 can be saved and later restored without having to redefine it.

1. Click the **UCS** tool on the **UCS** toolbar.

    Specify origin of UCS or [Face/Named/Object/ Previous/World/X/Y/Zaxis]<World>:

Use the Box tool on the Modeling toolbar and draw a rectangular box as shown.

5, 5, 2

5, 5, 0

0, 0, 0

Figure 14-14

Move the origin to the top surface. Use Osnap, Endpoint to define the upper corner.

Draw a circle on the top surface using the Circle tool.

New UCS

Figure 14-16

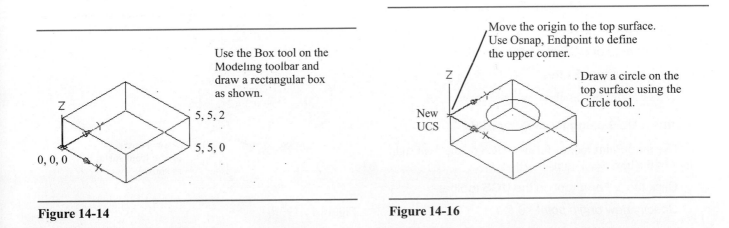

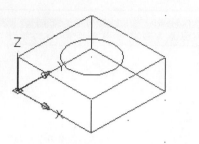

Use the World tool to return
the origin to its original location.

Figure 14-17

2. Type **NA**; press **Enter**.

Enter an option [Restore/Save/Delete/?]:

3. Type **s**; press **Enter**.

Enter name to save current UCS or [?]:

4. Type **top**; press **Enter**.

To return to the WCS

1. Click the **World** tool on the **UCS** toolbar.

The origin will be located on the original WCS axis.
See Figure 14-17.

To return to a saved UCS

1. Click the **UCS** tool on the **UCS** toolbar.

*Specify origin of UCS or [Face/Named/Object/
Previous/ World/X/Y/Zaxis] <World>:*

2. Type **NA**; press **Enter**.

Enter an option [Restore/Save/Delete/?]:

3. Type **r**; press **Enter**.

Enter name of UCS to restore or [?]:

4. Type **top**; press **Enter**.

The UCS defined as above will be restored.

To define a UCS using the 3 Point tool

Use the **3 Point** tool to define a UCS on the front right
surface of the box. See Figure 14-18.

1. Click the **3 Point** tool on the **UCS** toolbar.

Specify new origin point <0,0,0>:

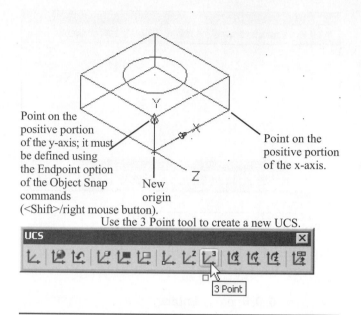

Point on the
positive portion
of the y-axis; it must
be defined using
the Endpoint option
of the Object Snap
commands
(<Shift>/right mouse button).

New
origin

Point on the
positive portion
of the x-axis.

Use the 3 Point tool to create a new UCS.

Figure 14-18

2. Click the lower corner of the box.

*Specify point on the positive portion of the X-axis
<x,x,x>:*

3. Click the lower far right corner of the box.

*Specify point on the positive portion of the Y-axis
<x,x,x>:*

The upper front corner of the box, just above the new
origin, will be specified, but it cannot simply be clicked. The
corner is not on the current XY plane.

4. Use the **Endpoint** option of the **Object Snap**
commands and click the corner.

5. Use the **Circle** command on the **Draw** toolbar
and draw a circle on the new UCS.

See Figure 14-19.

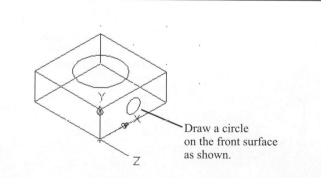

Draw a circle
on the front surface
as shown.

Figure 14-19

14-6 TO EDIT A SOLID MODEL

Solid models may be edited; that is, you can change their shapes after they have been created. Figure 14-20 shows the $5 \times 5 \times 2$ solid model created in the last section.

To change the length and width of a solid model

1. Click the box.

   A help box will appear. See Figure 14-21.

2. Close the help box.

   Blue arrowheads and boxes will appear on the base plane of the box. See Figure 14-22.

3. Locate the cursor on the **5,5,0** corner of the box, then click and hold the mouse button.

4. Move the cursor so that the corner has coordinate value **7,7,0.** Click the mouse.

5. Press the **<Esc>** key.

14-7 CREATING UCSs ON A PERSPECTIVE GRID

The procedures just defined for creating and saving UCSs are exactly the same for objects created on a perspective grid. Figure 14-23 shows the same $5 \times 5 \times 2$ solid model in Figure 14-14 drawn on a perspective grid. The box was drawn using the **Box** tool on the **Modeling** toolbar and an **acad3D** template.

The solid model will appear as solid gray object that is difficult to see.

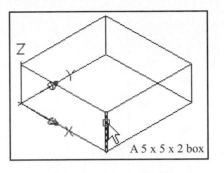

A 5 x 5 x 2 box

Figure 14-20

To change visual styles

1. Click the **View** pull-down menu.
   See Figure 14-24.

2. Select the **Visual Styles** option, then the **3DWireframe** option.

   The model will change to the wireframe format.
   Figure 14-25 shows the block displayed using each of the four options.

14-8 ROTATING A UCS AXIS

Figure 14-26 shows a $5 \times 5 \times 2$ box drawn on the WCS axis with its origin on the 0,0,0 point. It uses the **Conceptual** visual style. A new UCS can be created by rotating the coordinates about one of the major axes.

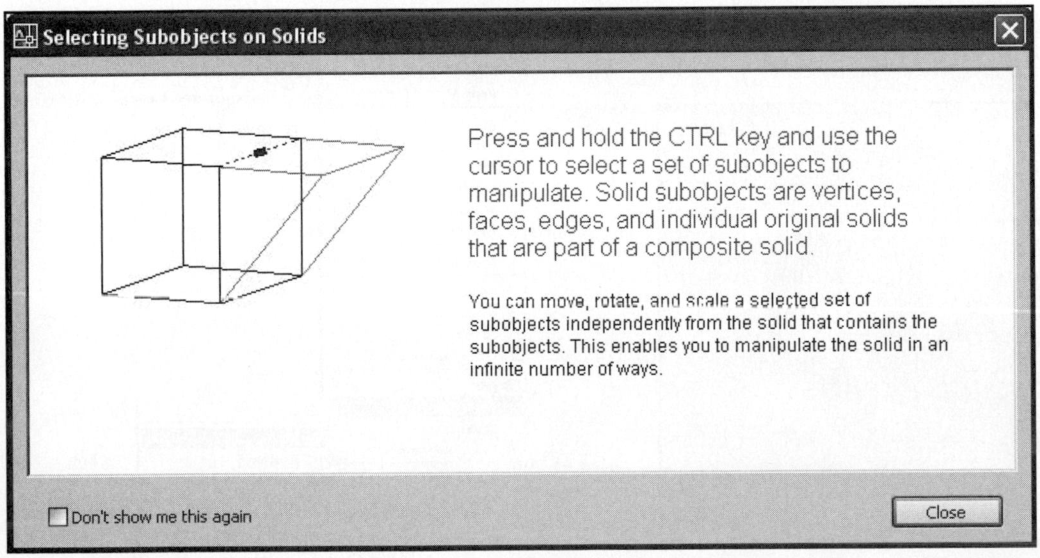

⚏ Selecting Subobjects on Solids   ☒

Press and hold the CTRL key and use the cursor to select a set of subobjects to manipulate. Solid subobjects are vertices, faces, edges, and individual original solids that are part of a composite solid.

You can move, rotate, and scale a selected set of subobjects independently from the solid that contains the subobjects. This enables you to manipulate the solid in an infinite number of ways.

☐ Don't show me this again   [Close]

Figure 14-21

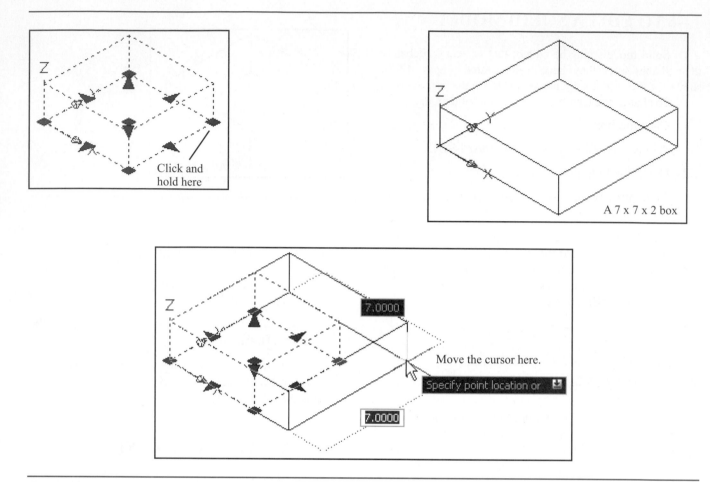

Figure 14-22

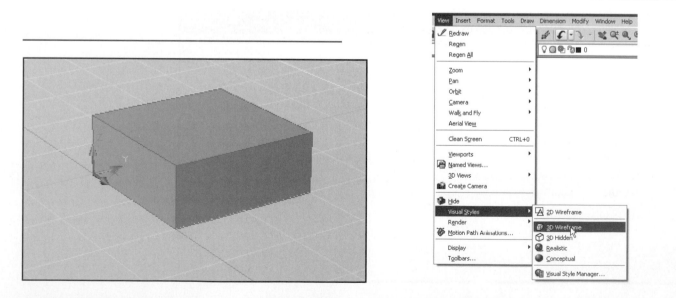

Figure 14-23

Figure 14-24

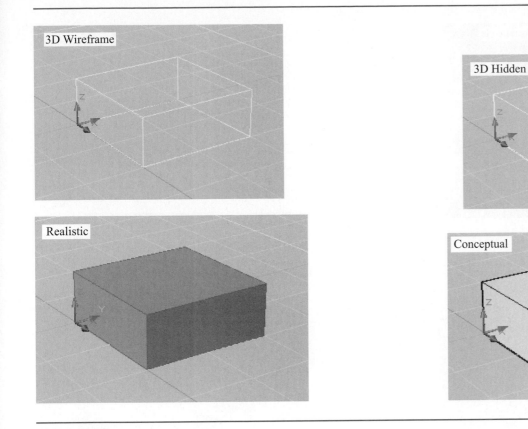

Figure 14-25

To rotate about the X-axis

See Figure 14-26.

1. Click the **X-axis** tool on the **UCS** toolbar.

   Specify rotation axis about the X-axis <90°>:

2. Press **Enter.**

This input accepts the 90° default value. The axis and grid will rotate 90° about the X axis. You can now draw on the new XY axis, which has the positive Y direction in the vertical direction.

3. Click the **Y-axis** tool on the **UCS** toolbar.

   Specify rotation axis about the Y-axis <90°>:

4. Press **Enter.**

The XZ axis and grid will have rotated 90°. You can now draw on the back plane of the box.

5. Click the **Z-axis** tool.

14-9 SAMPLE PROBLEM SP14-1

Draw an L-shaped bracket. The bracket will be used to demonstrate how to use UCSs.

1. Start a new drawing using the **acad3D** template with a **Perspective** grid.
2. Use the **Box** tool on the **Modeling** toolbar and draw a **5,5,2 box** using the **3DWireframe** visual style.

   See Figure 14-27.

3. Use the **Origin** tool on the **UCS** toolbar and move the origin to the top surface of the box as shown.
4. Use the new origin and draw a **2,5,2.5** box.
5. Use the **Union** tool on the **Modeling** toolbar and join the two boxes to form an L-shaped bracket.

14-10 VISUAL ERRORS

When working in 3D, it is important to remember that you cannot rely on visual inputs to locate shapes. What you see may be misleading. For example, Figure 14-28 shows a circle that appears to be drawn on the upper right surface of an L-shaped bracket. In fact, the circle is not located on the surface. This visual distortion is due to the line of sight of the view. The fact that the circle is on the back surface can be verified by changing the point of view from **SE Isometric** to **Top.**

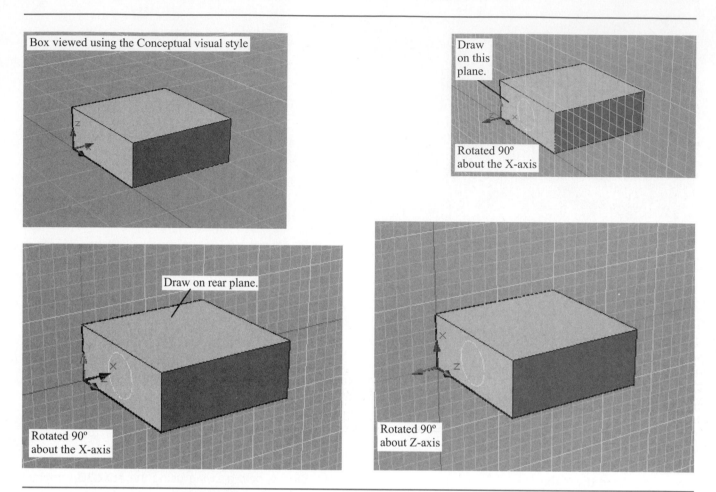

Figure 14-26

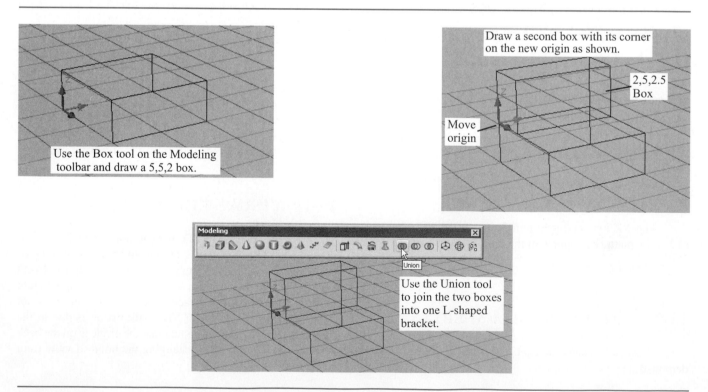

Figure 14-27

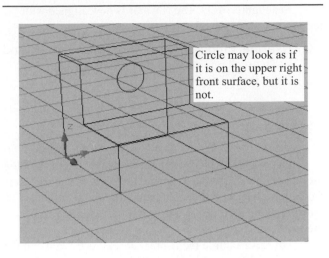

Circle may look as if it is on the upper right front surface, but it is not.

Figure 14-28

To change views

1. Click the **View** pull-down menu, then select the **3D** command, then the **Top** view as shown in Figure 14-29.

The **Top** view is shown in Figure 14-30. This view clearly shows that the circle is drawn on the back surface.

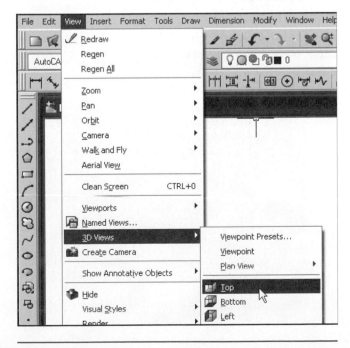

Figure 14-29

14-11 SAMPLE PROBLEM SP14-2

Figure 14-31 shows the same L-bracket drawn for Figure 14-27. This section will use different UCSs to draw 2D shapes on three of the L-bracket's surfaces.

To draw a circle on the upper front surface

1. Use the **3 Point** tool on the UCS toolbar and locate the origin and axis system as shown in Figure 14-32.

Remember, you can draw only on the XY plane. Use the **Object Snap** options to assure that the correct origin location is selected.

2. Use the **Circle** tool on the **Draw** toolbar and draw a circle on the front upper surface, as shown.

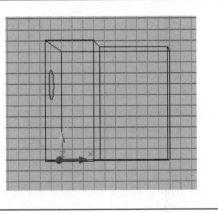

Figure 14-30

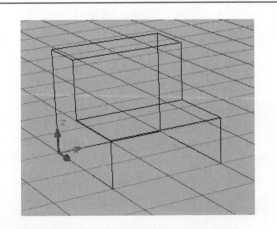

Figure 14-31

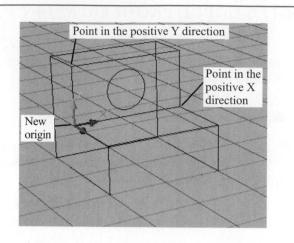

Figure 14-32

The purpose of the circle to help visually identify the surface it is on, so any location and size may be used.

To draw a rectangle on the top surface

1. Return to the WCS, then use the **Origin** tool on the **UCS** toolbar to move the axis to the top surface.

See Figure 14-33. The grid has been turned off for clarity.

2. Use the **Rectangle** tool on the **Draw** toolbar and draw a rectangle on the top surface.

To draw an ellipse on the left vertical surface

1. Return to the WCS, then use the **X** tool on the **UCS** toolbar and rotate the axis system **90°** about the X axis.

See Figure 14-34. The XY axis is now aligned with the left vertical axis.

2. Use the **Ellipse** tool on the **Draw** toolbar and draw an ellipse on the left vertical surface.
3. Save the L-bracket.

14-12 ORTHOGRAPHIC VIEWS

Once a 3D object has been created, orthographic views may be taken directly from the object. The screen is first split into four ports, each showing the 3D object. The viewpoint of three of the ports will be changed to create the front, top, and right-side views of the object.

For this example the L-bracket was redrawn using the same dimension and presented for Figure 14-14, but using an acad template. The SE Isometric view was used in Figure 14-35.

To create four viewports

1. Select the **View** pull-down menu.

See Figure 14-35.

2. Select **Viewports, 4 Viewports.**

The four viewports can also be created using the **Viewports** toolbar.

3. Select from the **Viewports** toolbar.

The **Viewports** dialog box will appear. See Figure 14-36.

4. Select **Four: Equal,** then **OK.**

See Figure 14-37. Note that the crosshairs are visible only in the active viewport. An arrow will appear if the cursor is moved into any other port.

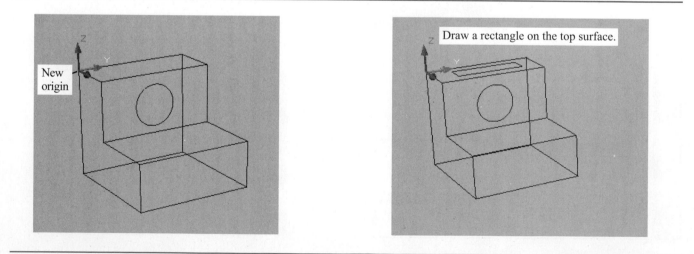

Figure 14-33

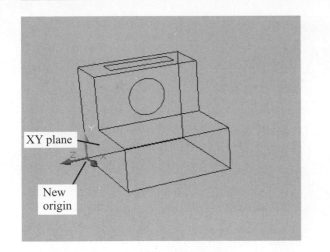

Figure 14-34

To create orthographic views

1. Move the cursor into the top left port and press the left mouse button.

   This will make this viewport the current viewport, and the crosshairs will appear in the port.

2. Select the **Top** tool from the **View** toolbar.

   The **View** dialog box will appear.

3. Make the **Top** view the current view.

   An oversized top view of the object will appear in the port.

4. Type **Zoom**; press **Enter**.

   All/Center/Dynamic/Extents/Left/Previous/Vmax/Window/<Scale(X/XP)>:

5. Type **1**; press **Enter**.

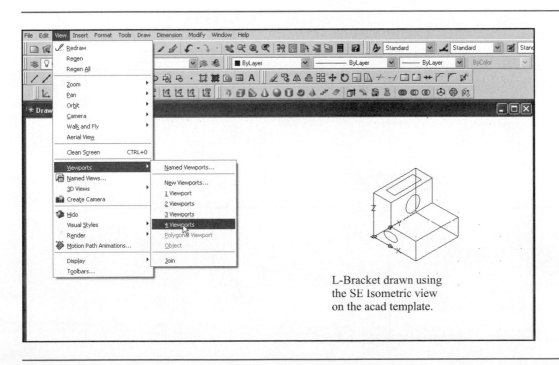

L-Bracket drawn using the SE Isometric view on the acad template.

Figure 14-35

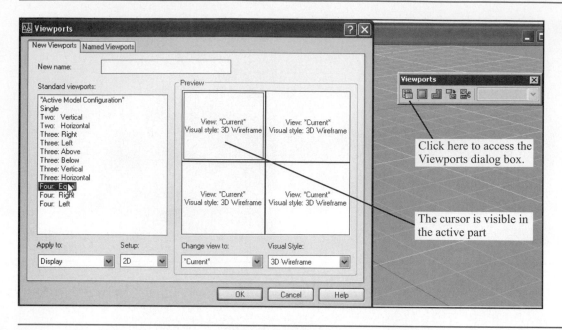

Figure 14-36

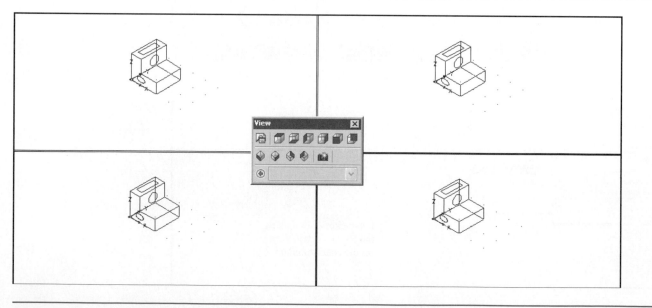

Figure 14-37

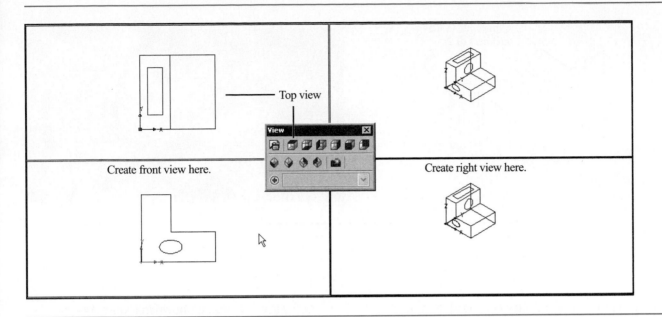

Figure 14-38

See Figure 14-38. Repeat the proceedure and make the lower left port the front view and the lower right port the right view. Figure 14-39 shows the completed orthographic views.

14-13 ELEV

The **Elev** (elevation) command is used to create 3D surfaces from 2D drawing commands. Figure 14-40 shows a line and a circle drawn as normal 2D entities and then

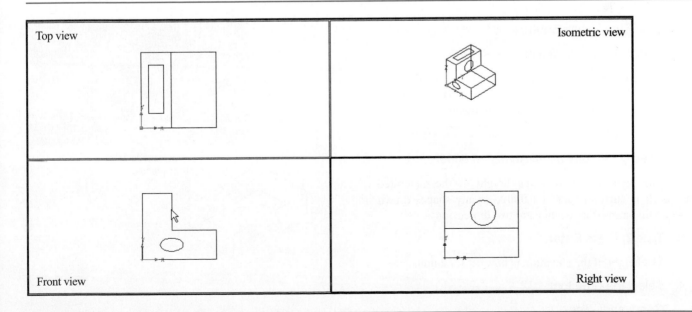

Figure 14-39

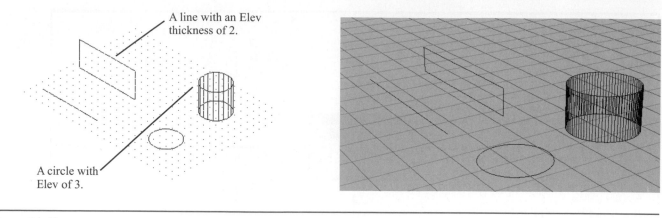

Figure 14-40

drawn a second time with the **Elev** command thickness set for 2 and 3, respectively. The figures were drawn using a 3D perspective grid background. The **Line** command generates a plane perpendicular to the plane of the original line, and the **Circle** command generates a cylinder perpendicular to the plane of the base circle.

To use the Elev command

There is no icon for the **Elev** command. It is accessed by typing **Elev** in response to a command prompt. In the following example, **Grid** and **Snap** have .5 spacing using decimal units, and there is an **SE Isometric** viewpoint. The same procedure may be applied using the acad3D template.

Command:

1. Type **Elev;** press **Enter.**

Specify new default elevation <0.0000>:

The current elevation is the base plane for the constructions. In this example the base plane is the current X,Y (WCS) plane, that is, the plane with the grid.

2. Press **Enter.**

Specify new default thickness <0.0000>:

The thickness defines the height of the generated planes. It is currently set at **0.0000,** so any shapes drawn have no thickness; that is, they are two dimensional.

3. Type **2;** press **Enter.**

The height of the elevation is now set for 2 units.

4. Select the **Line** tool.

Command: _line
From point:

5. Draw the box shape shown in Figure 14-41.

The shape shown in the figure is not a box but is four perpendicular planes. There is no top or bottom surface.

The default thickness value may also be defined using the **Properties** command on the **Modify** pull-down menu. The **Properties** dialog box will appear. See Figure 14-42. Change the thickness value to **2.0000.**

Figure 14-42 shows a hexagon and an arc drawn with the **Elev** command set at thickness values of 2 and 4, respectively.

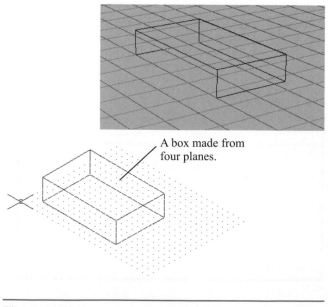

A box made from four planes.

Figure 14-41

Specify the default thickness value here.

Figure 14-42

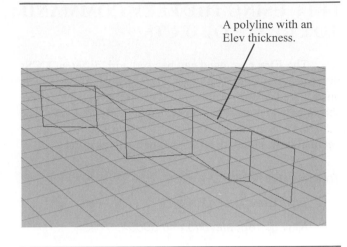

A polyline with an Elev thickness.

Figure 14-44

To draw a curve using Elev (See Figure 14-44.)

Use the **acad3D** template.

1. Type **Elev;** press **Enter.**

    Specify new default elevation <0.0000>:

2. Press **Enter.**

    Specify new default thickness <0.0000>:

3. Type **3.5;** press **Enter.**
4. Select the **Polyline** tool from the **Draw** toolbar.

    Command: _pline
    Specify start point:

5. Draw a polyline approximately like the one shown in Figure 14-44.
6. Select the **Edit Polyline** tool from the **Modify II** toolbar.

    Command: _pedit
    Select polyline:

7. Select the polyline.

    Enter an option [Close/Join/Width/Edit vertex/Fit/
    Spline Decurve/Ltype gen/Undo]:

8. Type **f;** press **Enter.**

    See Figure 14-45.

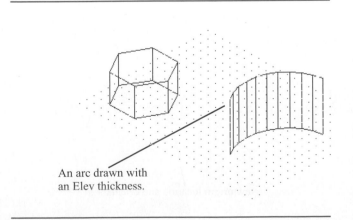

An arc drawn with an Elev thickness.

Figure 14-43

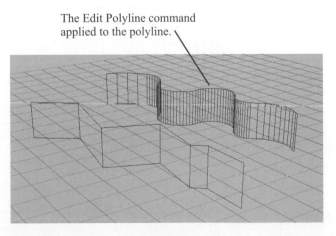

The Edit Polyline command applied to the polyline.

Figure 14-45

14-14 USING THE ELEV COMMAND TO CREATE OBJECTS

The **Elev** command can be used with different UCSs to create simulated 3D objects. The objects will actually be open-ended plane structures. The following procedure shows how to create the object shown in Figure 14-46. The drawing was created with **Grid** and **Snap** set to **.5** using decimal units and with an **SE Isometric** viewpoint.

To draw the box

1. Type **Elev**; press **Enter**.

   New current elevation <0.0000>:

2. Press **Enter**.

   New current thickness <0.0000>:

3. Type **6**; press **Enter**.

   Command:

4. Select the **Line** tool and draw a **10 x 10** box as shown in Figure 14-47.

To create a new UCS

1. Select the **Right** tool from the **UCS II** toolbar.
2. Select the **Origin** tool from the **UCS** toolbar.

   Specify new origin point <0,0,0>:

3. Select the lower right corner as shown in Figure 14-48.

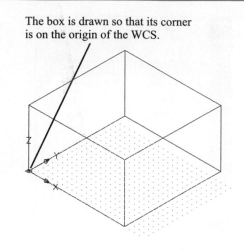

The box is drawn so that its corner is on the origin of the WCS.

Figure 14-47

To draw the right cylinder

This cylinder will be 6 units long, so there is no need to change the **Elev** thickness setting.

1. Select the **Circle** tool and draw a cylinder centered on the right face of the box.

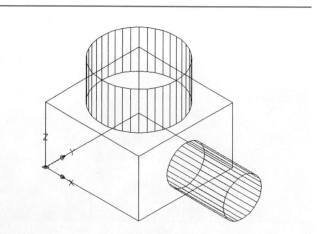

Figure 14-46

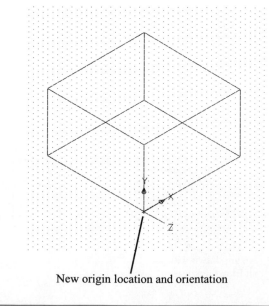

New origin location and orientation

Figure 14-48

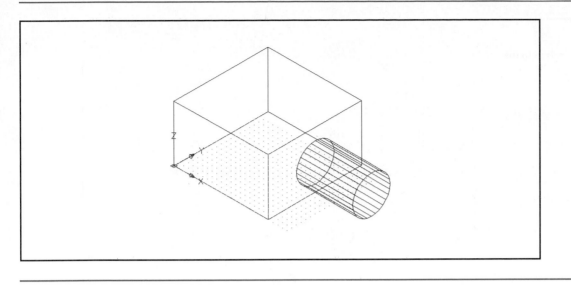

Figure 14-49

The coordinate value for the cylinder's center point is **5,3** and the radius value is **2.00.** See Figure 14-49.

To draw the top cylinder

First return the drawing to the original WCS X,Y axes, then change the location of the X,Y plane so the top cylinder can be drawn in the correct position.

1. Select the **World** tool from the **UCS** toolbar.
2. Type **Elev;** press **Enter.**

   Specify new default elevation <0.0000>:

3. Type **6;** press **Enter.**

   The value 6 is used because this is the thickness of the box.

   Specify new default thickness <6.0000>:

4. Type **4;** press **Enter.**

Note the shift in the grid pattern. The WCS origin icon is still located at the same place, but the grid origin is on the top surface. See Figure 14-50. Positions on the top surface may be located using the displayed coordinate values.

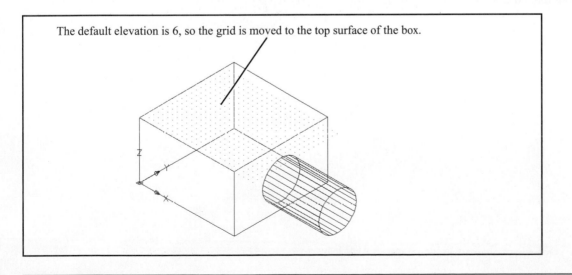

The default elevation is 6, so the grid is moved to the top surface of the box.

Figure 14-50

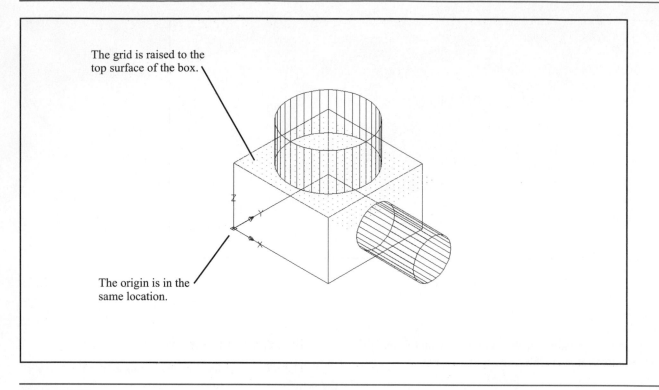

The grid is raised to the top surface of the box.

The origin is in the same location.

Figure 14-51

5. Select the **Circle** tool and draw the cylinder as shown in Figure 14-51.

   The cylinder's center point is at **5,5** and its radius = **2.**

To return the drawing to its original settings

1. Type **Elev; press Enter.**

   New current elevation <6.0000>:

2. Type **0; press Enter.**

   New current thickness <4.0000>:

3. Type **0; press Enter.**

   The object should look like the one shown in Figure 14-51.

14-15 EXERCISE PROBLEMS

Exercise problems EX14-1 through EX14-4 require you to draw 2D shapes on various surfaces of 3D objects created using the **Elev** command. All 2D shapes should be drawn at the center of the surfaces on which they appear. Either the acad or acad3D template may be used.

A. Draw the 2D shapes as shown.
B. Divide the screen into four viewports, and create front, top, and right-side orthographic views for each object.

EX14-1 INCHES

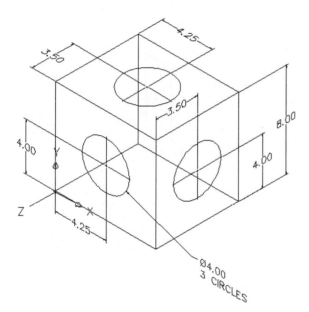

EX14-3 MILLIMETERS

Both cylinders are centered on the top surfaces as shown.

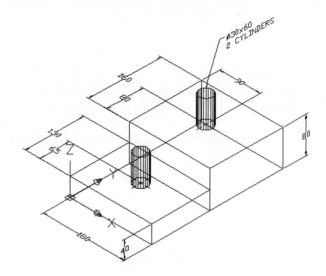

EX14-2 MILLIMETERS

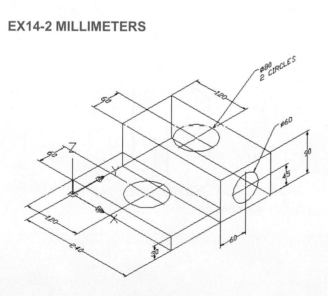

EX14-4 MILLIMETERS

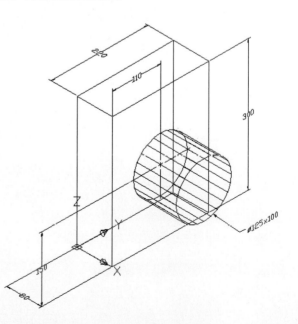

Redraw the figures presented in Exercise Problems EX14-5 through EX14-17 as wireframe models using the **Elev** command. Divide the screen into four viewports and create front, top, and right-side orthographic views and one isometric view as shown in Figure 14-38.

EX14-5 INCHES

Box a: X = 6, Y = 5, Z = 2
Box b: X = 4, Y = 4, Z = 4
Box c: X = 5, Y = 2, Z = 1

Hint: Consider a modified version of the NE Isometric viewpoint.

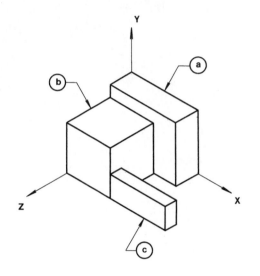

EX14-6 INCHES

Box a: X = 8, Y = 8, Z = 1
Box b: X = 6, Y = 6, Z = 2
Box c: X = 2, Y = 2, Z = 6

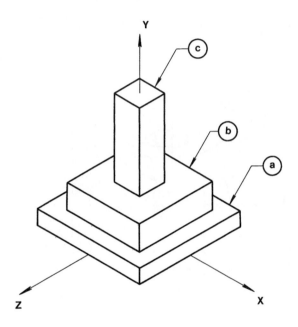

EX14-7 MILLIMETERS

Each box is 2 × 2 × 5.

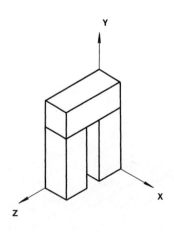

EX14-8 MILLIMETERS

Cylinder a: Ø10 × 30 LONG
Cylinder b: Ø20 × 8 LONG
Cylinder c: Ø35 × 18 LONG

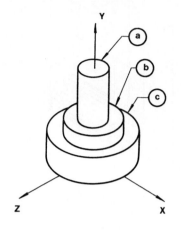

EX14-10 MILLIMETERS

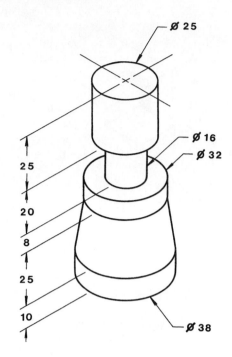

EX14-9 MILLIMETERS

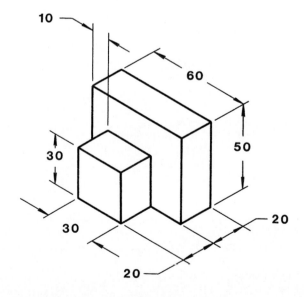

EX14-11 MILLIMETERS

Cylinders are 30 LONG.

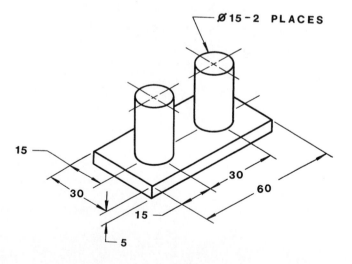

EX14-12 MILLIMETERS

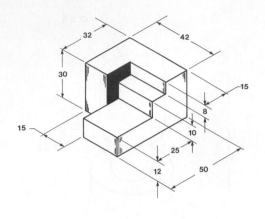

EX14-14 MILLIMETERS

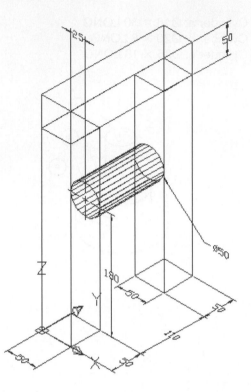

EX14-13 MILLIMETERS

The Ø12 cylinder is 20 LONG.
The Ø20 cylinder is 12 LONG.

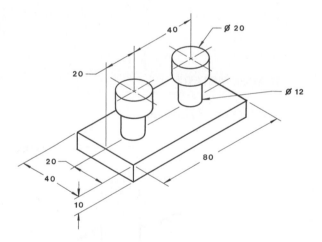

EX14-15 MILLIMETERS

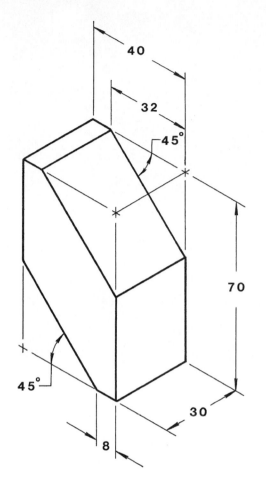

EX14-16 MILLIMETERS

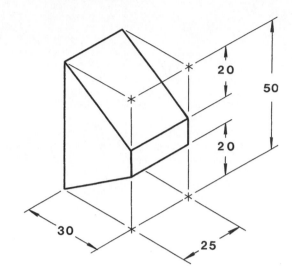

EX14-17 MILLIMETERS

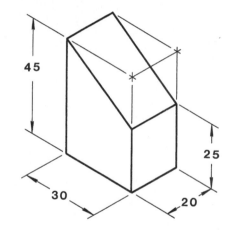

15

3D Surfaces

15-1 INTRODUCTION

This chapter explains how to draw 3D surfaces using AutoCAD. There are nine 3D surface options. To access these options type **3D** in response to a command prompt.

The options will appear as shown in Figure 15-1. In the example shown a dish was drawn.

Surface modeling is different from solid modeling. Surface modeling deals only with individual surfaces, whereas solid modeling deals with entire solid shapes. A box

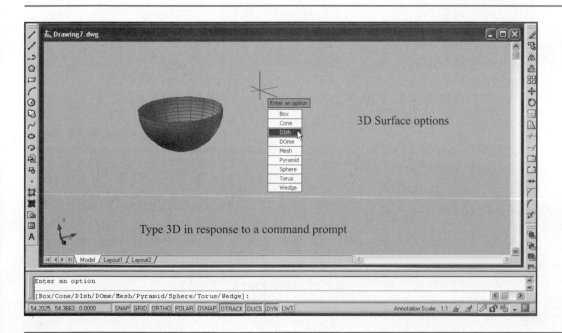

Figure 15-1

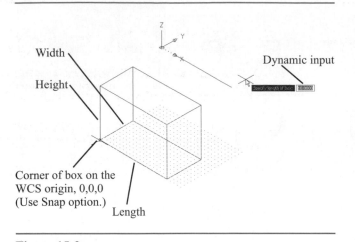

Figure 15-2

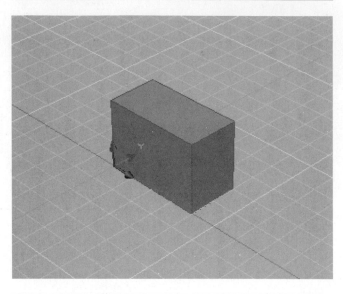

Figure 15-3

created as a surface model comprises six surfaces. A box created using solid modeling is a solid object with six sides. Surface models cannot be unioned or subtracted as can solid models.

Surface modeling is particularly well suited for drawing complex 3D meshes, such as might be found on a surface profile map, or for the transition area between an airfoil and the fuselage on an aircraft.

Other commands that can be used to draw 3D surfaces as well as solid 3D surfaces are **Revolved Surface, Tabulated Surface, Ruled Surface,** and **Edge Surface.** These commands will be explained in Chapter 16.

15-2 BOX

The **Box** command is used to draw a box comprising six surfaces. See Figure 15-2. The command sequence is as follows.

Figure 15-3 shows the same box drawn using an **acad3D** template and the **Realistic** visual style. The drawing procedure is the same.

To draw a box

1. Type **3D** in response to a command prompt; press **Enter** and select the **Box** option.
2. Select a corner point of the box.

In the example shown, the coordinate display was used to locate the corner point on the 0,0,0 WCS origin. The **Snap** option was on. The corner could have been defined using a 3-D coordinate value (X, Y, Z).

3. Type **0,0,0**; press **Enter.**

Specify length of box:

4. Type **10;** press **Enter** or use dynamic input to define the length.

The length can also be determined by moving the crosshairs and then picking a random point, using **Osnap** to align with an existing entity, or using the values displayed in the coordinate display box.

Specify width of box or [cube]:

5. Type **5;** press **Enter.**

If the box is a cube, enter a **C** response instead of a width value. AutoCAD will automatically take the length value and apply it to the width and height, skipping the width and height prompts in the sequence.

Specify height of box:

6. Type **7;** press **Enter.**

Specify rotation angle of box about the Z axis or [Reference]:

AutoCAD is now in a dynamic mode, meaning that as you move the crosshairs, the box will rotate about a Z axis projected out of the original corner point.

7. Type **0;** press **Enter.**

15-3 WEDGE

The wedge in this example will be drawn aligned with the box drawn in the proceding section. See Figure 15-4.

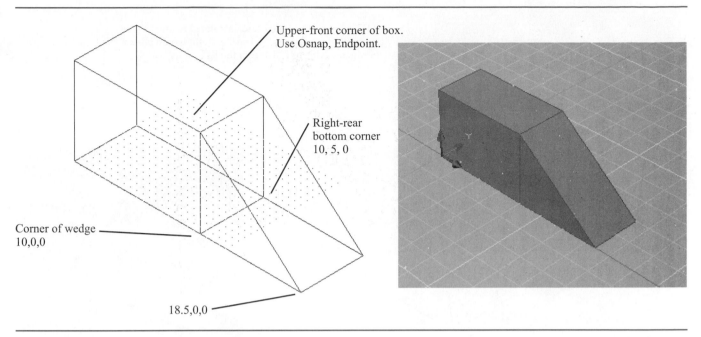

Upper-front corner of box.
Use Osnap, Endpoint.

Right-rear
bottom corner
10, 5, 0

Corner of wedge
10,0,0

18.5,0,0

Figure 15-4

To draw a wedge

1. Type **3D** in response to a command prompt; press **Enter** and select the **Wedge** option.

   Specify corner of wedge:

2. Use **Osnap, Endpoint** (press the **<Shift>** key and right mouse button simultaneously) and select the lower right corner of the box, or specify the coordinate point.

   Specify length of wedge:

3. Type **8.5**.

   Specify width of wedge:

4. Use **Osnap, Endpoint** and select the right rear bottom corner of the box or type **5**; press **Enter**.

   Specify height of wedge:

5. Use **Osnap, Endpoint** and select the upper front corner of the box as shown in Figure 15-4 or type **7**; press **Enter**.

   Specify rotation angle of wedge about the Z axis:

6. Type **0**; press **Enter**.

   The box and the wedge are not joined objects. Each has a surface that occupies the same space, but they are still two individual entities.

7. Clear the screen using the **Erase** tool.

15-4 PYRAMID

There are several different types of pyramids. Pyramids may have triangular, rectangular, or square bases. Any polygon can be used as the base of a pyramid, but the **Surfaces** toolbar commands are limited to triangular and rectangular bases. Polygons with more points can be used with the solid modeling options.

A tetrahedron is a pyramid made from three triangles. Pyramids whose apexes are located directly over the center point of their base polygons are called *right pyramids. Pyramids* whose apexes are not located directly over the center point of their base polygons are called *oblique pyramids.* Pyramids with their top sections removed are called *truncated pyramids,* and pyramids whose tops are edges and not point apexes are called *ridged pyramids.*

To draw a rectangular pyramid (See Figure 15-5.)

1. Type **3D** in response to a command prompt and select the **Pyramid** option.

   Specify first corner point for base pyramid:

2. Type **0,0**; press **Enter**.

   Specify second corner point for base of pyramid:

3. Type **8,0**; press **Enter**.

   Specify third corner point for base of pyramid:

4. Type **0,10**; press **Enter**.

NOTE: Coordinate values are calculated relative to the last point entered. The 8, 10 corner point would have a coordinate value of 0, 10 from the 8,0 corner.

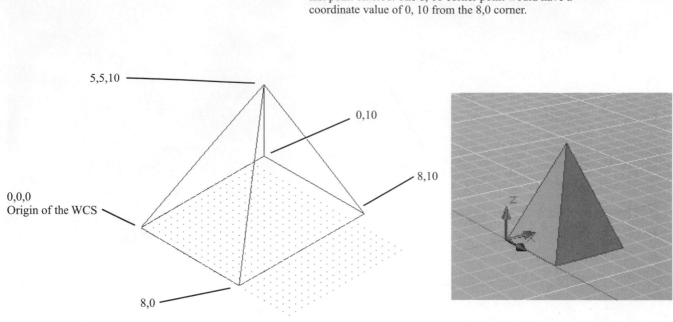

Figure 15-5

Note: Coordinate Values are calculated relative to the last point entered. The 8,10 corner point would have a coordinate value of 0,10 from the 8,0 corner.

> *Specify fourth corner point for base of pyramid or [Tetrahedron]:*

5. Type **−8,0** press **Enter.**

> *Specify apex point of pyramid or [Right/Top]:*

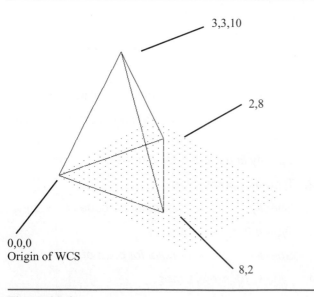

Figure 15-6

The apex point location requires a Z component. All the previous points have been in the X, Y plane, so only X and Y coordinates were necessary. The apex point can be located randomly; that is, you can simply move the crosshairs and select a point, but this procedure can be visually misleading. It is better to use either coordinate values or the **Osnap** command to locate a pyramid's apex.

6. Type **5,5,10;** press **Enter.**

> A negative Z value could have been entered.

To draw a tetrahedron (See Figure 15-6.)

1. Type **3D** in response to a command prompt and select the **Pyramid** option.

> *Specify first corner point for base of pyramid:*

2. Type **0,0;** press **Enter.**

> *Specify second corner point for base of pyramid:*

3. Type **8,2;** press **Enter.**

> *Specify third corner point for base of pyramid:*

4. Type **2,8;** press **Enter.**

> *Specify fourth corner point for base of pyramid or [Tetrahedron]:*

5. Type **t;** press **Enter.**

> *Specify apex point of tetrahedron or [Top]:*

6. Type **3,3,10;** press **Enter.**

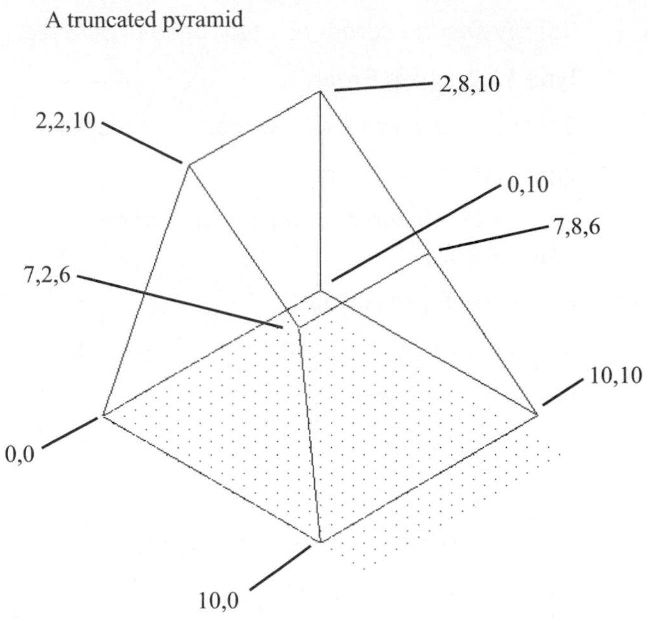

A truncated pyramid

2,8,10

2,2,10

0,10

7,8,6

7,2,6

0,0

10,10

10,0

Figure 15-7

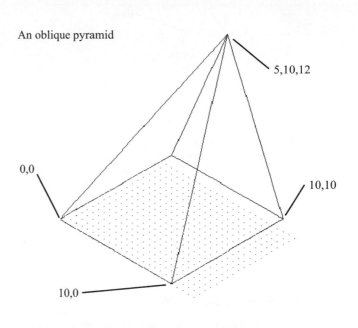

An oblique pyramid

5,10,12

0,0

10,10

10,0

Figure 15-8

To draw a truncated pyramid (See Figure 15-7.)

1. Select the **Pyramid** tool from the **Surfaces** toolbar.

   Command: ai_pyramid
   Specify first corner point for pyramid:

2. Type **0,0**; press **Enter**.

   Specify second corner point for the base of pyramid:

3. Type **10,0**; press **Enter**.

   Specify third corner point for base of pyramid:

4. Type **0,10**; press **Enter**.

   Specify fourth corner point for base of pyramid or [Tetrahedron]:

5. Type **−10,0**; press **Enter**.

   Specify apex point of pyramid or [Ridge/Top]:

6. Type **t**; press **Enter**.

   Specify first corner point for top of pyramid:

7. Type **2,2,10**; press **Enter**.

Remember that top points are not in the base plane and therefore their coordinate values require a Z component. Also note that a line will appear from the crosshairs to the corner point whose top point is being defined. The coordinate volues are relative to those corners.

Specify second corner point for top of pyramid:

8. Type **−3,2,6**; press **Enter**.

   Specify third corner point for top of pyramid:

9. Type **−3,−2,6**; press **Enter**.

   Specify fourth corner point for top of pyramid:

10. Type **2,−2,10**; press **Enter**.

To draw an oblique pyramid (See Figure 15-8.)

1. Select the **Pyramid** tool from the **Surfaces** toolbar.

   Command: ai_pyramid
   Specify first corner point for base of pyramid:

2. Type **0,0**; press **Enter**.

   Specify second corner point for base of pyramid:

3. Type **10,0**; press **Enter**.

   Specify third corner point for base of pyramid:

4. Type **0,10**; press **Enter**.

   Specify fourth corner point for base of pyramid or [Tetrahedron]:

5. Type **−10,0**; press **Enter**.

   Specify apex point of pyramid or [Ridge/Top]:

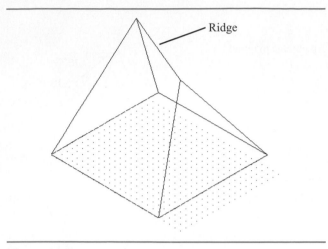

Ridge

Figure 15-9

6. Type **5,10,12**; press **Enter.**

This apex point is located over the center point of the rear line in the base polygon. Because the apex is not located over the base polygon's center point, the figure is classified as an oblique pyramid.

To draw a ridged pyramid (See Figure 15-9.)

1. Select the **Pyramid** tool from the **Surfaces** toolbar.

   Command: ai_pyramid
   Specify first corner point for base of pyramid:

2. Type **0,0**; press **Enter.**

   Specify second corner point for base of pyramid:

3. Type **10,0**; press **Enter.**

   Specify third corner point for base of pyramid:

4. Type **0,10**; press **Enter.**

   Specify fourth corner point for base of pyramid or [Tetrahedron]:

5. Type **−10,0**; press **Enter.**

   Specify apex point of pyramid or [Ridge/Top]:

6. Type **r**; press **Enter.**

   Specify first ridge end point of pyramid:

7. Type **3,5,10**; press **Enter.**

   Specify second ridge end point of pyramid:

8. Type **4,0,−3**; press **Enter.**

15-5 CONE (See Figure 15-10.)

To draw a cone

1. Type **3D** in response to a command prompt and select the **Cone** option.

   Specify center point for base of cone:

2. Type **5,5,0**; press **Enter.**

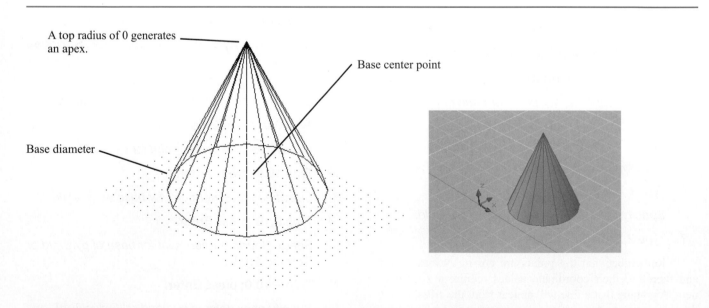

A top radius of 0 generates an apex.

Base center point

Base diameter

Figure 15-10

The cone's center point could also have been selected by locating a point using the crosshairs, then pressing the left mouse button.

Specify radius for base of cone or [Diameter]:

3. Type **4**; press **Enter**.

Specify radius for top of cone or [Diameter]<0>:

A top radius of 0 will create a cone with an apex. A radius value other than 0 will create a truncated cone.

4. Press **Enter**.

Specify height of cone:

5. Type **9**; press **Enter**.

Enter number of segments for surface of cone <16>:

The number of segments defines the number of facets used to show the cone. The more segments, the smaller each segment will be, and the resulting cone will appear smoother; however, a large number of segments will use more memory and will require more time to process. The default value of 16 is a good compromise between visual accuracy and drawing speed.

6. Press **Enter**.

To draw a truncated cone (See Figure 15-11.)

1. Type **3D** in response to a command prompt and select the **Cone** option.

Specify center point for base of cone:

2. Type **5,5**; press **Enter**.

Specify radius for base of cone or [Diameter]:

3. Type **6**; press **Enter**.

Specify radius for top of cone or [Diameter]<0>:

4. Type **3**; press **Enter**.

Specify height of cone:

5. Type **8**; press **Enter**.

Enter number of segments for surface of cone <16>:

6. Press **Enter**.

15-6 SPHERE (See Figure 15-12.)

To draw a sphere

In this example, the sphere will be drawn above the X, Y plane.

1. Type **3D** in response to a command prompt and select the **Sphere** option.

Specify center point of sphere:

2. Type **5,5,5**; press **Enter**.

Specify radius of sphere or [Diameter]:

3. Type **5**; press **Enter**.

Specify number of longitudinal segments for surface of sphere <16>:

4. Press **Enter**.

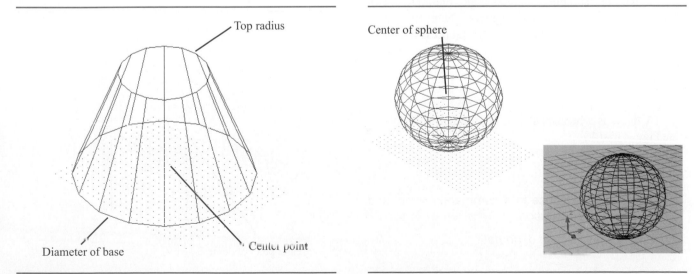

Top radius

Diameter of base

Center point

Center of sphere

Figure 15-11

Figure 15-12

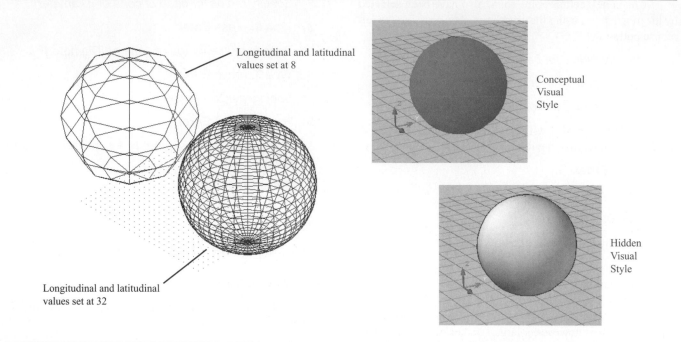

Longitudinal and latitudinal values set at 8

Conceptual Visual Style

Hidden Visual Style

Longitudinal and latitudinal values set at 32

Figure 15-13

Specify number of latitudinal segments for surface of sphere <16>:

5. Press **Enter.**

The number of longitudinal and latitudinal segments controls the number of facets that will show on the sphere. Longitudinal lines are equivalent to vertical lines. Latitudinal lines are equivalent to horizontal lines. The more segments defined, the smoother the resulting sphere, but more line segments increase the drawing time needed to generate the sphere.

Figure 15-13 shows two spheres: one drawn with 8 longitudinal and latitudinal segments, and one drawn with 32 longitudinal and latitudinal segments. Note the differences in the overall smoothness of the visual presentation.

15-7 DOME

A *dome* is a hemisphere, or the top half of a sphere. (See Figure 15-14.)

To draw a dome

1. Type **3D** in response to a command prompt and select the **Dome** option.

   Specify center point of dome:

2. Type **5,5**; press **Enter.**

   Specify radius of dome or [Diameter]:

3. Type **8**; press **Enter.**

   Enter number of longitudinal segments for surface of dome <16>:

4. Press **Enter.**

   Enter number of latitudinal segments for surface of dome <8>:

5. Press **Enter.**

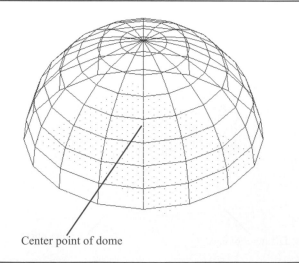

Center point of dome

Figure 15-14

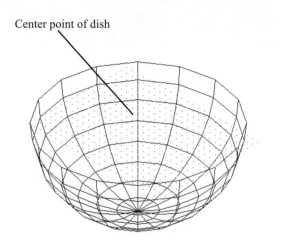

Center point of dish

Figure 15-15

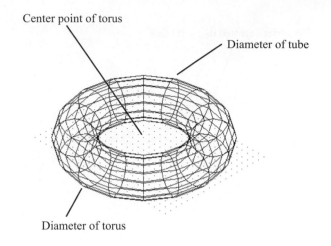

Center point of torus

Diameter of tube

Diameter of torus

Figure 15-16

15-8 DISH

A *dish* is a semisphere, or the bottom half of a sphere. (See Figure 15-15.)

To draw a dish

1. Type **3D** in response to a command prompt and select the **Dish** option.

   Specify center point of dish:

2. Type **5,5**; press **Enter.**

   Specify radius or dish or [Diameter]:

3. Type **7**; press **Enter.**

   Enter number of longitudinal segments for surface of dish <16>:

4. Press **Enter.**

   Enter number of latitudinal segments for surface of dish <8>:

5. Press **Enter.**

15-9 TORUS

A *torus* is a donutlike shape. See Figure 15-16.

To draw a torus

1. Type **3D** in response to a command prompt and select the **Torus** option.

   Specify center point of torus:

2. Type **5,5**; press **Enter.**

   Specify radius of torus or [Diameter]:

3. Type **6**; press **Enter.**

   Specify radius of tube or [Diameter]:

4. Type **1.5**; press **Enter.**

   The radius of a torus is the distance from the center point to the outside edge of the torus, as measured along the center plane.

   Enter number of segments around the tube circumference <16>:

5. Press **Enter.**

   Enter number of segments around the torus circumference <16>:

6. Press **Enter.**

A surface created using 3D face

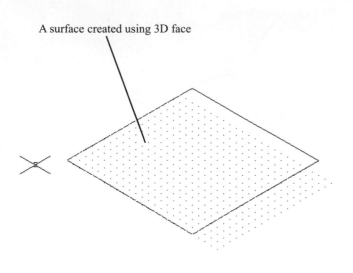

Figure 15-17

15-10 3D FACE

A 3D face is a plane. 3D faces may be created as individual planes or as adjoining groups of planes. A 3D face can be added to an existing 3D object to close an open area or to create a transition between existing objects.

To draw a single 3D face

In this example we will select points on a grid. Coordinate values can also be used to define the 3D face's corner points. See Figure 15-17.

1. Type **3Dface** in response to a command prompt.

   Command: _3dface
   Specify first point or [Invisible]:

2. Select the 0,0 point on the grid.

   Use the coordinate display at the lower left of the screen to verify the point location.

   Specify second point or [Invisible]:

3. Select **10,0.**

   Specify third point or [Invisible]<exit>:

4. Select **0,10.**

   Specify fourth point or [Invisible]<create three-sided face>:

5. Select −**10,0**.

   Specify third point or [Invisible]:

Adjoining surfaces created using 3D face

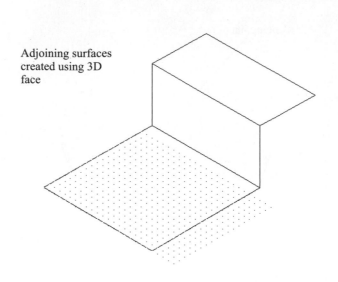

Figure 15-18

This point assumes that you will also be creating an adjacent plane, but because you are not, enter a null value.

6. Press **Enter.**

To draw adjoining 3D faces

In this example we will draw three adjoining planes, each 90° to the others. See Figure 15-18.

1. Type **3Dface** in response to a command prompt.

   Command: _3dface
   Specify first point or [Invisible]:

2. Select the **0,0** point on the grid.

   Specify second point or [Invisible]:

3. Select **10,0.**

   Specify third point or [Invisible]<exit>:

4. Select **0,10.**

   Specify fourth point or [Invisible]<create three-sided face>:

5. Select −**10,0.**

   The adjoining plane will be perpendicular to the line on the Y axis.

   Specify third point or [Invisible] <exit>:

6. Type **0,0,5;** press **Enter.**

   Specify fourth point or [Invisible] <create three-sided face>:

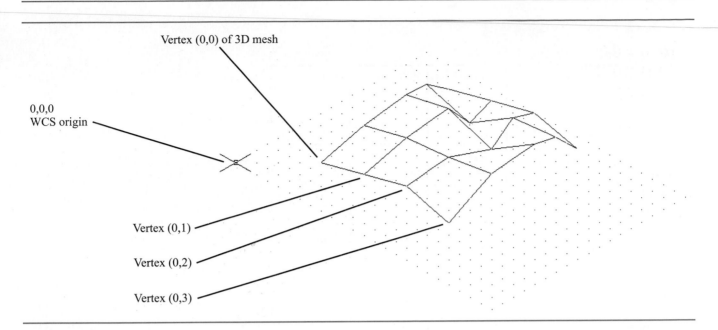

Figure 15-19

7. Type **10,0,0**; press **Enter**.

   Specify third point or [Invisible] <exit>:

8. Type **0,5,0**; press **Enter**.

   Specify fourth point or [Invisible] <create three-sided face>:

9. Type **−10,0,0**; press **Enter**.

   Specify third point or [Invisible] <exit>:

10. Press **Enter**.

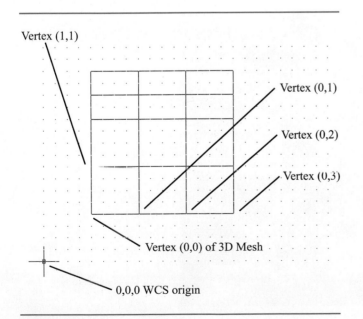

Figure 15-20

15-11 3D MESH

The **3D Mesh** command defines a surface by defining a series of points on the surface. Mesh surfaces are usually irregular surfaces, that is, not flat. (See Figure 15-19.)

The points used to define a mesh surface are located in terms of M,N axes. First, the number of points in both the M and N directions is defined, then each point is automatically assigned a vertex notation. The vertex notation 0,2 means first row, that is, the first line in the M direction, and the third column, third line in the N direction. The first line in both the M and N directions is labeled 0. Figure 15-20 is a top view of the finished 3D mesh.

To demonstrate how to create a 3d mesh, a 3 × 4 mesh will be created using the values listed in Figure 15-21.

To create a 3D mesh

1. Type **3d** in response to a command prompt, then select the **Mesh** option.

   Specify first corner of mesh:

2. Type **0,0**; press **Enter**.

   Specify second corner of mesh:

3. Type **4,0**; press **Enter**.

   Specify third corner of mesh:

4. Type **0,5**; press **Enter**.

   Specify fourth corner of mesh:

5. Type **−4,0**; press **Enter**.

   Enter mesh size in the M direction:

(0,0) = 0,0
(0,1) = 0,.25
(0,2) = 0,.50
(0,3) = 0,.38
(1,0) = 0,.25
(1,1) = 0,.38
(1,2) = 0,.38
(1,3) = 0,.50
(2,0) = 0,.50
(2,1) = 0,.25
(2,2) = 0,.50
(2,3) = 0,.63
(3,0) = 0,.63
(3,1) = 0,.38
(3,2) = 0,.63
(3,3) = 0, 50
(4,0) = 0,1.00
(4,1) = 0,.38
(4,2) = 0,.50
(4,3) = 0,.38

4,0	4,1	4,2	4,2
3,0	3,1	3,2	3,3
2,0	2,1	2,2	2,3
1,0	1,1	1,2	1,3
0,0	0,1	0,2	0,3

Figure 15-21

6. Type **5**; press **Enter.**

    Enter mesh size in the N direction:

7. Type **4**; press **Enter.**

    See Figure 15-22.

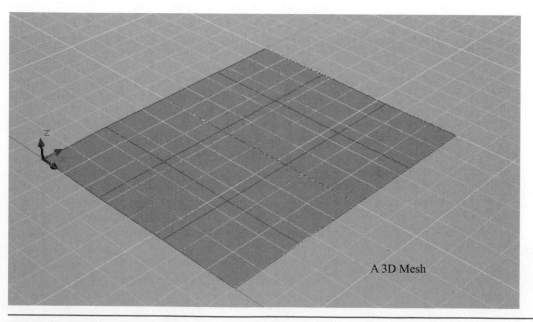

A 3D Mesh

Figure 15-22

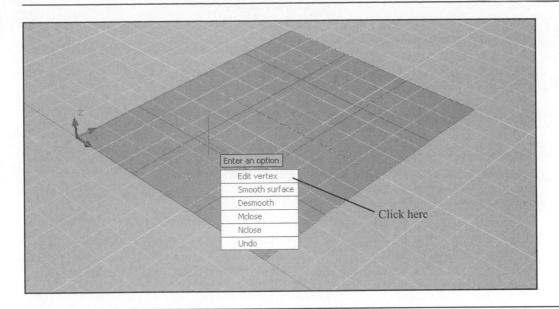

Figure 15-23

8. Type **pedit;** press **Enter.**

   Pedit is the polyedit command.
   Select polyline or [Multiple]:

9. Select the mesh grid.

   A list of options will appear. See Figure 15-23

10. Select the **Edit vertex** option.

    Current vertex (0,0);
    See Figure 15-24. The system will list the vertices in numerical order.

11. Press **Enter,** accepting the default option **Next.** In this example the 0,0 origin will not be changed.

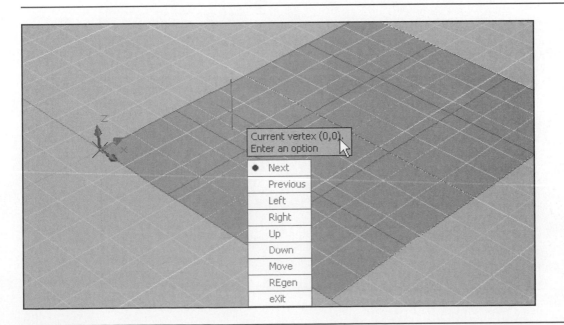

Figure 15-24

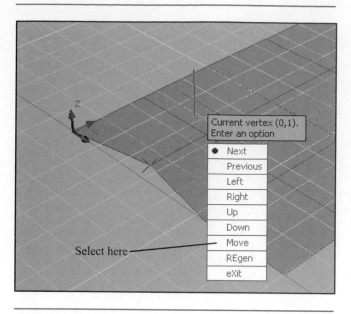

Figure 15-25

Current vertex (0,1):
See Figure 15-25.

12. Select the **Move** option.

   Specify new location for marked vertex:

13. Type **0,.25;** press **Enter.**
14. Press **Enter** again, accepting the default **Next** option.

   Current vertex (0,2):

15. Select the **Move** option.

   Specify new location for marked vertex:

16. Type **0,.5;** press **Enter.**
17. Proceed through all the vertex points, entering the values listed in Figure 15-21. The finished 3D mesh should look like the one shown in Figure 15-26.

To smooth a 3D mesh surface

1. Right-click the mouse.

   A listing of options will appear. See Figure 15-27.

2. Select the **Smooth surface** option; press **Enter.**

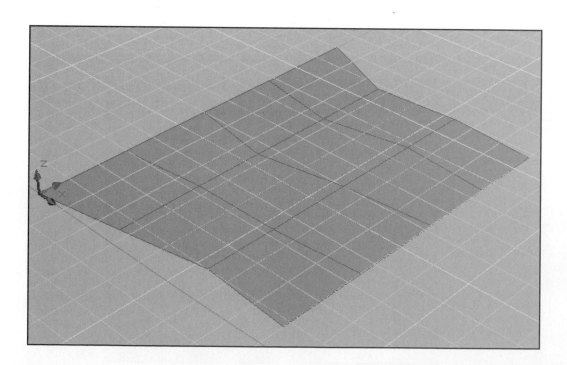

Figure 15-26

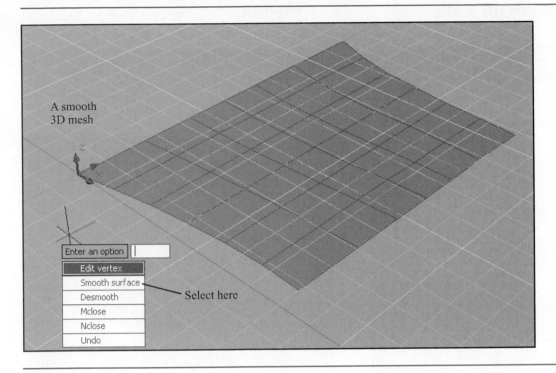

A smooth
3D mesh

Select here

Figure 15-27

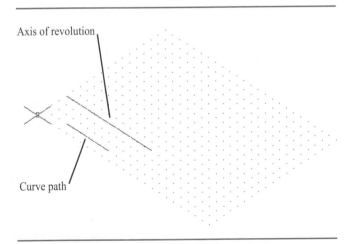

Axis of revolution

Curve path

Figure 15-28

15-12 EXERCISE PROBLEMS

EX15-1 INCHES

Draw the following box:
Corner point = 0,0
Length = 2.0
Width = 3.0
Height = 1.0
Rotation angle = 0°

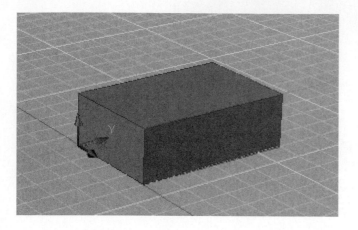

EX15-2 INCHES

Draw the following box:
Corner point = 2,2
Length = 8.0
Width = 4.0
Height = 6.0
Rotation angle = 0°

EX15-3 MILLIMETERS

Draw the following box:
Corner point = 0,0
Length = 60
Width = 60
Height = 40
Rotation angle = 0°

EX15-4 MILLIMETERS

Draw the following box:
Corner point = 5,5
Length = 100
Width = 50
Height = 60
Rotation angle = 0°

EX15-5 INCHES

Redraw the stacked boxes as shown. The box sizes are as follows:
1. $10 \times 10 \times 4$
2. $6 \times 6 \times 4$
3. $3 \times 3 \times 4$

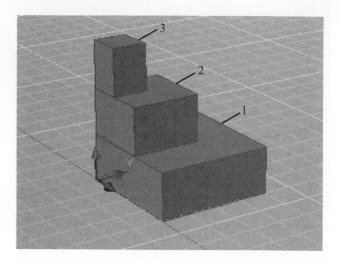

EX15-6 MILLIMETERS

Redraw the stacked boxes as shown so that their centerlines align.

The box sizes are as follows:

1. $100 \times 50 \times 20$
2. $50 \times 30 \times 20$
3. $20 \times 10 \times 20$

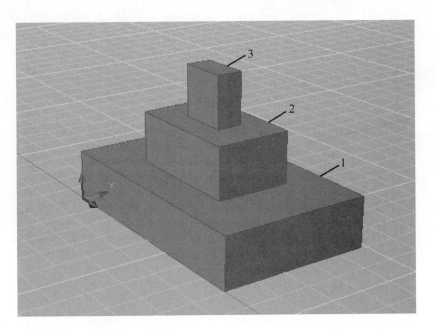

EX15-7 INCHES

Draw two wedges back to back as shown. Both wedges are the same size: $15 \times 10 \times 8$

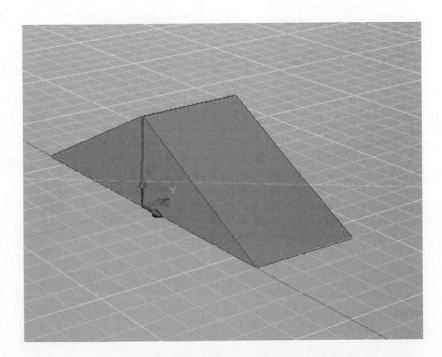

EX15-8 MILLIMETERS

Draw a box made from two wedges as shown. Both wedges are the same size: 60 × 30 × 40

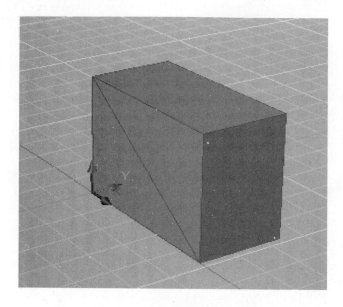

EX15-9 INCHES

Draw a pyramid that has an 8 × 8 square base and an apex located 10 above the center of the square.

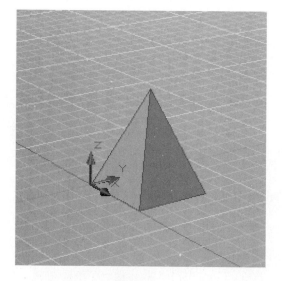

EX15-10 MILLIMETERS

Draw a tetrahedron pyramid based on the given coordinates relative to the 0,0,0 origin.

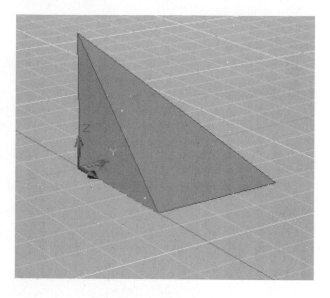

EX15-11 INCHES

Draw a pyramid with a 10 × 10 square base and an apex located 12 above the center of the square. The pyramid is to be centered about the 0,0 origin.

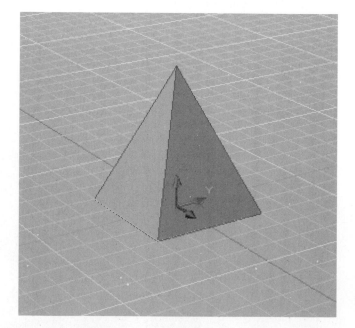

EX15-12 MILLIMETERS

Draw a pyramid based on the given coordinate values.

EX15-13 INCHES

Draw a pyramid based on the given coordinate values.

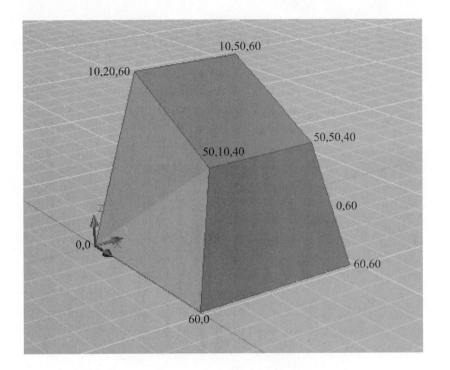

EX15-14 MILLIMETERS

Draw a pyramid based on the given coordinate values.

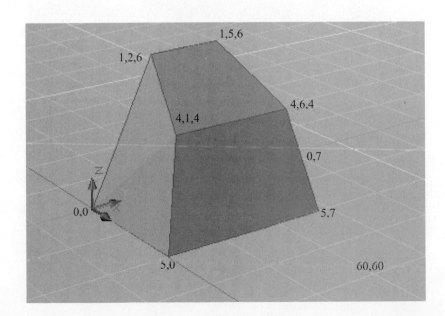

EX15-15 INCHES

Draw a pyramid based on the given coordinate values.

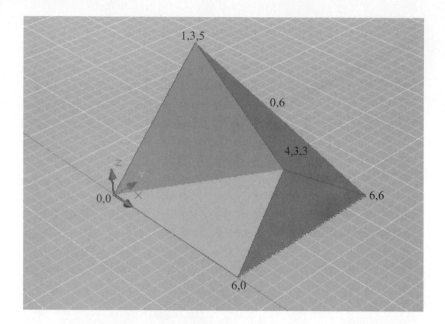

EX15-16 MILLIMETERS

Draw a pyramid based on the given coordinate values.

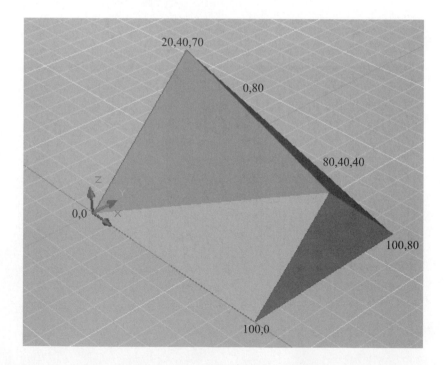

EX15-17 INCHES

Draw a cone with a Ø6.0 base and an apex 11.0 above
the center point of the base. Use 24 segments.

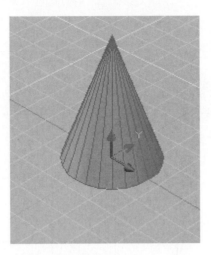

EX15-18 MILLIMETERS

Draw a cone with a Ø60 base, a Ø22 top, and a height
of 30. Use 24 segments.

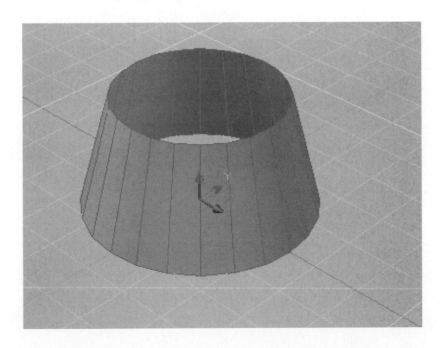

EX15-19 INCHES

Draw a Ø6.00 sphere.
Use 28 segments.

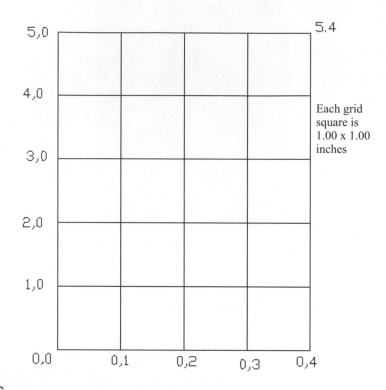

Each grid
square is
1.00 x 1.00
inches

EX15-20 MILLIMETERS

Draw a Ø80 sphere.
Use 20 segments.

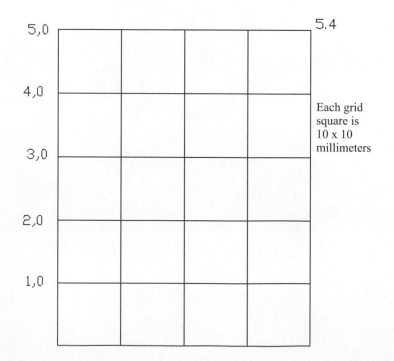

Each grid
square is
10 x 10
millimeters

EX15-21 INCHES

Draw a Ø4.0 dish.
Use 16 segments.

EX15-22 MILLIMETERS

Draw a Ø45.0 dish.
Use 24 segments.

EX15-23 INCHES

Draw a torus with a Ø5.5 and a tube radius of 0.50.
Use 24 segments.

EX15-24 INCHES

Draw a 3D mesh based on the given data. Each square
on the base grid is 1.00 ×1.00 inches. The given values are for height above the initial XY plane.
Grid Points
0,0 = 0
0,1 = .20
0,2 = .30
0,3 = .60
0,4 = .20
1,0 = .20
1,1 = .30
1,2 = .40
1,3 = .50
1,4 = .30
2,0 = .40
2,1 = .50
2,2 = .60
2,3 = .40
2,4 = .40
3,0 = .40
3,1 = .50
3,2 = 70
3,3 = .60
3,4 = .40
4,0 = .50
4,1 = .60
4,2 = .80
4,3 = .40

4,4 = .20
5,0 = .30
5,1 = .50
5,2 = .50
5,3 = .30
5,4 = .20

EX15-25 MILLIMETERS

Draw a 3D mesh based on the given data. Each square
on the base grid is 10 ×10 millimeters. The given values are for height above the initial XY plane.
Grid Points
0,0 =0
0,1 = 5
0,2 = 10
0,3 = 15
0,4 = 7
1,0 = 5
1,1 = 10
1,2 = 15
1,3 = 20
1,4 = 15
2,0 =10
2,1 = 15
2,2 = 20
2,3 = 25
2,4 = 15
3,0 = 30
3,1 = 20
3,2 = 15
3,3 = 15
3,4 = 10
4,0 =15
4,1 = 10
4,2 = 5
4,3 = 10
4,4 = 15
5,0 = 20
5,1 = 20
5,2 = 20
5,3 = 30
5,4 = 20

C H A P T E R 16

Modeling

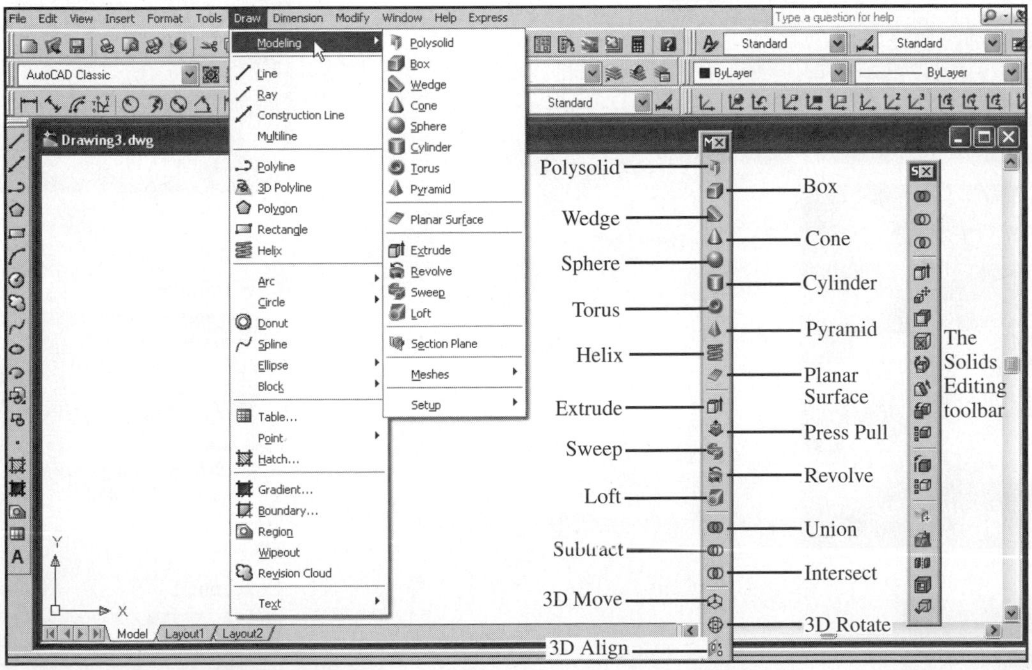

Figure 16-1

16-1 INTRODUCTION

This chapter introduces solid modeling. The solid modeling commands can be accessed using the **Modeling** toolbar, the **Solids Editing** toolbar or by using the **Modeling** option on the **Draw** pull-down menu or the **Solids Editing** option on the **Modify** pull-down menu. See Figure 16-1.

Solid modeling allows you to create objects as solid entities. Solid models differ from the surface models created in the last chapter in that solid models have density and are not merely joined surfaces.

Solid models are created by joining together, or *unioning*, basic primitive shapes: boxes, cylinders, wedges, and so on, or by defining a shape as a polyline and extruding it into a solid shape. Solid primitives may also be subtracted from one another. For example, to create a hole in a solid box, draw a solid cylinder, then subtract the cylinder from the box. The result will be an open volume in the shape of a hole.

The first part of the chapter deals with the individual tools of the **Solids** toolbar. The second part gives examples of how to create solid objects by joining primitive shapes and changing UCSs. The chapter ends with a discussion of the **Solids Editing** toolbar.

The **acad3D** template will be used throughout the chapter as a background grid. All the command sequences presented may also be applied using the **3D view** options of the **View** pull-down menu.

16-2 BOX

The **Box** tool on the **Solids** toolbar has two options: **Center** and **Corner. Corner** is the default option. The **Center** tool is used to draw a box by first locating its center point. The **Corner** tool is used to draw a box by first locating one of its corner points.

Set the drawing for **SE Isometric 3D View.**

To draw a box using the Corner option (See Figure 16-2.)

1. Select the **Box** tool from the **Modeling** toolbar.

   Command: _box
   Specify corner of box or [Center]:

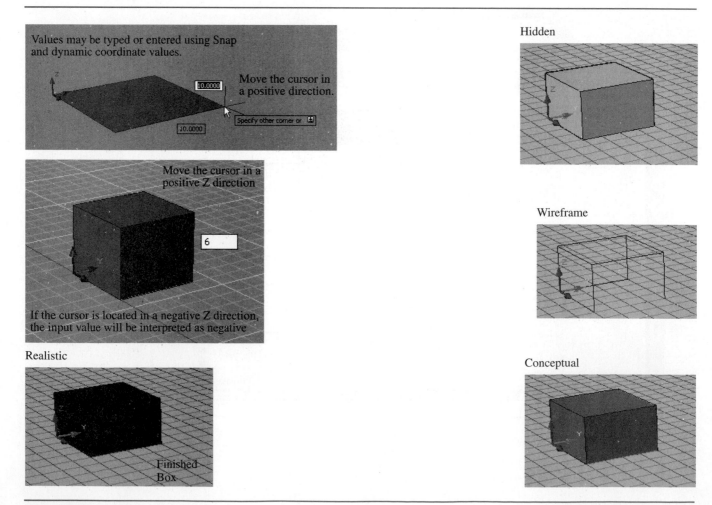

Figure 16-2

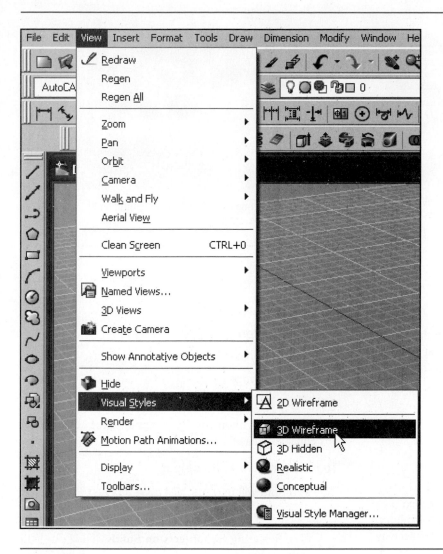

Figure 16-3

2. Type **0,0,0**; press **Enter.**

This means that the corner of the box will be located on the 0,0,0 point of the XY plane.

Specify other corner or [Cube/Length]:

3. Type **10,10,0**; press **Enter,** or use dynamic input to locate the corner.

Specify height or [2 points] <0.9396>:

4. Locate the cursor in a positive Z direction and type **6**; press **Enter.**

To change the visual style

Figure 16-2 shows the generated box using four different visual styles: realistic, wireframe, hidden, and conceptual. The **Visual Styles** command is located on the **View** pull-down menu. See Figure 16-3. To change the visual style of an object, click on the desired new style.

To use dynamic grips

Dynamic grips allow you to change the shape of an object in real time. See Figure 16-6.

1. Click the object A **Help** dialog box will appear.

2. Click **Close.**

Blue arrowheads will appear on the object. These arrowheads indicate the direction that the grip points may be moved.

3. Click and grab the arrowhead on the lower right edge.

Note how the object's shape is changed.

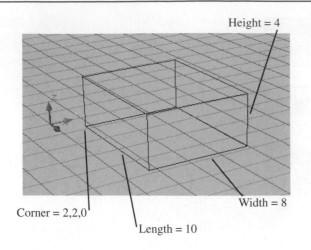

Height = 4

Corner = 2,2,0

Length = 10

Width = 8

Figure 16-4

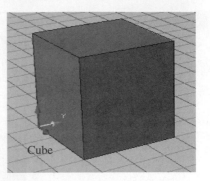

Cube

The cube option makes the length, width, and height of the box of equal length.

Figure 16-5

To draw a box from given dimensions

See Figure 16-4. Draw a box with a length of **10,** a width of **8,** and a height of **4,** with its corner at the **2,2,0** point. Use the cursor to define a positive X direction.

1. Select the **Box** tool from the **Modeling** toolbar.

   Command: _box
   Specify corner or [Center]:

2. Type **2,2,0**; press **Enter.**

   Specify corner or [Cube/Length]:

There are three different ways to define the next point for the box: enter coordinate values (10,0) as was explained in Section 15–2, activate the **Snap** command and use the dynamic screen coordinate display to select the point, or activate the **Ortho** command and enter a length value as will be defined in this example.

3. Type **L** and click the **Ortho** button at the bottom of the screen. Move the cursor to the positive **X** direction; press **Enter.**

   Specify length:

4. Type **10**; press **Enter.**

   Specify width:

5. Move the cursor to the positive **Y** direction and type **8**; press **Enter.**

   Specify height:

6. Move the cursor to a positive **Z** direction and type **4**; press **Enter.**

To draw a cube (See Figure 16-5.)

1. Select the **Corner** tool from the **Modeling** toolbar.

   Command: _box
   Specify corner of [Center]:

2. Press **Enter.**

   Specify corner or [Cube/Length]:

3. Type **c**; press **Enter.**

   Specify length:

4. Type **7.5**; press **Enter.**

AutoCAD automatically makes all three edges of the box equal lengths of 7.5.

To manipulate a solid object

This example will use the **10×8×4** box just drawn.

1. Click any edge of the box.

   The **Selecting Subobjects on Solids** dialog box will appear.

2. Read the dialog box; click **Close.**

   Blue arrowheads will appear on the XY plane of the box. See Figure 16-6.

3. Click and hold one of the arrowheads and drag the cursor away from the box.

   Note how the box changes shape.

16-3 SPHERE

The visual accuracy of solids depends on the settings of the **Isolines** command. The sphere shown in Figure 16-7 was drawn with an **Isolines** setting of **4.** For this section the **Isolines setting** will be changed to **18.**

To change the Isolines settings

1. Type **Isolines**; press **Enter.**

   There is no tool for the **Isolines** command.

   Enter new value for ISOLINES <4>:

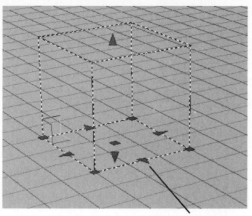

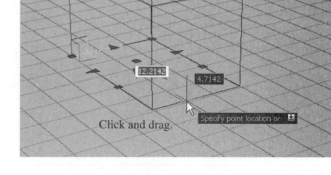

Click and drag here.

Click and drag.

Figure 16-6

2. Type **18**; press **Enter.**

Command:

The new value is now in place and will be used for all solids drawn until it is changed or a new drawing is started.

To draw a sphere (See Figure 16-7.)

1. Select **Sphere** from the **Modeling** toolbar.

Specify center of sphere or [3P/2P/Ttr]:

2. Type **5,5,0**; press **Enter.**

Specify radius or [Diameter]:

3. Type **4**; press **Enter.**

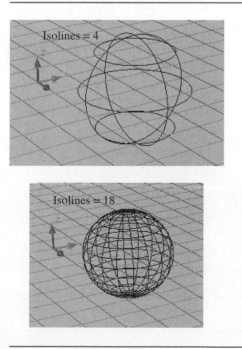

Figure 16-7

16-4 CYLINDER

There are two options associated with the **Cylinder** command: circular and elliptical. This means cylinders may be drawn with either elliptical or circular base planes. The base elliptical shape is drawn using the same procedure as was outlined for the **Ellipse** command in Chapter 2.

To draw a cylinder with a circular base

See Figure 16-8. Note that **Isolines** is still set for **18.**

1. Select the **Cylinder** tool from the **Modeling** toolbar.

Command: _cylinder
Specify center point or base or [3P/2P/Ttr/Elliptical] <0,0,0>:

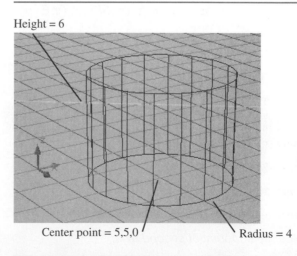

Height = 6

Center point = 5,5,0   Radius = 4

Figure 16-8

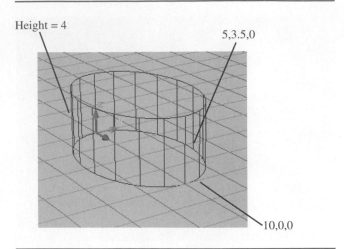

Figure 16-9

2. Type **5,5,0;** press **Enter.**

   Specify base radius or [Diameter]:

3. Type **4;** press **Enter.**

   Specify height or [2 Point/Axis endpoint] <0,0000>:

4. Locate the cursor in the positive Z direction and type **6;** press **Enter.**

To draw a cylinder with an elliptical base

See Figure 16-9. The following sequence uses coordinate value input. The same points could have been selected by moving the crosshairs and pressing the left mouse button.

1. Select the **Cylinder** tool from the **Modeling** toolbar.

   Command: _cylinder
   Specify center point or base or [3P/2P/Ttr/Elliptical]<0,0,0>:

2. Type **e;** press **Enter.**

   Specify endpoint of first axis or [center]:

3. Type **0,0,0;** press **Enter.**

   This input locates one end of the base axis on the 0,0,0 point of the XY plane.

   Specify other endpoint of first axis:

4. Type **10,0,0;** press **Enter.**

   This input locates the axis line along the X axis.

   Specify endpoint of second axis:

Note that the line dragging from the crosshairs has one end centered on the axis just defined.

5. Type **5,3.5,0;** press **Enter.**

   Specify height or [2 Point/Axis endpoint]

6. Type **4;** press **Enter.**

An elliptical base can also be defined by first defining a center point for the ellipse, then defining the length of the radii of the major and minor axes.

16-5 CONE

There are two options associated with the **Cone** tool: circular and elliptical. This means that cones can be drawn with either an elliptical or circular base plane. The base elliptical shape is drawn using the same procedure as was outlined for the **Ellipse** command in Chapter 2. The **Cone** command cannot be used to draw truncated cones, as could the **Cone** tool associated with the **Surfaces** toolbar. Solid truncated cones are created by subtracting the top portion of the cone.

To draw a cone with an elliptical base
(See Figure 16-10.)

1. Select the **Cone** tool from the **Modeling** toolbar.

   Command: _cone
   Specify center point or base or [3P/2P/Ttr/Elliptical] <0,0,0>:

2. Type **e;** press **Enter.**

   Specify endpoint of first axis or [Center]:

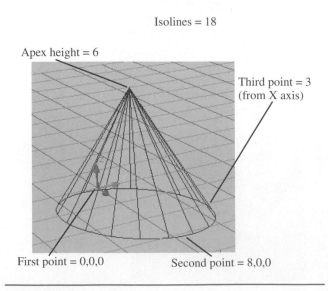

Figure 16-10

3. Type **0,0,0**; press **Enter.**

This input will locate one end of the ellipse axis at the 0,0,0 point of the XY plane.

Specify other endpoint of first axis:

4. Type **8,0,0**; press **Enter.**

Specify endpoint of second axis:

5. Type **3,0,0**; press **Enter.**

Specify height or [2 Point/Axis endpoint/Tap radius]:

6. Move the cursor to the positive Z direction and type **6**; press **Enter.**

A response of **A** to the Apex/<Height>: prompt allows you to select the height of the cone using the crosshairs.

To draw a cone with a circular base
(See Figure 16-11.)

1. Select the **Cone** tool from the **Modeling** toolbar.

Command: _cone
Specify center point [3P/2P/Ttr/Elliptical] <0,0,0>:

2. Type **5,5,0**; press **Enter.**

Specify base radius or [Diameter]:

3. Type **3**; press **Enter.**

Specify height or [2 Point/Axis endpoint/Top radius]:

4. Move the cursor to the positive Z direction and type **7**; press **Enter.**

Figure 16-12 shows a cone with a top radius of 2.00 and height of 7.00. To create a top radius on the cone type **T**

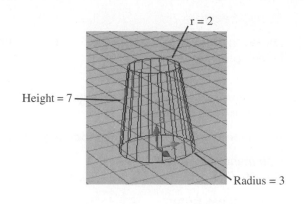

Figure 16-12

in response to the "Specify height" prompt and enter a radius value.

16-6 WEDGE

There are two options associated with the **Wedge** command: center and corner.

To draw a wedge by defining its corner point
(See Figure 16-13.)

1. Select the **Wedge** tool from the **Modeling** toolbar.

Command: _wedge
Specify first corner or [Center]:

2. Type **0,0,0**; press **Enter.**

This input will locate the corner of the wedge on the 0,0,0 point of the XY plane.

Specify other corner or [Cube/Length]:

The default response to this command defines the diagonal corner of the wedge's base.

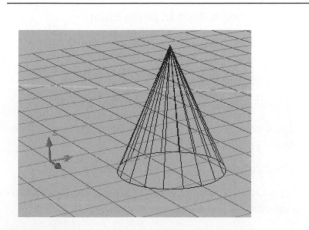

Figure 16-11

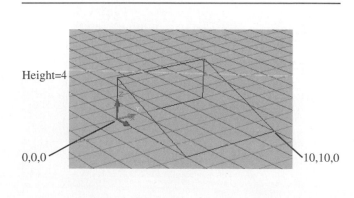

Figure 16-13

3. Type **10,10,0**; press **Enter**.

    Specify height:

4. Locate the cursor in the positive Z direction and type **4**; press **Enter**.

To draw a wedge by defining its center point (See Figure 16-14.)

1. Select the **Wedge** tool from the **Modeling** toolbar.

    Command: _wedge
    Specify first corner or [Center] <0,0,0>

2. Type **c**; press **Enter**.

    Specify center:

3. Type **5,5,0**; press **Enter**.

    Specify corner or [Cube/Length]:

4. Type **10,10,0**; press **Enter**.

    Specify height or [2 Point] <4.000>:

5. Locate the cursor in the positive Z direction and type **4**; press **Enter**.

The wedge shown in Figure 16-14 is centered about the XY plane; that is, part of the wedge is above the plane, and part is below. This is not easy to see even with the grid shown. The far right corner of the wedge is actually located below the

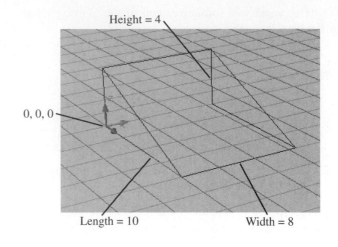

Figure 16-15

grid at the 10,10 point. Figure 16-14 also shows a side view of the same wedge with a line drawn on the XY plane added. Note how the line bisects the height line of the wedge.

To align a wedge with an existing wedge

This example illustrates how you can use different inputs to position a wedge. Figure 16-15 shows a **10 × 8 × 4** wedge. The problem is to draw another wedge with its back surface aligned with the back surface of the existing wedge.

1. Select the **Wedge** tool from the **Modeling** toolbar.

    Command: _wedge
    Specify first corner or [Center]:

2. Type **0,0,0**; press **Enter**.

    Specify other corner or [Cube/Length]:

3. Specify length: Type **L**; press **Enter**.

Move the cursor so that it is located in the negative X direction.

4. Type **10**; press **Enter**.

    Specify width:

5. Move the cursor so it is located in a positive **Y** direction. Type **8.00**; press **Enter**.

    Specify height or [2 Point] <4.0000>:

6. Use **Osnap, Endpoint** and select the Z-axis corner point of the existing wedge.

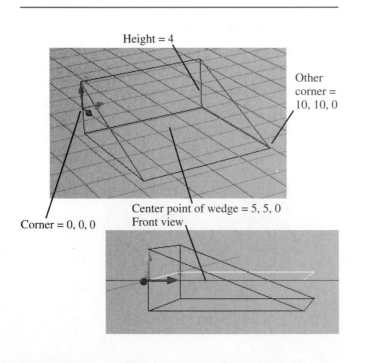

Figure 16-14

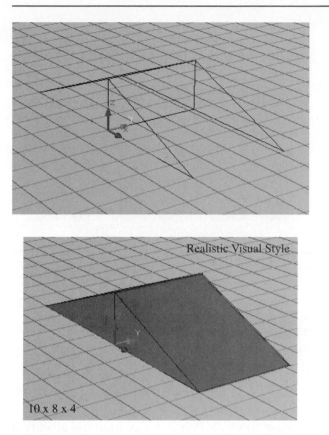

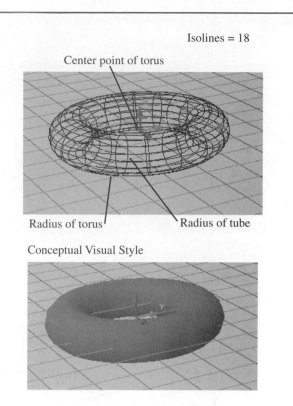

Isolines = 18

Center point of torus

Radius of torus   Radius of tube

Conceptual Visual Style

Figure 16-17

Figure 16-16

See Figure 16-16. The model could also have been constructed by using the **Copy** command to create a second wedge, the **Rotate** command to rotate the new wedge 180°, and the **Move** command to align the wedge with the existing wedge. Use **Osnap, Endpoint** to ensure exact alignment.

16-7 TORUS

A torus is a donutlike shape. See Figure 16-17.

To draw a torus

1. Select the **Torus** tool from the **Modeling** toolbar.

   Command: _torus
   Specify center point or [3P/2P/Ttr]:

2. Type **0,0,0;** press **Enter.**

This input will locate the center of the torus at the 0,0,0 point on the XY plane.

   Specify radius or [Diameter] <3.0000>:

3. Type **5;** press **Enter.**

   Specify tube radius or [2 Point/Diameter]:

4. Type **1.5;** press **Enter.**

The torus shown in Figure 16-17 was created with **Isolines** set at **18.**

16-8 EXTRUDE

The **Extrude** command is used to extend existing 2D shapes into 3D shapes. The **Extrude** tool can be applied only to a polyline.

To extrude a 2D Polyline

Figure 16-18 shows a hexagon drawn using the **Polygon** command. All shapes drawn using the **Polygon** command are automatically drawn as a polyline, so the hexagon can be extruded. How to draw a polygon is discussed in Chapter 2.

1. Select the **Extrude** tool from the **Modeling** toolbar.

   Command: _extrude
   Select objects to extrude:

2. Select the hexagon.

   Select objects:

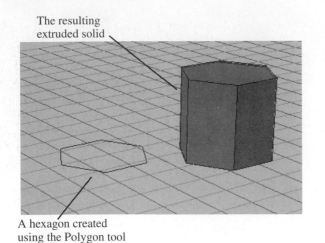

The resulting extruded solid

A hexagon created using the Polygon tool

Figure 16-18

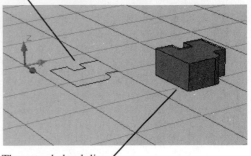

A shape that was created as individual line segments, then joined using the Edit Polyline tool to form a polyline.

The extruded polyline

Figure 16-20

3. Press **Enter.**

   Specify height of extrusion or [Direction/Path/ Taper angle] <4.0000>:

4. Locate the cursor in a positive Z direction, then type **6**; press **Enter.**

   Specify angle of taper for extrusion <0>:

5. Press **Enter.**

   Figure 16-19 shows a closed spline created using **Polyline, Polyline edit.** The **Extrude** command was applied specifying a height of **5** and a taper angle of **5°.**

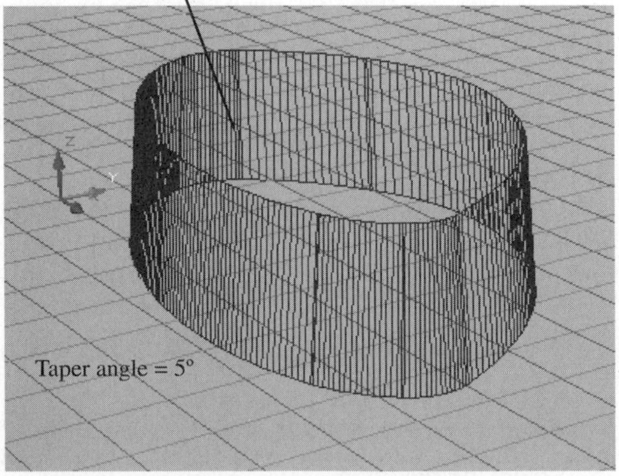

To create a polyline from line segments

   Figure 16-20 shows a 2D shape that was created using the **Line** and **Circle** tools from the **Draw** toolbar. The object must be converted to a polyline before it can be extruded.

1. Select the **Edit Polyline** tool from the **Modify II** toolbar.
   Select polyline or [Multiple]:

2. Select any one of the lines in the 2D shape.

   Object selected is not a polyline Do you want to turn it into one? <Y>:

3. Press **Enter.**

   Enter an option [Close/Join/Width/Edit vertex/Fit/ Spline/Decurve/Ltype gen/Undo]:

4. Select the **Join** option.

   The polyline will be defined by joining together all the line segments to form a polyline. A curve is considered to be a line segment.

   Select objects:

5. Window the entire object.

   Enter an option [Close/Join/Width/Edit vertex/Fit/ Spline/Decurve/Ltype gen/Undo]:

6. Right-click twice; press **Enter.**

7. Select the **Extrude** tool from the **Modeling** toolbar and create an extrusion **5** units high with a **0°** taper angle.

The Extrude command applied to an edited polyline.

Taper angle = 5°

Figure 16-19

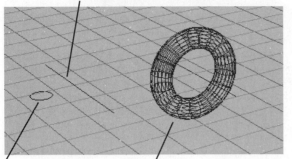

The axis of revolution

The object to be revolved

The resulting revolved solid

Figure 16-21

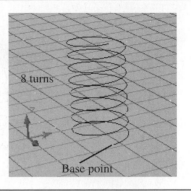

8 turns

Base point

Figure 16-22

16-9 REVOLVE

The **Revolve** tool on the **Solids** toolbar is used to create a solid 3D object by rotating a 2D shape around an axis of revolution. Figure 16-21 shows a torus created by rotating a circle around a straight line. The density of the resulting object is controlled by the **Isolines** command. In this example **Isolines** is set at **18.**

To create a revolved solid object

This procedure assumes that the curve path (2D shape) and the line that will be used as the axis of revolution already exist on the drawing.

1. Select the **Revolve** tool from the **Modeling** toolbar.

   Command: _revolve
   Current wire frame density: ISOLINES = 18
   Select objects to revolve:

2. Select the circle.

   Select objects to revolve:

3. Press **Enter.**

   Specify start point for axis of revolution or define axis by [Object/X/Y/Z] <object>:

4. Select one end of the line to be used as the axis of revolution. Use **Osnap, Endpoint** if necessary

   Specify axis endpoint:

5. Select the other end of the axis line.

   Specify angle of revolution or [Start angle] <360>:

6. Press **Enter.**

16-10 HELIX

See Figure 16-22.

1. Select the **Helix** tool from the **Modeling** toolbar.

   Number of turns = 3, Twist = CCW.

   Specify center point of base:

2. Select a point on the screen; press the left mouse button.

   Specify base radius or [Diameter] <1.0000>:

3. Type **2**; press **Enter.**

   Specify top radius or [Diameter] <2.0000>:

4. Press **Enter.**

   Specify helix height or [Axis endpoint/Turns/Turn height/tWist] <6.3885>:

5. Type **t**; press **Enter.**

   Enter number of turns <3.0000>:

6. Type **8**; press **Enter.**

   Specify helix height or [Axis endpoint/Turns/Turn height/tWist] <6.3885>:

7. Use the dynamic input option and select a helix height by moving the cursor.

8. Press the left mouse button.

16-11 POLYSOLID

See Figures 16-23 and 16-24.

1. Select the **Polysolid** tool from the **Modeling** toolbar.

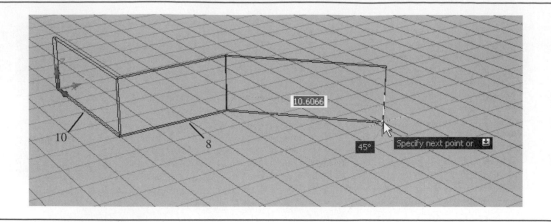

Figure 16-23

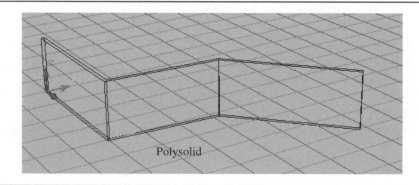

Figure 16-24

Polysolid Specify start point or [Object/Height/ Width/Justified] <Object>:

2. Type **0,0,0;** press **Enter.**

   Specify next point or [Arc/Close/Undo]:

3. Position the cursor in the positive X direction, then type **10;** press **Enter.**

   Specify next point or [Arc/Close/Undo]:

4. Position the cursor in the positive Y direction, then type **8;** press **Enter.**

   Specify next point or [Arc/Close/Undo]:

5. Position the cursor in the positive X direction, then use the dynamic inputs to select a point at **45°** from the X-axis.

6. Right-click the mouse and enter the polysolid.

16-12 LOFT

See Figure 16-25.

1. Use the **Circle** tool on the **Draw** toolbar and draw a **Ø10** circle centered about the 0,0,0 origin.

2. Use the **Circle** tool on the **Draw** toolbar and draw a **Ø5** circle centered about 0,0,12.

3. Select the **Loft** tool from the **Modeling** toolbar.

   Select the cross sections in lofting order:

4. Select the **Ø10** circle.

   Select the cross sections in lofting order:

5. Select the **Ø5** circle.

   Select the cross sections in lofting order:

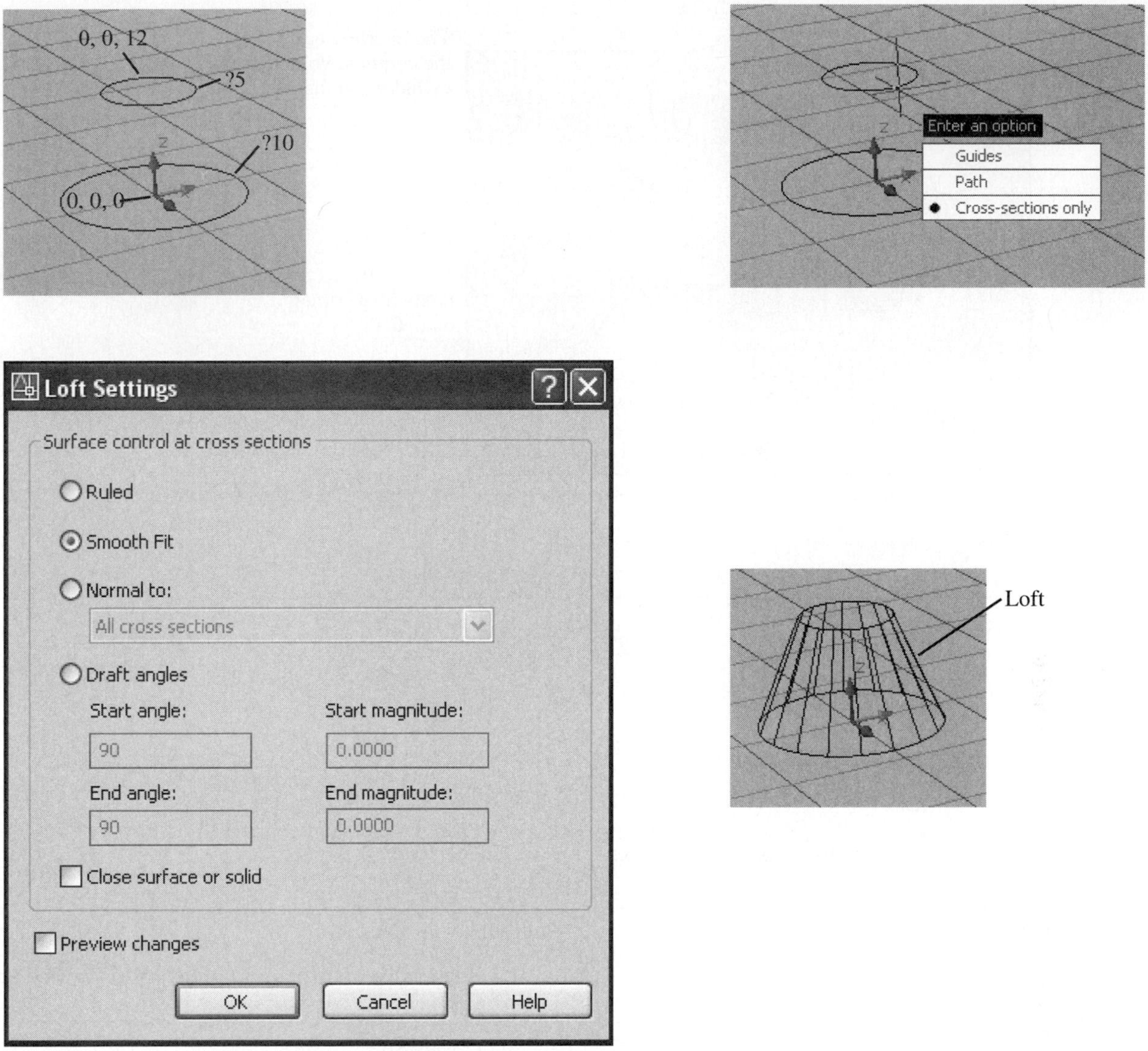

Figure 16-25

6. Press the right mouse button

    *Enter an option [Guides/Path/ Cross-sections only]
    <Cross-sections only>:*

7. Accept the **Cross-sections only** default option;
    press **Enter.**

    The **Loft Settings** dialog box will appear.

8. Select the **Smooth Fit** option (all other options
    should be open); click **OK.**

16-13 INTERSECT

The **Intersect** command is used to define a volume
common to two or more existing solid objects. Figure 16-26
shows a ⌀12 × 6 cylinder centered about the 0,0,0 point
and a 10 × 10 × 5 box with one of its corners located on
the 0,0,0 point. They were both drawn on the XY plane.
The following procedure will define the volume common
to both of them.

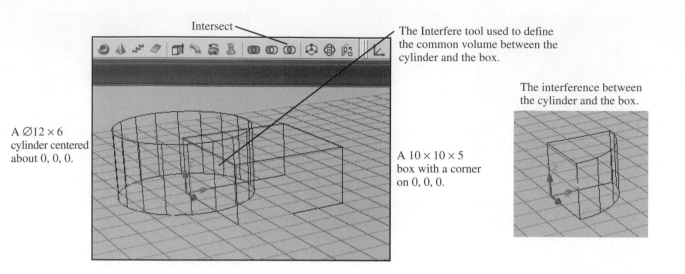

Intersect

The Interfere tool used to define the common volume between the cylinder and the box.

The interference between the cylinder and the box.

A ⌀12 × 6 cylinder centered about 0, 0, 0.

A 10 × 10 × 5 box with a corner on 0, 0, 0.

Figure 16-26

1. Select the **Intersect** tool from the **Modeling** toolbar.

   Command: _intersect
   Select objects:

2. Select the box.

   Select objects:

3. Select the cylinder.

   Select objects:

4. Press **Enter.**

   Figure 16-26 shows the resulting common volume.

16-14 UNION AND SUBTRACT

Solid objects may be combined to form more complex objects. Objects can be added together using the **Union** command and subtracted from each other using the **Subtract** command. A volume common to two or more objects may be defined using the **Intersect** command. The tools for these three commands are located on both the **Modeling** and **Solid Editing** toolbars or under the **Solid Editing** command located on the **Modify** pull-down menu. See Figure 16-27.

To Union two objects

Figure 16-28 shows two solid boxes drawn with adjoining surfaces.

1. Select the **Union** tool from the **Solid Editing** toolbar.

   Command: _union
   Select objects:

2. Select a box.

   Select objects:

3. Select the other box.

   Select objects:

4. Press **Enter.**

Note the changes in the boxes after they have been unioned. The solid object is no longer two boxes but an L-shaped object.

To Subtract an object

Figure 16-29 shows the L-shaped bracket formed in Figure 16-28. This exercise will add a hole to the front surface. Holes are created in solid objects by subtracting solid cylinders from the existing objects.

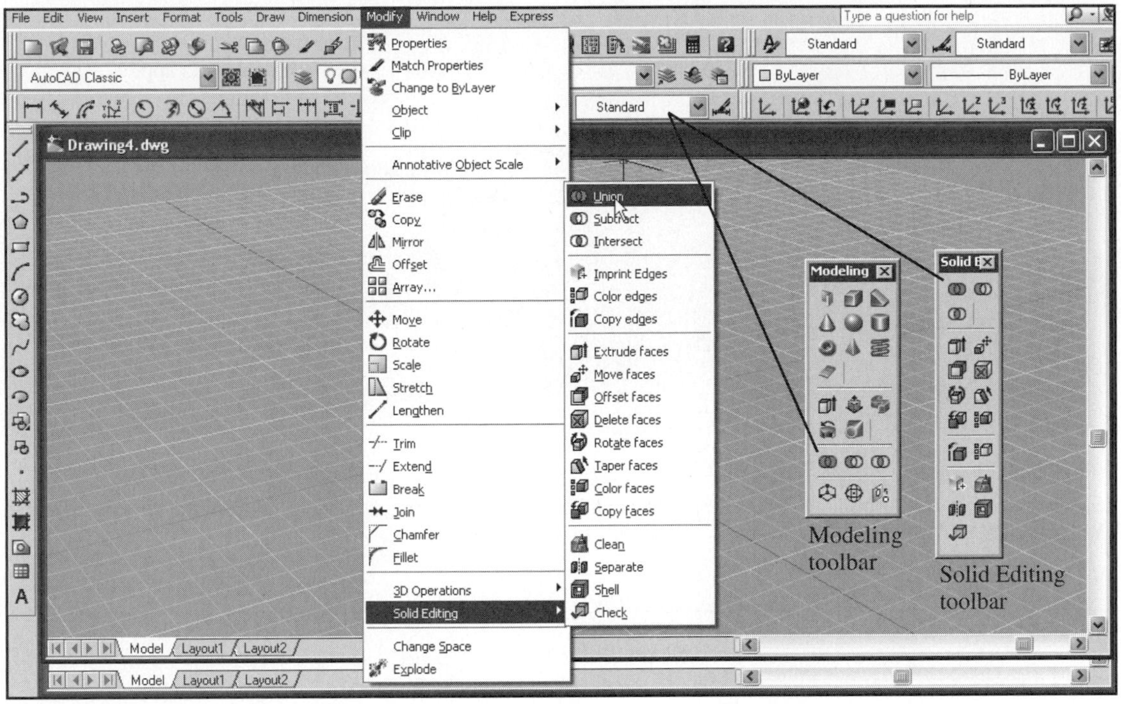

Figure 16-27

1. Select the **Cylinder** tool from the **Solids** toolbar.

   Command: _cylinder
   Specify center point of base or [3P/2P/Ttr/Elliptical]:

2. Type **5,5,0**; press **Enter.**

   Specify base radius or [2 Point/Axis endpoint] <5.0000>:

3. Type **3**: press **Enter.**

   Specify height or [2 Point/Axis endpoint] <10.0000>:

4. Type **5**; press **Enter.**

   The height of the cylinder was deliberately drawn higher than the top surface of the bracket to illustrate that the two heights need not be equal for the **Subtract** command. The only requirement is that the cylinder be equal to

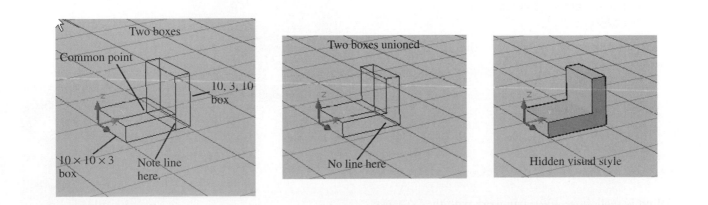

Figure 16-28

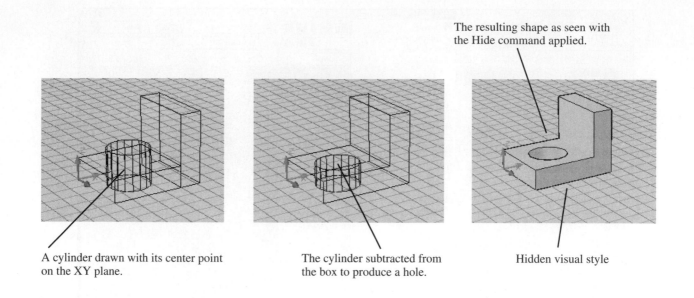

The resulting shape as seen with the Hide command applied.

A cylinder drawn with its center point on the XY plane.

The cylinder subtracted from the box to produce a hole.

Hidden visual style

Figure 16-29

or greater than the height of the box surface. See Figure 16-29.

5. Select the **Subtract** tool from the **Modeling** toolbar.

    Command: _subtract
    Select solids and regions to subtract from...
    Select objects:

    This prompt is asking you to define the main object, that is, the object you want to remain after the subtraction.

6. Select the L-shaped bracket.

    Select objects:

7. Press **Enter.**

    Select solids and regions to subtract...
    Select objects:

    This prompt is asking you to define the object you want removed by the subtraction.

8. Select the cylinder.

    Select objects:

9. Press **Enter.**

16-15 SOLID MODELING AND UCSs

In this section we will again work with the L-shaped bracket and add a hole to the upper surface. The procedure is to create a new UCS with its origin at the left intersection of the two perpendicular surfaces, then create and subtract a cylinder. See Figure 16-30.

1. Select the **Origin** tool from the **UCS** toolbar.

    Specify new origin point <0.0000>:

2. Use the **Endpoint** option of the **Object Snap** command (shift/right button) and select the corner as shown.

    The coordinate system will move to the new location.

3. Select the **X** tool on the **UCS** toolbar.

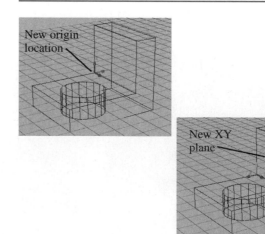

New origin location

New XY plane

Figure 16-30

Specify rotation angle about the X axis <90>:

4. Press **Enter.**

The axis will be rotated about the new X axis 90° and will be aligned with the front vertical surface of the L-bracket.

5. Select the **Cylinder** tool from the **Modeling** toolbar.

Specify center point of the base or [3P/2P/Ttr/ Elliptical]:

6. Type **5,3.5,0;** press **Enter.**

Specify base radius or [Diameter] <1.5000>:

7. Type **2.5000;** press **Enter.**

Specify height or [2 Point/Axis endpoint] <5.0000>:

8. Use the dynamic input options and move the cursor in the negative Z direction.

The cylinder height will increase as the cursor is moved.

9. Define the cylinder height at any distance greater than 3, the thickness of the L-bracket; press **Enter.**

See Figure 16-31.

10. Select the **Subtract** tool from the **Modeling** toolbar.

Command: _subtract

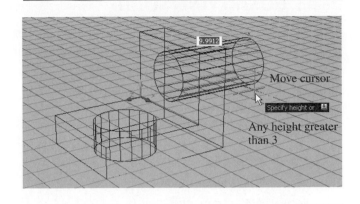

Figure 16-31

Select solids and regions to subtract from...
Select objects:

11. Select the L-shaped bracket.

Select objects:

12. Press **Enter.**

Select solids and regions to subtract...
Select objects:

13. Select the cylinder.

Select objects:

14. Press **Enter.**

Figure 16-32 shows the resulting solid object.

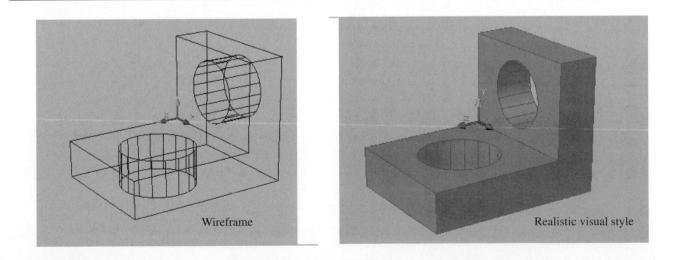

Figure 16-32

16-16 COMBINING SOLID OBJECTS

Figure 16-33 shows a dimensioned object. The following section explains how to create the object as a solid model. There are many different ways to create a solid model. The sequence presented here was selected to demonstrate several different input options.

To set up the drawing

Set up the drawing as follows:

Units = Decimal (millimeters)
Drawing Limits = 297,210
Grid = 10 × 10 parallel
Snap = 10
View = SE Isometric
Toolbars = Draw, Modify, Solids, Solid Editing, View, UCS, and **UCS II**

The object is relatively small, so use the **Zoom** tool to create a size that you find visually comfortable.

To draw the first box

The size specifications for this box are based on the given dimensions.

1. Select the **Box** tool from the **Solids** toolbar.

   Command: _box
   Specify first corner or [Center]:

2. Select the **0,0,0** point on the WCS by pressing **Enter.**

   Specify other or [Cube/Length]:

3. Type **L**; press **Enter.**

   Specify length:

4. Locate the cursor on the positive X axis and type **80**; press **Enter.**

   Specify width:

5. Locate the cursor in the positive Y direction and type **35**; press **Enter.**

   Specify height:

6. Locate the cursor in the positive Z direction and type **30**; press **Enter.**

   See Figure 16-34.

To create the internal open volume

The volume will be created by subtracting a second box from the first box.

1. Select the **Box** tool from the **Solids** toolbar.

   Command: _box
   Specify corner of box or [Center]:

2. Select the **0,10,0** point on the WCS.

   The point 0,10,0 was selected based on the given 10-millimeter dimension. The point can be selected using the crosshair because it is on the grid located on the XY plane.

   Specify corner or [Cube/Length]:

3. Type **L**; press **Enter.**

   Specify length:

4. Locate the cursor in the positive X direction and type **80**; press **Enter.**

   Specify width:

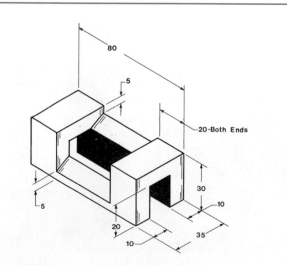

Figure 16-33

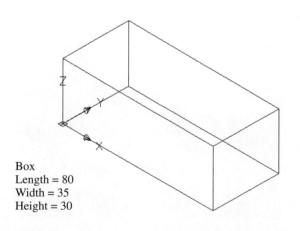

Box
Length = 80
Width = 35
Height = 30

Figure 16-34

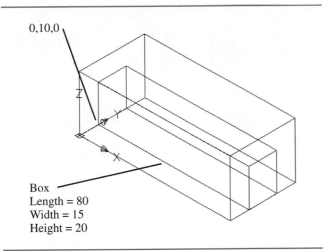

0,10,0

Box
Length = 80
Width = 15
Height = 20

Figure 16-35

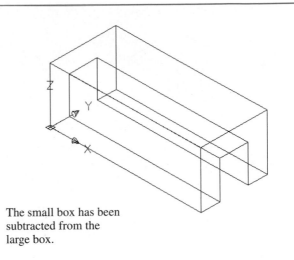

The small box has been
subtracted from the
large box.

Figure 16-36

5. Locate the cursor in the positive Y direction and type **15**; press **Enter.**

   Specify height:

6. Locate the cursor in the positive Z direction and type **20**; press **Enter.**

   Figure 16-35 shows the second box within the first box.

7. Select the **Subtract** tool from the **Solid Editing** toolbar.

   Command: _subtract
   Select solids and regions to subtract from...
   Select objects:

8. Select the first box.

   Select objects:

9. Press **Enter.**

   Select solids and regions to subtract...
   Select objects:

10. Press **Enter.**

    See Figure 16-36.

To create the cutout

   See Figure 16-37.

   The cutout will be created by drawing a box and wedge and subtracting them from the existing object.

To create a box

1. Move the origin to the top surface of the box.
2. Select the **Box** tool from the **Modeling** toolbar.

   Specify first corner or [Center]:

3. Type **20,0,0**; press **Enter.**

   Specify other corner or [Cube/Length]:

4. Position the cursor in the positive X and Y directions and type **40,35,0**; press **Enter.**

   Specify height or [2 Point]<0.0000>:

5. Position the cursor in the negative Z direction and type **5**; press **Enter.**

6. Use the **Subtract** tool from the **Modeling** toolbar and subtract the box from the existing object.

   See Figure 16-38.

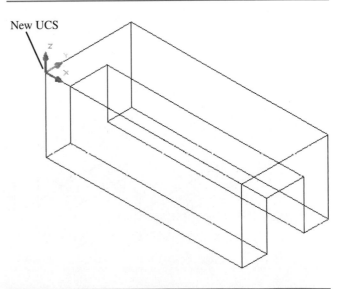

New UCS

Figure 16-37

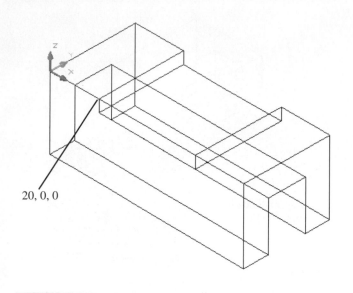

20, 0, 0

Figure 16-38

To create a wedge

See Figure 16-39.

1. Move the origin to the corner of the box cutout as shown.
2. Select the **Z** tool from the **UCS** toolbar and rotate the axis **90°** around the Z axis.

See Figure 16-40.

3. Select the **Wedge** tool from the **Modeling** toolbar.

Specify first corner or (Center):

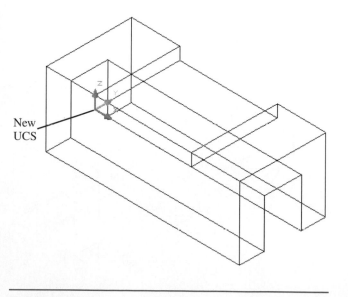

New UCS

Figure 16-39

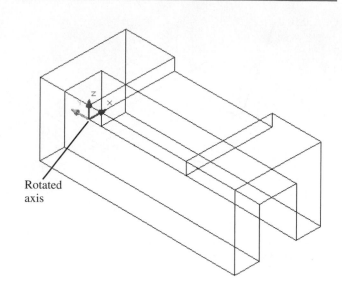

Rotated axis

Figure 16-40

4. Select the **0,0,0** point of the current UCS.

Specify other corner or [Cube/Length]:

5. Use the **Endpoint** option of the **Object Snap** commands and select the opposite diagonal corner of the box cutout as shown.

Specify height or [2 Point]<0.0000>:

6. Locate the cursor in the negative Z direction and type **20**; press **Enter.**

See Figure 16-41.

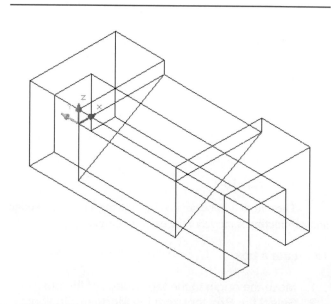

Figure 16-41

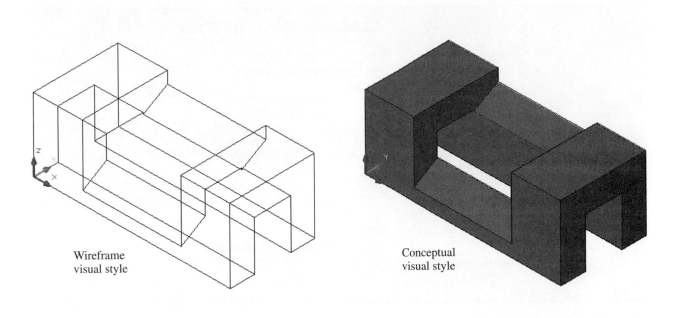

Wireframe
visual style

Conceptual
visual style

Figure 16-42

7. Subtract the wedge from the existing object.
8. Use the **World** tool from the **UCS** toolbar and return the axis system to the original location.

See Figure 16-42.

16-17 INTERSECTING SOLIDS

Figure 16-43 shows an incomplete 3D drawing of a cone and a cylinder. The problem is to complete the drawing in 3D and show the front, top, and right-side orthographic views of the intersecting objects. If this problem were to be done by hand on a drawing board, it would require extensive projection between views, as well as a high degree of precision in the line work. Done as a solid model, the problem is much simpler and serves to show the strength of solid modeling as a design tool.

To set up the drawing

Set up the drawing screen as follows.

Grid = 0.50 × 0.50
Snap = 0.50
Units = Decimal
View = SE Isometric
Toolbars = Modeling, Solid Editing, View, UCS, UCS II, Modify, and Modify II
Template = acad3D

See Figure 16-44. The objects are small, so use the **Zoom Window** command to create a comfortable visual size.

To draw the cone

1. Select the **Cone** tool from the **Solids** toolbar.

   Command: _cone
   Specify center point of base or [3P/2P/ Ttr/Elliptical]

2. Type **0,0,0**; press **Enter.**

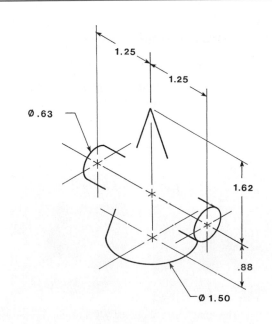

Figure 16-43

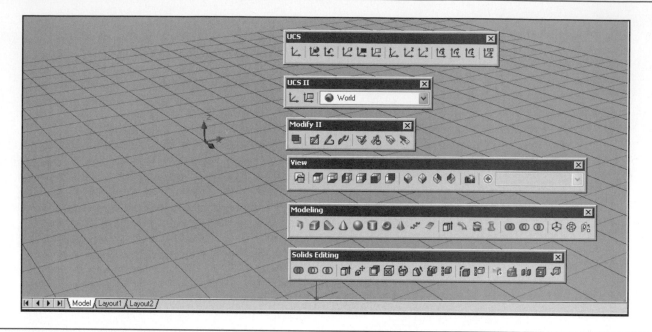

Figure 16-44

The center point of the cone will be located on the origin of the WCS.

Specify base radius or [Diameter] <0,0000>:

3. Type **d;** press **Enter.**

Specify diameter:

4. Type **1.50.**

Specify height or [2 Point/Axis endpoint/Top radius]<1.0000>:

5. Locate the cursor in the positive **Z** direction and type **2.50;** press **Enter.**

See Figure 16-45.

6. Select the **Visual Styles** tool on the **New** pull-down menu and change to **Conceptual.**

To draw the cylinder

1. Select the **Origin** UCS tool from the **UCS** toolbar.

Command: _ucs
Specify new origin point <0,0,0>:

2. Type **1.25,0,0;** press **Enter.**

This input locates the origin in the same plane as the end of the cylinder.

3. Select the **X Axis Rotate** UCS tool from the **UCS** toolbar and press **Enter,** accepting the **90°** default value, then select the **Y Axis Rotate** UCS tool and accept the **90°** default value.

See Figure 16-46.

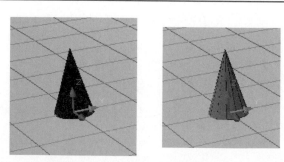

The cone drawn about the WCS origin

Figure 16-45

The Right UCS located at the new origin

Figure 16-46

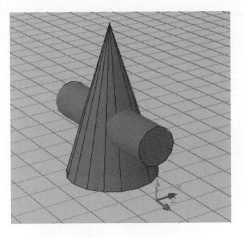

Figure 16-47

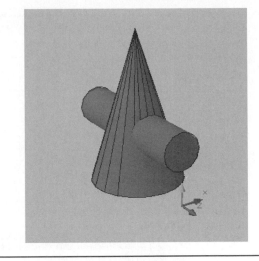

Figure 16-48

4. Select the **Cylinder** tool from the **Solids** toolbar.

   Command: _cylinder
   Specify center point of base or [3P/2P/Elliptical]:

5. Type **0,.88,0.**

   Specify base radius or [Diameter] <0.0000>:

6. Type **d**; press **Enter.**

   Specify diameter <1.0000>:

7. Type **.63**; press **Enter.**

   Specify height or [2Point/Axis endpoint] <1.0000>:

8. Locate the cursor in the negative Z direction and type **2.50**; press **Enter.**

   See Figure 16-47.

To complete the 3D drawing

1. Select the **Union** tool from the **Solid Editing** toolbar.

   Command: _union
   Select objects:

2. Select the cone.

   Select objects:

3. Select the cylinder.

   Select objects:

4. Press **Enter.**

   Command:

5. Select the **World** tool from the **UCS** toolbar.

6. Turn off **Grid.**

   See Figure 16-48.

To create the viewports for the orthographic views

1. Select the **View** pull-down menu.
2. Select **Viewports,** then **4 Viewports.**

   See Figures 16-49 and 16-50. The toolbars have been deleted from Figure 16-50 for clarity. Only the **View** toolbar is shown.

To create a top orthographic view

1. Move the cursor into the top left port and press the left mouse button.

   The crosshairs will appear in the port.

2. Select the **Top view** tool from the **View** toolbar.
3. Right-click the mouse and select the **Zoom** option.
4. Right-click the mouse again and select the **Parallel** option.

   The **Parallel** command assures that the axis lines remain parallel as they recede. A perspective axis converges to vanishing points as the lines recede.

5. Zoom and align as needed.
6. Right-click the mouse and select the **Exit** option.

   See Figure 16-51.

To create the front orthographic view

1. Move the cursor into the lower left port and press the left mouse button.
2. Select the **Front view** tool from the **View** toolbar and make it the current view.
3. Right-click the mouse and select the **Zoom** option.
4. Right-click the mouse again and select the **Parallel** option.

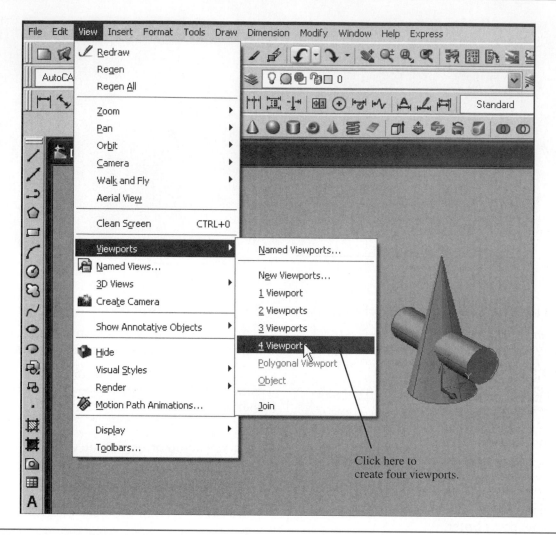

Click here to
create four viewports.

Figure 16-49

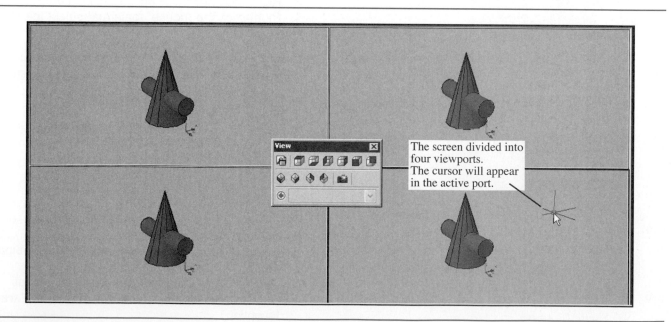

The screen divided into
four viewports.
The cursor will appear
in the active port.

Figure 16-50

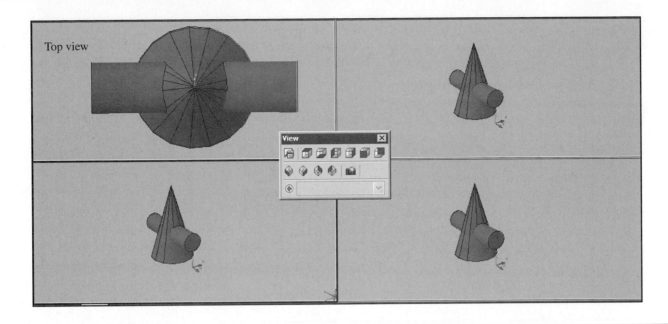

Figure 16-51

5. Zoom both top and front views so that they align.
6. Right-click and select the **Exit** option.

See Figure 16-52.

To create the right-side orthographic view

1. Move the cursor to the lower right port and press the left mouse button.

2. Select the **Right View** tool from the **View** toolbar and make it the current view.
3. Right-click the mouse and select the **Zoom** option.
4. Right-click the mouse again and select the **Parallel** option.

See Figure 16-53.

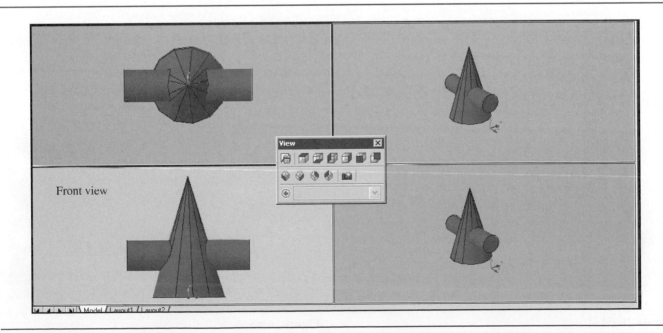

Figure 16-52

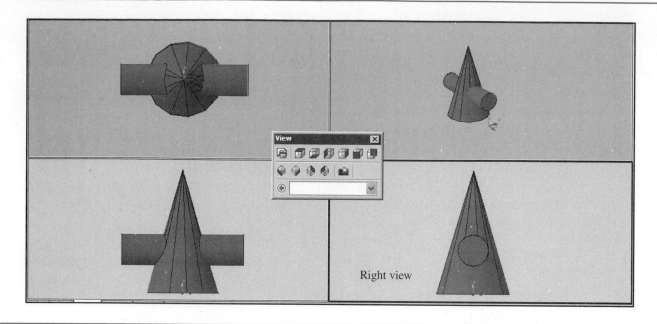

Figure 16-53

16-18 SOLID MODELS OF CASTINGS

Figure 16-54 shows a casting. Note that the object includes rounded edges. These rounded edges can be created on a solid model using the **Fillet** command found on the **Modify** toolbar. The **Fillet** command was explained in Chapter 2.

This example will be presented without specific dimensions and will use a generalized approach to creating the model. Decimal inches are used for all dimensions.

To draw the basic shape

The basic shape will first be drawn in 2D, then extruded into the 3D solid model.

1. Set up the drawing screen as needed.

In this example the **acad3D** template was used, and a top view was created using the **Top** tool from the **View** toolbar.

2. Draw the basic shape using the **Circle** tool and then the **Line** tool along with the **Osnap, Tangent** option.

See Figure 16-55. It is important to know the center point locations for the two circles in terms of their X,Y components.

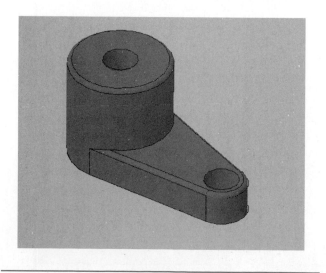

Figure 16-54

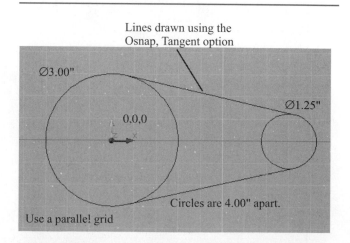

Lines drawn using the Osnap, Tangent option

Ø3.00"

0,0,0

Ø1.25"

Circles are 4.00" apart.

Use a parallel grid

Figure 16-55

In this example the center point location for the large circle is 4.5,5, and the location for the small circle is 11,5. The large circle's diameter equals 3.00 inches and the small circle's diameter equals 1.25 inches. The circles are 4.0 inches apart.

To create a polyline from the basic shape

Because only polylines can be extruded, some of the lines in the basic shape must be formed into a polyline. The large circle can be extruded, so it need not be included as part of the polyline; however, the polyline must be a closed area, and so it will need part of the circle. The needed circular segment can be created by drawing a second large circle directly over the existing circle and then using the **Trim** command to remove the excess portion. Remember that two lines can occupy the same space in AutoCAD drawings. If there is difficulty working with the two large circles, trim the circles, then add another larger circle, if needed.

1. After the shape shown in Figure 16-55 is drawn, right-click the mouse and select the **Zoom** option, then right-click the mouse again and select the **Parallel** option.
2. Right-click the mouse and select the **Exit** option.
3. Use the **Trim** tool to remove the excess portion of both the large and small circles.

Figure 16-56 shows the resulting shape that will be joined to form a polyline.

4. Select the **Edit Polyline** tool from the **Modify II** toolbar.

   Command: _pedit
   Select polyline or [Multiple]:

5. Select the remaining portion of the small circle.

   Object selected is not a polyline
   Do you want to turn it into one? <Y>

6. Press **Enter.**

   Enter an option [Close/Join/Width/Edit vertex/ Fit/Spline/Decurve/Ltype gen/Undo/eXit/ <X>:

7. Select the **Join** option.

   Select objects:

8. Window the object; press **Enter.**

   Select objects:

9. Press **Enter** twice.

There will be no visible changes to the lines, but they have been combined to form a single polyline.

To extrude the shape

1. Select the **SE Isometric** tool from the **View** toolbar and make it the current view.
2. Use the **Zoom** tool, if needed, to present the figure at a comfortable visual size.

   See Figure 16-57.

3. Type **Isolines** and change the value to **18.**
4. Select the **Extrude** tool from the **Solids** toolbar.

   Command: _extrude
   Select objects to extrude:

5. Select the polyline, and assign a height and **0°** taper.

   In this example a height of 1 was assigned. See Figure 16-58.

   Command:

6. Change the object to a **Wireframe** visual style.
7. Select the **Cylinder** tool from the **Solids** toolbar and create a cylinder whose diameter equals

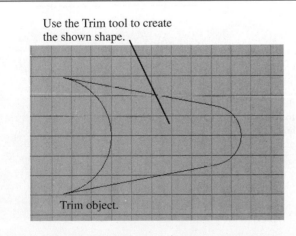

Use the Trim tool to create the shown shape.

Trim object.

Figure 16-56

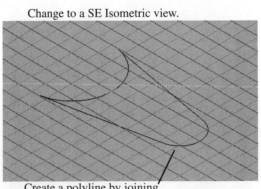

Change to a SE Isometric view.

Create a polyline by joining four line segments.

Figure 16-57

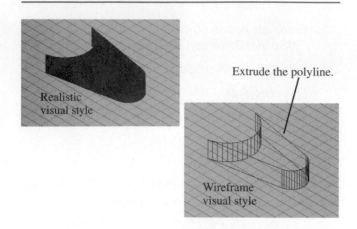

Figure 16-58

the original large-circle diameter (∅3.00) and whose height is 3, with a center point of **0,0,0.**

See Figure 16-59.

To add the holes

Create the holes by subtracting cylinders from the object.

1. Select the **Cylinder** tool from the **Solids** toolbar.

   Command: _cylinder
   Specify center point for base of cylinder or [Elliptical] <0,0,0>:

The ∅3.00 × 3.00 cylinder was created using the Cylinder tool on the Solids toolbar. Center point = 0,0,0.

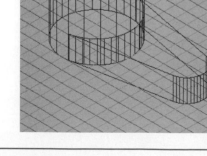

Figure 16-59

2. Type **0,0,0;** press **Enter.**

   This value came from the original circle's center point location.

3. Enter the appropriate diameter and height values. In this example a radius of **0.50** inch and a height of **3.5** were selected.

   Command:

4. Repeat the cylinder command.

   Specify center point for base of cylinder or [Elliptical] <0,0,0>:

5. Type **4,0,0.**
6. Draw a ∅**1.00 × 1.25** cylinder.

   See Figure 16-60.

7. Select the **Union** tool from the **Solid Editing** toolbar and join the large cylinder portion of the object to the polyline portion.
8. Select the **Subtract** tool from the **Solid Editing** toolbar and subtract the cylinders from the basic shape.

   See Figure 16-61.

To create the rounded edges

1. Select the **Fillet** tool from the **Modify** toolbar.

   Command: _fillet
   Current settings: Mode = TRIM, Radius=0.5000.
   Select first object or [Undo/Polyline/Radius/Trim/ Multiple]:

2. Type **r** and enter the appropriate radius value, if necessary.

   In this example a radius of **0.125** was used.

Use the Cylinder tool to create two holes in the model.

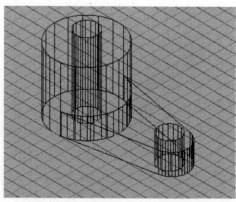

Figure 16-60

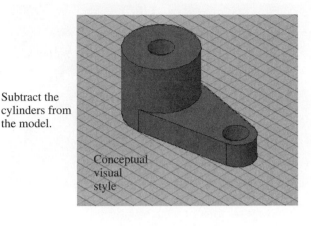

Subtract the cylinders from the model.

Conceptual visual style

Figure 16-61

3. Repeat the **Fillet** command and select the outside edge of the top surface of the large cylindrical portion of the object.

   Enter fillet radius <0.1250>:

4. Press **Enter.**

   Select first object or [Undo/Polyline/Radius/Trim/Multiple]:

5. Press **Enter.**

   See Figure 16-62.

6. Use the **Fillet** tool to create a fillet along the top edges of the object as shown in Figure 16-63.

   Note that the arc edge line between the large cylindrical portion of the object and the extended flat area cannot be filleted.

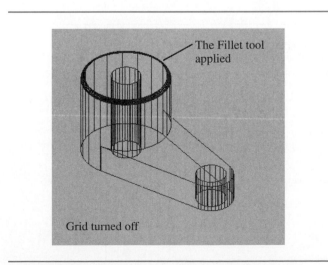

The Fillet tool applied

Grid turned off

Figure 16-62

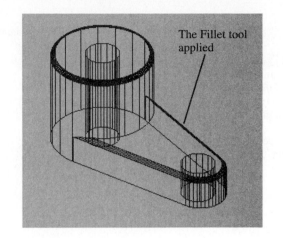

The Fillet tool applied

Figure 16-63

16-19 THREAD REPRESENTATIONS IN SOLID MODELS

This section explains how to draw thread representations for solid models. The procedure presented only represents a thread. It is not an actual detailed solid drawing of a thread. As with the thread representations presented in Chapter 11 for 2D drawings, 3D representations are acceptable for most applications.

1. Select the **Cylinder** tool from the **Solids** toolbar and draw a cylinder.

   In the example shown, a cylinder of diameter **3** and a height of **6** was drawn centered about the **0,0,0** point of the WCS. See Figure 16-64. It is presented on the **acad3D** template with a **Wireframe** visual style.

2. Draw a circle with a diameter equal to the diameter of the cylinder using center point **5,0,0.**

3. Select the **3D Array** command from the **Modify** pull-down menu, located under the **3D Operation** heading.

   Select object:

4. Select the circle.

   Enter the type of array [Rectangular or Polar array] <e>:

5. Type **r**; press **Enter.**

   Enter the number of rows (___)<1>:

6. Press **Enter.**

   Enter the number of columns: (|||)<1>:

7. Press **Enter.**

   Enter the number of levels:

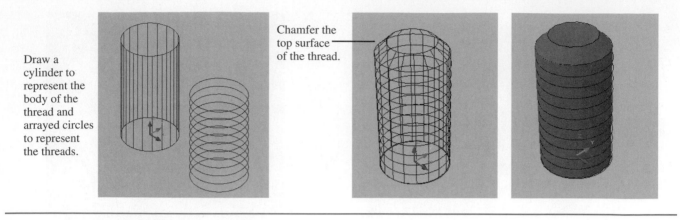

Draw a cylinder to represent the body of the thread and arrayed circles to represent the threads.

Chamfer the top surface of the thread.

Figure 16-64

8. Type **11**; press **Enter.**

The number 11 is used because the cylinder is 6 units high, and in this example circles representing threads will be spaced .5 apart. The top edge of the thread will be chamfered.

Specify the distance between levels: (...):

9. Type **.5**; press **Enter.**
10. Select the **Move** tool from the **Modify** toolbar and move the arrayed circles' center points to the **0,0,0** origin.
11. Select the **Chamfer** tool from the **Modify** toolbar and draw **a .5 × .5** chamfer around the top edge of the cylinder.

16-20 LIST

The **List** command is used to display database information for a drawn solid object. Figure 16-65 shows a list for the thread model created for Figure 16-64. There is no icon

for the **List** command. Type **list** in response to a command prompt and select the object.

16-21 MASSPROP

The **Massprop** command is used to display information about the structural characteristics of an object. Figure 16-66 shows the **Massprop** information for the object shown in Figure 16-64. To access the **Massprop** command, type **Massprop** in response to a command prompt and select the object.

16-22 FACE AND EDGE EDITING

AutoCAD 2008 has the capability to edit the faces and edges of existing solids. The **Solid Editing** command can be accessed using the **Solid Editing** toolbar or by using the **Solid Editing** option on the **Modify** pull-down menu.

The List command output

```
AutoCAD Text Window - Drawing2.dwg

Edit

Command: *Cancel*

Command: _.undo Current settings: Auto = On, Control = All, Combine = Yes
Enter the number of operations to undo or [Auto/Control/BEgin/End/Mark/Back]
<1>: 1 3DZOOM Regenerating model.

Command: list

Select objects: 1 found

Select objects:
                    3DSOLID    Layer: "0"
                               Space: Model space
                          Handle = c9
              History = Record
       Show History = No
          Solid type = Cylinder
          Position, X = 0.0000     Y = 0.0000     Z = 0.0000
              Radius: 1.5000
              Height: 6.0000

Command: |
```

Figure 16-65

The Massprop command output

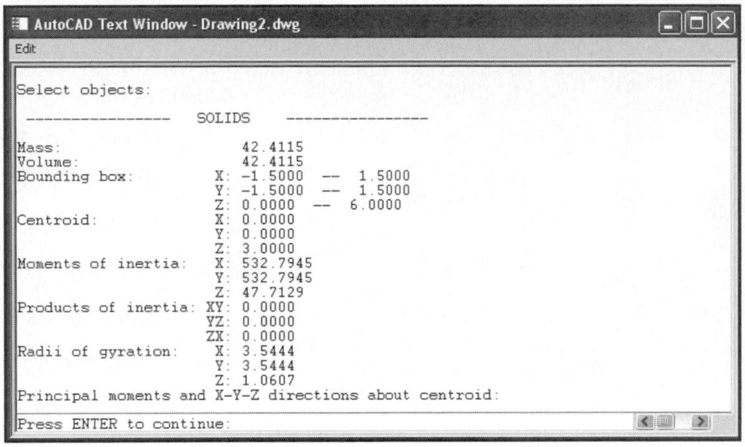
AutoCAD Text Window - Drawing2.dwg
Edit
Select objects:
---------------- SOLIDS ----------------
Mass: 42.4115
Volume: 42.4115
Bounding box: X: -1.5000 -- 1.5000
Y: -1.5000 -- 1.5000
Z: 0.0000 -- 6.0000
Centroid: X: 0.0000
Y: 0.0000
Z: 3.0000
Moments of inertia: X: 532.7945
Y: 532.7945
Z: 47.7129
Products of inertia: XY: 0.0000
YZ: 0.0000
ZX: 0.0000
Radii of gyration: X: 3.5444
Y: 3.5444
Z: 1.0607
Principal moments and X-Y-Z directions about centroid:
Press ENTER to continue:

Figure 16-66

See Figure 16-67. The following examples are based on a solid box of dimensions 4 × 2 × 3 located on a parallel axis system.

To extrude a face (See Figure 16-68.)

1. Select the **Extrude faces** tool from the **Solid Editing** toolbar.

   Select faces or [Undo/Remove]:

2. Select the right face by moving the cursor to the center of the face and left-clicking.

   Select faces or [Undo/Remove]:

3. Press the right mouse button and enter the face.

   Specify height of extrusion or [Path]:

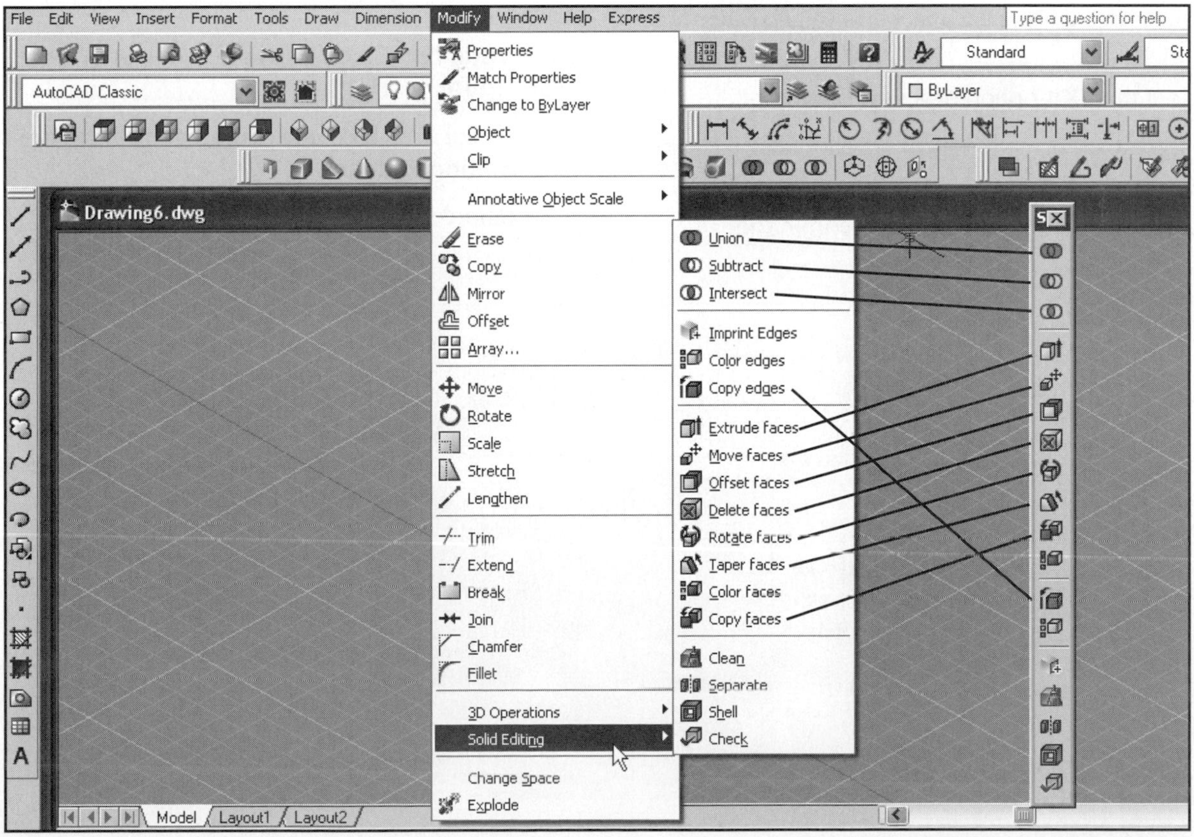
File Edit View Insert Format Tools Draw Dimension Modify Window Help Express
Type a question for help
Standard
AutoCAD Classic
ByLayer
Drawing6.dwg
Properties
Match Properties
Change to ByLayer
Object
Clip
Annotative Object Scale
Erase
Copy
Mirror
Offset
Array...
Move
Rotate
Scale
Stretch
Lengthen
Trim
Extend
Break
Join
Chamfer
Fillet
3D Operations
Solid Editing
Change Space
Explode
Union
Subtract
Intersect
Imprint Edges
Color edges
Copy edges
Extrude faces
Move faces
Offset faces
Delete faces
Rotate faces
Taper faces
Color faces
Copy faces
Clean
Separate
Shell
Check
Model
Layout1
Layout2

Figure 16-67

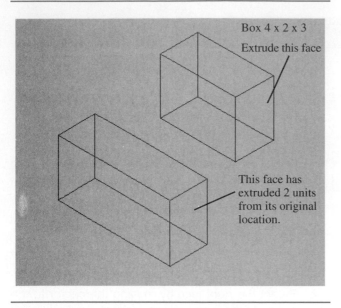

Figure 16-68

4. Type **2**; press **Enter.**

Specify angle of taper for extrusion <0>:

A warning dialog box will appear.

5. Press **Enter.**

A dialog box will appear under the heading **Enter face editing option.**

6. Select the **eXit** option or a new command.

A second set of options will appear under the same heading.

7. Select the **eXit** option or a new command.

To extrude a face along a path

See Figure 16-69. A line has been drawn from the right corner of the box 3″, 30°. This line will serve as the extrusion path.

1. Select the **Extrude faces** tool from the **Solid Editing** toolbar.

Select faces or [Undo/Remove]:

2. Click the right face.

Select faces or [Undo/Remove]:

3. Press the right mouse button and enter the face.

Specify height of extrusion or [Path]:

4. Type **p**; press **Enter.**

Select extrusion path:

5. Select the line.

A dialog box will appear under the heading **Enter face editing option.**

6. Select the **eXit** option or a new command.

A second set of options will appear under the same heading.

7. Select the **eXit** option or a new command.

To extrude two faces at the same time (See Figure 16-70.)

1. Select the **Extrude faces** tool from the **Solid Editing** toolbar.

Select faces or [Undo/Remove]:

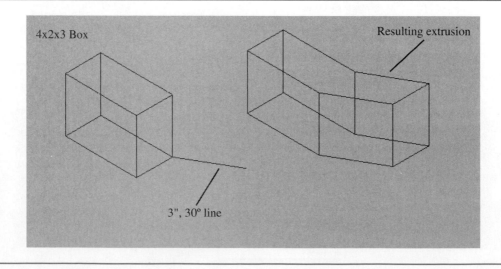

Figure 16-69

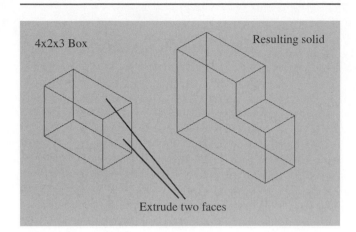

Figure 16-70

2. Select the right face, then select the top face.

Select faces or [Undo/Remove]:

3. Press the right mouse button and enter the face.

Specify height of extrusion or [Path]:

4. Type **2**; press **Enter.**

A dialog box will appear under the heading **Enter face editing option.**

5. Select the **eXit** option or a new command.

A second set of options will appear under the same heading.

6. Select the **eXit** option or a new command.

To move a face (See Figure 16-71.)

1. Select the **Move faces** tool from the **Solid Editing** toolbar.

Select faces or [Undo/Remove]:

2. Select the right face and enter it.

Specify a base point or displacement:

3. Select the corner of the box as shown.

Use the **Endpoint** option of the **Object snap** toolbar.

Specify a second point of displacement.

4. Select a second point along the X axis.
5. Press **Enter.**

A dialog box will appear under the heading **Enter face editing option.**

6. Select the **eXit** option or a new command.

A second set of options will appear under the same heading.

7. Select the **eXit** option or a new command.

To offset faces (See Figure 16-72.)

1. Select the **Offset Faces** tool from the **Solid Editing** toolbar.

Select faces or [Undo/Remove]:

2. Select the right and front faces and enter them.

Specify the offset distance:

3. Type **1.5**; press **Enter.**

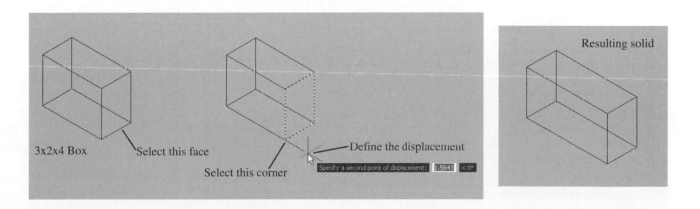

Figure 16-71

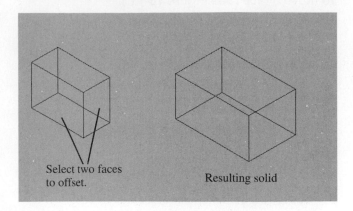

Select two faces to offset.   Resulting solid

Figure 16-72

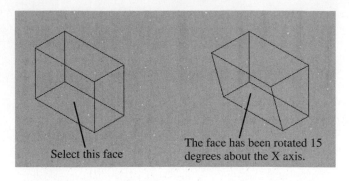

Select this face   The face has been rotated 15 degrees about the X axis.

Figure 16-73

A dialog box will appear under the heading **Enter face editing option.**

4. Select the **eXit** option or a new command.

A second set of options will appear under the same heading.

5. Select the **Exit** option or a new command.

To rotate a face (See Figure 16-73.)

1. Select the **Rotate faces** tool from the **Solid Editing** toolbar.

   Select faces or [Undo/Remove]:

2. Select the front face and enter it.

   Specify an axis point or [Axis by object/View/ Xaxis/Yaxis/Zaxis]<2 points>:

3. Type **x**; press **Enter.**

   Specify the origin of the rotation <0,0,0>:

4. Press **Enter.**

   Specify the rotation angle or [Reference]:

5. Type **15**; press **Enter.**

   An input of −**15** would have rotated the face in the opposite direction.

   A dialog box will appear under the heading **Enter face editing option.**

6. Select the **eXit** option or a new command.

   A second set of options will appear under the same heading.

7. Select the **eXit** option or a new command.

To taper a face

   See Figure 16-74.

1. Select the **Taper faces** tool from the **Solid Editing** toolbar.

   Select faces or [Undo/Remove]:

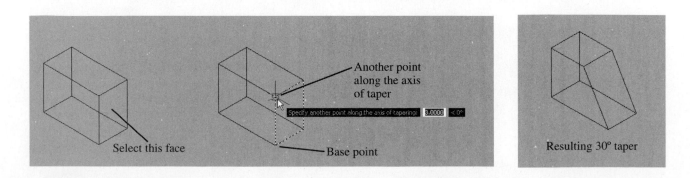

Select this face   Another point along the axis of taper   Base point   Resulting 30° taper

Specify another point along the axis of tapering: 3.0000 < 0°

Figure 16-74

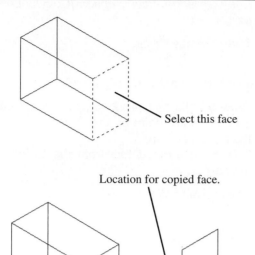

Select this face

Location for copied face.

Figure 16-75

2. Select the right front face.

   Select faces or [Undo/Remove]:

3. Press the right mouse button and enter the face.

   Specify the base point:

4. Select the lower right corner of the box.

   Specify another point along the axis of tapering:

5. Use **Osnap, Endpoint** and specify the top corner as shown in Figure 16-74.

   Specify the taper angle:

6. Type **30**; press **Enter.**

   [Extrude/Move/Rotate/Offset/Taper/Delete/Copy/coLor/Undo/eXit]<eXit>:

   A dialog box will appear under the heading **Enter face editing option.**

7. Select the **eXit** option or a new command.

   A second set of options will appear under the same heading.

8. Select the **eXit** option or a new command.

To copy a face

   See Figure 16-75.

1. Select the **Copy faces** tool from the **Solid Editing** toolbar.

   Select faces or [Undo/Remove]:

2. Select the right front face.

   Select faces or [Undo/Remove]:

3. Press the right mouse button and enter the face.

   Specify a base point or displacement:

4. Select the lower right corner of the box.

   Specify a second point of displacement:

5. Select the location for the copied face.

   [Extrude/Move/Rotate/Offset/Taper/Delete/Copy/coLor/Undo/eXit]<eXit>:

   A dialog box will appear under the heading **Enter face editing option.**

6. Select the **eXit** option or a new command.

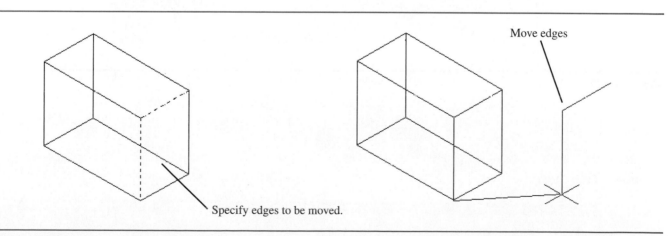

Specify edges to be moved.

Move edges

Figure 16-76

A second set of options will appear under the same heading.

7. Select the **eXit** option or a new command.

To copy edges (See Figure 16-76.)

1. Select the **Copy edges** tool from the **Solid Editing** toolbar.

   Select edges or [Undo/Remove]:

2. Select two edges.

   Select edges or [Undo/Remove]:

3. Press the right mouse button and enter the edges.

   Specify a base point or displacement:

4. Select the lower endpoint of the vertical line.

   Specify a second point of displacement:

5. Select a second point.

Enter an edge editing option [Copy/coLor/Undo/eXit] <eXit>:

6. Type **x**; press **Enter.**

To imprint an object (See Figure 16-77.)

1. Draw a solid box **(4×2×3)**, then a **Ø1.00** circle with its center point on the upper right corner of the box as shown.

2. Select the **Imprint** tool from the **Solid Editing** toolbar.

   Select a 3D solid

3. Select the box.

   Select an object to imprint:

4. Select the circle.

   Select the source object <N>:

5. Type **y**; press **Enter** twice.

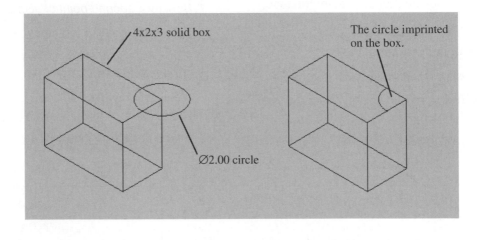

Figure 16-77

16-23 EXERCISE PROBLEMS

Draw the objects in Exercise Problems EX16-1 through EX16-45 as follows.

A. Draw each as a solid model.
B. Create front, top, and right-side orthographic views from the solid models.
C. Dimension the orthographic views.

EX16-1 INCHES

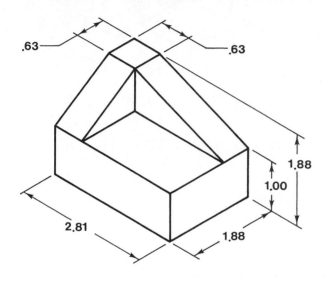

EX16-2 INCHES

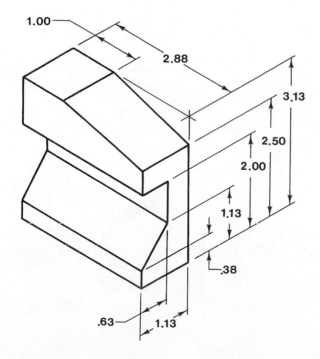

EX16-3 MILLIMETERS

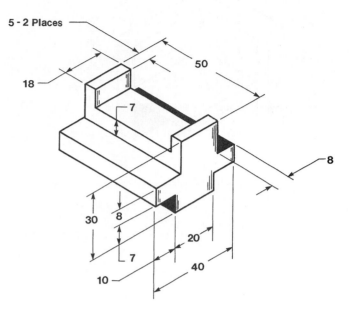

EX16-4 MILLIMETERS

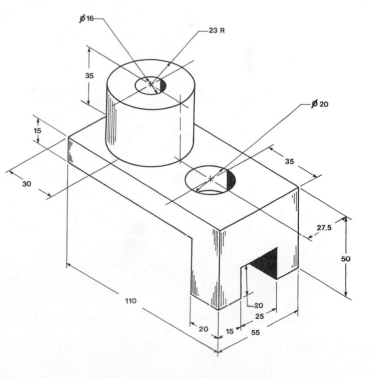

EX16-5 INCHES

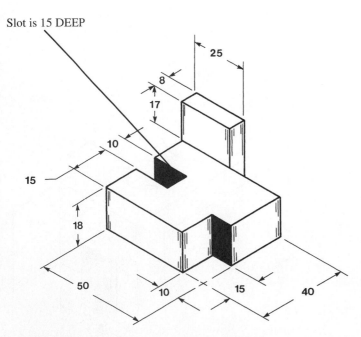

EX16-7 MILLIMETERS

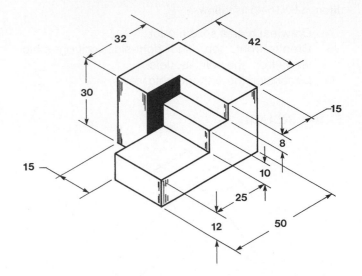

EX16-6 MILLIMETERS

Slot is 15 DEEP

EX16-8 MILLIMETERS

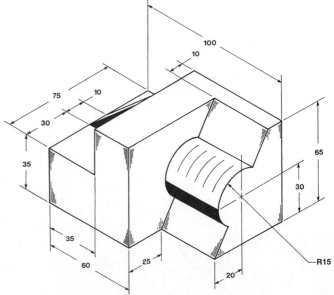

EX16-9 INCHES

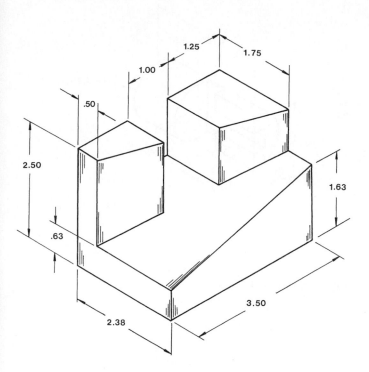

EX16-11 MILLIMETERS

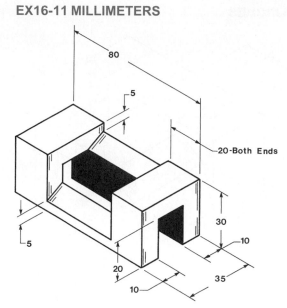

EX16-10 MILLIMETERS

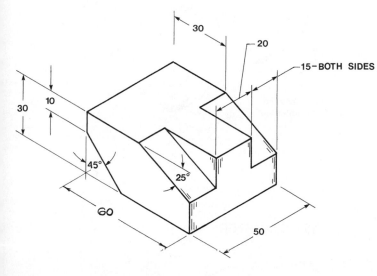

EX16-12 INCHES

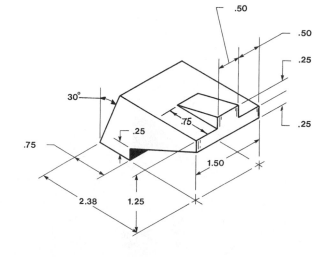

EX16-13 MILLIMETERS

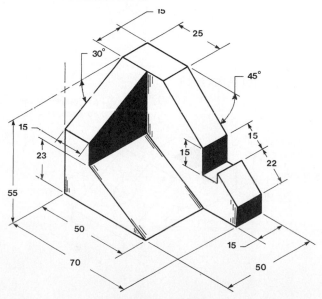

EX16-14 INCHES

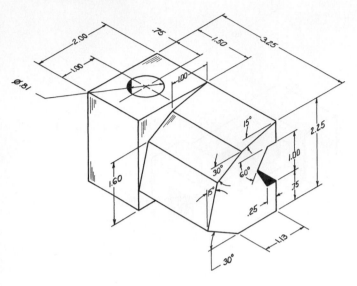

EX16-17 INCHES

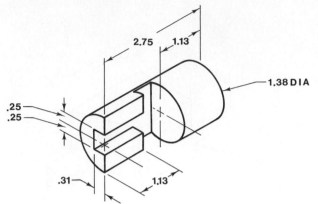

EX16-15 MILLIMETERS

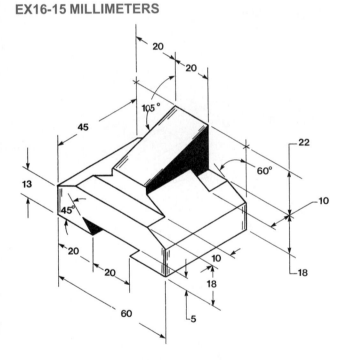

EX16-18 MILLIMETERS

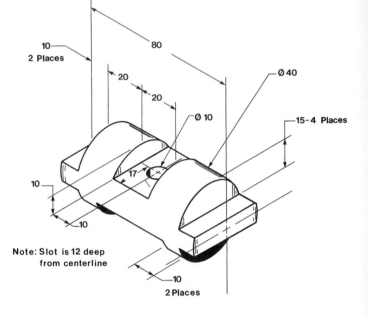

Note: Slot is 12 deep
from centerline

EX16-16 MILLIMETERS

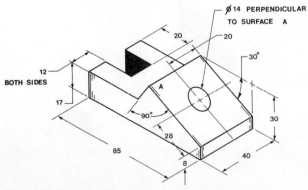

EX16-19 MILLIMETERS

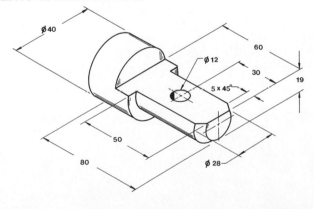

EX16-20 MILLIMETERS

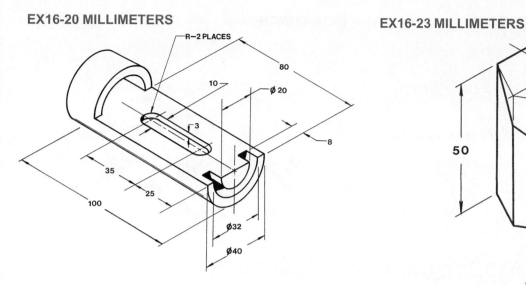

R−2 PLACES
80
10
Ø 20
3
8
35
25
100
Ø32
Ø40

EX16-23 MILLIMETERS

50
25

32 ACROSS
THE FLATS

EX16-21 INCHES

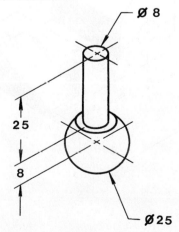

.69

Ø 1 $\frac{7}{8}$

EX16-24 MILLIMETERS

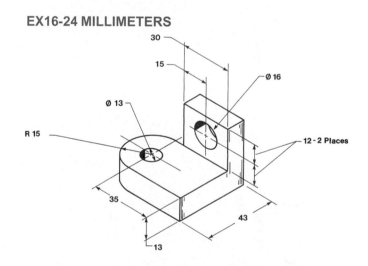

30
15
Ø 16
Ø 13
R 15
12 - 2 Places
35
43
13

EX16-22 MILLIMETERS

Ø 8

25

8

Ø25

EX16-25 MILLIMETERS

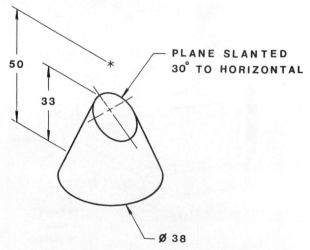

PLANE SLANTED
30° TO HORIZONTAL

50
33

Ø 38

EX16-26 MILLIMETERS (SCALE 2:1)

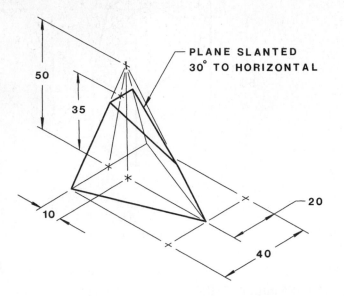

PLANE SLANTED
30° TO HORIZONTAL

50

35

10

20

40

EX16-28 INCHES

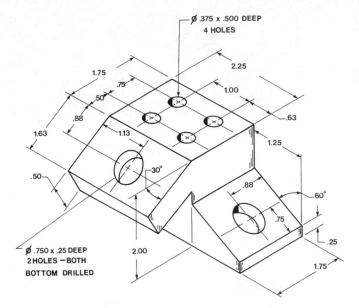

Ø .375 x .500 DEEP
4 HOLES

1.75

.75

.50

2.25

1.00

.88

.63

1.63

1.13

1.25

.50

30°

.88

60°

.75

.25

Ø .750 x .25 DEEP
2 HOLES—BOTH
BOTTOM DRILLED

2.00

1.75

EX16-27 MILLIMETERS

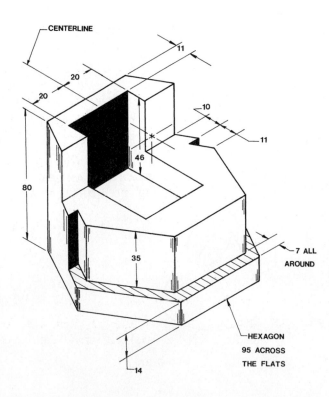

CENTERLINE

11

20

20

10

11

46

80

35

7 ALL
AROUND

35

HEXAGON
95 ACROSS
THE FLATS

14

EX16-29 MILLIMETERS

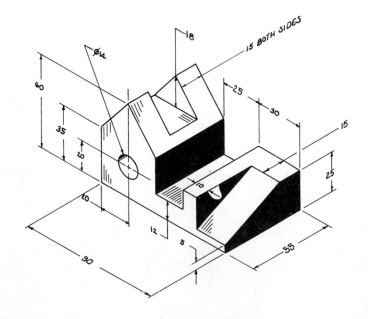

Ø 14

18

15 BOTH SIDES

25

30

15

60

35

20

10

25

20

12

8

10

90

55

EX16-30 INCHES (SCALE 2:1)

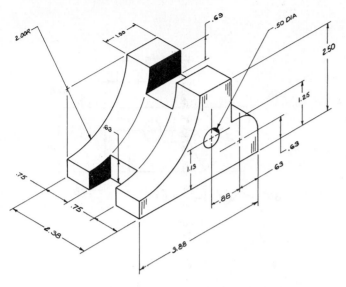

EX16-33 MILLIMETERS

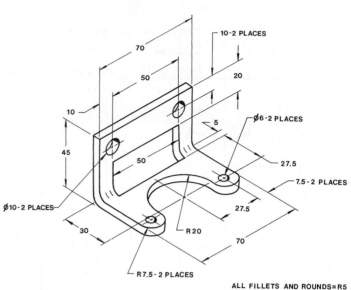

ALL FILLETS AND ROUNDS=R5
MATL 5 THK

EX16-31 MILLIMETERS

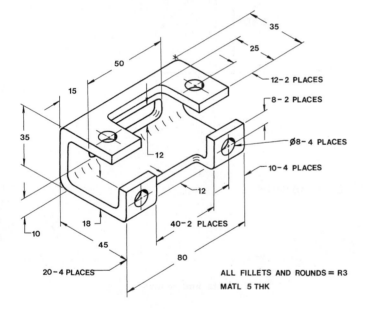

ALL FILLETS AND ROUNDS = R3
MATL 5 THK

EX16-34 MILLIMETERS

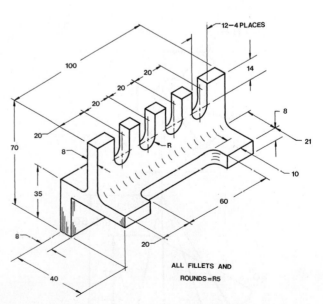

ALL FILLETS AND
ROUNDS=R5

EX16-32 MILLIMETERS (SCALE 2:1)

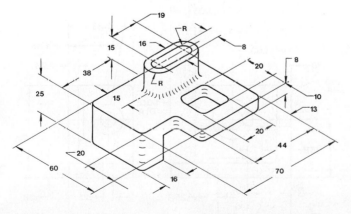

ALL FILLETS AND ROUNDS= R3

EX16-35 INCHES (SCALE 2:1)

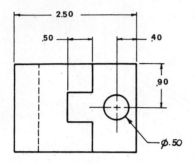

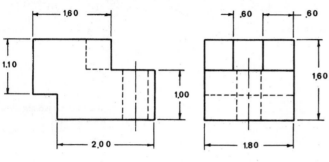

EX16-37 INCHES

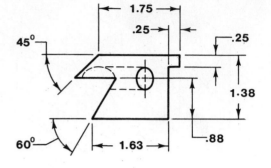

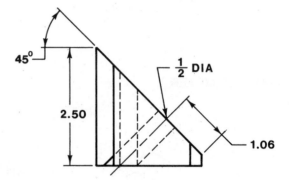

EX16-36 INCHES (SCALE 4:1)

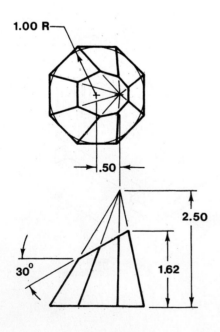

EX16-38 INCHES

TOP

All fillets and rounds = $\frac{1}{8}$ R

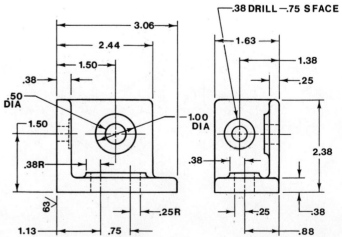

EX16-39 MILLIMETERS

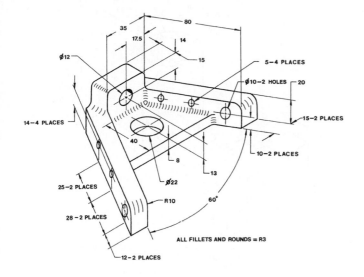

EX16-41 MILLIMETERS

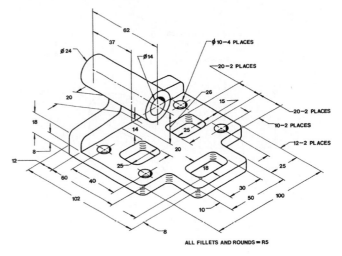

EX16-40 MILLIMETERS

EX16-42 MILLIMETERS

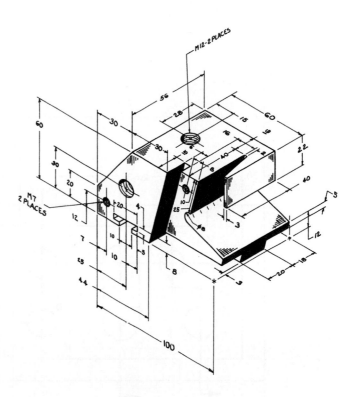

EX16-43 MILLIMETERS

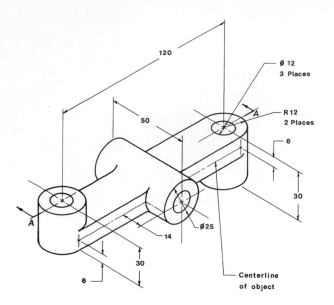

EX16-45 MILLIMETERS

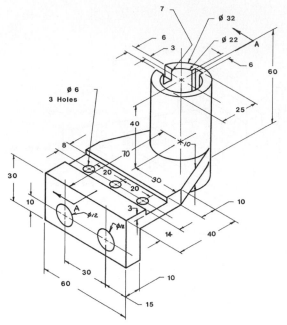

EX16-44 INCHES

Draw a solid model of the object, then create the three indicated sectional views from the model.

HOLE	X	Y	DIA
A	1.63	2.00	.44
B	1.13	1.00	.56
C	2.50	2.00 1.00	.63
D	3.88	2.00 1.00	.50

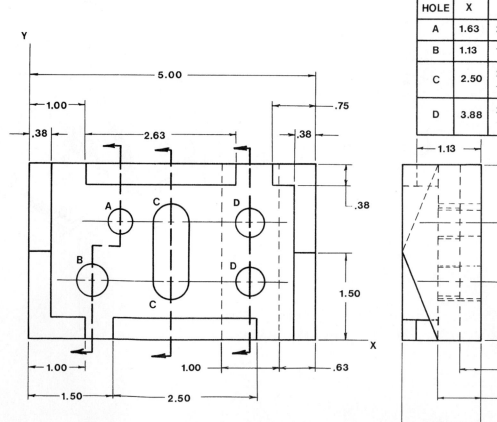
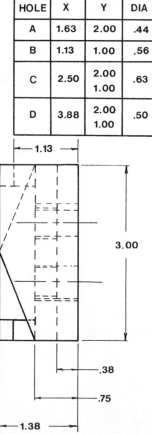

Redraw the assemblies in Exercise Problems EX16-46 through EX16-48 as solid models with the individual parts located in approximately the positions shown.

EX16-46 MILLIMETERS

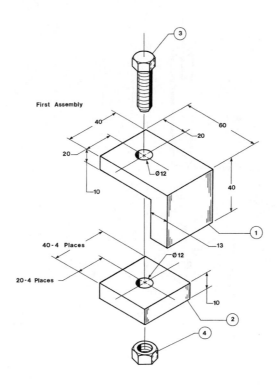

EX16-47 MILLIMETERS

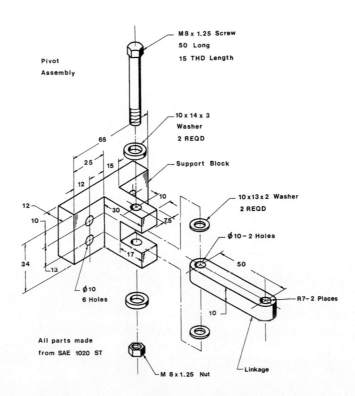

EX16-48 MILLIMETERS

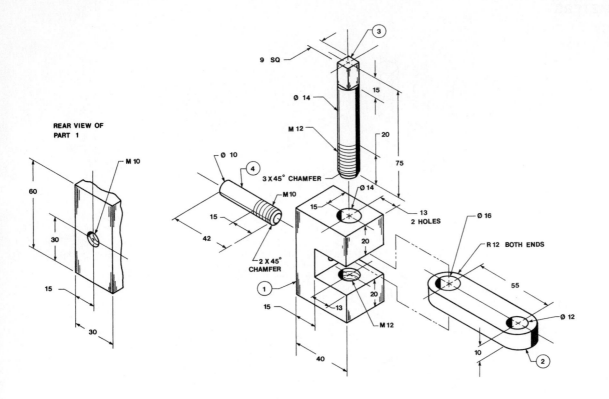

Prepare solid models and 3D orthographic views of the intersecting objects in Exercise Problems EX16-49 through EX16-54.

EX16-49 INCHES

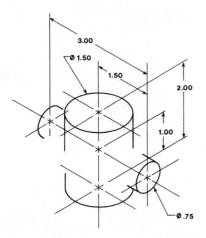

EX16-52 MILLIMETERS

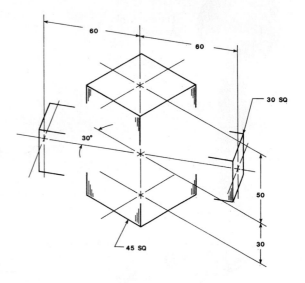

EX16-50 MILLIMETERS

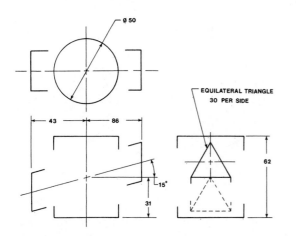

EX16-53 MILLIMETERS

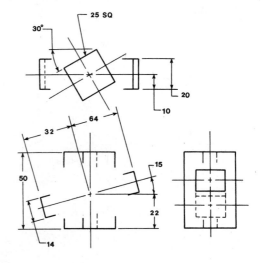

EX16-51 MILLIMETERS

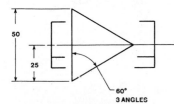

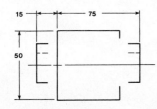

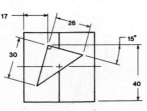

EX16-54 INCHES

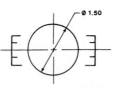

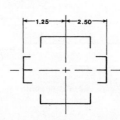

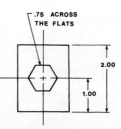

EX16-55 MILLIMETERS

Throttle Link
Assembly

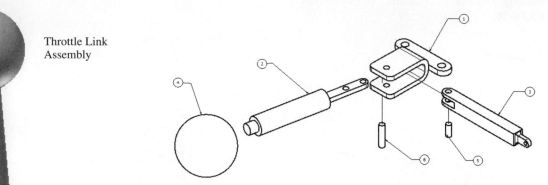

Parts List				
ITEM	PART NUMBER	DESCRIPTION	MATERIAL	QTY
1	ENG-A43	BOX,PIVOT	SAE1020	1
2	ENG-A44	POST,HANDLE	SAE1020	1
3	ENG-A45	LINK	SAE1020	1
4	AM300-1	HANDLE	STEEL	1
5	EK-132	POST-Ø6x14	STEEL	1
6	EK-131	POST-Ø6x26	STEEL	1

Pivot Box

Handle

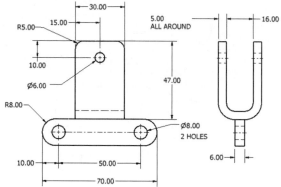

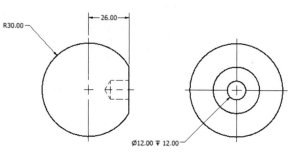

Handle Post

Link

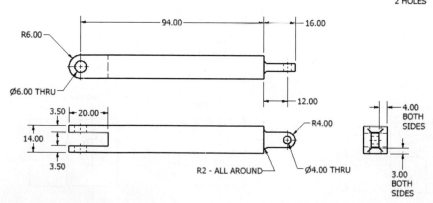

EX16-56

Adjustable Assembly

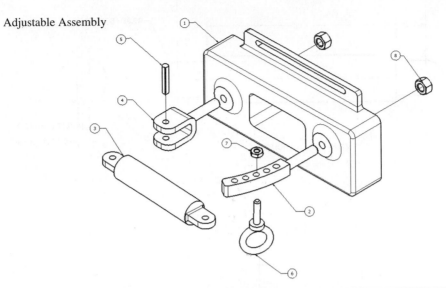

Parts List				
ITEM	PART NUMBER	DESCRIPTION	MATERIAL	QTY
1	ENG-311	BASE#4, CAST	Cast Iron	1
2	ENG-312	SUPPORT, ROUNDED	SAE 1040 STEEL	1
3	ENG-404	POST, ADJUSTABLE	Steel, Mild	1
4	BU-1964	YOKE	Cast Iron	1
5	ANSI B18.8.2 1/4x1.3120	Grooved pin, Type C - 1/4x1.312 ANSI B18.8.2	Steel, Mild	1
6	ANSI B18.15 - 1/4 - 20. Shoulder Pattern Type 2 - Style A	Forged Eyebolt	Steel, Mild	1
7	ANSI B18.6.3 - 1/4 - 20	Hex Machine Screw Nut	Steel, Mild	1
8	ANSI B18.2.2 - 3/8 - 16	Hex Nut	Steel, Mild	2

Cast Base #4

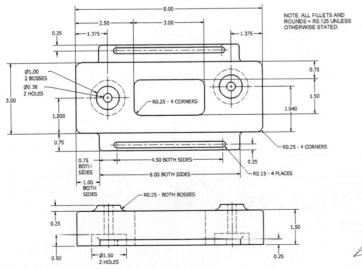

Rounded support

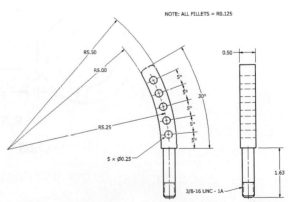

Adjustable Post

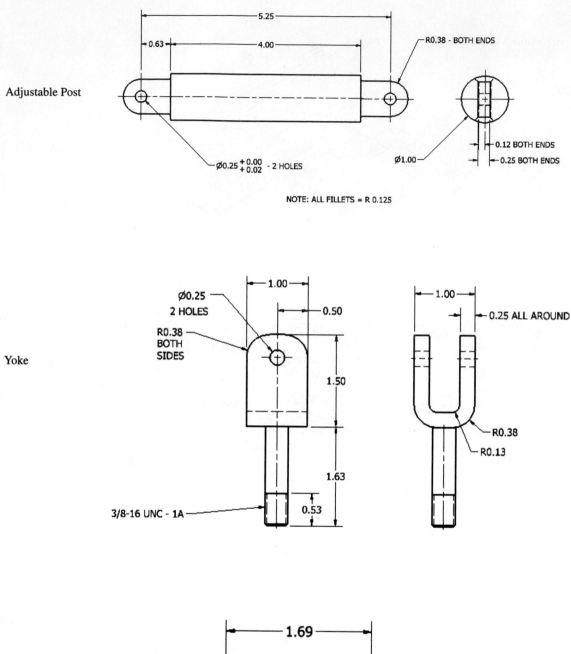

Yoke

Forged Eyebolt

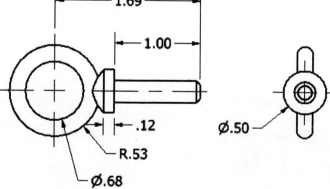

EX16-57

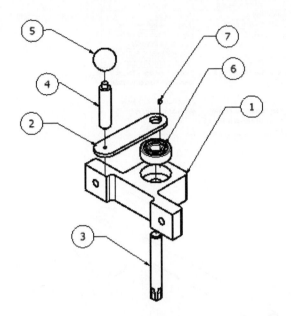

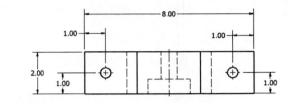

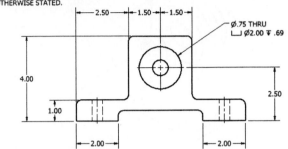

NOTE: ALL FILLETS AND ROUNDS R=0.250
UNLESS OTHERWISE STATED.

Ø.75 THRU
⌴ Ø2.00 ▼ .69

Parts List				
ITEM	PART NUMBER	DESCRIPTION	MATERIAL	QTY
1	EK131-1	SUPPORT	STEEL	1
2	EK131-2	LINK	STEEL	1
3	EK131-3	SHAFT,DRIVE	STEEL	1
4	EK131-4	POST, THREADED	STEEL	1
5	EK131-5	BALL	STEEL	1
6	BS 292 - BRM 3/4	Deep Groove Ball Bearings	STEEL,MILD	1
7	3/16x1/8x1/4	RECTANGULAR KEY	STEEL	1

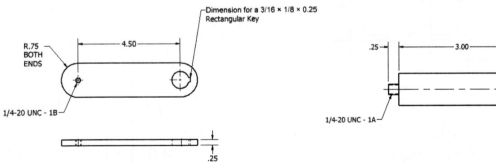

Dimension for a 3/16 × 1/8 × 0.25
Rectangular Key

R.75 BOTH ENDS

1/4-20 UNC - 1B

1/4-20 UNC - 1A

3/8-16 UNC - 1A

Ø.75

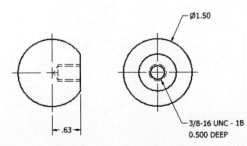

Ø1.50

3/8-16 UNC - 1B
0.500 DEEP

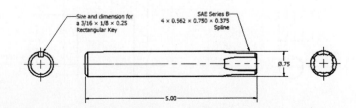

Size and dimension for
a 3/16 × 1/8 × 0.25
Rectangular Key

SAE Series B
4 × 0.562 × 0.750 × 0.375
Spline

Ø.75

EX16-58

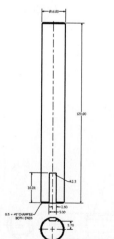

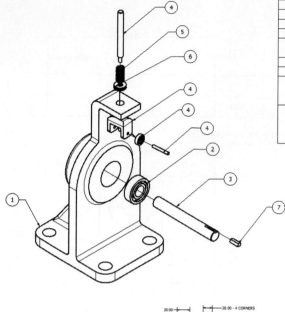

Parts List			
ITEM	QTY	PART NUMBER	DESCRIPTION
1	1	ENG-2008-A	BASE, CAST
2	1	DIN625 - SKF 6203	Single row ball bearings
3	1	SHF-4004-16	SHAFT: Ø16×120, WITH 2.3×5×16 KEYWAY
4	1		SUB-ASSEMBLY, FOLLOWER
5	1	SPR-C22	SPRING, COMPRESSION
6	1	GB 273.2-87 - 7/70 - 8 x 18 x 5	Rolling bearings - Thrust bearings - Plan of boundary dimensions
7	1	IS 2048 - 1983 - Specification for Parallel Keys and Keyways B 5 x 5 x 16	Specification for Parallel Keys and Keyways

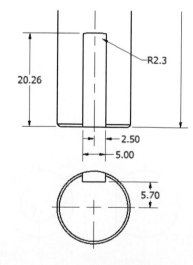

Parts List				
ITEM	PART NUMBER	DESCRIPTION	MATERIAL	QTY
1	AM-232	HOLDER	STEEL	1
2	AM-2S6	POST, FOLLOWER	STEEL	1
3	BS 1804-2 - 4 × 30	Parallel steel dowel pins - metric series	Steel, Mild	1
4	DIN625- SKF 634	Single row ball bearings	Steel, Mild	1

NOTE: ALL FILLETS AND ROUNDS = R5.0 UNLESS OTHERWISE STATED.

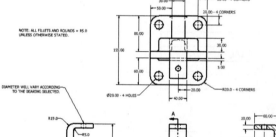

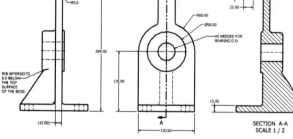

SECTION A-A
SCALE 1 / 2

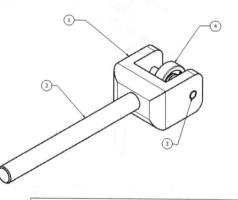

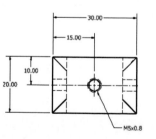

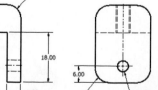

EX16-59

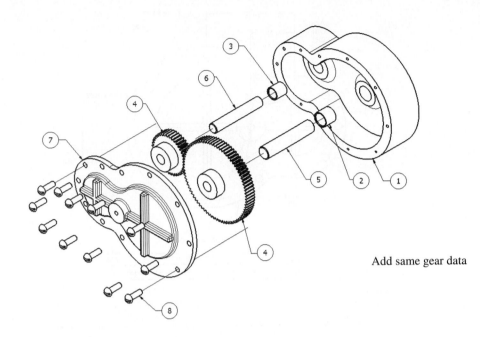

Add same gear data

Add same gear data

		Parts List			
ITEM	PART NUMBER	DESCRIPTION	MATERIAL	QTY
1	ENG-453-A	GEAR, HOUSING	CAST IRON	1
2	BU-1123	BUSHING Ø0.75	Delrin, Black	1
3	BU-1126	BUSHING Ø0.625	Delrin, Black	1
4	ASSEMBLY-6	GEAR ASSEEMBLY	STEEL	1
5	AM-314	SHAFT, GEAR Ø.625	STEEL	1
6	AM-315	SHAFT, GEAR Ø.0.500	STEEL	1
7	ENG -566-B	COVER, GEAR	CAST IRON	1
8	ANSI B18.6.2 - 1/4-20 UNC - 0.75	Slotted Round Head Cap Screw	Steel, Mild	12

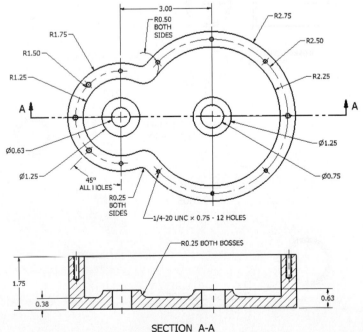

SECTION A-A
SCALE 3 / 4

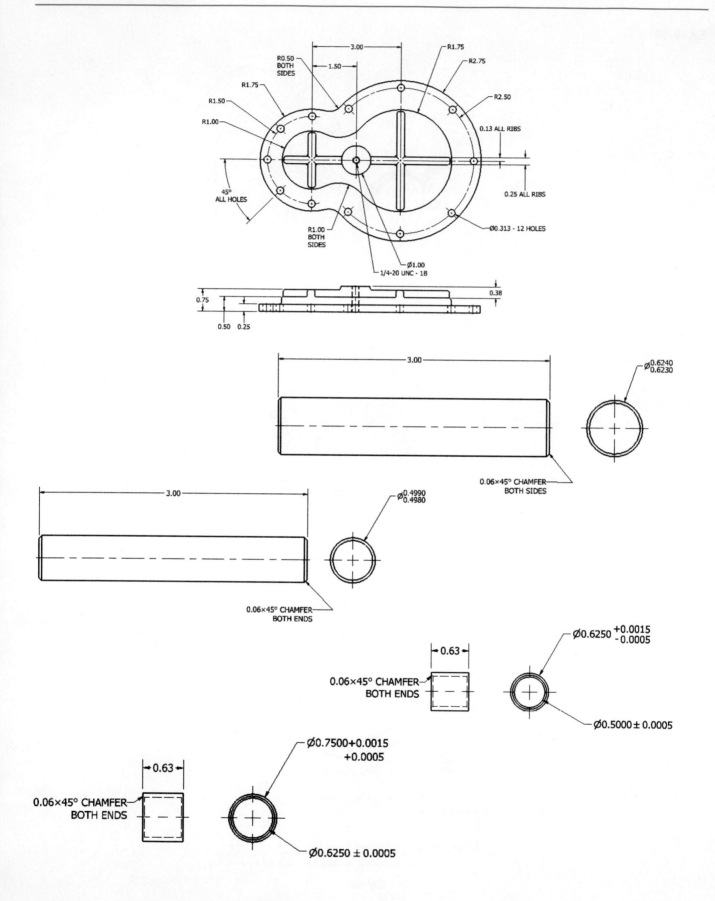

EX16-60

Design an access controller based on the information given. The controller works by moving an internal cylinder up and down within the base to align with output holes A and B. Liquids will enter the internal cylinder from the top, then exit the base through holes A and B. Include as many holes in the internal cylinder as necessary to create the following liquid exit combinations:

1. A open, B closed
2. A open, B open
3. A closed, B open

The internal cylinder is held in place by an alignment key and a stop button. The stop button is to be spring-loaded so that it will always be held in place. The internal cylinder will be moved by pulling out the stop button, repositioning the cylinder, then reinserting the stop button.

Prepare the following drawings.

A. Draw the objects as solid models.
B. Draw an assembly drawing.
C. Draw detail drawings of each nonstandard part. Include positional tolerances for all holes.
D. Prepare a parts list.

INTERNAL CYLINDER

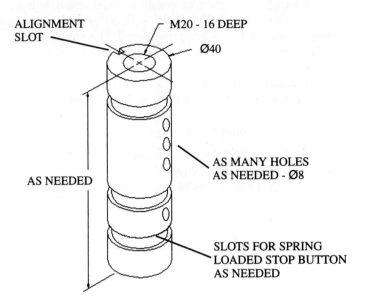

STOP BUTTON ASSEMBLY

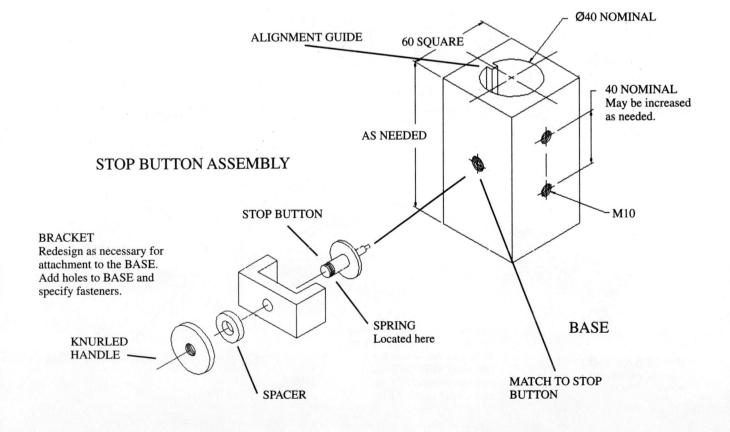

EX16-61

Design a hand-operated grinding wheel specifically for sharpening a chisel. The chisel is to be located on an adjustable rest while it is being sharpened. The mechanism should be able to be clamped to a table during operation using two thumbscrews.

A standard grinding wheel is Ø6.00 inch, is ½ inch thick, and has an internal mounting hole with a 50.00±0.3 millimeter bore.

Prepare the following drawings.

A. Draw the objects as solid models.
B. Draw an assembly drawing.
C. Draw detail drawings of each nonstandard part.
D. Prepare a parts list.

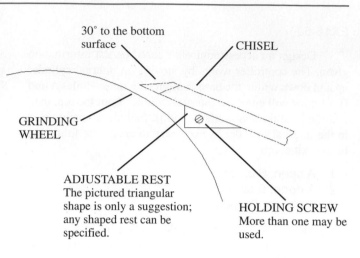

30° to the bottom surface

CHISEL

GRINDING WHEEL

ADJUSTABLE REST
The pictured triangular shape is only a suggestion; any shaped rest can be specified.

HOLDING SCREW
More than one may be used.

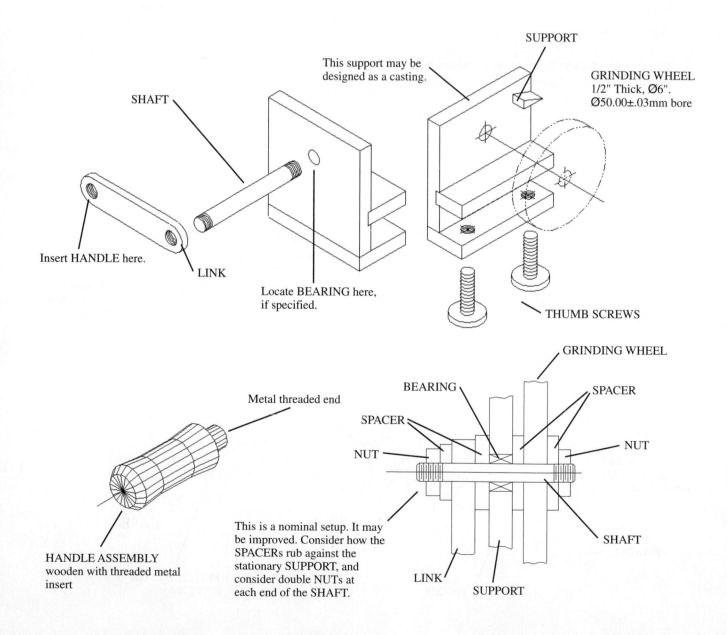

SHAFT

This support may be designed as a casting.

SUPPORT

GRINDING WHEEL
1/2" Thick, Ø6".
Ø50.00±.03mm bore

Insert HANDLE here.

LINK

Locate BEARING here, if specified.

THUMB SCREWS

GRINDING WHEEL

BEARING

SPACER

SPACER

NUT

NUT

SHAFT

Metal threaded end

HANDLE ASSEMBLY
wooden with threaded metal insert

This is a nominal setup. It may be improved. Consider how the SPACERs rub against the stationary SUPPORT, and consider double NUTs at each end of the SHAFT.

LINK

SUPPORT

EX16-62

A. Draw an assembly drawing of the given object.
B. Prepare a parts list.
C. Select appropriate fasteners to hold the object together.
D. Define the appropriate tolerances.

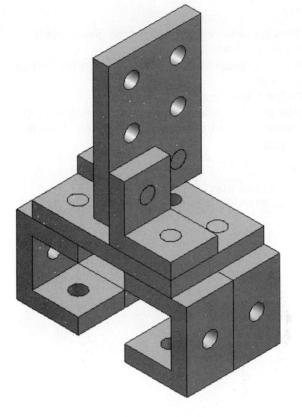

C-BRACKET
SAE 1020 STEEL, 4 REQD
Part Number: AM311-1

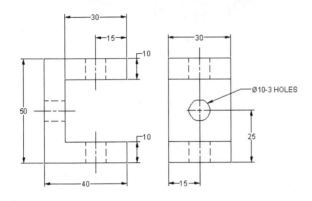

Ø10-3 HOLES

L-CLIP
SAE 1040 STEEL, 2 REQD
Part Number: AM312-4

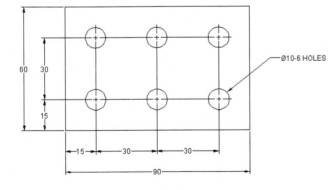

Ø10-6 HOLES

SUPPORT PLATE
SAE 1040 STEEL, 2 REQD
Part Number: AM312-3

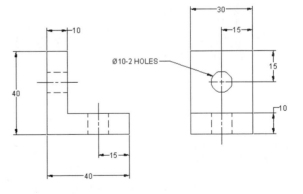

Ø10-2 HOLES

EX16-63

A. Draw an assembly drawing of the given object.
B. Prepare a parts list.
C. Select appropriate fasteners to hold the object together.
D. Define the appropriate tolerances.
E. Use phantom lines and define the motion of the center link and the rocker link, if the drive link rotates 360°.

ROCKER ASSEMBLY

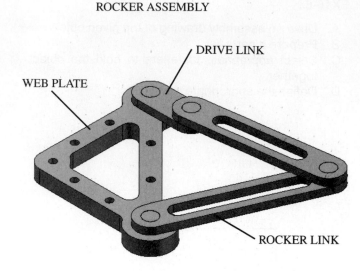

DRIVE LINK

WEB PLATE

ROCKER LINK

DRIVE LINK, SAE 1040 STEEL
Part Number: AM311-22A
5mm THK

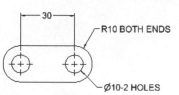

30
R10 BOTH ENDS
Ø10-2 HOLES

ALL FILLETS AND ROUNDS = R3

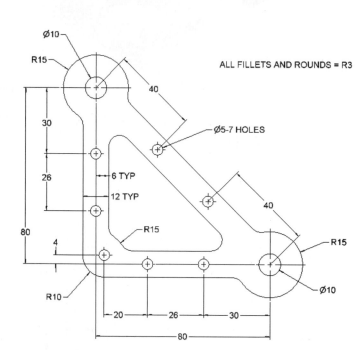

Ø10
R15
40
30
Ø5-7 HOLES
26
6 TYP
12 TYP
40
80
R15
4
R15
R10
Ø10
20 26 30
80

ROCKER LINK, SAE 1040 STEEL
Part Number: AM311-22C
5mm THK

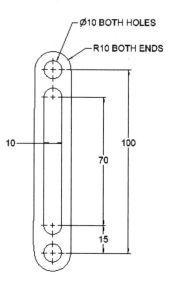

Ø10 BOTH HOLES
R10 BOTH ENDS
10
100
70
15

WEB PLATE, SAE 1040 STEEL
Part Number: AM311-22B
10mm THK

CENTER LINK, SAE 1040 STEEL
Part Number: AM311-22D
5mm THK

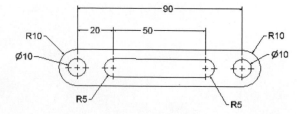

90
20 50
R10
Ø10
R10
Ø10
R5
R5

EX16-64

A. Draw an assembly drawing of the minivise.
B. Prepare a parts list.
C. Select the appropriate fasteners to hold the vise together.

D. Redesign the interface between the drive screw and the holder plate.

MINIVISE

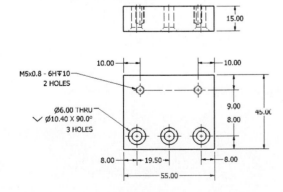

MINIVISE -ISO

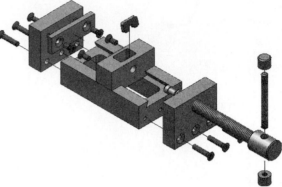

1. BASE
SAE 1040 STEEL

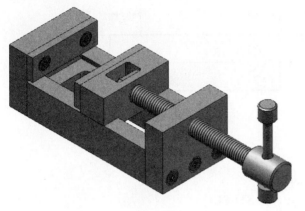

OBJECT IS SYMMETRICAL
ABOUT THIS CENTERLINE

M5x0.8 - 6H ▼8
6 HOLES

2. DRIVE SCREW
SAE 1040 STEEL

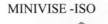
NOTE: ALL CHAMFERS = 1x45

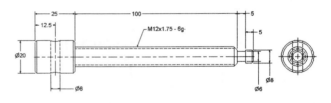

3. END PLATE
SAE 1040 STEEL

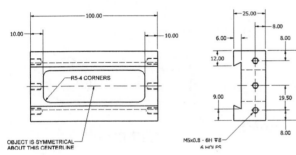

M5x0.8 - 6H▼10
2 HOLES

Ø6.00 THRU
∨ Ø10.40 X 90.0°
3 HOLES

4. SLIDER
SAE 1040 STEEL

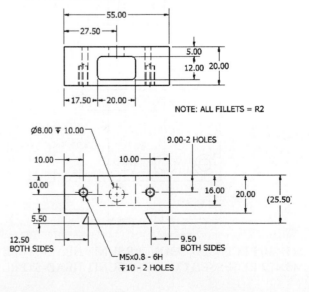

NOTE: ALL FILLETS = R2

Ø8.00 ▼ 10.00

9.00-2 HOLES

12.50
BOTH SIDES

9.50
BOTH SIDES

M5x0.8 - 6H
▼10 - 2 HOLES

5. HOLDER PLATE
   SAE 1040 STEEL

6. FACE PLATE
   SAE 1040 STEEL

7. HANDLE
   SAE 1040 STEEL

8. END CAP
   SAE 1040 STEEL

9. DRIVE PLATE
   SAE 1040 STEEL

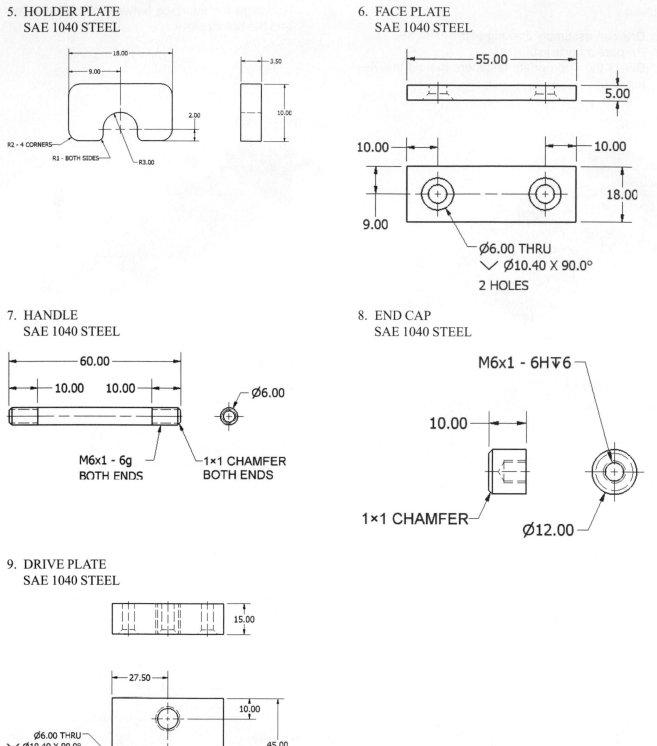

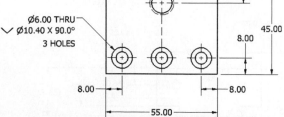

10. M5×10 RECESSED COUNTERSUNIT HEAD-STEEL
11. M5×22 RECESSED COUNTERSUNIT HEAD-STEEL

EX16-65

Create the following for the slider assembly shown below.

1. Select the appropriate fasteners.
2. Create an assembly drawing.
3. Create a parts list.
4. Specify the tolerance for the guide shaft/bearings interface.

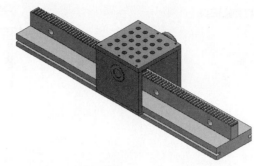

RACK ASSEMBLY

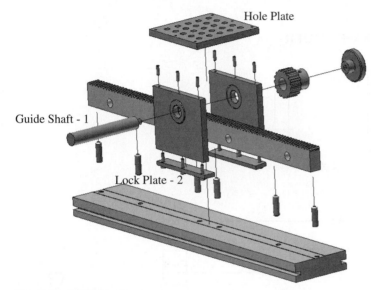

Hole Plate

Guide Shaft - 1

Lock Plate - 2

RACK ASSEMBLY - 1

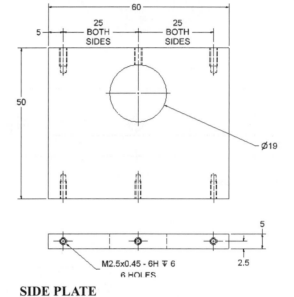

60

25 BOTH SIDES

25 BOTH SIDES

5

50

Ø19

M2.5x0.45 - 6H ▼ 6
6 HOLES

5

2.5

SIDE PLATE

1.25 MODULE, 20° PRESSURE ANGLE

Ø7-4 HOLES

17.17

C

282.8

4   50   75   75   50   4

10

M5x0.8 - 6H ▼ 8
6 HOLES

VIEW C-C
SCALE 1 : 1

5

C

BASED ON: ⊐PRECISION RACK R1M-2
BERG MANUFACTURING CO

RACK

Ø16
Ø10 THRU
M3.5x0.6 - 6H ▼ 4

10   10

PITCH Ø = 20.00
1.25 MODULE

PRECISION SPUR GEAR
BERG MANUFACTURING CO.
PART NUMBER: PBS86-16

GEAR 16

Ø19

Ø10

5

0.5X0.5 CHAMFER

BALL BEARING
BERG MANUFACTURING CO.
PART NUMBER: B11M-11

BEARING

EX16-65 CONTINUED

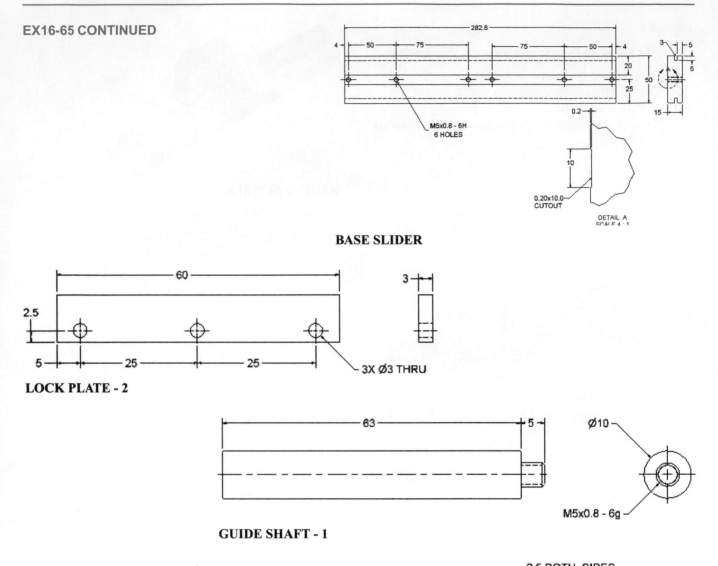

BASE SLIDER

LOCK PLATE - 2

3X Ø3 THRU

GUIDE SHAFT - 1

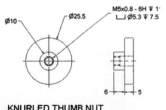

KNURLED THUMB NUT
BERG MANUFACTURING CO.
STOCK NUMBER: PD1M-15

THUMB NUT

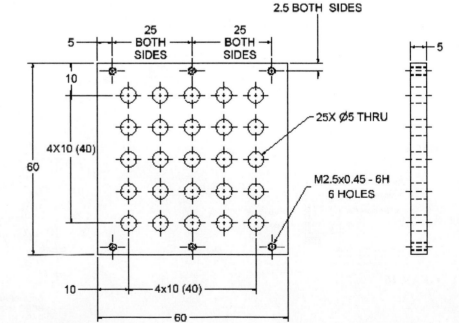

HOLE PLATE

Descriptive Geometry

17-1 INTRODUCTION

Descriptive geometry is the study of points, lines, and planes in space to determine their locations and true shapes. Classic descriptive geometry solutions involve using projection between both standard and auxiliary orthographic views. This chapter introduces these solutions and shows how they can be done using AutoCAD. The chapter also introduces some different approaches to the problems using AutoCAD's 3D capabilities.

17-2 ORTHOGRAPHIC PROJECTION

Figure 17-1 shows a point P located in space represented by the box. The front, top, and right-side orthographic views of the point are used to define the location of the point by relating the point's location to the sides of the box or to the orthographic planes. The point's locations in the orthographic views are related to each other using a projection rectangle. The front and side views are related

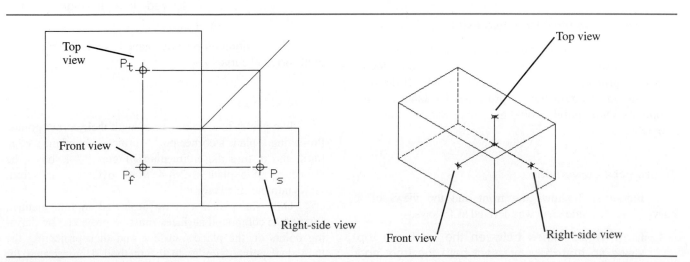

Figure 17-1

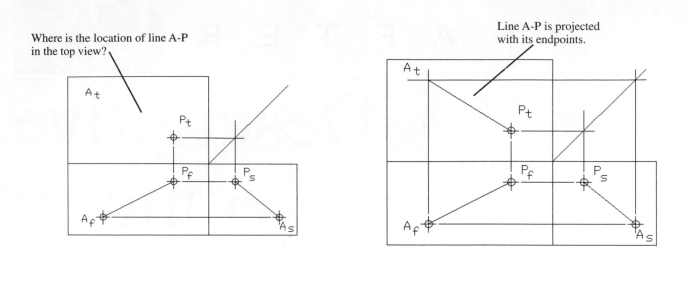

Figure 17-2

by a horizontal line, the top and front views by a vertical line, and the top and side views by a combination of a horizontal line and a vertical line.

Each orthographic view is two-dimensional and includes two pieces of locating information. For example, if the bottom surface of the box representing space were on an X,Y axis, the top view would locate the point using its X,Y coordinates, the front view the X,Z coordinates, and the side the Y,Z coordinates. This means that any two views will generate the X,Y,Z coordinate values for the point. This also means that if any two orthographic views are defined, the third orthographic view, or an auxiliary view, can be derived from the defined views.

To create a third view from two existing views

Figure 17-2 shows the front and side views of a line A-P. The location of the line in the top view can be determined by projecting the endpoints of the line. If the line were curved, it could not be projected by using just its endpoints. Other points would have to be defined and then projected.

To project a curved line

Figure 17-3 shows the front and top views of a curved line. The side view was derived as follows.

1. Draw vertical lines between the front and top views so that they intersect both the front and top views of the curved line.

In this example two vertical lines were drawn.

2. Label the intersections between the vertical lines and the front and top views of the curved line.

These intersections are defined points on the curved line common to both views of the line.

3. Project the points into the side view.
4. Use the **Polyline** tool from the **Draw** toolbar to join the endpoints and the projected points with a polyline.
5. Use the **Polyline Edit** tool from the **Modify II** toolbar, then the **Fit** option to change the straight polyline into a curved line through all the defined points.

The resulting elliptically shaped line is the side view of the original curved line.

To project a plane

Flat planes may be projected with their corner points. Projecting a plane's corner points projects the plane's edge lines, and in turn the entire plane. Figure 17-4 shows the projection of a plane 1-2-3-4 from given top and front views into the side view.

Only flat planes can be projected in this manner. Warped or compound surfaces must be projected by defining points on the plane's surface and then projecting the individual points, similar to the method demonstrated for curved lines.

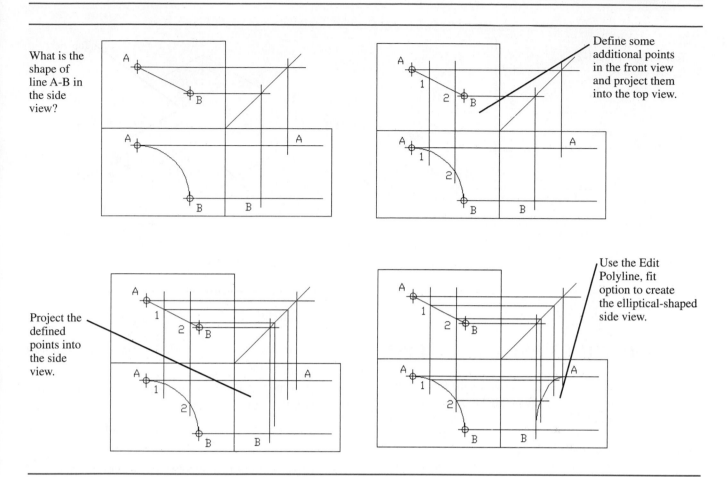

What is the shape of line A-B in the side view?

Define some additional points in the front view and project them into the top view.

Project the defined points into the side view.

Use the Edit Polyline, fit option to create the elliptical-shaped side view.

Figure 17-3

Project a plane by projecting its corner points.

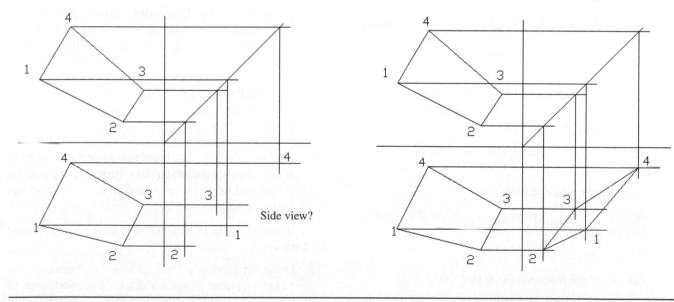

Side view?

Figure 17-4

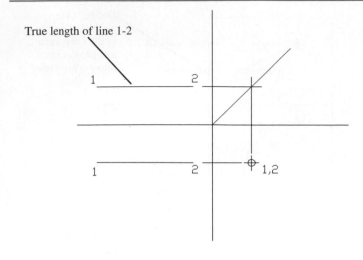

True length of line 1-2

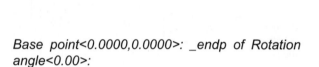

Top view

Front view

Side view

Line 1-2 in 3-D space

Figure 17-5

17-3 THE TRUE LENGTH OF A LINE

An orthographic view of a line shows the true length of the line only if the line is exactly parallel to the orthographic view's plane. Figure 17-5 shows three orthographic views of a line. Note that the side view is a point, or the end view of a line. If one of the views of a line is an end view, it means that a view 90° to the end view is a true-length view. In the example shown, both the front and top views are true-length views.

To determine the true length of a line

Figure 17-6 shows three views of line 1-2. None of the views is parallel to the front, top, or side orthographic plane, so none of the views is a true-length view. A true-length view can be created by defining an auxiliary orthographic plane parallel to one of the views of the line and then projecting the line into that view. The procedure is as follows.

1. Use the **Offset** tool from the **Modify** toolbar and create a line parallel to the side view at any distance from the side view.

This line is a reference plane line for an auxiliary view parallel to the side view of line 1-2.

2. Type **Snap**; press **Enter**.

   Specify snap spacing or [ON/OFF/Aspect/ Rotate/Style] <0.2500>:

3. Type **r**; press **Enter**.

   Specify base point <0.0000,0.0000>:

4. Use the **Osnap, Endpoint** option and select point **2** in the side view.

Base point<0.0000,0.0000>: _endp of Rotation angle<0.00>:

5. Use the **Osnap, Endpoint** option and select point **1** in the side view.

The crosshairs will align with the side view of line 1-2.

6. Draw lines perpendicular to the two endpoints of the auxiliary plane reference line.

7. Use the **Snap, Rotate** option and return the crosshairs to their normal horizontal and vertical orientation.

Set the base point value to 0.0000,0.0000 and the angle of rotation to 0.

8. Use the **Osnap, Endpoint** and **Perpendicular** options to draw lines from the endpoints of the front view of the line perpendicular to the reference plane line between the front and side views.

9. Click the **Distance** tool from the **Standard** toolbar.

10. Use the **Distance** tool to determine the lengths of the perpendicular lines drawn in step 8. Use **Osnap** to ensure accuracy.

11. Use the **Offset** tool from the **Modify** toolbar to draw lines parallel to the auxiliary reference plane line at distances equal to the distances determined in step 10.

In the example shown, the distances are represented by distances A and B.

12. Draw an auxiliary view of line 1-2 between the offset line and the lines drawn perpendicular to the auxiliary plane reference line as shown.

13. Erase any excess lines.

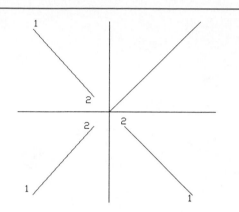

Orthographic views of line 1-2.
What is the true length of the line?

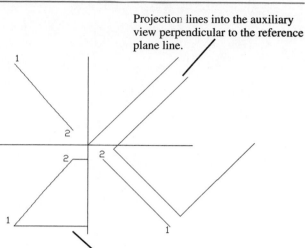

Projection lines into the auxiliary view perpendicular to the reference plane line.

Draw lines from the line's endpoints perpendicular to the vertical axis line. Use the Distance tool to determine the distances from the points to the vertical reference line.

This is the true length of line 1-2.

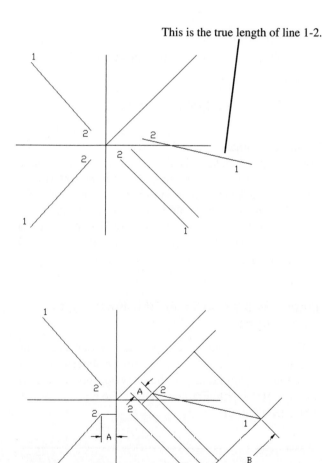

Use Snap, Rotate and align the crosshairs with the side view of line 1-2.

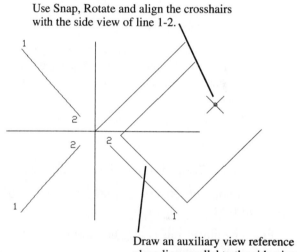

Draw an auxiliary view reference plane line parallel to the side view of line 1-2.

Use Offset to transfer the distances determined in the last step into the auxiliary view.

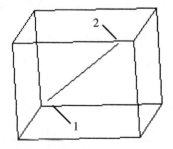

Line 1-2 in 3D space

Figure 17-6

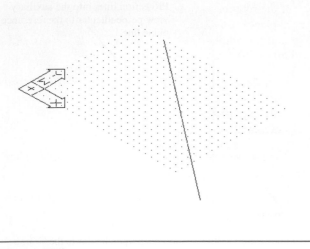

Figure 17-7

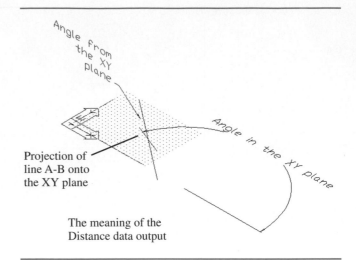

The meaning of the
Distance data output

Figure 17-9

To determine the properties of a line using AutoCAD

Figure 17-7 shows a line A-B in 3D space. The location of the line's endpoints and the line's true length can be determined using the **ID** (identification) and **Distance** commands.

1. Type **ID** in response to a command prompt.

   Point:

2. Use **Osnap, Endpoint** and select endpoint **A.**

   Point:_endp of X = 6.0000   Y = 5.0000 Z = 4.0000

These coordinate values are the location of point A relative to the WCS origin.

3. Press **Enter** and repeat the **ID** command.

   Point:

4. Use **Osnap, Endpoint** and select endpoint **B.**

   Point: _endp of X = 12.0000   Y = 2.0000 Z = −3.0000

5. Select the **Distance** tool.

Specify first point:

6. Use **Osnap, Endpoint** and select point **A.**

Specify second point:

7. Use **Osnap, Endpoint** and select point **B.**

Figure 17-8 shows the resulting value readout that will appear in the command area of the screen. The distance value, **9.6954,** is the true length of the line. The values listed as delta values are the changes in the individual component values. For example, the X value for point A is 6 and the X value for B is 12. The change in the values equals 6. Likewise, the Y value for point A is 5 and the Y value for B is 2. The delta value is −3 measured from point A to point B. Figure 17-9 shows the meaning of the two given angle values.

To determine the true length of a line using the rotation method

Figure 17-10 shows the front and top views of a line A-B. The true length of the line can be determined by rotating the line so that one of the views of the line is parallel to one of the orthographic planes. The other view will then be a true length.

Data generated by the Distance tool

```
 Specify second point: _endp of
Distance = 9.6954,   Angle in XY Plane = 333,   Angle from XY Plane = 314
Delta X = 6.0000,   Delta Y = −3.0000,   Delta Z = −7.0000
Command:
```

Figure 17-8

1. Select the **Rotate** tool from the **Modify** toolbar.

    Select objects:

2. Select the front view of line A-B.

    Select objects:

3. Press **Enter.**

    Specify base point:

4. Use **Osnap, Endpoint** and select point **A.**

    Specify rotation angle or [Reference]:

5. Rotate line A-B so that it is parallel to the reference plane line between the two views.

    This rotation can be done visually. The line will be smooth only when it is perfectly horizontal or vertical, or parallel to one of the orthographic view reference plane lines. If the rotation is to be done in another type of orthographic plane, use **Offset** to create a line parallel to the required orthographic view through one of the point's endpoints, then align the rotated line to the offset line.

6. Turn **Ortho** on and draw a horizontal line from point **B** in the top view.

7. Draw a vertical line from point **B** in the front view as shown in Figure 17-10.

    The intersection of the lines is the new location for point B.

8. Draw a line in the top view from point **A** to the new point **B.**

    This line is the true length of line A-B.

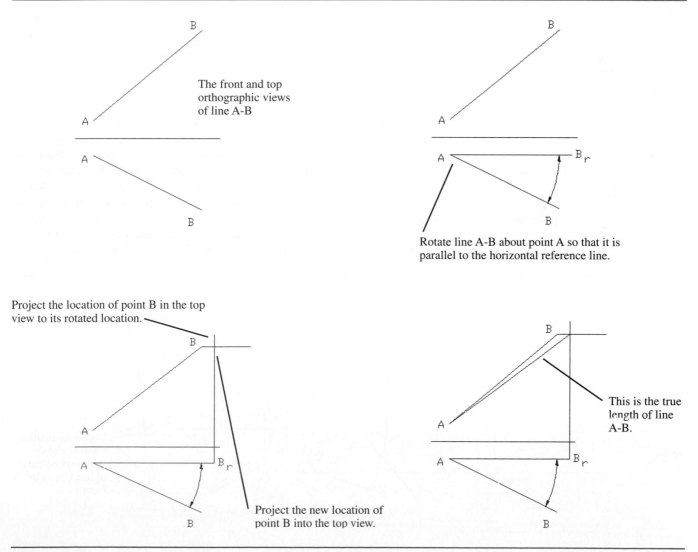

The front and top orthographic views of line A-B

Rotate line A-B about point A so that it is parallel to the horizontal reference line.

Project the location of point B in the top view to its rotated location.

Project the new location of point B into the top view.

This is the true length of line A-B.

Figure 17-10

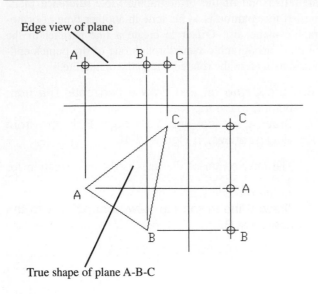

Edge view of plane

True shape of plane A-B-C

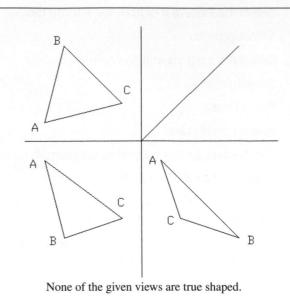

None of the given views are true shaped.

Figure 17-11

Figure 17-12

17-4 THE TRUE SHAPE OF A PLANE

Figure 17-11 shows three views of a plane A-B-C. The top and side views appear as single lines, or as edge views of the plane. This means that the front view is a view perpendicular to the plane: a view of the plane's true shape.

Figure 17-12 shows three views of a plane A-B-C. None of the views appears as a straight line, so none of the views is the true shape of the plane. All the views are distorted.

The true shape of the plane can be determined by creating an auxiliary orthographic view parallel to the plane. This is done by first creating an auxiliary view that defines an edge view of the plane (a straight line), and then taking a second auxiliary view perpendicular to the first.

To create a secondary auxiliary view of a plane

In this example, only the front and top views will be used. To define an edge view of the plane, we need a reference line that defines the true angle between the plane and the orthographic reference axis. None of the edge lines is parallel to the orthographic axis; there is no appropriate reference line. A plane contains an infinite number of lines, so a line can be drawn parallel to the orthographic axis in one of the views, then projected into the other view to yield a correct reference line. See Figure 17-13, Parts 1 to 3.

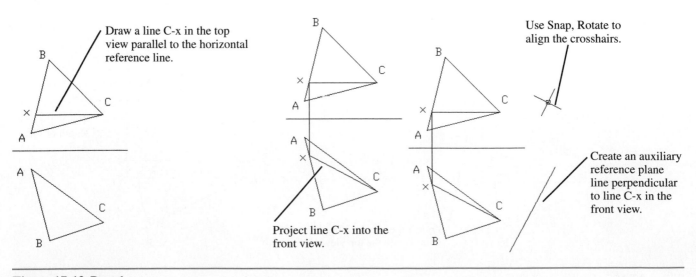

Draw a line C-x in the top view parallel to the horizontal reference line.

Project line C-x into the front view.

Use Snap, Rotate to align the crosshairs.

Create an auxiliary reference plane line perpendicular to line C-x in the front view.

Figure 17-13, Part 1

To create an edge view of a plane

1. Draw a horizontal line **C-x** in the top view.

   Any horizontal line within the plane can be used. It is simply convenient to select one end of the line on a known corner point.

2. Project the line into the front view.
3. Type **snap**; press **Enter**.

   Specify snap spacing or [ON/OFF/Aspect/ Rotate/Style/Type] <0.5000>:

4. Type **r**; press **Enter.**

   Specify base point <0.0000,0.0000>:

5. Use **Osnap, Endpoint** and select point **x.**

   Specify base point <0.0000,0.0000>: _endp of Specify rotation angle <0>:

6. Select point **C.**

   The crosshairs will rotate and align with line C-x.

7. Turn **Ortho** on and draw a line to the right of and perpendicular to line C-x.

   This is a reference plane line for the first auxiliary view.

8. Draw projection lines parallel to line C-x from points **A, B,** and **C** in the front view into the auxiliary view.
9. Turn **Ortho** off and draw a line from the corner point in the top view to the reference line between the front and top views.
10. Use the **Distance** tool to determine the lengths of the projected lines drawn in step 9.
11. Record the lengths and then use **Offset** to create lines parallel to the auxiliary view reference line in the auxiliary view plane.
12. Use the appropriate intersections to draw the edge view of plane A-B-C.

    The distances measured in step 9 determine where the corner points are located in the auxiliary view along the projected lines created in step 8.

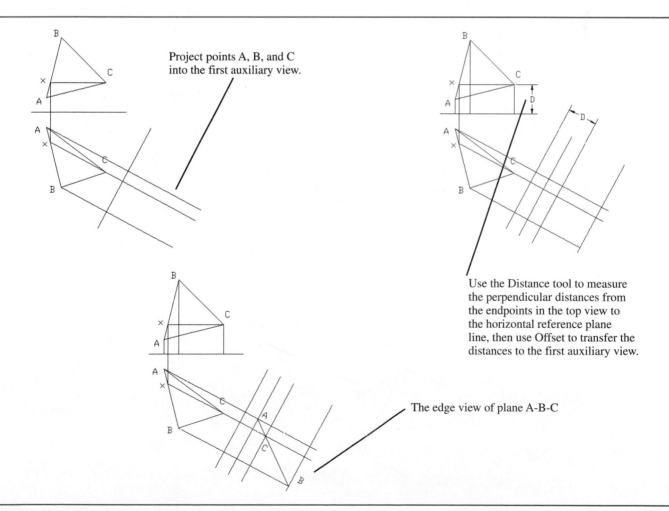

Project points A, B, and C into the first auxiliary view.

Use the Distance tool to measure the perpendicular distances from the endpoints in the top view to the horizontal reference plane line, then use Offset to transfer the distances to the first auxiliary view.

The edge view of plane A-B-C

Figure 17-13, Part 2

13. Erase the offset lines and trim the projection lines using the edge view as a cutting edge line.

To create a secondary auxiliary view perpendicular to the edge view of the plane

1. Use the **Snap, Rotate** command to reposition the crosshairs so that they are aligned with the edge view line.
2. Draw a line parallel to and above the edge view line.

This line is the reference plane line for the secondary auxiliary view.

3. Turn **Ortho** on and draw projection lines from the corner point of the plane in the first auxiliary view into the second auxiliary view.
4. Use the **Distance** command to determine the distance from the corner point in the front view to the first auxiliary view reference line.
5. Use the **Offset** command to transfer these distances to the secondary auxiliary view.
6. Draw lines between the appropriate intersections to create the secondary auxiliary view of plane A-B-C.
7. Erase any excess lines.

The secondary auxiliary view is the true shape of plane A-B-C.

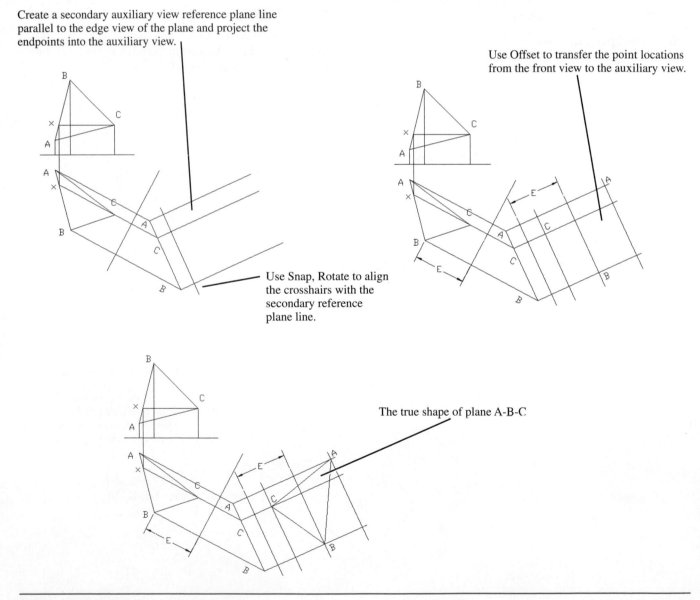

Create a secondary auxiliary view reference plane line parallel to the edge view of the plane and project the endpoints into the auxiliary view.

Use Snap, Rotate to align the crosshairs with the secondary reference plane line.

Use Offset to transfer the point locations from the front view to the auxiliary view.

The true shape of plane A-B-C

Figure 17-13, Part 3

To determine the true shape of a plane using AutoCAD

Figure 17-14 shows plane A-B-C in 3D space. The true shape of the plane can be determined by creating a UCS aligned with the plane and then taking a plan view of the UCS.

1. Select the **3 Point UCS** tool from the **UCS** toolbar.

   Specify new origin point <0,0,0>:

2. Use **Osnap, Endpoint** and select point **A.**

   Specify point on the positive portion of the X-axis <>:

3. Use **Osnap, Endpoint** and select point **B.**

   Specify point on the positive-Y portion of the UCS XY plane <>:

4. Use **Osnap, Endpoint** and select point **C.**

   The coordinate system will change, creating a UCS aligned with the plane. In the example shown, the UCS icon was located on the UCS's origin.

5. Select the **View** pull-down menu, **3D Views,** then **Plan View,** then **Current UCS.**

   The view shown is the true shape of plane A-B-C.

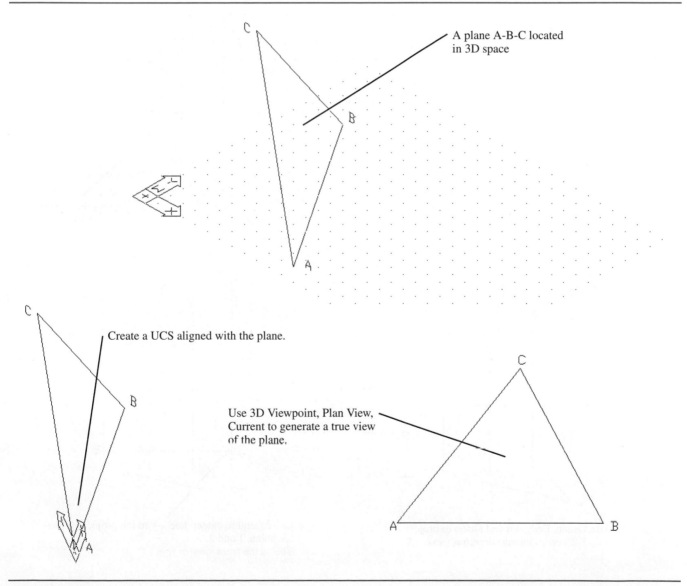

A plane A-B-C located in 3D space

Create a UCS aligned with the plane.

Use 3D Viewpoint, Plan View, Current to generate a true view of the plane.

Figure 17-14

17-5 LOCATING A LINE IN A PLANE

Figure 17-15 shows the front and top views of plane A-B-C and a line 1-2 in the top view. Line 1-2 may be located in the front view as follows.

1. Define the intersection points of the line with the edge lines of the plane as **x** and **y.**
2. Project points **x** and **y** into the front view.

This is done by drawing vertical lines from the intersection points in the top view into the front view through the appropriate edge lines. In this example point x is on edge line A-C, and point y is on A-B.

3. Draw a line between points **x** and **y** in the front view.
4. Draw vertical lines **(Ortho)** from the endpoints of the line into the front view.
5. Use the **Extend** command to extend line **x-y** to the projection lines from endpoints **1** and **2.**

Line 1-2 is now defined in the front view.

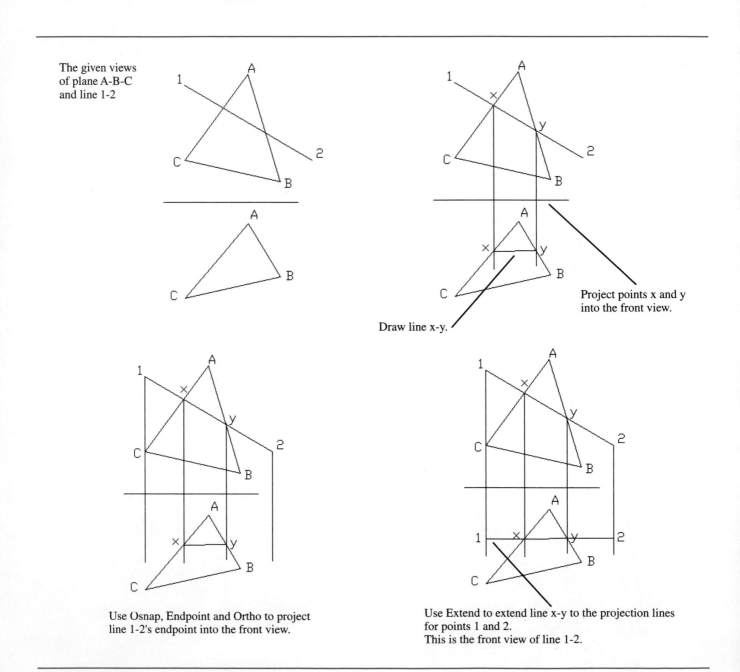

The given views of plane A-B-C and line 1-2

Project points x and y into the front view.

Draw line x-y.

Use Osnap, Endpoint and Ortho to project line 1-2's endpoint into the front view.

Use Extend to extend line x-y to the projection lines for points 1 and 2.
This is the front view of line 1-2.

Figure 17-15

To locate a line not in a plane

The line need not actually intersect the plane. Figure 17-16 shows line 1-2 in the top view, which does not intersect plane A-B-C. The line can be located in the front view as follows.

1. Project the endpoints into the front view.

At this point the exact location of the endpoints is unknown, so draw the line into the front view, past the front view of the plane.

2. Use the **Extend** command and extend line **1-2** into the plane so that it crosses at least two edge lines.

3. Label intersection points **x** and **y,** and project them into the front view.
4. Draw a line between points **x** and **y** in the front view.
5. Use the **Extend** command and extend line **x-y** in the front view to the projection line for endpoint **2.**
6. Use the endpoint projection lines to trim line **1-2** to its correct length in both the front and top views.
7. Erase any excess lines.

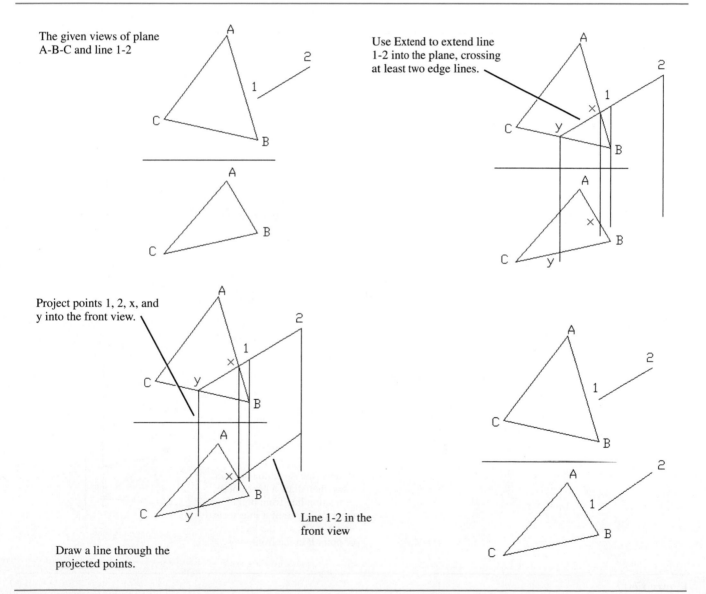

The given views of plane A-B-C and line 1-2

Use Extend to extend line 1-2 into the plane, crossing at least two edge lines.

Project points 1, 2, x, and y into the front view.

Line 1-2 in the front view

Draw a line through the projected points.

Figure 17-16

17-6   LOCATING A POINT IN A PLANE

Figure 17-17 shows front and top views of plane A-B-C. Point P is shown in the top view. The location of the point can be determined by drawing a line from one of the plane's corner points through point P and intersecting one of the plane's edge lines. This line can then be projected into the front view. It is known that the point's location in the front view is directly below the top view location, so the intersection of a vertical line from the point location in the top view with the projected line in the front view defines the point's location in the front view.

1. Draw a line from corner point **A** to point **P.**
2. Use the **Extend** command and extend line **A-P** to the edge of the plane.

3. Label the intersection point with edge line **x.**
4. Project line **A-x** into the front view.

   Use **Osnap** to ensure accuracy.

5. Project point **P** into the front view by drawing a vertical line from point P in the top view so that it intersects line A-x in the front view.

The intersection point is the location of point P in the front view.

The preceding method can also be used to determine the location of a point relative to a known plane. In Figure 17-18, a point is located to the right of plane A-B-C. The point can be located relative to the plane by drawing a line from the point to a corner point of the plane that intersects an edge line. The line can then be projected into the other views, and the point can be projected using the line. See Figure 17-18.

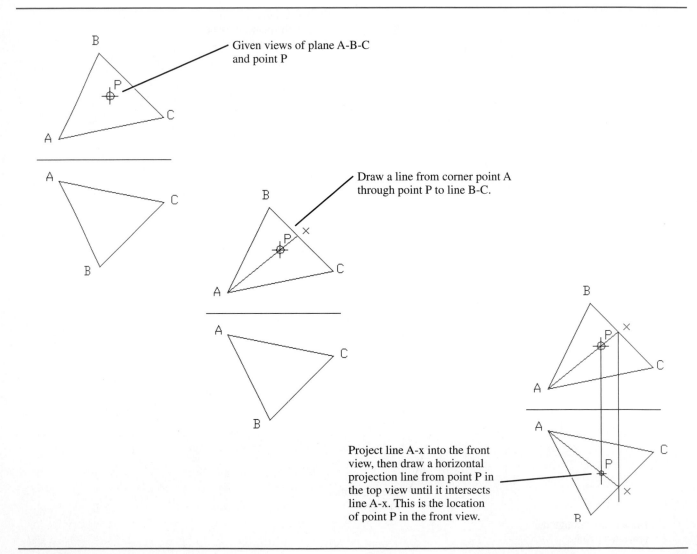

Given views of plane A-B-C and point P

Draw a line from corner point A through point P to line B-C.

Project line A-x into the front view, then draw a horizontal projection line from point P in the top view until it intersects line A-x. This is the location of point P in the front view.

Figure 17-17

Given views of plane A-B-C and a point

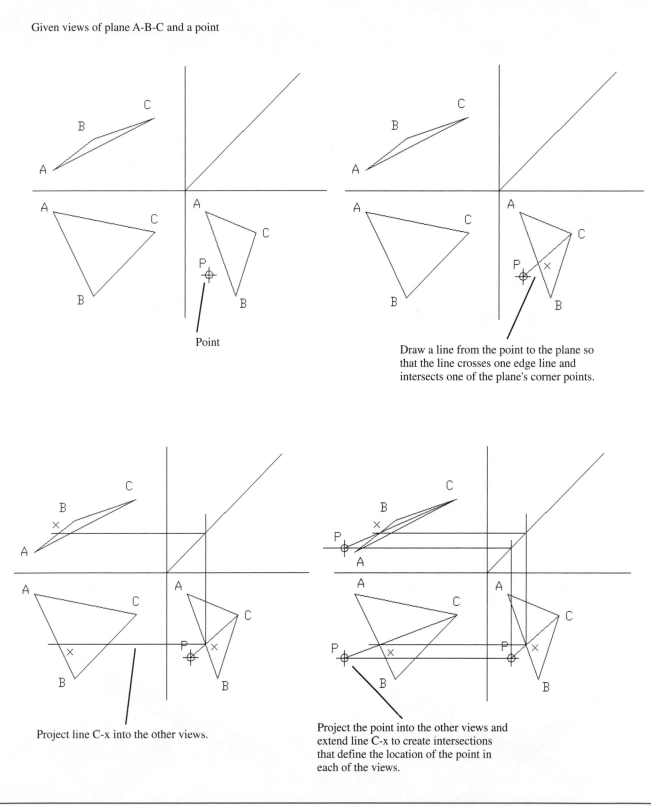

Point

Draw a line from the point to the plane so
that the line crosses one edge line and
intersects one of the plane's corner points.

Project line C-x into the other views.

Project the point into the other views and
extend line C-x to create intersections
that define the location of the point in
each of the views.

Figure 17-18

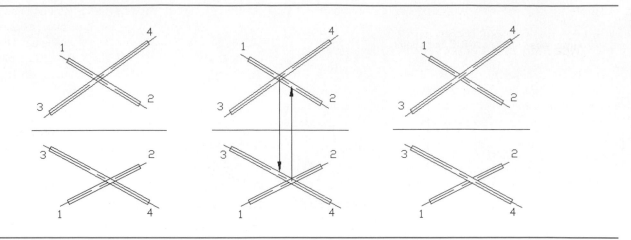

Figure 17-19

17-7 VISIBILITY OF A LINE

When working in 2D orthographic views it is sometimes difficult to determine the relative position of two lines with respect to each other. In the example presented here, solid bars were used rather than lines so it would be easier to see the visual distinction between the lines' relative locations.

Figure 17-19 shows two intersecting bars. Which bar is closer in the top view, and which is closer in the front view?

For the top view

If the intersection point of lines 1-2 and 3-4 in the top view is projected into the front view by drawing a vertical line downward, the projection line intersects line 3-4 before it intersects line 1-2. Remember that objects near the top of

the front view are higher than those near the bottom of the view. If line 3-4 is intersected first, it must be higher than line 1-2 at the intersection point in the top view.

For the front view

If the intersection of lines 1-2 and 3-4 in the front view is projected into the top view by drawing a vertical line upward, the projection line intersects line 1-2 first, indicating that line 1-2 is in front of line 3-4. Points near the bottom of the top view are closer than those located in the top portion of the view.

Figure 17-19 shows the resulting intersections with their visibility displayed.

Figure 17-20 shows two lines in 3D space. They were both drawn as solid cylinders so that their relative positions could be more easily seen.

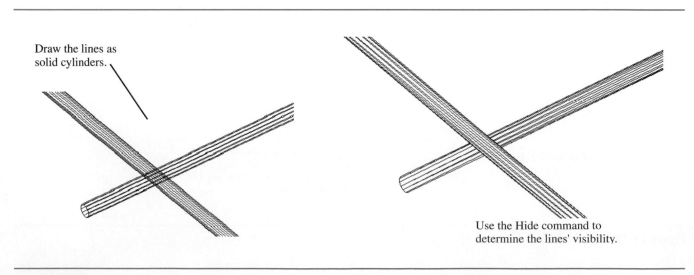

Draw the lines as solid cylinders.

Use the Hide command to determine the lines' visibility.

Figure 17-20

To determine the lines' visibility using AutoCAD

1. Zoom the intersecting area.
2. Type **hide;** press **Enter.**

It can be clearly seen that line 1-2 is above line 3-4 from this viewpoint. Different viewpoints would yield different results. This may be verified by returning to the original **Zoom** scale, then using the **Zoom** command again and creating a window that includes both lines in their entirety. Use the **3D Dynamic View** command to rotate the lines about the screen so you can see their relative positions and different viewpoints. If you are unsure about the lines' visibility at any orientation, use the **Hide** command to see the current visibility. See Figure 17-20. The **Regen** command will always return the lines to their original wireframe visibility.

17-8 PIERCING POINTS

Figure 17-21, Parts 1 and 2, shows front and top views of plane A-B-C and line 1-2. What is the position of the line relative to the plane? There are three possible results: the line pierces the plane, the line is parallel to the plane, or the line is within the same plane as the designated plane. There are two methods that can be used to determine whether the line pierces the plane: the cutting plane method or the auxiliary view method.

Cutting plane method

1. Complete line **1-2** across the top view.
2. Label the intersection points with the plane's edge lines as **x** and **y.**

Think of this line as representing the top edge of a plane that includes the line 1-2.

3. Project points **x** and **y** into the front view and draw line **x-y.**
4. Complete line **1-2** in the front view.

The intersection of lines 1-2 and x-y in the front view defines the piercing point between the line and the plane.

5. Project the piercing point into the top view.

The visibility of the line relative to the plane can be determined by using the method explained in Section 17-7

Given front and top views of plane A-B-C and line 1-2

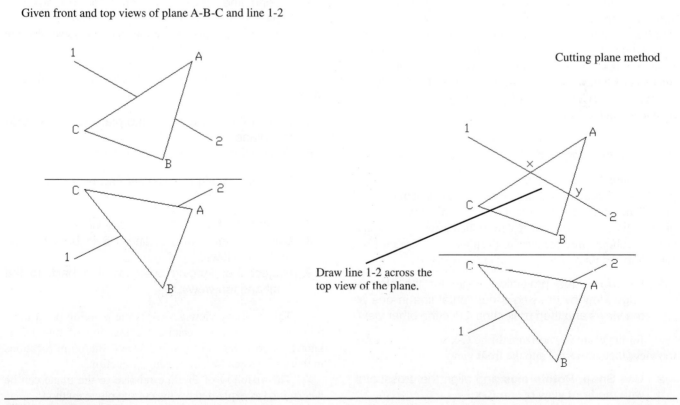

Draw line 1-2 across the top view of the plane.

Cutting plane method

Figure 17-21, Part 1

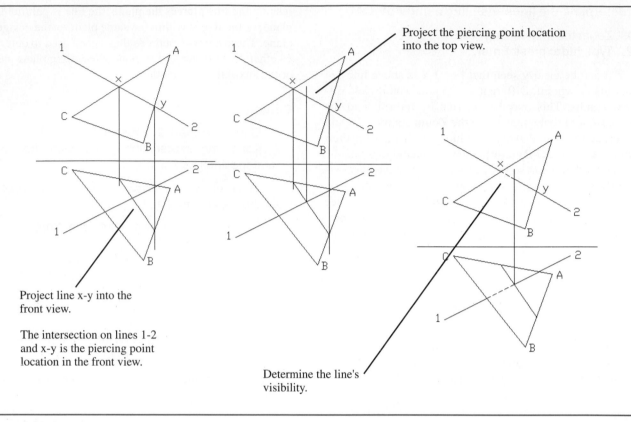

Project the piercing point location into the top view.

Project line x-y into the front view.

The intersection on lines 1-2 and x-y is the piercing point location in the front view.

Determine the line's visibility.

Figure 17-21, Part 2

for the visibility of lines. Note the intersection of lines A-C and 1-2 in the top view, intersection x. Follow the projection line from the intersection down to the front view. Line A-C is intersected before line 1-2, indicating that line A-C is higher than line 1-2. Likewise, the projection of the intersection between lines C-B and 1-2 in the front view strikes line C-B first, indicating that line C-B is closer to the viewer in the front view.

Auxiliary view method

Figure 17-22 shows the same front and top views of plane A-B-C and line 1-2 that were used to explain the cutting plane method in Figure 17-21. In this example, the piercing point location will be determined using an auxiliary view that shows the edge view of the plane. See Section 17-4 for a more detailed explanation of auxiliary views.

1. Determine the projection angle for the plane's edge view by drawing a horizontal line in one of the views and then projecting it into the other view.

In this example, horizontal line C-x was drawn in the top view, then projected into the front view.

2. Use **Snap, Rotate axis** and align the crosshairs with the front view of line **C-x**.

3. Extend line C-x into an area for the auxiliary view.
4. Draw a line perpendicular to the extension of line C-x.

The perpendicular line is the reference plane line for the auxiliary view.

5. Project the endpoints of the plane into the auxiliary plane.
6. Draw a line between the projected endpoints.

The result should be a straight line, or the edge view of plane A-B-C.

7. Project line **1-2** into the auxiliary view.
8. Complete the views of line 1-2 in both the top and front views.
9. Project the piercing point location back to the front and top views.

The auxiliary view generates the piercing point location on line 1-2, and because the location of line 1-2 is known in both the front and top views, the point locations in those views can be found using projection.

The visibility of the line relative to the plane can be determined as explained for the cutting plane method.

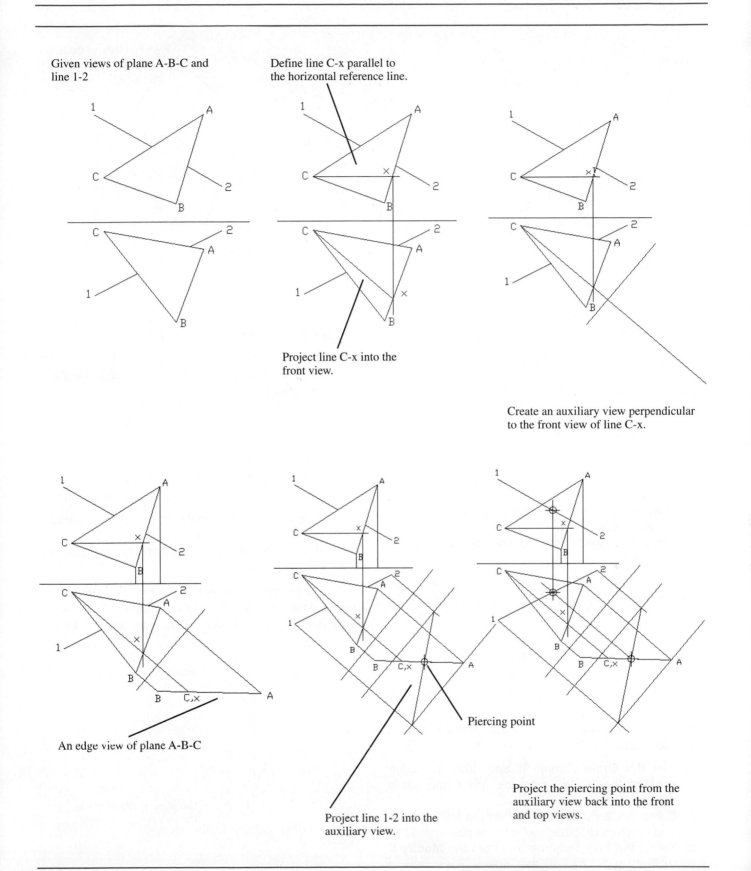

Given views of plane A-B-C and line 1-2

Define line C-x parallel to the horizontal reference line.

Project line C-x into the front view.

Create an auxiliary view perpendicular to the front view of line C-x.

An edge view of plane A-B-C

Project line 1-2 into the auxiliary view.

Piercing point

Project the piercing point from the auxiliary view back into the front and top views.

Figure 17-22

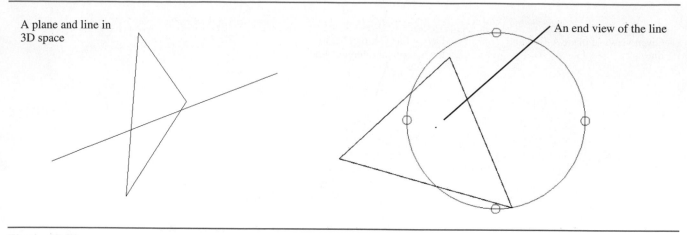

A plane and line in 3D space

An end view of the line

Figure 17-23

To determine if a plane and a line intersect using AutoCAD

Figure 17-23 shows a plane and a line in 3D. We can quickly determine if the line pierces the plane, but not the piercing point's exact location, by using the **3D Orbit** command. In this example, the plane and line are not solids; they are surface models.

1. Select the **View** pull-down menu, then **3D Orbit.**

A large green circle will appear on the drawing screen with its four quadrant points indicated by small circles.

2. Hold the left mouse button down and move the mouse around.
3. Move the objects around until an end view of the line is found.

The endpoint of line 1-2 appears as a point. If the point appears within plane A-B-C, then the line pierces the plane.

To determine the piercing point of a line and a plane using AutoCAD's solid modeling

In this example the corner points of the plane and the endpoints of the line are used to create solid models, then the Intersection command is used to determine the location of the piercing point relative to the WCS. See Figure 17-24.

1. Use the **Draw** command and draw the edge lines of the plane using the given coordinate values.
2. Select the **3 Point UCS** tool on the **UCS** toolbar and create a UCS aligned with the plane.
3. Select the **Edit Polyline** tool from the **Modify II** toolbar and change the drawn lines into a polyline.

The **Edit Polyline** command cannot be applied to planes that are not parallel to the current UCS.

4. Select the **Extrude** tool from the **Solids** toolbar and create a solid plane .00001 thick (height of extrusion).
5. Save the UCS and return to the WCS.
6. Select the **Solid Cylinder** command.
7. Draw a cylinder between the two known endpoints of the line of radius .00001.

Because the thickness and radius values are so small, the plane and line will appear as they did as surface models.

8. Zoom the approximate area of the piercing point.
9. Use the **Intersection** command to determine the piercing point.

The piercing point for two solid models is their common volume. It will appear as a small dot on the screen after the Intersection command has been applied. Be sure

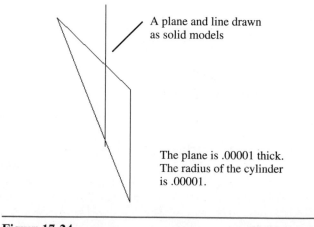

A plane and line drawn as solid models

The plane is .00001 thick. The radius of the cylinder is .00001.

Figure 17-24

```
AutoCAD Text Window - SecondaryView                          _ □ ×
Edit
Specify radius for base of cylinder or [Diameter]: .00001

Specify height of cylinder or [Center of other end]: 10

Command: '_pan
Press ESC or ENTER to exit, or right-click to display shortcut menu.

Command:  <Grid off>
Command: '_zoom
Specify corner of window, enter a scale factor (nX or nXP), or
[All/Center/Dynamic/Extents/Previous/Scale/Window] <real time>: _w
Specify first corner: Specify opposite corner:
Command:
Command:
Command: _intersect
Select objects: 1 found

Select objects: 1 found, 2 total

Select objects:


Command: list

Select objects: 1 found

Select objects:
                     3DSOLID   Layer: "0"
                               Space: Model space
                      Handle = 6B9
    Bounding Box: Lower Bound X =  2.0000   , Y =  2.0000   , Z =  2.3809
                  Upper Bound X =  2.0000   , Y =  2.0000   , Z =  2.3810

Command: |
```

Location of the intersection between the cylinder and the plane, the piercing point

Figure 17-25

to turn off **Grid** if it is on, so that the intersection volume can be seen.

10. Type **List**; press **Enter.**
11. Select the common volume (the dot that represents the common volume).

Figure 17-25 shows the resulting AutoCAD Text Window and the X,Y,Z coordinate values relative to the WCS. Two values are listed for the Y coordinate. Use the upper value to verify the location of the piercing point. If you were working with the plane's UCS, the resulting values would be relative to that UCS.

If you need to work with the piercing point, first use the **Point** command to define the point as an entity. Then you can use it as a selected object.

17-9 DISTANCE BETWEEN A LINE AND A POINT

Figure 17-26, Parts 1 and 2, shows the front and top views of a line 1-2 and point P. The shortest distance between the point and the line is determined by taking a secondary auxiliary view of the line, which will be a point view of the line. The shortest possible distance can then be measured directly between the end view and the point.

1. Use the **Offset** command to create a line parallel to the top view of line 1-2.

This line is the first auxiliary view reference plane line.

2. Project line **1-2** and point **P** into the first auxiliary view.

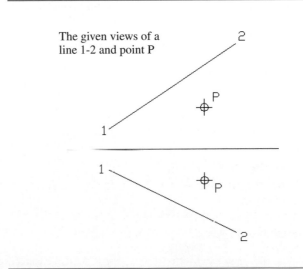

The given views of a line 1-2 and point P

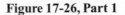

Figure 17-26, Part 1

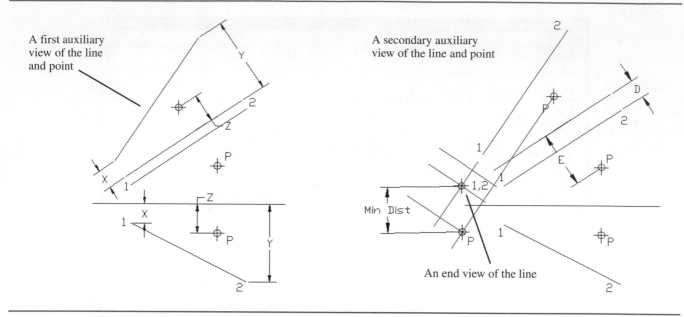

A first auxiliary view of the line and point

A secondary auxiliary view of the line and point

Min Dist

An end view of the line

Figure 17-26, Part 2

3. Use the **Snap, Rotate** option and align the crosshairs with the first auxiliary view of line 1-2.
4. Create a secondary auxiliary view that shows the end view of the line.

The distance between the edge view of the line and the point in the secondary auxiliary view is the minimum distance between the point and the line. See Section 17-4 for a more detailed explanation of auxiliary views.

To determine the minimum distance between a line and a point using 3D AutoCAD

Figure 17-27 shows line 1-2 and point P. The minimum distance between the point and the line is determined as follows.

1. Select the **3 Point UCS** tool from the **UCS** toolbar.

   Specify new origin point <0,0,0>:

2. Use **Osnap, Endpoint** and select point **1**.

   Specify point on the positive portion of the X-axis:

3. Use **Osnap, Endpoint** and select point **2**.

   Specify point on the positive-Y portion of the UCS XY plane:

4. Use **Osnap, Nearest** and select point **P**.
5. Select the **View** pull-down menu, then **3D views, Plan View, Current UCS**.
6. Select the **Distance** tool.

   Specify first point:

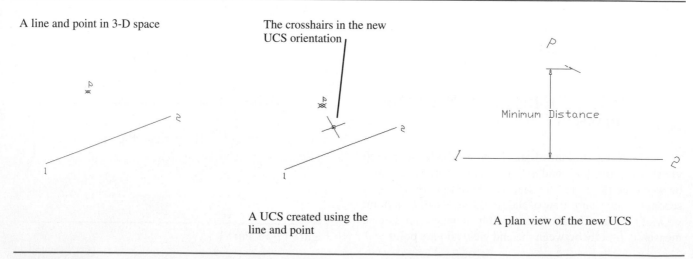

A line and point in 3-D space

The crosshairs in the new UCS orientation

A UCS created using the line and point

A plan view of the new UCS

Minimum Distance

Figure 17-27

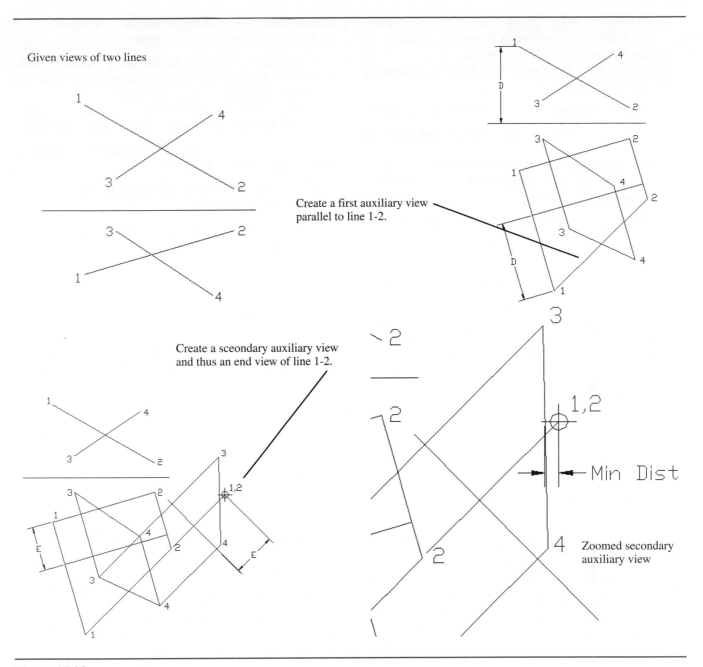

Given views of two lines

Create a first auxiliary view
parallel to line 1-2.

Create a sceondary auxiliary view
and thus an end view of line 1-2.

Min Dist

Zoomed secondary
auxiliary view

Figure 17-28

7. Use **Osnap, Nearest** and select point **P.**

   Second extension line origin:

8. Use **Osnap, Perpendicular** and select line **1-2.**

17-10 DISTANCE BETWEEN TWO LINES

Figure 17-28 shows the front and top views of lines 1-2 and 3-4.

The minimum distance between the lines can be determined by using a secondary auxiliary view to create an end view of one of the lines. The minimum distance can then be measured by drawing a line from the point end view of the line perpendicular to the other line.

The secondary auxiliary view method

1. Use the **Snap, Rotate** command to align the crosshairs with the front view of line 1-2.
2. Draw a reference plane line for the first auxiliary view parallel to the front view of line 1-2, then project both lines' endpoints into the auxiliary view from the front view.
3. Use the **Osnap, Rotate** command and align the crosshairs with the first auxiliary view of line 1-2.

4. Create the edge plane line for the secondary auxiliary view by drawing a line perpendicular to the first auxiliary view of line 1-2.
5. Project both lines' endpoints into the secondary auxiliary view.

Line 1-2 should appear as a point or end view of the line.

6. Draw a line from the end view perpendicular to the view of line 3-4.
7. Use the **Linear, Dimension** command to measure the length of the perpendicular line, or use the **Distance** command.

The distance found in step 7 is the minimum distance between the two lines.

Using AutoCAD's 3D capabilities

Figure 17-29 shows two lines in 3D space. The minimum distance between them may be determined by rotating the lines until one appears as an end view, then measuring the perpendicular distance between the end view and the other line. The **Osnap** commands cannot be applied in this situation. The resulting distance is an approximation. It is, however, a very close approximation.

1. Select the **View** pull-down menu, then **3D Orbit**.
2. Hold down the left mouse button and rotate the lines until one appears as a point.
3. Zoom the area around the end view and draw a line from the end view perpendicular to the line.

Draw the line by eye; that is, select the endpoints on the screen. Do not use **Osnap.**

4. Use the **Distance** command to determine the length of the line.

This distance is the approximate minimum distance between the two lines.

17-11 ANGLE BETWEEN A PLANE AND A LINE

Figure 17-30 shows front and top views of plane A-B-C and a line 1-2. The true angle between them can be

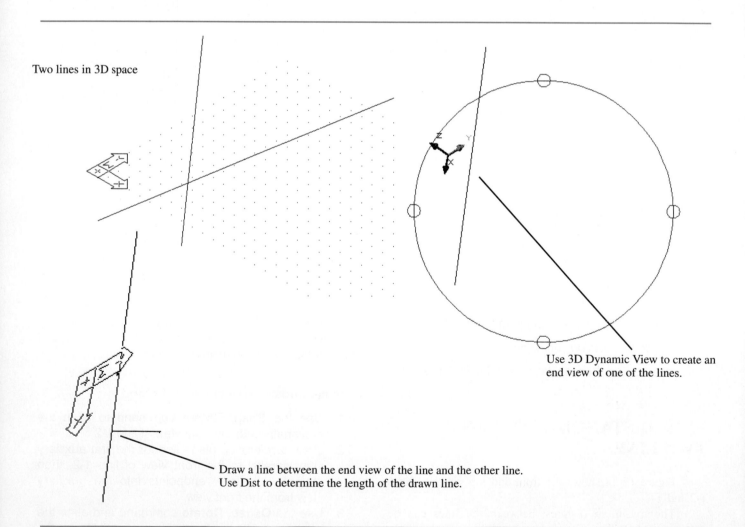

Two lines in 3D space

Use 3D Dynamic View to create an end view of one of the lines.

Draw a line between the end view of the line and the other line.
Use Dist to determine the length of the drawn line.

Figure 17-29

determined by taking several auxiliary views so that the plane appears as an edge view and the line appears as a true length. See Section 17-6 for additional explanation of how to create the end view of a plane.

The auxiliary view method

1. Create a first auxiliary view that shows plane A-B-C as an end view (a straight line).

This view shows the edge view of the plane but not the true length of the line, so the angle between them is not a true angle.

2. Draw a secondary auxiliary view of the line and plane that shows the true shape of the plane.
3. Draw a third auxiliary view so that its edge plane line is parallel to the secondary auxiliary view of line 1-2.

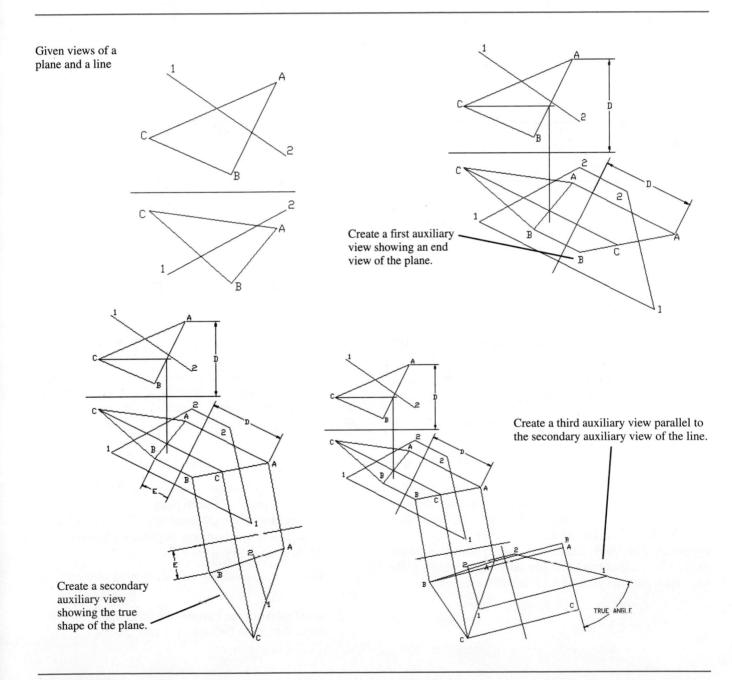

Given views of a plane and a line

Create a first auxiliary view showing an end view of the plane.

Create a secondary auxiliary view showing the true shape of the plane.

Create a third auxiliary view parallel to the secondary auxiliary view of the line.

TRUE ANGLE

Figure 17-30

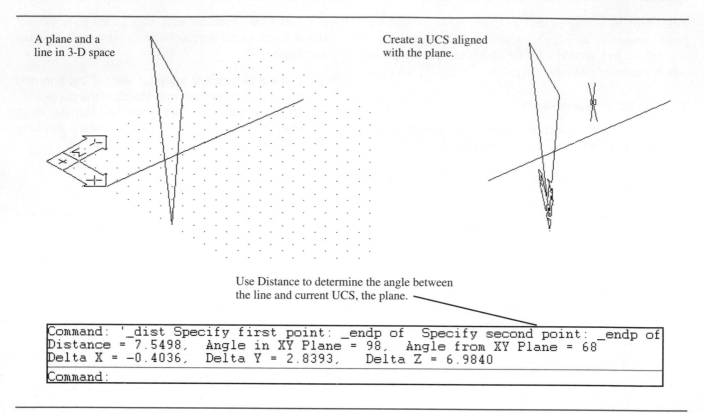

A plane and a line in 3-D space

Create a UCS aligned with the plane.

Use Distance to determine the angle between the line and current UCS, the plane.

```
Command: '_dist Specify first point: _endp of  Specify second point: _endp of
Distance = 7.5498,   Angle in XY Plane = 98,   Angle from XY Plane = 68
Delta X = -0.4036,   Delta Y = 2.8393,   Delta Z = 6.9840
Command:
```

Figure 17-31

Using 3D AutoCAD

Figure 17-31 shows plane A-B-C and line 1-2 in 3D space. The angle between them can be determined by first defining a UCS aligned with the plane, then using the Dist command.

1. Select the **3 Point UCS** tool from the **UCS** toolbar and define a UCS aligned with plane A-B-C.

   Use **Osnap, Endpoint** to ensure accuracy.

2. Select the **Distance** tool.
3. Use **Osnap, Endpoint** and select points 1 and 2.

The value listed in the command box at the bottom of the screen for "Angle from XY plane" is the angle between the line and the plane. In this example, the angle from the XY plane is defined as 9.65°. See Figure 17-31. If the endpoints had been selected in the opposite sequence, the resulting angular value would have been 350.35°, or the complementary angle of the 9.65° initially determined.

17-12 ANGLE BETWEEN TWO PLANES

Figure 17-32 shows front and top views of two intersecting planes, A-B-C and B-C-D. The true angle between the planes can be determined by using a secondary auxiliary view positioned to show an edge view of each plane. This can be done only by aligning the secondary view so that it shows an end view of the line common to the two planes.

The auxiliary view method

1. Create a first auxiliary view aligned with line B-C, the line common to both planes.
2. Create a secondary auxiliary view perpendicular to the projection line aligned with line B-C in the first auxiliary view.

The resulting auxiliary view shows end views of both planes.

3. Use the **Angular Dimension** command to determine the angle between the two planes.

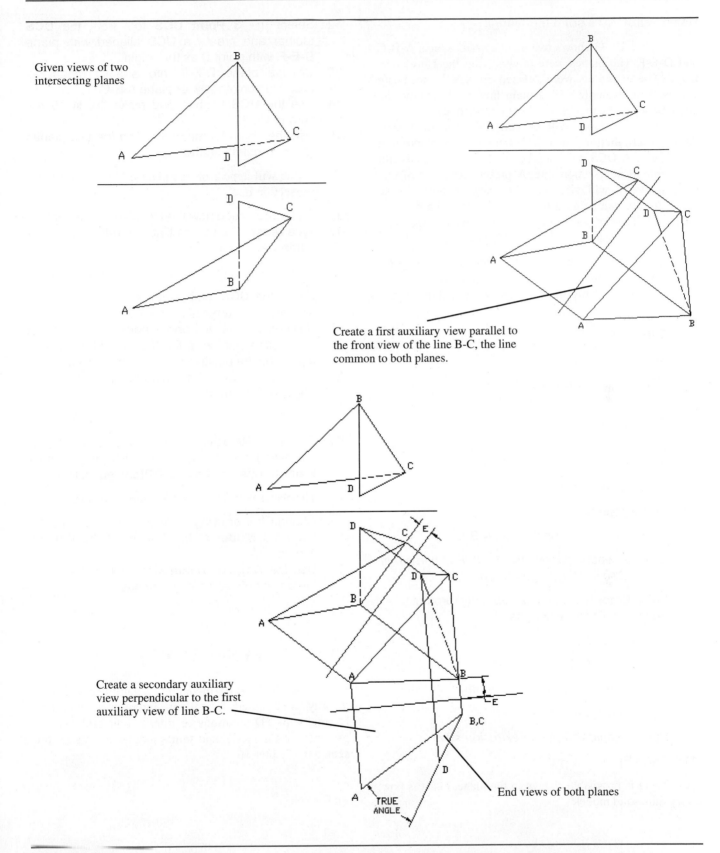

Given views of two
intersecting planes

Create a first auxiliary view parallel to
the front view of the line B-C, the line
common to both planes.

Create a secondary auxiliary
view perpendicular to the first
auxiliary view of line B-C.

End views of both planes

TRUE
ANGLE

Figure 17-32

Using AutoCAD's solid modeling

Figure 17-33 shows two planes in 3D space: A-B-C and D-E-F. The planes were drawn using the **Line** command. The endpoints were defined by X,Y,Z coordinate values. This example will explain how to determine the angle between the planes using solid modeling.

The two planes will be changed to solid models .00001 thick, then unioned to determine their intersection line, if any. A UCS will then be created and aligned with the common intersection line. A preset viewpoint of this UCS set to 0° will yield an edge view of both planes, which in turn can be used to determine the angle between the planes. The **Origin** icon will appear in most of the illustrations to help clarify the different UCSs used.

1. Use the **UCS** toolbar and create a UCS aligned with plane **A-B-C,** with point **A** as the origin.
2. Select **Edit Polyline** from the **Modify II** toolbar.

   Command: _pedit
   Select polyline:

3. Select line **A-B.**

   Object selected is not a polyline.
   Do you want to turn it into one? <Y>

4. Press **Enter.**

   Enter an option [Close/Join/Width/Edit vertext/Fit/ Spline/Decurve/Ltype gen/ Undo]:

5. Type **j;** press **Enter.**

   Select objects:

6. Select the three lines in plane A-B-C.

   Enter an option [Open/Join/Width/Edit vertex/Fit/ Spline/Decurve/Ltype gen/Undo]:

7. Press **Enter** to return to a command prompt.
8. Select the **Extrude** command.

   Select objects:

9. Select plane **A-B-C.**

   Specify height of extrusion or [Path]:

10. Type **.00001;** press **Enter.**

    Specify angle of taper for extrusion <0>:

11. Press **Enter.**

There is no visible change in the plane, but it is now a very thin, solid model.

12. Select the **3 Point UCS** tool from the **UCS** toolbar and create a UCS aligned with plane **D-E-F,** with point **D** as the origin.
13. Change plane D-E-F into a solid model as explained previously for plane A-B-C.
14. Use the **UCS** toolbar, and make the **WCS** the current UCS.
15. Use the **Union** command to join the two planes (solid models) together.

A line will appear on the planes. This is their common intersection line.

16. Type **hide** to see how the two planes intersect.
17. Type **regen,** then press **Enter** to return to a wireframe view of the planes.

In this example, the intersection line is labeled **1-2.**

18. Select the **UCS** toolbar and create a third UCS aligned with intersection line 1-2.
19. Select point **1** as the origin, point **2** as a point on the positive portion of the X axis, and point **D** as a point on the positive portion of the X axis.
20. Select the **View** pull-down menu, then **3D Views, Viewpoint Presets.**

The **Viewpoint Presets** dialog box will appear.

21. Select the **Relative to UCS** button at the top of the dialog box, then set both **Viewing Angles From X Axis** and **From XY Plane** equal to **0°.**

The planes will appear as two intersecting lines.

22. Change the drawing to paper space by clicking the word **Model** at the bottom of the drawing screen.
23. Use the **Angular Dimension** tool to determine the angle between the two planes.

17-13 INTERSECTION

AutoCAD's solid-modeling capabilities are particularly well suited to solving intersection problems. Examples of 2D solutions to intersection problems are presented in Chapter 5, and solid-model solutions are presented in Chapter 16.

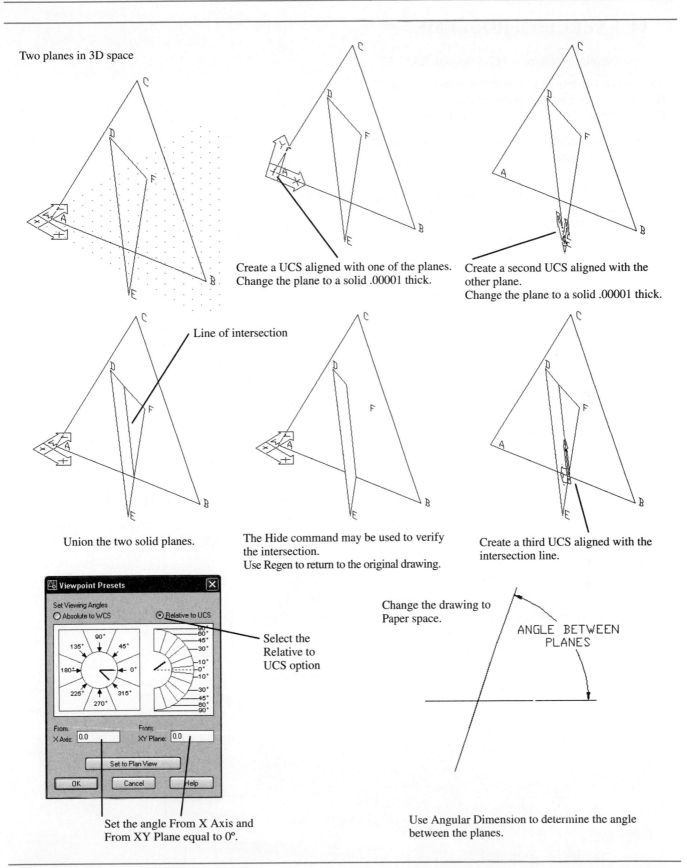

Two planes in 3D space

Create a UCS aligned with one of the planes.
Change the plane to a solid .00001 thick.

Create a second UCS aligned with the other plane.
Change the plane to a solid .00001 thick.

Line of intersection

Union the two solid planes.

The Hide command may be used to verify the intersection.
Use Regen to return to the original drawing.

Create a third UCS aligned with the intersection line.

Select the Relative to UCS option

Change the drawing to Paper space.

ANGLE BETWEEN PLANES

Set the angle From X Axis and From XY Plane equal to 0°.

Use Angular Dimension to determine the angle between the planes.

Figure 17-33

17-14 EXERCISE PROBLEMS

For Exercise Problems EX17-1 through EX17-12:

A. Locate the line or plane in the missing view.
B. Determine the true length of the line.
C. Determine the true shape of the plane.

The grid spacing for the dot background is either .5 inch or 10 millimeters.

EX17-1

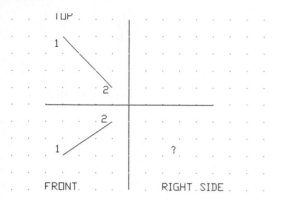

EX17-2

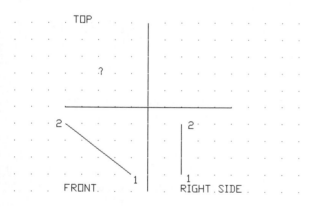

EX17-3

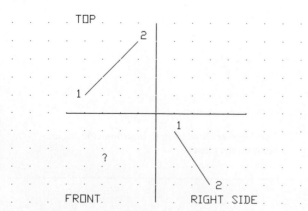

EX17-4

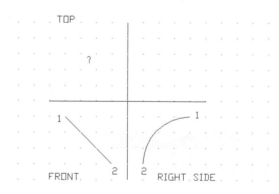

EX17-5

EX17-6

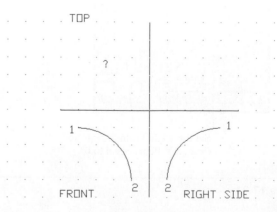

EX17-7

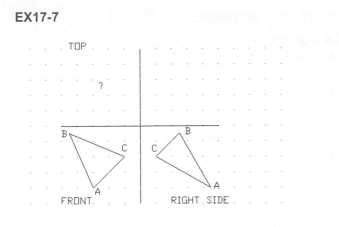

EX17-10 INCHES, SCALE 4:1

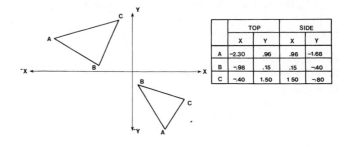

	TOP		SIDE	
	X	Y	X	Y
A	-2.30	.96	.96	-1.68
B	-.98	.15	.15	-.40
C	-.40	1.50	1 50	-.80

EX17-8

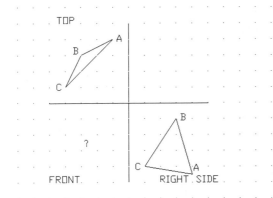

EX17-11 MILLIMETERS, SCALE 3:1

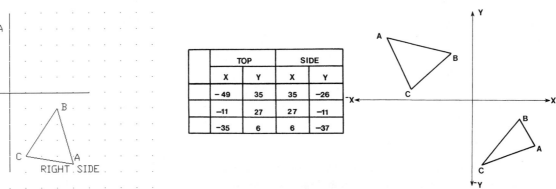

	TOP		SIDE	
	X	Y	X	Y
	- 49	35	35	-26
	-11	27	27	-11
	-35	6	6	-37

EX17-9

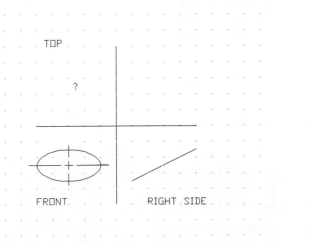

EX17-12 INCHES, SCALE 4:1

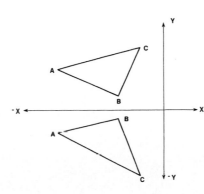

	FRONT		TOP	
	X	Y	X	Y
A	-2.80	-.60	-2.80	1.05
B	-1.17	-.20	-1.17	.35
C	-.61	-1.62	-.61	1.62

Find the true shape of the triangular planes in Exercise Problems EX17-13 through EX17-16. The three corner points of the planes are defined as A, B, and C with the following coordinate values.

EX17-13 INCHES

A = 2,6,−5
B = 3,2,−4
C = 6,5,1

EX17-14 INCHES

A = 0,7,3
B = 8,0,−1
C = 5,5,−6

EX17-15 MILLIMETERS

A = 50,50,125
B = 175,50,−100
C = 150,125,25

EX17-16 MILLIMETERS

A = 20,200,80
B = 200,0,−30
C = 120,120,−160

Determine the length of each of the lines in Exercise Problems EX17-17 through EX17-20 and the minimum distance between the two lines. The lines are defined as A-B and C-D.

EX17-17 INCHES

A = 1,1,1
B = 8,8,−3

C = 3,4,5
D = 9,6,−7

EX17-18 INCHES

A = 2,6,−4
B = 7,1,6

C = 0,1,3
D = 8,6,−2

EX17-19 MILLIMETERS

A = 30,20,20
B = 200,200,−70

C = 60,100,130
D = 200,150,−170

EX17-20 MILLIMETERS

A = 50,150,−100
B = 180,20,150

C = 0,20,70
D = 200,150,−60

Determine the minimum distance between the points and planes in Exercise Problems EX17-21 to EX17-24.

EX17-21 INCHES

Plane:

A = 1,6,4
B = 3,3,−4.5
C = 6,5,0

Point:

P = 4,3,6

EX17-22 INCHES

Plane:

A = 0,7,3
B = 7,1,−1
C = 5,5,−6

Point:

P = 2,3,5

EX17-23 MILLIMETERS

Plane:

A = 55,150,120
B = 75,50,−100
C = 140,130,20

Point:

P = 100,60,130

EX17-24 MILLIMETERS

Plane:

A = 20,180,85
B = 200,0,−35
C = 120,130,−150

Point:

P = 50,75,130

Determine the true shape of each plane in Exercise Problems EX17-25 through EX17-30 and the perpendicular distance from the point to the plane.

EX17-25 INCHES

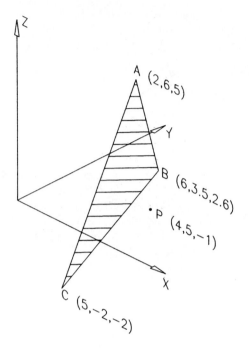

EX17-27 MILLIMETERS

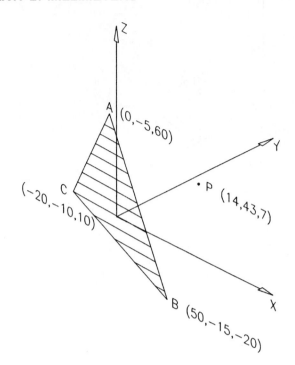

EX17-26 MILLIMETERS

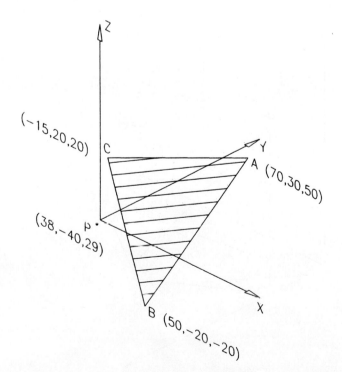

EX17-28 MILLIMETERS

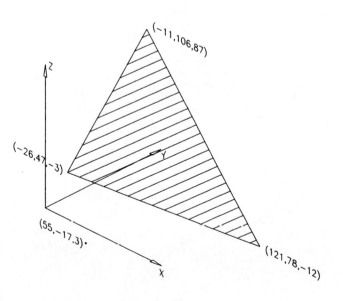

EX17-29 INCHES

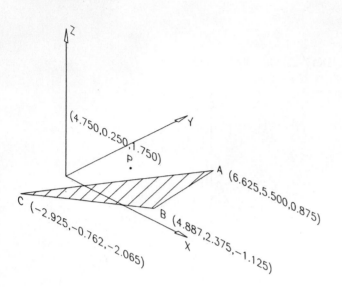

Determine the intersection line and angle of intersection between the planes in Exercise Problems EX17-31 through EX17-34.

EX17-31 INCHES

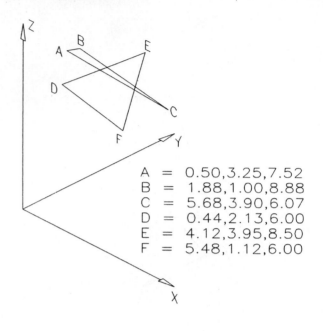

```
A  =  0.50,3.25,7.52
B  =  1.88,1.00,8.88
C  =  5.68,3.90,6.07
D  =  0.44,2.13,6.00
E  =  4.12,3.95,8.50
F  =  5.48,1.12,6.00
```

EX17-30 MILLIMETERS

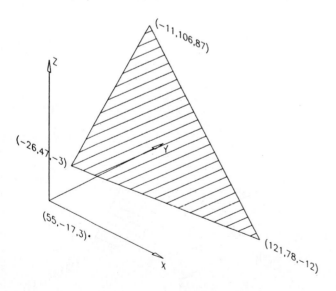

EX17-32 MILLIMETERS

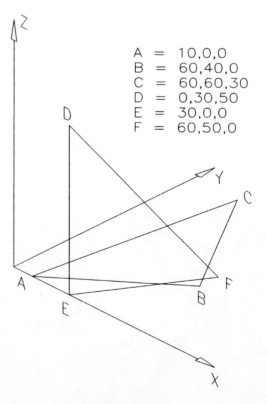

```
A  =  10,0,0
B  =  60,40,0
C  =  60,60,30
D  =  0,30,50
E  =  30,0,0
F  =  60,50,0
```

EX17-33 MILLIMETERS

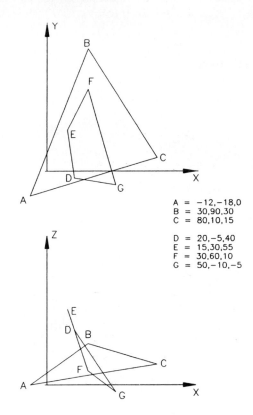

A = −12,−18,0
B = 30,90,30
C = 80,10,15

D = 20,−5,40
E = 15,30,55
F = 30,60,10
G = 50,−10,−5

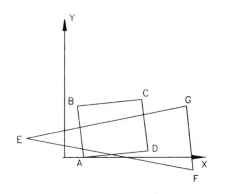

Determine the minimum distance between the given lines in Exercise Problems EX17-35 and EX17-36.

EX17-35 MILLIMETERS

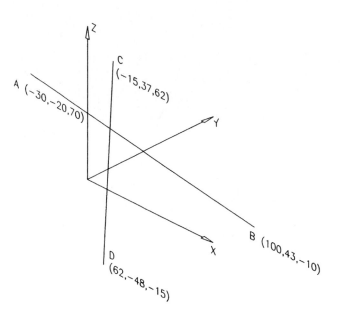

EX17-34 MILLIMETERS

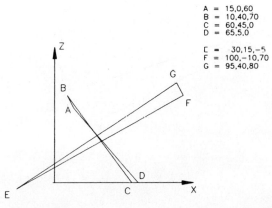

A = 15,0,60
B = 10,40,70
C = 60,45,0
D = 65,5,0

E = 30,15,−5
F = 100,−10,70
G = 95,40,80

EX17-36 MILLIMETERS

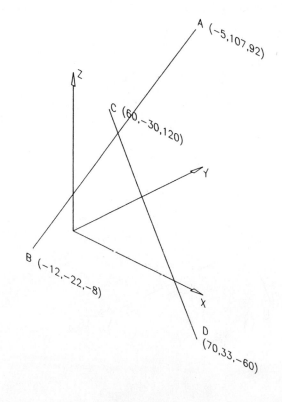

Determine the minimum distance between the given line and the point in Exercise Problems EX17-37 through EX17-40.

EX17-37 MILLIMETERS

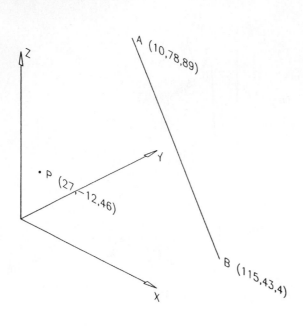

EX17-39 INCHES

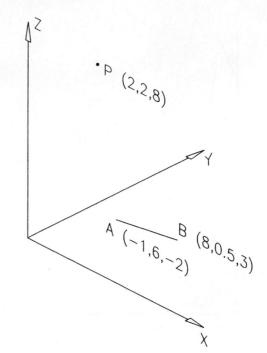

EX17-38 INCHES

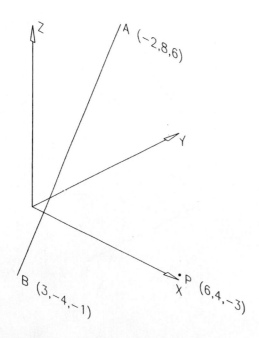

EX17-40 MILLIMETERS

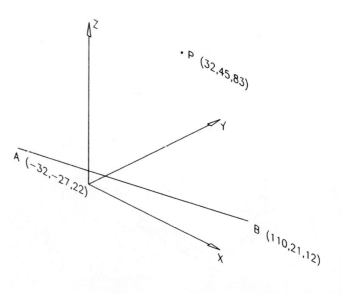

In Exercise Problems EX17-41 through EX17-45, determine the point on the plane where the line intersects. Label the point and specify its X,Y,Z coordinates relative to the WCS.

EX17-41 MILLIMETERS

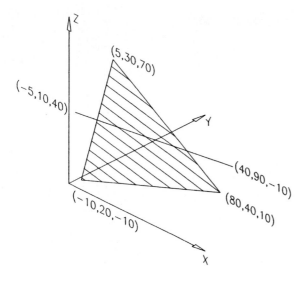

EX17-43 MILLIMETERS

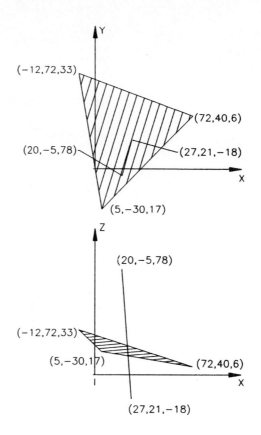

EX17-42 MILLIMETERS

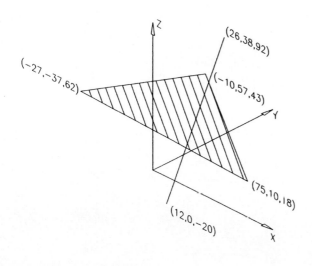

EX17-44 INCHES

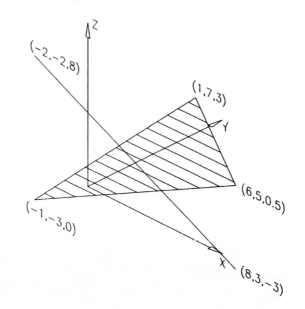

EX17-45 INCHES

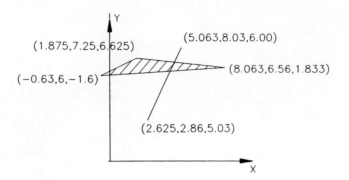

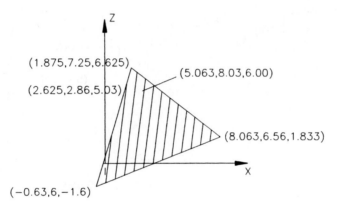

EX17-46

A pole 19.7 feet high is supported by three guy wires. Wire 1 is attached to the pole 1.00 foot from the top, wire 2 is attached 2.00 feet from the top, and wire 3 is attached 2 feet 6 inches from the top. The location of the pole and the locations of the ground positions of the three guy wires are as shown. What is the true length of each of the guy wires?

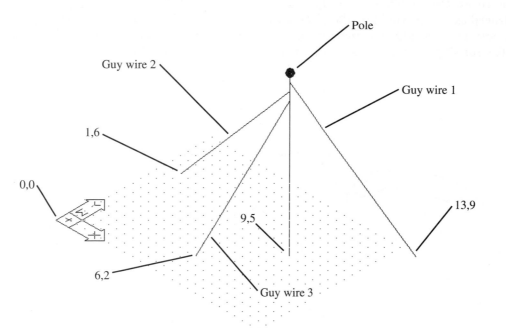

EX17-47

A work platform is to be added to the cowling of a helicopter. It is to fold down, giving access to the transmission area and supplying a place for the mechanic to sit while working. The platform is to be 24 by 24 inches.

Two support wires will be required to support the mechanic's weight and tools. The wires are to attach to the platform 1 inch from each of the outside edges and attach to the cowling 6 inches to the left and right of the top edge of the platform opening as shown.

What is the true length of the support wires?

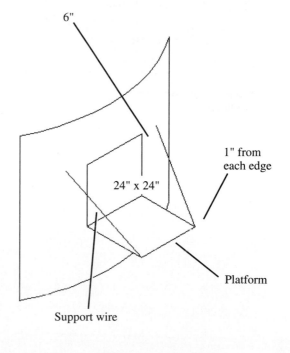

EX17-48

A satellite for scientific experiments involving the solar wind is to be made in an approximately spherical shape. The diameter of the sphere is 20 cm, but the top and bottom portions have to be truncated so that they are 19 cm apart. See below.

The satellite is to be made from 6 rows of 8 flat panels with straight edges. The satellite is symmetrical about its central horizontal axis. The symmetry allows the satellite to be made from just three different-sized panels.

What is the true shape of each of the three different panels?

Solar Wind Satellite

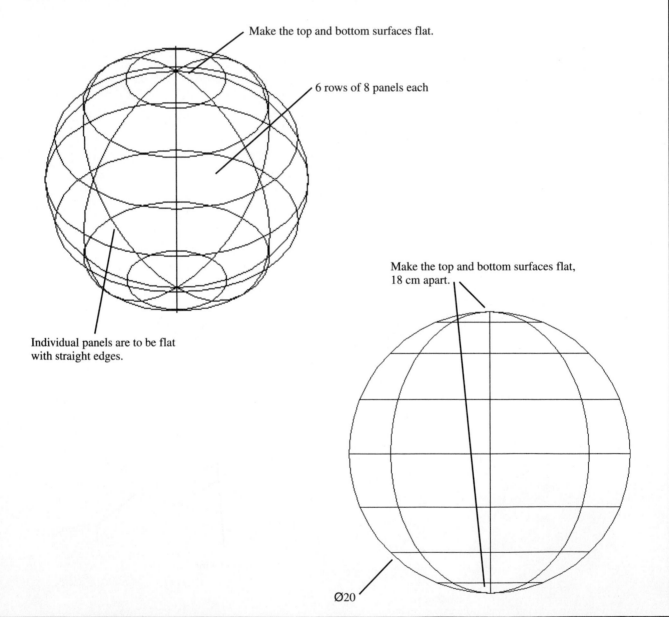

Make the top and bottom surfaces flat.

6 rows of 8 panels each

Individual panels are to be flat with straight edges.

Make the top and bottom surfaces flat, 18 cm apart.

Ø20

Appendix

Wire and Sheet Metal Gauges

Gauge			Thickness	Gauge	Thickness
000	000		0.5800	18	0.0403
00	000		0.5165	19	0.0359
0	000		0.4600	20	0.0320
	000		0.4096	21	0.0285
	00		0.3648	22	0.0253
	0		0.3249	23	0.0226
	1		0.2893	24	0.0201
	2		0.2576	25	0.0179
	3		0.2294	26	0.0159
	4		0.2043	27	0.0142
	5		0.1819	28	0.0126
	6		0.1620	29	0.0113
	7		0.1443	30	0.0100
	8		0.1285	31	0.0089
	9		0.1144	32	0.0080
	10		0.1019	33	0.0071
	11		0.0907	34	0.0063
	12		0.0808	35	0.0056
	13		0.0720	36	0.0050
	14		0.0641	37	0.0045
	15		0.0571	38	0.0040
	16		0.0508	39	0.0035
	17		0.0453	40	0.0031

Figure A-1

American Standard Clearance Locational Fits

Nominal Size Range Inches Over — To	Limits of Clearance	Class LC1 Standard Limits Hole H6	Shaft h5	Limits of Clearance	Class LC2 Standard Limits Hole H7	Shaft h6	Limits of Clearance	Class LC3 Standard Limits Hole H8	Shaft h7	Limits of Clearance	Class LC4 Standard Limits Hole H10	Shaft h9
0 — 0.12	0 / 0.45	+0.25 / 0	0 / -0.2	0 / 0.65	+0.4 / 0	0 / -0.25	0 / 1	+0.6 / 0	0 / -0.4	0 / 2.6	+1.6 / 0	0 / -1.0
0.12 — 0.24	0 / 0.5	+0.3 / 0	0 / -0.2	0 / 0.8	+0.5 / 0	0 / -0.3	0 / 1.2	+0.7 / 0	0 / -0.5	0 / 3.0	+1.8 / 0	0 / -1.2
0.24 — 0.40	0 / 0.65	+0.4 / 0	0 / -0.25	0 / 1.0	+0.6 / 0	0 / -0.4	0 / 1.5	+0.9 / 0	0 / -0.6	0 / 3.6	+2.2 / 0	0 / -1.4
0.40 — 0.71	0 / 0.7	+0.4 / 0	0 / -0.3	0 / 1.1	+0.7 / 0	0 / -0.4	0 / 1.7	+1.0 / 0	0 / -0.7	0 / 4.4	+2.8 / 0	0 / -1.6
0.71 — 1.19	0 / 0.9	+0.5 / 0	0 / -0.4	0 / 1.3	+0.8 / 0	0 / -0.5	0 / 2	+1.2 / 0	0 / -0.8	0 / 5.5	+3.5 / 0	0 / -2.0
1.19 — 1.97	0 / 1.0	+0.6 / 0	0 / -0.4	0 / 1.6	+1.0 / 0	0 / -0.6	0 / 2.6	+1.6 / 0	0 / -1.0	0 / 6.5	+4.0 / 0	0 / -2.5

Figure A-2A

Nominal Size Range Inches Over — To	Limits of Clearance	Class LC5 Standard Limits Hole H7	Shaft g6	Limits of Clearance	Class LC6 Standard Limits Hole H9	Shaft f8	Limits of Clearance	Class LC7 Standard Limits Hole H10	Shaft e9	Limits of Clearance	Class LC8 Standard Limits Hole H10	Shaft d9
0 — 0.12	0.1 / 0.75	+0..4 / 0	-0.1 / -0.35	0.3 / 1.9	+1.0 / 0	-0.3 / -0.9	0.6 / 3.2	+1.6 / 0	-0.6 / -1.6	1.0 / 3.6	+1.6 / 0	-1.0 / -2.0
0.12 — 0.24	0.15 / 0.95	+0.5 / 0	-0.15 / -0.45	0.4 / 2.3	+1.2 / 0	-0.4 / -1.1	0.8 / 3.8	+1.8 / 0	-0.8 / -2.0	1.2 / 4.2	+1.8 / 0	-1.2 / -2.4
0.24 — 0.40	0.2 / 1.2	+0.6 / 0	-0.2 / -0.6	0.5 / 2.8	+1.4 / 0	-0.5 / -1.4	1.0 / 4.6	+2.2 / 0	-1.0 / -2.4	1.6 / 5.2	+2.2 / 0	-1.6 / -3.0
0.40 — 0.71	0.25 / 1.35	+0.7 / 0	-0.25 / -0.65	0.6 / 3.2	+1.6 / 0	-0.6 / -1.6	1.2 / 5.6	+2.8 / 0	-1.2 / -2.8	2.0 / 6.4	+2.8 / 0	-2.0 / -3.6
0.71 — 1.19	0.3 / 1.6	+0.8 / 0	-0.3 / -0.8	0.8 / 4.0	+2.0 / 0	-0.8 / -2.0	1.6 / 7.1	+3.5 / 0	-1.6 / -3.6	2.5 / 8.0	+3.5 / 0	-2.5 / -4.5
1.19 — 1.97	0.4 / 2.0	+1.0 / 0	-0.4 / -1.0	1.0 / 5.1	+2.5 / 0	-1.0 / -2.6	2.0 / 8.5	+4.0 / 0	-2.0 / -4.5	3.0 / 9.5	+4.0 / 0	-3.0 / -5.5

Figure A-2B

American Standard Running and Sliding Fits
(Hole Basis)

Nominal Size Range Inches		Class RC1				Class RC2				Class RC3				Class RC4	
	Limits of Clearance	Standard Limits		Limits of Clearance	Standard Limits			Limits of Clearance	Standard Limits			Limits of Clearance	Standard Limits		
Over To		Hole H5	Shaft g4	Limits of Clearance	Hole H6	Shaft g5		Limits of Clearance	Hole H7	Shaft f6		Limits of Clearance	Hole H8	Shaft f7
0 - 0.12	0.1 0.45	+0.2 0	-0.1 -0.25	0.1 0.55	+0.25 0	-0.1 -0.3		0.3 0.95	+0.4 0	-0.3 -0.55		0.3 1.3	+0.6 0	-0.3 -0.7
0.12 - 0.24	0.15 0.5	+0.2 0	-0.15 -0.3	0.15 0.65	+0.3 0	-0.15 -0.35		0.4 1.12	+0.5 0	-0.4 -0.7		0.4 1.5	+0.7 0	-0.4 -0.0
0.24 - 0.40	0.2 0.6	+0.25 0	-0.2 -0.35	0.2 0.85	+0.4 0	-0.2 -0.45		0.5 1.5	+0.6 0	-0.5 -0.9		0.5 2.0	+0.9 0	-0.5 -1.1
0.40 - 0.71	0.25 0.75	+0.3 0	-0.25 -0.45	0.25 0.95	+0.4 0	-0.25 -0.55		0.6 1.7	+0.7 0	-0.6 -1.0		0.6 2.3	+1.0 0	-0.6 -1.3
0.71 - 1.19	0.3 0.95	+0.4 0	-0.3 -0.55	0.3 1.2	+0.5 0	-0.3 -0.7		0.8 2.1	+0.8 0	-0.8 -1.3		0.8 2.8	+1.2 0	-0.8 -1.6
1.19 - 1.97	0.4 1.1	+0.4 0	-0.4 -0.7	0.4 1.4	+0.6 0	-0.4 -0.8		1.0 2.6	+1.0 0	-1.0 -1.6		1.0 3.6	+1.6 0	-1.0 -2.0

Figure A-3A

Nominal Size Range Inches		Class RC5				Class RC6				Class RC7				Class RC8	
	Limits of Clearance	Standard Limits		Limits of Clearance	Standard Limits			Limits of Clearance	Standard Limits			Limits of Clearance	Standard Limits		
Over To		Hole H8	Shaft e7	Limits of Clearance	Hole H9	Shaft e8		Limits of Clearance	Hole H9	Shaft d8		Limits of Clearance	Hole H10	Shaft c9
0 - 0.12	0.6 1.6	+0.6 0	-0.6 -1.0	0.6 2.2	+1.0 0	-0.6 -1.2		1.0 2.6	+1.0 0	-1.0 -1.6		2.5 5.1	+1.6 0	-2.5 -3.5
0.12 - 0.24	0.8 2.0	+0.7 0	-0.8 -1.3	0.8 2.7	+1.2 0	-0.8 -1.5		1.2 3.1	+1.2 0	-1.2 -1.9		2.8 5.8	+1.8 0	-2.8 -4.0
0.24 - 0.40	1.0 2.5	+0.9 0	-1.0 -1.6	1.0 3.3	+1.4 0	-1.0 -1.9		1.6 3.9	+1.4 0	-1.6 -2.5		3.0 6.6	+2.2 0	-3.0 -4.4
0.40 - 0.71	1.2 2.9	+1.0 0	-1.2 -1.9	1.2 3.8	+1.6 0	-1.2 -2.2		2.0 4.6	+1.6 0	-2.0 -3.0		3.5 7.9	+2.8 0	-3.5 -5.1
0.71 - 1.19	1.6 3.6	+1.2 0	-1.6 -2.4	1.6 4.8	+2.0 0	-1.6 -2.8		2.5 5.7	+2.0 0	-2.5 -3.7		4.5 10.0	+3.5 0	-4.5 -6.5
1.19 - 1.97	2.0 4.6	+1.6 0	-2.0 -3.0	2.0 6.1	+2.5 0	-2.0 -3.6		3.0 7.1	+2.5 0	-3.0 -4.6		5.0 11.5	+4.0 0	-5.0 -7.5

Figure A-3B

American Standard Transition Locational Fits

Nominal Size Range Inches Over — To	Class LT1 Fit	Standard Limits Hole H7	Shaft js6	Class LT2 Fit	Standard Limits Hole H8	Shaft js7	Class LT3 Fit	Standard Limits Hole H7	Shaft k6
0 — 0.12	−0.10 / +0.50	+0.4 / 0	+0.10 / −0.10	−0.2 / +0.8	+0.6 / 0	+0.2 / −0.2			
0.12 — 0.24	−0.15 / −0.65	+0.5 / 0	+0.15 / −0.15	−0.25 / +0.95	+0.7 / 0	+0.25 / −0.25			
0.24 — 0.40	−0.2 / +0.5	+0.6 / 0	+0.2 / −0.2	−0.3 / +1.2	+0.9 / 0	+0.3 / −0.3	−0.5 / +0.5	+0.6 / 0	+0.5 / +0.1
0.40 — 0.71	−0.2 / +0.9	+0.7 / 0	+0.2 / −0.2	−0.35 / +1.35	+1.0 / 0	+0.35 / −0.35	−0.5 / +0.6	+0.7 / 0	+0.5 / +0.1
0.71 — 1.19	−0.25 / +1.05	+0.8 / 0	+0.25 / −0.25	−0.4 / +1.6	+1.2 / 0	+0.4 / −0.4	−0.6 / +0.7	+0.8 / 0	+0.6 / +0.1
1.19 — 1.97	−0.3 / +1.3	+1.0 / 0	+0.3 / −0.3	−0.5 / +2.1	+1.6 / 0	+0.5 / −0.5	+0.7 / +0.1	+1.0 / 0	+0.7 / +0.1

Figure A-4A

Nominal Size Range Inches Over — To	Class LT4 Fit	Standard Limits Hole H8	Shaft k7	Class LT5 Fit	Standard Limits Hole H7	Shaft n6	Class LT6 Fit	Standard Limits Hole H7	Shaft n7
0 — 0.12				−0.5 / +0.15	+0.4 / 0	+0.5 / +0.25	−0.65 / +0.15	+0.4 / 0	+0.65 / +0.25
0.12 — 0.24				−0.6 / +0.2	+0.5 / 0	+0.6 / +0.3	−0.8 / +0.2	+0.5 / 0	+0.8 / +0.3
0.24 — 0.40	−0.7 / +0.8	+0.9 / 0	+0.7 / +0.1	−0.8 / +0.2	+0.6 / 0	+0.8 / +0.4	−1.0 / +0.2	+0.6 / 0	+1.0 / +0.4
0.40 — 0.71	−0.8 / +0.9	+1.0 / 0	+0.8 / +0.1	−0.9 / +0.2	+0.7 / 0	+0.9 / +0.5	−1.2 / +0.2	+0.7 / 0	+1.2 / +0.5
0.71 — 1.19	−0.9 / +1.1	+1.2 / 0	+0.9 / +0.1	−1.1 / +0.2	+0.8 / 0	+1.1 / +0.6	−1.4 / +0.2	+0.8 / 0	+1.4 / +0.6
1.19 — 1.97	−1.1 / +1.5	+1.6 / 0	+1.1 / +0.1	−1.3 / +0.3	+1.0 / 0	+1.3 / +0.7	−1.7 / +0.3	+1.0 / 0	+1.7 / +0.7

Figure A-4B

American Standard Interference Locational Fits

Nominal Size Range Inches		Limits of Interference	Class LN1 Standard Limits		Limits of Interference	Class LN2 Standard Limits		Limits of Interference	Class LN3 Standard Limits	
Over	To		Hole H6	Shaft n5		Hole H7	Shaft p6		Hole H7	Shaft r6
0	- 0.12	0 / 0.45	+0.25 / 0	+0.45 / +0.25	0 / 0.65	+0.4 / 0	+0.63 / +0.4	0.1 / 0.75	+0.4 / 0	+0.75 / +0.5
0.12	- 0.24	0 / 0.5	+0.3 / 0	+0.5 / +0.3	0 / 0.8	+0.5 / 0	+0.8 / +0.5	0.1 / 0.9	+0.5 / 0	+0.9 / +0.6
0.24	- 0.40	0 / 0.65	+0.4 / 0	+0.65 / +0.4	0 / 1.0	+0.6 / 0	+1.0 / +0.6	0.2 / 1.2	+0.6 / 0	+1.2 / +0.8
0.40	- 0.71	0 / 0.8	+0.4 / 0	+0.8 / +0.4	0 / 1.1	+0.7 / 0	+1.1 / +0.7	0.3 / 1.4	+0.7 / 0	+1.4 / +1.0
0.71	- 1.19	0 / 1.0	+0.5 / 0	+1.0 / +0.5	0 / 1.3	+0.8 / 0	+1.3 / +0.8	0.4 / 1.7	+0.8 / 0	+1.7 / +1.2
1.19	- 1.97	0 / 1.1	+0.6 / 0	+1.1 / +0.6	0 / 1.6	+1.0 / 0	+1.6 / +1.0	0.4 / 2.0	+1.0 / 0	+2.0 / +1.4

Figure A-5

American Standard Force and Shrink Fits

Nominal Size Range Inches		Limits of Interference	Class FN 1 Standard Limits		Limits of Interference	Class FN 2 Standard Limits		Limits of Interference	Class FN 3 Standard Limits		Limits of Interference	Class FN 4 Standard Limits	
Over	To		Hole	Shaft		Hole	Shaft		Hole	Shaft		Hole	Shaft
0	- 0.12	0.05 / 0.5	+0.25 / 0	+0.5 / +0.3	0.2 / 0.85	+0.4 / 0	+0.85 / +0.6				0.3 / 0.95	+0.4 / 0	+0.95 / +0.7
0.12	- 0.24	0.1 / 0.6	+0.3 / 0	+0.6 / +0.4	0.2 / 1.0	+0.5 / 0	+1.0 / +0.7				0.4 / 1.2	+0.5 / 0	+1.2 / +0.9
0.24	- 0.40	0.1 / 0.75	+0.4 / 0	+0.75 / +0.5	0.4 / 1.4	+0.6 / 0	+1.4 / +1.0				0.6 / 1.6	+0.6 / 0	+1.6 / +1.2
0.40	- 0.56	0.1 / 0.8	+0.4 / 0	+0.8 / +0.5	0.5 / 1.6	+0.7 / 0	+1.6 / +1.2				0.7 / 1.8	+0.7 / 0	+1.8 / +1.4
0.56	- 0.71	0.2 / 0.9	+0.4 / 0	+0.9 / +0.6	0.5 / 1.6	+0.7 / 0	+1.6 / +1.2				0.7 / 1.8	+0.7 / 0	+1.8 / +1.4
0.71	- 0.95	0.2 / 1.1	+0.5 / 0	+1.1 / +0.7	0.6 / 1.9	+0.8 / 0	+1.9 / +1.4				0.8 / 2.1	+0.8 / 0	+2.1 / +1.6
0.95	- 1.19	0.3 / 1.2	+0.5 / 0	+1.2 / +0.8	0.6 / 1.9	+0.8 / 0	+1.9 / +1.4	0.8 / 2.1	+0.8 / 0	+2.1 / +1.6	1.0 / 2.3	+0.8 / 0	+2.1 / +1.8
1.19	- 1.58	0.3 / 1.3	+0.6 / 0	+1.3 / +0.9	0.8 / 2.4	+1.0 / 0	+2.4 / +1.8	1.0 / 2.6	+1.0 / 0	+2.6 / +2.0	1.5 / 3.1	+1.0 / 0	+3.1 / +2.5
1.58	- 1.97	0.4 / 1.4	+0.6 / 0	+1.4 / +1.0	0.8 / 2.4	+1.0 / 0	+2.4 / +1.8	1.2 / 2.8	+1.0 / 0	+2.8 / +2.2	1.8 / 3.4	+1.0 / 0	+3.4 / +2.8

Figure A-6

Preferred Clearance Fits — Cylindrical Fits
(Hole Basis; ANSI B4.2)

Basic Size		Loose Running			Free Running			Close Running			Sliding			Locational Clear.		
		Hole H11	Shaft c11	Fit	Hole H9	Shaft d9	Fit	Hole H8	Shaft f7	Fit	Hole H7	Shaft g6	Fit	Hole H7	Shaft h6	Fit
4	Max	4.075	3.930	0.220	4.030	3.970	0.090	4.018	3.990	0.040	4.012	3.996	0.024	4.012	4.000	0.020
	Min	4.000	3.855	0.070	4.000	3.940	0.030	4.000	3.978	0.010	4.000	3.988	0.004	4.000	3.992	0.000
5	Max	5.075	4.930	0.220	5.030	4.970	0.090	5.018	4.990	0.040	5.012	4.996	0.024	5.012	5.000	0.020
	Min	5.000	4.855	0.070	5.000	4.940	0.030	5.000	4.978	0.010	5.000	4.988	0.004	5.000	4.992	0.000
6	Max	6.075	5.930	0.220	6.030	5.970	0.090	6.018	5.990	0.040	6.012	5.996	0.024	6.012	6.000	0.020
	Min	6.000	5.885	0.070	6.000	5.940	0.030	6.000	5.978	0.010	6.000	5.988	0.004	6.000	5.992	0.000
8	Max	8.090	7.920	0.260	8.036	7.960	0.112	8.022	7.987	0.050	8.015	7.995	0.029	8.015	8.000	0.024
	Min	8.000	7.830	0.080	8.000	7.924	0.040	8.000	7.972	0.013	8.000	7.986	0.005	8.000	7.991	0.000
10	Max	10.090	9.920	0.260	10.036	9.960	0.112	10.022	9.987	0.050	10.015	9.995	0.029	10.015	10.000	0.024
	Min	10.000	9.830	0.080	10.000	9.924	0.040	10.000	9.972	0.013	10.000	9.986	0.005	10.000	9.991	0.000
12	Max	12.110	11.905	0.315	12.043	11.950	0.136	12.027	11.984	0.061	12.018	11.994	0.035	12.018	12.000	0.029
	Min	12.000	11.795	0.095	12.000	11.907	0.050	12.000	11.966	0.016	12.000	11.983	0.006	12.000	11.989	0.000
16	Max	16.110	15.905	0.315	16.043	15.950	0.136	16.027	15.984	0.061	16.018	15.994	0.035	16.018	16.000	0.029
	Min	16.000	15.795	0.095	16.000	15.907	0.050	16.000	15.966	0.016	16.000	15.983	0.006	16.000	15.989	0.000
20	Max	20.130	19.890	0.370	20.052	19.935	0.169	20.033	19.980	0.074	20.021	19.993	0.041	20.021	20.000	0.034
	Min	20.000	19.760	0.110	20.000	19.883	0.065	20.000	19.959	0.020	20.000	19.980	0.007	20.000	19.987	0.000
25	Max	25.130	24.890	0.370	25.052	24.935	0.169	25.033	24.980	0.074	25.021	24.993	0.041	25.021	25.000	0.034
	Min	25.000	24.760	0.110	25.000	24.883	0.065	25.000	24.959	0.020	25.000	24.980	0.007	25.000	24.987	0.000
30	Max	30.130	29.890	0.370	30.052	29.935	0.169	30.033	29.980	0.074	30.021	29.993	0.041	30.021	30.000	0.034
	Min	30.000	29.760	0.110	30.000	29.883	0.065	30.000	29.959	0.020	30.000	29.980	0.007	30.000	29.987	0.000

Figure A-7

Preferred Transition and Interference Fits — Cylindrical Fits
(Hole Basis; ANSI B4.2)

Basic Size		Locational Trans.			Locational Trans.			Locational Inter.			Medium Drive			Force		
		Hole H7	Shaft k6	Fit	Hole H7	Shaft n6	Fit	Hole H7	Shaft p6	Fit	Hole H7	Shaft s6	Fit	Hole H7	Shaft u6	Fit
4	Max	4.012	4.009	0.011	4.012	4.016	0.004	4.012	4.020	0.000	4.012	4.027	-0.007	4.012	4.031	-0.011
	Min	4.000	4.001	-0.009	4.000	4.008	-0.016	4.000	4.012	-0.020	4.000	4.019	-0.027	4.000	4.023	-0.031
5	Max	5.012	5.009	0.011	5.012	5.016	0.004	5.012	5.020	0.000	5.012	5.027	-0.007	5.012	5.031	-0.011
	Min	5.000	5.001	-0.009	5.000	5.008	-0.016	5.000	5.012	-0.020	5.000	5.019	-0.027	5.000	5.023	-0.031
6	Max	6.012	6.009	0.011	6.012	6.016	0.004	6.012	6.020	0.000	6.012	6.027	-0.007	6.012	6.031	-0.011
	Min	6.000	6.001	-0.009	6.000	6.008	-0.016	6.000	6.012	-0.020	6.000	6.019	-0.027	6.000	6.023	-0.031
8	Max	8.015	8.010	0.014	8.015	8.019	0.005	8.015	8.024	0.000	8.015	8.032	-0.008	8.015	8.037	-0.013
	Min	8.000	8.001	-0.010	8.000	8.010	-0.019	8.000	8.015	-0.024	8.000	8.023	-0.032	8.000	8.028	-0.037
10	Max	10.015	10.010	0.014	10.015	10.019	0.005	10.015	10.024	0.000	10.015	10.032	-0.008	10.015	10.037	-0.013
	Min	10.000	10.001	-0.010	10.000	10.010	-0.019	10.000	10.015	-0.024	10.000	10.023	-0.032	10.000	10.028	-0.037
12	Max	12.018	12.012	0.017	12.018	12.023	0.006	12.018	12.029	0.000	12.018	12.039	-0.010	12.018	12.044	-0.015
	Min	12.000	12.001	-0.012	12.000	12.012	-0.023	12.000	12.018	-0.029	12.000	12.028	-0.039	12.000	12.033	-0.044
16	Max	16.018	16.012	0.017	16.018	16.023	0.006	16.018	16.029	0.000	16.018	16.039	-0.010	16.018	16.044	-0.015
	Min	16.000	16.001	-0.012	16.000	16.012	-0.023	16.000	16.018	-0.029	16.000	16.028	-0.039	16.000	16.033	-0.044
20	Max	20.021	20.015	0.019	20.021	20.028	0.006	20.021	20.035	-0.001	20.021	20.048	-0.014	20.021	20.054	-0.020
	Min	20.000	20.002	-0.015	20.000	20.015	-0.028	20.000	20.022	-0.035	20.000	20.035	-0.048	20.000	20.041	-0.054
25	Max	25.021	25.015	0.019	25.021	25.028	0.006	25.021	25.035	-0.001	25.021	25.048	-0.014	25.021	25.061	-0.027
	Min	25.000	25.002	-0.015	25.000	25.015	-0.028	25.000	25.022	-0.035	25.000	25.035	-0.048	25.000	25.048	-0.061
30	Max	30.021	30.015	0.019	30.021	30.028	0.006	30.021	30.035	-0.001	30.021	30.048	-0.014	30.021	30.061	-0.027
	Min	30.000	30.002	-0.015	30.000	30.015	-0.028	30.000	30.022	-0.035	30.000	30.035	-0.048	30.000	30.048	-0.061

Figure A-8

Preferred Clearance Fits — Cylindrical Fits
(Shaft Basis; ANSI B4.2)

Basic Size		Loose Running			Free Running			Close Running			Sliding			Locational Clear.		
		Hole C11	Shaft h11	Fit	Hole D9	Shaft h9	Fit	Hole F8	Shaft h7	Fit	Hole G7	Shaft h6	Fit	Hole H7	Shaft h6	Fit
4	Max	4.145	4.000	0.220	4.060	4.000	0.090	4.028	4.000	0.040	4.016	4.000	0.024	4.012	4.000	0.020
	Min	4.070	3.925	0.070	4.030	3.970	0.030	4.010	3.988	0.010	4.004	3.992	0.004	4.000	3.992	0.000
5	Max	5.145	5.000	0.220	5.060	5.000	0.090	5.028	5.000	0.040	5.016	5.000	0.024	5.012	5.000	0.020
	Min	5.070	4.925	0.070	5.030	4.970	0.030	5.010	4.988	0.010	5.004	4.992	0.004	5.000	4.992	0.000
6	Max	6.145	6.000	0.220	6.060	6.000	0.090	6.028	6.000	0.040	6.016	6.000	0.024	6.012	6.000	0.020
	Min	6.070	5.925	0.070	6.030	5.970	0.030	6.010	5.988	0.010	6.004	5.992	0.004	6.000	5.992	0.000
8	Max	8.170	8.000	0.260	8.076	8.000	0.112	8.035	8.000	0.050	8.020	8.000	0.029	8.015	8.000	0.024
	Min	8.080	7.910	0.080	8.040	7.964	0.040	8.013	7.985	0.013	8.005	7.991	0.005	8.000	7.991	0.000
10	Max	10.170	10.000	0.260	10.076	10.000	0.112	10.035	10.000	0.050	10.020	10.000	0.029	10.015	10.000	0.024
	Min	10.080	9.910	0.080	10.040	9.964	0.040	10.013	9.985	0.013	10.005	9.991	0.005	10.000	9.991	0.000
12	Max	12.205	12.000	0.315	12.093	12.000	0.136	12.043	12.000	0.061	12.024	12.000	0.035	12.018	12.000	0.029
	Min	12.095	11.890	0.095	12.050	11.957	0.050	12.016	11.982	0.016	12.006	11.989	0.006	12.000	11.989	0.000
16	Max	16.205	16.000	0.315	16.093	16.000	0.136	16.043	16.000	0.061	16.024	16.000	0.035	16.018	16.000	0.029
	Min	16.095	15.890	0.095	16.050	15.957	0.050	16.016	15.982	0.016	06.006	15.989	0.006	16.000	15.989	0.000
20	Max	20.240	20.000	0.370	20.117	20.000	0.169	20.053	20.000	0.074	20.028	20.000	0.041	20.021	20.000	0.034
	Min	20.110	19.870	0.110	20.065	19.948	0.065	20.020	19.979	0.020	20.007	19.987	0.007	20.000	19.987	0.000
25	Max	25.240	25.000	0.370	25.117	25.000	0.169	25.053	25.000	0.074	25.028	25.000	0.041	25.021	25.000	0.034
	Min	25.110	24.870	0.110	25.065	24.948	0.065	25.020	24.979	0.020	25.007	24.987	0.007	25.000	24.987	0.000
30	Max	30.240	30.000	0.370	30.117	30.000	0.169	30.053	30.000	0.074	30.028	30.000	0.041	30.021	30.000	0.034
	Min	30.110	29.870	0.110	30.065	29.948	0.065	30.020	29.979	0.020	30.007	29.987	0.007	30.000	29.987	0.000

Figure A-9

Preferred Transition and Interference Fits — Cylindrical Fits
(Shaft Basis; ANSI B4.2)

Basic Size		Locational Trans.			Locational Trans.			Locational Inter.			Medium Drive			Force		
		Hole K7	Shaft h6	Fit	Hole N7	Shaft h6	Fit	Hole F7	Shaft h6	Fit	Hole S7	Shaft h6	Fit	Hole U7	Shaft h6	Fit
4	Max	4.003	4.000	0.011	3.996	4.000	0.004	3.992	4.000	0.000	3.985	4.000	-0.007	3.981	4.000	-0.011
	Min	3.991	3.992	-0.009	3.984	3.992	-0.016	3.980	3.992	-0.020	3.973	3.992	-0.027	3.969	3.992	-0.031
5	Max	5.003	5.000	0.011	4.996	5.000	0.004	4.992	5.000	0.000	4.985	5.000	-0.007	4.981	5.000	-0.011
	Min	4.991	4.992	-0.009	4.984	4.992	-0.016	4.980	4.992	-0.020	4.973	4.992	-0.027	4.969	4.992	-0.031
6	Max	6.003	6.000	0.011	5.996	6.000	0.004	5.992	6.000	0.000	5.985	6.000	-0.007	5.981	6.000	-0.011
	Min	5.991	5.992	-0.009	5.984	5.992	-0.016	5.980	5.992	-0.020	5.973	5.992	-0.027	5.969	5.992	-0.031
8	Max	8.005	8.000	0.014	7.996	8.000	0.005	7.991	8.000	0.000	7.983	8.000	-0.008	7.978	8.000	-0.013
	Min	7.990	7.991	-0.010	7.981	7.991	-0.019	7.976	7.991	-0.024	7.968	7.991	-0.032	7.963	7.991	-0.037
10	Max	10.005	10.000	0.014	9.996	10.000	0.005	9.991	10.000	0.000	9.983	10.000	-0.008	9.978	10.000	-0.013
	Min	9.990	9.991	-0.010	9.981	9.991	-0.019	9.976	9.991	-0.024	9.968	9.991	-0.032	9.963	9.991	-0.037
12	Max	12.006	12.000	0.017	11.995	12.000	0.006	11.989	12.000	0.000	11.979	12.000	-0.010	11.974	12.000	-0.015
	Min	11.988	11.989	-0.012	11.977	11.989	-0.023	11.971	11.989	-0.029	11.961	11.989	-0.039	11.956	11.989	-0.044
16	Max	16.006	16.000	0.017	15.995	16.000	0.006	15.989	16.000	0.000	15.979	16.000	-0.010	15.974	16.000	-0.015
	Min	15.988	15.989	-0.012	15.977	15.989	-0.023	15.971	15.989	-0.029	15.961	15.989	-0.039	15.956	15.989	-0.044
20	Max	20.006	20.000	0.019	19.993	20.000	0.006	19.986	20.000	-0.001	19.973	20.000	-0.014	19.967	20.000	-0.020
	Min	19.985	19.987	-0.015	19.972	19.987	-0.028	19.965	19.987	-0.035	19.952	19.987	-0.048	19.946	19.987	-0.054
25	Max	25.006	25.000	0.019	24.993	25.000	0.006	24.986	25.000	-0.001	24.973	25.000	-0.014	24.960	25.000	-0.027
	Min	24.985	24.987	-0.015	24.972	24.987	-0.028	24.965	24.987	-0.035	24.952	24.987	-0.048	24.939	24.987	-0.061
30	Max	30.006	30.000	0.019	29.993	30.000	0.006	29.986	30.000	-0.001	29.973	30.000	-0.014	29.960	30.000	-0.027
	Min	29.985	29.987	-0.015	29.972	29.987	-0.028	29.987	29.987	-0.035	29.952	29.987	-0.048	29.939	29.987	-0.061

Figure A-10

American National Standard Type A Plain Washers
(ANSI B18.22.1-1965, R1975)

Nominal Washer Size		Series	Inside Diameter			Outside Diameter			Thickness		
			Basic	Tolerance		Basic	Tolerance		Basic	Max.	Min.
				Plus	Minus		Plus	Minus			
#6	.138		.156	.008	.005	.375	.015	.005	.049	.065	.036
#8	.164		.188	.008	.005	.438	.015	.005	.049	.065	.036
#10	.190		.219	.008	.005	.500	.015	.005	.049	.065	.036
1/4	.250	N	.281	.015	.005	.625	.015	.005	.065	.080	.051
1/4	.250	W	.312	.015	.005	.734	.015	.007	.065	.080	.051
5/16	.312	N	.344	.015	.005	.688	.015	.007	.065	.080	.051
5/16	.312	W	.375	.015	.005	.875	.030	.007	.083	.104	.064
3/8	.375	N	.406	.015	.005	.812	.015	.007	.065	.080	.051
3/8	.375	W	.438	.015	.005	1.000	.030	.007	.083	.104	.064
7/16	.438	N	.469	.015	.005	.922	.015	.007	.065	.080	.051
7/16	.438	W	.500	.015	.005	1.250	.030	.007	.083	.104	.064
1/2	.500	N	.531	.015	.005	1.062	.030	.007	.095	.121	.074
1/2	.500	W	.562	.015	.005	1.375	.030	.007	.109	.132	.086
9/16	.562	N	.594	.015	.005	1.156	.030	.007	.095	.121	.074
9/16	.562	W	.625	.015	.005	1.469	.030	.007	.109	.132	.086
5/8	.625	N	.656	.030	.007	1.312	.030	.007	.095	.121	.074
5/8	.625	W	.688	.030	.007	1.750	.030	.007	.134	.160	.108
3/4	.750	N	.812	.030	.007	1.469	.030	.007	.134	.160	.108
3/4	.750	W	.812	.030	.007	2.000	.030	.007	.148	.177	.122
7/8	.875	N	.938	.030	.007	1.750	.030	.007	.134	.160	.108
7/8	.875	W	.938	.030	.007	2.250	.030	.007	.165	.192	.136
1	1.000	N	1.062	.030	.007	2.000	.030	.007	.134	.160	.108
1	1.000	W	1.062	.030	.007	2.500	.030	.007	.165	.192	.136
1 1/8	1.125	N	1.250	.030	.007	2.250	.030	.007	.134	.160	.108
1 1/8	1.125	W	1.250	.030	.007	2.750	.030	.007	.165	.192	.136
1 1/4	1.250	N	1.375	.030	.007	2.500	.030	.007	.165	.192	.136
1 1/4	1.250	W	1.375	.030	.007	3.000	.030	.007	.165	.192	.136
1 3/8	1.375	N	1.500	.030	.007	2.750	.030	.007	.165	.192	.136
1 3/8	1.375	W	1.500	.045	.010	3.250	.045	.010	.180	.213	.153
1 1/2	1.500	N	1.625	.030	.007	3.000	.030	.007	.165	.192	.136
1 1/2	1.500	W	1.625	.045	.010	3.500	.045	.010	.180	.213	.153
1 5/8	1.625		1.750	.045	.010	3.750	.045	.010	.180	.213	.153
1 3/4	1.750		1.875	.045	.010	4.000	.045	.010	.180	.213	.153
1 7/8	1.875		2.000	.045	.010	4.250	.045	.010	.180	.213	.153
2	2.000		2.125	.045	.010	4.500	.045	.010	.180	.213	.153
2 1/4	2.250		2.375	.045	.010	4.750	.045	.010	.220	.248	.193
2 1/2	2.500		2.625	.045	.010	5.000	.045	.010	.238	.280	.210
2 3/4	2.750		2.875	.065	.010	5.250	.065	.010	.259	.310	.228
3	3.000		3.125	.065	.010	5.500	.065	.010	.284	.327	.249

Figure A-11

American National Standard Helical Spring Lock Washers (ANSI B18.21.1-1972)

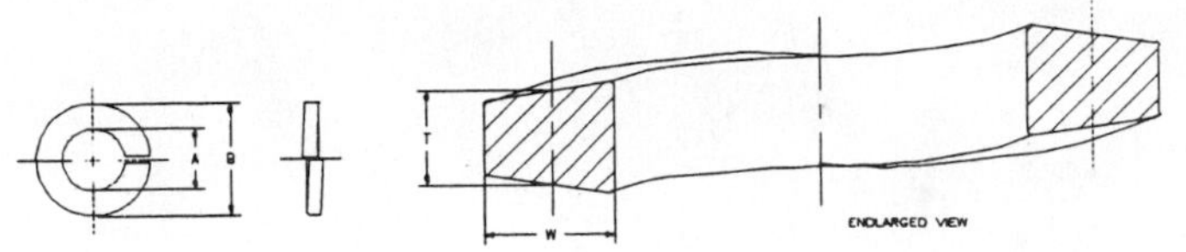

Nominal Washer Size		Inside Diameter, A		Regular			Heavy			Extra Duty		
		Max	Min	O.D., B Max	Section Width	Section Thickness	O.D., B Max	Section Width	Section Thickness	O.D., B Max	Section Width	Section Thickness
#2	.086	.094	.088	.172	.035	.020	.182	.040	.025	.208	.053	.027
#3	.099	.107	.101	.195	.040	.025	.209	.047	.031	.239	.062	.034
#4	.112	.120	.114	.209	.040	.025	.223	.047	.031	.253	.062	.034
#5	.125	.133	.127	.236	.047	.031	.253	.055	.040	.300	.079	.045
#6	.138	.148	.141	.250	.047	.031	.266	.055	.040	.314	.079	.045
#8	.164	.174	.167	.293	.055	.040	.307	.062	.047	.375	.096	.057
#10	.190	.200	.193	.334	.062	.047	.350	.070	.056	.434	.112	.068
#12	.216	.227	.220	.377	.070	.056	.391	.077	.063	.497	.130	.080
1/4	.250	.262	.254	.489	.109	.062	.491	.110	.077	.535	.132	.084
5/16	.312	.326	.317	.586	.125	.078	.596	.130	.097	.622	.143	.108
3/8	.375	.390	.380	.683	.141	.094	.691	.145	.115	.741	.170	.123
7/16	.438	.455	.443	.779	.156	.109	.787	.160	.133	.839	.186	.143
1/2	.500	.518	.506	.873	.171	.125	.883	.176	.151	.939	.204	.162
9/16	.562	.582	.570	.971	.188	.141	.981	.193	.170	1.041	.223	.182
5/8	.625	.650	.635	1.079	.203	.156	1.093	.210	.189	1.157	.242	.202
11/16	.688	.713	.698	1.176	.219	.172	1.192	.227	.207	1.258	.260	.221
3/4	.750	.775	.760	1.271	.234	.188	1.291	.244	.226	1.361	.279	.241
13/16	.812	.843	.824	1.367	.250	.203	1.391	.262	.246	1.463	.298	.261
7/8	.875	.905	.887	1.464	.266	.219	1.494	.281	.266	1.576	.322	.285
15/16	.938	.970	.950	1.560	.281	.234	1.594	.298	.284	1.688	.345	.308
1	1.000	1.042	1.017	1.661	.297	.250	1.705	.319	.306	1.799	.366	.330
1 1/16	1.062	1.107	1.080	1.756	.312	.266	1.808	.338	.326	1.910	.389	.352
1 1/8	1.125	1.172	1.144	1.853	.328	.281	1.909	.356	.345	2.019	.411	.375
1 3/16	1.188	1.237	1.208	1.950	.344	.297	2.008	.373	.364	2.124	.341	.396
1 1/4	1.250	1.302	1.271	2.045	.359	.312	2.113	.393	.384	2.231	.452	.417
1 5/16	1.312	1.366	1.334	2.141	.375	.328	2.211	.410	.403	2.335	.472	.438
1 3/8	1.375	1.432	1.398	2.239	.391	.344	2.311	.427	.422	2.439	.491	.458
1 7/16	1.438	1.497	1.462	2.334	.406	.359	2.406	.442	.440	2.540	.509	.478
1 1/2	1.500	1.561	1.525	2.430	.422	.375	2.502	.458	.458	2.638	.526	.496

Figure A-12

American National Standard Internal-External Tooth Lock Washers (ANSI B18.21.1-1972)

Size	A Inside Diameter Max.	Min.	B Outside Diameter Max.	Min.	C Thickness Max.	Min.	Size	A Inside Diameter Max.	Min.	B Outside Diameter Max.	Min.	C Thickness Max.	Min.
#4	.123	.115	.475	.460	.021	.021	5/16	.332	.320	.900	.865	.040	.032
	.123	.115	.510	.495	.021	.017		.332	.320	.985	.965	.045	.037
			.610	.580				.332	.320	1.070	1.045	.050	.042
										1.155	1.130		
#6	.150	.141	.510	.495	.028	.023	3/8	.398	.384	.985	.965	.045	.037
			.610	.580				.398	.384	1.070	1.045	.050	.042
			.690	.670						1.155	1.130		
										1.260	1.220		
#8	.176	.168	.610	.580	.034	.028	7/16	.464	.448	1.070	1.045	.050	.042
			.690	.670						1.155	1.130		
			.760	.740				.464	.448	1.260	1.220	.055	.047
										1.315	1.290		
#10	.204	.195	.610	.580	.034	.028	1/2	.530	.512	1.260	1.220	.055	.047
	.204	.195	.690	.670	.040	.032				1.315	1.290		
			.760	.740				.530	.512	1.410	1.380	.060	.052
			.900	.880				.530	.512	1.620	1.590	.067	.059
#12	.231	.221	.690	.670	.040	.032	9/16	.596	.576	1.315	1.290	.055	.047
	.231	.221	.760	.725				.596	.576	1.430	1.380	.060	.052
			.900	.880				.596	.576	1.620	1.590	.067	.059
			.985	.965	.045	.037				1.830	1.797		
1/4	.267	.256	.760	.725	.040	.032	5/8	.663	.640	1.410	1.380	.060	.052
	.267	.256	.900	.880				.663	.640	1.620	1.590	.067	.059
			.985	.965	.045	.037				1.830	1.797		
			1.070	1.045						1.975	1.935		

Figure A-13

British Standard Bright Metal Washers - Metric Series (BS 4320:1968)

NORMAL DIAMETER SIZES											
Nominal Size of Bolt or Screw	Inside Diameter			Outside Diameter			Thickness					
							Form A (Normal Range)			Form B (Light Range)		
	Nom.	Max.	Min.	Nom.	Max.	Min.	Nom.	Max.	Min.	Nom.	Max.	Min.
M 1.0	1.1	1.25	1.1	2.5	2.5	2.3	.3	.4	.2			
M 1.2	1.3	1.45	1.3	3.0	3.0	2.8	.3	.4	.2			
M 1.4	1.5	1.65	1.5	3.0	3.0	2.8	.3	.4	.2			
M 1.6	1.7	1.85	1.7	4.0	4.0	3.7	.3	.4	.2			
M 2.0	2.2	2.35	2.2	5.0	5.0	4.7	.3	.4	.2			
M 2.2	2.4	2.55	2.4	5.0	5.0	4.7	.5	.6	.4			
M 2.5	2.7	2.85	2.7	6.5	6.5	6.2	.5	.6	.4			
M 3	3.2	3.4	3.2	7	7	6.7	.5	.6	.4			
M 3.5	3.7	3.9	3.7	7	7	6.7	.5	.6	.4			
M 4	4.3	4.5	4.3	9	9	8.7	.8	.9	.7			
M 4.5	4.8	5.0	4.8	9	9	8.7	.8	.9	.7			
M 5	5.3	5.5	5.3	10	10	9.7	1.0	1.1	.9			
M 6	6.4	6.7	6.4	12.5	12.5	12.1	1.6	1.8	1.4	.8	.9	.7
M 7	7.4	7.7	7.4	14	14	13.6	1.6	1.8	1.4	.8	.9	.7
M 8	8.4	8.7	8.4	17	17	16.6	1.6	1.8	1.4	1.0	1.1	.9
M 10	10.5	10.9	10.5	21	21	20.5	2.0	2.2	1.8	1.25	1.45	1.05
M 12	13.0	13.4	13.0	24	24	23.5	2.5	2.7	2.3	1.6	1.80	1.40
M 14	15.0	15.4	15.0	28	28	27.5	2.5	2.7	2.3	1.6	1.8	1.4
M 16	17.0	17.4	17.0	30	30	29.5	3.0	3.3	2.7	2.0	2.2	1.8
M 18	19.0	19.5	19.0	34	34	33.2	3.0	3.3	2.7	2.0	2.2	1.8
M 20	21	21.5	21	37	37	36.2	3.0	3.3	2.7	2.0	2.2	1.8
M 22	23	23.5	23	39	39	38.2	3.0	3.3	2.7	2.0	2.2	1.8
M 24	25	25.5	25	44	44	43.2	4.0	4.3	3.7	2.5	2.7	2.3
M 27	28	28.5	28	50	50	49.2	4.0	4.3	3.7	2.5	2.7	2.3
M 30	31	31.6	31	56	56	55.0	4.0	4.3	3.7	2.5	2.7	2.3
M 33	34	34.6	34	60	60	59.0	5.0	5.6	4.4	3.0	3.3	2.7
M 36	37	37..6	37	66	66	65.0	5.0	5.6	4.4	3.0	3.3	2.7
M 39	40	40.6	40	72	72	71.0	6.0	6.6	5.4	3.0	3.3	2.7

Figure A-14

American National Standard and Unified Standard Square Bolts (ANSI B18.2.1-1972)

SQUARE BOLTS

Nominal Size or Basic Product Diameter		Body Diam., E	Width Across Flats, F			Width Across Corners, G		Height, H			Radius of Fillet, R
		Max.	Basic	Max.	Min.	Max.	Min.	Basic	Max.	Min.	Max.
1/4	.2500	.260	3/8	.375	.362	.530	.498	11/64	.188	.156	.03
5/16	.3125	.324	1/2	.500	.484	.707	.665	13/64	.220	.186	.03
3/8	.3750	.388	9/16	.562	.544	.795	.747	1/4	.268	.232	.03
7/16	.4375	.452	5/8	.625	.603	.884	.828	19/64	.316	.278	.03
1/2	.5000	.515	3/4	.750	.725	1.061	.995	21/64	.348	.308	.03
5/8	.6250	.642	15/16	.938	.906	1.326	1.244	37/64	.444	.400	.06
3/4	.7500	.768	1 1/8	1.125	1.088	1.591	1.494	1/2	.524	.476	.06
7/8	.8750	.895	1 5/16	1.312	1.269	1.856	1.742	19/32	.620	.568	.06
1	1.0000	1.022	1 1/2	1.500	1.450	2.121	1.991	21/32	.684	.628	.09
1 1/8	1.1250	1.149	1 11/16	1.688	1.631	2.386	2.239	3/4	.780	.720	.09
1 1/4	1.2500	1.277	1 7/8	1.875	1.812	2.652	2.489	27/32	.876	.812	.09
1 3/8	1.3750	1.404	2 1/16	2.062	1.994	2.917	2.738	29/32	.940	.872	.09
1 1/2	1.5000	1.531	2 1/4	2.250	2.175	3.182	2.986	1	1.036	.964	.09

Figure A-15

American National Standard and Unified Standard Hex Head Screws
(ANSI B18.2.1-1972)

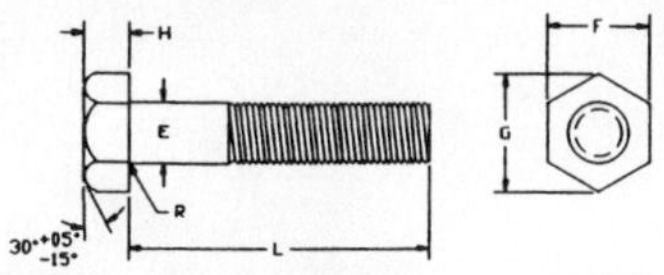

Nominal Size or Basic Diam.	Body Diam., E	Width Across Flats, F			Width Across Corners, G		Height, H			Radius of Fillet, R	
	Max.	Basic	Max.	Min.	Max.	Min.	Basic	Max.	Min.	Max.	Min.
HEX BOLTS											
1/4	.2500	.260	7/16	.438	.425	.505	.484	11/64	.188	.150	.03	.01
5/16	.3125	.324	1/2	.500	.484	.577	.552	7/32	.235	.195	.03	.01
3/8	.3750	.388	9/16	.562	.544	.650	.620	1/4	.268	.226	.03	.01
7/16	.4375	.452	5/8	.625	.603	.722	.687	19/64	.316	.272	.03	.01
1/2	.5000	.515	3/4	.750	.725	.866	.826	11/32	.364	.302	.03	.01
5/8	.6250	.642	15/16	.938	.906	1.083	1.033	27/64	.444	.378	.06	.02
3/4	.7500	.768	1 1/8	1.125	1.088	1.299	1.240	1/2	.524	.455	.06	.02
7/8	.8750	.895	1 5/16	1.312	1.269	1.516	1.447	37/64	.604	.531	.06	.02
1	1.0000	1.022	1 1/2	1.500	1.450	1.732	1.653	43/64	.700	.591	.09	.03
1 1/8	1.1250	1.149	1 11/16	1.688	1.631	1.949	1.859	3/4	.780	.658	.09	.03
1 1/4	1.2500	1.277	1 7/8	1.875	1.812	2.165	2.066	27/32	.876	.749	.09	.03
1 3/8	1.3750	1.404	2 1/16	2.062	1.994	2.382	2.273	29/32	.940	.810	.09	.03
1 1/2	1.5000	1.531	2 1/4	2.250	2.175	2.598	2.480	1	1.036	.902	.09	.03
1 3/4	1.7500	1.785	2 5/8	2.625	2.538	3.031	2.893	1 5/32	1.196	1.054	.12	.04
2	2.0000	2.039	3	3.000	2.900	3.464	3.306	1 11/32	1.388	1.175	.12	.04
2 1/4	2.2500	2.305	3 3/8	3.375	3.262	3.897	3.719	1 1/2	1.548	1.327	.19	.06
2 1/2	2.5000	2.559	3 3/4	3.750	3.625	4.330	4.133	1 21/32	1.708	1.479	.19	.06
2 3/4	2.7500	2.827	4 1/8	4.125	3.988	4.763	4.546	1 13/16	1.869	.1632	.19	.06
3	3.0000	3.081	4 1/2	4.500	4.350	5.196	4.959	2	2.060	1.815	.19	.06
3 1/4	3.2500	3.335	4 7/8	4.875	4.712	5.629	5.372	2 3/16	2.251	1.936	.19	.06
3 1/2	3.5000	3.589	5 1/4	5.250	5.075	6.062	5.786	2 5/16	2.380	2.057	.19	.06
3 3/4	3.7500	3.858	5 5/8	5.625	5.437	6.495	6.198	2 1/2	2.572	2.241	.19	.06
4	4.0000	4.111	6	6.000	5.800	6.982	6.612	2 11/16	2.764	2.424	.19	.06

Figure A-16

Coarse-Thread Series, UNC, UNRC, and NC — Basic Dimensions

Sizes	Basic Major Diam., D	Thds. per Inch, n	Basic Pitch Diam., E	Minor Diameter		Lead Angle at Basic P.D.		Area of Minor Diam. at D-2h	Tensile Stress Area
				Ext. Thds., Ks	Int. Thds., Kn	Deg.	Min.		
	Inches		Inches	Inches	Inches			Sq. In.	Sq. In.
1 (.073)	.0730	64	.0629	.0538	.0561	4	31	.00218	.00263
2 (.086)	.0860	56	.0744	.0641	.0667	4	22	.00310	.00370
3 (.099)	.0990	48	.0855	.0734	.0764	4	26	.00406	.00487
4 (.112)	.1120	40	.0958	.0813	.0849	4	45	.00496	.00604
5 (.125)	.1250	40	.1088	.0943	.0979	4	11	.00672	.00796
6 (.138)	.1380	32	.1177	.0997	.1042	4	50	.00745	.00909
8 (.164)	1.640	32	.1437	.1257	.1302	3	58	.01196	.0140
10 (.190)	.1900	24	.1629	.1389	.1449	4	39	.01450	.0175
12 (.216)	.2160	24	.1889	.1649	.1709	4	1	.0206	.0242
1/4	.2500	20	.2175	.1887	.1959	4	11	.0269	.0318
5/16	.3125	18	.2764	.2443	.2524	3	40	.0454	.0524
3/8	.3750	16	.3344	.2983	.3073	3	24	.0678	.0775
7/16	.4375	14	.3911	.3499	.3602	3	20	.0933	.1063
1/2	.5000	13	.4500	.4056	.4167	3	7	.1257	.1419
9/16	.5625	12	.5084	.4603	.4723	2	59	.162	.182
5/8	.6250	11	.5660	.5135	.5266	2	56	.202	.226
3/4	.7500	10	.6850	.6273	.6417	2	40	.302	.334
7/8	.8750	9	.8028	.7387	.7547	2	31	.419	.462
1	1.0000	8	.9188	.8466	.8647	2	29	.551	.606
1 1/8	1.1250	7	1.032	.9497	.9704	2	31	.693	.763
1 1/4	1.2500	7	1.572	1.0747	1.0954	2	15	.890	.969
1 3/8	1.3750	6	1.2667	1.1705	1.1946	2	24	1.054	1.155
1 1/2	1.5000	6	1.3917	1.2955	1.3196	2	11	1.294	1.405

Figure A-17

Fine-Thread Series, UNC, UNRC, and NC — Basic Dimensions

Sizes	Basic Major Diam., D	Thds. per Inch, n	Basic Pitch Diam., E	Minor Diameter		Lead Angle at Basic P.D.		Area of Minor Diam. at D-2h	Tensile Stress Area
				Ext. Thds., Ks	Int. Thds., Kn	Deg.	Min.		
	Inches		Inches	Inches	Inches	Deg.	Min.	Sq. In.	Sq. In.
1 (.073)	.0730	72	.0640	.0560	.0580	3	57	.00237	.00278
2 (.086)	.860	64	.0759	.0668	.0691	3	45	.00339	.00394
3 (.099)	.990	56	.0874	.0771	.0797	3	43	.00451	.00523
4 (.112)	.1120	48	.0985	.0864	.0894	3	51	.00566	.00661
5 (.125)	.1250	44	.1102	.0971	.1004	3	45	.00716	.00830
6 (.138)	.1380	40	.1218	.1073	.1109	3	44	.00874	.01015
8 (.164)	.1640	36	.1460	.1299	.1339	3	28	.01285	.01474
10 (.190)	.1900	32	.1697	.1517	.1562	3	21	.0175	.0200
12 (.216)	.2160	28	.1928	.1722	.1773	3	22	.0226	.0258
1/4	.2500	28	.2268	.2062	.2113	2	52	.0326	.0364
5/16	.3125	24	.2854	.2614	.2674	2	40	.0524	.0580
3/8	.3750	24	.3479	.3239	.3299	2	11	.0809	.0878
7/16	.4375	20	.4050	.3762	.3834	2	15	.1090	.1187
1/2	.5000	20	.4675	.4387	.4459	1	57	.1486	.1599
9/16	.5625	18	.5264	.4943	.5024	1	55	.189	.203
5/8	.6250	18	.5889	.5568	.5649	1	43	.240	.256
3/4	.7500	16	.7094	.6733	.6823	1	36	.351	.373
7/8	.8750	14	.8286	.7874	.7977	1	34	.480	.509
1	1.0000	12	.9459	.8978	.9098	1	36	.625	.663
1 1/8	1.1250	12	1.0709	1.0228	1.0348	1	25	.812	.856
1 1/4	1.2500	12	1.1959	1.1478	1.1598	1	16	1.024	1.073
1 3/8	1.3750	12	1.3209	1.2728	1.2848	1	9	1.260	1.315
1 1/2	1.5000	12	1.4459	1.3978	1.4098	1	3	1.521	1.581

Figure A-18

American National Standard General-Purpose Acme Screw Thread Form—
Basic Dimensions (ANSI B1.5-1977)

Thds. per Inch	Pitch	Height of Thread (Basic)	Total Height of Thread	Thread Thickness (Basic)	Width of Flat	
					Crest of Internal Thread (Basic)	Root of Internal Thread
16	.06250	.03125	.0362	.03125	.0232	.0206
14	.07143	.03571	.0407	.03571	.0265	.0239
12	.08333	.04167	.0467	.04167	.0309	.0283
10	.10000	.05000	.0600	.05000	.0371	.0319
8	.12500	.06250	.0725	.06250	.0463	.0411
6	.16667	.08333	.0933	.08333	.0618	.0566
5	.20000	.10000	.1100	.10000	.0741	.0689
4	.25000	.12500	.1350	.12500	.0927	.0875
3	.33333	.16667	.1767	.16667	.1236	.1184
2 1/2	.40000	.20000	.2100	.20000	.1483	.1431
2	.50000	.25000	.2600	.25000	.1853	.1802
1 1/2	.66667	.33333	.3433	.33333	.2471	.2419
1 1/3	.75000	.37500	.3850	.37500	.2780	.2728
1	1.0000	.50000	.5100	.50000	.3707	.3655

Figure A-19

60°   Stub Threads

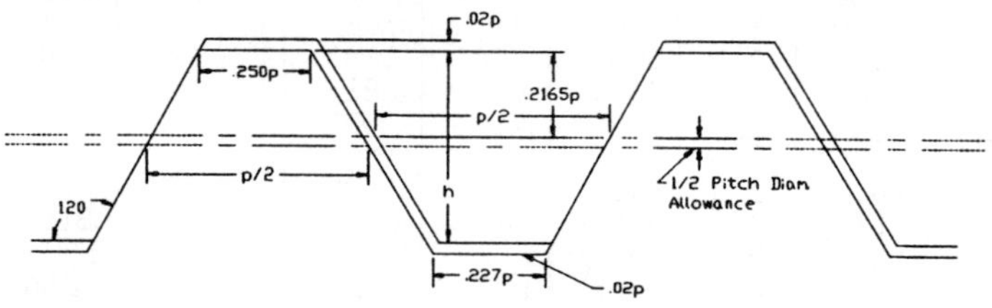

Threads per Inch	Pitch, Inch	Depth of Thread (Basic)	Total Depth of Thread	Thickness (Basic)	Width of Flat at Crest of Screw (Basic)	Width of Flat at Root of Screw
16	.06250	.0271	.0283	.0313	.0156	.0142
14	.07143	.0309	.0324	.0357	.0179	.0162
12	.08333	.0361	.0378	.0417	.0208	.0189
10	.10000	.0433	.0453	.0500	.0250	.0227
9	.11111	.0481	.0503	.0556	.0278	.0252
8	.12500	.0541	.0566	.0625	.0313	.0284
7	.14286	.0619	.0648	.0714	.0357	.0324
6	.16667	.0722	.0755	.0833	.0417	.0378
5	.20000	.0866	.0906	.1000	.0500	.0454
4	.25000	.1083	.1133	.1250	.0625	.0567

Figure A-20

American National Standard Slotted 100° Flat Countersunk
Head Machine Screws (ANSI B18.6.3-1972, R1977)

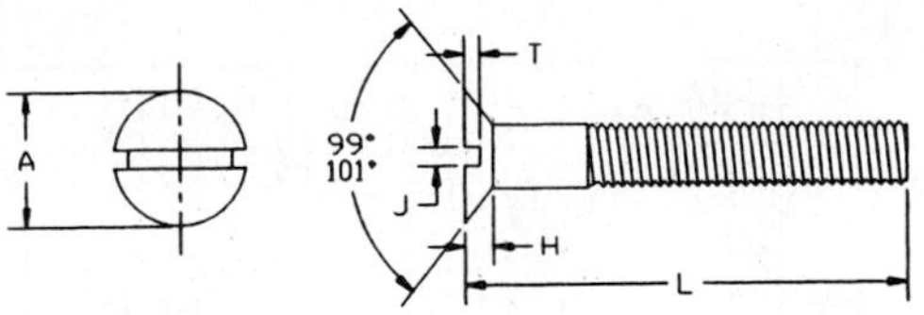

Nominal Size or Basic Screw Diam.		Head Diam., A		Head Height, H	Slot Width, J		Slot Depth, T	
		Max., Edge Sharp	Min., Edge Rounded or Flat	Ref.	Max.	Min.	Max.	Min.
0000	.0210	.043	.037	.009	.008	.005	.008	.004
000	.0340	.064	.058	.014	.012	.008	.011	.007
00	.0470	.093	.085	.020	.017	.010	.013	.008
0	.0600	.119	.096	.026	.023	.016	.013	.008
1	.0730	.146	.120	.031	.026	.019	.016	.010
2	.0860	.172	.143	.037	.031	.023	.019	.012
3	.0990	.199	.167	.043	.035	.027	.022	.014
4	.1120	.225	.191	.049	.039	.031	.024	.017
6	.1380	.279	.238	.060	.048	.039	.030	.022
8	.1640	.332	.285	.072	.054	.045	.036	.027
10	.1900	.385	.333	.083	.060	.050	.042	.031
1/4	.2500	.507	.442	.110	.075	.064	.055	.042
5/16	.3125	.635	.556	.138	.084	.072	.069	.053
3/8	.3750	.762	.670	.165	.094	.081	.083	.065

Figure A-21

American National Standard Slotted Truss Head Machine Screws
(ANSI B18.6.3-1972, R1977)

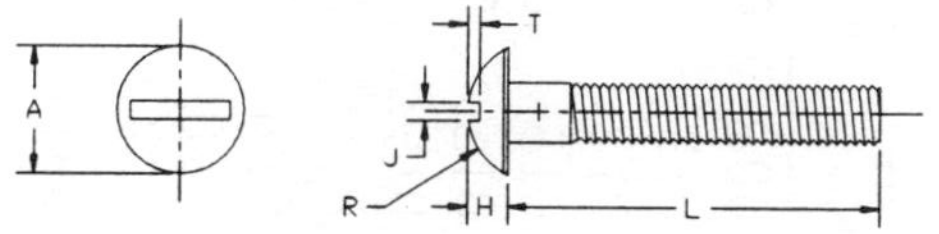

Nominal Size or Basic Screw Diam.		Head Diam., A		Head Height, H		Head Radius, R	Slot Width, J		Slot Depth, T	
		Max.	Min.	Max.	Min.	Max.	Max.	Min.	Max.	Min.
0000	.0210	.049	.043	.014	.010	.032	.009	.005	.009	.005
000	.0340	.077	.071	.022	.018	.051	.013	.009	.013	.009
00	.0470	.106	.098	.030	.024	.070	.017	.010	.018	.012
0	.0600	.131	.119	.037	.029	.087	.023	.016	.022	.014
1	.0730	.164	.149	.045	.037	.107	.026	.019	.027	.018
2	.0860	.194	.180	.053	.044	.129	.031	.023	.031	.022
3	.0990	.226	.211	.061	.051	.151	.035	.027	.036	.026
4	.1120	.257	.241	.069	.059	.169	.039	.031	.040	.030
5	.1250	.289	.272	.078	.066	.191	.043	.035	.045	.034
6	.1380	.321	.303	.086	.074	.211	.048	.039	.050	.037
8	.1640	.384	.364	.102	.088	.254	.054	.045	.058	.045
10	.1900	.448	.425	.118	.103	.283	.060	.050	.068	.053
12	.2160	.511	.487	.134	.118	.336	.067	.056	.077	.061
1/4	.2500	.573	.546	.150	.133	.375	.075	.064	.087	.070
5/16	.3125	.698	.666	.183	.162	.457	.084	.072	.106	.085
3/8	.3750	.823	.787	.215	.191	.538	.094	.081	.124	.100
7/16	.4375	.948	.907	.248	.221	.619	.094	.081	.142	.116
1/2	.5000	1.073	1.028	.280	.250	.701	.106	.091	.161	.131
9/16	.5625	1.198	1.149	.312	.279	.783	.118	.102	.179	.146
5/8	.6250	1.323	1.269	.345	.309	.863	.133	.116	.196	.162
3/4	.7500	1.573	1.511	.410	.368	1.024	.149	.131	.234	.182

Figure A-22

American National Standard Plain and Slotted Hexagon Head Machine Screws (ANSI B18.6.3-1972, R1977)

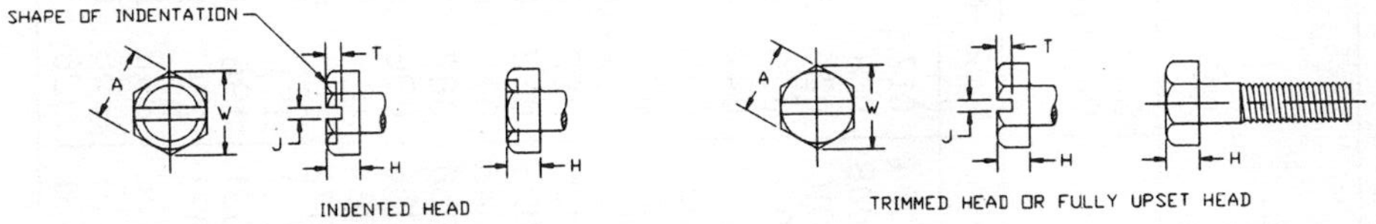

Nominal Size or Basic Screw Diam.	Regular Head Width Across Flats A Max.	Min.	Regular Head Across Corn. W Min.	Large Head Width Across Flats A Max.	Min.	Large Head Across Corn. W Min.	Head Height H Max.	Min.	Slot Width J Max.	Min.	Slot Depth T Max.	Min.
1 .0730	.125	.120	.134				.044	.036				
2 .0860	.125	.120	.134				.050	.040				
3 .0990	.188	.181	.202				.055	.044				
4 .1120	.188	.181	.202	.219	.213	.238	.060	.049	.039	.031	.036	.025
5 .1250	.188	.181	.202	.250	.244	.272	.070	.058	.043	.035	.042	.030
6 .1380	.250	.244	.272				.093	.080	.048	.039	.046	.033
8 .1640	.250	.244	.272	.312	.305	.340	.110	.096	.054	.045	.066	.052
10 .1900	.312	.305	.340				.120	.105	.060	.050	.072	.057
12 .2160	.312	.305	.340	.375	.367	.409	.155	.139	.067	.056	.093	.077
1/4 .2500	.375	.367	.409	.438	.428	.477	.190	.172	.075	.064	.101	.083
5/16 .3125	.500	.489	.545				.230	.208	.084	.072	.122	.100
3/8 .3750	.562	.551	.614				.295	.270	.094	.081	.156	.131

SHAPE OF INDENTATION

INDENTED HEAD

TRIMMED HEAD OR FULLY UPSET HEAD

Figure A-23

Slotted Round Head Machine Screws
(ANSI B18.6.3-1972, R1977 Appendix)

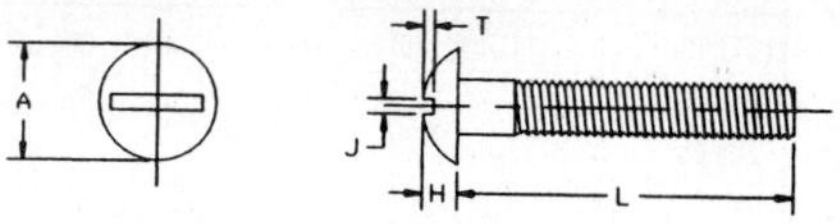

Nominal Size or Basic Screw Diam.		Head Diameter, A		Head Height, H		Slot Width, J		Slot Depth, T	
		Max.	Min.	Max.	Min.	Max.	Min.	Max.	Min.
0000	.0210	.041	.035	.022	.016	.008	.004	.017	.013
000	.0340	.062	.056	.031	.025	.012	.008	.018	.012
00	.0470	.089	.080	.045	.036	.017	.010	.026	.018
0	.0600	.113	.099	.053	.043	.023	.016	.039	.029
1	.0730	.138	.122	.061	.051	.026	.019	.044	.033
2	.0860	.162	.146	.069	.059	.031	.023	.048	.037
3	.0990	.187	.169	.078	.067	.035	.027	.053	.040
4	.1120	.211	.193	.086	.075	.039	.031	.058	.044
5	.1250	.236	.217	.095	.083	.043	.035	.063	.047
6	.1380	.260	.240	.103	.091	.048	.039	.068	.051
8	.1640	.309	.287	.120	.107	.054	.045	.077	.058
10	.1900	.359	.334	.137	.123	.060	.050	.087	.065
12	.2160	.408	.382	.153	.139	.067	.056	.096	.073
1/4	.2500	.472	.443	.175	.160	.075	.064	.109	.082
5/16	.3125	.590	.557	.216	.198	.084	.072	.132	.099
3/8	.3750	.708	.670	.256	.237	.094	.081	.155	.117
7/16	.4375	.750	.707	.328	.307	.094	.081	.196	.148
1/2	.5000	.813	.766	.355	.332	.106	.091	.211	.159
9/16	.5625	.938	.887	.410	.385	.118	.102	.242	.183
5/8	.6250	1.000	.944	.438	.411	.133	.116	.258	.195
3/4	.7500	1.250	1.185	.547	.516	.149	.131	.320	.242

Figure A-24

AMERICAN NATIONAL STANDARD SQUARE HEAD SETSCREWS (ANSI B18.6.2)

Nominal Size of Basic Screw	Diameter	Width Across Flats		Width Across Corners		Head Height		Neck Relief Diameter		Max Neck Relief Fillet Radius	Min Neck Relief Width	Min Head Radius
		Max.	Min.	Max.	Min.	Max.	Min.	Max.	Min.			
10	0.1900	0.188	0.180	0.265	0.247	0.148	0.134	0.145	0.140	0.027	0.083	0.48
1/4	0.2500	0.250	0.241	0.354	0.331	0.196	0.178	0.185	0.170	0.032	0.100	0.62
5/16	0.3125	0.312	0.302	0.442	0.415	0.245	0.224	0.240	0.225	0.036	0.111	0.78
3/8	0.3750	0.375	0.362	0.530	0.497	0.293	0.270	0.294	0.279	0.041	0.125	0.94
7/16	0.4375	0.438	0.423	0.619	0.581	0.341	0.315	0.345	0.330	0.046	0.143	1.09
1/2	0.5000	0.500	0.484	0.707	0.665	0.389	0.361	0.400	0.385	0.050	0.154	1.25
9/16	0.5625	0.562	0.545	0.795	0.748	0.437	0.407	0.454	0.439	0.054	0.167	1.41
5/8	0.6250	0.625	0.606	0.884	0.833	0.485	0.452	0.507	0.492	0.059	0.182	1.56
3/4	0.7500	0.750	0.729	1.060	1.001	0.582	0.544	0.620	0.605	0.065	0.200	1.88
7/8	0.8750	0.875	0.852	1.237	1.170	0.678	0.635	0.731	0.716	0.072	0.222	2.19
1	1.0000	1.000	0.974	1.414	1.337	0.774	0.726	0.838	0.823	0.081	0.250	2.50
1 1/8	1.1250	1.125	1.096	1.591	1.505	0.870	0.817	0.939	0.914	0.092	0.283	2.81
1 1/4	1.2500	1.250	1.219	1.768	1.674	0.966	0.908	1.064	1.039	0.092	0.283	3.12
1 3/8	1.3750	1.375	1.342	1.945	1.843	1.063	1.000	1.159	1.134	0.109	0.333	3.44
1 1/2	1.5000	1.500	1.464	2.121	2.010	1.159	1.091	1.284	1.259	0.109	0.333	3.75

Figure A-25

AMERICAN NATIONAL STANDARD SQUARE HEAD SETSCREWS

(ANSI B18.6.2)

Nominal Size or Basic Screw Diameter		Cup and Flat Point Diameters		Dog and Half-Dog Point Diameters		Point Length				Oval Point Radius +0.031 −0.000	Cone Point Angle 90° ± 2° for these Nominal Lengths or Longer, 118° ± 2° for Shorter Screws
						Dog		Half-Dog			
		Max.	Min.	Max.	Min.	Max.	Min.	Max.	Min.		
10	0.1900	0.102	0.088	0.127	0.120	0.095	0.085	0.050	0.040	0.142	1/4
1/4	0.2500	0.132	0.118	0.156	0.149	0.130	0.120	0.068	0.058	0.188	5/16
5/16	0.3125	0.172	0.156	0.203	0.195	0.161	0.151	0.083	0.073	0.234	3/8
3/8	0.3750	0.212	0.194	0.250	0.241	0.193	0.183	0.099	0.089	0.281	7/16
7/16	0.4375	0.252	0.232	0.297	0.287	0.224	0.214	0.114	0.104	0.328	1/2
1/2	0.5000	0.291	0.270	0.344	0.334	0.255	0.245	0.130	0.120	0.375	9/16
9/16	0.5625	0.332	0.309	0.391	0.379	0.287	0.275	0.146	0.134	0.422	5/8
5/8	0.6250	0.371	0.347	0.469	0.456	0.321	0.305	0.164	0.148	0.469	3/4
3/4	0.7500	0.450	0.425	0.562	0.549	0.383	0.367	0.196	0.180	0.562	7/8
7/8	0.8750	0.530	0.502	0.656	0.642	0.446	0.430	0.227	0.221	0.656	1
1	1.0000	0.609	0.579	0.750	0.734	0.510	0.490	0.260	0.240	0.750	1 1/8
1 1/8	1.1250	0.689	0.655	0.844	0.826	0.572	0.552	0.291	0.271	0.844	1 1/4
1 1/4	1.2500	0.767	0.733	0.938	0.920	0.635	0.615	0.323	0.303	0.938	1 1/2
1 3/8	1.3750	0.848	0.808	1.031	1.011	0.698	0.678	0354	0.334	1.031	1 5/8
1 1/2	1.5000	0.926	0.886	1.125	1.105	0.760	0.740	0.385	0.365	1.125	1 3/4

Figure A-26

American National Standard Slotted Headless
Setscrews (ANSI B18.6.2)

Nominal Size or Basic Screw Diameter		Crown Radius Basic	Slot Width		Slot Depth		Cup and Flat Point Diameters		Dog Point Diameters		Point Length				Oval Point Radius Basic	Cone Point Angle 90°±2° For These Nominal Lengths or Longer 118°±2° For Shorter
											Dog		Half-Dog			
			MAX	MIN	MAX	MIN	MAX	MIN	MAX	MIN	MAX	MIN	MAX	MIN		
0	0.0600	0.060	0.014	0.0010	0.020	0.016	0.033	0.027	0.040	0.037	0.032	0.028	0.017	0.013	0.045	5/64
1	0.0730	0.073	0.016	0.012	0.020	0.016	0.040	0.033	0.049	0.045	0.040	0.036	0.021	0.017	0.055	3/32
2	0.0860	0.086	0.018	0.014	0.025	0.019	0.047	0.039	0.057	0.053	0.046	0.042	0.024	0.020	0.064	7/64
3	0.0990	0.099	0.020	0.016	0.028	0.022	0.054	0.045	0.066	0.062	0.052	0.048	0.027	0.023	0.074	1/8
4	0.1120	0.112	0.024	0.018	0.031	0.025	0.061	0.051	0.075	0.070	0.058	0.054	0.030	0.026	0.084	5/32
5	0.1250	0.125	0.026	0.020	0.036	0.026	0.067	0.057	0.083	0.078	0.063	0.057	0.033	0.027	0.094	3/16
6	0.1380	0.138	0.028	0.022	0.040	0.030	0.074	0.064	0.092	0.087	0.073	0.067	0.038	0.032	0.104	3/16
8	0.1640	0.164	0.032	0.026	0.046	0.036	0.087	0.076	0.109	0.103	0.083	0.077	0.043	0.037	0.123	1/4
10	0.1900	0.190	0.035	0.029	0.053	0.043	0.102	0.088	0.127	0.120	0.095	0.085	0.050	0.040	0.142	1/4
12	0.2160	0.216	0.042	0.035	0.061	0.051	0.115	0.101	0.144	0.137	0.115	0.105	0.060	0.050	0.162	5/16
1/4	0.2500	0.250	0.049	0.041	0.068	0.058	0.132	0.118	0.156	0.149	0.130	0.120	0.068	0.058	0.188	5/16
5/16	0.3125	0.312	0.055	0.047	0.083	0.073	0.172	0.156	0.203	0.195	0.161	0.151	0.083	0.073	0.234	3/8
3/8	0.3750	0.375	0.068	0.060	0.099	0.089	0.212	0.194	0.250	0.241	0.193	0.183	0.099	0.089	0.281	7/16
7/16	0.4375	0.438	0.076	0.068	0.114	0.104	0.252	0.232	0.297	0.287	0.224	0.214	0.114	0.104	0.328	1/2
1/2	0.5000	0.500	0.086	0.078	0.130	0.120	0.291	0.270	0.344	0.334	0.255	0.245	0.130	0.120	0.375	9/16
9/16	0.5625	0.562	0.096	0.088	0.146	0.136	0.332	0.309	0.391	0.379	0.287	0.275	0.146	0.134	0.422	5/8
5/8	0.6250	0.625	0.107	0.097	0.161	0.151	0.371	0.347	0.469	0.456	0.321	0.305	0.164	0.148	0.469	3/4
3/4	0.7500	0.750	0.134	0.124	0.193	0.183	0.450	0.425	0.562	0.549	0.383	0.367	0.196	0.180	0.562	7/8

Figure A-27

Lengths for Threaded Fasteners

DIAMETER	.250	.313	.375	.438	.500	.563	.625	.750	.875	1.000	1.250	1.500	1.750	2.000	2.500	3.000	3.500	4.000
5(.125)	●	●	●	●	●	●	●	●	●	●		●						
6(.138)	●	●	●	●	●	●	●	●	●	●	●	●		●				
8(.164)	●	●	●	●	●	●	●	●	●	●	●	●		●				
10(.190)	●	●	●	●	●	●	●	●	●	●	●	●	●	●				
12(.216)	●	●	●	●	●	●	●	●	●	●		●		●				
.250	●	●	●	●	●	●	●	●	●	●	●	●	●	●	●			
.313	●	●	●	●	●	●	●	●	●	●	●	●	●	●	●			
.375	●	●	●	●	●	●	●	●	●	●	●	●		●	●	●		
.438	●	●	●	●	●	●	●	●	●	●	●	●		●	●	●		
.500	●	●	●	●	●	●	●	●	●	●	●	●		●	●	●	●	●
.563			●	●	●	●	●	●	●	●		●		●	●	●	●	●
.625			●	●	●	●	●	●	●	●		●		●		●		●
.750					●	●	●	●	●	●		●		●		●	●	●
.875							●	●	●	●		●		●		●	●	●
1.000												●		●	●	●		●

LENGTHS

Figure A-28

Lengths for Metric Threaded Fasteners

DIAMETER	4	5	8	10	12	16	20	24	30	36	40	45	50	60	70
1.6	●	●	●												
2	●	●	●												
2.5	●	●	●	●	●										
3		●	●	●	●										
4			●	●	●	●	●								
5			●	●	●	●	●	●							
6				●	●	●	●	●							
8					●	●	●	●	●	●	●				
10						●	●	●	●	●	●	●	●	●	
12						●	●	●	●	●	●	●	●	●	●
16							●	●	●	●	●	●	●	●	●
20								●	●	●	●	●	●	●	●
24									●	●	●	●	●	●	●
30										●	●	●	●	●	●

LENGTHS

Figure A-29

American National Standard Square and Hexagon Machine Screw Nuts
(ANSI B18.6.3-1972, R1977)

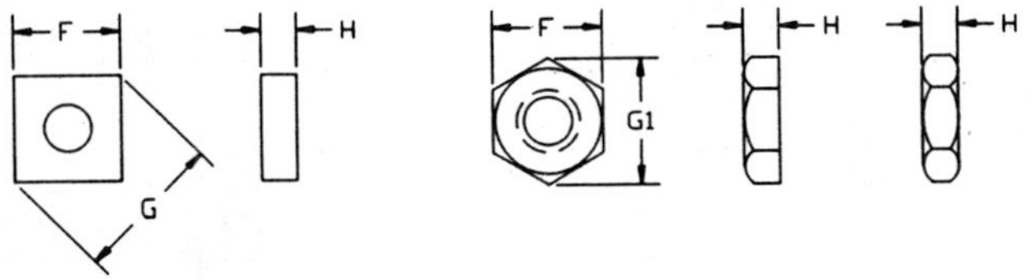

Nom. Size	Basic Diam.	Basic F	Max. F	Min. F	Max. G	Min. G	Max. G1	Min. G1	Max. H	Min. H
0	.0600	5/32	.156	.150	.221	.206	.180	.171	.050	.043
1	.0730	5/32	.156	.150	.221	.206	.180	.171	.050	.043
2	.0860	3/16	.188	.180	.265	.247	.217	.205	.066	.057
3	.0990	3/16	.188	.180	.265	.247	.217	.205	.066	.057
4	.1120	1/4	.250	.241	.354	.331	.289	.275	.098	.087
5	.1250	5/16	.312	.302	.442	.415	.361	.344	.114	.102
6	.1380	5/16	.312	.302	.442	.415	.361	.344	.114	.102
8	.1640	11/32	.344	.332	.486	.456	.397	.378	.130	.117
10	.1900	3/8	.375	.362	.530	.497	.433	.413	.130	.117
12	.2160	7/16	.438	.423	.619	.581	.505	.482	.161	.148
1/4	.2500	7/16	.438	.423	.619	.581	.505	.482	.193	.178
5/16	.3125	9/16	.562	.545	.795	.748	.650	.621	.225	.208
3/8	.3750	5/8	.625	.607	.884	.833	.722	.692	.257	.239

Figure A-30

743

Standard Twist Drill Sizes (Inches)

SIZE	DIAMETER	SIZE	DIAMETER	SIZE	DIAMETER	SIZE	DIAMETER
40	.098	19	.166	C	.242	U	.368
39	.0995	18	.1695	D	.246	3/8	.375
38	.1015	11/64	.1719	1/4(E)	.250	V	.377
37	.104	17	.173	F	.257	W	.386
36	.1065	16	.177	G	.261	25/64	.3906
7/64	.1094	15	.180	17/64	.2656	X	.397
35	.110	14	.182	H	.266	Y	.404
34	.111	13	.185	I	.272	13/32	.4062
33	.113	3/16	.1875	J	.277	Z	.413
32	.116	12	.189	K	.281	27/64	.4219
31	.120	11	.191	9/32	.2812	7/16	.4375
1/8	.125	10	.1935	L	.290	29/64	.4531
30	.1285	9	.196	M	.295	15/32	.4688
29	.136	8	.199	19/64	.2969	31/64	.4844
28	.1405	7	.201	N	.302	1/2	.5000
9/64	.1406	13/64	.2031	5/16	.3125	9/16	.5625
27	.144	6	.204	O	.316	5/8	.625
26	.147	5	.2055	P	.323	11/16	.6875
25	.1495	4	.209	21/64	.3281	3/4	.750
24	.152	3	.213	Q	.332	13/16	.8125
23	.154	7/32	.2188	R	.339	7/8	.875
5/32	.1562	2	.221	11/32	.3438	15/16	.9375
22	.157	1	.228	S	.348		
21	.159	A	.234	T	.358		
20	.161	B	.238	23/64	.3594		

NOTES FOR TWIST DRILL SIZES – INCHES
1. This is only a partial list of standard drill sizes.
2. Whenever possible, specify holes sizes that correspond to standard drill sizes.
3. Drill sizes are available in 1/64 increments between .5000 and 1.2500.
4. Drill sizes are available in 1/32 increments between 1.2500 and 1.500.

Figure A-31

Standard Twist Drill Sizes (Millimeters)

0.40	2.05	5.10	8.60	15.25	30.00
0.42	2.10	5.20	8.70	15.50	30.50
0.45	2.15	5.30	8.80	15.75	31.00
0.48	2.20	5.40	8.90	16.00	31.50
0.50	2.25	5.50	9.00	16.25	32.00
0.55	2.30	5.60	9.10	16.50	32.50
0.60	2.35	5.70	9.20	16.75	33.00
0.65	2.40	5.80	9.30	17.00	33.50
0.70	2.45	5.90	9.40	17.25	34.00
0.75	2.50	6.00	9.50	17.50	34.50
0.80	2.60	6.10	9.60	17.75	35.00
0.85	2.70	6.20	9.70	18.00	35.50
0.90	2.80	6.30	9.80	18.50	36.00
0.95	2.90	6.40	9.90	19.00	36.50
1.00	3.00	6.50	10.00	19.50	37.00
1.05	3.10	6.60	10.20	20.00	37.50
1.10	3.20	6.70	10.50	20.50	38.00
1.15	3.30	6.80	10.80	21.00	40.00
1.20	3.40	6.90	11.00	21.50	42.00
1.25	3.50	7.00	11.20	22.00	44.00
1.30	3.60	7.10	11.50	22.50	46.00
1.35	3.70	7.20	11.80	23.00	48.00
1.40	3.80	7.30	12.00	23.50	50.00
1.45	3.90	7.40	12.20	24.00	
1.50	4.00	7.50	12.50	24.50	
1.55	4.10	7.60	12.80	25.00	
1.60	4.20	7.70	13.00	25.50	
1.65	4.30	7.80	13.20	26.00	
1.70	4.40	7.90	13.50	26.50	
1.75	4.50	8.00	13.80	27.00	
1.80	4.60	8.10	14.00	27.50	
1.85	4.70	8.20	14.25	28.00	
1.90	4.80	8.30	14.50	28.50	
1.95	4.90	8.40	14.75	29.00	
2.00	5.00	8.50	15.00	29.50	

Figure A-32

Index